Exercise to Prevent and Manage Chronic Disease Across the Lifespan

Exercise to Prevent and Manage Chronic Disease Across the Lifespan

Jack Feehan

Institute for Health and Sport, Victoria University, Melbourne, VIC, Australia
First Year College, Victoria University, Melbourne, VIC, Australia
The Australian Institute for Musculoskeletal Science, Western Health, Victoria University and the
University of Melbourne, Melbourne, VIC, Australia
Osteopathy group, College of Health and Biomedicine, Victoria University, Melbourne, VIC, Australia

Nicholas Tripodi

Institute for Health and Sport, Victoria University, Melbourne, VIC, Australia
First Year College, Victoria University, Melbourne, VIC, Australia
The Australian Institute for Musculoskeletal Science, Western Health, Victoria University and the
University of Melbourne, Melbourne, VIC, Australia
Osteopathy group, College of Health and Biomedicine, Victoria University, Melbourne, VIC, Australia

Vasso Apostolopoulos

Institute for Health and Sport, Victoria University, Melbourne, VIC, Australia
The Australian Institute for Musculoskeletal Science, Western Health, Victoria University and the
University of Melbourne, Melbourne, VIC, Australia

ACADEMIC PRESS
An imprint of Elsevier

Academic Press is an imprint of Elsevier
125 London Wall, London EC2Y 5AS, United Kingdom
525 B Street, Suite 1650, San Diego, CA 92101, United States
50 Hampshire Street, 5th Floor, Cambridge, MA 02139, United States
The Boulevard, Langford Lane, Kidlington, Oxford OX5 1GB, United Kingdom

ISBN 978-0-323-89843-0

For information on all Academic Press publications
visit our website at https://www.elsevier.com/books-and-journals

Publisher: Stacy Masucci
Senior Acquisitions Editor: Elizabeth Brown
Editorial Project Manager: Susan Ikeda
Production Project Manager: Punithavathy Govindaradjane
Cover Designer: Christian J. Bilbow

Typeset by STRAIVE, India

Contents

12. Exercise in the management of multiple sclerosis

*Narges Dargahi,
Melina Haritopoulou-Sinanidou, and
Vasso Apostolopoulos*

13. Exercise and menopause

*Serene Hilary, Habiba Ali,
Leila Cheikh Ismail, Ayesha S. Al Dhaheri,
and Lily Stojanovska*

14. Exercise for chronic pain

*Della Buttigieg, Nick Efthimiou, and
Alison Sim*

27. Tai Chi exercise to improve balance and prevent falls among older people with dementia

Yolanda Barrado-Martín, Remco Polman, and Samuel R. Nyman

28. Osteosarcopenia and exercise

Troy Walker, Jordan Dixon, Ian Haryono, and Jesse Zanker

29. Exercise and older adults receiving home care services

Elissa Burton and Anne-Marie Hill

Contributors

Numbers in parenthesis indicate the pages on which the authors' contributions begin.

Ayesha S. Al Dhaheri (175), Department of Nutrition and Health, College of Medicine and Health Sciences, United Arab Emirates University, Al-Ain, United Arab Emirates

Habiba Ali (175), Department of Nutrition and Health, College of Medicine and Health Sciences, United Arab Emirates University, Al-Ain, United Arab Emirates

Saud Almaslmani (301), Department of Orthopedic Surgery, Western Health, Melbourne, VIC, Australia; Department of Surgery, Faculty of Medicine in Al-Qunfudhah, Umm Al-Qura University, Makkah, Saudi Arabia

Vasso Apostolopoulos (3, 7, 49, 163), Institute for Health and Sport, Victoria University; The Australian Institute for Musculoskeletal Science, Western Health, Victoria University and the University of Melbourne, Melbourne, VIC, Australia

Yolanda Barrado-Martín (363), Centre for Ageing Population Studies, Research Department of Primary Care & Population Health, London, United Kingdom

Rezaul Begg (423), Institute for Health and Sport, Victoria University, Melbourne, VIC, Australia

Erin Bicknell (317), Physiotherapy, Western Health, St Albans, VIC, Australia

Kelly Bower (317), Physiotherapy, University of Melbourne, Parkville, VIC, Australia

Peter Brukner (97), La Trobe Sport and Exercise Medicine Research Centre, La Trobe University, Melbourne, VIC, Australia

Elissa Burton (391), Curtin School of Allied Health; enAble Institute, Curtin University, Perth, WA, Australia

Della Buttigieg (193), Melbourne, VIC, Australia

Claire J. Cadwallader (403, 437), The Turner Institute for Brain and Mental Health, School of Psychological Sciences, Monash University, Melbourne, VIC, Australia

Celso R.F. Carvalho (117), Department of Physical Therapy, School of Medicine, University of São Paulo, São Paulo, Brazil

Trevor T-J. Chong (403, 437), The Turner Institute for Brain and Mental Health, School of Psychological Sciences, Monash University; Department of Neurology, Alfred Health; Department of Clinical Neurosciences, St Vincent's Hospital, Melbourne, VIC, Australia

Sheri R. Colberg (79, 141), Old Dominion University, Norfolk, VA, United States

Daniel Corcoran (289), Osteopathy group, College of Health and Biomedicine, Victoria University, Melbourne, VIC, Australia

Maximilian de Courten (23), Institute for Health and Sport; Mitchell Institute for Education and Health Policy, Victoria University, Melbourne, VIC, Australia

James P. Coxon (403, 437), The Turner Institute for Brain and Mental Health, School of Psychological Sciences, Monash University, Melbourne, VIC, Australia

Dylan Curtin (403, 437), The Turner Institute for Brain and Mental Health, School of Psychological Sciences, Monash University, Melbourne, VIC, Australia

Narges Dargahi (163), Institute for Health and Sport, Victoria University, Melbourne, VIC, Australia

Jordan Dixon (373), Australian Institute for Musculoskeletal Science, The University of Melbourne and Western Health, St. Albans, VIC, Australia

Nick Efthimiou (193), Sydney, NSW, Australia

Luke Ellis (289), Osteopathy group, College of Health and Biomedicine, Victoria University, Melbourne, VIC, Australia

Jack Feehan (3, 289), Institute for Health and Sport, Victoria University; First Year College, Victoria University; The Australian Institute for Musculoskeletal Science, Western Health, Victoria University and the University of Melbourne; Osteopathy group, College of Health and Biomedicine, Victoria University, Melbourne, VIC, Australia

Natalie Fini (317), Physiotherapy, University of Melbourne, Parkville, VIC, Australia

Michael Fleischmann (233), School of Public Health, University of Technology Sydney, Sydney, NSW; Torrens University Australia, Melbourne, VIC, Australia

Monique E. Francois (141), University of Wollongong; Illawarra Health and Medical Research Institute, Wollongong, NSW, Australia

Daniel Friedman (97), Alfred Health, Melbourne, VIC, Australia

Jenna B. Gillen (141), University of Toronto, Toronto, ON, Canada

Catherine Giuliano (261), Institute for Health and Sport, Victoria University, Melbourne, VIC, Australia

Mansueto Gomes-Neto (131), Federal University of Bahia, Salvador, BA, Brazil

Melina Haritopoulou-Sinanidou (163), School of Biomedical Sciences, University of Queensland, Brisbane, QLD, Australia

Katherine Harkin (49), Institute for Health and Sport; First Year College, Victoria University, Melbourne, VIC, Australia

Esther Hartman (65), University of Groningen, University Medical Center Groningen, Center for Human Movement Sciences, Groningen, The Netherlands

Ian Haryono (373), Australian Institute for Musculoskeletal Science, The University of Melbourne and Western Health, St. Albans, VIC, Australia

Joshua J. Hendrikse (403, 437), The Turner Institute for Brain and Mental Health, School of Psychological Sciences, Monash University, Melbourne, VIC, Australia

Serene Hilary (175), Department of Nutrition and Health, College of Medicine and Health Sciences, United Arab Emirates University, Al-Ain, United Arab Emirates

Anne-Marie Hill (391), School of Allied Health, WA Centre for Health & Ageing, University of Western Australia, Perth, WA, Australia

Joel Hiney (289), Osteopathy group, College of Health and Biomedicine, Victoria University, Melbourne, VIC, Australia

Maja Husaric (349), First Year College, Victoria University, Melbourne, VIC, Australia

Susan Irvine (49, 413), First Year College, Victoria University, Melbourne, VIC, Australia

Leila Cheikh Ismail (175), Department of Clinical Nutrition and Dietetics, College of Health Sciences, Research Institute of Medical and Health Sciences (RIMHS), University of Sharjah, Sharjah, United Arab Emirates; Nuffield Department of Women's & Reproductive Health, University of Oxford, Oxford, United Kingdom

Liam Johnson (317), Physiotherapy, University of Melbourne, Parkville; School of Behavioural and Health Sciences, Australian Catholic University, Melbourne; Physiotherapy Department, Epworth HealthCare, Melbourne, VIC, Australia

Amy Lawton (221, 349), College of Health & Biomedicine, Victoria University, Melbourne, VIC, Australia

Itamar Levinger (261), Institute for Health and Sport, Victoria University, Melbourne; Australian Institute for Musculoskeletal Science, University of Melbourne and Western Health, St Albans, VIC, Australia

Leandro Teixeira Paranhos Lopes (329), Brazil University, Fernandópolis, SP, Brazil

Camilo Luís Monteiro Lourenço (131, 329), Federal University of Santa Catarina, Florianópolis, SC, Brazil

Peter Malliaras (233), Department of Physiotherapy, School of Primary and Allied Health Care, Monash University, Melbourne, VIC, Australia

Vanessa M. McDonald (117), National Health and Medical Research Council, Centre of Excellence in Treatable Traits; Hunter Medical Research Institute; School of Nursing and Midwifery, University of Newcastle; Department of Respiratory and Sleep Medicine, John Hunter Hospital, Newcastle, NSW, Australia

Luke C. McIlvenna (273), Institute for Health and Sport, Victoria University, Melbourne, VIC, Australia

Rebecca F. McLoughlin (117), National Health and Medical Research Council, Centre of Excellence in Treatable Traits; Hunter Medical Research Institute; School of Nursing and Midwifery, University of Newcastle, Newcastle, NSW, Australia

Anna Meijer (65), Vrije Universiteit Amsterdam, Clinical Neuropsychology Section, Amsterdam; Universiteit Leiden, Clinical Neurodevelopmental Sciences, Leiden, The Netherlands

Jack Mest (233), University of Canberra Health Clinics, Faculty of Health, University of Canberra, Canberra, ACT, Australia

Melanie Moore (335), Faculty of Health; Prehabilitation, Activity, Cancer, Exercise, and Survivorship (PACES) Research Group; Faculty of Health Clinics, University of Canberra, Canberra, ACT, Australia

Alba Moreno-Asso (273), Institute for Health and Sport, Victoria University; Australian Institute for Musculoskeletal Science, Victoria University and Western Health, Melbourne, VIC, Australia

Michael Musker (243), The University of South Australia; South Australian Health & Medical Research Institute; The University of Adelaide; Flinders University, Adelaide, SA, Australia

Hanatsu Nagano (423), Institute for Health and Sport, Victoria University, Melbourne, VIC, Australia

Samuel R. Nyman (363), Bournemouth University Clinical Research Unit, Department of Medical Science and Public Health, Bournemouth University, Poole, United Kingdom

Alexandra G. Parker (73, 251), Institute for Health and Sport, Victoria University, Melbourne, VIC, Australia

Brigitte Marie Pascal (49), College of Arts and Education, Victoria University, Melbourne, VIC, Australia

Michaela C. Pascoe (73, 251), Institute for Health and Sport, Victoria University, Melbourne, VIC, Australia

Rhiannon K. Patten (273), Institute for Health and Sport, Victoria University, Melbourne, VIC, Australia

Bojana Klepac Pogrmilovic (23), Institute for Health and Sport; Mitchell Institute for Education and Health Policy, Victoria University, Melbourne, VIC, Australia

Remco Polman (363), School of Exercise & Nutrition Sciences, Queensland University of Technology, Brisbane, QLD, Australia

Supa Pudkasam (7), Institute for Health and Sport, Victoria University, Melbourne, VIC, Australia; Faculty of Nursing Science, Assumption University, Bangkok, Thailand

Leonardo Roever (131, 329), Federal University of Uberlândia, Uberlândia, MG, Brazil

Catherine Said (317), Physiotherapy, University of Melbourne, Parkville; Physiotherapy, Western Health, St Albans; Australian Institute of Musculoskeletal Science, St Albans, VIC, Australia

Alison Sim (193), School of Medicine, Faculty of Medicine and Health, The University of Sydney, Sydney, NSW, Australia

William Anthony Sparrow (423), Institute for Health and Sport, Victoria University, Melbourne, VIC, Australia

Lily Stojanovska (175), Department of Nutrition and Health, College of Medicine and Health Sciences, United Arab Emirates University, Al-Ain, United Arab Emirates; Institute for Health and Sport, Victoria University, Melbourne, VIC, Australia

Julie C. Stout (403, 437), The Turner Institute for Brain and Mental Health, School of Psychological Sciences, Monash University, Melbourne, VIC, Australia

Kathy Tangalakis (49, 413), First Year College, Victoria University, Melbourne, VIC, Australia

Eleanor M. Taylor (403, 437), The Turner Institute for Brain and Mental Health, School of Psychological Sciences, Monash University, Melbourne, VIC, Australia

Melanie Thewlis (23), Institute for Health and Sport; Mitchell Institute for Education and Health Policy, Victoria University, Melbourne, VIC, Australia

Kellie Toohey (335), Faculty of Health; Prehabilitation, Activity, Cancer, Exercise, and Survivorship (PACES) Research Group; Faculty of Health Clinics, University of Canberra, Canberra, ACT, Australia

Phong Tran (301), Department of Orthopedic Surgery, Western Health, Melbourne; Victoria University, Footscray; Australian Institute for Musculoskeletal Science, The University of Melbourne and Western Health, St. Albans; Swinburne University, Melbourne, VIC, Australia

Nicholas Tripodi (3, 289), Institute for Health and Sport, Victoria University; First Year College, Victoria University; The Australian Institute for Musculoskeletal Science, Western Health, Victoria University and the University of Melbourne; Osteopathy group, College of Health and Biomedicine, Victoria University, Melbourne, VIC, Australia

Paola D. Urroz (117), National Health and Medical Research Council, Centre of Excellence in Treatable Traits; Hunter Medical Research Institute; School of Nursing and Midwifery, University of Newcastle, Newcastle, NSW, Australia

Patrick Vallance (233), Department of Physiotherapy, School of Primary and Allied Health Care, Monash University, Melbourne, VIC, Australia

Brett Vaughan (233), Department of Medical Education, Melbourne Medical School, University of Melbourne, Melbourne, VIC; School of Public Health, University of Technology Sydney, Sydney, NSW; Faculty of Health, Southern Cross University, Lismore, NSW, Australia

Lot Verburgh (65), Rivas Zorggroep, Gorinchem, The Netherlands

Troy Walker (373), Global Obesity Centre (GLOBE), Institute for Health Transformation, Deakin University, Geelong, VIC, Australia

Mary Woessner (261), Institute for Health and Sport, Victoria University, Melbourne, VIC, Australia

Douglas Wong (221), College of Health & Biomedicine, Victoria University, Melbourne, VIC, Australia

Breanna Wright (221), College of Health & Biomedicine, Victoria University, Melbourne, VIC, Australia

Jane E. Yardley (79), University of Alberta, Camrose, AB, Canada

Hugo Ribeiro Zanetti (131, 329), University Center IMEPAC, Araguari; Federal University of Triângulo Mineiro, Uberaba, MG, Brazil

Jesse Zanker (373), Australian Institute for Musculoskeletal Science, The University of Melbourne and Western Health, St. Albans; Department of Medicine—Western Health, The University of Melbourne, St. Albans; Institute for Health Transformation, Deakin University, Geelong, VIC, Australia

Acknowledgments

A text of the size and scale of this book is a work that requires significant collaboration and input, and there are many people worthy of our thanks who have contributed to this book.

We were lucky to receive enormous support from international leaders across a number of fields. More than 50 authors contributed to chapters, throughout a time of significant challenges for researchers and academics globally. Without their contributions, the book would not have come to fruition, and so we offer our sincere gratitude.

We thank Victoria University and the Institute for Health and Sport for their ongoing support throughout the process of publishing this volume. We also thank Elsevier and their staff for their ongoing and tireless editorial support throughout the publication process. We thank Myfanwy Thewlis for the production of the cover art.

Part I

Introduction

Chapter 1

Introduction

Jack Feehan[a,b,c,d], Nicholas Tripodi[a,b,c,d], and Vasso Apostolopoulos[a,c]

[a]*Institute for Health and Sport, Victoria University, Melbourne, VIC, Australia* [b]*First Year College, Victoria University, Melbourne, VIC, Australia*
[c]*The Australian Institute for Musculoskeletal Science, Western Health, Victoria University and the University of Melbourne, Melbourne, VIC, Australia*
[d]*Osteopathy group, College of Health and Biomedicine, Victoria University, Melbourne, VIC, Australia*

Medical communities around the world are under strain from the huge burden of various preventable and chronic diseases. Even the most established, well-resourced healthcare systems are encountering increasing difficulty to fund the treatments and procedures required for treating a growing, aging, and unhealthy population of the new age than in years gone by. While rapid advancement in medical science has provided new and powerful treatments for diseases, these are often costly and rarely provide a complete cure. There is, however, an intervention that offers preventative, protective, and therapeutic benefits in most disease states, which is cost-effective and can be accessible to all.

Exercise can appear an impenetrable area to many in clinical services. A host of conflicting guidelines, perceptions of risk, challenges in patient adherence and accessibility, and a lack of training in prescription in medical specialties have led to widespread underutilization. While a slow cultural shift is occurring, it remains the case across almost all medical specialties. However, incorporating exercise programs into standard care provides widespread benefits in most scenarios, both directly to the pathology being managed and indirectly through secondary exercise effects. There is no other treatment that so routinely improves patient quality of life and disease outcomes while reducing costs no matter the disease state.

In this book, the beneficial effects of exercise are explored in a range of common disease states with a focus on the outcomes and prescription alongside background on the mechanisms underlying the changes where known. While it is obviously not possible to specifically cover all disease states, we have aimed to include the most common and prevalent conditions—indeed, almost 100% of the population will experience at least one of the included diseases in their lifetime. We hope that it will act as a guide to inform prescription and encourage clinicians to recommend exercise to all patients in a way that will improve their outcomes.

Throughout the book, there are many recommendations, guidelines, and considerations discussed, but we recommend a general approach for all as a first principle:

Some is better than none, more is better than some.

Exercise to Prevent and Manage Chronic Disease Across the Lifespan. https://doi.org/10.1016/B978-0-323-89843-0.00001-5

Part II

General exercise

Chapter 2

Exercise and immunity

Supa Pudkasam[a,b] and Vasso Apostolopoulos[a,c]

[a]*Institute for Health and Sport, Victoria University, Melbourne, VIC, Australia* [b]*Faculty of Nursing Science, Assumption University, Bangkok, Thailand*
[c]*The Australian Institute for Musculoskeletal Science, Western Health, Victoria University and the University of Melbourne, Melbourne, VIC, Australia*

Introduction

Exercise is well recognized as a powerful driver of antiinflammatory responses [1]. Both acute high-intensity exercise as well as moderate-intensity training improve immune function and induce antiinflammatory responses in healthy individuals [1,2]. However, there is conflicting evidence on biological mechanisms which may affect human immunity; for example, intensive training in elite athletes can result in their susceptibility to viral infections [3]. Exercise-induced immune regulation involves both innate and adaptive immune systems as well as inflammatory responses [4]. However, the response to exercise in some settings is very similar to an inflammatory response depending on the mode, intensity, and duration of exercise [5]. Herein, we discuss how exercise (acute or long-term) and physical activity can alter immune function and inflammatory processes.

Effects of exercise on inflammatory markers

Exercise-induced inflammatory responses involve a complex array of mechanisms and are associated with tissue damage and oxidative stress [6]. Previous studies have shown inflammatory responses such as cytokines and neutrophil activation following strenuous exercise in muscle [7] and adipose tissue [8]. Whether exercise-induced inflammatory responses are beneficial or deleterious for muscle tissue and body systems is currently unknown [7]. However, there is evidence to suggest that exercise training in myocardial infarction-induced rats can increase peroxisome proliferator-activated receptors (PPARs) in the myocardium, which are known to regulate inflammatory responses [9]. There are several mechanisms explaining the reduction of the chronic inflammatory process following exercise training such as immune functions to reduce pro-inflammatory cells and muscle cells secreting antiinflammatory myokines [10].

The alteration of inflammatory responses following exercise depends on the modality, intensity, and duration of activity. Extreme or strenuous exercise can initiate inflammation as a response by immune cells and activation of inflammatory biomarkers, however, this response is likely to repair exercise-induced muscle or organ damage [5]. The association between regular exercise and antiinflammatory responses may extend to the benefits of exercise on health. Exercise can mitigate the incidence of low-grade inflammatory-related chronic diseases such as cardiovascular diseases and type 2 diabetes [11]. Antiinflammatory responses to exercise also involve the reduction of adipokines and body fat mass [12]. Additionally, regular moderate exercise can reduce pro-inflammatory cytokines such as tumor necrosis factor-alpha (TNF-α) and interleukin 6 (IL-6) [13]. Additionally, muscle contraction during exercise can release antagonists to the IL-1 and TNF-receptors which further decreases inflammation [12]. Furthermore, a review study has found that regular exercise prevents inflammation in the central nervous system via antimicroglial activation through the IL-10 and cluster of differentiation-200 (CD200)/CD200 receptor pathways [14].

Exercise and C-reactive protein

C-reactive protein (CRP) is a nonspecific biomarker of systemic inflammation and is a commonly used indicator of chronic inflammatory diseases [15]. Increases in the level of CRP highlight low-grade systemic inflammation, often accompanied by a local inflammatory response and associated cytokine production [16]. Early studies showed an inverse relationship between CRP concentration and physical activity in adults suggesting a reduction of inflammation following exercise [17]. More recent studies on the effect of exercise on CRP concentrations have been less clear, however. A single bout of resistance exercise at moderate to vigorous intensity was shown to increase CRP, IL-6, and leukocyte counts after 3–24 h in

Exercise to Prevent and Manage Chronic Disease Across the Lifespan. https://doi.org/10.1016/B978-0-323-89843-0.00033-7

middle-aged sedentary overweight males [18], and a 4-month moderate to vigorous exercise (60%–80% VO₂max) training program was associated with both CRP reduction and weight loss in middle-aged adults [15]. Conversely, however, another study showed that one bout of exhaustive cycling increased serum IL-6 but not CRP after 2-h in young male athletes [19].

Exercise and fibrinogen

Fibrinogen is produced by the liver and is secreted into the peripheral circulation, forming an important part of the coagulation cascade [20]. Importantly fibrinogen activation increases erythrocyte sedimentation rate (ESR) which is another widely used nonspecific indicator of inflammation [21]. Broadly, it has been shown that increasing exercise is associated with decreasing fibrinogen levels, particularly in healthy individuals [13,22]. However other studies have shown no effect, with no change in fibrinogen and fibrin D-dimer levels after 3 months of moderate to vigorous aerobic training (75%–85% maximal heart rate) in middle-aged healthy males [23], nor did a single bout of exercise influence plasma fibrinogen [24]. Additionally, 8-weeks of exhaustive exercise training in young to middle-aged males also led to no change to fibrinogen or coagulation physiology [25]. The nuance of these findings, as well as the ultimate clinical result of them, requires further evaluation.

Exercise and toll-like receptors

Toll-like receptors (TLRs) are expressed on immune cells which recognize internal and external pathogens [26]. TLR-4 binds to bacterial lipopolysaccharide which activates inflammatory mediates, TNF-α, IL-1, IL-6, IL-12, and nitric oxide (NO). Long-term regular exercise training reduces TLR-4 expression by immune cells, which promotes antiinflammatory responses [27]. A recent systematic review concluded that TLR responses to exercise are heterogeneous based on the modality of exercise training. Upregulation of TLR is likely to occur after one bout of resistance exercise, whereas acute aerobic exercise may not alter TLR [26]. In obese mice, running on a treadmill over a 16 week period, inhibited mRNA expression of TLR4 [28]. Likewise, aerobic training in animals induces downregulation of TLR2 and TLR4 expression together with reduction in nuclear factor kappa beta (NF-κB) activation. On the contrary, antiinflammatory functions of TLR following exercise cannot be detected in humans [29]. As only a few studies have investigated the relationship between exercise and TLR and have demonstrated conflicting results, further investigations in the area are required [30].

Exercise and oxidative stress

Oxidative stress is a condition that occurs as a result of an imbalance between pro-oxidative factors (radicals or free radicals) and antioxidative reactions and is associated with chronic inflammatory diseases such as cardiovascular disease and cancer [31]. Most radicals generated in the body systems derive from nitrogen or oxygen [32]. Reactive oxygen or nitrogen species (RONS) are the intermediates in oxygen or nitrogen-derived radical reactions [32]. Healthy cells produce free radicals and reactive oxygen species (ROS) by normal metabolic processes, however are neutralized by enzymes such as superoxide dismutase, catalase, and glutathione peroxidase [33]. Exercise can enhance free radical generation and oxidative stress by increasing oxidative phosphorylation which depends on the intensity and duration of the activity [32,33]. Over the years, there has been growing evidence that prolonged and/or vigorous exercise increases biomarkers of oxidative stress [34]. As such, prolonged aerobic exercise (about an hour) at moderate intensity can increase lipid peroxidation and its biomarker, pentane [35]. Vigorous exercise training, especially in athletes, could increase RONS production, faster than they can be cleared, and as a consequence, increased RONS levels can induce contractile dysfunction of muscles [36]. Exercise at high intensity produces greater ROS concentration than exercise at moderate intensity [37]. It is asserted that in mitochondria, phospholipase A2 (PLA2)-dependent processes, nicotinamide adenine dinucleotide phosphate (NADPH) oxidase and xanthine oxidase contribute to muscle contraction induced ROS during exercise, however, the primary site of this in muscle fibers remains unclear [38]. The contraction-induced ROS following prolong and strenuous exercise would facilitate to muscle fatigue [38]. Malondialdehyde (MDA), a biomarker of lipid peroxidation, an indicator of pathological condition [39] is commonly increased after acute exhaustive exercise e.g., running sprint test [40]. Additionally, oxidized glutathione is also used for the measurement of exhaustive exercise-induced oxidative stress [41].

On the contrary, long-term aerobic and resistance training has a likelihood of antioxidant improvement [36]. Many research studies demonstrate the adaptive hormesis against exercise-induced oxidant production by the role of endogenous antioxidant system [35]. Interestingly, ROS responding to exercise can facilitate training adaptation by upregulation of endogenous antioxidants [42]. For example, exercise training adaptation can detoxify oxidative stress reactions via antioxidant enzymes such as catalase, glutathione reductase, and glutathione peroxidase [43]. It has been argued that endogenous and exogenous antioxidants can mitigate the effects of oxidative stress, therefore dietary antioxidants such as vitamin C, vitamin D, methionine, and cysteine are very popular for athletes for their performance recovery after heavy training season [42]. Nevertheless, 10 weeks of strenuous weight training in young adults who received a daily high dose of vitamin C and E did not increase gene expression of heat shock protein (HSP) and some endogenous antioxidants such as superoxide dismutases (SOD) and glutathione peroxidases (GPx) [44].

Effects of exercise on immune cells

Many studies reveal the effects of exercise on the immune system across both human and animal studies. Regular exercise at a moderate workload is helpful for boosting immunity in general people and those who are living with chronic diseases, however, prolonged exhaustive exercise such as training for competition in athletes can increase susceptibility to respiratory tract infection [45]. Innate immunity which incorporates adaptive immunity is a complex system that protects the body from viral infection and cancer. Each exercise bout can activate mucosal defenses, enhance the functions of immunoglobulins, antiinflammatory cytokines, and influx of CD8+ T cells, natural killer cells [4], and induce leukocytosis [46]. Likewise, the long-term effects of exercise training can prevent low-grade chronic inflammation, influenza and enhance the effectiveness of the influenza vaccine. Therefore, moderate exercise intensity and duration should be recommended for most to improve the robustness of their immune system.

T cells, Treg cells, B cells, NK cells, MDSC

Exercise has a dual effect on circulating lymphocyte numbers, initially increasing them during training, before inducing lymphocytopenia afterward [46,47]. It has been hypothesized that these changes to lymphocyte mobilization can enhance immunosurveillance [47]. It appears that T-lymphocytes are particularly impacted by exercise likely explained by a significant redistribution or migration of T cells between lymphoid tissues and organs. For example, the pathway of naïve T cells is to preferentially migrate to secondary lymphoid organs such as the lymph nodes, spleen, and the Peyer's patches through specialized blood vessels known as high endothelial venules, while antigen-experienced (memory) T-cells more commonly transit to bone marrow, skin and lamina propria [47]. The mechanism driving exercise-induced mobilization of lymphocytes from the vascular endothelium to circulation involves the elevation of catecholamines [48]. It has been shown that both CD4+ T helper cells and regulatory T cells (Treg) in the circulation increase during and immediately after exhaustive exercise. However, both T cell subsets decrease to a level lower than baseline 2-h postexercise [46]. Moderate aerobic exercise for 4-weeks in mice induces an immunosuppressive phenotype of CD4$^+$CD25$^+$Foxp3$^+$ Treg cells which suppress CD4$^+$CD25$^-$ T cell proliferation and Th2 cytokine levels, associated with inflammation. An exercise-mediated increase in the immunosuppressive activity of Treg cells may play a role in the exercise-associated mitigation of airway inflammation [49]. In contrast, highly intensive exercise training over a one-week period in athletes reduces CD8+ Treg mobilization and induces lymphocytopenia potentially impairing host immunosurveillance [50].

Antibody secreting B-cells increase during acute exhaustive exercise. An experimental study found that short bouts of heavy workload cycling increases immature (CD27−CD10+) B cell levels, but did not affect effector or memory cells [48]. There have been many studies showing the effect of exercise and acquired immune cells. This exercise-induced adaptive immune response likely stems from the secretion of IL-10 and creatine kinase by the musculoskeletal system, which activates T, B, and NK cells [51].

NK cells, an innate immune cell that expresses CD65 but not the lymphocyte marker CD3, are characterized by their high cytotoxic capacity. Acute physical activity has been reported to increase NK cell number immediately after exercise; however, this is followed by a decrease for up to 24 h. This mobilization of NK cells from the spleen to the circulation following exercise is the result of epinephrine-β adrenergic action [52].

Myeloid-derived suppressor cells (MDSCs) are immature myeloid cells that facilitate cancer growth and metastasis by suppressing both innate and adaptive immune responses [53]. In animal studies, the decrease of MDSCs in the spleen has been reported after aerobic exercise training which in turn may have anticancer effects [54,55].

Dendritic cells, macrophages, monocytes

Dendritic cells (DCs) take an important role in presenting small antigenic fragments to adaptive immune cells, priming a response against pathogens. Exercise has been shown to induce DC (plasmacytoid and myeloid) mobilization into the blood ha immediately after heavy exercise in young healthy adults, likely leading to improved immunosurveillance and adaptive response against viral infections and cancer [56].

Another key innate immune cell in the priming of adaptive responses which is benefited from exercise is monocytes. The intermediate (CD14^{++}CD16^{+}) and nonclassical (CD14^{Low}CD16^{++}) subsets are more mature pro-inflammatory monocytes that produce the inflammatory IL-1β and TNF-α in response to tissue damage [57]. Once monocytes enter a tissue they differentiate into macrophages, which can be categorized as either M1 pro-inflammatory or M2 antiinflammatory cells. Acute moderate aerobic exercise stimulates monocyte CCR2 expression and macrophage polarization response [57]. Moderate exercise also increases macrophage phagocytic capacity, and exercise-induced cortisol can stimulate their secretion of the immunomodulatory cytokines IL-6 and IL-10. Macrophage recruitment and polarization are likely associated with muscle stress and overload, therefore macrophage exercise effects are likely limited physical activity involving muscle contraction and overload [58]. Exercise training has also been shown to induce a switch from M1 pro-inflammatory to M2 antiinflammatory polarization [28]. A study in animals showed that although the expression of CD163 mRNA, a marker of M2 antiinflammatory macrophages, was decreased in high fat diet-induced obese mice, exercise training increased its expression in fat tissue, suggesting a further antiinflammatory effect. On the contrary, exercise training can attenuate the expression of CD11c mRNA, an M1 pro-inflammatory macrophage in these obese mice [28]. Exercise training can also hinder the infiltration of M1 macrophages into fat tissue, mitigating inflammatoryresponse [28].

Neutrophils, eosinophils, mast cells, and platelet-activating factors

Neutrophils have an important role in frontline defense against infectious microorganisms. Acute intensive aerobic exercise in young healthy adults has been shown to suppress neutrophil activity including, impaired chemotaxis and polarization [59]. In addition, although neutrophil numbers appear unaltered after endurance exercise in middle-aged healthy males, a significant delay of neutrophil apoptosis was shown. This delay of apoptosis may be accompanied by granulocyte-colony-stimulating factor (G-CSF) and some exercise-associated physiological changes such as decreased glutathione levels and altered intracellular calcium signaling [60].

Eosinophils are highly associated with the inflammatory disease; for example, the accumulation of eosinophils in the lung results in chronic inflammatory eosinophilic pneumonia [61]. Exercise-induced muscle damage increases eosinophil number 24 h after strenuous exercise in older adults [62]. More recently, a case was reported of a young female presenting with eosinophilic fasciitis, painful limb sclerosis with inflammation of the deep fascia, and symmetric skin stiffness including tissue and blood eosinophilia, following a bout of strenuous exercise [63]. This has also been shown in postpubertal strenuous exercise, where absolute eosinophil counts are increased, potentially preceding the onset of respiratory tract inflammation such as in exercise-induced asthma among adolescents [64].

Mast cells are innate immune cells that play an important function in allergic reactions via the release of pro-inflammatory prostaglandins, histamine, and proteases such as tryptase, chymase [65]. Acute exercise has been shown to induce anaphylaxis with associated tachycardia, bronchospasm, and urticaria due to mast cell degranulation and release of histamine, tryptase, and prostaglandins [66]. Interstitial histamines in muscular tissue have also been shown to increase after exercise due to the induction of the enzyme histidine decarboxylase and/or mast cell degranulation. This anaphylactoid reaction may enhance muscle recovery and influence skeletomuscular blood flow (vasodilation) during exercise [67]. Additionally, postmarathon-induced hypotension is related to increased mast cell degranulation and tryptase secretion which mediate vasodilation [68].

Effects of exercise on cytokines and chemokines

Cytokines are polypeptides secreted by a number of immune cells, [69] including helper T cells and macrophages [70]. Cytokines are classified further as chemokines if they demonstrate chemotactic activities [70]. Others are classified by their original biological structure such as tumor necrosis factor (TNF) [71]. Cytokine production is a local response to physical or chemical stress such as inflammation or infection [72], which is initially a protective mechanism against tissue injury [73]. Importantly, strenuous exercise has a significant effect on cytokine secretion as a result of muscle overload and microtrauma, which underpins many of the physiological effects of physical activity [72].

Generally, cytokines are classified as either pro-inflammatory or antiinflammatory. Pro-inflammatory cytokines (IL-1, TNF, IL-8, MCP-1) promote inflammation with manifestations such as fever, tissue damage, infection, and sepsis. Conversely, antiinflammatory cytokines (IL-4, IL-10, IL-13) are known to suppress pro-inflammatory cytokine functions, reducing inflammation, promoting tissue healing, and alleviating certain chronic diseases [71].

Exercise-induced inflammation shows similar changes to those observed in pathological tissue injury and infections [74]. A well-regulated inflammatory response and balance between pro-and antiinflammatory cytokines are important for muscle repair following exercise [75]. Pro-and antiinflammatory cytokines are altered by physical activity or exercise if the activity is vigorous enough to induce inflammation [74]. Strenuous and prolonged bouts of exercise can induce temporary respiratory tract infection due to suppressed immune function which involves an increase in IL-1 receptor, alongside increased IL-6 and IL-10 [2]. On the contrary, some studies suggest that short-term exercise induces an acute stress response which can enhance immune protection against pathogens (both innate and adaptive immune functions) and cancer [76]. The heterogeneity of inflammatory outcomes and exercise training, however, is also influenced by the variety of exercise modalities, duration, and population groups [77]. Therefore, the discussion of the "advantage" and "disadvantage" coming from alteration in cytokines following exercise requires consideration of exercise types, intensity, and duration as well as demographics.

Inflammatory response following a single bout of exercise

Following a bout of heavy and prolonged exercise, there is a clear increase in pro-inflammatory cytokines [78], antiinflammatory cytokines [79], natural cytokine inhibitors, and chemokine production [72]. The increase in pro-inflammatory markers such as TNF-α, IL-6, following heavy bouts of exercise makes them potentially unsuitable for people living with chronic diseases such as heart failure due to deleterious effects [78]. In healthy adult males, incrementally increasing cycling load to exhaustion increases endotoxin, lactate, and TNF-α in the peripheral blood immediately after exercise, with antiinflammatory IL-10, increasing an hour following [79]. It has also been shown that increases in IL-10 after strenuous exercise are associated with aerobic fitness [79]. Many cytokines (both pro-and antiinflammatory) such as IL-1 receptor antagonist (IL-1ra), IL-6, IL-8, and IL-10 are secreted into the blood after endurance exercise such as marathon running, likely due to neutrophilia second to muscle damage and immune suppression [73]. IL-6 is known to be significantly increased in parallel with exercise intensity, duration, and muscle recruitment. Additionally, IL-6 mRNA expression is upregulated in contracting muscle during exercise [80]. TheIL-6 response to endurance exercise results from increased heat stress and increased cellular metabolism, therefore IL-6 could be an indicator of physical exercise-induced muscle damage [73]. IL-6 in the circulation secreted by contracting muscle cells can later induce the secretion of antiinflammatory cytokines (IL-1ra, IL-10), and inhibit TNF-α production [80]. After one bout of endurance exercise (running) antiinflammatory cytokines (IL-1ra, IL-10, IL-15) peak at 1–2 h [81]. For chemokines responding to acute exercise, not only does prolonged strenuous exercise induces IL-8 and MCP1 secretion, short bouts of strenuous exercise can also increase both these cytokines in the circulation [73].

It is well described that a single short bout of heavy and prolonged exhaustive exercise can stimulate pro-inflammatory cytokines and chemokine inducing muscular and systemic inflammation following exercise. However, there is a compensatory response with late production of antiinflammatory cytokines which promotes tissue healing following exercise [73].

Inflammatory cytokine response to endurance exercise training

It is well-known that regular moderate physical activity or exercise can mitigate cellular aging processes and is likely to reduce pro-inflammatory markers such as IL-6, CRP, TNF-α [82]. Sedentary men in middle-age tend to have higher serum CRP levels which are one indicator of cardiovascular disease risk [82]. Following 16 weeks of aerobic and resistance exercise at moderate intensity, three times a week this population showed reduced IL-6, CRP, TNF-α, and serum triglycerides [82]. Two weeks of heavy cross-country skiing also immediately increased CRP and TNF-α in sedentary men, however, CRP was significantly reduced (less than baseline) during recovery [83]. Meanwhile, a CRP reduction was detected 12 weeks postaerobic and resistance training in sedentary young and older men which may help mitigate the risk of cardiovascular disease [84]. A recent systematic review and meta-analysis found that exercise training can reduce IL-6, TNF-α, and CRP, whereas it increased adiponectin in postmenopausal women [85]. An randomized controlled trial found that 12 weeks of moderate aerobic training (30 min walking on the treadmill at 60% heart rate reserve), resistance training (30 min, 4 sets of 8–12 repetition at 10 RM intensity), and combination training, all reduced TNF-α in overweight men and women [86]. Short durations of moderate exercise training however, did not affect inflammatory biomarkers. Both moderate endurance training (65% VO_2 max cycling on ergometer) and sprint interval training (Wingate test) in younger adults also did not alter serum IL-6, IL-10, and CRP levels 2 weeks after training [87].

Aerobic training (20 min cycling or jogging) combined with 30 min of resistance training three-times per week for 4-months induced antiinflammatory effects in stable chronic heart failure patients. The exercise program reduced pro-inflammatory cytokines including the soluble TNF receptors (sTNFR1 and sTNFR2,) but did not alter IL-6 and TNF-α [88]. Middle-aged adults of both sexes with obesity (body mass index 31–36 kg/m^2) and type 2 diabetes mellitus show a reduction in insulin resistance index, leptin, IL-2, IL-4, IL-6, and TNF-α after 3 months of mild (at 55%–65% maximum heart rate) and moderate (at 65%–75% maximum heart rate) treadmill-based aerobic exercise training [89].

In women living with breast cancer following primary treatment, a randomized controlled study showed antiinflammatory protection against breast cancer relapse through the reduction of TNF-α secreted by NK cells and T cells at 16 weeks after supervised machine-based weight training sessions [90].

In summary, long durations of aerobic exercise, resistance training, and/or a combination of these at mild to moderate intensity, provides beneficial effects to antiinflammatory processes due to the decrease in pro-inflammatory (IL-6, TNF-α, CRP) mediators in both sedentary healthy people and those living with chronic illness. Enhancement of antiinflammatory functions and a reduction in pro-inflammatory cytokines are good indicators of reducing the risk of many chronic diseases. Additionally, the balance of pro-inflammatory and antiinflammatory cytokines following moderate aerobic exercise can reduce chronic inflammation and improve clinical symptoms in patients with chronic diseases such as cardiovascular diseases, type 2 diabetes, and cancer.

However, some studies did not show changes in inflammatory cytokines or biomarkers after endurance exercise training in healthy adults. Therefore, it is very challenging to identify the exercise program-related factors affecting inflammatory cytokines. More specifically, suitable exercise programs for specific groups of the population which can alter inflammatory cytokines or biomarkers are required.

The good, the bad, and the ugly

The inconsistency of immune and inflammatory responses to exercise depends on the intensity and duration of physical activity. It also depends on the heterogeneity of the population groups [77]. Regular exercise at moderate intensity generally strengthens the immune system, whereas prolonged exercise with vigorous levels can suppress immune functions [91].

Inflammatory cytokines responding to exercise are likely to have the same effect as the phenomenon of exercise-induced immunity alteration. One bout of strenuous and prolonged exercise such as marathon running, could increase physical stress, muscle damage and stimulate the inflammatory process (increasing pro-inflammatory cytokines and chemokines); as a consequence, it may have negative effects on tissue and body systems [72]. On the other hand, regular long-term aerobic and/or weight training at mild to moderate levels can enhance antiinflammatory responses which can reduce the chronic inflammatory process in both healthy people and chronic illness patients [86,88].

Different exercise regimes and intensities

Studies in exercise immunology have found the positive effects of a single and multiple bouts of exercise training in both aerobic and resistant training modes on immune biomarkers such as leukocytes, T cells, and B cells [92]. However, vigorous training in high-performance athletes exhibits adverse effects on the immune system [93].

An early study reported the reduction in upper respiratory tract infection symptoms after 15 weeks of aerobic training (brisk walking) at moderate intensity (60% of heart rate reserve) five times per week among sedentary female adults, possibly due to the increase in NK cell activity [94]. Further, an animal study demonstrated that an increase in IgA levels in submandibular glands in Wistar rats after voluntary exercise for 3 weeks which has been implied for strengthening immunity against upper respiratory tract infections following exercise [95]. On the other hand, extreme exercise intensity mitigates immune function. IgA in saliva significantly decreased after one heavy bout of exercise such as marathon running, with many athletes reporting URI symptoms up to 2 weeks after a marathon [96]. This topic discusses the heterogeneity of immune response to exercise regimes as evidenced by empirical research studies.

Single bout exercise and immunity

A single bout of vigorous exercise can place stress on skeletal muscle tissue by inducing immediate pro-inflammatory responses which are regulated by both innate (neutrophils and macrophages) and adaptive immune cells (T cells) [97]. Immune biomarkers such as leukocytes and macrophage infiltration into muscle cells are dependent on the intensity of exercise. Approximately 2 h following exercise, the tissue repair and antiinflammatory phase of postexercise recovery begins, involving the secretion of the antiinflammatory cytokine, IL-10, and stimulation of M2 macrophages and regulatory T cells [97].

The effect of a single bout exercise on leukocytes is dependent on the intensity and duration. A randomized crossover trial reported a higher leukocyte count after heavy aerobic exercise (running on treadmill at 80% VO_2 peak) when compared to low aerobic exercise (walking on treadmill at 40% VO_2 peak) among healthy young men [98]. Prolonged aerobic exercise has been reported for an increase of leukocyte counts up to fivefold. Even a few minutes of exercise can leukocytes increase by twofold in the bloodstream [91]. In contrast, leukocytes in the circulation and its functions are significantly reduced following prolonged heavy exercise (more than 1.5 h) because of increased free radicals [2]. The alteration in some substances in plasma such as inflammatory cytokines is reported after vigorous exercise which can affect leukocyte function [2].

As mentioned above, a single bout of exercise can stimulate innate immune cells (neutrophils and lymphocytes increase up to fivefold after acute exercise). Additionally, some studies report the functionality of neutrophils increase (phagocytosis) immediately after exercise [91]. NK cells, especially those of the $CD3^-CD16^+CD56^+$ phenotype, are the most sensitive innate immune cells to a single bout exercise and this response is associated with immune surveillance [99]. NK cells in the circulation seem to increase by a short single bout but not in prolonged single bout exercise [99]. Aerobic exercise for 30 min increases the number of NK cells in the circulation due to elevated heart rate and plasma catecholamines [100]. Increasing NK cells after exercise may also be caused by stress signals from muscle injury [101]. In addition, exercise-induced NK cells activation can facilitate the mechanism of cancer protection by the immune system [100]. Humoral factors of the innate mucosal immune system such as lysosome and lactoferrin which have prophylactic roles and antimicrobial functions have been reported to be elevated during and after exercise [102]. Single bouts of exhaustive exercise by running on a treadmill have been found to increase the number of monocyte-derived DC 4-fold, which in turn facilitates T cell activation [103]. Nevertheless, prolonged and heavy single bout exercise such as 2 h heavy cycling (at 90% of ventilatory threshold) has been reported to suppress NK cell counts, neutrophil phagocytic activities and total lymphocyte count at 2–8 h following exercise in young adult cyclists [104]. This may anticipate susceptibility to upper respiratory infections following high intensity exercise [104].

Lymphocytosis especially mobilization of T cells (CD3 + CD4 + and CD3 + CD8 +) is detected during and after 1-h of a single bout of exhaustive cycling in well-trained male adults. Increased T cells are noted during and immediately after exercise which returns to baseline levels at 1-h recovery [105]. In this regard, mobilization of T cells may be affected by sympathetic action influencing blood circulation and shear pressure on blood vessel walls [106]. Some studies have shown the activation in T cells by mitogen or antigen stimulation as evidenced by expression of CD25, CD69, CD45RO, CD45RA, and the HLA-DR antigen after acute exercise [93]. There may be another immune response leading to lymphocytosis during acute exercise such as cytokine releasing [93]. However, the suppression of lymphocytes has been observed if exercise duration is longer than 1 h [107]. The decrease of T cells after acute exercise is highly associated with the decrease in Th1 cells [93] and CD3 + IFN-gamma + cells [105]. For B cells or humoral immune functions, the concentration of antibodies in serum or mucosa has not been shown to change after a bout exercise. However, the concentration of IgM response to mitogen stimulation has been found for increasing after acute exercise [93].

Exercise training and immunity

It is evident that long periods of exercise training can alter immune responses in both athletes and generally active people [108], however, the alteration is based on the intensity of exercise training [108]. Well-trained endurance athletes are likely to have lower numbers of resting leukocyte, neutrophil, and monocyte than sedentary people [91]. T-helper cells (CD4 +) and T-suppressor cells (CD8 +) are slightly decreased after intensified training in male athletes [109]. The most focusing of studies is concerned with excessive training in athletes for competition which in turn declines the function of immunity associated with susceptibility of upper airway inflammatory [110]. Likewise, studies investigating regular exercise training on immunity improvement in general adults are required for identifying the mechanism of immunological response [91].

Moderate-intensity exercise training seems to enhance innate immune functions. The improvement of monocytes (CD14+/CD16+), and phagocytosis of monocytes and neutrophils, including the oxidative killing of bacteria were found after 10 weeks of moderate (70% of maximal heart rate) and high-intensity (more than 90% of maximal heart rate) exercise in adults [111]. High-intensity interval (80%–90% VO_2 reserve) walking has shown similar effects in older adults with rheumatoid arthritis, improving the neutrophil number, monocyte phagocytosis and cytokine concentrations, as well as joint swelling and ESR after 10 weeks [112]. In contrast, long periods of high-intensity exercise training significantly reduce the numbers of leukocytes and neutrophils in young and elderly athletes [113]. Moreover, 3 weeks of moderate-intensity treadmill running with mice increased antimicrobial mediators such as pro-inflammatory cytokines and lipopolysaccharide-stimulated nitric oxide which are secreted by macrophages [114].

A previous study reported higher transcriptional changes in blood leucocytes in young swimming athletes during their regular training period compared to sedentary volunteers [115]. The alteration of leucocyte transcription can be anticipated for the improvement of innate immune functions in well-trained athletes [115]. On the other hand, heavy training may suppress innate immunity in athletes. Elite rowers have a 50% lower concentration of salivary lactoferrin and lysozyme in the resting stage than sedentary people after 5 months of heavy training season [116]. This condition in elite rowers can be anticipated for vulnerability to upper airway infection [116]. A longitudinal observation in sedentary adults training for a few weeks or months of physical activity noted no changes in T cell and B cell activity whereas these adaptive immune cells are likely to be altered by intense training in fit athletes [93]. Chronic exercise training in male cyclists leads to the reduction of Th1 cells in the circulation responding to a single bout of exercise [105]. The results suggest that sporters who perform heavy load training illustrate the decline in T cell-mediated immune functions which is likely correlated to elevated stress hormone levels such as cortisol [93].

Active elderly people who regularly perform prolonged exercise demonstrate lymphocytosis stimulated with anti-CD 3 monoclonal antibody [117]. In addition, physical activity in aging can increase the level of CD4+ T cells [117]. Vigorous exercise in athletes with a long training history is associated with higher differentiation of the CD4+ T cells [113].

The positive benefits of exercise

The overall mortality rate has been reported to decrease after long periods of regular exercise in humans [118]. It is generally accepted that exercise provides positive, immune-boosting benefits in the immune system for the general population if they perform exercise or physical activity regularly at moderate intensity [119]. Strengthening of immunity following exercise is thought to be due to antiinflammatory responses which prevent noncommunicable chronic diseases such as type 2 diabetes, cardiovascular diseases, neurodegeneration, and cancers [120]. Exercise training stimulates IL-6 secretion from skeletal muscle and it further induces antiinflammatory cytokines such as IL-1ra and IL-10 [121]. IL-6 also disturbs the production of TNF-α [121]. As a result, exercise can mitigate the process of chronic low-grade inflammation [121]. Additionally, exercise prevents chronic diseases possibly due to reduced expression of TLR on monocytes as well as an increase in PGC-1α gene expression in skeletal muscle which can reduce chronic pro-inflammatory processes [122].

Exercise is well-known as adjuvant therapy for people living with cancer through the regulation of immunity and inflammatory process [123]. A previous systematic review confirmed that NK cytotoxic activity and lymphocyte proliferation increased following exercise training among patients diagnosed with heterogeneous cancers [124]. A previous systematic review assured that exercise training can improve quality of life but did not alter immune functions in those with HIV infection [125]. Interestingly, the exercise demonstrates protective immune functions against viral respiratory tract infection; for example, after H1N1 influenza A virus infection in 2009, there was evidence that physical activity before the pandemic can reduce the incidence of influenza and diminish the duration and severity by self-reported infected people [126].

The potential negative effects of exercise: Too much of a good thing?

Long duration bouts of heavy exercise and strenuous training for the competition can induce immunosuppression in elite athletes [11]. Approximately 5%–7% of well-trained athletes reported URI symptoms associated with exhaustion [45]. Infection following high-intensity training in athletes can be intensified by precipitating factors such as close exposure to pathogens while traveling, insufficient sleeping, malnutrition, and stress [127]. The environment can also affect exercise immunology, with evidence showing that both cold and hot conditions can alter immune responses to exercise [118].

Even though some studies have reported heavy exercise training-induced immunosuppression, there are no studies that have reported associations between exercise and confirmed severe illness incidence in athletes [2]. However, a research study assures that exercise training load is associated with the frequency of illness reported in athletes [128]. Interestingly, exercise-induced pro-inflammatory can be attenuated by antioxidant supplementation and increased carbohydrate consumption in athletes during their training period [2]. Additionally, sufficient fluid intake before exercise can attenuate inflammatory responding to heavy exercise training [73]. So it is necessary to bring an issue of adaptive training modality for over-training syndrome prevention in highly-trained athletes [129].

Some ugly effects of exercise on the immune system and inflammatory cytokines

Immune suppression often occurs after prolonged and continued heavy exercise without sufficient food consumption, which may be a precursor for infection among elite athletes [130]. Some athletes can suffer from respiratory tract infection after periods of heavy training and competition, as a result of temporary immunosuppression [131]. The infection phenomena can be explained by "open window" theory, which suggests that repetition of extreme exercise with incomplete immune recovery could bring a potential risk of infections in elite athletes [131].

Exercise-induced inflammation may be harmful to patients with chronic disease. As it is evident that chronic heart failure is associated with some pro-inflammatory cytokines such as TNF-α and IL-6, people are warned that too much exercise, which can enhance the said cytokine level, maybe a precipitating factor for exacerbated heart failure [78]. Prolonged endurance exercise during the menstrual period in sedentary females can induce systemic inflammation, as evidenced by a significant increase of IL-6 following 60 min of cycling at moderate intensity [132]. The challenge of exercise-induced inflammation remains to be explored in elite athletes and people living with chronic diseases. Likewise, potential countermeasures against systemic inflammatory process following exercise need to be more investigated [6].

Clinical trial summaries and exercise recommendations for boosting the immune system

A clinical trial points out that the patients with chronic insomnia (sleeping disturbance) can reduce the ratio of CD4+ and CD8+, which indicates antiinflammatory response and increase their sleeping time after a four-month exercise training regimen at ventilatory threshold intensity continuously for 50 min, three times per week [133]. A randomized controlled study in a wide range of adults living with human immunodeficiency virus/acquired immune deficiency syndrome (HIV/AIDS) reveals an increase of CD4+ T cell count after 8 weeks of moderate endurance training (walking on treadmill at 60%–79% of maximal heart rate, 45–60 min, three times per week) [134]. Improvement of CD4+ count is a positive indicator of antiviral therapy in people with HIV/AIDS [135]. On the other hand, a pre-post experimental design implementing 16-weeks of supervised aerobic and resistance exercise at moderate intensity 60 min per session, twice a week in middle-aged adults infected with human immunodeficiency virus (HIV) reported that the program can increase aerobic fitness but and quality of life (QoL), but no change in CD4+ cell counts [136]. Furthermore, a randomized-controlled trial implementing 15 weeks of stationary cycling at moderate intensity (70%–75% VO_2 peak), 15–35 min, three times per week has demonstrated an increase in innate immune functions by enhancing NK cell cytotoxic activity in female breast cancer survivors which can be beneficial for cancer recurrence prevention [137] (Fig. 1).

Performing regular exercise at a moderate intensity ranging 30–60 min per session and three times weekly appears to be beneficial for boosting immune function in sedentary individuals and those people who have clinical health problems. Regarding the consistency of results in empirical studies on exercise immunology, regular exercise or physical activity at moderate intensity for 150 min per week is recommended to strengthen immune function [126]. The immune response to exercise in sedentary people, athletes, and people living with chronic illness is presented in Table 1.

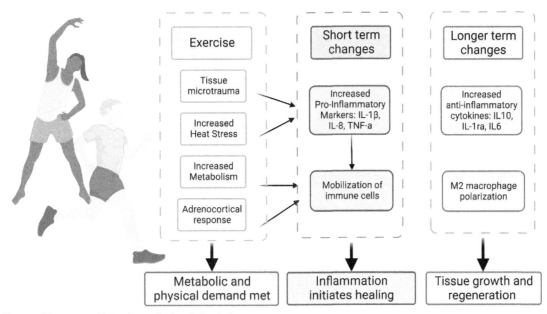

FIG. 1 Short- and long-term effects of exercise in relation to immune responses.

TABLE 1 The effects of exercise on immune functions in sedentary people, athletes, and some people with chronic diseases.

Study	Study design	Exercise modality	Participants	Immune response
Acute exercise				
Neves et al. [98]	RCT with crossover design	Running on treadmill at 80% of VO_2 peak for approximately 30 min	Sedentary young men	Increase in leukocyte count
Kakanis et al. [104]	Pre-postexperimental design	Two hours heavy cycling at 90% of ventilatory threshold	Young adult cyclists	Suppression of NK cells, neutrophil phagolytic activity, and leukocyte count
Lancaster et al. [105]	Pre-postexperimental design	Exhaustive cycling	Young male cyclists (moderate to well-trained)	Lymphocytosis especially mobilization of T cells (CD3+CD4+ andCD3+CD8+) Reduction of type 1 helper T cells (CD3+ IFN-Y) in blood circulation
Exercise training				
Bartlett et al. [111]	RCT	10 weeks of moderate (70% of maximal heart rate) and high intensity (more than 90% of maximal heart rate)	Sedentary adults	Increase of monocytes (percentage of CD14+/CD16+) and phagocytosis of monocytes and neutrophils
Bartlett et al. [117]	RCT pilot study	10 weeks walking at 80%–90% VO_2 reserve (30 min/three times per week)	Older adults with rheumatoid arthritis	Improvement of neutrophils monocyte antibacterial phagocytosis and cytokine concentration
Liu et al. [115]	A cohort study	Regular swim training	Young swimming athletes	Higher transcriptional changes in gene level of blood leucocytes
Nieman et al. [94]	RCT	15 weeks brisk walking a moderate intensity (60% of heart rate reserve) 45 min/five times per week	Mildly obese women	Reduction in upper respiratory tract infection (URI) symptoms
West et al. [116]	A prospective cohort study	5 months of heavy training	Elite rowers	50% lower concentration of salivary lactoferrin and lysozyme in the resting stage
Clinical trial				
Passos et al. [133]	Pre-postexperimental design	4 months exercise training at individual ventilatory threshold intensity continuously 50 min, three times per week	The patients with chronic insomnia in middle age	Reduce the ratio of CD4+ and CD8+ which indicates an antiinflammatory response
Ezema et al. [134]	RCT	8 weeks moderate walking on treadmill a 60%–79% of maximal heart rate, 45–60 min, three times per week)	Wider age of adults with HIV/AIDS	Increase of CD4+ T cell count
Fairey et al. [137]	RCT	15 weeks cycling on ergometer 70%–75% VO_2 peak, 15–35 min, three times per week	Oder female breast cancer survivors	Enhancing NK cell cytotoxic activity

Abbreviations: *CD*: a cluster of differentiation; *HIV/AIDS*: human immunodeficiency virus/acquired immune deficiency syndrome; *NK cells*: natural killer cells; *RCT*: randomized controlled trial; *VO2*: oxygen consumption.

Conclusion and Future Prospects

Immune and inflammatory responses to exercise depend on the intensity and duration of exercise activity. Regular exercise training at moderate intensity has been recommended for boosting immunity and antiinflammatory responses in the general population and those living with chronic disease. Evidence reveals that regular aerobic exercise and/or resistance training, which is not too strenuous enhances innate and adaptive immune functions. Likewise, this exercise training modality induces an antiinflammatory effect by mitigating pro-inflammatory biomarkers, while increasing antiinflammatory biomarkers.

A well-regulated inflammatory response is advantageous for muscle recovery after heavy exercise. Although pro-inflammatory cytokines such as IL-6 are often detected immediately after a single heavy bout of exercise, antiinflammatory cytokines such as IL-10 are significantly increased after 1 h. Likewise, long-term aerobic exercise can change the phenotype of macrophages from M1 pro-inflammatory to M2 antiinflammatory.

On the contrary, deleterious effects of exercise on immunity and inflammation are highlighted in a strenuous, exhaustive, or prolonged exercise training regime. The susceptibility of a viral infection such as upper respiratory tract infection is often detected among those during their heavy training and competition season. The increase of pro-inflammatory cytokines (IL-6, IL-8, TNF-α) is likely to occur after acute heavy exercise or after a short duration (a couple of weeks) in intense sports and exercise. Such pro-inflammatory cytokines can induce chronic inflammatory processes which may further develop chronic health problems; for example, cardiovascular disease, type 2 diabetes, and cancers. Therefore, coaches and athletes alike should pay close attention to adverse effects on immune and inflammatory during and after heavy periods of training and competition in athletes.

Currently, there is strong evidence supporting the recommendation of moderate endurance exercise for boosting immunity and balancing inflammatory responses in the general population and in people living with chronic illnesses. However, heavy and prolonged training in elite athletes can cause potential issues. Additionally, more studies regarding the mechanisms of immune responses to exercise in people with chronic inflammatory diseases need to be conducted. Exercise and health coaches should be aware of adverse effects on immunity and inflammatory responses to exercise in well-trained athletes and people with chronic inflammatory diseases. Some countermeasures to mitigate a deleterious response to imbalanced exercise in such people should be implemented such as antioxidant supplementation, sufficient nutrition, and fluid intake as well as appropriate climatic conditions, where possible.

References

[1] Brown WM, et al. A systematic review of the acute effects of exercise on immune and inflammatory indices in untrained adults. Sports Med-Open 2015;1(1):1–10.
[2] Gleeson M. Immune function in sport and exercise. J Appl Physiol 2007;103(2):693–9.
[3] Martin SA, Pence BD, Woods JA. Exercise and respiratory tract viral infections. Exerc Sport Sci Rev 2009;37(4):157.
[4] Ranasinghe C, Ozemek C, Arena R. Exercise and well-being during COVID 19—time to boost your immunity. Expert Rev Anti Infect Ther 2020;18(12):1195–200.
[5] Shek PN, Shephard RJ. Physical exercise as a human model of limited inflammatory response. Can J Physiol Pharmacol 1998;76(5):589–97.
[6] Suzuki K, et al. Characterization and modulation of systemic inflammatory response to exhaustive exercise in relation to oxidative stress. Antioxidants 2020;9(5):401.
[7] Toumi H, Best T. The inflammatory response: friend or enemy for muscle injury? Br J Sports Med 2003;37(4):284–6.
[8] Rosa JC, et al. Exhaustive exercise increases inflammatory response via toll like receptor-4 and NF-κBp65 pathway in rat adipose tissue. J Cell Physiol 2011;226(6):1604–7.
[9] Santos MHH, et al. Previous exercise training increases levels of PPAR-α in long-term post-myocardial infarction in rats, which is correlated with better inflammatory response. Clinics 2016;71(3):163–8.
[10] You T, et al. Effects of exercise training on chronic inflammation in obesity. Sports Med 2013;43(4):243–56.
[11] Gleeson M, Bishop N, Walsh N. Exercise Immunology. Routledge; 2013.
[12] Apostolopoulos V, et al. Physical and immunological aspects of exercise in chronic diseases. Immunotherapy 2014;6(10):1145–57.
[13] Metsios GS, Moe RH, Kitas GD. Exercise and inflammation. Best Pract Res Clin Rheumatol 2020;34, 101504.
[14] Mee-Inta O, Zhao Z-W, Kuo Y-M. Physical exercise inhibits inflammation and microglial activation. Cell 2019;8(7):691.
[15] Church TS, et al. Exercise without weight loss does not reduce C-reactive protein: the INFLAME study. Med Sci Sports Exerc 2010;42(4):708.
[16] Petersen AMW, Pedersen BK. The antiinflammatory effect of exercise. J Appl Physiol 2005;98(4):1154–62.
[17] Ford ES. Does exercise reduce inflammation? Physical activity and C-reactive protein among US adults. Epidemiology 2002;561–8.
[18] Mendham AE, et al. Effects of mode and intensity on the acute exercise-induced IL-6 and CRP responses in a sedentary, overweight population. Eur J Appl Physiol 2011;111(6):1035–45.
[19] Czarkowska-Paczek B, et al. Lack of relationship between interleukin-6 and CRP levels in healthy male athletes. Immunol Lett 2005;99(1):136–40.

[20] Pieters M, Wolberg AS. Fibrinogen and fibrin: an illustrated review. Res Pract Thromb Haemost 2019;3(2):161–72.

[21] Grechin C, et al. Inflammatory marker alteration in response to systemic therapies in psoriasis. Exp Ther Med 2020;20(1):42–6.

[22] Gomez-Marcos MA, et al. Relationship between physical activity and plasma fibrinogen concentrations in adults without chronic diseases. PLoS One 2014;9(2), e87954.

[23] Bizheh N, Jaafari M. Effects of regular aerobic exercise on cardiorespiratory fitness and levels of fibrinogen, fibrin D-dimer and uric acid in healthy and inactive middle aged men. J Shahrekord Univ Med Sci 2012;14(3).

[24] El-Sayed MS. Fibrinogen levels and exercise. Sports Med 1996;21(6):402–8.

[25] Sackett JR, Farrell DP, Nagelkirk PR. Hemostatic adaptations to high intensity interval training in healthy adult men. Int J Sports Med 2020;41 (12):867–72.

[26] Favere K, et al. A systematic literature review on the effects of exercise on human toll-like receptor expression. Exerc Immunol Rev 2021;27:84–124.

[27] Chen C-W, et al. Long-term aerobic exercise training-induced antiinflammatory response and mechanisms: focusing on the toll-like receptor 4 signaling pathway. Chin J Physiol 2020;63(6):250.

[28] Kawanishi N, et al. Exercise training inhibits inflammation in adipose tissue via both suppression of macrophage infiltration and acceleration of phenotypic switching from M1 to M2 macrophages in high-fat-diet-induced obese mice. Exerc Immunol Rev 2010;16.

[29] Rada I, et al. Toll like receptor expression induced by exercise in obesity and metabolic syndrome: a systematic review. Exerc Immunol Rev 2018;24.

[30] Cavalcante PAM, et al. Aerobic but not resistance exercise can induce inflammatory pathways via toll-like 2 and 4: a systematic review. Sports Med-Open 2017;3(1):1–18.

[31] Lichtenberg D, Pinchuk I. Oxidative stress, the term and the concept. Biochem Biophys Res Commun 2015;461(3):441–4.

[32] Fisher-Wellman K, Bloomer RJ. Acute exercise and oxidative stress: a 30 year history. Dyn Med 2009;8(1):1–25.

[33] Urso ML, Clarkson PM. Oxidative stress, exercise, and antioxidant supplementation. Toxicology 2003;189(1–2):41–54.

[34] Powers SK, et al. Exercise-induced oxidative stress: friend or foe? J Sport Health Sci 2020;9(5):415–25.

[35] Powers SK, Radak Z, Ji LL. Exercise-induced oxidative stress: past, present and future. J Physiol 2016;594(18):5081–92.

[36] Magherini F, et al. Oxidative stress in exercise training: the involvement of inflammation and peripheral signals. Free Radic Res 2019;53 (11–12):1155–65.

[37] Radak Z, et al. Exercise, oxidants, and antioxidants change the shape of the bell-shaped hormesis curve. Redox Biol 2017;12:285–90.

[38] Powers SK, Nelson WB, Hudson MB. Exercise-induced oxidative stress in humans: cause and consequences. Free Radic Biol Med 2011;51(5):942–50.

[39] Tsikas D. Assessment of lipid peroxidation by measuring malondialdehyde (MDA) and relatives in biological samples: analytical and biological challenges. Anal Biochem 2017;524:13–30.

[40] Spirlandeli A, Deminice R, Jordao A. Plasma malondialdehyde as biomarker of lipid peroxidation: effects of acute exercise. Int J Sports Med 2014;35(01):14–8.

[41] Sastre J, et al. Exhaustive physical exercise causes oxidation of glutathione status in blood: prevention by antioxidant administration. Am J Physiol Regul Integr Comp Physiol 1992;263(5):R992–5.

[42] McLeay Y, et al. Dietary thiols in exercise: oxidative stress defence, exercise performance, and adaptation. J Int Soc Sports Nutr 2017;14(1):1–8.

[43] Martínez-Noguera FJ, et al. Differences between professional and amateur cyclists in endogenous antioxidant system profile. Antioxidants 2021;10(2):282.

[44] Cumming KT, et al. Vitamin C and E supplementation does not affect heat shock proteins or endogenous antioxidants in trained skeletal muscles during 12 weeks of strength training. BMC Nutr 2017;3(1):1–8.

[45] Simpson RJ, et al. Can exercise affect immune function to increase susceptibility to infection? Exerc Immunol Rev 2020;26:8–22.

[46] Gleeson M, Bishop NC. The T cell and NK cell immune response to exercise. Ann Transplant 2005;10(4):44.

[47] Kruger K, Mooren F. T cell homing and exercise. Exerc Immunol Rev 2007;13:37–54.

[48] Turner JE, et al. Exercise-induced B cell mobilisation: preliminary evidence for an influx of immature cells into the bloodstream. Physiol Behav 2016;164:376–82.

[49] Lowder T, et al. Repeated bouts of aerobic exercise enhance regulatory T cell responses in a murine asthma model. Brain Behav Immun 2010;24(1):153–9.

[50] Witard OC, et al. High-intensity training reduces CD8+ T-cell redistribution in response to exercise. Med Sci Sports Exerc 2012;44(9):1689–97.

[51] Jee Y-S. Acquired immunity and moderate physical exercise: 5th series of scientific evidence. J Exerc Rehabil 2021;17(1):2.

[52] Zimmer P, et al. Exercise induced alterations in NK-cell cytotoxicity-methodological issues and future perspectives. Exerc Immunol Rev 2017;23.

[53] Garritson J, et al. Exercise reduces proportions of tumor resident myeloid-derived suppressor cells. FASEB J 2019;33(S1):lb13.

[54] Garritson J. Effects of Exercise on Myeloid-Derived Suppressor Cell-Related Tumor Progression and Metastasis. University of Northern Colorado; 2020.

[55] Hirsch AL, et al. Voluntary physical exercise decreases MDSC (myeloid derived suppressor cell) populations in the spleen and circulation in a rat model of mammary adenocarcinoma. Am Assoc Immnol 2018.

[56] Brown FF, et al. Acute aerobic exercise induces a preferential mobilisation of plasmacytoid dendritic cells into the peripheral blood in man. Physiol Behav 2018;194:191–8.

[57] Blanks AM, et al. Impact of physical activity on monocyte subset CCR2 expression and macrophage polarization following moderate intensity exercise. Brain Behav Immun-Health 2020;2, 100033.

[58] Silveira LS, et al. Macrophage polarization: implications on metabolic diseases and the role of exercise. Crit Rev Eukaryot Gene Expr 2016;26(2).

[59] Gavrieli R, et al. The effect of aerobic exercise on neutrophil functions. Med Sci Sports Exerc 2008;40(9):1623–8.

[60] Mooren FC, et al. Exercise delays neutrophil apoptosis by a G-CSF-dependent mechanism. J Appl Physiol 2012;113(7):1082–90.

[61] Suzuki Y, Suda T. Long-term management and persistent impairment of pulmonary function in chronic eosinophilic pneumonia: a review of the previous literature. Allergol Int 2018;67(3):334–40.

[62] Clifford T, et al. The effects of a high-protein diet on markers of muscle damage following exercise in active older adults: a randomized, controlled trial. Int J Sport Nutr Exerc Metab 2020;1(aop):1–7.

[63] Al-Ghamdi HS. Vigorous exercise-induced unilateral eosinophilic fasciitis: rare and easily misdiagnosed subtype. Int J Clin Exp Pathol 2020;13(7):1739.

[64] Pal S, Chaki B, Bandyopadhyay A. High intensity exercise induced alteration of hematological profile in sedentary post-pubertal boys and girls: a comparative study. Indian J Physiol Pharmacol 2021;64(3):207–14.

[65] Bagher M, et al. Mast cells and mast cell tryptase enhance migration of human lung fibroblasts through protease-activated receptor 2. Cell Commun Signal 2018;16(1):1–13.

[66] Giannetti MP. Exercise-induced anaphylaxis: literature review and recent updates. Curr Allergy Asthma Rep 2018;18(12):1–8.

[67] Romero SA, et al. Mast cell degranulation and de novo histamine formation contribute to sustained postexercise vasodilation in humans. J Appl Physiol 2017;122(3):603–10.

[68] Parsons I, Stacey M, Woods D. 4 Histamine, mast cell tryptase and postexercise hypotension in healthy and collapsed marathon runners. Brit Med J 2021.

[69] Jaffer U, Wade RG, Gourlay T. Cytokines in the systemic inflammatory response syndrome: a review. HSR Proc Intensive Care Cardiovasc Anesth 2010;2(3):161–75.

[70] Zhang J-M, An J. Cytokines, inflammation, and pain. Int Anesthesiol Clin 2007;45(2):27–37.

[71] Dinarello CA. Proinflammatory cytokines. Chest 2000;118(2):503–8.

[72] Pedersen BK. Exercise and cytokines. Immunol Cell Biol 2000;78(5):532–5.

[73] Suzuki K. Cytokine response to exercise and its modulation. Antioxidants 2018;7(1):17.

[74] Moldoveanu AI, Shephard RJ, Shek PN. The cytokine response to physical activity and training. Sports Med 2001;31(2):115–44.

[75] Peake JM, et al. Muscle damage and inflammation during recovery from exercise. J Appl Physiol 2017;122(3):559–70.

[76] Dhabhar FS. Effects of stress on immune function: the good, the bad, and the beautiful. Immunol Res 2014;58(2):193–210.

[77] Khosravi N, et al. Exercise training, circulating cytokine levels and immune function in cancer survivors: a meta-analysis. Brain Behav Immun 2019;81:92–104.

[78] Niebauer J. Effects of exercise training on inflammatory markers in patients with heart failure. Heart Fail Rev 2008;13(1):39–49.

[79] Antunes BM, et al. Anti-inflammatory response to acute exercise is related with intensity and physical fitness. J Cell Biochem 2019;120(4):5333–42.

[80] Nielsen AR, Pedersen BK. The biological roles of exercise-induced cytokines: IL-6, IL-8, and IL-15. Appl Physiol Nutr Metab 2007;32(5):833–9.

[81] Reichel T, et al. Reliability and suitability of physiological exercise response and recovery markers. Sci Rep 2020;10(1):1–11.

[82] Libardi CA, et al. Effect of resistance, endurance, and concurrent training on TNF-α, IL-6, and CRP. Med Sci Sports Exerc 2012;44(1):50–6.

[83] Andersson J, et al. Effects of heavy endurance physical exercise on inflammatory markers in non-athletes. Atherosclerosis 2010;209(2):601–5.

[84] Stewart LK, et al. The influence of exercise training on inflammatory cytokines and C-reactive protein. Med Sci Sports Exerc 2007;39(10):1714.

[85] Khalafi M, Malandish A, Rosenkranz SK. The impact of exercise training on inflammatory markers in postmenopausal women: a systemic review and meta-analysis. Exp Gerontol 2021;, 111398.

[86] Ho SS, et al. Effects of chronic exercise training on inflammatory markers in Australian overweight and obese individuals in a randomized controlled trial. Inflammation 2013;36(3):625–32.

[87] Hovanloo F, Arefirad T, Ahmadizad S. Effects of sprint interval and continuous endurance training on serum levels of inflammatory biomarkers. J Diabetes Metab Disord 2013;12(1):1–5.

[88] Conraads V, et al. Combined endurance/resistance training reduces plasma TNF-α receptor levels in patients with chronic heart failure and coronary artery disease. Eur Heart J 2002;23(23):1854–60.

[89] Abd El-Kader S, Gari A, El-Den AS. Impact of moderate versus mild aerobic exercise training on inflammatory cytokines in obese type 2 diabetic patients: a randomized clinical trial. Afr Health Sci 2013;13(4):857–63.

[90] Hagstrom AD, et al. The effect of resistance training on markers of immune function and inflammation in previously sedentary women recovering from breast cancer: a randomized controlled trial. Breast Cancer Res Treat 2016;155(3):471–82.

[91] Simpson RJ, et al. Exercise and the regulation of immune functions. In: Progress in molecular biology and translational science. Elsevier; 2015. p. 355–80.

[92] Szlezak AM, et al. Establishing a dose-response relationship between acute resistance-exercise and the immune system: protocol for a systematic review. Immunol Lett 2016;180:54–65.

[93] Walsh NP, et al. Position statement part one: immune function and exercise; 2011.

[94] Nieman D, et al. The effects of moderate exercise training on natural killer cells and acute upper respiratory tract infections. Int J Sports Med 1990;11(6):467–73.

[95] Kurimoto Y, et al. Voluntary exercise increases IgA concentration and polymeric Ig receptor expression in the rat submandibular gland. Biosci Biotechnol Biochem 2016;80(12):2490–6.

[96] Campbell JP, Turner JE. There is Limited Existing Evidence to Support the Common Assumption that Strenuous Endurance Exercise Bouts Impair Immune Competency. Taylor & Francis; 2018.

[97] Jones B, Hoyne G. The role of the innate and adaptive immunity in exercise induced muscle damage and repair. J Clin Cell Immunol 2017;8(1).

[98] Neves PRDS, et al. Acute effects of high- and low-intensity exercise bouts on leukocyte counts. J Exerc Sci Fit 2015;13(1):24–8.

[99] Timmons BW, Cieslak T. Human natural killer cell subsets and acute exercise: a brief review. Exerc Immunol Rev 2008;14(905):8–23.

[100] Idorn M, Hojman P. Exercise-dependent regulation of NK cells in cancer protection. Trends Mol Med 2016;22(7):565–77.

[101] Inkabi SE, et al. Exercise immunology: involved components and varieties in different types of physical exercise. Scientect J Life Sci 2017;1(1):31–5.

[102] West NP, et al. Antimicrobial peptides and proteins, exercise and innate mucosal immunity. FEMS Microbiol Immunol 2006;48(3):293–304.

[103] LaVoy EC, et al. A single bout of dynamic exercise by healthy adults enhances the generation of monocyte-derived-dendritic cells. Cell Immunol 2015;295(1):52–9.

[104] Kakanis MW, et al. The open window of susceptibility to infection after acute exercise in healthy young male elite athletes. Exerc Immunol Rev 2010;16:119–37.

[105] Lancaster GI, et al. Effects of acute exhaustive exercise and chronic exercise training on type 1 and type 2 T lymphocytes. Exerc Immunol Rev 2004;10(91):106.

[106] Sapp RM, Hagberg JM. CrossTalk opposing view: acute exercise does not elicit damage to the endothelial layer of systemic blood vessels in healthy individuals. J Physiol 2018;596(4):541–4.

[107] Siedlik JA, et al. Acute bouts of exercise induce a suppressive effect on lymphocyte proliferation in human subjects: a meta-analysis. Brain Behav Immun 2016;56:343–51.

[108] Mackinnon LT. Chronic exercise training effects on immune function. Med Sci Sports Exerc 2000;32(7 Suppl):S369–76.

[109] Greenham G, et al. Biomarkers of physiological responses to periods of intensified, non-resistance-based exercise training in well-trained male athletes: a systematic review and Meta-analysis. Sports Med 2018;48(11):2517–48.

[110] Bermon S. Airway inflammation and upper respiratory tract infection in athletes: is there a link. Exerc Immunol Rev 2007;13(6):14.

[111] Bartlett DB, et al. Neutrophil and monocyte bactericidal responses to 10 weeks of low-volume high-intensity interval or moderate-intensity continuous training in sedentary adults. Oxid Med Cell Longev 2017;2017.

[112] Bartlett DB, et al. Ten weeks of high-intensity interval walk training is associated with reduced disease activity and improved innate immune function in older adults with rheumatoid arthritis: a pilot study. Arthritis Res Ther 2018;20(1):127.

[113] Moro-García MA, et al. Frequent participation in high volume exercise throughout life is associated with a more differentiated adaptive immune response. Brain Behav Immun 2014;39:61–74.

[114] Kizaki T, et al. Adaptation of macrophages to exercise training improves innate immunity. Biochem Biophys Res Commun 2008;372(1):152–6.

[115] Liu D, et al. Immune adaptation to chronic intense exercise training: new microarray evidence. BMC Genomics 2017;18(1):29.

[116] West NP, et al. The effect of exercise on innate mucosal immunity. Br J Sports Med 2010;44(4):227–31.

[117] Malaguarnera L, et al. Acquired immunity: immunosenescence and physical activity. Eur Rev Aging Phys Act 2008;5(2):61.

[118] Malm C. Exercise immunology. Sports Med 2004;34(9):555–66.

[119] Simpson RJ, et al. Chapter fifteen—exercise and the regulation of immune functions. In: Bouchard C, editor. Progress in Molecular Biology and Translational Science. Academic Press; 2015. p. 355–80.

[120] Scheffer DL, Latini A. Exercise-induced immune system response: anti-inflammatory status on peripheral and central organs. Biochim Biophys Acta (BBA) - Mol Basis Dis 2020;1866(10), 165823.

[121] Gleeson M. Immune Function in Sport and Exercise. Elsevier Health Sciences; 2006.

[122] Handschin C, Spiegelman BM. The role of exercise and PGC1α in inflammation and chronic disease. Nature 2008;454(7203):463–9.

[123] Hojman P. Exercise protects from cancer through regulation of immune function and inflammation. Biochem Soc Trans 2017;45(4):905–11.

[124] Kruijsen-Jaarsma M, et al. Effects of exercise on immune function in patients with cancer: a systematic review. Exerc Immunol Rev 2013;19.

[125] Ibeneme SC, et al. Impact of physical exercises on immune function, bone mineral density, and quality of life in people living with HIV/AIDS: a systematic review with meta-analysis. BMC Infect Dis 2019;19(1):1–18.

[126] Laddu DR, et al. Physical activity for immunity protection: inoculating populations with healthy living medicine in preparation for the next pandemic. Prog Cardiovasc Dis 2021;64:102.

[127] Nieman DC. Current perspective on exercise immunology. Curr Sports Med Rep 2003;2(5):239–42.

[128] Foster C. Monitoring training in athletes with reference to overtraining syndrome. Med Sci Sports Exerc 1998;30(7):1164–8.

[129] Hug M, et al. Training modalities: over-reaching and over-training in athletes, including a study of the role of hormones. Best Pract Res Clin Endocrinol Metab 2003;17(2):191–209.

[130] Gleeson M. Immune system adaptation in elite athletes. Curr Opin Clin Nutr Metab Care 2006;9(6).

[131] Ronsen O, et al. Leukocyte counts and lymphocyte responsiveness associated with repeated bouts of strenuous endurance exercise. J Appl Physiol 2001;91(1):425–34.

[132] Hayashida H, et al. Exercise-Induced Inflammation during Different Phases of the Menstrual Cycle; 2016.

[133] Passos GS, et al. Exercise improves immune function, Antidepressive response, and sleep quality in patients with chronic primary insomnia. Biomed Res Int 2014;2014, 498961.

[134] Ezema C, et al. Effect of aerobic exercise training on cardiovascular parameters and CD4 cell count of people living with human immunodeficiency virus/acquired immune deficiency syndrome: a randomized controlled trial. Niger J Clin Pract 2014;17(5):543–8.

[135] Baker JV, et al. CD4+ count and risk of non-AIDS diseases following initial treatment for HIV infection. AIDS (London, England) 2008;22(7):841.

[136] Rojas R, Schlicht W, Hautzinger M. Effects of exercise training on quality of life, psychological well-being, immune status, and cardiopulmonary fitness in an HIV-1 positive population. J Sport Exerc Psychol 2003;25(4):440–55.

[137] Fairey AS, et al. Randomized controlled trial of exercise and blood immune function in postmenopausal breast cancer survivors. J Appl Physiol 2005;98(4):1534–40.

Chapter 3

Physical activity interventions for culturally and linguistically diverse populations: A critical review

Melanie Thewlis[a,b], Maximilian de Courten[a,b], and Bojana Klepac Pogrmilovic[a,b]

[a]Institute for Health and Sport, Victoria University, Melbourne, VIC, Australia [b]Mitchell Institute for Education and Health Policy, Victoria University, Melbourne, VIC, Australia

Introduction

Regular physical activity has numerous health benefits, including decreasing the risk of atherosclerotic cardiovascular disease [1–4], type 2 diabetes [5,6], functional disability [7], and some cancers [8,9]. It also improves well-being [10], sleep [11], and mental health [12], and promotes social relationships [13]. In older adults, exercise has been shown to decrease the risk of falls [14,15], muscular atrophy [16], and osteoporotic changes [17], and to delay cognitive decline [18]. However, the World Health Organization estimates that 25% of adults fail to meet moderate-vigorous activity recommendations and that physical inactivity is a leading worldwide cause of mortality [19]. Physical activity (PA) interventions that seek to make lasting changes in the lifestyle habits of inactive adults have been a growing research area since at least 1996 [20]. Successful PA interventions have led to positive improvements for participants in clinical measures such as blood pressure [21–23], adiposity measures [24–26], cholesterol [23], and blood glucose [27].

This chapter reviews PA interventions for culturally and linguistically diverse (CALD) populations. For the purpose of this chapter, we define a CALD population as migrants or their descendants who differ by culture and/or language from the mainstream culture in the country in which they live [28–31]. Indigenous populations and African Americans are not defined by these variables and hence not included in this review, as is common in the literature [28,30–32].CALD populations face a double challenge when engaging in PA interventions. The first challenge is that people from CALD backgrounds living in the United States (US), United Kingdom (UK), and Australia are generally less physically active than the rest of the population [30,33]. They carry a higher burden of lifestyle disease and have generally poorer health outcomes [34,35]. They may face difficulties in accessing health care due to language or cultural barriers, socio-economic conditions, or stigma and marginalization [35–39]. The migration experience itself can be a source of ongoing stress [40].

Understanding the reasons for widespread physical under-activity in CALD communities is a complex task. A systematic review of the literature on physical activity participation in CALD populations found 44 correlates of participation across four broad domains: acculturation, demographic, psychosocial, and environmental/organizational. Factors ranged from walkability of local neighborhoods to cultural norms about exercise, length of time since migration, or employment status. The underlying reasons for the lack of PA are unique to each individual and correlate such as "cultural norms" may have diametrically opposite impacts in different communities [28]. The second challenge is that PA intervention studies may not adequately reflect the cultural diversity of the country in which the study takes place. A 2011review of PA interventions found that participants were more likely to be white, women, tertiary educated, and middle class, when compared to nonparticipants [41]. Given the unique experiences of people with CALD backgrounds, PA interventions that are shown to be successful in a typical study cohort may not be readily transferable to CALD populations [42].

In recognition of these challenges, there is a growing research interest in PA interventions that are culturally adapted to target unique CALD communities. This research is given both urgency and further impetus by the policy attention and financial investment that has been directed toward sport and exercise as sites for cultural integration and acculturation, and as tools for the promotion of migrant and refugee well-being [43]. Since 2010, there has been a steady increase in the sophistication and scope of studies in this area, from focus groups and surveys to single group pretest-post-test trials, to larger-scale randomized controlled trials(RCT) and clinical controlled trials (CCT). At the same time, systematic and

Exercise to Prevent and Manage Chronic Disease Across the Lifespan. https://doi.org/10.1016/B978-0-323-89843-0.00032-5

critical reviews have exposed various issues of research in the field such as a heavy bias toward studies on women; limited use of theoretical frameworks for understanding migration and acculturation; complexities of terminology and definition of CALD communities, particularly related to refugees and people experiencing forced migration; the vast diversity of the migrant experience; and the frequent absence of postcolonialist, genuinely participatory research methods [28,42,43].

Given that CALD communities may be vulnerable populations with significant barriers to self-advocacy [28], it is vital to critically review the literature on interventions in CALD groups, both in terms of the clinical efficacy of different intervention approaches, and also to understand features of the field as a whole. By this, we mean what populations are being studied, where, and by whom; the ethics and power dynamics of that relationship; how the problems and causes of physical inactivity are understood and addressed; and what theories of behavior change and acculturation are being operationalized.

A review of this nature can inform policy and practice in designing PA programs at the community or legislative level. It can identify directions for future research and guide study design toward practices grounded in a sound theoretical framework, clinical evidence, and equity. We hope that for practitioners, this chapter will provide both an overview of promising developments and an aid to the critical reading of the literature.

The aims of this chapter are to:

- Provide an integrated analysis of data obtained from RCTs, CCTs, and multiarm interventional trials (MAIs) of physical activity interventions for CALD populations; and
- Critically examine the practices, theories, and ethics that shape the studies under review.

There is an earlier systematic review that also examined PA interventions in CALD populations, and there is overlap with the studies included in this review [29]. However, the earlier paper included single group pretest posttest design studies, which are excluded from this review. Further, this chapter expands on previous work by including results from more recent studies; providing a more detailed analysis of intervention practices; by assessing outcomes in terms of a novel framework of knowledge and beliefs, behaviors, and clinical measures; and by exploring questions of theory and ethics informed by critical literature.

The chapter first describes the methodology used for the search and review of the literature. The results of the data synthesis and analysis are presented according to the characteristics of the study participants, interventions, cultural adaptations, measures, and outcomes. Critical lenses identified from other literature reviews of the field are employed in a discussion of study design, cultural adaptation, community engagement, ethics, and individualization. Finally, further research and practice recommendations are outlined.

Methods

This review aims to provide a critical assessment of the literature on exercise and physical activity interventions for CALD populations.

Problem identification

Problem identification was undertaken through an initial review of the existing literature on PA in CALD populations. By focusing on systematic and narrative reviews of the field, we were able to identify current research topics of interest including barriers to and enablers of PA participation, methodologies for developing PA interventions, and qualitative, quantitative, and mixed-methods studies trialing PA interventions. Given the emerging maturity of the field, a decision was made to include only randomized controlled trials (RCTs), clinical controlled trials (CCTs), and multiarm interventional studies (MAIs).

Search strategy

The search was conducted in PubMed, Scopus, and Embase in March 2021. The search was performed through titles, abstracts, and keywords. The headline search terms were migrant, physical activity, and intervention. For each of these search terms, synonyms or related terms were also included (Table 1), connected by appropriate Boolean operators.

Study selection and inclusion criteria

Studies selected for inclusion were required to include physical activity, sport, or exercise component intervention component, cultural adaptation of the intervention for a target CALD population, and a minimum of 50% CALD participants.

TABLE 1 Key words included in the search of the literature.

1	2	3
Migrant	Physical activity	Intervention
CALD	Sport	Strategy
Minority	Exercise	Treatment
Newly arrived	Training	Therapy
Refugee		Program
Asylum seeker		Best practice
Culturally diverse		Management
Linguistically diverse		Maintain
Immigrant		Adherence
Foreign-born		Participate
Ethnically diverse		Uptake

They also needed to be RCTs, CCTs, or MAIs, published in a peer-reviewed journal, between 1990 and 2021, with the full text available in English. We included only studies with adult participants aged 18 years or older who are identified as CALD or who are experiencing, or have experienced, migration. The outcome reporting of the intervention needed to include a formal evaluation or physical health outcome or behaviors, which could include program adherence.

We excluded: single group, qualitative, and mixed-methods studies; studies including solely African American participants; studies focusing on participants that are Indigenous to the country such as Aboriginal Australians, Indigenous Canadians, and Native Americans; studies in which ethnicity of participants was used only as a control variable; literature reviews, dissertations, conference proceedings, and opinion articles; studies which did not provide detailed specification as to the type of ethnicity of participants other than "white," "black" and/or "others"; and studies in which only psychological or social outcomes were formally evaluated.

Critical appraisal

The studies were assessed using the Critical Appraisal Skills Programme Randomized Controlled Trial (CASPRCT) Checklist. The checklist includes 11 questions, however, the final two were not relevant for this purpose as they are designed to assist practitioners in applying study results locally. Question four includes three subquestions, and each subquestion was scored as a single point. Hence, the highest possible CASPRCT Checklist score was 11. Although the appraisal tool was developed for assessing RCTs, only one question specifically addresses randomization, and the remainder address topics such as participant selection, blinding, and results reporting that is also relevant to CCTs and MAIs. We, therefore, decided it was an acceptable tool for making comparisons across study designs.

Data extraction and analysis

The studies were analyzed by organizing the data from the publications into an Excel spreadsheet matrix. Where multiple publications described a single study the data from all publications were integrated. The data extracted into the matrix included general information about:

1. The study, including design, research purpose, date, and location;
2. The participants, including their age, gender, health status, and cultural background;
3. The intervention, including mode of recruitment, type of physical activity, duration, cultural adaptations, evaluation measures, results, and retention rates;
4. Other important components such as limitations, underpinning theory, and ethical process; and
5. The CASPRCT assessment of the study.

Results

Thirty publications describing 23 unique studies were included in this review (Table 2). The earliest publication was from 2003 and the most recent from 2021.

Study design and critical appraisal

Sixty-one percent ($n = 14$) of studies were RCTs with attention or care as usual control group. There was one study that used an RCT with waitlist design. Twenty-two percent ($n = 5$) of studies used a MAI design. Thirteen per cent ($n = 3$) were CCTs, in which assignment to intervention or control groups was not randomized. Each of these three studies used a different type of control method—attention control, waitlist control, and one in which the control group received no intervention aside from the measurement procedures. The minimum CASPRCT score was four and the maximum was nine. The mean CASPRCT score was seven.

Study location

Sixty-five per cent ($n = 15$) of studies included in this review took place in the United States. The remainder were located in South Korea, northern Europe, and Australia. All studies took place in high-income countries.

Participants' characteristics

All studies specified the number of participants, which ranged from 33 to 571. The mean number of participants was 168 and the total number across all studies was 3869. Forty-eight per cent ($n = 11$) of studies included only women. Four per cent ($n = 1$)of studies included only men. Of the studies that included both men and women, the mean percentage of female participants was 59.28%. No studies reported participants of any gender other than male or female.

All studies recorded the mean age of participants, which ranged from 26.5 to 73.9 years old. The mean age across all participants was 52.67 years old. Twenty-six per cent ($n = 6$) of studies focused on older adults. The mean age for participants in single-gender studies was 42.49 years and 63.41 years in mixed-gender studies.

Thirty per cent ($n = 7$) of studies included participants from Latin American immigrant communities in the United States, variously defined as Hispanic, Latino, or Latina, or of Latin American origin. Table 3 shows the country/region/culture of origin of participants as described in the studies, and the number of studies that include that population.

The health status of participants could in general be described as at risk. Where reported, the mean Body Mass Index(BMI) ranged from 27.1 to 32.86, indicating that on average, participants were overweight or obese. Thirty-nine per cent ($n = 9$)of studies reported that most or all participants had a history of or abnormal clinical markers indicating high risk for at least one disease, including Atherosclerotic Cardiovascular Disease (ASCVD), stroke, cognitive impairment, hypertension, prediabetes, gestational diabetes, and low mental well-being. Twenty-six per cent ($n = 6$) of studies specified that participants needed to be physically inactive to meet inclusion criteria. Seventeen per cent ($n = 4$) of studies excluded participants who had been diagnosed with diabetes.

Interventions

Duration

The duration of interventions ranged from a single 1-h session to 6 months. The total duration of the studies from initial assessment to final follow-up ranged from 2 weeks to 24 months. The mean intervention duration was 13.13 weeks. The mean duration of interventions targeted at older adults was 6.17 weeks while the mean duration of interventions targeted at adults under 65 years old was 15.59 weeks. The mean duration of single-gender studies was longer at 16.58 weeks than the 9.36 weeks for mixed-gender studies.

Components of the interventions

Ninety-one per cent of the studies ($n = 21$) included an education component in the intervention. These were a combination of one or more lectures, interactive group discussions, and practical classes on various healthy lifestyle topics and specific disease prevention strategies. Seventy-eight per cent ($n = 18$)of studies had in-person classes, 9% ($n = 2$) used a print-based

TABLE 2 Summary of key results.

Author(s)	Study design	CASP score	Number (n =)	Number of females	Mean age	Country/region/culture of origin and current residence	Theory	Duration Intervention components Cultural adaptation Outcome measures Results
Reijneveld, Westhoff, and Hopman-Rock [44]	RCT	8	n = 126	F = 98	54.8	Turkish immigrants living in the Netherlands	None reported	• 6-Week, 8 × 2-h group health education and exercise sessions (adapted Health and Vitality program) • Control received 6-week existing older adults' program through welfare services • Feasibility for adaptation was conducted with 12 Turkish women in three focus groups. Program education adapted to older Turks' culture and knowledge, same-gender groups, run by Turkish peer educators, using the Turkish language • Assessed by program adherence, Voorrips questionnaire, SF-12, MHI-5 • Participants in the intervention group showed an improvement in mental health (effect size: 0.38 SD (95% CI 0.03–0.73), P = 0.03)
Chiang and Sun [45]	MIA	6	n = 128	F = 83	73.4	Chinese immigrants living in the United States	Trans-theoretical model	• 8-Week walking program using Jitramontree's protocol translated to Chinese and modified to emphasize the Chinese cultural values of authority, family members' involvement, harmony, and balance. Program delivered by weekly phone calls • Control group translation only • Blood pressure, PA state of readiness to change, self-reported PA • No significant differences between control and intervention arms
Borschmann, Moore, Russell, Ledgerwood, Renehan, Lin, Brown, and Sison [46]	RCT	7	n = 121	F = 76	69.9	Macedonian and Polish immigrants living in Australia	Trans-theoretical model	• 7-Week, 3 × 1-h interactive lectures with personal goal setting, led by exercise physiologist, to discuss and overcome barriers to PA • Control group 1 × 1h general health promotion lecture • Sessions were based on adult learning and self-management principles • Stages of change Questionnaire, Human Activity Profile Adjusted Activity Score 14, pedometers, BMI, fitness measures • No significant between-group difference

Continued

TABLE 2 Summary of key results—cont'd

Author(s)	Study design	CASP score	Number (n =)	Number of females	Mean age	Country/region/ culture of origin and current residence	Theory	Duration Intervention components Cultural adaptation Outcome measures Results
Sundquist, Hagströmer, Johansson, and Sundquist [47]	RCT	7	n = 243	F = 243	43	First-generation refugees from the Middle East or Latin America, living in Sweden	Social-cognitive theory	• 28-Week group exercise sessions with integrated group counseling, 1 × session individual counseling • Control received a written prescription for exercise and 1 × individual counseling • Written materials translated to Arabic and Spanish. Spanish and Arabic speakers or translators present in classes. Group sessions addressed cultural barriers to exercise • Aerobic capacity and fitness level • The intervention group increased both measures relative to control at 6-month
Qi, Resnick, Smeltzer, and Bausell [48]	RCT	7	n = 83	F = 63	64.08	Foreign-born Mandarin-speaking Asians living in the United States	Self-efficacy theory	• 1-h group presentation and discussion by a nurse researcher on general information on cerebrovascular diseases • Attention control • Delivered in Mandarin in a community setting • Osteoporosis knowledge tests, self-efficacy for exercise and medicine adherence, time, energy spent, and outcome expectations for exercise • Significant improvements in knowledge on osteoporosis, self-efficacy to exercise, time and energy spent in moderate exercise at 2-week follow-up

Andersen, Burton, and Anderssen [49]	RCT	7	$n=150$	$F=0$	35.7	Pakistani immigrants or their children living in Norway	Self-efficacy	• 5-Month program, $2 \times 1h$/week group exercise plus $2 \times$ group lectures, $1 \times$ individual counseling, $1 \times$ phone call counseling • Control group offered organized exercise (once a week for 4 months), one group lecture, and written material after the end of the trial • Focus groups before the study identified floor ball as a familiar sport. Education addressed lack of knowledge about vigorous exercise, benefits of exercise, lack of social context for exercise, problem-solving, self-efficacy
Andersen, Høstmark, and Anderssen [50]		7						• PA habits (accelerometer), cardiorespiratory fitness (oxygen consumption), and Metabolic Syndrome risk factors (waist circumference, serum TG concentration, HDLc concentration, systolic blood pressure or diastolic blood pressure or fasting plasma glucose concentration) • The intervention group increased the total PA level from baseline to the 6-month follow-up by a mean of 36 CPM (95% CI=4–70; $P=0.02$), an increase of 10% (95% CI=2–17), and time spent in MVPA by an average of 7.3 min•day-1 (95% CI=0.8–13.7; $P=0.03$), an increase of 21% (95% CI=10–31). The intervention group reduced the sedentary time by a mean of 0.7h•day-1 (95% CI=−0.3 to −1.1; $P=0.001$), a reduction of 9% (95% CI=1.5–16)
Gademan, Deutekom, Hosper, and Stronks [51]	CCT	4	$n=514$	$F=514$	45	Turkish, Surinamese, Moroccan, and other immigrants living in the Netherlands	Social-cognitive theory	• EoP consisted of an intake followed by 18 sessions of supervised physical activity and a final evaluation. Training sessions were held once a week. EoP offered Fitness, Aquarobics, Aerobics, and Dancing • Control group received care as usual • Personal coaching during sessions on social support, attitudes toward sports, ways of coping with (negative) feedback from the community, increasing awareness of the positive effects of exercise, empowering the participant with respect to the continuation of the healthy (physical) behavior. Training sessions were held in their neighborhood in a supportive environment under the supervision of a female coach and financial incentive was available • Self-reported physical activity, weight, height, fat percentage, waist circumference, and physical fitness level • Total PA did not change. Small positive impact on PA during leisure time at 6-month and 12-month and PA during household activities at 12-month ($P<0.05$)

Continued

TABLE 2 Summary of key results—cont'd

Author(s)	Study design	CASP score	Number ($n=$)	Number of females	Mean age	Country/region/culture of origin and current residence	Theory	Duration Intervention components Cultural adaptation Outcome measures Results
Marcus, Dunsiger, Pekmezi, Larsen, Bock, Gans, Marquez, Morrow, and Tilkemeier [52]	RCT	8	$n=266$	$F=266$	41.61	Spanish speakers who self-identify as Latino or Hispanic, living in the United States	Social-cognitive theory	• 6-Month mail delivered, print-based physical activity intervention for Latinas. Participants received pedometers and PA logs to encourage self-monitoring of PA behavior • Control received a wellness contact control condition • Translated to Spanish, culturally and linguistically adapted, individually tailored print materials. Addressed PA barriers identified by Latinas in focus groups and in the literature review • Change in MVPA measured by 7-day PAR interview and accelerometers • Increases in min/week of MVPA measured by the 7-day PAR were significantly greater in the intervention group compared to the control group (mean difference=41.36, SE=7.93, $P<0.01$). Corroborated by accelerometer readings (rho=0.44, $P<0.01$)
Hawkins, Chasan-Taber, Marcus, Stanek, Braun, Ciccolo, and Markenson [53]	RCT	8	$n=260$	$F=260$	26.5	Hispanic women living the United States	Trans-theoretical model	• 12-Week individually tailored, motivationally matched exercise intervention • One in-person session with a health educator to complete the tailoring questionnaire and set behavioral goals. Weekly goal to increase time spent in moderate-intensity physical activity by 10% each week • Self-selected activities such as dancing, walking, and yard work. Digital pedometer and activity diary for self-monitoring • Monthly follow-up questionnaires and individually tailored reports via mail
Nobles, Marcus, Stanek, Braun, Whitcomb, Solomon, Manson, Markenson, and Chasan-Taber [54]		8						• Tip sheets mailed weekly for the first 4 weeks and then every other week • Control arm received comparison health and wellness intervention. Both arms received booster telephone calls at 2-, 6-, and 10-week • Used research findings on the social, cultural, economic, and environmental resources and challenges faced by women of diverse backgrounds. Intervention materials were written at the sixth-grade reading level. • Pregnancy Physical Activity Questionnaire • Intervention group had significantly greater increases in sports or exercise activity (0.3 vs 5.3 MET hours/week; $P<0.001$), smaller declines in total activity (−42.7 vs −2.1 MET hours/week; $P=0.02$), and activities of moderate to vigorous intensity (−30.6 vs −10.6 MET hours/week; $P=0.05$), and was more likely to achieve recommended guidelines for physical activity (odds ratio=2.12; 95% confidence interval=1.45, 3.10)

	Design	Quality	Sample		Age	Population	Theory	Description / Findings
Khare, Cursio, Locklin, Bates, and Loo [55]	RCT	7	180	F = 180	50.9	Spanish speaking immigrants living in the United States	Social-cognitive theory	• 12-Week nutrition and physical activity lifestyle change intervention. (IWP). • Control and intervention groups received CVD risk factor screening and educational handouts. • Curriculum translated and back-translated into Spanish. Orientation with family members. • Physical activity questionnaire (CHAMPS), BMI, blood pressure, lipid counts, blood glucose levels
Lee, Chae, Wilbur, Miller, Lee, and Jin [56]	MAI	7	80	F = 80	52.43	Korean-Chinese migrant workers living in South Korea	Acculturation	• Intervention received phone counseling and text messages for 12-Week, then three sets of acculturation workshops during weeks 13–24 • Control and intervention instructed to carry out stretching program 3 × 6-min/day and 5-days/week • Individual phone and SMS counseling, three acculturation workshops: "How to apply make-up," "Communication skills in everyday life," and "How to choose healthy foods" • Musculoskeletal fitness, symptoms, and acculturative stress
Lee, Chae, Cho, Kim, and Yoo [57]		7						• Acculturative stress was significantly reduced only in control group. Significant improvements in flexibility of the back and work-related musculoskeletal disorder at 24-weeks but no significant between-group difference
Kandula, Dave, De Chavez, Bharucha, Patel, Seguil, Kumar, Baker, Spring, and Siddique [58]	RCT	8	63	F = 39	50	South Asian immigrants living in the United States	Community-based participatory research	• 6 Classes on physical activity, healthful diet, weight, and stress management. Telephone support calls to focus on self-reflection, behavior goals, and problem-solving. Four heart health melas (festivals) over 12-month period. Pedometers and daily steps tracking training were provided • Control received translated print education materials about ASCVD and healthy behaviors. Control received the intervention after study completion • The community partner and academic partner collaboratively chose a research focus after working together to understand community needs and the socio-cultural context of South Asian ASCVD risk behaviors. Community advisory board involved in study design, focus groups, review of study materials, and questionnaires. Melas incorporated culturally salient activities • MVPA levels, weight, waist circumference, blood sugar, hemoglobin A1c, lipids, blood pressure, psychosocial outcomes • Intervention group showed significant weight loss (−1.5 kg, P-value = 0.04) and greater sex-adjusted decrease in hemoglobin A1c (−0.43%, P-value <0.01) at 6-month

Continued

TABLE 2 Summary of key results—cont'd

Author(s)	Study design	CASP score	Number (n =)	Number of females	Mean age	Country/region/ culture of origin and current residence	Theory	Duration / Intervention components / Cultural adaptation / Outcome measures / Results
Martin, Signorile, Kahn, Perkins, Ahn, and Perry [59]	MAI	7	231	F = 231	36.8	English speaking Latinas living in the United States	Social-cognitive theory	• 12-Week program, labs available for resistance training Mon-Sat. Participants committed to do routine at least 1/week • EC group: three evaluative conditioning trials at 3-week intervals (pairing positive words with images of exercise on a computer) • NC: dummy task containing unrelated target images paired with neutral words • Evaluative conditioning • Exercise adherence, Rosenberg Self-Esteem scale, Exercise Self-efficacy scale, and Body Esteem scale • Functional variables: 30s Arm Curl, 30s Chair Stand, usual and maximum gait speeds, reactive balance, and single-leg stand test • Body composition variables: BMI, percent body fat, lean body mass, waist, and hip circumference
Chee, Kim, Chu, Tsai, Ji, Zhang, Chee, and Im [60] Chee, Ji, Kim, Park, Zhang, Chee, Tsai, and Im [61] Chee, Kim, Tsai, and Im [62] Chee, Kim, Tsai, Liu, and Im [63]	RCT	8	n = 33	F = 33	41.9	Korean American or Chinese American women living in the United States	Self-efficacy theory	• 3-Month web-based physical activity program inc. social media site with chat functionality, interactive education sessions, and online resources. Individual and group coaching & support using Fitbits from culturally matched peers and health care providers • Control group told to maintain their usual information searches through existing resources • Program developed in English, Mandarin Chinese, and Korean, bilingual interventionists; education modules adapted to include cultural context, e.g., benefits of PA, PA in daily schedule in cultural context, general and cultural-specific exercises, cultural beliefs about the value of PA • Kaiser Physical Activity Survey; Questions on Attitudes Toward Physical Activity; Subjective Norm, Perceived Behavioral Control, and Behavioral Intention; Physical Activity Assessment Inventory; Modified Barriers to Health Activities Scale; The Center for Epidemiologic Studies Depression Scale-Korean; Acculturation Stress Scale; Social Readjustment Rating Scale; Discrimination Scale • Depression and total physical activity scores did not significantly improve over time for both groups

Author	Design	Weeks	N	F	Age	Population	Framework	Findings
Yeh, Heo, Suchday, Wong, Poon, Liu, and Wylie-Rosett [64]	RCT	6	60	F = 34	60.9	Chinese immigrants living in the United States	CBPR	• Diabetes Prevention Program lifestyle intervention of 12×2-h bi-weekly core sessions and 6 monthly postcore sessions on topics such as healthy eating, physical activity, stress reduction, and problem-solving skills, use of self-monitoring tools (e.g., pedometer) • Control consisted of quarterly mailing of diabetes prevention information • Conducted in Chinese at a community site that could accommodate an exercise program. Conveniently located meeting place, and informal buddy system • The DPP curriculum was adapted based on feedback from three focus groups of Chinese participants with prediabetes and one advisory group meeting of physicians and community leaders. Adaptations: more information about Asian diabetes risk disparity, following each intervention with a physical activity session, inviting family members to attend sessions, culturally and linguistically tailoring • Weight, BMI, hemoglobin bA1c concentration • The participant attrition rate was <5% (2 out of 60) at 12 months. Significantly greater percent weight loss in the intervention group (-3.5 vs -0.1%; $P = 0.0001$) at 6-month, largely maintained at 12-month (-3.3 vs 0.3%, $P = 0.0003$)
Lee, Cho, Wilbur, Kim, Park, Lee, and Lee [65]	MAI	5	132	F = 132	56.2	Korean-Chinese women living in South Korea	Acculturation theory	• All participants had two face-to-face meetings with a nurse interventionist and received a walking manual, a pedometer, a walking step goal (3000 daily step increase on previous 1 week of pedometer count), and a walking step diary • Time difference of 1 year between recruitment and assessment of two groups. Enhanced intervention group also received 12 motivation text messages in the first 12 weeks and 6 mobile phone cartoon illustrations to help cultural adaptation • Walking adherence (10,000+ daily steps measured with a pedometer, average daily step count), blood pressure, blood lipid tests, BMI, weight, waist-hip ratio, 10year CVD risk • No difference between groups except BMI, with greater loss in enhanced intervention group ($t = -2.010$, $P = 0.046$). Significant changes over time in both groups: decreased 10-year CVD risk (at week 12: $t = -2.245$, $P = 0.027$; at week 24: $t = -2.534$, $P = 0.013$), lower systolic blood pressure (at week 12: $t = -4.704$, $P < 0.001$; at week 24: $t = -6.371$, $P < 0.001$), lower fasting glucose (at week 12: $t = -4.734$, $P < 0.001$; at week 24: $t = -4.027$, $P < 0.001$), decreased BMI (at week 12: $t = -4.701$, $P < 0.001$; at week 24: $t = -5.693$, $P < 0.001$), and decreased WHR (at week 12: $t = -3.821$, $P < 0.001$; at week 24: $t = -5.975$, $P < 0.001$)

Continued

TABLE 2 Summary of key results—cont'd

Author(s)	Study design	CASP score	Number (n =)	Number of females	Mean age	Country/region/ culture of origin and current residence	Theory	Duration / Intervention components / Cultural adaptation / Outcome measures / Results
Piedra, Andrade, Hernandez, Boughton, Trejo, and Sarkisian [66]	RCT	7	571	F = 441	73.12	Older Hispanic adults living in the United States	Not reported	• EF 4-week core program of 1-h exercise class plus 1-h/week group discussion sessions either attribution retraining (treatment) or generic health education (control) • Culturally tailored attribution retraining or health education discussion groups, led by a bilingual health educator, low cost • Program adherence; diary of activities and problems encountered; cognitive function (3MS) • 80% attendance rate for both groups for the 4-week core program. Statistically significant improvement in 3MS cognitive functioning scores for both groups at 1-year ($P = 0.001$); no at 2-year. No significant difference in cognitive function between the treatment and control groups over time
Siddiqui, Kurbasic, Lindblad, Nilsson, and Bennet [67]	RCT	8	96	F = 51	47.9	Iraqi immigrants living in Sweden	None reported	• Seven group sessions on a healthy diet and physical activity inc. One cooking class • Intervention received a step-counter with written advice on achieving 10,000 steps per day • Control received "treatment as usual," i.e., 3 × written advice on PA levels and adopting healthy eating habits
Siddiqui, Koivula, Kurbasic, Lindblad, Nilsson, and Bennet [68]		8						• Culturally modified from DPP to include gender-specific groups led by Arabic speaking health coaches with experience and knowledge of the Middle-Eastern lifestyle, culture, and food habits and with experience in patient education and counseling. Additionally: identification of social and cultural barriers to lifestyle change, cooking classes, and economic support for training clothes and admittance to the local PA centers • Reduction in body weight by at least 5%, moderate-intensity PA of at least 30-min/day, 5-days/week, reduction in mean caloric intake; cardio-metabolic bio-markers such as FPG, 2-h glucose, Hemoglobin A1C, systolic blood pressure, diastolic blood pressure, plasma triglycerides and LDL-cholesterol, insulin sensitivity index, disposition index and/or HDL-cholesterol • The mean insulin sensitivity index increased significantly at follow-up in the intervention group compared to the control group (10.9% per month, $P = 0.005$). The intervention group also reached a significant reduction in body weight (0.4% per month, $P = 0.004$), body mass index (0.4% per month, $P = 0.004$), and LDL-cholesterol (2.1% per month, $P = 0.036$) compared to the control group. In total, 14.3% in the intervention group reached the goal to lose $\geq$5% of body weight vs none in the control group

| An, Nahm, Shaughnessy, Storr, Han, and Lee [69] | CCT | 5 | 73 | F = 54 | 71.49 | Korean-Americans living in the United States | Self-efficacy theory | • 8-Week stroke prevention program focused on stroke knowledge and lifestyle changes inc. healthy eating and physical activity. 1-h/week session inc. 30 min lecture and 30 min group discussion
• Control received assessment only
• Course material translated to Korean, community discussion groups
• IPAQ-SF, Exercise Self-Efficacy Scale, Exercise Benefits/Barriers Scale; pedometers worn for step counts 1-week pre and posttrial
• Average attendance 80.4, Likert scale satisfaction 3.84/4; physical activity showed marginal effects assessed by the IPAQ-SF and step counts ($P = 0.075$)
  Pretest steps: control: 6142.78 [+ or −] 2224.20 intervention: 6919.54 [+ or −] 2835.18
  Posttest step: control: 6514.68 [+ or −] 2396.39 intervention: 7918.75 [+ or −] 2992.42
• The between-group difference was not statistically significant |
| D'Alonzo, Smith, and Dicker [70] | CCT | 7 | 76 | F = 76 | 29.8 | Spanish speaking immigrant Latinas living in the United States | Community-based participatory research | • A 12-week program of 2 x 1h/week classes in low-impact aerobic exercise/Latin dance, and strength and flexibility activities, a short lecture on different topics during cool down
• Waitlist control
• Classes at neighborhood centers or charter schools, in Spanish, using culturally appropriate music. Babysitting provided
• During each cooldown session, the promoter led a brief discussion on one of a number of scripted topics
• Daily PA levels; aerobic fitness, muscle strength, and flexibility; BMI and percentage of body fat
• Significant improvements were measured in aerobic fitness, muscle strength and flexibility, and daily PA levels
• No significant change in BMI or body fat percentage
  G1 and G2—statistically significant improvements in PA ($P \leq 10-10$, 95% CI [0.21, 0.32]) and flexibility ($P < 10-10$, 95% CI [2.95, 4.59])
• The intervention led to higher improvements in muscle strength ($P < 10-10$, CI [17.58, 23.16]) in G2 as compared with G1 ($P = 0.01$, CI [1.05, 8.94]) and greater improvements in aerobic fitness ($P < 10-10$, CI [4.03, 7.11]) among participants in G1 as compared with G2 ($P \leq 10-10$, CI [3.39, 4.81]) |

Continued

TABLE 2 Summary of key results—cont'd

Author(s)	Study design	CASP score	Number (n =)	Number of females	Mean age	Country/region/ culture of origin and current residence	Theory	Duration Intervention components Cultural adaptation Outcome measures Results
Menkin, McCreath, Song, Carrillo, Reyes, Trejo, Choi, Willis, Jimenez, Ma, Chang, Liu, Kwon, Kotick, and Sarkisian [62,71]	RCT	9	233	F = 161	73.9	Latino, Korean, and Chinese immigrants living in the United States	Social-cognitive theory	• 1-month, 8-session, culturally tailored behavioral intervention to black, Latino, Chinese, and Korean-American seniors • Control and intervention received the same frequency of contact inc. reminder phone calls to wear pedometers, and same incentives (pedometer and honoraria) • Survey instruments were forward- and back-translated into Spanish, Korean, and Chinese. The curriculum content combined aspects of social-cognitive theory and attribution theory to motivate change in walking behavior. Sessions 6 and 7 were culturally tailored to each racial/ethnic group to enhance relevance and impact, using insight gained from the collaboration with racial/ethnic-specific community action boards and 12 previously conducted focus groups • Daily steps (weekly pedometer use) systolic and diastolic blood pressure, BMI, blood lipids; Medical Outcomes Study 12-item Short Form, 9-item Patient Health Questionnaire • Small improvements in daily steps not sustained 2-month after the intervention
Towfighi, Cheng, Hill, Barry, Lee, Valle, Mittman, Ayala-Rivera, Moreno, Espinosa, Dombish, Wang, Ochoa, Chu, Atkins, and Vickrey [72]	RCT	9	100	F = 38	58	English or Spanish speaking Hispanics living in the United States	Lifestyle Redesign theory	• 6-Week occupational therapist-led group lifestyle intervention. 2-h session in English or Spanish including presentations on diet and physical activity, peer exchange, personal exploration with goal setting, self-management tools (pedometers, food, and physical activity logs), and activities including walking at a park, stretching, strengthening, and yoga • Control received usual care, including primary care, vascular neurology, therapies, and optional DPP sessions • Delivery in Spanish or English • BMI, fruit/vegetable intake, physical activity, waist circumference, smoking, blood pressure, high-density lipoprotein, low-density lipoprotein, triglyceride, total cholesterol, glycosylated hemoglobin levels, quality of care, and perceptions of care • No significant changes in outcomes at 6 months. Effect sizes for all outcomes were small (<0.2)

| Daniel, Marquez, Ingram, and Fogg [73] | MAI | 6 | 50 | F = 50 | 50 | South Asian Indian women living in the United States | Social-cognitive theory | • 12-Week program
• G1 and G2: 6 × 1h motivational workshops G1: 6 × 30-min dance intervention based on YouTube video of a Bollywood dance in Hindi following each workshop G2: 6 × 20-min motivational phone calls in Hindi, Punjabi, or English, starting 1-week after the first workshop
• Materials provided in English, then discussion points were addressed speaking Hindi, Punjabi, or English; motivational materials given cultural context; Bollywood dance routine
• Height, weight, BMI, waist circumference, blood pressure, finger prick hemoglobin A1C, FBS, cholesterol level, and a 12-lead electrocardiogram, 2-min step test, average daily steps (pedometer)
• Both interventional arms were found to be equally effective, except for the LDL level, which worsened for the phone call group at 24 weeks. Significant differences were seen in body weight ($P = 0.024$), waist circumference ($P = 0.001$), systolic blood pressure ($P = 0.000$), triglyceride ($P = 0.000$), cholesterol ($P = 0.001$), blood sugar level ($P = 0.000$), and average daily steps across both intervention groups over time ($P = 0.000$) |

3MS, modified minimental state; 7-Day PAR, 7 day physical activity recall; ASCVD, atherosclerotic cardiovascular disease; BMI, body mass index; CCT, clinical controlled trial; CHAMPS, Community Healthy Activities Model Program for Seniors; CI, confidence interval; CPM, average counts per minute per day (accelerometer data); DPP, Diabetes Prevention Program; EC, evaluative conditioning; EF, EnhanceFitness program; EoP, exercise on prescription; FBS, fasting blood sugar; FPG, fasting plasma glucose; HDLc, high-density lipid cholesterol; IPAQ-SF, 7-item International Physical Activity Questionnaire-Short Form; IWP, Illinois WISEWOMAN program; LDL, low-density lipoprotein; MAI, multi-arm interventional trial; MET, metabolic equivalent of task; MHI-5, Mental health calculated from five items of the Short Form 36; MVPA, moderate to vigorous physical activity; NC, neutral conditioning; PA, physical activity; RCT, randomized controlled trial; SD, standard deviation; SF-12, Short Form 12; TG, triglycerides.

TABLE 3 Country/region/culture of origin of study participants and number of students that include that population.

Country/region/culture of origin	Number of studies[a]
Latin America/Hispanic[b]	8
China	5
Korea	3
Turkey	3
South Asia[c]	2
Korean Chinese[d]	2
Iraq	2
Pakistan	1
Macedonia	1
Poland	1
Suriname	1
Morocco	1
Syria	1
Lebanon	1
Chile	1
Middle East	1

[a]Several studies include multiple countries of origin so the number totals to more than 23.
[b]Seven studies took place in the United States and one study in Sweden.
[c]Includes participants from India, Pakistan, Bangladesh, Sri Lanka, Nepal, and Bhutan.
[d]Korean-Chinese are Chinese citizens of Korean descent, whose ancestors migrated to China in the 1700–1800s, and who now make up 60% of migrant workers in Korea.

education program delivered by mail, and 4% ($n = 1$) delivered motivational messages and acculturation information via text message.

Forty-three per cent ($n = 10$) of studies included individualized counseling or goal setting in at least one intervention arm. Thirteen per cent ($n = 3$)of studies delivered individual support via phone calls and 30% ($n = 7$) via in-person meetings. A further 13% ($n = 3$) delivered retention reminders via phone calls.

Thirty-five per cent ($n = 8$) of studies provided exercise classes or made exercise facilities available to participants in at least one intervention arm. Thirty-nine per cent ($n = 9$) of studies included only an education component. Fifty-seven per cent ($n = 13$) of studies included two components out of education, individual counseling, and exercise classes. Four per cent ($n = 1$) of studies included all three components.

Three of the 6 studies that targeted older adults included only an education component, as opposed to 6 of the 17 studies that targeted adults under 65 years old. Two of the 12 single-gender studies included only an education component, as opposed to 7 of the 11 mixed-gender studies.

Physical activities

The studies assessed and/or promoted a broad range of physical activities. Fifty-seven per cent ($n = 13$) of studies promoted walking and 52% ($n = 12$) of studies tracked participants' daily steps. Some studies assessed and promoted active living habits, such as doing yard work or parking further from a destination. More formal activities included floorball, dance, aquarobics, aerobics, weightlifting, stretching, indoor cycling, treadmill walking, stair climbing, strengthening, yoga, and tai chi.

Cultural adaptations

Sixty-one per cent ($n = 15$) of studies translated intervention materials and/or conducted education components in the participants' first language. Four per cent ($n = 1$) of studies adapted reading materials to be in English but at a sixth-grade reading level.

Additionally, in some cases, the education component was adjusted to address various themes, including cultural beliefs about or barriers to exercise (39%, $n = 9$), the role of community and peer support (35%, $n = 8$), self-efficacy, and motivation (30%, $n = 7$), the role of family (17%, $n = 4$), and cultural competency in the host culture (9%, $n = 2$).

Twenty-six per cent ($n = 6$) of studies reported using focus groups or pilot studies to develop cultural adaptations. Seventeen per cent ($n = 4$) of studies selected the exercise activity for cultural reasons (floorball, Latin dance, Bollywood dance, yoga).

Further adaptations included 57% of studies ($n = 13$) that either included only a single-gender or grouped participants by gender with a gender-matched educator; 13% ($n = 3$) of studies that had interventions delivered at least partly by peer educators, 4% ($n = 1$)of studies that trialed evaluative conditioning, and 4% ($n = 1$) of studies that provided babysitting.

Recruitment and retention

Ninety-six per cent ($n = 22$) of studies reported on the recruitment methods used. Fifty-two per cent ($n = 12$) of studies worked with community centers to distribute flyers. For 30% of studies ($n = 7$), participants were referred by general practitioners or other medical services. Twenty-six per cent ($n = 6$) of studies worked with churches or mosques to distribute flyers. Twenty-two per cent ($n = 5$) of studies used snowball or word of mouth. Four per cent ($n = 1$) of studies recruited from online communities, 4% ($n = 1$) of studies from welfare services, 4% ($n = 1$) of studies used ads in local newspapers and radio, and 4% ($n = 1$) of studies used mail and telephone outreach based on an existing population survey. There was a wide variance in retention rates. The minimum retention rate at final follow-up was 23% and the maximum was 100%. The mean retention rate was 78%.

Assessments and measures

The studies used a wide range of assessment measures. We categorized the measures into three domains:

1. Eighty-seven per cent ($n = 20$) of studies assessed physical activity behaviors, as measured by program adherence, questionnaires, activity diaries, accelerometers, and pedometers.
2. Seventy-four per cent ($n = 17$) of studies assessed clinical markers, including blood pressure, BMI, blood lipids and glucose, waist circumference, percentage body fat, and performance in strength, flexibility, and aerobic fitness tests.
3. Forty-three per cent ($n = 10$) of studies assessed motivation, knowledge, and beliefs about exercise, as measured by questionnaires and knowledge tests.

Theoretical/conceptual frameworks

Eighty-seven per cent ($n = 20$) studies reported one or more theoretical frameworks that guided the intervention design, as follows. Thirty-five per cent ($n = 8$) of studies drew on social-cognitive theory, 22% ($n = 5$) of studies employed self-efficacy theory, 22% ($n = 5$)of studies used the trans-theoretical model of behavior change, 13% ($n = 3$) of studies used a community-based participatory research model (CBPR), and 9% ($n = 2$) used acculturation theory. Other theories used were Leininger's Culture Care Theory, Lifestyle Redesign theory, and a social-ecological model.

Ethics procedures

All studies reported that the study had been granted ethics approval by an institutional review board. Ninety-one per cent ($n = 21$) of studies stated that informed consent was obtained from the participants. Four per cent ($n = 1$) of studies reported that verbal consent was obtained as many participants were not literate in any language.

Fifty-two per cent ($n = 12$) of studies either excluded participants on the basis of preexisting disease or injury or reported that participants were screened to ensure they could safely commence exercise using tools such as the Physical Activity Readiness Questionnaire (PAR-Q). Thirteen per cent ($n = 3$) of studies offered exercise classes but did not report that they conducted a check for medical contraindications for increasing physical activity. Four per cent ($n = 1$) of studies reported referring participants who presented with risk factors for exercise to seek assistance from a medical practitioner. Four per cent ($n = 1$) of studies reported on whether participants had sustained injuries or other harms as a result of the intervention; this was also the only study to report on the cost of delivering the intervention.

TABLE 4 Outcomes of the studies and number/percentage.

Outcome	Percentage of studies (%)	Number of studies (n =)
Statistically significant change in participants' physical activity behaviors	35	8
Statistically significant improvement in clinical markers	43	10
Statistically significant improvement in participants' knowledge, motivation, or beliefs about exercise	22	5
Change in a single domain	35	8
Change in two domains	26	6
Change in all three domains	4	1
No improvement	35	8

Thirteen per cent ($n = 3$) of studies used a study design in which all participants eventually received the intervention and 4% ($n = 1$)of studies mailed the intervention materials to the control group after trial completion.

Outcomes

As stated, we have categorized the outcome measures into three domains: physical activity behaviors; clinical markers; and motivation, knowledge, and beliefs about exercise. As shown in Table 4, statistically significant change/improvement in the first domain was reported in 35% ($n = 8$) of studies, in the second domain in 43% ($n = 10$) of studies, and in the third domain in22% ($n = 5$) of studies.

Studies that recorded a statistically significant outcome in at least one domain had younger participants, with a mean age of 48.26 years vs 60.66 years for studies that did not record statistically significant outcomes. The baseline mean BMI of participants in studies that recorded a statistically significant outcome was also lower: 28.88 vs 30.85. These studies also had a higher average retention rate than unsuccessful studies:80.07% vs 75.67%.

Of the studies that assessed clinical markers such as blood pressure, the ones that achieved statistically significant change in those markers were of longer duration, at 17.18 weeks compared to 14.43 weeks on average.

Studies that performed a physical safety review or excluded participants with disease or injury were more likely than average to report a successful outcome. Ten out of 13 studies that excluded participants on the basis of risk factors for beginning exercise were successful as opposed to six out of 10 studies that did not.

Nine of the 12 single-gender studies reported a statistically significant outcome as opposed to 6 of the 11 mixed-gender studies.

Discussion

Characteristics of interventions with positive outcomes

We found that studies that achieved statistically significant outcomes tended to have one or more of the following characteristics:

- Be of longer duration (15 weeks vs 9.5 weeks);
- Use translated education materials and/or bilingual instructors;
- Promote culturally relevant exercise such as dance or yoga;
- Use pedometers in combination with individualized goal setting for daily step tracking; and
- Comprise at least two components out of education, individualized counseling, and exercise classes.

All six studies that reported using focus groups achieved statistically significant positive outcomes such as improvement in physical activity behavior. There are several reasons why using focus groups as a method could have contributed to a successful intervention design and implementation. Focus groups have been reported as a tool to build relationships with key stakeholders and potential intervention participants [74]. Focus groups that employ open-ended questions can invite novel responses that might not have been predicted by theory or the assumptions of researchers. They can allow researchers to uncover crucial barriers to behavior change and aspects of interventions that may be unacceptable to participants [75].

Furthermore, there was evidence that the use of phone calls to offer individual counseling and reminders to encourage retention may lead to good outcomes. Five out of six studies that made use of phone calls in some form achieved a statistically significant outcome as compared to 11 out of 17 studies that did not. Single-gender studies were more likely than mixed-gender studies to achieve statistically significant outcomes. However, the participants in single-gender studies were over 20 years younger on average than for studies that included mixed genders. The duration of single-gender studies was also longer, and they were more likely to include additional components beyond group education, such as individual counseling or exercise classes.

Challenges of study design—Randomization, blinding, recruitment, and outcome measures

The critical appraisal scores as measured by CASPRCT were in general low. One reason for this is that it is difficult to double or even single blind interventions of this nature. Some studies attempted to provide a single blind using an attention control or care as usual treatment for the control group.

Another reason for low CASPRCT scores was that many studies, even those described as RCTs, did not employ random assignment of participants to the intervention and control groups. Several studies raised concerns about the potential for cross-interaction of groups in small communities. To counter this risk, some studies delivered the intervention at one community site and the control at another. Other studies used a waitlist control. One study reported it would be too challenging to recruit a large enough cohort to run the intervention and control groups simultaneously and used a waitlist control for this reason. Another study used participant-selected waitlist control as a tool to make the intervention more convenient for participants and therefore boost retention rates. In general, studies that did not use true randomization provided reasonable explanations of why a different method of group assignment was used.

Among studies that reported a recruitment rate, it was generally low. Several studies reported that it took longer than expected to recruit sufficient participants; several studies ultimately went ahead with fewer participants than researchers had aimed to recruit on the basis of providing sufficient statistical power; and one study changed recruitment methods after the trial had begun. Where data was available, participants tended to be younger, less overweight, and fitter than the target CALD population average.

It was noticeable that there was an overall lack of standardization of measures across the studies under review, which make the comparison of outcomes challenging. The reliance on questionnaires or PA diaries introduces the risk of social desirability bias. Many studies made use of states of change or self-efficacy theory in study design, and some used these models as part of their outcome measures. The authors of one study noted that there are inherent problems with this approach, as it is difficult to recruit people who have low motivation to change [46]. There is therefore a ceiling effect on how much motivation can be increased by the intervention when motivation is already high at baseline. In addition, the authors hypothesized that as people learn more about how much PA is recommended for maintaining health, their confidence to incorporate it into everyday life diminishes [46].

Overall, the complexity and variability of intervention design make it difficult to assess which elements contribute to the success of the intervention or lack thereof. There seems to be a tendency to include as many elements as possible such as education, counseling, classes, step tracking, and group discussions, with very few studies isolating those various elements to determine their individual impact.

Participants

On average, older adults received shorter interventions and were more likely to receive interventions with only an education component and no individual counseling or exercise classes. This is despite the additional barriers older adults may face to achieving adequate PA levels, such as health problems or fear of falling [76].

Men were also underserved in the studies selected for review, as we found only one study that included only men. Additionally, the mixed-gender studies were of shorter duration and more likely to only include a group education component. This is possibly because women are perceived to have a greater need for PA interventions. Women are in general less physically active and report greater barriers to exercise than men. In addition, normative female gender roles can be a barrier to exercise [77–79]. However, the large proportion of women-only studies in the field has been described as problematic, as both men and women in CALD groups are typically not active enough to maintain health benefits, longevity is typically lower for men, and focusing on CALD women in isolation may disrupt cultural gender balances between women and their families [28].

Theoretical and conceptual frameworks

The majority of studies drew on Albert Bandura's social-cognitive theory(SCT) or its subcomponent, self-efficacy theory (SET) [80]. Self-efficacy, defined by Bandura as one's belief in one's ability to succeed in specific situations or accomplish a task [81], is a widely applied theory used to understand health behavior and facilitate behavioral modification, such as the increase of PA. Apart from self-efficacy, other SCT constructs include outcome expectations, goal setting and self-monitoring, and impediments and facilitators of behavior [82]. Studies have found that PA interventions aimed at improving the self-perception of exercise self-efficacy can have positive effects on confidence and the ability to initiate and maintain PA behavior [83,84] and that self-efficacy is a predictor of the adoption and maintenance of PA behaviors in adults [85–87].

The second most used theory was the trans-theoretical model (TTM) for behavior change, which proposes six stages of change: precontemplation, contemplation, preparation, action, maintenance, and termination [88]. Applied research based on this model attempts to tailor interventions to the stage of change of the participant.

Accordingly, studies that used SCT, SET, or TTM as their theoretical foundation developed education and individual counseling components that targeted motivation, self-belief, knowledge about exercise, and individual barriers to PA. Far fewer studies had a theoretical basis that guided the cultural adaptation of the intervention. Two studies, both in South Korea, made use of the theory of acculturation [89] and the documented link between increased acculturation and higher PA levels among immigrants [90]. These were the only studies in the review that focused on the cultural adaptation component on increasing participants' knowledge and skills for life in the host country, rather than making reference to practices and beliefs of the participants' culture of origin. This is a potentially interesting path to explore; however, in these two studies, the cultural adaptation component was delivered solely by a series of weekly text messages, and showed no significant advantage over the standard intervention alone except for a moderate decrease in BMI in the intervention group in one of the studies.

Only one study drew on the social-ecological model (SEM) for health promotion, which makes the model underutilized. The SEM extends motivation and behavioral theories to consider additional factors that influence health behaviors. It provides an overarching framework for analyzing the interrelationships between individuals and the social, physical, and policy environment [91]. Therefore, given a large number of correlates of PA behaviors, the SEM could have been a helpful tool in identifying barriers to PA beyond individual low motivation and designing cultural adaptations to overcome those barriers. An example of such a constraint from one of the studies under review is the intersection of crowded living conditions for older Turkish migrants in the Netherlands and their perception that exercising in front of younger family members is undignified [44].

Finally, three of the studies used the CBPR model. CBPR approaches are shaped by a commitment to building an equitable partnership between researchers and the people for whom the research is intended to be of use. The aim is to reduce inequities identified by community members themselves and engage in team building, sharing of resources, and the mutual exchange of expertise to inform study design and implementation [92]. CBPR has deepened understanding of the complex factors that influence health behaviors, led to new ideas and innovations, and is an expanding field of practice [93]. It has been shown to be an effective strategy for generating environmental changes to promote increased physical activity [94]. Researchers have found that CBPR projects lead to more sustainable public health interventions due to the careful tailoring of interventions to participants' needs, well-articulated division of responsibility, the potential to make use of community resources, and the strengthening of community skills, experience, and knowledge [95–98].

CBPR approaches vary considerably; public health interventions have made use of frameworks including action research, feminist action research, community coalitions, community advisory boards, consumer research, and participatory research teams. Methodologies include focus groups, interviews, photovoice, brainstorm sessions, and concept mapping, workshops, observations, field notes, and surveys [99]. It's possible to embed a CBPR approach within an SEM theoretical framework [100].CBPR is not without challenges; these include time constraints, finding appropriate community partners, gaining trust, reconciling differences within the community, addressing power imbalances, the risk of sampling bias, and sustainability [99]. Despite this, the successful scaling up and transfer of CBPR-based projects has been demonstrated, given sufficient research support and funding [101].

The studies in this review that used CBPR all achieved high retention rates, ranging from 83.2% (with all dropouts attributable to pregnancy) to 100% [58,64,70]. They also all had significant positive outcomes for clinical measures, although not all achieved significant improvements in PA levels. Looking more broadly, all studies that included some type of community input such as using focus groups to guide the intervention development had statistically significant positive outcomes. Given the general lack of theoretical/conceptual frameworks for developing cultural adaptations, CBPR seems to offer a promising approach for researchers to be able to understand the barriers and enablers to PA for specific CALD communities, and thereby adapt the content, delivery methods, and study design of PA interventions to better serve the participants.

Cultural adaptation

In the reviewed studies, the descriptions of how educational materials were culturally adapted tended to be of a very general nature and somewhat lacking in detail. Cultural adaptation and a focus on motivation and self-efficacy were often conflated, and the depth and quality of cultural adaptation varied.

Studies in which researchers were fluent in the CALD's community's language and had a good understanding of the specific cultural context were able to carry out more meaningful cultural adaptations. In contrast, several studies that trialed existing programs with a post hoc cultural adaptation found the programs to be no more effective than a straightforward translation of the program into the target language. Partnerships with community organizations and leaders were typically critical to a successful recruitment process. Furthermore, several studies pointed to the important role of peer educators, social connections, and buddy systems in achieving successful program adherence and retention rates.

As already mentioned, studies that reported engaging the community through focus groups had a statistically significant positive outcome, as did studies in which culturally relevant PA was used. This suggests that cultural adaptation can extend beyond translation and encompass the community's own perception of health priorities, concerns, and goals. Overall, the literature displayed limited willingness to consider the external environment, for example, unsafe neighborhoods, or specific personal conditions, for example, lack of space for exercise at home, as reasons for physical inactivity. These factors might have been given more prominence through the use of the SEM or CBPR model.

Ethical considerations

With regards to the ethics procedures, we found some practices that should be considered for improvement in future studies:

- Only one study explicitly reported on injury rates, which contributed to the overall low CASPRCT scores of the studies under review.
- While the majority of studies excluded participants on the basis of preexisting disease, injury, or risk factors for undertaking exercise, only one study reported that participants with risk factors were referred to a medical care provider after the participants' initial eligibility intake assessment.
- Two studies failed to report whether informed consent was obtained.
- Only one study reported modifying the consent process to reflect the low level of overall literacy among participants.
- Only one study reported signing a formal memorandum of understanding with a community partner that set out the roles, obligations, and processes of engagement.

Three of the studies delivered the intervention to the control group, either using waitlist control or after the study assessments had been completed. Two of these used a CBPR model, and of those one explicitly reported that the community partner organization had raised ethical concerns about delivering the intervention to only half the participants. In response to these concerns, the study design was modified to ensure all participants would receive the intervention. All three of these studies achieved high retention rates (ranging from 83.2% to 100%) and statistically significant outcomes.

Recommendations for future research

We recommend that future research engages older adults, men, and people with existing diseases or injuries. Studies should transparently report on informed consent processes, injury rates, and cost-effectiveness. The use of phone calls, pedometers, and automated individual tailoring tools such as online programs warrant further investigation. CBPR seems to be a promising approach for developing interventions that respond to the health goals, challenges, and capabilities of CALD populations. Future interventions should consider engaging long-term community partners to assist in study design and cultural adaptation of the intervention. Future research should report in more detail on the methodology for developing cultural adaptations and their specific content. Further, more studies using an MAI design that compares the effectiveness of various components of an intervention (e.g., education, counseling, exercise classes) are needed on CALD populations.

Recommendations for practitioners

Practitioners can adapt successful approaches from interventions by working with individuals to better understand their health goals and concerns, current activity levels, limiting beliefs or lack of knowledge about PA, and barriers such as time, cost, safety, or social environment issues. Practitioners should consider including their patients in the decision-making process about the treatment. Together with patients, they can develop achievable, individualized goals to gradually increase

PA levels. This can include increasing daily step counts, encouraging active transport, and focusing on culturally relevant PA/sport. Achieving change in PA behavior is a long-term 'project' and change in clinical markers may be visible several months after behavioral change. Therefore, follow-up phone calls to offer active support and the recruitment of family and/or peer support may be beneficial to prevent relapse.

Limitations

This review has several limitations. Even though we followed a structured search strategy and reviewed three databases, some rigorous methods generally applied in systematic literature reviews were not followed. For example, only one author screened, selected, and assessed the studies. Further, the majority of the included studies were conducted in the US with Hispanic communities. Therefore, the findings of this review may not be applicable in other contexts. The decision to only include papers published in English is another limitation and a review that is able to include publications in other languages might not be so US-centric. The critical appraisal tool used (CASPRCT) was developed to assess RCTs, while this review also included CCTs and MAIs. This may have led to a lower appraisal score for those studies. Finally, this review excluded some types of publications such as dissertations and gray literature, which may have led to the omission of relevant studies.

Conclusion

This review aimed to identify characteristics of PA interventions for CALD communities. Data from 23 RCTs, CCTs, and MAIs were analyzed. The studies were assessed based on several components, including how they carried out cultural adaptation; which components out of education, individualized counseling, and exercise classes they comprised; and whether they achieved statistically significant results across three domains—knowledge and beliefs about exercise, PA behaviors, and clinical markers. The synthesized data showed that interventions were more likely to achieve positive outcomes when there was meaningful and equitable engagement with CALD community partners at the stages of problem identification, setting health priorities, activity selection, study design, and recruitment process. There was good evidence that the use of phone calls to deliver individual counseling and motivate retention led to good outcomes. Studies that achieved statistically significant outcomes tended to be of longer duration; use translated material and bilingual instructors; promote culturally relevant activities; use pedometers for daily step tracking; and comprise at least two components out of three—education, individualized counseling, and exercise classes.

The review also remarked on commonly encountered challenges in recruitment, randomization, blinding, and measurement. Many studies excluded participants with preexisting diseases, which carries implications for the generalizability of findings to clinical practice. Many of these challenges are applicable to PA interventions not limited to those targeting CALD communities, although recruitment, randomization, and blinding can be especially difficult for interventions delivered to small, close-knit communities. We were guided by critical literature to draw attention to commonly encountered problems in this research field. Descriptions of cultural adaptations tended to lack depth and detail that would allow reproducibility. Active equitable engagement practices such as obtaining verbal consent where literacy is low or incorporating community feedback on study design were not widespread, and ethics reporting could be significantly improved.

References

[1] Wing RR. Long-term effects of a lifestyle intervention on weight and cardiovascular risk factors in individuals with type 2 diabetes mellitus: four-year results of the look AHEAD trial. Arch Intern Med 2010;170(17):1566–75.

[2] Appel LJ, et al. Effects of comprehensive lifestyle modification on blood pressure control: main results of the PREMIER clinical trial. JAMA 2003;289(16):2083–93.

[3] Elmer PJ, et al. Effects of comprehensive lifestyle modification on diet, weight, physical fitness, and blood pressure control: 18-month results of a randomized trial. Ann Intern Med 2006;144(7):485–95.

[4] Mann S, Beedie C, Jimenez A. Differential effects of aerobic exercise, resistance training and combined exercise modalities on cholesterol and the lipid profile: review, synthesis and recommendations. Sports Med 2014;44(2):211–21.

[5] Knowler WC, et al. Reduction in the incidence of type 2 diabetes with lifestyle intervention or metformin. N Engl J Med 2002;346(6):393–403.

[6] Knowler WC, et al. 10-year follow-up of diabetes incidence and weight loss in the diabetes prevention program outcomes study. Lancet 2009;374 (9702):1677–86.

[7] Bauman A, et al. Updating the evidence for physical activity: summative reviews of the epidemiological evidence, prevalence, and interventions to promote "active aging". Gerontologist 2016;56(Suppl. 2):S268–80.

[8] Lee IM, et al. Effect of physical inactivity on major non-communicable diseases worldwide: an analysis of burden of disease and life expectancy. Lancet 2012;380(9838):219–29.

[9] Lee IM. Exercise and physical health: cancer and immune function. Res Q Exerc Sport 1995;66(4):286–91.

[10] Penedo FJ, Dahn JR. Exercise and well-being: a review of mental and physical health benefits associated with physical activity. Curr Opin Psychiatry 2005;18(2):189–93.

[11] Das P, Horton R. Rethinking our approach to physical activity. Lancet 2012;380(9838):189–90.

[12] Brosse AL, et al. Exercise and the treatment of clinical depression in adults. Sports Med 2002;32(12):741–60.

[13] Di Bartolomeo G, Papa S. The effects of physical activity on social interactions: the case of trust and trustworthiness. J Sports Econ 2017;20(1):50–71.

[14] Messier SP, et al. Long-term exercise and its effect on balance in older, osteoarthritic adults: results from the fitness, arthritis, and seniors trial (FAST). J Am Geriatr Soc 2000;48(2):131–8.

[15] Steinberg M, et al. A sustainable programme to prevent falls and near falls in community dwelling older people: results of a randomised trial. J Epidemiol Community Health 2000;54(3):227–32.

[16] Cartee GD, et al. Exercise promotes healthy aging of skeletal muscle. Cell Metab 2016;23(6):1034–47.

[17] Gregg EW, et al. Physical activity and osteoporotic fracture risk in older women. Study of osteoporotic fractures research group. Ann Intern Med 1998;129(2):81–8.

[18] Panza GA, et al. Can exercise improve cognitive symptoms of Alzheimer's disease? J Am Geriatr Soc 2018;66(3):487–95.

[19] W.H. Organization. Physical activity; 2020 [cited 2021 24/07/2021].

[20] Varela AR, et al. Mapping the historical development of physical activity and health research: a structured literature review and citation network analysis. Prev Med 2018;111:466–72.

[21] Bento VF, et al. Impact of physical activity interventions on blood pressure in Brazilian populations. Arq Bras Cardiol 2015;105(3):301–8.

[22] Cornelissen VA, Smart NA. Exercise training for blood pressure: a systematic review and meta-analysis. J Am Heart Assoc 2013;2(1), e004473.

[23] Arija V, et al. Effectiveness of a physical activity program on cardiovascular disease risk in adult primary health-care users: the "Pas-a-Pas" community intervention trial. BMC Public Health 2017;17(1):576.

[24] Baker A, Sirois-Leclerc H, Tulloch H. The impact of long-term physical activity interventions for overweight/obese postmenopausal women on adiposity indicators, physical capacity, and mental health outcomes: a systematic review. J Obes 2016;2016:6169890.

[25] Chin S-H, Kahathuduwa CN, Binks M. Physical activity and obesity: what we know and what we need to know. Obes Rev 2016;17(12):1226–44.

[26] Jakicic JM, et al. Time-based physical activity interventions for weight loss: a randomized trial. Med Sci Sports Exerc 2015;47(5):1061–9.

[27] Quiles NN, Piao L, Ortiz A. The effects of exercise on lipid profile and blood glucose levels in people living with HIV: a systematic review of randomized controlled trials. AIDS Care 2020;32(7):882–9.

[28] O'Driscoll T, et al. A systematic literature review of sport and physical activity participation in culturally and linguistically diverse (CALD) migrant populations. J Immigr Minor Health 2014;16(3):515–30.

[29] El Masri A, Kolt GS, George ES. Physical activity interventions among culturally and linguistically diverse populations: a systematic review. Ethn Health 2019;1–21.

[30] Caperchione CM, Kolt GS, Mummery WK. Physical activity in culturally and linguistically diverse migrant groups to Western society: a review of barriers, enablers and experiences. Sports Med 2009;39:167.

[31] Australia. Commonwealth Interdepartmental Committee on Multicultural, A. In: I. Australia. Department of and A. Multicultural, editor. The guide : implementing the standards for statistics on cultural and language diversity / a publication of the Commonwealth Interdepartmental Committee on Multicultural Affairs, a committee managed by DIMA. Canberra: Commonwealth Interdepartmental Committee on Multicultural Affairs; 2001.

[32] Montayre J, et al. What makes community-based physical activity programs for culturally and linguistically diverse older adults effective? A systematic review. Australas J Ageing 2020;39(4):331–40.

[33] Statistics, A.B.o. Migrants and participation in sport and physical activity; 2006. Canberra, Australia.

[34] Michael J, Aylen T, Ogrin R. Development of a translation standard to support the improvement of health literacy and provide consistent high-quality information. Aust Health Rev 2013;37(4):547–51.

[35] Nelson A. Unequal treatment: confronting racial and ethnic disparities in health care. J Natl Med Assoc 2002;94(8):666–8.

[36] Henderson S, Kendall E. Culturally and linguistically diverse peoples' knowledge of accessibility and utilisation of health services: exploring the need for improvement in health service delivery. Aust J Prim Health 2011;17(2):195–201.

[37] Derose KP, Escarce JJ, Lurie N. Immigrants and health care: sources of vulnerability. Health Aff 2007;26(5):1258–68.

[38] Harrison R, et al. Beyond translation: engaging with culturally and linguistically diverse consumers. Health Expect 2020;23(1):159–68.

[39] White J, et al. What is needed in culturally competent healthcare systems? A qualitative exploration of culturally diverse patients and professional interpreters in an Australian healthcare setting. BMC Public Health 2019;19(1):1096.

[40] Berry JW, Kim U. Acculturation and mental health. In: Health and cross-cultural psychology: toward applications. Thousand Oaks, CA, US: Sage Publications, Inc; 1988. p. 207–36.

[41] Waters LA, et al. Who participates in physical activity intervention trials? J Phys Act Health 2011;8(1):85–103.

[42] Caperchione CM, Kolt GS, Mummery WK. Examining physical activity service provision to culturally and linguistically diverse (CALD) communities in Australia: a qualitative evaluation. PLoS One 2013;8(4), e62777.

[43] Spaaij R, et al. Sport, refugees, and forced migration: a critical review of the literature. Front Sports Act Living 2019;1:47.

[44] Reijneveld SA, Westhoff MH, Hopman-Rock M. Promotion of health and physical activity improves the mental health of elderly immigrants: results of a group randomised controlled trial among Turkish immigrants in the Netherlands aged 45 and over. J Epidemiol Community Health 2003;57(6):405–11.

[45] Chiang CY, Sun FK. The effects of a walking program on older Chinese American immigrants with hypertension: a pretest and posttest quasi-experimental design. Public Health Nurs 2009;26(3):240–8.

[46] Borschmann K, Moore K, Russell M, Ledgerwood K, Renehan E, Lin X, Brown C, Sison J. Overcoming barriers to physical activity among culturally and linguistically diverse older adults: a randomised controlled trial. Australas J Ageing 2010;29(2):77–80.

[47] Sundquist J, Hagströmer M, Johansson SE, Sundquist K. Effect of a primary health-care-based controlled trial for cardiorespiratory fitness in refugee women. BMC Fam Pract 2010;11:55.

[48] Qi BB, Resnick B, Smeltzer SC, Bausell B. Self-efficacy program to prevent osteoporosis among Chinese immigrants: a randomized controlled trial. Nurs Res 2011;60(6):393–404.

[49] Andersen E, Burton NW, Anderssen SA. Physical activity levels six months after a randomised controlled physical activity intervention for Pakistani immigrant men living in Norway. Int J Behav Nutr Phys Act 2012;9:47.

[50] Andersen E, Høstmark AT, Anderssen SA. Effect of a physical activity intervention on the metabolic syndrome in Pakistani immigrant men: a randomized controlled trial. J Immigr Minor Health 2012;14(5):738–46.

[51] Gademan MGJ, Deutekom M, Hosper K, Stronks K. The effect of exercise on prescription on physical activity and wellbeing in a multi-ethnic female population: a controlled trial. BMC Public Health 2012;12(1):758.

[52] Marcus BH, Dunsiger SI, Pekmezi DW, Larsen BA, Bock BC, Gans KM, et al. The Seamos Saludables study: a randomized controlled physical activity trial of Latinas. Am J Prev Med 2013;45(5):598–605.

[53] Hawkins M, Chasan-Taber L, Marcus B, Stanek E, Braun B, Ciccolo J, et al. Impact of an exercise intervention on physical activity during pregnancy: the behaviors affecting baby and you study. Am J Public Health 2014;104(10):e74–81.

[54] Nobles C, Marcus BH, Stanek 3rd EJ, Braun B, Whitcomb BW, Solomon CG, Manson JE, Markenson G, Chasan-Taber L. Effect of an exercise intervention on gestational diabetes mellitus: a randomized controlled trial. Obstet Gynecol 2015;125(5):1195–204.

[55] Khare MM, Cursio JF, Locklin CA, Bates NJ, Loo RK. Lifestyle intervention and cardiovascular disease risk reduction in low-income hispanic immigrant women participating in the Illinois WISEWOMAN program. J Community Health 2014;39(4):737–46.

[56] Lee H, Chae D, Wilbur J, Miller A, Lee K, Jin H. Effects of a 12 week self-managed stretching program among Korean-Chinese female migrant workers in Korea: a randomized trial. Jpn J Nurs Sci 2014;11(2):121–34.

[57] Lee H, Chae D, Cho S, Kim J, Yoo R. Influence of a community-based stretching intervention on the health outcomes among Korean-Chinese female migrant workers in South Korea: a randomized prospective trial. Jpn J Nurs Sci 2017;14(4):277–87.

[58] Kandula NR, Dave S, De Chavez PJ, Bharucha H, Patel Y, Seguil P, Kumar S, Baker DW, Spring B, Siddique J. Translating a heart disease lifestyle intervention into the community: the South Asian heart lifestyle intervention (SAHELI) study; a randomized control trial. BMC Public Health 2015;15:1064.

[59] Martin L, Signorile JF, Kahn BE, Perkins AW, Ahn S, Perry AC. Improving exercise adherence and physical measures in English-speaking Latina women. J Racial Ethn Health Disparities 2015;2(4):517–26.

[60] Chee W, et al. The effect of a culturally tailored web-based physical activity promotion program on Asian American midlife women's depressive symptoms. Asian Pac Isl Nurs J 2016;1:162–73.

[61] Chee W, Ji X, Kim S, Park S, Zhang J, Chee E, Tsai HM, Im EO. Recruitment and retention of Asian Americans in web-based physical activity promotion programs: a discussion paper. Comput Inform Nurs 2019;37(9):455–62.

[62] Chee W, Kim S, Tsai HM, Im EO. Decreasing sleep-related symptoms through increasing physical activity among Asian American midlife women. Menopause 2019;26(2):152–61.

[63] Chee W, Kim S, Tsai HM, Liu J, Im EO. Effect of an online physical activity promotion program and cardiovascular symptoms among Asian American women at midlife. Comput Inform Nurs 2020;26:198–207.

[64] Yeh MC, Heo M, Suchday S, Wong A, Poon E, Liu G, Wylie-Rosett J. Translation of the diabetes prevention program for diabetes risk reduction in Chinese immigrants in New York City. Diabet Med 2016;33(4):547–51.

[65] Lee H, Cho S, Wilbur J, Kim J, Park CG, Lee YM, Lee H. Effects of culturally adaptive walking intervention on cardiovascular disease risks for middle-aged Korean-Chinese female migrant workers. Arch Environ Occup Health 2017;72(6):317–27.

[66] Piedra LM, et al. The influence of exercise on cognitive function in older hispanic/latino adults: results from the "¡Caminemos!" study. Gerontologist 2017;57(6):1072–83.

[67] Siddiqui F, Kurbasic A, Lindblad U, Nilsson PM, Bennet L. Effects of a culturally adapted lifestyle intervention on cardio-metabolic outcomes: a randomized controlled trial in Iraqi immigrants to Sweden at high risk for type 2 diabetes. Metabolism 2017;66:1–13.

[68] Siddiqui F, Koivula RW, Kurbasic A, Lindblad U, Nilsson PM, Bennet L. Physical activity in a randomized culturally adapted lifestyle intervention. Am J Prev Med 2018;55(2):187–96.

[69] An M, Nahm ES, Shaughnessy M, Storr CL, Han HR, Lee J. A pilot primary stroke prevention program for elderly Korean Americans. J Neurosci Nurs 2018;50(6):327–33.

[70] D'Alonzo KT, Smith BA, Dicker LH. Outcomes of a culturally tailored partially randomized patient preference controlled trial to increase physical activity among low-income immigrant Latinas. J Transcult Nurs 2018;29(4):335–45.

[71] Menkin JA, McCreath HE, Song SY, Carrillo CA, Reyes CE, Trejo L, Choi SE, Willis P, Jimenez E, Ma S, Chang E, Liu H, Kwon I, Kotick J, Sarkisian CA. "Worth the walk": culturally tailored stroke risk factor reduction intervention in community senior centers. J Am Heart Assoc 2019; 8(6), e011088.

[72] Towfighi AC, Hill EM, Barry VA, Lee F, Valle M, Mittman NP, Ayala-Rivera B, Moreno M, Espinosa L, Dombish A, Wang H, Ochoa D, Chu D, Atkins A, Vickrey M, Barbara G. Results of a pilot trial of a lifestyle intervention for stroke survivors: healthy eating and lifestyle after stroke. J Stroke Cerebrovasc Dis 2020;29(12):105323.

[73] Daniel M, Marquez D, Ingram D, Fogg L. Group dance and motivational coaching for walking: a physical activity program for South Asian Indian immigrant women residing in the United States. J Phys Act Health 2021;18:262–71.

[74] Burn NL, Weston M, Atkinson G, Weston KL. Using focus groups and interviews to inform the design of a workplace exercise programme: an example from a high-intensity interval training intervention. J Occup Environ Med 2021;63(2):e63–74.

[75] Yardley L, et al. The person-based approach to intervention development: application to digital health-related behavior change interventions. J Med Internet Res 2015;17(1):e30.

[76] Mathews AE, et al. Older adults' perceived physical activity enablers and barriers: a multicultural perspective. J Aging Phys Act 2010;18(2):119–40.

[77] Azevedo MR, et al. Gender differences in leisure-time physical activity. Int J Public Health 2007;52(1):8–15.

[78] Edwards ES, Sackett SC. Psychosocial variables related to why women are less active than men and related health implications. Clin Med Insights Womens Health 2016;9(Suppl. 1):47–56.

[79] Segar M, et al. Fitting fitness into women's lives: effects of a gender-tailored physical activity intervention. Womens Health Issues 2002;12(6):338–47.

[80] Bandura A. Social foundations of thought and action: a social cognitive theory; 1986. p. xiii. 617-xiii, 617.

[81] Bandura A. Self-efficacy: the exercise of control. New York, NY, US: W H Freeman/Times Books/ Henry Holt & Co; 1997. p. ix. 604-ix, 604.

[82] McAuley E, Szabo A, Gothe N, Olson EA. Self-efficacy: implications for physical activity, function, and functional limitations in older adults. Am J Lifestyle Med 2011;5(4). https://doi.org/10.1177/1559827610392704.

[83] Lee LL, Arthur A, Avis M. Using self-efficacy theory to develop interventions that help older people overcome psychological barriers to physical activity: a discussion paper. Int J Nurs Stud 2008;45(11):1690–9.

[84] Dutton GR, et al. Relationship between self-efficacy and physical activity among patients with type 2 diabetes. J Behav Med 2009;32(3):270–7.

[85] Sharma M, Sargent L, Stacy R. Predictors of leisure-time physical activity among African American women. Am J Health Behav 2005;29(4):352–9.

[86] Rovniak LS, et al. Social cognitive determinants of physical activity in young adults: a prospective structural equation analysis. Ann Behav Med 2002;24(2):149–56.

[87] Bauman AE, et al. Correlates of physical activity: why are some people physically active and others not? The Lancet 2012;380(9838):258–71.

[88] Prochaska JO, Velicer WF. The transtheoretical model of health behavior change. Am J Health Promot 1997;12(1):38–48.

[89] Berry JW. Theories and models of acculturation. In: Schwartz SJ, Unger JB, editors. The Oxford handbook of acculturation and health. New York, NY, US: Oxford University Press; 2017. p. 15–28.

[90] Gerber M, Barker D, Pühse U. Acculturation and physical activity among immigrants: a systematic review. J Public Health 2012;20(3):313–41.

[91] Stokols D. Translating social ecological theory into guidelines for community health promotion. Am J Health Promot 1996;10(4):282–98.

[92] Jull J, Giles A, Graham ID. Community-based participatory research and integrated knowledge translation: advancing the co-creation of knowledge. Implement Sci 2017;12(1):150.

[93] Horowitz CR, Robinson M, Seifer S. Community-based participatory research from the margin to the mainstream: are researchers prepared? Circulation 2009;119(19):2633–42.

[94] Suminski RR, et al. Neighborhoods on the move: a community-based participatory research approach to promoting physical activity. Prog Community Health Partnersh 2009;3(1):19–29.

[95] Jagosh J, et al. Uncovering the benefits of participatory research: implications of a realist review for health research and practice. Milbank Q 2012;90(2):311–46.

[96] Rifkin SB. Examining the links between community participation and health outcomes: a review of the literature. Health Policy Plan 2014;29 Suppl 2(Suppl. 2):ii98–106.

[97] Shediac-Rizkallah MC, Bone LR. Planning for the sustainability of community-based health programs: conceptual frameworks and future directions for research, practice and policy. Health Educ Res 1998;13(1):87–108.

[98] Wiltsey Stirman S, et al. The sustainability of new programs and innovations: a review of the empirical literature and recommendations for future research. Implement Sci 2012;7:17.

[99] Quirk H, Speake H. Doncaster community engagement & physical activity literature review: a review of community engagement and community-based participatory research approaches in the development of community-based physical activity interventions; 2018.

[100] Bammann K, et al. Promoting physical activity among older adults using community-based participatory research with an adapted PRECEDE-PROCEED model approach: the AEQUIPA/OUTDOOR ACTIVE project. Am J Health Promot 2021;35(3):409–20.

[101] Herbert-Maul A, Abu-Omar K, Frahsa A, Streber A, Reimers AK. Transferring a community-based participatory research project to promote physical activity among socially disadvantaged women—experiences from 15 years of BIG. Front Public Health 2020;8(546).

Chapter 4

Physical activity adherence: Worldwide trends, barriers and facilitators and tools to improve it

Katherine Harkin[a,b], Brigitte Marie Pascal[c], Susan Irvine[b], Kathy Tangalakis[b], and Vasso Apostolopoulos[a,d]

[a]Institute for Health and Sport, Victoria University, Melbourne, VIC, Australia [b]First Year College, Victoria University, Melbourne, VIC, Australia [c]College of Arts and Education, Victoria University, Melbourne, VIC, Australia [d]The Australian Institute for Musculoskeletal Science, Western Health, Victoria University and the University of Melbourne, Melbourne, VIC, Australia

Introduction

The World Health Organization (WHO) estimated that in 2008, 36 million deaths worldwide were due to noncommunicable diseases (NCD), with the four most common causes being cardiovascular disease (48%), cancer (21%), chronic respiratory diseases (12%) and diabetes (3.5%) [1]. Within 10 years, this has increased to 41 million, which constituted 70% of total global mortality [2]. There are four common modifiable behavioral risk factors that collectively influence these four main NCDs: tobacco use, unhealthy diet, alcohol use, and physical inactivity [1]. The global impact of NCDs has become so significant that in 2011 a political declaration was embraced by the United Nations General Assembly requiring member states to act in order to reduce their prevalence [3].

Physical activity recommendations

The *2020 WHO Guidelines on Physical Activity and Sedentary Behavior* estimates that if the worldwide population were more physically active, between 4 and 5 million deaths could be averted annually [4]. Literature has consistently confirmed that sufficient levels of physical activity can, in part, treat and prevent many of the NCDs by improving associated risk factors such as adiposity, hypertension as well as mental health, and quality of life [5]. In addition to decreasing the incidence of disease and mortality rates, increasing levels of physical activity can reduce national health care costs. In the year 2013 alone, it was estimated that physical inactivity placed a global financial burden on health care services to the cost of $54 billion International Dollars (INT) [5,6].

In 2010, the WHO published a set of evidence-based global physical activity guidelines [7] that have informed the recommendations implemented by various organizations and governments worldwide [8–11]. These guidelines were updated in 2020 to include added recommendations for children under the age of 5, people with certain chronic conditions, and pregnant and postpartum women [4]. However, much of the available evidence is based upon the earlier 2010 WHO guidelines [7], outlined in Fig. 1.

Global physical inactivity trends

The establishment of physical activity recommendations is important to help guide communities in improving their collective health outcomes. However, if effective strategies do not successfully engage people to become physically active, then the benefits are lost. Obtaining physical inactivity data and their respective demographic components (age, sex, income level, etc.) in specific regions throughout the world enables researchers to establish relationships that can then be further investigated to help identify key determinants of physical inactivity.

Retrieval of data on physical inactivity levels is commonly done through large national databases, which largely rely upon surveys utilizing self-report of activity levels. These methods of data collection produce results that are prone to overestimation in part due to the phenomenon of the Hawthorne effect [12]. The Hawthorne effect is a change in behavior of

Exercise to Prevent and Manage Chronic Disease Across the Lifespan. https://doi.org/10.1016/B978-0-323-89843-0.00015-5

Children 5-17 years of age

> 60 minutes Moderate to Vigorous Physical Activity (MVPA) daily

• *Amounts > 60 minutes provide additional health benefits*
•*Include bouts of vigorous intensity*
•*At least 3 times per week*
•*Activity should mostly be aerobic and include aspects of strengthening muscle and bone*

Adults 18-64 years of age

> 150 minutes Moderate Intensity Physical Activity or 75 minutes Vigorous Intensity Physical Activity or an equal combination of both per week

•*Increase to 300 minutes MIPA or 150 minutes VIPA or an equal combination of both per week to provide additional health benefits*
•*Aerobic bouts at least 10 minutes in duration Include aspects of strengthening major muscle groups at least twice per week*

Older adults 65 years of age and above

> 150 minutes Moderate Intensity Physical Activity or 75 minutes Vigorous Intensity Physical Activity or an equal combination of both per week

•*Increase to 300 minutes MIPA or 150 minutes VIPA or an equal combination of both per week to provide additional health benefits*
•*Aerobic bouts at least 10 minutes in duration Include aspects of strengthening major muscle groups at least twice per week*
•*Include aspects that address balance and prevent falls at least three times per week for those with poor mobility*
•*Older adults with health related limitations should be as active as they are capable*

FIG. 1 Physical activity recommendations. The World Health Organization (WHO) recommendations of physical activity, intensity, and duration for children 5–17 years of age, adults 18–64 years, and older adults 65 years of age and above. *(Adapted from W.H. Organisation. Global recommendations on physical activity for health. 2010;1–60.)*

participants in a study, occurring as a result of their awareness of being observed. Thus, participants can be prone to mis-reporting their reported levels of physical activity within a study and to a lesser extent within national census population surveys [12].

The alternative to this method of data collection is the use of digital monitors (mobile apps and wearable monitors), which can provide objective, detailed information such as step count, time inactivity, and heart rate, and can be accessed remotely via a digital platform [13]. Importantly, they are less prone to recall bias and can impact motivation [14]. However there are limitations—cohorts with poor digital literacy, such as the elderly and lower socioeconomic populations [5], inaccurate physical activity measurements due to the poor positioning of the monitor e.g. worn on the leg or wrist [4], and different time point measurements such as daily, weekly, or monthly [15].

Adults

One of the largest observational studies into worldwide trends of adult physical inactivity levels compiled self-reported data from 1.9 million participants aged 18 years and older from 168 countries, between 2001 and 2016 [16]. The assessment of "inactivity" was based upon individuals not achieving the WHO 2010 physical activity recommendations [7] of 150 min of aerobic MIPA or 75 min of aerobic VIPA per week or an equivalent combination of the two. The researchers reported the global prevalence of insufficient physical activity for adults in 2001 was on average 28.5% after which it decreased to 23.3% in 2010, but rose again to 27.5% in 2016 [16].

When comparing prevalence between regions, there was twice as much insufficient physical activity in high-income countries when compared to low-income countries [16]. However, it was mentioned that there was more data available from higher-income countries than from low-income countries, which may have influenced the results. Furthermore, there were large discrepancies between some countries within the same regions.

Further analysis of the demographic data indicates that in 2016, women had higher levels of insufficient physical activity (32%) compared to men (23%), except for East and Southeast Asia, where inactivity levels in women were marginally less than men. Furthermore, women from poorer countries do less leisure-time and low-intensity physical activity compared to women in wealthy countries [16,17].

Adolescents and children

WHO researchers who analyzed global adult physical inactivity levels [16], also conducted a pooled analysis of cross-sectional survey data of the insufficient physical activity in adolescents aged 11–17 years between 2001 and 2016 [18]. Data was used from 298 school-based surveys of 1.6 million students from 146 countries. The status of "inactivity" was defined as failure to achieve the 2010 WHO physical activity recommendations for children and adolescents [7] of 60 min of MVPA daily. It was reported that, in 2001, 80.1% of boys and 85.1% of girls were insufficiently physically active. The levels improved for boys in 2016 to 77.6%, however, there was no significant change for girls. Interestingly, there were 27 countries that had insufficient physical activity levels of 90% or more for girls, whereas there were only two countries where inactivity of 90% or more occurred for boys, with the highest prevalence in both sexes occurring in the Philippines and South Korea [18].

There was no significant difference between low-income countries (84.9% average of all adolescents aged 11–17), low-middle-income countries (79.3%), upper-middle-income countries (83.9%), and high-income countries (79.4%) in 2016.

The International Children's Accelerometry Database (ICAD) is an open-access repository of data regarding physical activity levels of young people aged 3–18 years of age produced by researchers from Europe, North and South America, Brazil, and Australia between 2008 and 2010. ICAD originally contained data obtained from over 45,000 children worldwide using Acti-Graph accelerometers [15], however now includes longitudinal data as well as data obtained from nonaccelerometry methods.

Since its inception, the ICAD data has been used by numerous researchers worldwide to help answer important questions regarding child health and physical activity. One of the most comprehensive and most cited papers to include ICAD data was in 2015 [15], which investigated the physical activity data from 27,637 children between 3 and 18 years of age, showing associations between age, sex, weight status, and country. From the age of 2, levels of physical activity progressively increase, until age 5, where they slowly decline until 18 years of age, among all countries. One specific analysis within this paper showed that only 9% of males and 1.9% of females in the 5–17 age range achieved the WHO recommendations of 60 min of MVPA per day globally. However, the levels of physical activity varied greatly between countries for different age groups [15].

Another comprehensive study of physical activity levels studied almost half a million adolescents aged 11, 13, and 15 years of age in 32 countries from Europe and North America [19]. The data was derived from the Health Behavior in School-Aged Children (HBSC) study (now a collaborative section of the WHO) which uses self-reported surveys collected between 2002 and 2010. The authors found that when the three age groups were combined, on average, 23% of boys and 14% of girls achieved the WHO recommendations of ≥60 min per day of MVPA, higher than those in the previous studies [15]. For example, the accelerometer study [15] reported physical activity levels in Norwegian boys (aged 9–10) to be 13% and in the HBSC study [19] it was 19.5% (data from 2010, in 11, 13, and 15-year-old males). However, the studies are hard to compare as sample sizes, regions, ages, and timeframes differ. However, irrespective of the demographic differences the prevalence reported is not dissimilar, and the physical activity level decline from approximately 5 years of age through adolescence was consistent with earlier reports.

Determinants of physical activity

While cross-sectional studies can determine the prevalence of physical inactivity, studies that have an etiological component, enable causative relationships to be established. Many of these studies have revealed a variety of determining factors that influence physical activity in certain populations. These factors are either barriers or facilitators to behavior change in individuals, and so, it is important to understand the theories that underpin them.

The theory of planned behavior

Previously called the Theory of Reasoned Action in 1980, this behavioral theory focuses on the ability to explain and predict an individual's likelihood to make a change in health behavior at a particular location and time frame. This theory is based on the assumption that the ability to exert self-control is dependent upon intention, motivation, and capability [20].

The theory of planned behavior (TPB) begins with a definition of the *goal*, the *action* required to achieve that goal, the *context*, and the expected *time* frame to achieve that goal, with identification of an individual's intention to help predict their likelihood of success.

The originator of this theory, Icek Ajzen a social psychologist, describes the three factors that influence the behavioral intention (the strength of the motivating factors that drive engagement of the new behavior) are [20]:

1. *Attitude to the behavior:* the preexisting beliefs surrounding the particular behavioral change and the possible positive outcomes associated.
2. *Subjective norm:* the individual perception of external beliefs or attitudes of others or their community (social norms) to their potential behavioral change
3. *Perceived behavioral control:* an individual's perception of their self-efficacy at engaging in the behavioral change which includes their sense of control over any facilitating or limiting factors

Some limitations of this theory are that many influences on an individual's access to resources are required to engage in the change regardless of intention, considering any unpredictable economic, environmental, or emotional factors such as fear, threat, or previous experiences. The theory also does not account for stagnation or reversal of decision making nor any indication of timeframes between intent and behavioral action.

The "Great Live and Move Challenge" (GLMC) is a TPB-based intervention that was used in a large prospective cluster randomized controlled trial (RCT) aiming to promote physical activity levels among children aged 7–11 years of age [21]. The GLMC is a multilevel intervention requiring preparation and input from various sources such as family, school, and community. The challenge is separated into two main active phases—a motivational phase designed to encourage, educate, scaffold, goal setting, and plan with children in how to engage in physical activity, and a volitional phase where the provision of family, school, and community-based events facilitates the engagement in physical activity for children. This design is founded in the TPB model where predetermined attitudes and subjective norms are identified and potentially influenced in the motivational phase and then confidence and efficacy in engaging in physical activity are carried out during the volitional phase.

Social cognitive theory

The social cognitive theory of behavioral change takes into consideration an individual's past experiences, immediate social influences, expectations, and reinforcement which contribute to behavioral change. The context of this theory also allows for explanation and facilitation of long-term maintenance of behavioral change rather than just initiation of change. This theory consists of six constructs [22]:

1. Understanding of the associated benefits and potential risks
2. Perceived self-efficacy of an individual's capability of behavioral modification is based on the assumption that they possess the relevant knowledge and skills to achieve it
3. Expectation of outcomes, costs, and benefits as a result of behavioral modification and if that facilitates further successful engagement or not
4. Reinforcement feedback and support whether positive or negative from others or the environment around them or internally
5. Health-related goals
6. Local barriers or facilitators

Some limitations apply with this theory such as lack of clarity around which of the person, environment, and behavior are more influential, and the assumption that a change in environment automatically triggers a change in the individual. Additionally, by focusing more on social influences, there is a disregard for internal influences such as motivation, emotion, or biological predispositions of the individual.

An RCT that used an online Facebook-based portal to deliver information on a physical activity intervention to a group of young (21–39 years of age) cancer survivors was based on social cognitive theory [23]. The intervention involved directly on the type and frequency of the activity regimen, advice on goal setting, and possible barriers that may arise. Most importantly, the online environment also allowed for social support between participants which encouraged further engagement in the regime.

Self-determination theory

Self-determination theory (SDT) is a framework focusing on the intrinsic motivational components that underpin and drive human development and wellness. It is proposed within this theory that the three basic psychological needs of competence, autonomy, and relatedness need to be facilitated in order for development to occur [24]. In the context of health behavior changes, these three psychological needs must be supported within the intervention for the intrinsic motivators to drive the behavioral change [25]. In comparison to other behavioral change models which focus on extrinsic motivators such as financial reward or instruction from a health professional, SDT focuses on intrinsic motivators such as enjoyment and satisfaction within oneself, which are believed to be stronger drivers for initiation and ongoing behavior change [26].

Transtheoretical model

The transtheoretical model (TTM) was originally proposed in 1982 based upon the earlier research of the Stages-of-Change model and "self-changers" in the context of smoking cessation [27] and the idea that an individual undergoes various stages of "readiness" when making health behavioral changes. It is based on five main stages (Fig. 2): precontemplation, where the individual is not thinking about change; contemplation, where the individual is now genuinely considering in favor of or not in favor of change; toward preparation, where scheduling, arrangement, and commitment are in place. The successful achievement of these preliminary stages allows the person to take action to create the desired behavioral change. If accomplished, action advances to the final stage of maintenance, where the individual, with effort, strives to maintain long-lasting change [28].

An individual can enter the process at various stages and can either stay stagnant, relapse or progress depending on the presence and/or effectiveness of an intervention [29]. This model has been useful in recognizing what stage an individual may be in, and tailoring interventions to target progression from one stage to the next, which has been shown to be more effective than nontailored interventions [30]. Despite evidence supporting this model, some dispute its validity to all types of behavioral change processes. While the original conception of the TTM was around the cessation of unhealthy behavior (smoking), other situations require adherence and adoption of new behaviors such as increasing and sustaining physical activity levels. Others suggest that the stages may be more akin to a continuum so rather than "processes of change" and "self-efficacy" being determinants of stage progression, they are actually secondary outcomes [31]. Thus, actual changes in behavior (increase in physical activity levels) become the determining factor for progression through the later stages and thus causing a change in decision making and self-efficacy.

In 2002, an RCT focused on using a TTM based intervention to promote and encourage participation in physical activity in a group of young individuals [32]. The study was conducted over a 6-month period and the intervention group received two sets of mail-out pamphlets that had information and questions centered on motivating the participants to engage in physical activity. The first pamphlet was mailed to the intervention group participants at the commencement of the study and contained questions and prompts that were related to the first three stages of change, and the second sent 3 months later contained questions and prompts relating to the later two stages. This intervention showed a modest improvement in physical activity compared to a control group [32].

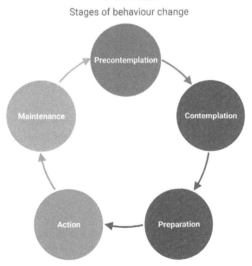

The transtheoretical model

Stages of behaviour change

Precontemplation

Contemplation

Maintenance

Preparation

Action

FIG. 2 Stages of change.

Barriers and facilitators

Many of the barriers and facilitators to physical activity commonly identified in the literature fall into four domains: accessibility/environmental, biophysical, aspirational/psychological, and social/cultural. Bauman et al. [33] describe the use of an "ecological model" to help consider factors from various domains that may be more influential in certain cohorts depending on their specific demographics. This helps to build a framework that comprehensively explains which determinants affect certain populations, and therefore informs local interventions and policies aimed at increasing physical activity.

Data on important barriers and facilitators of physical activity in healthy adult populations are shown in Table 1.

While the majority of this data was derived from developed Western countries common trends are apparent. In studies that looked at older adults, factors such as fear of injury, social engagement, and chronic conditions featured more heavily, which is likely due to the increased likelihood of comorbidities and social isolation which are more prevalent in these cohorts.

When considering barriers to physical activity in childhood cohorts, it is important to consider that there is often a flow-on effect so that barriers that restrict adults can inherently become barriers for children. In one study, it was reported that children from a higher SES background were more likely to meet physical activity guidelines, likely due to financial and transport factors enabling children to participate more frequently in activities [19]. Additionally, it was suggested that less active modes of transport and more sedentary-based leisure activities in communities may contribute to activity level plateau or decrease -conclusions which are supported in other studies [35].

Another study analyzed data from ICAD and made correlations between physical activity levels of children and weather features in their respective countries [37]. It found that higher levels of physical activity were associated with temperatures between 0 and 20 degrees Celsius with activity levels decreasing in temperatures outside of that range. The exception to this trend was in Melbourne, Australia, and Northern Europe where children appeared to maintain activity levels outside those temperatures suggesting acclimatization to these temperatures or activity infrastructure which facilitates ongoing activity throughout the seasons.

A comprehensive European study drew data from the 2013 Eurobarometer survey of 27,919 participants and compared influencing factors to sedentary behavior according to age group: young (15–25 years), adult (26–44 years), middle-aged (45–64 years), and older (65 years+) [38]. They found that in the young, adult, and middle-aged groups, occupation was the largest barrier or facilitator of sedentary behavior. This was suggested to occur because the type of occupation can determine financial, geographical, educational, and biological accessibility to physical activity resources either directly or indirectly.

Physical inactivity levels are more prevalent in higher-income countries compared to lower [16]. This difference has been suggested to be due to more common sedentary work behaviors and motorized transportation methods in higher-income countries. Another study published in 2016, investigated the correlation between urban environmental features of 14 cities and 10 countries (from both low and high-income levels), and adult physical activity levels [39]. They found that there were three main factors that correlated with high physical activity levels: Net residential density, whereby the higher density leads to more patronage in local shops and services hence contributing to the "walkability" of an environment, public transport density, which is interdependent with residential density and leisure park density, to cater for leisure activity. It was calculated that by improving these environmental attributes, it would enable residents to achieve 45%–59% of the WHO recommended guidelines.

Differences in physical activity levels between females and males are similar in both youth and adult populations. As highlighted by global data, females are more physically inactive than males [16,18], however, females in lower-income countries have less leisure-time physical activity, potentially due to affordability, safety, and accessibility factors [16,17]. In some regions of the world, other social and cultural factors that limit participation in leisure-time physical activity [40].

Barriers and facilitators of physical activity vary widely across minority groups. A study on the motivation to exercise in paraplegic patients revealed that avoidance of ill-health was the strongest facilitator, closely followed by physical fitness improvement [41]. Additionally, it was noted that participants also preferred exercise programs that incorporated a competitive and enjoyment factor, suggesting strong intrinsic motivators in these minority groups. Similarly, other research identifies ill-health as a biophysical barrier to exercise while fear of further ill-health was simultaneously a strong motivator for other groups such as hemodialysis [42], severe mental illness [43], cardiac [44], chronic obstructive pulmonary disease [45] and cancer [46] patients.

The multifactorial nature of the determinants of physical activity is incorporated by the WHO in their strategy [5,47]. They focus on Active Societies, Active Environments, Active People, and Active Systems—highlighting the importance of addressing multiple determinants to ensure improvements in overall health outcomes.

TABLE 1 Determinants of physical activity (PA) in adult populations.

	Accessibility/environmental	Biophysical	Aspirational/psychological	Social/cultural
Adults				
Kelly et al. [34] Systematic Review of 19 studies Adults 40–64 years from mostly developed countries worldwide Funded by NICE	*Barriers to PA:* Time limitations, poor weather, poor environmental safety, lack of access to recreational facilities, financial costs, transport limitations, technological illiteracy *Facilitators to PA:* Availabilities to facilities	*Barriers to PA:* Physical ailments, chronic conditions *Facilitators to PA:* Health benefits, fear of ill-health that restricted lifestyle, weight loss	*Barriers to PA:* Lack of knowledge of benefits, poor motivation, low confidence *Facilitators to PA:* Enjoyment, integration of behaviors into lifestyle, greater self-esteem and confidence as motivators, body image	*Barriers to PA:* Social concerns, language and cultural barriers, low SES *Facilitators to PA:* Social support
Older adults				
Gomes et al. [35] Cross-sectional analysis Adults 55 years + 19,298 participants Survey of Health, Aging, and Retirement in Europe (SHARE)		*Barriers to PA:* Increasing age, Physical limitations	*Barriers to PA:* Depression, poor sense of life meaning, memory loss	*Barriers to PA:* Lack of social support
Rosenkranz et al. (2013) Review of 29 studies Adults 65 years + International populations	*Barriers to PA:* Time limitations, financial costs, poor weather, lack of transport, poor accessibility to facilities *Facilitators to PA:* Good transport and facilities	*Barriers to PA:* Ill-health, physical inability *Facilitators to PA:* Perceived physical health benefits,	*Barriers to PA:* Fears of injury, Low confidence, or self-efficacy *Facilitators to PA:* Enjoyment, knowledge of benefits, self-confidence, recommendation by health professional	*Barriers to PA:* Lifestyles not conducive to PA, social and cultural limitations, lack of social support *Facilitators to PA:* Social support
Franco et al. [36] Systematic review of qualitative literature 132 studies including 5987 individuals from 24 countries Adults 60yoa+	*Barriers to PA:* Time limitations, poor transport, poor accessibility of facilities, unsafe environment, poor weather, financial limitations	*Barriers to PA:* Physical ailments, comorbidities *Facilitators to PA:* Experienced physical health benefits	*Barriers to PA:* Perceived incapability to complete the activity, fear of injury, poor motivation *Facilitators to PA:* Enjoyment of activity, sense of belonging, instruction by health professional, sense of competency, sense of independence, improved mental health, developing a habitual pattern	*Barriers to PA:* Social awkwardness, cultural factors, lack of social support *Facilitators to PA:* Social interaction, social support

Methods to improve physical activity levels

Once factors affecting physical activity within a population group are determined, innovative approaches to enhancing facilitators or overcoming barriers can be developed. There has been much research conducted in trialing various novel interventions over the last few decades however the challenge will be to not just elicit a change but a sustainable and feasible one over time.

Motivational interviewing

Founded in self-determination theory, Motivational Interviewing (MI) is a counseling technique used to encourage behavioral change in a particular health area. It was developed by William R. Miller in the early 1980s, originating from counseling alcohol-addicted individuals. A state of "cognitive dissonance" was noted in these individuals where they continued with excessive consumption of alcohol despite awareness around the negative consequences of it [48,49].

Motivational interviewing takes a patient-centered approach and involves collaboration between practitioner and patient through open-ended questions posed to help uncover reasons for the patient's ambivalence to change. Goals and strategies are then set, preferably by the patient, however, ideas can be posed by the practitioner if the patient fails to suggest any. It is vital that the practitioner expresses empathy and recognition to foster an environment where the patient's sense of autonomy is maintained. Additionally, it is important that the practitioner avoid confrontation or argument around resistance that may be demonstrated by the patient, instead of using encouragement to achieve the identified goals and aims [50]. This process is thought to satisfy an individual's feelings of competency, autonomy, and relatedness which, according to self-determination theory, are the essential factors for successful behavior change to occur.

Motivational interviewing has been suggested to have a greater impact in cases of cessation of negative behavior such as gambling, substance use, and alcohol addiction whereas there is only a small-to-moderate impact, of short duration (<6 months), in cases where there is an adoption or increase in beneficial behaviors such as physical activity [51]. A systematic review and meta-analysis published in 2014 looked at the effect of motivational interviewing on physical activity levels in patients with chronic conditions including fibromyalgia, obesity, hypertension, cardiovascular disease, and multiple sclerosis, finding evidence of small-to-moderate positive effects of motivational interviewing on physical activity levels [52]. Despite positive early results, there remains a lack of high-quality evidence about the impact of motivational interviewing on physical activity adherence.

Environmental urban changes

The use of ecological models to help explain multifactorial determinants of physical activity highlights the dual impacts of environmental and behavioral factors. However, this is not reflected in the literature as the majority of studies are dedicated to behavioral interventions [33]. A systematic review into the environmental determinants of physical activity looked at studies with a prospective longitudinal cohort and natural experimental design. It was shown that infrastructure that facilitated more active transportation methods such as walking and cycling paths, public transportation, and greater accessibility to facilities, had the greatest positive impact on physical activity. To a lesser extent, esthetically pleasing neighborhood features and safety also positively influenced physical activity levels [53].

Type of exercise intervention

There has been much interest on whether group or individual physical activities have a greater impact on adherence levels. The physical activity type, (group or individual and supervised or nonsupervised) can have a large impact on adherence, however, this is significantly influenced by individual preference and situational factors.

A mixed-methods study that investigated patients' adherence to a prescribed exercise program found that the development of a therapeutic alliance between patient and professional was a strong motivator [54]. The study cohort were adults (45 years +) who had knee pain caused by osteoarthritis who were assigned to one of three physiotherapy-based exercise programs with varying degrees of individualizing, supervision, and feedback for a period of 12 weeks. They found that adherence levels at all of the time points (3, 6, 9, and 18 months) decreased over time irrespective of which program they followed with no significant differences between groups. However, in interviews at two-time points (immediately, and 12 months postintervention) highlighted that among other common reported facilitators, such as physical benefits, a desire to prevent a recurrence, and reattain previous activity levels, the therapeutic alliance was integral to building their confidence, knowledge, and motivation to continue exercise in the longer term. These results could be subject to selection bias as participants who agreed to participate (postintervention $n = 30$, 12 months postintervention $n = 22$) were only a small subset of the overall cohort ($n = 526$) from the original quantitative study. A European study into the effect of "green" or outdoor exercise on adherence, yielded interesting findings [55]. The study compared three different groups who engaged in "green exercise" and found that while extrinsic social and environmental factors such as the environment, family, and friends were integral in initiating and maintaining exercise, intrinsic factors such as enjoyment and physical benefits remained superior motivators.

Educational

Community-Wide Intervention (CWI) approaches to increasing physical activity levels have been studied to a moderate degree over the last 20 years. CWI includes widespread community educational programs on the benefits of physical activity as well as investment in local infrastructure to improve accessibility, which is known to be a common barrier. A 2015 Cochrane review [56] showed that there were no significant improvements in physical activity levels using community-wide physical activity promotion interventions. The authors included studies that were conducted in higher-income (25 studies) and lower-income (eight studies) countries and encompassed 267 different communities. The authors highlight the presence of methodological flaws in many studies such as subjective measurement and selection bias which could explain the poor results, however, even the highest-quality studies still failed to report any significant improvement in community activity levels.

However, some argue that the median duration of 12 months of the CWI intervention in the review was not long enough for widespread community changes to occur. Interestingly, the authors of one of the studies extended their original 1-year duration study [57] to investigate the effectiveness of a CWI which promoted physical activity in middle-aged and older adults (both male and female of age range 40–79 years) in a number of communities in Unnan, Japan to include 3- and 5-year follow up [58]. The study randomized 12 communities to either intervention (9) or control (3) with further randomization of the intervention communities to one of three subgroups which consisted of aerobic, resistance and flexibility, or aerobic, resistance and flexibility physical activity types [58]. The intervention was the promotion of the particular intervention via social marketing, including information, education, and support delivery of the particular physical activity. Community members were followed up at 1, 3, and 5 years and activity levels were measured via self-report questionnaires. The results showed that a longer duration of CWI resulted in higher impacts on physical activity levels across a community with 1.6% change seen at 3 years and 4.6% at 5 years [58].

Exercise referral schemes

In an attempt to increase physical activity levels in the UK in the early 1990s, an exercise referral scheme (ERS) was designed, where a primary care health professional could refer an individual they deemed "inactive" to an external exercise provider [59]. More recently reviewed guidelines published by the UK-based National Institute for Health and Care Excellence (NICE) in 2014, stated that only inactive or sedentary individuals who are at risk of or currently diagnosed with a chronic health condition were eligible to be part of the referral Scheme [59]. ERS is founded on the premise that the main reason sedentary individuals are inactive is that they lack the health literacy to make healthy behavioral choices [60]. Research has shown an increase of 45% (self-reported) in an individual's physical activity levels following advice from a primary healthcare professional [61]. This provides health care providers with the unique opportunity to educate and encourage engagement in physical activity for these at-risk individuals. Over time, the practice of ERS has become prevalent in other countries across Europe, the USA, Canada, New Zealand, and Australia.

Since their inception, there has been doubt regarding the effectiveness of ERS in increasing physical activity levels in the short and long term. One of the earliest systematic reviews into attendance levels in ERS was conducted in 2005 and despite reviewing only nine pieces of literature, they found that approximately 80% of referred participants did not complete their prescribed program [62]. A more extensive systematic review conducted in 2015 reported on "uptake" and "adherence" levels finding wide variation in reported uptake rates – from 85% to 35%. It was also noted that completion rates also varied from 25% to 86%, however, there was significant heterogeneity between studies that may explain the differences [63]. Factors that facilitated adherence to ERS, reported by participants, were social support, the variety and tailoring of activity, and formation of habitual exercise postintervention [59,64]. Barriers were mainly environmental such as timing, cost, and location of physical activity setting.

Financial reward

Several studies have used a financial incentive as an extrinsic motivator to initiate an increase in physical activity levels. A study recruited Canadian adult female breast cancer survivors (a cohort with an inactivity rate of up to 70%) to form groups of 8–12 and design their own physical activity regime to undertake for a 12-month period [65]. Data including physical activity levels (via accelerometer), quality of life, and motivational outcome measures were taken at baseline, 6 months, and 12 months. The groups then submitted their designed regime to the researchers and if successful, received up to $2000 budget to spend on materials related to the regime with an additional $500 to the group if they had increased physical activity levels at 6-month follow up [65]. Developing one's own exercise regime incorporates the psychological needs of autonomy and relatedness, important constructs of SDT. Competence was also fostered, via feedback received within the group and activity level data that was reported back to members from the researchers [65]. The project showed that intrinsic motivational levels (via the Behavioral Regulation Exercise Questionnaire) and physical activity levels improved between baseline and 6 months significantly, however, was not sustained at the 12-month period.

A systematic review and meta-analysis of literature between 2012 and 2018 identified a dose-relationship between the magnitude of financial incentive and physical activity level [66]. However, contrary to a similar review conducted by the same authors in 2013 [67], the improved physical activity levels are sustained even after the incentive is removed. Arguably though, only three out of 11 studies reviewed in 2013 [67] reported postintervention (financial incentive removed) data, whereas 18 out of the 23 studies had in the 2020 review [66]. However, only four of the 18 studies in the 2020 review showed a sustained elevation in physical activity levels despite meta-analysis revealing a significant increase of 514 steps observed 3–6 months after the removal of incentives [66]. The majority of studies included focused on sedentary populations and older age groups who may be more responsive to extrinsic motivators and yield larger effects from baseline which is also reported in other similar reviews on sedentary populations with chronic conditions [68].

Digital monitors

Research into digital wearable monitors to track physical activity levels has become more mainstream over the last few decades as technology has improved. Monitors such as applications on mobile devices, step count trackers, fitness bands, and smartwatches have all been used, however, accelerometers are most commonly used as they are affordable, user friendly, and provide objective measurements via a remote digital platform [69]. In addition to being a valuable method of data collection, digital monitors can act as an effective motivational tool.

Many systematic reviews have shown a small-to-moderate positive impact of digital monitors on physical activity levels in healthy adults [70,71]; however, systematic reviews that focus on more specific subpopulations seem to report more significant positive effects. A review into the effect of digital monitors on physical activity levels of individuals with osteoarthritis found that there was a positive influence for up to 12 months and that interventions with more realistic, autonomous goals, and which involved interaction and feedback from peers, had a greater impact [72]. These interventions are grounded in the social cognitive theory highlighting the impact of social engagement on behavior change. In a study on the influence of FitBit step count monitors on adherence to exercise in overweight, postmenopausal women over 16 weeks, it was reported that the FitBit significantly improved adherence in the intervention group compared to the control by giving real-time feedback and encouragement via achievement of short-term goals [73].

Social media

The influence of social factors on mental and physical health is well recognized [74] and these influences have informed the social components present in many of the behavioral change models outlined previously. Therefore, investigation of interventions that utilize more modern online platforms facilitating the social dynamics is an integral approach for behavioral change. There are numerous studies investigating the efficacy of various platforms in facilitating change in physical activity levels with conflicting results.

A small Saudi Arabian study investigated the use of the popular social media application "Instagram" as a motivational method to influence adherence to a four-week home exercise program [75]. Female undergraduate students ($n = 58$) aged between 18 and 25 years were assigned to either the intervention (received the Instagram application and home exercise program) or control group (received the home exercise program only). The application included educational information, as well as reminder alarms and videos, sent to participants in the intervention group at various intervals. At the completion of the four-week period, all participants were required to fill out an Exercise Motivation Inventory-2 (EMI) questionnaire which includes 51 items regarding adherence and motivational factors on a 5-point Likert scale. Results showed that

4% of the control group were adherent to the prescribed exercise program and 17% of the intervention group were adherent. Interestingly the results on motivational factors revealed that extrinsic motivators such as: "positive health," "ill-health avoidance" and "weight management and nimbleness" were more prevalent than intrinsic factors [75].

Contrary to these findings, a larger RCT from the US using a similar social media intervention (a Facebook group) on a similar cohort of participants found that there was no difference between perceived social support or adherence levels between intervention and control groups over a 12 week period [76]. However, participants did report their perceived satisfaction in the usage of the Facebook group, which suggests that it could be a feasible platform for other interventions grounded in social cognitive theory.

These conflicting results highlight that there are different aspects of the online social media platforms that are more influential in facilitating behavioral change than others. A unique four-arm RCT conducted in 2014 aimed to determine which features of an online social media platform were more influential in increasing physical activity levels [77]. The trial compared the social support aspect, with participants giving each other online support, to the social comparison component, where performance levels were compared with others. The authors found that social comparison was a more motivating factor than social support, however, a combination of the two provided the greatest benefit. This suggests that while a social media platform can be a feasible method of influencing physical activity levels, various components such as educational information, adherence tracking, social support, and social comparison factors need to be included for maximal benefit.

Digital games

As previously described, digital technology has been pivotal in providing both an objective form of collating physical activity data, while simultaneously being a motivating factor in adherence. However, for certain populations who may be faced with certain barriers such as safety, accessibility, high-injury risk, or family commitment [5], physical activity may only be achievable indoors. In this cohort, research has focused on the impact of emerging technology-based "exergaming" platforms such as Nintendo, Xbox, and PlayStation [78] on physical activity participation.

A 2018 systematic review of 22 articles looked at the impact of technology-based exercise programs of older adults with an average age range of 67–86 years of age and found that they were comparable and commonly superior in adherence, to supervised or home-based unsupervised exercise programs [79]. Much of the literature reviewed indicated that older adults commonly cite poor weather, fear of injury/falls, transport challenges, financial restriction, and emotional apprehension as barriers to physical activity (particularly outside the home). However, eight studies on "exergames" in the home reported elevated levels of enjoyment, and improved intrinsic motivation to continue. Where individuals have control over the design of their program (autonomy), the intensity or frequency, and can receive instant feedback on performance (competence), two of the three essential psychological needs that form the foundation of the SDT of behavioral change. However, one major limitation highlighted was that of technological literacy in older adults, particularly in programs that didn't have sufficiently detailed explanations [79].

Another limitation with technology-based programs is an individual preference for the group-based exercise environment. A 2019 systematic review of the use of digital games in influencing physical activity behaviors in patients with various cardiovascular conditions found that one of the main reasons for loss of interest was a desire for social, group exercise programs [80]. While high enjoyment levels of average range 79%–93% were reported and adherence levels were also high at between 70% and 100%, higher values were evident in the studies that contained human interaction whether in the context of supervision or social groups, support the social cognitive theory of behavioral change.

Conclusion

Physical inactivity is not just a major modifiable risk factor for noncommunicable diseases but also a public health issue in itself, which significantly impacts global morbidity and mortality and the financial health burden worldwide. Many organizations have trialed innovative technological, financial, and socially orientated interventions aimed at increasing physical activity levels with variable success. The diversity of individuals, communities, and nations is the reason why health strategies and promotional methods should be tailored to the needs of a particular individual or group. Awareness of the common barriers and facilitators in play combined with an understanding of the behavioral models that underpin behavior modification can enable the design and implementation of effective strategies and incentives to enhance adherence to physical activity, and through that global health.

References

[1] W.H.Organisation. Global action plan: For the prevention and control of noncommunicable diseases. vols. 2013–2020; 2013. p. 1–55.

[2] Nugent R, et al. Investing in non-communicable disease prevention and management to advance the sustainable development goals. Lancet 2018;391 (10134):2029–35.

[3] Kontis V, et al. Contribution of six risk factors to achieving the 25×25 non-communicable disease mortality reduction target: a modelling study. Lancet 2014;384(9941):427–37.

[4] W.H.Organisation. WHO guidelines on physical activity and sedentary behaviour. Geneva: World Health Organisation; 2020.

[5] W.H.Organisation. Global action plan on physical activity 2018–2030: More active people for a healthier world; 2018. p. 1–104.

[6] Ding D, et al. The economic burden of physical inactivity: a global analysis of major non-communicable diseases. Lancet 2016;388(10051):1311–24.

[7] W.H.Organisation. Global recommendations on physical activity for health; 2010. p. 1–60.

[8] Secretary U.S. Department of Health and Human Services. Physical activity guidelines for Americans. 2nd ed. Washington D.C.: U.S Department of Health and Human Services; 2018. p. 1–118.

[9] Care, D.o.H.a.S. UK chief medical officers' physical activity guidelines; 2019.

[10] Ross R, et al. Canadian 24-hour movement guidelines for adults aged 18-64 years and adults aged 65 years or older: an integration of physical activity, sedentary behaviour, and sleep. Appl Physiol Nutr Metab 2020;45(10):S57–S102.

[11] Australian Institute of Health and Welfare. Physical activity across the life stages. Australian Institute of Health and Welfare; 2018.

[12] McCambridge J, Witton J, Elbourne DR. Systematic review of the Hawthorne effect: new concepts are needed to study research participation effects. J Clin Epidemiol 2014;67(3):267–77.

[13] Gresham G, et al. Wearable activity monitors in oncology trials: current use of an emerging technology. Contemp Clin Trials 2018;64:13–21.

[14] Wilde LJ, et al. Apps and wearables for monitoring physical activity and sedentary behaviour: a qualitative systematic review protocol on barriers and facilitators. Digital Health 2018;4.

[15] Cooper AR, et al. Objectively measured physical activity and sedentary time in youth: the international children's accelerometry database (ICAD). Int J Behav Nutr Phys Act 2015;12:113.

[16] Guthold R, et al. Worldwide trends in insufficient physical activity from 2001 to 2016: a pooled analysis of 358 population-based surveys with 1·9 million participants. Lancet Glob Health 2018;6(10):e1077–86.

[17] Hallal PC, et al. Global physical activity levels: surveillance progress, pitfalls, and prospects. Lancet 2012;380(9838):247–57.

[18] Guthold R, et al. Global trends in insufficient physical activity among adolescents: a pooled analysis of 298 population-based surveys with 1·6 million participants. Lancet Child Adolesc Health 2020;4(1):23–35.

[19] Kalman M, et al. Secular trends in moderate-to-vigorous physical activity in 32 countries from 2002 to 2010: a cross-national perspective. Eur J Public Health 2015;25(suppl_2):37–40.

[20] Ajzen I. The theory of planned behavior: frequently asked questions. Human Behav Emerg Technol 2020;2(4):314–24.

[21] Cousson-Gélie F, et al. The "great live and move challenge": a program to promote physical activity among children aged 7-11 years. Design and implementation of a cluster-RCT. BMC Public Health 2019;19(1).

[22] Stacey FG, et al. A systematic review and meta-analysis of social cognitive theory-based physical activity and/or nutrition behavior change interventions for cancer survivors. J Cancer Surviv 2015;9(2):305.

[23] Valle CG, et al. A randomized trial of a Facebook-based physical activity intervention for young adult cancer survivors. J Cancer Surviv 2013; 7(3):355.

[24] Ryan RM, Deci EL. Intrinsic and extrinsic motivation from a self-determination theory perspective: definitions, theory, practices, and future directions. Contemp Educ Psychol 2020;61.

[25] Pudkasam S, et al. Physical activity and breast cancer survivors: importance of adherence, motivational interviewing and psychological health. Maturitas 2018;116:66–72.

[26] Teixeira Pedro J, et al. Exercise, physical activity, and self-determination theory: a systematic review. Int J Behav Nutr Phys Act 2012;9(1):78.

[27] Prochaska JO, DiClemente CC. Transtheoretical therapy: toward a more integrative model of change. Psychother Theory Res Pract 1982;19 (3):276–88.

[28] DiClemente CC, Prochaska JO. Toward a comprehensive, transtheoretical model of change: Stages of change and addictive behaviors. In: Miller WR, Heather N, editors. Treating addictive behaviors. 2nd ed. New York, NY: Plenum Press; 1998. p. 3–24.

[29] Prochaska JO. Decision making in the transtheoretical model of behavior change. Med Decis Making 2008;28(6):845–9.

[30] Adams J, White M. Are activity promotion interventions based on the transtheoretcal model effective? A critical review. Br J Sports Med 2003; 37(2):106–14.

[31] Nigg CR, et al. Temporal sequencing of physical activity change constructs within the transtheoretical model. Psychol Sport Exerc 2019;45.

[32] Woods C, Mutrie N, Scott M. Physical activity intervention: a transtheoretical model-based intervention designed to help sedentary young adults become active. Health Educ Res 2002;17(4):451–60.

[33] Bauman AE, et al. Correlates of physical activity: why are some people physically active and others not? Lancet 2012;380(9838):258–71.

[34] Kelly S, et al. Barriers and facilitators to the uptake and maintenance of healthy behaviours by people at mid-life: a rapid systematic review. PLoS One 2016;1–26.

[35] Gomes M, et al. Physical inactivity among older adults across Europe based on the SHARE database. Age Ageing 2017;46(1):71–7.

[36] Franco MR, et al. Older people's perspectives on participation in physical activity: a systematic review and thematic synthesis of qualitative literature. Br J Sports Med 2015;49(19):1268–76.

[37] Harrison F, et al. Weather and children's physical activity; how and why do relationships vary between countries? Int J Behav Nutr Phys Act 2017; 14(1).

[38] Buck C, et al. Factors influencing sedentary behaviour: a system based analysis using Bayesian networks within DEDIPAC. PLoS One 2019; 14(01):1–18.

[39] Sallis JF, et al. Physical activity in relation to urban environments in 14 cities worldwide: a cross-sectional study. Lancet 2016;387(10034):2207–17.

[40] Abbasi IN. Socio-cultural barriers to attaining recommended levels of physical activity among females: a review of literature. Quest (00336297) 2014;66(4):448–67.

[41] Ferri-Caruana A, et al. Motivation to physical exercise in manual wheelchair users with paraplegia. Top Spinal Cord Inj Rehabil 2020;26(1):1–10.

[42] Hornik B, Dulawa J. Frailty, quality of life, anxiety, and other factors affecting adherence to physical activity recommendations by hemodialysis patients. Int J Environ Res Public Health 2019;16(10).

[43] Firth J, et al. Motivating factors and barriers towards exercise in severe mental illness: a systematic review and meta-analysis. Psychol Med 2016;46 (14):2869–81.

[44] Resurrección DM, et al. Barriers for nonparticipation and dropout of women in cardiac rehabilitation programs: a systematic review. J Womens Health 2002;26(8):849–59. 2017.

[45] Thorpe O, Johnston K, Kumar S. Barriers and enablers to physical activity participation in patients with COPD: a systematic review. J Cardiopulm Rehabil Prev 2012;32(6):359–69.

[46] Morrison KS, et al. What are the barriers and enablers to physical activity participation in women with ovarian Cancer? A rapid review of the literature. Semin Oncol Nurs 2020;36(5):1–10.

[47] World Health Organisation. ACTIVE: a technical package for increasing physical activity. Geneva: World Health Organisation; 2018. https://apps.who.int/iris/handle/10665/275415. License: CC BY-NC-SA 3.0 IGO.

[48] Miller WR. Motivational interviewing with problem drinkers. Behav Psychother 1983;11:147–72.

[49] Miller WR, Baca LM. Two-year follow-up of bibliotherapy and therapist-directed controlled drinking training for problem drinkers. Behav Ther 1983;14(3):441–8.

[50] Miller WR, Rollnick S. Motivational interviewing: preparing people for change. 2nd ed. Guilford Press; 2002.

[51] Frost H, et al. Effectiveness of motivational interviewing on adult behaviour change in health and social care settings: a systematic review of reviews. PLoS One 2018;13(10):1–39.

[52] O'Halloran PD, et al. Motivational interviewing to increase physical activity in people with chronic health conditions: a systematic review and meta-analysis. Clin Rehabil 2014;28(12):1159–71.

[53] Kärmeniemi M, et al. The built environment as a determinant of physical activity: a systematic review of longitudinal studies and natural experiments. Ann Behav Med 2018;52(3):239–51.

[54] Moore AJ, et al. Therapeutic alliance facilitates adherence to physiotherapy-led exercise and physical activity for older adults with knee pain: a longitudinal qualitative study. J Physiother 2020;66(1):45–53.

[55] Fraser M, Munoz S-A, MacRury S. What motivates participants to adhere to green exercise? Int J Environ Res Public Health 2019;16(10).

[56] Baker PRA, et al. Community wide interventions for increasing physical activity. Cochrane Database Syst Rev 2015;1, CD008366.

[57] Kamada M, et al. A community-wide campaign to promote physical activity in middle-aged and elderly people: A cluster RCT. Springer Nature; 2013.

[58] Kamada M, et al. Community-wide intervention and population-level physical activity: a 5-year cluster randomized trial. Int J Epidemiol 2018; 47(2):642–53.

[59] Morgan F, et al. Adherence to exercise referral schemes by participants—what do providers and commissioners need to know? A systematic review of barriers and facilitators. BMC Public Health 2016;16:227.

[60] Albert FA, et al. Functionality of physical activity referral schemes (PARS): a systematic review. Front Public Health 2020;8:1–13.

[61] Orrow G, et al. Effectiveness of physical activity promotion based in primary care: systematic review and meta-analysis of randomised controlled trials. BMJ Br Med J 2012;344(7850):1–17.

[62] Gidlow C, et al. Attendance of exercise referral schemes in the UK: a systematic review. Health Educ J 2005;64(2):168–86.

[63] Campbell F, et al. A systematic review and economic evaluation of exercise referral schemes in primary care: a short report. Health Technol Assess (Winch Eng) 2015;19(60):1–110.

[64] Eynon M, et al. Assessing the psychosocial factors associated with adherence to exercise referral schemes: a systematic review. Scand J Med Sci Sports 2019;29(5):638–50.

[65] Caperchione CM, et al. A preliminary trial examining a 'real world' approach for increasing physical activity among breast cancer survivors: findings from project MOVE. BMC Cancer 2019;19(1):272.

[66] Mitchell MS, et al. Financial incentives for physical activity in adults: systematic review and meta-analysis. Br J Sports Med 2020;54 (21):1259–68.

[67] Mitchell MS, et al. Financial incentives for exercise adherence in adults: systematic review and Meta-analysis. Am J Prev Med 2013;45(5):658–67.

[68] Gong Y, et al. Financial incentives for objectively-measured physical activity or weight loss in adults with chronic health conditions: a meta-analysis. PLoS One 2018;13(9):1–16.

[69] Phillips SM, et al. Wearable technology and physical activity in chronic disease: opportunities and challenges. Am J Prev Med 2018;54(1):144–50.

[70] Foster C, et al. Remote and web 2.0 interventions for promoting physical activity. Cochrane Database Syst Rev 2013;2013(2).

[71] Aalbers T, Baars MAE, Rikkert MGMO. Characteristics of effective internet-mediated interventions to change lifestyle in people aged 50 and older: a systematic review. Ageing Res Rev 2011;10(4):487–97.

[72] Berry A, Muir S, McCabe C. Digital behaviour change interventions for osteoarthritis—a systematic literature review. Front Public Health 2015;4.

[73] Cadmus-Bertram L, et al. Use of the Fitbit to measure adherence to a physical activity intervention among overweight or obese, postmenopausal women: self-monitoring trajectory during 16 weeks. J Med Internet Res 2015;3(4), e96.

[74] Berkman LF, et al. From social integration to health: Durkheim in the new millennium. Soc Sci Med 2000;51(6):843–57.

[75] Al-Eisa E, et al. Effect of motivation by "Instagram" on adherence to physical activity among female college students. Biomed Res Int 2016;2016:1546013.

[76] Cavallo DN, et al. A social media–based physical activity intervention: A RCT. Am J Prev Med 2012;43(5):527–32.

[77] Zhang J, et al. Support or competition? How online social networks increase physical activity: a RCT. Prev Med Rep 2016;4:453–8.

[78] Chaddha A, et al. Technology to help promote physical activity. Am J Cardiol 2017;119(1):149–52.

[79] Valenzuela T, et al. Adherence to technology-based exercise programs in older adults: a systematic review. J Geriatr Phys Ther 2001;41(1):49–61. 2018.

[80] Radhakrishnan K, et al. Role of digital games in self-Management of Cardiovascular Diseases: a scoping review. Games Health J 2019;8(2):65–73.

Part III

Children and early teens (5-15)

Chapter 5

The effects of physical exercise on the brain and neurocognitive functioning during childhood

Anna Meijer[a,b], Lot Verburgh[c], and Esther Hartman[d]

[a]*Vrije Universiteit Amsterdam, Clinical Neuropsychology Section, Amsterdam, The Netherlands* [b]*Universiteit Leiden, Clinical Neurodevelopmental Sciences, Leiden, The Netherlands* [c]*Rivas Zorggroep, Gorinchem, The Netherlands* [d]*University of Groningen, University Medical Center Groningen, Center for Human Movement Sciences, Groningen, The Netherlands*

Introduction

Physical exercise during youth benefits physical and cognitive functioning. Physical exercise diminishes the risk for cardiovascular illness, obesity, and type 2 diabetes, while improving bone density, and cardiorespiratory fitness [1]. Additionally, regular physical exercise is associated with improved cognitive functioning and even enhanced academic achievement in children and adolescents [2–6].

Today, the most common explanation for this relationship is a strong association between physical exercise and the developing brain. The physical exercise-induced effects in brain structure and/or functioning are thought to lead to enhanced cognitive functioning, which in turn leads to improved academic achievement during childhood (Fig. 1). Further details of this relationship and an overview of the literature are provided in this chapter. In particular, the chapter focuses on the beneficial effects of physical exercise on brain structure and neurophysiological functioning, neurocognitive functioning, and academic achievement in children. Furthermore, we aim to introduce practical implications resulting from the current knowledge.

Among the various domains of neurocognitive functioning, executive functions are considered as most responsive to the effect of physical exercise [1]. Executive functions are cognitive processes that are necessary for the control of behavior. Executive functioning encompasses a subset of cognitive skills that we use for guided behavior towards a goal [2] and includes processes such as interference control, inhibition, and working memory. These processes enable us to solve problems, plan, and reason [3,4]. Lower-level, more basic neurocognitive functions, such as information processing speed and attention are considered fundamental functions for these executive functions. Executive functions are crucial for learning in new situations and more broadly success throughout life, including vocation, quality of life, and academic achievement [2]. To be able to read, spell and do mathematics, a child needs the ability to inhibit automatic behavior, to move among approaches, and to update their working memory [5]. In children, academic achievement (i.e., mathematics and language performance) is extremely important, as it provides a crucial foundation for the child's future development [2].

Research has shown a number of neural mechanisms that may explain the beneficial effects of physical exercise on neurocognitive functioning. Within this chapter, physical exercise is defined as moderate to vigorous activity, which encompasses activities that cause a noticeable increase in heart rate and breathing rates, such as jogging or cycling. A single bout of physical exercise has been reported to directly improve cerebral blood flow [6] and to trigger the upregulation of neurotransmitters that facilitate cognitive processes, such as dopamine and epinephrine [7,8]. In the scientific literature, these immediate effects resulting from short-term physical exercise, are referred to as the acute effects of physical exercise.

Regular or chronic physical exercise comprises of several workout sessions per week and is thought to trigger several important neural mechanisms. Regular physical exercise appears to raise the levels of neurotrophic factors such as brain-derived neurotrophic factors and nerve growth factors. These neurotrophic factors are small proteins in the brain that boost neural blood vessel formation (angiogenesis) and the development of new neurons (neurogenesis) [7,9]. Neuroimaging

Exercise to Prevent and Manage Chronic Disease Across the Lifespan. https://doi.org/10.1016/B978-0-323-89843-0.00011-8

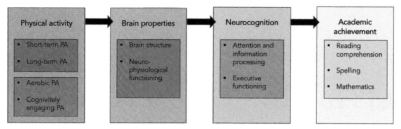

FIG. 1 Conceptual model of the impact of exercise on childhood development and academic outcomes.

studies have shown that physical exercise can indeed impact both brain structure as brain functioning [10]. These effects of regular or long-term physical exercise are referred to as chronic effects.

The current chapter provides an overview of the effects of physical exercise on brain structure neurophysiological functioning, neurocognitive functioning, and academic achievement in children, adolescents, and young adults. Randomized controlled trials (RCTs) and cross-over trials provide evidence for causal relationships between physical exercise and outcomes. Therefore, we have selected meta-analyses and systematic reviews from the last decade that provide an overview of the causal effects of physical exercise on brain structure and neurophysiological functioning, neurocognitive functioning, or academic achievement in children, adolescents, and young adults [11–15]. Together, these publications provide an extensive overview of effects in a wide age range from 5 to 35 years of age.

When possible, we've made a distinction between the effects of single bout exercise interventions (examining acute effects) and the effects of regular exercise interventions (examining chronic effects). Acute effects are investigated by crossover designs and chronic effects are investigated by RCTs. Both research designs provide the opportunity to study causal relationships. Furthermore, because of the likely different effects of physical exercise in children with and without clinical disorders, a distinction is made between these populations in the findings below.

The impact of physical exercise on brain structure and neurophysiological functioning

The impact of physical exercise on brain structure and neurophysiological functioning in children is difficult to narrow down to a single factor. Differences in brain properties in children are mostly assessed by non-invasive techniques such as electroencephalography (EEG) which measures neurophysiological functioning by detecting electrical activity in the brain using electrodes on to the scalp. Another commonly used technique is magnetic resonance imaging (MRI), which is an imaging technique using a magnetic field and radio waves to produce detailed images of the brain (structural MRI). A well-established MRI technique called diffusion tensor imaging (DTI) makes it possible to display the connections between brain structures, i.e., white matter tracts and white matter integrity. Besides the structural images, MRI can also indirectly measure neurophysiological function by detecting changes in blood flow associated with cognitive performance (functional MRI, fMRI).

Brain structure

Recently, a meta-analysis on the effects of physical exercise on brain structure in typically developing children and clinical pediatric populations (5–12 years old) was performed [13]. Studies describing the effects of regular physical exercise programs on brain structure using DTI to assess white matter tracts and white matter integrity. The specific studies reported mixed results, such as both increases and decreases in white matter integrity in response to long-term physical exercise interventions compared to sedentary controls. No overall evidence was found for the effects of physical exercise on brain structure in children. However, this was based on a very limited number of studies with heterogeneous samples of healthy, obese, and deaf children. Hence, it is still unknown whether there is a causal effect of physical exercise on brain structure and whether these results are generalizable to a general pediatric population.

Neurophysiological functioning

The same study also reported the effects of short-term and regular physical exercise on neurophysiological functioning by appraising the literature on the acute effects of physical exercise that used EEG to measure neurophysiological functioning [13]. Meta-analyses revealed small and positive effects of short-term physical exercise on neurophysiological functioning.

Results indicated for example, that greater P3 amplitudes during cognitive goal-directed tasks after a short bout of exercise compared to a period of rest. Greater P3 amplitudes are associated with orienting attention to novel stimuli [16]. Interestingly, the acute effects of physical exercise on neurophysiological functioning were observed in studies that included typically developing children, but also in studies that focused on children with attention-deficit/hyperactivity disorder (ADHD). This disorder is the most widely recognized mental disorder affecting children with symptoms including distractedness, hyperactivity, and impulsivity. Separate meta-analyses for children with ADHD and without ADHD did not reach significance, likely due to the limited number of studies in these analyses. Taken together, the results suggest that short-term physical exercise has acute effects on neurophysiological functioning in children, but no firm conclusions could be made for specific populations.

Experiments on the effects of long-term physical exercise on the neurophysiological function used both EEG and MRI techniques and observed brain activity during a period of rest or during cognitive tasks. Meta-analyses revealed small and positive effects for long-term physical exercise on neurophysiological functioning [13]. Furthermore, additional analyses distinguishing between typically developing children and children from clinical populations revealed a small effect for typically developing children and a large effect for children with ADHD or obesity. In both typically developing children as for the studied clinical groups, results indicated altered brain activity after exercise interventions during cognitive tasks or a period of rest and an improved state of alertness. Additionally, the observed differences in brain activity were associated with improved neurocognitive functioning [13]. Hence, the authors conclude that the observed chronic effects in neurophysiological functioning could be indicated as beneficial. These results suggest that long-term physical exercise has a beneficial effect on neurophysiological functioning in both typically developing children as children with ADHD or obesity. In summary, no evidence was found for the effects of physical exercise on brain structure in children. However, both short-term and long-term physical exercise seems to be beneficial for neurophysiological functioning in children.

Effects of physical exercise on neurocognitive functioning

Basic neurocognitive functioning

Recently, the effects of physical exercise interventions on cognitive functions in healthy children (6–12 years old) and young adults (12–30 years old), were described [11,12]. Moderate positive effects were found for both short- and long-term physical exercise on neurocognitive functions such as information processing speed and attention. This is interesting as performance on these two basic neurocognitive functions are considered prerequisites for more complex executive functions to emerge.

Executive functions

A moderate positive overall acute effect of short-term physical exercise on executive functions has been found in children, adolescents, and young adults [2]. Most of the included studies in the meta-analysis examined the acute effects of short-term exercise on inhibition, and results indicated at least a temporary enhancement of inhibitory control [13]. This is an important finding as inhibitory control (the ability to stop or tune out irrelevant or distracting stimuli) is an essential cognitive function in daily life. Children with ADHD exhibit deficits in inhibitory control. This deficiency is responsible for several adverse developmental outcomes such as impaired cognitive performance, disruptive behavior, and impaired social skills, which in turn could lead to poor academic achievement [17]. Interestingly, a meta-analysis on functional outcomes (among executive functions) in children with ADHD (6.7–13.9 years old), reported a positive effect of physical exercise on executive functioning. Due to the limited number of studies, no separate analyses could be performed for acute and chronic exercise [18].

Recent meta-analyses on the acute effects of physical exercise on executive functioning in children showed positive effects on inhibition in children [11] and adolescents and young adults [12], whereas no clear effects were found on working memory, cognitive flexibility, and planning. Interestingly, another meta-analysis showed similar findings with again, significant effects of short-term exercise were also found on inhibitory control, but not on working memory [15]. With long-term physical exercise, positive effects were found on working memory and cognitive flexibility in both children as well as in adolescents and young adults [12]. It is interesting to see that results in children, adolescents, and young adults are largely in agreement, despite the rapid growth of the prefrontal cortex in adolescence [19]. Body mass index (BMI) was shown to be a moderator of the effect of long-term physical exercise on executive functioning in one meta-analysis, indicating that exercise interventions had a greater impact on executive functions in a population with higher BMI [15]. This is an interesting finding as previous studies have shown that obese children perform poorer on tasks measuring executive functioning.

In summary, both short-term and long-term physical exercise may have beneficial effects on basic neurocognitive functioning and on the more complex executive functions. Effects of long-term physical exercise are somewhat smaller and show more contrasting findings across studies, however, there may be several explanations for this. First, performing a randomized controlled trial in these age groups with high compliance of the participants, for a longer period of time, is highly difficult. Second, the effects of long-term physical exercise may have another underlying mechanism that takes longer to express and which is possibly harder to measure [12]. Third, the brain of preadolescent children and adolescents develops very rapidly during these age stages. This complicates measurements as a lot of interference might have taken place during the experiment, which can affect executive functioning. Examples are social environment and academic development of participants, or changes in behavior of daily physical exercise, which is known to decrease during adolescence.

Effects of physical exercise on academic achievement

Meta-analyses that specifically focus on the effects of short-term physical exercise on academic achievement in children and adolescents did not find any overall significant effect of physical exercise on academic achievement [11]. A closer look at the subdomains of academic achievement revealed a small positive effect of physical exercise on spelling skills, but no effect on reading and mathematics [11]. In adolescents and young adults, only 1 out of 44 studies focused on the acute effects of physical exercise on academic achievement [12]. Therefore, no meta-analysis was performed on acute effects on academic achievement in these target groups.

Meta-analysis concerning the chronic effects of long-term physical exercise interventions in children has shown a small to moderate positive overall effect on academic achievement. Additional analyses making a further distinction between specific domains of academic performance did not reveal significant effects for spelling, reading, and mathematics separately [11]. In adolescents and young adults, meta-analysis revealed a moderately positive overall effect on academic performance. Further distinction between the subdomains of academic performance showed a moderate effect for language, but no significant effects on grade point average and mathematics [12].

In summary, the role of short-term physical exercise in stimulating academic achievement in children seems to be modest, and in adolescents and young adults more research is warranted. Long-term physical exercise may improve academic achievement, although the results on specific subdomains are still inconsistent. This conclusion is in accordance with supporting past outcomes showing that the viability of physical exercise interventions on performance at school remains yet uncertain [20,21].

Practical implications

Duration of physical exercise interventions

The effect of duration on children's neurocognitive functioning has been evaluated [11]. The duration of acute physical exercise interventions varied from 5 to 60 min per session and different types of exercises were involved, for example, adapted physical education lessons, physical exercise in the classroom, running on a treadmill, or cycling on an ergometer within a lab. The length of the session (in minutes) had no significant impact on the effects of short-term physical exercise, yet more research is required to find the ideal length of short-term interventions.

In the appraised literature, the duration of the long-term physical exercise interventions varied from 6 weeks to 16 months, with a frequency from two to five times a week. Most studies conducted a school-based intervention during recess or after school hours. In several studies adapted physical education lessons were delivered or physical exercise in the classroom was provided [11]. The duration of the interventions (in weeks) had no significant influence on the effects of the programs [11].

Type of physical exercise interventions

The effects of different types of physical exercise interventions such as aerobic exercise and cognitively engaging physical exercise have also been studied. The focus of the aerobic exercise is to target moderate-to-vigorous intensity and to increase the heart rate and breathing rate, such as playing tag or running. Cognitively engaging physical exercise refers to exercise that requires a high amount of attention and cognitive effort, such as adapted versions of games in which complex rules were included. It was shown that long-term physical exercise programs that focused on cognitively engaging physical exercise showed moderate to large effects, while aerobic exercise showed small to moderate effects on neurocognitive functioning in children. This is most likely explained by the notion that physical exercise around the lactate threshold leads to immediate releases of catecholamines and neurotransmitters, leading to elevated arousal, and subsequently enhancing cognitive performance [14]. In addition, long-term physical exercise seems to have a widespread effect on academic achievement, suggesting that both quantitative as well as qualitative aspects of physical exercise should be considered when aiming to stimulate academic achievement [22]. Although both aerobic and cognitively engaging physical exercise showed beneficial effects, the results indicated that long-term physical exercise with high cognitive engagement has a greater effect on cognitive functioning in children compared to aerobic physical exercise.

There are different ways to develop cognitively engaging interventions. Cognitive engagement or exertion may be stimulated by including new, difficult, or changing assignments instead of repetitive tasks that have been profoundly learned with regular practice [23]. As long as motor skills, like balancing, running, jumping, throwing, and catching, are still challenging and not fully automated, these activities yield opportunities to exercise in a cognitively engaging manner. In addition, team games that require cooperation and anticipation of behavior offer opportunities to apply motor skills in a competitive and strategic way. Such team sports require adequate motor skills in combination with neurocognitive functioning, like attention, inhibition, and fast decision making [1]. Finally, incorporating physical exercise into the teaching of academic class content—physically active learning—is a promising new method to improve academic achievement in particular. It was found to be an effective way to promote acute effects on attention [24], and greater academic achievement in children in the long term [25,26]. Taken together, when aiming to stimulate neurocognitive functioning and academic achievement in children both quantitative and qualitative aspects of physical exercise should be considered.

Clinical populations

It is highly possible that the mechanisms responsible for the effects of physical exercise on brain structure and neurophysiological function, and thereby on neurocognitive functioning and academic performance, rely upon the pathophysiology of clinical disorders. The majority of studies concerning the impact of physical exercise focuses on children with ADHD or obesity. Hence, we could only draw conclusions for this limited number of specific groups.

Children with ADHD seem to respond similarly to physical exercise as compared to typically developing children. More specifically, studies that focused on the effects of physical exercise on neurophysiological functioning showed beneficial effects of both acute and chronic physical exercise in children with ADHD [12]. Furthermore, improved executive functioning was also observed as a result of physical exercise. The results concerning neurophysiological functioning and executive functioning suggest that physical exercise is a potentially effective treatment of ADHD. Stimulant medication is today's most common treatment of ADHD, which aims to alleviate symptoms such as hyperactivity and impulsive behavior. This type of stimulant medication is based on the upregulation of dopamine and norepinephrine [27], as both neurotransmitters that are implicated in the pathophysiology of ADHD [18]. Interestingly, this neurotransmitter upregulation affected by ADHD medication is similar to the same neurotransmitter upregulation caused by acute physical activity.

Furthermore, children with obesity seem to benefit from long-term physical exercise. Beneficial effects of long-term physical exercise have been demonstrated in both neurophysiological and executive functioning domains [13,15]. Children with obesity have an altered brain structure and functioning compared to leaner children, which can lead to cognitive dysfunction [28,29]. This difference may be explained by vasoactive effects on cerebral arteries and neurotoxicity by hyperinsulinemia [30,31]. Hence, in addition to the compelling physical advantages of physical exercise, children with obesity are likely to experience cognitive benefits from physical activity.

Taken together, the results suggest that physical exercise could be used as a treatment to improve neurophysiological functioning and thereby neurocognitive functioning and academic performance in pediatric clinical populations. In this chapter, we highlighted the effects of physical exercise for children with ADHD and children with obesity. Hence, it is important to note that this does not necessarily mean that these results may be generalized to other populations and more research is needed to better understand the impact of physical exercise in different pediatric clinical groups.

Future research directions

When designing future studies, we highly recommend that researchers should focus on RCT's and/or cross-over designs, which can demonstrate exercise as a causative factor of neurological function. Furthermore, the context should be taken into account, specifically the target population (e.g., fitness, clinical populations) and the type and intensity of exercise interventions (aerobic versus cognitively engaging exercise, etc.). Physical exercise-induced effects on brain structure or neurophysiological functioning, supplemented by changes in neurocognitive functioning may give more knowledge into the physiological mechanisms underlying physical exercise. Therefore, future studies should also include both structural and functional imaging techniques (e.g., MRI and EEG) in combination with behavioral outcome assessment.

Moreover, more research is needed to find the optimal duration for acute and chronic interventions. Nevertheless, for acute interventions, aerobic exercises seem to be most effective, whereas for chronic interventions, a combination of aerobic and cognitively engaging exercises seems to be most promising. This means that for a (possible) temporary boost of neurocognitive functioning, a short bout of moderate to vigorous exercise will be sufficient. This can be put into easily practice at schools during a recess of 15–20 min. For chronic effects on brain functioning and neurocognitive functioning, regular and frequent bouts of aerobic and cognitively engaging exercise (e.g., soccer training) will be more effective.

Primary and secondary schools are ideal environments to increase the level of physical exercise and to implement exercise interventions. Physical exercise could and should be encouraged before, during, and/or after school activities. Previous programs to expand the quantity of physical exercise at primary schools do not show interference with a performance at academic courses. This chapter highlights the importance of physical exercise during development. The current increase of sedentary behavior in youth underlines the need for new and innovative strategies to provide an adequate amount of physical exercise during the day.

Summary—Physical activity recommendations

- Only a short bout of moderate to vigorous exercise may already boost cognitive functioning, the precise duration does not seem to matter.
- Short exercise interventions (e.g., running or playing tag for 10 min) may be incorporated during school days.
- For chronic effects on brain functioning and neurocognitive functioning, regular and frequent bouts of aerobic and cognitively engaging exercise (e.g., soccer training or games with challenging tasks) may be effective for durable impact.

References

[1] Best JR. Effects of physical activity on Children's executive function: contributions of experimental research on aerobic exercise. Dev Rev 2010; 30(4):331–551.

[2] Diamond A. Executive functions. Annu Rev Psychol 2013;64:135–68.

[3] Collins A, Koechlin E. Reasoning, learning, and creativity: frontal lobe function and human decision-making. PLoS Biol 2012;10(3), e1001293.

[4] Zelazo PD, Müller U. Executive function in typical and atypical development. In: Goswami U, editor. Blackwell handbook of childhood cognitive development. Oxford: Wiley-Blackwell; 2002. p. 445–69.

[5] Best JR, Miller PH, Naglieri JA. Relations between executive function and academic achievement from ages 5 to 17 in a large, representative national sample. Learn Individ Differ 2011;21(4):327–36.

[6] Querido JS, Sheel AW. Regulation of cerebral blood flow during exercise. Sports Med 2007;37(9):765–82.

[7] Dishman RK, et al. Neurobiology of exercise. Obesity 2006;14(3):345–56.

[8] McAuley E, Kramer AF, Colcombe SJ. Cardiovascular fitness and neurocognitive function in older adults: a brief review. Brain Behav Immun 2004;18(3):214–20.

[9] Swain RA, et al. Prolonged exercise induces angiogenesis and increases cerebral blood volume in primary motor cortex of the rat. Neuroscience 2003;117(4):1037–46.

[10] Chaddock L, et al. A review of the relation of aerobic fitness and physical activity to brain structure and function in children. J Int Neuropsychol Soc 2011;17(6):975–85.

[11] de Greeff JW, et al. Effects of physical activity on executive functions, attention and academic performance in preadolescent children: a meta-analysis. J Sci Med Sport 2018;21(5):501–7.

[12] Haverkamp BF, et al. Effects of physical activity interventions on cognitive outcomes and academic performance in adolescents and young adults: a meta-analysis. J Sports Sci 2020;1–24.

[13] Meijer A, et al. The effects of physical activity on brain structure and neurophysiological functioning in children: a systematic review and Meta-analysis. Dev Cogn Neurosci 2020; 100828.

[14] Verburgh L, et al. Physical exercise and executive functions in preadolescent children, adolescents and young adults: a meta-analysis. Br J Sports Med 2014;48(12):973–9.

[15] Xue Y, Yang Y, Huang T. Effects of chronic exercise interventions on executive function among children and adolescents: a systematic review with meta-analysis. Br J Sports Med 2019;53(22):1397–404.

[16] Polich J. Updating P300: an integrative theory of P3a and P3b. Clin Neurophysiol 2007;118(10):2128–48.

[17] Scheres A, et al. Executive functioning in boys with ADHD: primarily an inhibition deficit? Arch Clin Neuropsychol 2004;19(4):569–94.

[18] Vysniauske R, et al. The effects of physical exercise on functional outcomes in the treatment of ADHD: a meta-analysis. J Atten Disord 2016; 1087054715627489.

[19] Lebel C, et al. Microstructural maturation of the human brain from childhood to adulthood. Neuroimage 2008;40(3):1044–55.

[20] Donnelly JE, et al. Physical activity, fitness, cognitive function, and academic achievement in children: a systematic review. Med Sci Sports Exerc 2016;48(6):1197.

[21] Singh AS, et al. Effects of physical activity interventions on cognitive and academic performance in children and adolescents: a novel combination of a systematic review and recommendations from an expert panel. Br J Sports Med 2018;53(10):640–7.

[22] De Bruijn AGM, et al. Effects of aerobic and cognitively-engaging physical activity on academic skills: a cluster randomized controlled trial. J Sports Sci 2020;1–12.

[23] Tomporowski PD, Horvat MA, McCullick BA. Role of contextual interference and mental engagement on learning. Nova Science Publishers; 2010.

[24] Mullender-Wijnsma MJ, et al. Moderate-to-vigorous physically active academic lessons and academic engagement in children with and without a social disadvantage: a within subject experimental design. BMC Public Health 2015;15(1):1–9.

[25] Mullender-Wijnsma MJ, et al. Physically active math and language lessons improve academic achievement: a cluster randomized controlled trial. Pediatrics 2016;137(3).

[26] Vetter M, et al. Effectiveness of active learning that combines physical activity and math in schoolchildren: a systematic review. J Sch Health 2020; 90(4):306–18.

[27] Ng QX, et al. Managing childhood and adolescent attention-deficit/hyperactivity disorder (ADHD) with exercise: a systematic review. Complement Ther Med 2017;34:123–8.

[28] Yau PL, et al. Obesity and metabolic syndrome and functional and structural brain impairments in adolescence. Pediatrics 2012;130(4):e856–64.

[29] Yeo BT, et al. Functional specialization and flexibility in human association cortex. Cereb Cortex 2015;25(10):3654–72.

[30] Bruce AS, Martin LE, Savage CR. Neural correlates of pediatric obesity. Prev Med 2011;52:S29–35.

[31] Raji CA, et al. Brain structure and obesity. Hum Brain Mapp 2010;31(3):353–64.

Chapter 6

Physical activity for young people with mental illness

Michaela C. Pascoe and Alexandra G. Parker
Institute for Health and Sport, Victoria University, Melbourne, VIC, Australia

Adolescents and mental illness

Over 25% of the worldwide population is aged 10–24 years [1–3] and most major mental disorders emerge during this period [4,5]. Every year, as many a 25% of young people are diagnosed with substance use or mental disorder [6], making mental disorders the number one cause of disability globally [7]. The common mental disorders are anxiety, depressive, and substance use disorders [5,8]. This means young people will commonly experience a mental disorder, or know someone who does, as they transition from adolescence to adulthood.

The development of mental disorders results in significant social and vocational functioning impairments for individuals, including social, employment abilities, and educational opportunities and therefore it is important to provide early treatment for mental disorders to reduce symptoms and functional disability [9]. Furthermore, many long-term lifelong health-related behaviors are formed and established during adolescence and early adulthood [10], highlighting this critical timeframe for intervention.

Evidence-based treatments for mental disorders such as psychotherapies and psychotropic medications have modest effects in young people [e.g., see 11,12]. As a key example of this, over half of the young people with depression do not respond to the best guideline-recommended treatment [11,13]. Furthermore, psychotropic medications can also have side effects, which young people can find intolerable and which can lead to further, serious health complications, such as the increased risk for metabolic syndrome [14].

Physical activity, as defined by the World Health Organization, includes sports or planned exercise as well as movement for recreation or in leisure-time, transport, household, and occupational domains [15]. Physical activity is a nonstigmatizing approach that has few reported side effects [16] and has been reported as helpful for mental health promotion and for the treatment of mental health problems, by young people [17].

Young people and physical activity

In addition to being the developmental stage when the onset of most mental health problems are likely to occur [4], adolescence is also a period during which people disengage from regular physical activity and sports [18,19]. A lack of sufficient physical activity results in an increased risk of poor physical health, including higher rates of diabetes, cardiovascular disease, and premature death among young people [20,21]. Cohort and large cross-sectional studies report that young people who are less physically active are at greater risk of experiencing mental disorders such as depression [22,23] and that young people who experience depression are far more likely to be physically inactive compared to the general population [24].

This chapter will focus on physical activity and exercise as an intervention for young people, defined as aged between 12 and 25 years, with a mental disorder diagnosed by a clinician or reaching a defined cut-off score on a scale indicating a disorder. We only review controlled trials. This chapter aims to provide a comprehensive evidence overview for physical activity and exercise as an innervation for mental disorders affecting young people as well as to explore opportunities for translating interventions into clinical practice.

Physical activity or exercise as a treatment for depression among young people.

There is ample evidence showing that physical activity or exercise can effectively treat or diminish depression symptoms in young people. Two meta-analyses including adolescents, children [25] and young people (12–25 years)

Exercise to Prevent and Manage Chronic Disease Across the Lifespan. https://doi.org/10.1016/B978-0-323-89843-0.00002-7

[26] show a moderate [25] to large [26] positive effect of physical activity or exercise as an intervention approach for depression symptoms or diagnosed depression (i.e., major depressive disorder).

A number of randomized controlled trials (RCTs) have demonstrated that moderate-to-vigorous intensity exercise is effective for decreasing depression symptoms. In a population of 64 depressed inpatients not taking antidepressant medication and aged 13–18 (mean age = 15.9 years), 6 weeks of both vigorous-intensity cardiovascular training and a muscular training exercise using a whole-body vibration device decreased depression symptoms compared to treatment as usual (TAU) [27]. A study comparing swimming (likely to be moderate-to-vigorous intensity) and a no-intervention control group, reported that 12–15 weeks of swimming resulted in decreases in the severity of symptoms of depression in male university students that were 19–22 years old [28]. One study compared 6 weeks of aerobic or swimming, both of which were moderate-to-vigorous intensity interventions, as well as a physical education intervention, (intensity not stated), in 75 depressed female university students, according to the Beck Depression Inventory. Both the physical activity interventions decreased depression symptoms from pre- to postintervention, with aerobic exercise decreasing depression symptoms more than physical education [29]. In another study, 8 weeks of aerobic exercise (moderate-to-vigorous intensity intervention) was compared with either a cognitive behavioral therapy (CBT) group or to an unguided group meeting, in 46 help-seeking university students (22% female, mean age 21 years), with a depression diagnosis and Beck Depression Inventory-II score between 13 and 28. In this study, both aerobic exercise and CBT decreased depression compared to the control group, however, aerobic exercise did not decrease negative thoughts or dysfunctional attitudes compared to CBT or no-intervention [30].

In an additional study, a 6-week-long moderate-intensity intervention (walking laps of a 15-m swimming pool) decreased depression symptoms by 50% from baseline, which was significant compared to no intervention [31], in 24 females in high school (mean age 16.9 years) with severe depression. However, a no-treatment control group does not control for nonspecific intervention factors such as participant expectancy, contact with intervention personnel, and social interaction, all of which might improve depression symptoms.

Two additional studies assesses the impact of a moderate-intensity intervention compared to a vigorous intensity intervention and found that both reduced depression symptoms [32,33].

In one of these studies, which comprised of 30 help-seeking adolescents diagnosed with moderate depression (12–18 years, 58% female) 12 weeks of both the vigorous (aerobic exercise) and moderate intervention (stretching exercises) improved, psychosocial functioning, social adjustment and mood states (anger, fatigue, and tension) from pre-post intervention [32]. Six months later, remission rates were 100% for the vigorous-intensity group and 70% for a moderate-intensity group the and by 12-months, remission rates were 100% and 88%, respectively. There were however significant dropouts in both groups during the follow-up period and therefore it is difficult to interpret these findings, highlighting the need for studies exploring the medium and long-term effects of exercise on mental health outcomes. A further limitation is that these studies did not include a nonexercise control group [32,33].

The second study compared a 6-week long moderate intensity, vigorous intensity, and light intensity intervention in 30 males who were not seeking help for their mental health (university students/staff with a mean age of 25.4 years), and found that both moderate and vigorous-intensity exercise interventions reduced depression symptoms pre-to-post intervention. No change in symptoms was reported in the light intensity group [33] However, no significant differences were found when comparing all three groups, likely as the trial was underpowered.

In 54 moderately depressed female university students with a mean age of 25.8 years, 10 weeks of either high or moderate aerobic exercise, or stretching decreased depression symptoms, with no differences between the three groups. The authors reported that after controlling for baseline depression symptoms that high-intensity exercise appeared to be more effective than moderate-intensity and stretching [16].

A number of RCTs have demonstrated that *light-to-moderate intensity* exercise is effective for decreasing depression symptoms. In one study, both a six-week-long preferred intensity intervention (light-to-moderate intensity) as an addition to treatment as usual (TAU), and TAU, reduced depression symptoms in 87 help-seeking adolescents (mean age = 15.4 years) with severe depression. Furthermore, exercise as an adjunct to TAU decreased depression symptoms compared to TAU at the 6-month follow-up but did not impact the quality of life [34].

In 40 female adolescents with mild depression (mean age = 16.0 years) a 12-week-long dance program (likely light-to-moderate intensity) decreased distress and psychological symptoms compared to no intervention [35].

In 176 people with subthreshold mild–moderate depression and/or depression (mean age 17.6 years), a 6-week long intervention encouraging self-selected exercise decreased depression symptoms, compared to psychoeducation, but did not impact social and occupational functioning, or substance use [36].

Finally, among 28 mildly depressed young people, 5 weeks of light-to-moderate yoga practice, reduced depression symptoms at postintervention, compared to a wait-list [3]. Exercise and physical activity interventions may improve depression via biological and psychosocial pathways. Biological processes likely include influencing neurogenesis, inflammatory and oxidative stress responses, the modulation of monoamines such as serotonin, and the regulation of the HPA axis as well as bio-rhythms such as sleep, all of which can be dysregulated in depression [37]. From a psychological perspective, behavioral activation provides an opportunity to improve self-efficacy and physical activity and exercise can provide a distraction from rumination or negative thoughts as well as opportunities for social interaction [38].

Physical activity or exercise as a treatment for anxiety in young people

Compared to depression, fewer studies have assessed the impact of physical activity interventions for anxiety in young people, however, the limited research to date does suggest that exercise is beneficial in terms of reducing symptoms of anxiety related to specific anxiety disorders.

In one study involving 176 subthreshold or mild-moderately depressed/and/or anxious young people (mean age 17.6 years) (also discussed above in the context of depression symptoms), an intervention encouraging self-selected exercise did not reduce anxiety symptoms, compared to psychoeducation [36].

In 38 university students with anxiety (mean age, 21.8 years) a light-intensity intervention (Collective Rehabilitation Training) in addition to counseling reduced anxiety symptoms, compared to counseling alone [39]. This study engaged participants in collective outdoor games, the aim of which was to increase team cooperation [39]. Earlier research shows that team games/sports may improve psychosocial health due to the social nature of participation [40]. Therefore, it may be worth considering group-based, outdoor activities when designing interventions for young people with anxiety symptoms.

Conversely, neither 6 weeks of light intensity aerobic exercise or vigorous-intensity resistance training delivered indoors, reduced anxiety symptoms compared to a waitlist control group in 30 women, with generalized anxiety disorder (GAD) and with a mean age of 23.5 years. Regarding remission, this same study showed that the vigorous-intensity, but not the light-intensity intervention reduced remission rates (60% for vigorous, 40% for light, and 30% for the control group) and irritability [41], however, there was no effect of worry.

Physical activity or exercise as a treatment for eating disorders in young people.

There is limited evidence indicating that exercise interventions are effective for eating disorders among young people. One study involving 53 adolescents diagnosed with eating disorders (mean age = 16.5 years) found that an 8-week long yoga intervention (likely light-to-moderate intensity) decreased eating disorder symptoms and food preoccupation, compared to a waitlist control group [42].

A second study, involving 64 people with bulimia nervosa (mean age = 22.5 years) reported that 32 weeks of moderate-to-vigorous intensity aerobic exercise reduces laxative use and, at follow-up, drives for thinness and bulimic symptoms, compared to CBT [43].

Finally, a third study involving 22 outpatients with restrictive anorexia nervosa and a mean age of 14.5 years, found no effect 12 weeks of moderate-to-vigorous weight training on quality of life, compared to no intervention [44].

Physical activity or exercise as a treatment for psychosis in young people

There is limited evidence indicating the benefit of exercise for psychosis. One study assigned140 females with psychosis and a mean age of 24.6 years, to 12 weeks of yoga (hatha), stationary cycling (moderate-intensity aerobic exercise), or a wait-list control group. In both exercise conditions, quality of life improved and clinical symptoms of psychosis reduced at postintervention, negative symptoms improved in the yoga group [45].

In a non-randomized, comparative pilot study involving 31 people with first-episode psychosis and aged 18–35 years, 10 weeks of aerobic and resistance activities selected according to participant preference was seen to reduce symptom scores, compared to TAU. Negative symptoms decreased 33% and general symptoms decreased by approximately 25% in the exercise group. Psychosocial functioning also improved following exercise, as did some domains of cognition, while positive symptoms decreased in both groups [46].

In 28 people with first-episode psychosis (mean age = 20.7 years) 12 weeks of moderate-to-vigorous intensity aerobic and resistance training did not improve self-esteem, sleep quality, occupational, social, and psychological functioning compared to TAU [47].

In 15 individuals with first-episode individuals with first-episode schizophrenia (mean age = 21 years), a 10-week-long intervention of cognitive training plus an exercise (unspecified intensity) improved cognitive, family relationships, and school/work functioning, as well as independent living skills, compared to TAU—however, it is unclear if participants were allocated randomly to interventions [48].

Practice recommendations for clinicians in youth mental health services

Clinicians, including allied health professionals and general practitioners, have a significant contribution with regard to how physical activity is perceived, in assisting young people to make informed choices and to increase engagement in physical activity for mental health benefits. Recent guidance indicates that "some movement is better than none" and greater mental health benefits are associated with leisure-time activities (Teychanne et al., 2020. Clinicians can assist young people to set achievable goals, brainstorm activities, monitor the benefits associated with physical activity engagement, as well as to draw on existing social supports, to assist in including physical activity as part of the treatment of mental disorders [49].

The delivery of individualized or tailored physical activities based on individual preferences is likely important with regard to the efficacy of exercise on mental health. Self-determination theory states that autonomy is a basic psychological need, fundamental to positive mental health [50,51]. Previous work shows participants have a greater tolerance to higher intensity exercise when intensity is self-selected, rather than prescribed [52]. Therefore, individual preference is an important consideration in terms of adherence to exercise and mental health outcomes. Self-selection of exercise type and intensity may also lead to a greater sense of mastery of the activity and therefore contribute to increases in self-efficacy. Indeed, previous work shows that reductions in depression symptoms corresponded with increased self-efficacy, suggesting that self-efficacy may contribute to the impact of exercise on depression [53]. This hypothesis is consistent with previous studies reporting a bidirectional relationship between low self-efficacy and elevated depression symptoms among young people [54] and suggests that exercise programs should be achievable and aim to increase young people's self-efficacy. Self-efficacy is also important in determining adherence to exercise as research with 8th-grade girls reports that self-efficacy was indicated related to physical activity engagement, mediated through perceived barriers to engagement [55].

One study discussed here that reduced anxiety engaged participants in team games delivered outdoors, the aim of which was to increase team cooperation [39], while two studies that did not decrease anxiety were delivered indoors [42,56]. This is consistent with other research reporting that exercise outdoors predicted lower somatic anxiety, while indoor exercise predicted higher somatic anxiety [57]. Further, a study that did not reduce anxiety was individually delivered, rather than group-based [42]. Previous work reports that team sports/games might improve psychosocial well-being beyond effects attributable to exercise, due to the social element [40]. Therefore, it may be beneficial to consider outdoor, group-based activities when designing and implementing programs for young people with anxiety symptoms; however, this requires further examination.

Conclusions

The evidence presented in this chapter overall indicates that exercise improves symptoms among young people with a diagnosed mental disorder. Limited evidence reports that light-to-moderate exercise delivered outdoors and in group formats reduces anxiety symptoms. Light and moderate-to-vigorous exercise decrease depression, perhaps via improvements in self-efficacy. A smaller group of studies similarly demonstrate benefits for eating disorder symptoms and psychosis. Few studies have included young people's preferences for exercise type and intensity, which could be important to increase engagement, adherence, and perhaps effectiveness, compared to prescribed exercise. Future studies might also include long-term follow-up assessment time points as well as include an active control group that does not engage in exercise, in order to improve the quality of the research in this field. Research should clearly detail exercise interventions including the type, intensity, frequency, and duration of exercise, supported with appropriate quantifiable measures (e.g., heart rate, rate of perceived exertion). The findings from the chapter are can inform the design of future physical activity studies, building on the evidence in depression and expanding to include quality research on other mental disorders. While there is certainly scope for further development of the evidence base, there is sufficient evidence to demonstrate that physical activity is an acceptable, low-risk intervention for young people with emerging or diagnosed mental disorders and has an important role to play in the youth mental health care (Fig. 1).

Recommendations for prescribers

1. Some is better than none, more is better than some.

2. Self-selection of activity by the individual likely leads to beneficial results, increased self-efficacy and adherence.

3. In anxious patients, outdoor team sports may have greater effects.

4. Greater benefits to leisure time activities

5. Draw on social support networks to engage and encourage young people to exercise

---

FIG. 1   Recommendations for exercise in youth mental health clinical settings.

References

[1] WHO. Global Health Risks: Mortality and Burden Of Disease Attributable to Selected Major Risks. Geneva: World Health Organisation; 2009.

[2] United Nations. World Population Monitoring. New York: Department of Economic and Social Affairs; 2012.

[3] Woolery A, et al. A yoga intervention for young adults with elevated symptoms of depression. Altern Ther Health Med 2004;10(2):60–3.

[4] Kessler RC, et al. Age of onset of mental disorders: a review of recent literature. Curr Opin Psychiatry 2007;20:359–64.

[5] Kessler RC, et al. Lifetime prevalence and age-of-onset distributions of DSM-IV disorders in the National Comorbidity Survey Replication. Arch Gen Psychiatry 2005;62(6):593–602.

[6] Patel V, et al. Mental health of young people: a global public-health challenge. The Lancet 2007;369:1302–13.

[7] Gore FM, et al. Global burden of disease in young people aged 10-24 years: a systematic analysis. The Lancet 2011;377:2093–102.

[8] Australian Institute of Health and Welfare. Young Australians: Their Health and Wellbeing 2011, in Cat. no. PHE 140. Canberra: AIHW; 2011.

[9] McGorry PD, et al. Cultures for mental health care of young people: an Australian blueprint for reform. Lancet Psychiatry 2014;1(7):559–68.

[10] Sawyer SM, et al. Adolescence: a foundation for future health. The Lancet 2012;379:1630–40.

[11] Hetrick SE, et al. Newer generation antidepressants for depressive disorders in children and adolescents. Cochrane Database Syst Rev 2012;2012 (Issue 11). p. Art. No.: CD004851.

[12] Weisz JR, McCarty CA, Valeri SM. Effects of psychotherapy for depression in children and adolescents: a meta-analysis. Psychol Bull 2006;132 (1):132–49.

[13] March J, et al. Fluoxetine, cognitive-behavioral therapy, and their combination for adolescents with depression: treatment for adolescents with depression study (TADS) randomized controlled trial. JAMA 2004;292(7):807–20.

[14] Curtis J, Newall HD, Samaras K. The heart of the matter: cardiometabolic care in youth with psychosis. Early Interv Psychiatry 2012;6:347–53.

[15] WHO. Global Recommendations on Physical Activity for Health. Geneva, Switzerland: World Health Organisation; 2010.

[16] Chu IH, et al. Effect of exercise intensity on depressive symptom in women. Ment Health Phys Act 2009;2:37–43.

[17] Jorm AF, Wright A. Beliefs of young people and their parents about the effectiveness of interventions for mental disorders. Aust N Z J Psychiatry 2007;41:656–66.

[18] Baldursdottir B, et al. Age-related differences in physical activity and depressive symptoms among 10-19-year-old adolescents: a population based study. Psychol Sport Exerc 2017;28:91–9.

[19] Zimmermann-Sloutskis D, et al. Physical activity levels and determinants of change in young adults: a longitudinal panel study. Int J Behav Nutr Phys Act 2010;7(2):1–13.

[20] Lee IM, et al. Effect of physical inactivity on major non-communicable diseases worldwide: an analysis of burden of disease and life expectancy. Lancet 2012;380:219–29.

[21] Commission NMH. A contributing life, the 2012 National Report Card on mental health and suicide prevention. Sydney: NHMRC; 2012.

[22] Belair MA, et al. Relationship between leisure time physical activity, sedentary behaviour and symptoms of depression and anxiety: evidence from a population-based sample of Canadian adolescents. BMJ Open 2018;8(e021119):1–8.

[23] Strohle A. Physical activity, exercise, depression and anxiety disorders. J Neural Transm 2009;116:777–84.

[24] Mangerud WL, et al. Physical activity in adolescents with psychiatric disorders and in the general population. Child Adolesc Psychiatry Ment Health 2014;8(2):1–10.

[25] Carter T, et al. The effect of exercise on depressive symptoms in adolescents: a systematic review and meta-analysis. J Am Acad Child Adolesc Psychiatry 2016;55:580–90.

[26] Bailey A, et al. Treating depression with physical activity in adolescents and young adults: a systematic review and meta-analysis of randomised controlled trials. Psychol Med 2017;10:1–20.

[27] Wunram HL, et al. Whole body vibration added to treatment as usual is effective in adolescents with depression: a partly randomized, three-armed clinical trial in inpatients. Eur Child Adolesc Psychiatry 2018;27(5):645–62.

[28] Yavari A. The effect of swimming in reduction of depression in university male students. Res J Biol Sci 2008.

[29] Nourbakhsh P. The effects of physical activity on the level of depression in female students of SHAHID CHAMRAN university in AHVAZ; 2004.

[30] Sadeghi K, et al. A comparative study of the efficacy of cognitive group therapy and aerobic exercise in the treatment of depression among the students. Glob J Health Sci 2016;8(10):54171.

[31] Roshan VD, Pourasghar M, Mohammadian Z. The efficacy of intermittent walking in water on the rate of MHPG sulfate and the severity of depression. Iran J Psychiatry Behav Sci 2011;5(2):26–31.

[32] Hughes CW, et al. Depressed adolescents treated with exercise (DATE): a pilot randomized controlled trial to test feasibility and establish preliminary effect sizes. Ment Health Phys Act 2013;6(2):119–31.

[33] Balchin R, et al. Sweating away depression? The impact of intensive exercise on depression. J Affect Disord 2016;200:218–21.

[34] Carter T, et al. Preferred intensity exercise for adolescents receiving treatment for depression: a pragmatic randomised controlled trial. BMC Psychiatry 2015;15:247.

[35] Jeong YJ, Hong SC. Dance movement therapt improves emotional responses and modulates neurohormones in adolescents with mild depression. Int J Neurosci 2005;115:1711–20.

[36] Parker AG, et al. The effectiveness of simple psychological and physical activity interventions for high prevalence mental health problems in young people: a factorial randomised controlled trial. J Affect Disord 2016;196:200–9.

[37] Schuch FB, et al. Neurobiological effects of exercise on major depressive disorder: a systematic review. Neurosci Biobehav Rev 2016;61:1–11.

[38] Salmon P. Effects of physical exercise on anxiety, depression, and sensitivity to stress: a unifying theory. Clin Psychol Rev 2001;21(1):33–61.

[39] Yang WL, et al. Collective rehabilitation training conductive to improve psychotherapy of college students with anxiety disorder. Int J Clin Exp Med 2015;8(6):9949–54.

[40] Eime RM, et al. A systematic review of the psychological and social benefits of participation in sport for children and adolescents: informing development of a conceptual model of health through sport. Int J Behav Nutr Phys Act 2013;10:98.

[41] Herring MP, et al. Feasibility of exercise training for the short-term treatment of generalized anxiety disorder: a randomized controlled trial. Psychother Psychosom 2011;81(1):21–8.

[42] Carei TR, et al. Randomized controlled clinical trial of yoga in the treatment of eating disorders. J Adolesc Health 2010;46(4):346–51.

[43] Sundgot-Borgen J, et al. The effect of exercise, cognitive therapy, and nutritional counseling in treating bulimia nervosa. Med Sci Sports Exerc 2002;34(2):190–5.

[44] del Valle MF, et al. Does resistance training improve the functional capacity and well being of very young anorexic patients? A randomized controlled trial. J Adolesc Health 2010;46(4):352–8.

[45] Lin J, et al. Aerobic exercise and yoga improve neurocognitive function in women with early psychosis. NPJ Schizophr 2015;1:15047.

[46] Firth J, et al. Exercise as an intervention for first-episode psychosis: a feasibility study. Early Interv Psychiatry 2016.

[47] Curtis J, et al. Evaluating an individualized lifestyle and life skills intervention to prevent antipsychotic-induced weight gain in first-episode psychosis. Early Interv Psychiatry 2016;10(3):267–76.

[48] Ventura J, Gretchen-Doorly D, Subotnik KL, Vinogradov S, Nahum M, Nuechterlein KH. Combining cognitive training and exercise to improve cognition and functional outcomes in the early course of schizophrenia: a pilot study. In: Abstracts for the 14th International Congress on Schizophrenia Research (ICOSR). Schizophrenia Bulletin; 2013. p. S1–S358.

[49] Ekkekakis P, Murri MB. Exercise as antidepressant treatment: time for the transition from trials to clinic? Gen Hosp Psychiatry 2017;49:A1–5.

[50] Craft LL, et al. Psychosocial correlates of exercise in women with self-reported depressive symptoms. J Phys Act Health 2008;5;469–80.

[51] Ryan RM, Deci EL. Self-determination theory and the facilitation of intrinsic motivation, social development, and well-being. Am Psychol 2000;55(1):68–78.

[52] Ekkekakis P, Parfitt G, Petruzzello SJ. The pleasure and displeasure people feel when they exercise at different intensities decennial update and Progress towards a tripartite rationale for exercise intensity prescription. Sports Med 2011;41(8):641–71.

[53] Brown SW, et al. Aerobic exercise in the psychological treatment of adolescents. Percept Mot Skills 1992;74(2):555–60.

[54] Tak YR, et al. The prospective associations between self-efficacy and depressive symptoms from early to middle adolescence: a cross-lagged model. J Youth Adolesc 2017;46(4):744–56.

[55] Dishman RK, et al. Social-cognitive correlates of physical activity in a multi-ethnic cohort of middle-school girls: two-year prospective study. J Pediatr Psychol 2010;35(2):188–98.

[56] Herring MP, et al. Effects of short-term exercise training on signs and symptoms of generalized anxiety disorder. Ment Health Phys Act 2011;4(2):71–7.

[57] Lawton E, et al. The relationship between the physical activity environment, nature relatedness, anxiety, and the psychological well-being benefits of regular exercisers. Front Psychol 2017;8:1058.

Chapter 7

Type 1 diabetes

Jane E. Yardley[a] and Sheri R. Colberg[b]

[a]*University of Alberta, Camrose, AB, Canada* [b]*Old Dominion University, Norfolk, VA, United States*

Introduction

As of 2021, the International Diabetes Federation (IDF) estimated that there were over 537 million people living with diabetes worldwide, with numbers expected to reach 643 million by 2030 and 783 million by 2045 [1]. The majority of these cases (approximately 90%) are type 2 diabetes, a condition in which insulin production is unable to compensate fully for the body's level of cellular insulin resistance, resulting in chronic hyperglycemia (elevated blood glucose levels). It is most common in older adults and often associated with obesity, physical inactivity, other lifestyle choices, genetics, and inappropriate diet [1]. Type 2 diabetes, gestational diabetes, where hyperglycemia first presents during pregnancy, and prediabetes (a glucose-intolerant state) are discussed in another chapter.

Roughly 10% of people with diabetes have type 1 diabetes (T1D). It is the most common chronic disease of childhood, although it can develop at any age and most people living with it are adults. The underlying causes are likely a genetic predisposition combined with an environmental trigger, such as a virus or toxin [1], leading to the destruction of the insulin-producing beta cells of the pancreas through an autoimmune response [1]. As a result, inadequate insulin is available to manage blood glucose levels, and insulin replacement is necessary. Since some individuals continue to produce a certain amount of insulin (measured by C-peptide in the blood) after diagnosis [2,3], the amount of insulin required by each person is highly variable.

Methods of insulin delivery in type 1 diabetes

The methods of insulin administration also vary. Some individuals choose to administer their insulin using multiple daily injections (MDI), which generally involves the daily injection of basal insulin lasting anywhere from 12 to 42 h depending on the type, dosing, and unique metabolic differences [4]. Basal insulins are typically supplemented with injections of faster (bolus) insulin to manage blood glucose fluctuations associated with meals and snacks or to correct hyperglycemia. Others use continuous subcutaneous insulin infusion (CSII) that involves using an insulin pump primed with rapid-acting insulin given to cover basal and bolus insulin requirements throughout each day. Both of these methods of insulin delivery typically require user decision-making based on glucose levels measured either by fingerstick blood glucose monitoring or, more recently, continuous glucose monitoring (CGM), which provides real-time measurement of interstitial blood glucose levels. When CGM is combined with CSII, users may have the option of a hybrid closed-loop system where an algorithm in the pump can adjust insulin delivery rates based on the data provided by the CGM, relatively independent of most user input [5]. These technologies will be discussed in more detail later in this chapter.

Managing insulin therapy can be very complex, and many individuals with T1D will experience hypoglycemia (low blood glucose levels) or hyperglycemia on an almost daily basis [6]. If not addressed promptly, hypoglycemia can become debilitating and may require assistance from others to correct through fast-acting carbohydrate consumption or glucagon administration [7]. Frequent hyperglycemia and, consequently, elevated A1C levels (a measure of average blood glucose over the last 2–3 months), on the other hand, may increase long-term risks of cardiovascular disease [8], retinopathy [9,10], nephropathy [11], and neuropathy [12]. When severe, these complications can result in blindness, kidney failure potentially requiring transplant, and foot ulcers and infections that may lead to amputation. Many of these conditions have been linked to systemic inflammation [13–16], which is a known result of chronic hyperglycemia in individuals with T1D [17]. There is also evidence to indicate that T1D may lead to lower overall bone quality and strength [18], and a faster loss of muscle mass [19], bone density [20], and/or bone quality [20–22] with aging, compared to individuals without diabetes, leading to an increased risk of frailty and a higher risk of fracture [23] in older adults with T1D, compared to adults without diabetes.

Exercise to Prevent and Manage Chronic Disease Across the Lifespan. https://doi.org/10.1016/B978-0-323-89843-0.00025-8

Effects of physical activity and exercise training

Cross-sectional and longitudinal studies of individuals with T1D have shown that higher physical activity levels are associated with increased longevity [24,25] and a decreased risk of health complications [24,26–30]. Being more frequently and/or more vigorously active is associated with having fewer cardiovascular risk factors (e.g., high body mass index, dyslipidemia, hypertension) [27], lower incident cardiovascular disease [24], and fewer incident cardiovascular events [28], along with lower risk and/or slower progression of retinopathy [27,30], nephropathy [26,29] and neuropathy [26]. Higher physical activity levels have also been correlated with better quality of life among adolescents and young adults with T1D [31–33]. Several (but not all) [34] cross-sectional or longitudinal studies also show an overall improvement in A1C measures with higher levels of physical activity and exercise [27,35–38], although notably one study using accelerometry only found improvements with vigorous and not moderate exercise [39]. Finally, higher levels of physical activity have been associated with greater bone mass [40] and bone mineral density [41] in this population.

In intervention studies involving exercise in T1D, the impact of regular exercise on A1C is unclear, with several studies showing no change over the course of training [42–48] and some studies showing an improvement [49–52]. A 2014 meta-analysis of intervention studies suggested that very few studies were of adequate intensity and/or duration to allow for changes in A1C [53]. It should be noted, however, that small studies involving interventions that used either resistance exercise [51,54,55], or high-intensity intermittent exercise (HIIE) [51], have generally seen beneficial effects on overall blood glucose management.

Aerobic training

In spite of unclear effects on blood glucose measured by A1C, aerobic exercise interventions involving individuals with T1D have produced many positive results for the participants. Not surprisingly, aerobic exercise interventions have led to improvements in aerobic fitness [43,45,47,56–58], increases in capillary density [59], enhanced endothelial function [47,59,60] (potentially due to decreases in systemic inflammation) [61], and improved lipid levels [47,56–58,62], along with a decrease in insulin dosage [47,57,62,63] and/or insulin resistance [45,57,62]. There have also been measured improvements in body composition (decreased waist circumference [64] and increased bone density [65]) measured after a period of supervised aerobic training.

High-intensity interval exercise training

Studies of HIIE training have found many similar benefits to aerobic training in individuals with T1D. HIIE appears to improve several markers of cardiovascular health, including an increase in aerobic capacity [51,60,66], a decrease in aortic stiffness (measured by pulse wave velocity) [66], and better endothelial function as a result of HIIE training [60,66]. It is unclear whether or not these improvements in fitness and endothelial function are greater with HIIE than with aerobic training in this population, with one study finding training outcomes to be similar [66] and another finding that HIIE provided superior results [60]. This type of training may also improve body composition (i.e., reduced fat mass and increased muscle mass) [51] and quality of life [67] in adults with T1D.

Resistance training

The limited number of resistance exercise interventions involving individuals with T1D suggests an increase in muscle strength [51,54] and better body composition (decreased fat mass and increased muscle mass) [51]. A small study with eight male participants with T1D [54] found that resistance exercise on its own may lower A1C and improve lipid profiles. In addition to increases in muscle strength [48,55] and improved lipid profiles [48,55], combined aerobic (or HIIE) and resistance exercise programs have produced additional benefits, including higher aerobic fitness [48,55], lower A1C [51,55], and a decrease in insulin needs [48,51].

Impact of sedentary behavior

Despite the many known benefits of exercise training in this population, a large proportion of individuals with T1D fail to achieve the recommended amount of weekly activity [68]. While this may also be true for the general population without diabetes [69], both youth [70,71] and adults [68] with T1D achieve fewer weekly minutes of moderate to vigorous physical activity and spend more time on sedentary behaviors than their counterparts without diabetes. In addition, women with T1D are less active than men with the condition [27,72].

Given that sedentary behavior, in general, may have a negative impact on blood glucose levels in individuals with T1D [72,73], simply decreasing sedentary time may have substantial health benefits for this population. The only study to date to examine this, however, found a decrease in body fat mass, but no changes in lean body mass or A1C as a result of performing three short bouts (6 × 1-min bouts of the resistance band activity, referred to as "exercise snacks") of exercise daily to break up sedentary time [74]. More research in this area is needed.

Insulin kinetics and risk of hypoglycemia with exercise

Whereas endogenous insulin has a half-life of approximately 5 min, even the fastest-acting insulins used by individuals with T1D take much longer than that after administration to hit their peak (see Table 1), and several hours or more to be cleared out of the body [75]. Their onset, peak, and duration of action vary by type but also by the dosage given at any one time (i.e., larger subcutaneous droplets take longer to be fully absorbed), with increased capillary blood flow during exercise potentially increasing absorption rates of faster-acting insulins [76]. Insulin levels decrease relatively rapidly at the start of exercise in individuals without diabetes. For individuals with T1D, given the peripheral delivery of replacement insulin and absorption kinetics, circulating insulin levels are frequently elevated during activity, which greatly increases their risk of hypoglycemia.

Insulin management for aerobic exercise

The risk of hypoglycemia is greatest when the exercise is aerobic in nature and performed within 2–3 h after a meal [77], as the blood glucose arising from that meal will have mostly been removed but insulin levels may still be elevated. If exercise is planned, insulin dosing for meals may be decreased a couple of hours before exercise, or individuals using a pump may lower basal insulin delivery prior to and during exercise. When activity is spontaneous or insulin levels cannot be lowered appropriately, carbohydrate intake will often be necessary to maintain blood glucose levels in a safe range during moderate aerobic activities. Lesser amounts may be needed to compensate for either light or vigorous activity compared to moderate [78], and the duration of the activity will also dictate the required carbohydrate intake for glycemic management.

There are two main options for adjusting insulin dosage prior to exercise when individuals with T1D are using an MDI regimen. Where exercise routines are well-established but exercise may not be happening daily, a decrease in the basal insulin injection of 20% the night before exercise (or morning of depending on the individual's regular management schedule) may be implemented [79]. If exercise is less predictable, a more suitable option may be to decrease the amount of insulin administered with the meal or snack consumed before exercise by 25%–75% [79], depending on the planned exercise duration and intensity.

Insulin pumps tend to offer a little bit more flexibility in that adjustments to basal insulin delivery only need to be made within hours of the exercise session. Recent studies have shown that making adjustments at the beginning of, or even up to 40 min prior to aerobic exercise [80], may be insufficient to prevent hypoglycemia, and current recommendations are that

TABLE 1 Approximate action of current insulins and insulin analogs.

Insulin	Onset	Peak	Duration
Afrezza (inhaled)	10–12 min	35–45 min	1.5–3 h
Fiasp (Aspart), Lyumjev (Lispro)	15–20 min	1–2 h	3–6 h
Humalog, Ademlog (Lispro), NovoLog (Aspart), Apidra (Glulisine)	20–30 min	1.5–3 h	5–7 h
Humulin R, Novolin R, Velosulin R (Regular)	30–45 min	2–5 h	5–8 h
Humulin N, Novolin N, ReliOn (NPH)	1–2 h	2–12 h	14–24 h
Lantus, Basaglar, Semglee, Rezvoglar, Toujeo* (Glargine)	1.5 h (*6 h since U-300)	None	20–24 h
Levemir (Detemir)	1–2 h	8–10 h	Up to 24 h
Tresiba (Degludec)	1–4 h	None	Over 24 h

basal rates be adjusted downward 90 min in advance of most aerobic activity. Reductions in basal insulin between 50% and 80% are recommended for aerobic exercise [79]. Full suspension of insulin delivery is possible without adverse effects [81], although not recommended for exercise sessions lasting longer than 60 min [79] without reconnecting and replacing some of the missed basal doses.

Insulin management for anaerobic exercise

In contrast to aerobic exercise, when short, very intense exercise (anaerobic—fueled without oxygen being required) is performed by individuals with T1D, blood glucose levels are more likely to remain stable or increase. These types of activities, which include sprints, plyometrics, and resistance exercise, can only be maintained for very short periods of time in all but highly trained individuals. The increase in blood glucose with maximal and near-maximal effort is often attributed to epinephrine-induced increases in hepatic glycogenolysis [82] and may be potentiated by decreased muscle glucose uptake due to high intracellular glucose-6-phosphate from muscle glycogenolysis [83]. Several studies have shown that these types of activities, when introduced in short bursts during an otherwise aerobic exercise session to produce HIIE, can delay declines in blood glucose during exercise and reduce the amount by which blood glucose decreases compared to aerobic exercise only [84–88]. Some additional insulin may be needed in the immediate postexercise period to correct hyperglycemia [89]. As these types of activities depend more on muscle and liver glycogen as a fuel source [82], postexercise hypoglycemia risk may be heightened as glycogen stores are being replenished [90], particularly if exercise is performed later in the day [91–93]. Modest reductions of 10%–30% of evening basal doses may prevent overnight hypoglycemia [94].

Recommended insulin adjustments prior to exercise involving an anaerobic component are frequently the same as those for aerobic exercise [79], even though they often do not need to be as aggressive. They should, however, still take place in advance of exercise, as HIIE on its own may be insufficient to maintain blood glucose levels during exercise [81,95]. It may be more important to decrease basal insulin by 20% overnight postexercise after anaerobic activity to prevent later declines in blood glucose as the muscles and liver replace their glycogen stores. It is also recommended that a postexercise meal or snack containing 1.0–1.2 g of carbohydrate per kg of body mass be consumed to assist in replenishing glycogen, which could also be accompanied by a reduced (50%) insulin bolus if the activities took place in the afternoon or evening [79].

Other exercise-related insulin management considerations

In the case of frequent, spontaneous, or long duration (i.e., over 60–90 min) activity, insulin adjustments may either not be possible, or may not be sufficient. During these instances, it is often possible to maintain blood glucose levels by consuming carbohydrates at a rate of between 0.5 and 1.0 g/kg of body mass, for each hour of activity where no insulin adjustments have been made [79]. If insulin has been adjusted in advance, carbohydrate supplements of 10–20 g/h may be sufficient, depending on the starting blood glucose levels of the individual, exercise timing, and the rate of change of their blood glucose during the activity [64]. For exercise sessions between 60 and 150 min in duration, 30–60 g of carbohydrate per hour are recommended for individuals with and without diabetes to maintain performance. Where exercise extends beyond 150 min, quantities of up to 90 g/h may be required [79].

The size of the insulin adjustment and/or the amount of carbohydrate required to maintain safe blood glucose levels are extremely variable, will depend on the type, intensity, and duration of activity (as shown in Fig. 1), and often need to be determined by trial and error. If the individual about to undertake exercise or physical activity has experienced a bout of hypoglycemia within the previous 24 h, it may be necessary to increase the size of the insulin reduction or the amount of carbohydrate consumed [96]. For a more comprehensive review of recommendations for insulin reductions and carbohydrate intake, the 2017 Consensus Statement by Riddell et al. [79] can be consulted.

Hypoglycemia prevention and treatment

Exercise participation significantly increases the risk of hypoglycemia, usually defined as blood glucose levels below $65 \, mg \, dL^{-1}$ ($3.6 \, mmol \, L^{-1}$). Hypoglycemia may occur in the presence of the following factors, in isolation or combination:

- Too much circulating insulin
- Too little carbohydrate (or other food) intake
- Missed meals
- Excessive or poorly planned exercise
- Late-day exercise

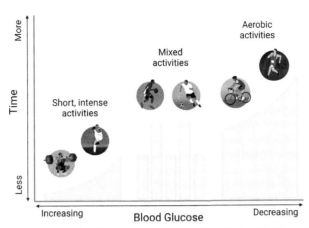

FIG. 1 Blood glucose effects of different types of exercise. *(Reprinted by permission from Colberg SR. The athlete's guide to diabetes. Champaign, IL: Human Kinetics; 2020. p. 30.)*

Hypoglycemia can occur either during exercise or hours to days later [97,98]. Late-onset postexercise hypoglycemia generally occurs within hours following moderate-to-high-intensity exercise that lasts longer than 30 min, frequently occurring during the overnight hours unless insulin adjustments are made [99]. Such hypoglycemia is likely the result of increased insulin sensitivity, heightened blood glucose use, and glycogen repletion via gluconeogenesis [100,101]. Individuals should monitor blood glucose before and periodically after exercise to assess glucose responses and be particularly vigilant after late-day exercise.

Hypoglycemia symptoms and awareness

The symptoms of hypoglycemia can be adrenergic or neuroglycopenic. As blood glucose decreases, glucose-raising hormones (i.e., glucagon, epinephrine, growth hormone, and cortisol) are released to help increase circulating blood glucose levels [102]. Adrenergic symptoms like shakiness, weakness, sweating, nervousness, anxiety, and tingling of the mouth and fingers, result primarily from epinephrine release. As blood glucose delivery to the brain decreases, neuroglycopenic symptoms including headache, visual disturbances, mental dullness, confusion, extreme fatigue, amnesia, seizures, or coma may occur.

It is well established that people with long-standing T1D have an altered counterregulatory hormone response to hypoglycemia, specifically with lower epinephrine release and greatly lower or absence release of glucagon [103]. Related to these hormonal deficiencies, some people with T1D lose their ability to sense hypoglycemic symptoms (termed *hypoglycemia unawareness*), which is essentially the onset of neuroglycopenic symptoms before the appearance of adrenergic ones [104]. Instituting tight management of blood glucose may lower the threshold so that symptoms do not occur until blood glucose drops quite low, whereas individuals whose blood glucose has been inadequately managed may have symptoms associated with low blood glucose at much higher levels.

With prior exercise or prior hypoglycemia, some individuals with T1D may experience *hypoglycemia-associated autonomic failure* or HAAF. In this case, their lack of adrenergic response leading to hypoglycemia unawareness may be related to having a prior hypoglycemic episode in the previous 24–48 h, especially one that was severe or long-lasting, or having exercised in that time period [105–107]. Both prior exercise and prior hypoglycemia may blunt the normal adrenergic response to the next bout of exercise or hypoglycemia and set individuals up for potentially dangerous hypoglycemic episodes [108,109].

Hypoglycemia treatment

Treating hypoglycemia consists of testing blood glucose to confirm hypoglycemia and consuming 4–20 g of carbohydrate (preferably glucose, but also sucrose, lactose, or juice containing fructose) that contains minimal or no fat [110]. Commercial products (glucose tablets, gels, or liquids) allow the consumption of precise amounts of carbohydrate, and glucose increases blood glucose levels most rapidly compared to other simple sugars or carbohydrate sources. Individuals should wait about 15 or 20 min to allow the symptoms to resolve and then recheck blood glucose levels to determine if additional carbohydrates or other food sources are necessary.

If the individual becomes unconscious because of hypoglycemia, glucagon should be administered. Both mini-doses of glucagon via subcutaneous injection with a pen and via nasal glucagon sprays are now available, potentially for prevention of hypoglycemia during exercise and for mild to moderate hypoglycemia treatment [111,112], and both delivery methods appear to work equally well in resolving hypoglycemia in conscious individuals with T1D [113]. Although prior full doses of glucagon that were injected frequently caused nausea and vomiting after the individual regained consciousness, the use of mini-doses for less severe hypoglycemia avoids the majority of those symptoms. If glucagon is not available and an individual is unconscious or unable to self-treat hypoglycemia, emergency medical services should be called immediately.

Additional considerations related to prandial state

Some of the variability seen in blood glucose outcomes to exercise noted in the research literature may also be related to the prandial state of the participants. When individuals with T1D perform exercise after an overnight fast, distinctly different patterns of blood glucose responses have been observed. Exercise performed in the fasted state can often cause an increase in blood glucose levels compared to the same exercise performed postprandially. This phenomenon has been found in small studies of aerobic exercise [114,115], resistance exercise [116], and HIIE [117] where the same participants performed an exercise protocol in a fed and fasted state on different days. While the phenomenon is currently not fully understood, the divergent blood glucose responses may be due to the impact of diurnal variations in growth hormone and cortisol levels on insulin sensitivity and fuel selection [118–120]. For any individual with T1D fearful of exercising due to the hypoglycemia risk, performing exercise in a fasting state may be a potential solution [66]. However, depending on the activity type, intensity, and duration undertaken, additional insulin may be needed to correct or prevent exercise-induced hyperglycemia related to fasting metabolic conditions [116,117].

Exercise recommendations for individuals with T1D of all ages

In general, exercise training recommendations for adults (Table 2) and youth with T1D differ very little from those for everyone else at the same stage of life. The main difference from the latest exercise guidelines (at least in the United States [121]) is that people with diabetes are recommended to not allow more than two consecutive days to lapse between bouts of activity due to the loss of insulin sensitivity over time when no activity is undertaken [122]. Regular moderate to vigorous aerobic activity is recommended, along with two to three nonconsecutive days of resistance training. Balance training is recommended for anyone over 40 years old, and flexibility exercises should be performed frequently by everyone with T1D. Children and adolescents with T1D can follow the recommendations for physical activity for youth without diabetes.

Infants, toddlers, and preschoolers (0–4 Years)

Infants' activity is based upon their interactions with caregivers who provide the infants with opportunities to explore movement and their surroundings and supports development. Toddlers should participate in at least 30 min of physical activity daily, often with unstructured play lasting longer and sedentary time limited to no more than 60 min at one time (except during sleep) [123]. They will likely engage in indoor and outdoor activities that promote the development of their large muscles and gross motor skills. Preschoolers should have at least 60 min of structured physical activity daily, along with several hours of unstructured play each day to help them develop motor skills and enhance socialization [123]. For every 30 min of physical activity, the young child with T1D may require additional carbohydrate intake in order to prevent hypoglycemia as it is difficult to adjust insulin in advance of the activity. In general, 5–10 g for every 30–60 min of activity will be needed, depending upon the child's initial blood glucose concentration and intensity of the exercise. Blood glucose levels should be checked frequently as these young children are unable to convey symptoms of hypoglycemia.

Young children (5–11 Years) and adolescents (12–17 Years)

All children (5–11 years) and youth (12–17 years) should undertake at least 60 min of moderate- to vigorous-intensity mixed aerobic and anaerobic activity daily (420 min/week) [123]. This recommendation appears appropriate for youth with T1D, although it can clearly increase hypoglycemia risk. These recommendations are particularly suitable for children and adolescents with T1D since physical activity patterns in youth track into adulthood. In addition, health benefits are similar to people without diabetes and cardiovascular disease is the major cause of early mortality and morbidity in this population.

To summarize, the recommendations for daily exercise in all children and adolescents with T1D are expected to include:

TABLE 2 Recommended aerobic, resistance, flexibility, and balance training for adults with T1D.

Training method	Mode	Intensity	Frequency	Duration	Progression	Important considerations
Aerobic	Walking, jogging, cycling, swimming, aquatic activities, rowing, dancing	40%–59% of VO$_2$R (moderate) RPE 11–12; or 60%–89% of VO$_2$R (vigorous)' RPE 14–17	3–7 days/week, with no more than 2 consecutive days between bouts of activity	Minimum of 150–300 min/week of moderate activity or 75–150 min of vigorous activity, or a combination thereof	Rate of progression depends on many factors including baseline fitness, age, weight, health status, and individual goals; gradual progression of both intensity and volume is recommended	Note: All special considerations apply to all these training methods. • Avoid or take precautions for exercise undertaken during insulin peak time. • Be aware of any signs and symptoms for vascular and neurological complications. • Include appropriate warm-ups and cooldowns. • Use proper footwear and inspect feet daily. • Avoid extreme environmental temperatures. • Avoid or postpone exercise when blood glucose is not well managed. • Maintain adequate hydration. • Monitor glucose and follow guidelines to prevent hypoglycemia and hyperglycemia.
Resistance	Free weights, machines, elastic bands, or bodyweight as resistance; do 8–10 exercises involving the major muscle groups	Moderate at 50%–69% of 1-RM or vigorous at 70%–85% of 1-RM	2–3 days/week, but never on consecutive days	10–15 repetitions per set, 1–3 sets per type of specific exercise	As tolerated Increase resistance first, followed by a greater number of sets, and then increased training frequency	
Flexibility and Balance Training	Static, dynamic, or PNF stretching; balance exercises; yoga and tai chi may be appropriate for a range of motion, balance, and strength	Stretch to the point of tightness or slight discomfort	≥2–3 days/week or more; usually done with when muscles and joints are warmed up; lower body and core resistance exercises may double as balance training	10–30 s per exercise of each muscle group; 2–4 repetitions of each; balance exercises can be practiced daily or as often as possible	As tolerated; may increase range of stretch as long as not painful; balance training should be done carefully to minimize the risk of falls	

Note: 1-RM, 1-repetition maximum; PNF, proprioceptive neuromotor facilitation; RPE, rating of perceived exertion; VO$_2$R, VO$_2$ reserve.

- At least 60 min of accumulated physical activity every day
- Vigorous-intensity aerobic activities at least 3 days/week
- Activities that strengthen muscle and bone at least 3 days/week each

Children and youth should be physically active daily as part of play, games, sports, transportation, recreation, physical education, or planned exercise in the context of family, school, and community activities (e.g., volunteer, employment). This should be achieved above and beyond the incidental physical activities accumulated in the course of daily living. Reducing sedentary time is convincingly associated with a favorable cardiovascular profile, and several expert panels recommend limiting leisure screen time to <2 h/day [123].

For school-aged children and youth who are physically inactive, doing amounts below the recommended levels (i.e., 30 min/day) likely provides some health benefits, compared with being completely sedentary, given that sedentary youth with T1D tend to have higher A1C levels compared with active youth [124]. For sedentary youth, it may be appropriate to start with smaller amounts of physical activity and gradually increase duration, frequency, and intensity as a stepping stone to meeting the guidelines.

These guidelines for daily physical activity, which should include vigorous aerobic and muscle-strengthening activities, may be considered ambitious for young persons with T1D, given their potential fear of hypoglycemia and sometimes sedentary nature. A large sample of children and adolescents with T1D (23,251 youth ages 3–18 years), found that ~45% of this cohort was generally sedentary (i.e., less than 30 min of continuous physical activity per week, excluding school sports), with only 37% of the cohort having regular physical activity for 30 min 1–2 times/week [125]. However, this study also found that A1C levels were lower in patients with higher frequencies of physical activity and that blood lipid profile was more favorable compared to those who were sedentary. Interestingly, multiple regression analysis revealed that regular physical activity was one of the most important factors influencing A1C levels [37,125].

Adults (18–64 Years)

To achieve health benefits, adults ages 18–64 years should accumulate at least 150–300 min of moderate or 75–150 min of vigorous-intensity aerobic physical activity per week [123,126]. It is also beneficial to add muscle- and bone-strengthening activities that use major muscle groups, at least 2 days/week (see Table 2). Although the total energy expenditure should be ~1000 kcal/week of physical activity, health benefits occur with energy expenditures as low as 500 kcal/week, with additional benefits occurring at higher levels [126].

Adults with T1D can meet recommended guidelines through planned exercise sessions, transportation, recreation, sports, or occupational demands in the context of family, work, volunteer, and community activities, above and beyond the incidental physical activities accumulated in the course of daily living. The potential benefits far exceed the potential risks associated with physical activity, even in people with T1D [127,128]. The regular activity appears to lower cardiovascular mortality risk at all levels of glycemic management. For those who are initially physically inactive, doing amounts below the recommended levels can provide some health benefits. For these adults, it is appropriate to start with smaller amounts of physical activity and then gradually increase the duration, frequency, and intensity as a stepping stone to meeting these guidelines.

Older adults (≥65 Years)

To achieve health benefits and improve functional abilities, adults aged 65 years and older should accumulate at least 150–300 min of moderate- to vigorous-intensity aerobic physical activity per week, as recommended for younger adults, if possible [123]. It is beneficial to add muscle- and bone-strengthening activities that use the major muscle groups at least 2 days/week.

While the exercise guidelines for adults also apply to older adults, there are some additional ones that apply only to older adults (with or without T1D) [123]:

- When older adults cannot do 150 min of moderate-intensity aerobic activity a week because of chronic conditions, they should be as physically active as their abilities and conditions allow.
- Recommend exercises that maintain or improve balance, particularly if they are at risk of falling.
- Determine their level of effort for physical activity relative to their level of fitness.
- Those with chronic conditions should understand whether and how their conditions affect their ability to do regular physical activity safely.

Older adults can meet these guidelines through the same means as their younger counterparts (i.e., increased activities of daily living) to reduce the risk of comorbid disease and premature death, maintain functional independence and mobility, as well as improve fitness, body composition, bone health, cognitive function, and indicators of mental health. These guidelines may be appropriate for older adults with frailty or other comorbid conditions; however, individuals with health issues should consult a health professional to understand the types and amounts of physical activity appropriate for them based on their exercise capacity and specific health risks or limitations.

Exercise safety considerations and precautions for all individuals with T1D

When beginning a regular exercise training program, most people with T1D can anticipate significant and meaningful improvements, assuming the following exercise training considerations and precautions are followed:

- Most individuals with T1D will benefit from frequent monitoring of glucose levels before, possibly during, and after exercise [79].
- Adjustments to insulin doses and food intake may be necessary to keep blood glucose levels in normal or near-normal ranges during and following all physical activities.
- Individuals should not begin the exercise with blood glucose >250 mg dL^{-1} (13.9 mmol L^{-1}) if moderate or higher levels of blood or urinary ketones are present.
- Caution should be used for exercising with blood glucose >300 mg dL^{-1} (16.7 mmol L^{-1}) without excessive ketones; individuals should attempt to stay hydrated and only begin activity if feeling well [79,122].
- Medical clearance (and exercise testing) prior to starting activities more vigorous than brisk walking may be recommended for individuals with any signs or symptoms of cardiovascular disease, longer duration of T1D, older age, or other diabetes-related complications [122].
- Both resting and exercise heart rates may be impaired in individuals with cardiac autonomic neuropathy, which can result in a higher than typical resting heart rate (usually 100 beats per minute or higher) and a blunted peak heart rate. If the heart rate is to be used to prescribe exercise intensity for such individuals, maximal heart rate should be measured rather than estimated, or ratings of perceived exertion can be used instead [129].
- Peripheral neuropathy and poor circulation may increase the risk of a foot injury and delay healing. Individuals should inspect their feet regularly for any signs of trauma and avoid weight-bearing exercise with unhealed plantar ulcers [122].

Individuals with health complications arising from their diabetes should also be aware that certain types of activities may be contraindicated due to their condition, or that special testing/preparation may be required (Table 3). In the presence of known macrovascular diseases, such as coronary artery disease and peripheral artery disease, preexercise screening should follow the guidelines set by the American College of Sports Medicine [128]. Those with cardiovascular autonomic neuropathy should consult a physician and undergo an exercise stress test prior to undertaking a physical activity program, but will also generally benefit from performing moderate aerobic exercise [130] provided that it is performed with adequate warm-up, and cool-down, and avoiding extremes of temperature.

As for microvascular complications, if an individual with T1D has severe nonproliferative or unstable proliferative retinopathy, activities that increase intraocular pressure (e.g., high-intensity activities, heavy weight lifting) or activities that involve jumping or jarring should be avoided [122]. Vigorous activities may also not be appropriate for individuals with advanced diabetic kidney disease, although moderate-intensity activities can be very beneficial [131]. Individuals with peripheral neuropathy, especially if they are lacking sensation in their legs and feet, should ensure that they use proper footwear and perform regular inspections of their feet for injuries, wounds, and infections [130]. Where wounds or injuries contraindicate weight-bearing activities, seated exercises using free weights or resistance bands should be encouraged [122].

Setting individualized goals for physical activity

Individuals with T1D may have varying goals for starting and maintaining a physically active lifestyle. Many are active and avid athletes—both recreational and competitive—and may have improved performance as a training goal, while others may be aiming for increased cardiorespiratory fitness or muscular strength and endurance [132]. Finally, many may be focused on weight loss, gain, or maintenance. All of these goals may require different insulin adjustments and dietary regimens.

TABLE 3 Special precautions for diabetes-related complications.

Health complication	Precaution
Autonomic (Central) neuropathy	Be aware of an increased likelihood of hypoglycemia, abnormal BP responses, and impaired thermoregulation, as well as elevated resting and blunted maximal HR. Use of RPE is suggested to monitor exercise intensity, and take steps to prevent dehydration and hyper/hypothermia.
Peripheral neuropathy	Limit participation in exercise that may cause trauma to the feet, such as prolonged hiking, jogging, or walking on uneven surfaces. Non-weight-bearing exercises (e.g., cycling, chair exercises, swimming) may be more appropriate, although aquatic exercise is not recommended with unhealed ulcers. Check feet daily for signs of trauma and redness and keep them clean and dry. Choose shoes carefully for proper fit and wear socks that keep feet dry. Avoid activities requiring a great deal of balance.
Diabetic retinopathy	With unstable proliferative and severe stages of retinopathy, avoid vigorous, high-intensity activities that involve breath-holding (e.g., weight lifting and isometrics) or overhead lifting. Avoid activities that lower the head (e.g., yoga, gymnastics) or that risk jarring the head. In the absence of stress test HR, the use of RPE is recommended (10–12 on a 20 scale). If an individual has proliferative retinopathy and has recently undergone photocoagulation or surgical treatment or is not properly treated, exercise is contraindicated. Consult an ophthalmologist for specific restrictions and limitations.
Diabetic kidney disease	Avoid exercise that causes excessive increases in BP (e.g., weight lifting, high-intensity aerobic exercise) and refrain from breath-holding. High BP is common, and lower intensity exercise may be necessary to manage BP responses and fatigue. Light to moderate exercise may be possible during dialysis treatments as long as electrolytes are managed.
Hypertension	Avoid heavy weight lifting or breath-holding. Perform dynamic exercises using large muscle groups, such as walking and cycling at a low to moderate intensity. Follow BP guidelines. In the absence of a maximal HR determined with an exercise stress test, use of RPE is recommended (10–12 on a 6–20 scale).
General precautions	Maintain hydration by drinking fluids before, during, and after exercise. Avoid exercising during the peak heat of the day or in direct sunlight. Carry rapid-acting carbohydrate sources during all activity. Have glucagon available to treat severe hypoglycemia.

Note: *BP*, blood pressure; *HR*, heart rate; *RPE*, rating of perceived exertion.

If individuals with T1D are actively engaged in regular exercise and are managing their blood glucose effectively, they can achieve elite-level performance. Good glycemic management, as measured by A1C, is associated with better performance, while suboptimal management is associated with deterioration in athletic performance [133].

Athletic goals aside, weight loss is often an incentive to be more active since many adults with T1D may become overweight or obese, contributing to increases in insulin resistance more characteristic of type 2 diabetes and prediabetes [134]. Moderate weight loss improves glucose management and decreases insulin resistance in adults with T1D, and dietary improvements and increased physical activity together are more effective than either alone in achieving moderate weight loss [135,136]. Increased visceral, or deep abdominal, body fat decreases peripheral insulin sensitivity. Exercise results in preferential mobilization of visceral body fat, likely contributing to the metabolic improvements even in the absence of significant weight loss [137]. Regular exercise is one of the strongest predictors of the success of maintenance of weight loss and long-term weight maintenance [138]. At the onset of T1D, most individuals have lost some body weight and may engage in exercise training, along with appropriate insulin therapy and dietary intake, to manage body composition during the weight regain [139].

Barriers to exercise participation and overcoming obstacles

All individuals with T1D undertaking physical activity must manage their blood glucose levels before, during, and afterward. Their primary barrier to being active is fear of hypoglycemia and loss of effective management of blood glucose levels [75]. This fear arises due in large part to the kinetics of insulin that has to be replaced through subcutaneous administration (injected or pumped) or inhalation.

Barriers

The promotion of regular exercise among individuals with any type of diabetes is important. However, many of the other barriers to exercise are similar among people with and without diabetes, but those with diabetes need appropriate education about exercise effects on diabetes management and complications. In people with T1D, other nondiabetes barriers to participation may include lack of time and work-related factors, limited access to facilities, lack of motivation; embarrassment and body image, and weather challenges [140]. The built environment can impact a person's ability and willingness to be regularly active [141]. The availability of facilities and pleasant places to walk is frequently an important predictor of regular physical activity [141]. Focusing on creating more exercise-friendly environments can certainly promote greater physical activity participation. In addition, the use of technology may promote greater adherence by assisting people in overcoming some of these barriers related to glycemic monitoring and diabetes management.

Promotion of self-efficacy

Exercise interventions should focus on self-efficacy (a person's belief in their ability to succeed), enjoyment, problem-solving and goal-setting, social-environmental support, and cultural nuances. Greater levels of activity are frequently related to higher levels of self-efficacy, which reflect confidence in the ability to be more active [142,143]. Health issues like obesity [144] and knee and hip osteoarthritis [145] may be barriers as they negatively impact self-efficacy.

Goal setting

Developing realistic goals, selecting appropriate types of activity, progressing slowly (to avoid injury or burnout), and getting supportive feedback can increase confidence [143,146,147]. For example, goals that are vague, overly ambitious, or long-term may not provide enough self-motivation to maintain short-term goals. Having defined strategies is helpful for exercise maintenance. Have individuals identify exercise benefits they find personally motivating and set physical activity goals that are not too vague, ambitious, or distant; instead, choose goals that are specific, measurable, attainable, realistic, and time-bound (that is, SMART). Establishing a routine to help exercise become more habitual can help, along with identifying available social support systems. Positive feedback and assistance with troubleshooting how to overcome any specific obstacles to being active that may arise from their diabetes (such as fear of hypoglycemia) will also be useful.

Counseling, supervision, and cultural sensitivity

Counseling performed by health care professionals may also be a meaningful and effective source of support [148]. Likewise, supervision of exercise sessions by qualified fitness professionals improves compliance and glycemic management [149]. Even in the absence of supervision, home-based exercise like HIIE may be a viable alternative for assisting some highly motivated T1D individuals with previous exercise experience to become more active without fear of hypoglycemia [150]. In general, individual cultural practices and beliefs may also influence the adoption of physical activity programs, and providing culturally appropriate suggestions including yoga, tai chi, and dancing may promote greater adoption.

Tracking activity

Finally, the use of objective measures like step counters (pedometers), accelerometers, and other physical activity trackers may aid individuals in reaching daily movement goals. Pedometers promote physical activity by increasing awareness of daily movement, providing motivation and visual feedback, and encouraging conversation and support among pedometer users, family, and friends [151]. Pedometers are most suitable for walking-based activities, but fail to detect changes in type, intensity, or patterns of activity. Accelerometers may alternately be used to detect activity, but are more expensive than pedometers and often require the use of sophisticated software. Some GPS-based devices allow for heart rate monitoring as well which can be related to glycemic changes and may improve management [152].

Use of diabetes technology with exercise

In the past several decades, technologies related to diabetes management have been developed and continually improved, including blood glucose monitors providing single point-in-time readings, continuous or intermittently scanned glucose monitors and insulin pumps. Their availability has led to the development of several hybrid closed-loop insulin delivery systems as well, although exercise and meals typically still have to be accounted for by the users [153–155].

Blood glucose monitors

Frequent self-monitoring of blood glucose (SMBG) has long been considered an important tool in the management of glycemic responses to exercise, food, stress, and more. These tools provide readings that are single points in time. Prior to exercise it may be helpful to take at least two glucose measurements, spaced 15–30 min apart, to determine any directional changes that may suggest which regimen adjustments may be needed. During longer activities, it may be useful to monitor blood glucose every 30 min to determine appropriate carbohydrate intake and insulin dosing changes. Blood glucose monitoring following activities will help detect low and high blood glucose that may need corrective measures.

Continuous glucose monitoring (CGM)

Use of CGM systems can reveal high and low blood glucose values and other fluctuations that may occur even when an individual has A1C levels in target ranges [156]. The percentage of time in the target glycemic range (typically 70–180 mg/dL) has emerged as one of the strongest indicators of good glycemic management in individuals with T1D [157,158]. Given that exaggerated fluctuations in blood glucose can cause health problems both short-term (hypoglycemia, hyperglycemia) and long-term (e.g., heart disease, neuropathy, diabetic kidney disease, and more), an international consensus panel has recommended that most individuals with T1D should aim to spend at least 17 h a day (that is, more than 70% of their time) in a blood glucose range of 70–180 mg dL^{-1} (3.9–10.0 mmol L^{-1}) [158]. This so-called time in range is typically measured via continuous glucose monitor (CGM) or intermittently scanned monitor readings in individuals who are wearing those devices. Individuals not using CGM can simply strive to have their self-monitored fingerstick glucose values in those ranges.

As for use during exercise, in most studies, CGM measures are reasonably accurate in tracking glycemic changes during activities whether readings are given every 5 min by continuous or intermittently scanned (isCGM) systems. For instance, the accuracy of the no-calibration Dexcom G6 CGM was not significantly impacted by aerobic, resistance, or HIIE in a recent study [159], although other research has shown a lower accuracy during aerobic activities [160]. However, all readings tend to overestimate blood glucose levels if hypoglycemia or hyperglycemia is developing rapidly due to the typical 6–20-min time delay in equilibrium between interstitial fluid and capillary glucose [161–163]. If hypoglycemia is suspected during exercise, individuals should confirm levels with a fingerstick blood glucose measurement. During vigorous exercise, CGM may fail to capture hyperglycemia if the changes are rapid.

Real-time CGM use may help reduce the fear of exercise-associated hypoglycemia when directional arrows and alerts are used to at least determine trends [164]. It may also be easier to use in situations in which SMBG is less convenient (sleep), impractical (e.g., cycling road racing), or impossible, such as during scuba diving [165]. Since glucose changes during exercise can be rapid, increasing the low-glucose-alert alarms to a higher value may confer additional protection against exercise-induced hypoglycemia, without promoting false alarms [166].

Insulin pumps

Insulin pump devices offer a number of advantages and disadvantages for the active individual with T1D. In general, pump users can adjust both the bolus insulin and the basal rate infusion before, during, and after exercise, thereby offering more flexibility in insulin dosing compared to fixed insulin injections. The insulin pump can also be suspended or disconnected during activities. One major advantage is that basal insulin reductions following exercise can occur automatically during discrete hours during sleep (i.e., bedtime to 3 a.m.) to help prevent nocturnal hypoglycemia [167]. Calculators can estimate the amount of onboard insulin during the activity and can help prevent excessive insulin dosing [168,169]. In contrast, such pumps may interfere or be damaged during contact sports, and hyperglycemia and ketosis can develop rapidly if the pump is disconnected or if the infusion set is blocked. Indeed, postexercise hyperglycemia often occurs if the pump is removed (or if the infusion is reduced to zero) for exercise lasting more than an hour [170]. Insulin infusion sets may become displaced in conditions of heavy perspiration or water exposure (e.g., swimming) and skin irritation may result at the site of infusion unless precautions are taken.

Conclusion

Undertaking physical activity with T1D can be complicated, and regimen changes to maintain normal or near-normal blood glucose levels will likely be necessary in most cases. All physical activity promotes blood glucose uptake into active muscles, and insulin use increases the risk of developing hypoglycemia or hyperglycemia both during or following activities. Therefore, altered regimens may be necessary for successful participation, and the use of select diabetes technologies may improve glycemic management. In addition, certain precautions may be necessary for individuals to exercise safely and effectively with diabetes-related health complications.

References

[1] International Diabetes Federation. IDF diabetes atlas. 10th ed. Brussels: International Diabetes Federation; 2021.

[2] Williams GM, et al. Beta cell function and ongoing autoimmunity in long-standing, childhood onset type 1 diabetes. Diabetologia 2016;59 (12):2722–6.

[3] Oram RA, et al. Most people with long-duration type 1 diabetes in a large population-based study are insulin microsecretors. Diabetes Care 2015; 38(2):323–8.

[4] Rodbard HW, Rodbard D. Biosynthetic human insulin and insulin analogs. Am J Ther 2020;27(1):e42–51.

[5] Cengiz E. Automated insulin delivery in children with type 1 diabetes. Endocrinol Metab Clin North Am 2020;49(1):157–66.

[6] Renard E, et al. The SAGE study: global observational analysis of glycaemic control, hypoglycaemia and diabetes management in T1DM. Diabetes Metab Res Rev 2020;37:e3430.

[7] Thieu VT, et al. Treatment and prevention of severe hypoglycaemia in people with diabetes: current and new formulations of glucagon. Diabetes Obes Metab 2020;22(4):469–79.

[8] de Ferranti SD, et al. Type 1 diabetes mellitus and cardiovascular disease: a scientific statement from the American Heart Association and American Diabetes Association. Circulation 2014;130(13):1110–30.

[9] Lind M, et al. HbA1c level as a risk factor for retinopathy and nephropathy in children and adults with type 1 diabetes: Swedish population based cohort study. BMJ 2019;366:l4894.

[10] Hainsworth DP, et al. Risk factors for retinopathy in type 1 diabetes: the DCCT/EDIC study. Diabetes Care 2019;42(5):875–82.

[11] Perkins BA, et al. Risk factors for kidney disease in type 1 diabetes. Diabetes Care 2019;42(5):883–90.

[12] Pinto MV, et al. HbA1c variability and long-term glycemic control are linked to peripheral neuropathy in patients with type 1 diabetes. Diabetol Metab Syndr 2020;12:85.

[13] Verges B. Cardiovascular disease in type 1 diabetes: a review of epidemiological data and underlying mechanisms. Diabetes Metab 2020;46 (6):442–9.

[14] Lin J, et al. Inflammation and progressive nephropathy in type 1 diabetes in the diabetes control and complications trial. Diabetes Care 2008;31 (12):2338–43.

[15] Rajab HA, et al. The predictive role of markers of inflammation and endothelial dysfunction on the course of diabetic retinopathy in type 1 diabetes. J Diabetes Complications 2015;29(1):108–14.

[16] Schram MT, et al. Markers of inflammation are cross-sectionally associated with microvascular complications and cardiovascular disease in type 1 diabetes- -the EURODIAB prospective complications study. Diabetologia 2005;48(2):370–8.

[17] Gordin D, et al. Acute hyperglycaemia induces an inflammatory response in young patients with type 1 diabetes. Ann Med 2008;40(8):627–33.

[18] Novak D, et al. Altered cortical bone strength and lean mass in young women with long-duration (19 years) type 1 diabetes. Sci Rep 2020;10 (1):22367.

[19] Mori H, et al. Advanced glycation end-products are a risk for muscle weakness in Japanese patients with type 1 diabetes. J Diab Investig 2017;8 (3):377–82.

[20] Halper-Stromberg E, et al. Bone mineral density across the lifespan in patients with type 1 diabetes. J Clin Endocrinol Metab 2020;105(3).

[21] Alhuzaim ON, et al. Bone mineral density in patients with longstanding type 1 diabetes: results from the Canadian study of longevity in type 1 diabetes. J Diabetes Complications 2019;33(11):107324.

[22] Kuroda T, et al. Quadrant analysis of quantitative computed tomography scans of the femoral neck reveals superior region-specific weakness in young and middle-aged men with type 1 diabetes mellitus. J Clin Densitom 2018;21(2):172–8.

[23] Vilaca T, et al. The risk of hip and non-vertebral fractures in type 1 and type 2 diabetes: a systematic review and meta-analysis update. Bone 2020;137:115457.

[24] Tielemans SM, et al. Association of physical activity with all-cause mortality and incident and prevalent cardiovascular disease among patients with type 1 diabetes: the EURODIAB prospective complications study. Diabetologia 2013;56(1):82–91.

[25] Moy CS, et al. Insulin-dependent diabetes mellitus, physical activity, and death. Am J Epidemiol 1993;137(1):74–81.

[26] Kriska AM, et al. The association of physical activity and diabetic complications in individuals with insulin-dependent diabetes mellitus: the epidemiology of diabetes complications study- -VII. J Clin Epidemiol 1991;44(11):1207–14.

[27] Bohn B, et al. Impact of physical activity on glycemic control and prevalence of cardiovascular risk factors in adults with type 1 diabetes: a cross-sectional multicenter study of 18,028 patients. Diabetes Care 2015;38(8):1536–43.

[28] Tikkanen-Dolenc H, et al. Frequent and intensive physical activity reduces risk of cardiovascular events in type 1 diabetes. Diabetologia 2017;60 (3):574–80.

[29] Waden J, et al. Leisure-time physical activity and development and progression of diabetic nephropathy in type 1 diabetes: the FinnDiane study. Diabetologia 2015;58(5):929–36.

[30] Tikkanen-Dolenc H, et al. Frequent physical activity is associated with reduced risk of severe diabetic retinopathy in type 1 diabetes. Acta Diabetol 2020;57(5):527–34.

[31] Lukacs A, et al. Health-related quality of life of adolescents with type 1 diabetes in the context of resilience. Pediatr Diabetes 2018;19(8):1481–6.

[32] Mozzillo E, et al. Unhealthy lifestyle habits and diabetes-specific health-related quality of life in youths with type 1 diabetes. Acta Diabetol 2017;54 (12):1073–80.

[33] Anderson BJ, et al. Factors associated with diabetes-specific health-related quality of life in youth with type 1 diabetes: the global TEENs study. Diabetes Care 2017;40(8):1002–9.

[34] Ligtenberg PC, et al. No effect of long-term physical activity on the glycemic control in type 1 diabetes patients: a cross-sectional study. Neth J Med 1999;55(2):59–63.

[35] Waden J, et al. Leisure time physical activity is associated with poor glycemic control in type 1 diabetic women: the FinnDiane study. Diabetes Care 2005;28(4):777–82.

[36] Miculis CP, De Campos W, da Silva Boguszweski MC. Correlation between glycemic control and physical activity level in adolescents and children with type 1 diabetes. J Phys Act Health 2015;12(2):232–7.

[37] Herbst A, et al. Effects of regular physical activity on control of glycemia in pediatric patients with type 1 diabetes mellitus. Arch Pediatr Adolesc Med 2006;160(6):573–7.

[38] Cuenca-Garcia M, et al. How does physical activity and fitness influence glycaemic control in young people with type 1 diabetes? Diabet Med 2012;29(10):e369–76.

[39] Carral F, et al. Intense physical activity is associated with better metabolic control in patients with type 1 diabetes. Diabetes Res Clin Pract 2013;101 (1):45–9.

[40] Leao AAP, et al. Bone mass and dietary intake in children and adolescents with type 1 diabetes mellitus. J Diabetes Complications 2020;34 (6):107573.

[41] Joshi A, et al. A study of bone mineral density and its determinants in type 1 diabetes mellitus. J Osteoporos 2013;2013:397814.

[42] Zinman B, Zuniga-Guajardo S, Kelly D. Comparison of the acute and long-term effects of exercise on glucose control in type I diabetes. Diabetes Care 1984;7(6):515–9.

[43] Wallberg-Henriksson H, et al. Long-term physical training in female type 1 (insulin-dependent) diabetic patients: absence of significant effect on glycaemic control and lipoprotein levels. Diabetologia 1986;29(1):53–7.

[44] Rowland TW, et al. Glycemic control with physical training in insulin-dependent diabetes mellitus. Am J Dis Child 1985;139(3):307–10.

[45] Landt KW, et al. Effects of exercise training on insulin sensitivity in adolescents with type I diabetes. Diabetes Care 1985;8(5):461–5.

[46] Huttunen NP, et al. Effect of once-a-week training program on physical fitness and metabolic control in children with IDDM. Diabetes Care 1989;12 (10):737–40.

[47] Fuchsjager-Mayrl G, et al. Exercise training improves vascular endothelial function in patients with type 1 diabetes. Diabetes Care 2002;25 (10):1795–801.

[48] D'Hooge R, et al. Influence of combined aerobic and resistance training on metabolic control, cardiovascular fitness and quality of life in adolescents with type 1 diabetes: a randomized controlled trial. Clin Rehabil 2011;25(4):349–59.

[49] Stratton R, et al. Improved glycemic control after supervised 8-wk exercise program in insulin-dependent diabetic adolescents. Diabetes Care 1987;10(5):589–93.

[50] Peterson CM, et al. Changes in basement membrane thickening and pulse volume concomitant with improved glucose control and exercise in patients with insulin-dependent diabetes mellitus. Diabetes Care 1980;3(5):586–9.

[51] Farinha JB, et al. Glycemic, inflammatory and oxidative stress responses to different high-intensity training protocols in type 1 diabetes: a randomized clinical trial. J Diabetes Complications 2018;32(12):1124–32.

[52] Dahl-Jorgensen K, et al. The effect of exercise on diabetic control and hemoglobin A1 (HbA1) in children. Acta Paediatr Scand Suppl 1980;283:53–6.

[53] Yardley JE, et al. A systematic review and meta-analysis of exercise interventions in adults with type 1 diabetes. Diabetes Res Clin Pract 2014;106 (3):393–400.

[54] Durak EP, Jovanovic-Peterson L, Peterson CM. Randomized crossover study of effect of resistance training on glycemic control, muscular strength, and cholesterol in type I diabetic men. Diabetes Care 1990;13(10):1039–43.

[55] Mosher PE, et al. Aerobic circuit exercise training: effect on adolescents with well-controlled insulin-dependent diabetes mellitus. Arch Phys Med Rehabil 1998;79(6):652–7.

[56] Laaksonen DE, et al. Aerobic exercise and the lipid profile in type 1 diabetic men: a randomized controlled trial. Med Sci Sports Exerc 2000;32 (9):1541–8.

[57] Yki-Jarvinen H, DeFronzo RA, Koivisto VA. Normalization of insulin sensitivity in type I diabetic subjects by physical training during insulin pump therapy. Diabetes Care 1984;7(6):520–7.

[58] Rigla M, et al. Effect of physical exercise on lipoprotein(a) and low-density lipoprotein modifications in type 1 and type 2 diabetic patients. Metabolism 2000;49(5):640–7.

[59] de Moraes R, et al. Effects of non-supervised low intensity aerobic excise training on the microvascular endothelial function of patients with type 1 diabetes: a non-pharmacological interventional study. BMC Cardiovasc Disord 2016;16:23.

[60] Boff W, et al. Superior effects of high-intensity interval vs. moderate-intensity continuous training on endothelial function and cardiorespiratory fitness in patients with type 1 diabetes: a randomized controlled trial. Front Physiol 2019;10:450.

[61] Scheffer DDL, Latini A. Exercise-induced immune system response: anti-inflammatory status on peripheral and central organs. Biochim Biophys Acta Mol Basis Dis 1866;2020(10):165823.

[62] Wallberg-Henriksson H, et al. Increased peripheral insulin sensitivity and muscle mitochondrial enzymes but unchanged blood glucose control in type I diabetics after physical training. Diabetes 1982;31(12):1044–50.

[63] Wrobel M, et al. Aerobic as well as resistance exercises are good for patients with type 1 diabetes. Diabetes Res Clin Pract 2018;144:93–101.

[64] Ramalho AC, et al. The effect of resistance versus aerobic training on metabolic control in patients with type-1 diabetes mellitus. Diabetes Res Clin Pract 2006;72(3):271–6.

[65] Maggio AB, et al. Physical activity increases bone mineral density in children with type 1 diabetes. Med Sci Sports Exerc 2012;44(7):1206–11.

[66] Scott SN, et al. High-intensity interval training improves aerobic capacity without a detrimental decline in blood glucose in people with type 1 diabetes. J Clin Endocrinol Metab 2019;104(2):604–12.

[67] Minnebeck K, et al. Four weeks of high-intensity interval training (HIIT) improve the cardiometabolic risk profile of overweight patients with type 1 diabetes mellitus (T1DM). Eur J Sport Sci 2020;1–11.

[68] Plotnikoff RC, et al. Factors associated with physical activity in Canadian adults with diabetes. Med Sci Sports Exerc 2006;38(8):1526–34.

[69] World Health Organization. Phyiscal Activity. [cited 2021 January 11, 2021]; Available from: https://www.who.int/news-room/fact-sheets/detail/physical-activity; 2020.

[70] Czenczek-Lewandowska E, et al. Sedentary behaviors in children and adolescents with type 1 diabetes, depending on the insulin therapy used. Medicine (Baltimore) 2019;98(19), e15625.

[71] de Lima VA, et al. Physical activity levels of adolescents with type 1 diabetes physical activity in T1D. Pediatr Exerc Sci 2017;29(2):213–9.

[72] Aman J, et al. Associations between physical activity, sedentary behavior, and glycemic control in a large cohort of adolescents with type 1 diabetes: the Hvidoere study group on childhood diabetes. Pediatr Diabetes 2009;10(4):234–9.

[73] Adamo M, et al. Active subjects with autoimmune type 1 diabetes have better metabolic profiles than sedentary controls. Cell Transplant 2017;26(1):23–32.

[74] Hasan R, et al. Can short bouts of exercise ("exercise snacks") improve body composition in adolescents with type 1 diabetes? A feasibility study. Horm Res Paediatr 2019;92(4):245–53.

[75] Brazeau AS, et al. Barriers to physical activity among patients with type 1 diabetes. Diabetes Care 2008;31(11):2108–9.

[76] Frank S, et al. Modeling the acute effects of exercise on insulin kinetics in type 1 diabetes. J Pharmacokinet Pharmacodyn 2018;45(6):829–45. https://doi.org/10.1007/s10928-018-9611-z [Epub 2018 Nov 3].

[77] Rabasa-Lhoret R, et al. Guidelines for premeal insulin dose reduction for postprandial exercise of different intensities and durations in type 1 diabetic subjects treated intensively with a basal-bolus insulin regimen (ultralente-lispro). Diabetes Care 2001;24(4):625–30.

[78] Shetty VB, et al. Effect of exercise intensity on glucose requirements to maintain Euglycemia during exercise in type 1 diabetes. J Clin Endocrinol Metab 2016;101(3):972–80.

[79] Riddell MC, et al. Exercise management in type 1 diabetes: a consensus statement. Lancet Diabetes Endocrinol 2017;5(5):377–90.

[80] Roy-Fleming A, et al. Timing of insulin basal rate reduction to reduce hypoglycemia during late post-prandial exercise in adults with type 1 diabetes using insulin pump therapy: a randomized crossover trial. Diabetes Metab 2019;45(3):294–300.

[81] Zaharieva DP, et al. No disadvantage to insulin pump off vs pump on during intermittent high-intensity exercise in adults with type 1 diabetes. Can J Diabetes 2020;44(2):162–8.

[82] Marliss EB, Vranic M. Intense exercise has unique effects on both insulin release and its roles in glucoregulation: implications for diabetes. Diabetes 2002;51(Suppl 1):S271–83.

[83] Fahey AJ, et al. The effect of a short sprint on postexercise whole-body glucose production and utilization rates in individuals with type 1 diabetes mellitus. J Clin Endocrinol Metab 2012;97(11):4193–200.

[84] Campbell MD, et al. Simulated games activity vs continuous running exercise: a novel comparison of the glycemic and metabolic responses in T1DM patients. Scand J Med Sci Sports 2015;25(2):216–22.

[85] Dube MC, Lavoie C, Weisnagel SJ. Glucose or intermittent high-intensity exercise in glargine/glulisine users with T1DM. Med Sci Sports Exerc 2013;45(1):3–7.

[86] Guelfi KJ, Jones TW, Fournier PA. The decline in blood glucose levels is less with intermittent high-intensity compared with moderate exercise in individuals with type 1 diabetes. Diabetes Care 2005;28(6):1289–94.

[87] Iscoe KE, Riddell MC. Continuous moderate-intensity exercise with or without intermittent high-intensity work: effects on acute and late glycaemia in athletes with type 1 diabetes mellitus. Diabet Med 2011;28(7):824–32.

[88] Moser O, et al. Effects of high-intensity interval exercise versus moderate continuous exercise on glucose homeostasis and hormone response in patients with type 1 diabetes mellitus using novel ultra-long-acting insulin. PLoS One 2015;10(8), e0136489.

[89] Aronson R, et al. Optimal insulin correction factor in post-high-intensity exercise hyperglycemia in adults with type 1 diabetes: the FIT study. Diabetes Care 2019;42(1):10–6. https://doi.org/10.2337/dc18-1475.

[90] Rempel M, et al. Vigorous intervals and hypoglycemia in type 1 diabetes: a randomized cross over trial. Sci Rep 2018;8(1):15879.

[91] Metcalf KM, et al. Effects of moderate-to-vigorous intensity physical activity on overnight and next-day hypoglycemia in active adolescents with type 1 diabetes. Diabetes Care 2014;37(5):1272–8.

[92] Yardley JE, et al. Resistance versus aerobic exercise: acute effects on glycemia in type 1 diabetes. Diabetes Care 2013;36(3):537–42.

[93] Maran A, et al. Continuous glucose monitoring reveals delayed nocturnal hypoglycemia after intermittent high-intensity exercise in nontrained patients with type 1 diabetes. Diabetes Technol Ther 2010;12(10):763–8.

[94] Lee AS, et al. High-intensity interval exercise and hypoglycaemia minimisation in adults with type 1 diabetes: a randomised cross-over trial. J Diabetes Complications 2020;34(3):107514.

[95] Zaharieva D, et al. The effects of basal insulin suspension at the start of exercise on blood glucose levels during continuous versus circuit-based exercise in individuals with type 1 diabetes on continuous subcutaneous insulin infusion. Diabetes Technol Ther 2017;19(6):370–8.

[96] Galassetti P, et al. Effect of antecedent hypoglycemia on counterregulatory responses to subsequent euglycemic exercise in type 1 diabetes. Diabetes 2003;52(7):1761–9.

[97] Davey RJ, et al. The effect of midday moderate-intensity exercise on postexercise hypoglycemia risk in individuals with type 1 diabetes. J Clin Endocrinol Metab 2013;98(7):2908–14.

[98] McMahon SK, et al. Glucose requirements to maintain euglycemia after moderate-intensity afternoon exercise in adolescents with type 1 diabetes are increased in a biphasic manner. J Clin Endocrinol Metab 2007;92(3):963–8.

[99] Campbell MD, et al. Insulin therapy and dietary adjustments to normalize glycemia and prevent nocturnal hypoglycemia after evening exercise in type 1 diabetes: a randomized controlled trial. BMJ Open Diabetes Res Care 2015;3(1), e000085.

[100] Suh SH, Paik IY, Jacobs K. Regulation of blood glucose homeostasis during prolonged exercise. Mol Cells 2007;23(3):272–9.

[101] Wahren J, Ekberg K. Splanchnic regulation of glucose production. Annu Rev Nutr 2007;27:329–45.

[102] Adolfsson P, et al. Hormonal response during physical exercise of different intensities in adolescents with type 1 diabetes and healthy controls. Pediatr Diabetes 2012;13(8):587–96. https://doi.org/10.1111/j.1399-5448.2012.00889.x.

[103] Bisgaard Bengtsen M, Møller N. Mini-review: glucagon responses in type 1 diabetes—a matter of complexity. Physiol Rep 2021;9(16):e15009. https://doi.org/10.14814/phy2.15009. 34405569. In this issue.

[104] Sejling AS, et al. Hypoglycemia-associated changes in the electroencephalogram in patients with type 1 diabetes and normal hypoglycemia awareness or unawareness. Diabetes 2015;64(5):1760–9.

[105] Mao Y, et al. The hypoglycemia associated autonomic failure triggered by exercise in the patients with "brittle" diabetes and the strategy for prevention. Endocr J 2019;66(9):753–62.

[106] Cryer PE. Hypoglycemia-associated autonomic failure in diabetes. Handb Clin Neurol 2013;117:295–307.

[107] Galassetti P, et al. Effect of differing antecedent hypoglycemia on counterregulatory responses to exercise in type 1 diabetes. Am J Physiol Endocrinol Metab 2006;290(6):E1109–17.

[108] Sandoval DA, et al. Acute, same-day effects of antecedent exercise on counterregulatory responses to subsequent hypoglycemia in type 1 diabetes mellitus. Am J Physiol Endocrinol Metab 2006;290(6):E1331–8.

[109] Sandoval DA, et al. Effects of low and moderate antecedent exercise on counterregulatory responses to subsequent hypoglycemia in type 1 diabetes. Diabetes 2004;53(7):1798–806.

[110] Carlson JN, et al. Dietary sugars versus glucose tablets for first-aid treatment of symptomatic hypoglycaemia in awake patients with diabetes: a systematic review and meta-analysis. Emerg Med J 2017;34(2):100–6. https://doi.org/10.1136/emermed-2015-205637 [Epub 2016 Sep 19].

[111] Rickels MR, et al. Mini-dose glucagon as a novel approach to prevent exercise-induced hypoglycemia in type 1 diabetes. Diabetes Care 2018;41(9):1909–16.

[112] Seaquist ER, et al. Prospective study evaluating the use of nasal glucagon for the treatment of moderate to severe hypoglycaemia in adults with type 1 diabetes in a real-world setting. Diabetes Obes Metab 2018;20(5):1316–20.

[113] Pontiroli A, Tagliabue E. Intranasal versus injectable glucagon for hypoglycemia in type 1 diabetes: systematic review and meta-analysis. Acta Diabetol 2020;57(6):743–9. https://doi.org/10.1007/s00592-020-01483-y.

[114] Ruegemer JJ, et al. Differences between prebreakfast and late afternoon glycemic responses to exercise in IDDM patients. Diabetes Care 1990;13(2):104–10.

[115] Yamanouchi K, et al. The effect of walking before and after breakfast on blood glucose levels in patients with type 1 diabetes treated with intensive insulin therapy. Diabetes Res Clin Pract 2002;58(1):11–8.

[116] Toghi-Eshghi SR, Yardley JE. Morning (fasting) vs afternoon resistance exercise in individuals with type 1 diabetes: a randomized crossover study. J Clin Endocrinol Metab 2019;104(11):5217–24.

[117] Yardley JE. Fasting may Alter blood glucose responses to high-intensity interval exercise in adults with type 1 diabetes: a randomized, acute crossover study. Can J Diabetes 2020;44(8):727–33.

[118] Yardley JE, et al. Performing resistance exercise before versus after aerobic exercise influences growth hormone secretion in type 1 diabetes. Appl Physiol Nutr Metab 2014;39(2):262–5.

[119] Vendelbo MH, et al. GH signaling in human adipose and muscle tissue during 'feast and famine': amplification of exercise stimulation following fasting compared to glucose administration. Eur J Endocrinol 2015;173(3):283–90.

[120] Vieira AF, et al. Effects of aerobic exercise performed in fasted v. fed state on fat and carbohydrate metabolism in adults: a systematic review and meta-analysis. Br J Nutr 2016;116(7):1153–64.

[121] Committee, P.A.G.A. Physical activity guidelines advisory Committee scientific report. U.S. Department of Health and Human Services; 2018. ed. 2018: Washington, D.C.

[122] Colberg SR, et al. Physical activity/exercise and diabetes: a position statement of the American Diabetes Association. Diabetes Care 2016;39(11):2065–79.

[123] Piercy KL, et al. The physical activity guidelines for Americans. JAMA 2018;320(19):2020–8.

[124] MacMillan F, et al. A systematic review of physical activity and sedentary behavior intervention studies in youth with type 1 diabetes: study characteristics, intervention design, and efficacy. Pediatr Diabetes 2014;15(3):175–89.

[125] Herbst A, et al. Impact of physical activity on cardiovascular risk factors in children with type 1 diabetes: a multicenter study of 23,251 patients. Diabetes Care 2007;30(8):2098–100.

[126] Garber CE, et al. American College of Sports Medicine position stand. Quantity and quality of exercise for developing and maintaining cardio-respiratory, musculoskeletal, and neuromotor fitness in apparently healthy adults: guidance for prescribing exercise. Med Sci Sports Exerc 2011;43(7):1334–59.

[127] Saint-Maurice PF, et al. Association of Leisure-Time Physical Activity across the adult life course with all-cause and cause-specific mortality. JAMA Netw Open 2019;2(3). https://doi.org/10.1001/jamanetworkopen.2019.0355, e190355.

[128] Riebe D, et al. Updating ACSM's recommendations for exercise Preparticipation health screening. Med Sci Sports Exerc 2015;47(11):2473–9.

[129] Colberg SR, Swain DP, Vinik AI. Use of heart rate reserve and rating of perceived exertion to prescribe exercise intensity in diabetic autonomic neuropathy. Diabetes Care 2003;26(4):986–90.

[130] Colberg SR, Vinik AI. Exercising with peripheral or autonomic neuropathy: what health care providers and diabetic patients need to know. Phys Sportsmed 2014;42(1):15–23.

[131] Pongrac Barlovic D, Tikkanen-Dolenc H, Groop PH. Physical activity in the prevention of development and progression of kidney disease in type 1 diabetes. Curr Diab Rep 2019;19(7):41.

[132] Colberg S. The athlete's guide to diabetes: expert advice for 165 sports and activities. Champaign, IL: Human Kinetics; 2020. p. 382.

[133] Baldi JC, et al. Glycemic status affects cardiopulmonary exercise response in athletes with type I diabetes. Med Sci Sports Exerc 2010;42(8): 1454–9.

[134] Mottalib A, et al. Weight Management in Patients with type 1 diabetes and obesity. Curr Diab Rep 2017;17(10):92.

[135] Wu T, et al. Long-term effectiveness of diet-plus-exercise interventions vs. diet-only interventions for weight loss: a meta-analysis. Obes Rev 2009;10(3):313–23.

[136] Catenacci VA, Wyatt HR. The role of physical activity in producing and maintaining weight loss. Nat Clin Pract Endocrinol Metab 2007;3(7): 518–29.

[137] O'Leary VB, et al. Exercise-induced reversal of insulin resistance in obese elderly is associated with reduced visceral fat. J Appl Physiol 2006;100 (5):1584–9.

[138] Catenacci VA, et al. Physical activity patterns using accelerometry in the National Weight Control Registry. Obesity (Silver Spring) 2011;19 (6):1163–70.

[139] Rosenfalck AM, et al. Body composition in adults with type 1 diabetes at onset and during the first year of insulin therapy. Diabet Med 2002;19 (5):417–23.

[140] Lascar N, et al. Attitudes and barriers to exercise in adults with type 1 diabetes (T1DM) and how best to address them: a qualitative study. PLoS One 2014;9(9), e108019.

[141] Karmeniemi M, et al. The built environment as a determinant of physical activity: a systematic review of longitudinal studies and natural experiments. Ann Behav Med 2018;52(3):239–51. https://doi.org/10.1093/abm/kax043.

[142] Delahanty LM, Conroy MB, Nathan DM. Psychological predictors of physical activity in the diabetes prevention program. J Am Diet Assoc 2006;106(5):698–705.

[143] Dutton GR, et al. Relationship between self-efficacy and physical activity among patients with type 2 diabetes. J Behav Med 2009;32(3):270–7.

[144] Graham D, Edwards A. The psychological burden of obesity: the potential harmful impact of health promotion and education programmes targeting obese individuals. Int J Health PromoT Educ 2013;13(3):124–33.

[145] Hutton I, et al. What is associated with being active in arthritis? Analysis from the obstacles to action study. Intern Med J 2010;40(7):512–20.

[146] Aljasem LI, et al. The impact of barriers and self-efficacy on self-care behaviors in type 2 diabetes. Diabetes Educ 2001;27(3):393–404.

[147] McAuley E, Blissmer B. Self-efficacy determinants and consequences of physical activity. Exerc Sport Sci Rev 2000;28(2):85–8.

[148] Armit CM, et al. Randomized trial of three strategies to promote physical activity in general practice. Prev Med 2009;48(2):156–63.

[149] Balducci S, et al. The Italian diabetes and exercise study. Diabetes 2008;57(Suppl. 1):A306–7.

[150] Scott SN, et al. Home-based high-intensity interval training reduces barriers to exercise in people with type 1 diabetes. Exp Physiol 2020;105 (4):571–8.

[151] Lauzon N, et al. Participant experiences in a workplace pedometer-based physical activity program. J Phys Act Health 2008;5(5):675–87.

[152] Gawrecki A, et al. Assessment of safety and glycemic control during football tournament in children and adolescents with type 1 diabetes-results of goaldiab study. Pediatr Exerc Sci 2019;1–7.

[153] Zaharieva DP, et al. Glucose control during physical activity and exercise using closed loop Technology in Adults and Adolescents with type 1 diabetes. Can J Diabetes 2020;44(8):740–9.

[154] Ekhlaspour L, et al. Closed loop control in adolescents and children during winter sports: use of the tandem control-IQ AP system. Pediatr Diabetes 2019;20(6):759–68.

[155] Jayawardene DC, et al. Closed-loop insulin delivery for adults with type 1 diabetes undertaking high-intensity interval exercise versus moderate-intensity exercise: a randomized. Diabetes Technol Ther 2017;19(6):340–8. https://doi.org/10.1089/dia.2016.0461 [Epub 2017 Jun 2].

[156] Kushner PR, Kruger DF. The changing landscape of glycemic targets: focus on continuous glucose monitoring. Clin Diab 2020;38(4):348–56.

[157] Kalra S, et al. Individualizing time-in-range goals in Management of Diabetes Mellitus and Role of insulin: Clinical insights from a multinational panel. Diabetes Ther 2020.

[158] Battelino T, et al. Clinical targets for continuous glucose monitoring data interpretation: recommendations from the international consensus on time in range. Diabetes Care 2019;42(8):1593–603. https://doi.org/10.2337/dci19-0028 [Epub 2019 Jun 8].

[159] Guillot FH, et al. Accuracy of the Dexcom G6 glucose sensor during aerobic, resistance, and interval exercise in adults with type 1 diabetes. Biosensors (Basel) 2020;10(10).

[160] Biagi L, et al. Accuracy of continuous glucose monitoring before, during, and after aerobic and anaerobic exercise in patients with type 1 diabetes mellitus. Biosensors (Basel) 2018;8(1). https://doi.org/10.3390/bios8010022, bios8010022.

[161] Zaharieva DP, et al. Lag time remains with newer real-time continuous glucose monitoring technology during aerobic exercise in adults living with type 1 diabetes. Diabetes Technol Ther 2019;21(6):313–21. https://doi.org/10.1089/dia.2018.0364 [Epub 2019 May 6].

[162] Iscoe KE, Davey RJ, Fournier PA. Is the response of continuous glucose monitors to physiological changes in blood glucose levels affected by sensor life? Diabetes Technol Ther 2012;14(2):135–42.

[163] Moser O, Yardley JE, Bracken RM. Interstitial glucose and physical exercise in type 1 diabetes: integrative physiology, technology, and the gap in-between. Nutrients 2018;10(1):nu10010093. https://doi.org/10.3390/nu10010093 [pii].

[164] Davey RJ, et al. The effect of short-term use of the Guardian RT continuous glucose monitoring system on fear of hypoglycaemia in patients with type 1 diabetes mellitus. Prim Care Diabetes 2012;6(1):35–9.

[165] Adolfsson P, Ornhagen H, Jendle J. The benefits of continuous glucose monitoring and a glucose monitoring schedule in individuals with type 1 diabetes during recreational diving. J Diabetes Sci Technol 2008;2(5):778–84.

[166] Iscoe KE, Davey RJ, Fournier PA. Increasing the low-glucose alarm of a continuous glucose monitoring system prevents exercise-induced hypoglycemia without triggering any false alarms. Diabetes Care 2011;34(6), e109.

[167] Taplin CE, et al. Preventing postexercise nocturnal hypoglycemia in children with type 1 diabetes. J Pediatr 2010;157(5):784–788.e1.

[168] Walsh J, Roberts R, Bailey T. Guidelines for optimal bolus calculator settings in adults. J Diabetes Sci Technol 2011;5(1):129–35.

[169] Walsh J, Roberts R, Heinemann L. Confusion regarding duration of insulin action: a potential source for major insulin dose errors by bolus calculators. J Diabetes Sci Technol 2014;8(1):170–8.

[170] Delvecchio M, et al. Effects of moderate-severe exercise on blood glucose in type 1 diabetic adolescents treated with insulin pump or glargine insulin. J Endocrinol Invest 2009;32(6):519–24.

Chapter 8

Exercise across the lifespan: Exercise and obesity

Daniel Friedman[a] and Peter Brukner[b]

[a]Alfred Health, Melbourne, VIC, Australia [b]La Trobe Sport and Exercise Medicine Research Centre, La Trobe University, Melbourne, VIC, Australia

Introduction

What is obesity?

In 2010, the Scottish Intercollegiate Guideline defined obesity as "a disease process characterized by excessive body fat accumulation with multiple organ-specific consequences" [1]. These consequences significantly impair health and are associated with a greater risk of morbidity and mortality due to cardiovascular disease, type 2 diabetes mellitus, and cancer, among other noncommunicable diseases. The medical definitions of "overweight" and "obesity" have traditionally been based on body mass index (BMI), as shown in Table 1. BMI is calculated using the following formula: BMI (kg/m^2) = Weight (kg)/Height^2 (m).

For example, if you are the average Australian male who is 178 cm tall and weighs 85 kg, your BMI is $85/1.78^2 = 26.8$ (and you would be considered overweight).

Contemporary medical use of BMI for classifying overweight and obese individuals was first introduced by American physiologist Ancel Keys in 1972 [2]. At the time, Keys and colleagues intended for BMI to be used for population-based epidemiological studies, rather than measurement of an individual—which it has come to be routinely used today. BMI was created as a score (an index), rather than an objective measure of actual adipose tissue mass, and cannot define excessive adipose accumulation. Measurement and classification using BMI have their limitations [3], especially in individuals with high muscle-to-fat ratios (e.g., athletes) and those with non-Caucasian ancestry. BMI weight categories can vary based on populations and subpopulations, for example, in Europe, Asia and Africa compared to Australia, the USA, Canada, and England. Definitions of obesity and BMI classifications are also different for children and adolescents. For children under 5 years of age, obesity is defined as a weight-for-height greater than 3 standard deviations above the World Health Organization (WHO) Child Growth Standards median. For children aged between 5 and 19 years, obesity is defined as greater than 2 standard deviations above the WHO Growth Reference median [4].

Multiple alternative methods of measuring obesity are shown in Box 1.

Not all fat is created equal

Obese individuals vary in sites of their adipose tissue distribution, their metabolic profile, and subsequent degree of associated cardiometabolic risk. There are two different general sites of fat accumulation: the relatively harmless subcutaneous fat—that is found under the skin on arms, legs, and buttocks (gluteofemoral); and the more dangerous visceral fat found deep within the abdominal cavity, filling spaces between and around abdominal organs.

Accumulation of visceral adipose tissue is strongly associated with obesity-related health consequences such as cardiovascular disease and type-2 diabetes mellitus [7], which can be independent of age, overall obesity, or the amount of subcutaneous adipose tissue. Visceral adipose tissue and its adipose-tissue resident macrophages produce more pro-inflammatory cytokines, such as tumor necrosis factor-alpha (TNF-alpha) and interleukin-6 (IL-6), and less of the antiinflammatory adiponectin. These cytokine changes induce insulin resistance and play a major role in the pathogenesis of endothelial dysfunction and subsequent atherosclerosis leading to cardiovascular disease and other disease processes [7].

The rate of visceral adipose tissue accumulation also differs according to the individual's sex and ethnic background. For example, visceral accumulation is more prominent in Caucasian men, African-American women, and Asian Indian and Japanese men and women [8].

Exercise to Prevent and Manage Chronic Disease Across the Lifespan. https://doi.org/10.1016/B978-0-323-89843-0.00004-0

TABLE 1 World Health Organization BMI categories for adults (aged 19 and over).

Weight category	BMI
Underweight	<18.5
Normal	18.5–24.9
Overweight	25.0–29.9
Obesity I	30.0–34.9
Obesity II	35.0–39.9
Extreme obesity III	>40.0

Box 1 More accurate measures of adiposity

Body fat percentage is more precise than BMI for measuring obesity. There are multiple ways to calculate body fat percentage.

DXA (dual-energy X-ray absorptiometry) scan
This total body scan is used more commonly to measure bone density but also measures fat and muscle mass. It is the most accurate but also commonly the most expensive method.

Body impedance analysis devices (BIA)
Bioelectrical impedance analysis (BIA) measures the body's resistance to a light electrical current. These are cheaper and quicker than DEXA scans but are not as accurate.

Skin calipers
In skinfold testing, skin calipers are used to pinch the skin and the subcutaneous fat (fat underneath the skin), pull the skinfold away from the underlying muscle, and measure its thickness. Skin calipers can be used at a single site or a sum of skinfolds can be performed. This is commonly used by fitness personnel working with sporting teams, and as long as the same person performing the measurements is reasonably accurate at detecting changes in body fat.

Bathroom scales
They may be old-fashioned, but they are still a good indication of progressive weight loss (or gain).

Waist measurement
Waist circumference is a good indicator of visceral fat, which usually poses more health risks than fat elsewhere. The WH ratio may be more accurate than BMI at predicting percentage body fat [5].
  Waist measurement can be used in multiple ways:

Waist circumference
Waist circumference is measured by placing a tape measure directly on the skin, halfway between the lowest rib and the top of the hip bone, roughly in line with the umbilicus (Table 2).

Waist-to-hip ratio (WHR)
Measure waist circumference as above. Measure hip circumference at the widest point, usually at the top of the hip bone. Divide waist measurement by hip measurement.
  Excess abdominal fat distribution is indicated by a WHR greater than 0.85 for women and 0.9 for men [6].

Waist to height (WH) ratio
WH ratio may be more accurate than BMI at predicting percentage body fat. Divide waist measurement by height (ensure both measurements are in the same units—i.e., height is in centimeters rather than meters).
  A ratio of more than 0.53 in men and 0.54 in women is a predictor of whole-body obesity. Greater than 0.59 in both sexes is a predictor of abdominal obesity [5].

TABLE 2 Waist circumference thresholds indicate increased risk of disease [5].

Gender	Waist circumference for:	
	Increased disease risk	High disease risk
Female	≥80 cm	≥88 cm
Male	≥94 cm	≥102 cm for men

"Metabolic obesity"—visceral adipose tissue accumulation in either lean or obese individuals—may identify those at risk for cardiovascular disease more accurately than current definitions of obesity [9].

Is obesity a "disease"?

The question of whether obesity should be classified as a "disease" or a "risk factor" is a source of considerable controversy. The Oxford Dictionary defines disease as "a disorder of structure or function…especially one that produces specific symptoms…and is not simply a direct result of physical injury" [10]. Obesity, a process in which excess adipose tissue accumulates due to abnormal biological regulation to a point at which health is impaired, meets this definition. However, others argue obesity should be considered no different from hypertension or hypercholesterolemia—all variables associated with an increased risk of disease, such as cardiovascular disease [10]. Differentiating disease from risk factors is complex, especially when it comes to chronic diseases, which fall on a spectrum from imperfect absolutes of health to illness.

Despite ongoing debate over the past 40 years, obesity has been recognized as a disease many times by various organizations. It was perhaps first recognized internationally as a disease by the WHO in 1948, included in the International Classification of Diseases [11]. More recently, The American Medical Association classified obesity as a disease in 2013 despite opposition from its own Council on Science and Public Health (CSPH). The American Heart Association, American College of Cardiology, and The Obesity Society released joint guidelines in 2013 stating that doctors should consider obesity a disease and more actively treat obese patients for weight loss [12]. In 2015, the Asia Oceania Association for the Study of Obesity differentiated "obesity" from "obesity disease": a pathological state caused by obesity and requiring clinical intervention [13]. Additionally in 2017, the World Obesity Federation argued that "obesity is a chronic, relapsing, progressive, disease process" that requires intervention. Regardless of whether obesity is defined as a disease or risk factor, obesity clearly poses a significant threat to public health in all countries, having replaced more traditional concerns such as malnutrition and infectious disease.

How obese are we?

According to the WHO, in 2016, more than 1.9 billion adults aged 18 years and older were overweight. Of these, over 650 million were obese. Additionally, 39% of adults aged 18 years and over were overweight, and 13% were obese [4]. In 2018, 67% of Australians aged 18 and over were overweight or obese. Of these, 31% were obese [14]. Forty-one million children under the age of 5, and over 340 million children and adolescents aged 5–19 were overweight or obese in 2016. The prevalence of overweight and obese children and adolescents aged 5–19 has risen dramatically from 4% in 1975 to over 18% in 2016. The rise has occurred similarly among both boys and girls—in 2016, 18% of girls and 19% of boys were overweight [4]. In 2017–18, 25% of Australian children and adolescents aged 2–17 were overweight or obese. Of these, 8.2% were obese [14].

While just under 1% of children and adolescents aged 5–19 were obese in 1975, more than 124 million children and adolescents (6% of girls and 8% of boys) were obese in 2016 [4].

Most of the world's population live in countries where overweight and obesity now kill more people than underweight. Worldwide obesity has steadily increased over the past 40 years (Fig. 1) and has nearly tripled since 1975.

Among OECD countries, USA and Mexico have the highest rates of obesity (Fig. 2), followed by New Zealand, Hungary, Australia, the UK, and Canada [15]. More adults are obese than children, but the rate of increase is higher among children. In particular, there has been a tripling of obesity in youth and young adults in developing, middle-income countries, such as China, Brazil, and Indonesia. This is a particularly worrisome trend because overweight children are at higher risk for the early onset of diseases such as type 2 diabetes, hypertension, and chronic kidney disease. Many low- and middle-income countries currently face a double burden of disease: a high prevalence of both malnutrition and obesity. As income increases, the burden of obesity shifts to the poor and to rural areas.

Obesity rates are steadily increasing in most Western countries, as shown in Fig. 3. If the obesity pandemic continues on its current trajectory, almost half of the world's adult population will be overweight or obese by 2030 [16].

A heavy and costly disease burden

Obesity is responsible for 5% of all deaths per year worldwide. In 2015, 8.4% of the disease burden in Australia was due to overweight and obesity, following only tobacco use as the leading factor contributing to the nationwide disease burden. In Australia, overweight and obesity are estimated to contribute to [17]:

- 54% of the type 2 diabetes burden
- 44% of the osteoarthritis burden
- 40% of the chronic kidney disease burden

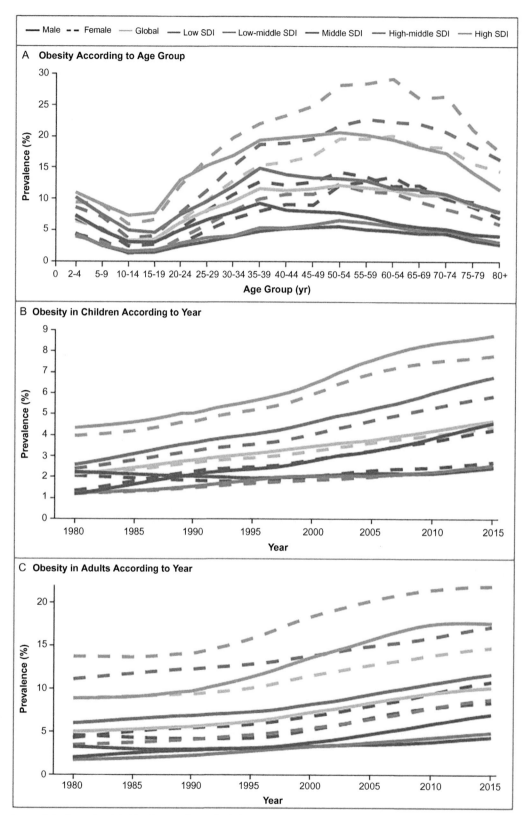

FIG. 1 Prevalence of obesity at the global level, according to Sociodemographic Index (SDI). Shown is the age-specific prevalence of obesity at the global level and according to SDI quintile in 2015 (A) and age-standardized prevalence trends at the global level and according to SDI quintile from 1980 through 2015 among children (B) and adults (C). *From GBD 2015 Obesity Collaborators. Health effects of overweight and obesity in 195 countries over 25 years. N Engl J Med 2017;377:13–27.*

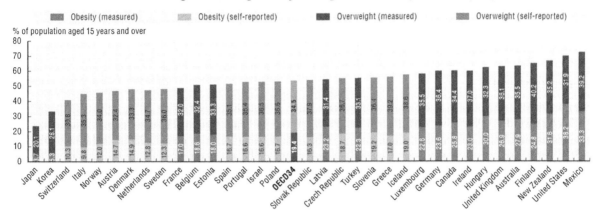

4.14. Overweight including obesity among adults, 2015 (or nearest year)

Source: OECD Health Statistics 2017.

StatLink http://dx.doi.org/10.1787/888933602956

FIG. 2 Proportion of obese people aged 15 and over, by selected OECD countries, 2016 or nearest year. *Source: Organization for Economic Cooperation and Development (OECD) 2017 . Health at a glance 2015: OECD indicators. Paris: OECDO.O.Update. Organization for Economic Cooperation and Development (OECD). 2017; O.Indicators. Health at a Glance 2011. Paris:OECD Indicators, OECD Publishing;2015 15:2016. doi: https://doi org/ 101787/health_glance-2015-en Accessed February, 2015.*

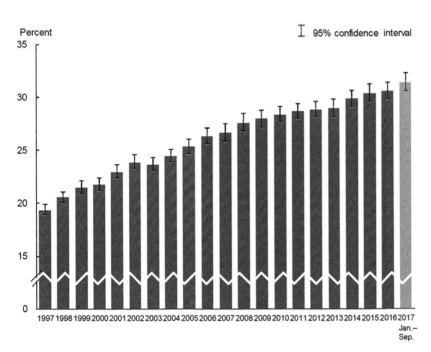

FIG. 3 Prevalence of obesity among adults aged 20 and over United States, 1997-September 2017. *Source: NCHS, National Health Interview Survey 1997-September 2017, Sample Adult Core component Martinez ME, Cohen RA, Zammitti EP. Health insurance coverage: early release of estimates from the National Health Interview Survey, January–September 2016. National Center for Health Statistics. Available at: http://wwwcdcgov/ nchs/nhishtm; 2017.*

- 25% of the coronary heart disease burden
- 24% of the asthma burden
- 21% of the stroke burden

The total annual direct cost of overweight and obesity in Australia in 2005 was $21 billion [18]. However, the burden of obesity extends well beyond healthcare dollars. Indirect costs that are not often considered include productivity losses due to premature mortality, disability, absenteeism, presenteeism (employees who come into work but have compromised productivity due to ill-health) as well as informal care and other nonmedical costs (see Box 2).

Obesity's global economic impact amounts to roughly $2 trillion annually, or 2.8% of global GDP—nearly equivalent to the global impact of smoking [16].

Box 2 How much does obesity cost per person?

In Australia, the average annual cost of government subsidies for the overweight and obese is roughly $3917 per person [18].

What causes obesity?

A common view is that obesity is the result of physical inactivity and the consumption of ultra-processed, energy-dense foods. In other words, if you move too little and eat too much, you will gain weight. This is largely based on the claim that weight gain is simply the consequence of an energy imbalance between calories consumed and calories expended. At a superficial level, this may be correct, however, it fails to recognize obesity's complex and multifactorial etiology. Maintenance and changes in body weight are regulated by the interaction of a number of processes, encompassing homoeostatic, environmental, and behavioral factors. Many factors that are emerging with globalization, urbanization, and technological development are also interacting to drive the rapidly increasing rates of obesity.

Risk factors for obesity [19]
- Individual
  - Genetics / pre- and perinatal exposures
  - Energy intake in excess of energy needs
  - Calorie-dense, nutrient-poor food choices (e.g., sugar-sweetened beverages)
  - Physical inactivity / sedentary behavior
  - Poor sleep
  - Certain medical conditions (e.g., Cushing's disease)
  - Mental health conditions (e.g., depression, stress)
  - Specific medication (e.g., steroids)
- Socioeconomic
  - Low education
  - Poverty
- Environmental
  - Lack of access to physical activity resources / low walkability neighborhoods
  - Food deserts (i.e., geographic areas with little to no ready access to healthy food, such as fresh produce/grocery)
  - Obesogens (e.g., endocrine-disrupting chemicals)

Genetics

Despite recent advances in genetic mapping and molecular biology, little is known about the genetics of obesity. Genome-wide association studies in which obese populations are compared with lean populations have explained less than 5% of the heritability of obesity [20]. These studies have found over 200 genetic variants involved in different biological pathways (central nervous system, digestion, adipocyte differentiation, insulin signaling, lipid metabolism, muscle and liver biology, gut microbiota) that are associated with polygenic obesity [21].

Multiple genetic factors responsible for rare forms of syndromal obesity have been identified, however, genes for common obesity have been harder to find. Pre- and perinatal exposures have been implicated in a child's risk of obesity. Parental obesity [22], smoking [23], endocrine-disrupting chemicals [24], weight gain during gestation [25], and gestational diabetes [26] are believed to contribute to hereditary obesity risk.

Energy balance

The first law of thermodynamics states that energy cannot be created or destroyed. In the nutrition world, this is referred to as "energy in must equal energy out." The energy balance equation: Calories in = Calories out.
or in other words:

$$\text{Weight gain (energy balance)} = \text{Energy in (food)} - \text{Energy out (exercise)}$$

This theory remains popular despite substantial evidence that losing weight as a result of either reducing caloric (food) intake or increasing energy output (exercise) is not effective in the medium to long term due to the body's hormonal response to caloric restriction and weight loss. Caloric restriction will usually result in short-term weight loss, however, to maintain that weight loss, individuals must adhere to behaviors that counteract physiological adaptations and other factors favoring weight regain. In response to calorie restriction, levels of leptin (the satiety hormone) decrease, and levels of ghrelin (the hunger hormone) increase [27]. Individuals in a calorie-restricted state experience more hunger, and feelings of hunger remain increased even after eating [28].

The other major response to caloric restriction and weight loss is compensatory changes in energy expenditure, which tend to favor weight gain. Depending on an individual's level of physical activity, between 50% and 80% of the energy expended each day is from essential metabolic processes (basal metabolic rate). Consumption and digestion of food also use energy and produce heat (the thermic effect of food). This thermogenesis accounts for about 10% of daily energy expenditure. Physical activity accounts for 20%–40% of total energy expenditure, depending on an individual's level of activity. Weight loss from calorie restriction leads to a decrease in total energy expenditure, resting energy expenditure (basal metabolic rate), and nonresting energy expenditure (thermic effect of eating and physical activity. Mechanisms involved in the decline of total energy expenditure following weight loss are likely linked to a reduction of body mass and enhanced metabolic efficiency [29]. If less total mass must be moved during physical activity, the same activity will have less energy cost, resulting in a decrease in nonresting energy expenditure, if levels of physical activity remain constant.

These hormonally driven responses to reduced caloric intake and initial weight loss prevent further weight loss. When the individual who is dieting eventually "gives up" on their reduced-calorie diet, weight is slowly regained. With the individual's BMR set at a lower rate, he or she is more likely to then not only regain all the initial weight loss but perhaps gain additional weight once normal caloric intake is resumed [27].

Diet

Perhaps deserving to be at the top of the list, the type and amount of food and drink people consume directly impacts their adiposity. Changes in caloric quantity and quality of the food supply combined with a global industrialized food system that produces and markets convenient, highly-processed foods which have been engineered to bypass normal appetite control are driving increased consumption [30]. Ultra-processed food usually has a poor nutrient profile–high in salt, added sugar, and unhealthy fats, and low in dietary fiber and micronutrients. The processing involves altering physical and structural characteristics, removal of water, flavor enhancers, colors, and other cosmetic additives, which makes the food hyperpalatable and addictive [31].

A growing body of evidence is demonstrating that ubiquitous access to and consumption of ultra-processed food is consistently associated with weight gain and obesity [31]. A notable 2019 randomized controlled trial showed that compared to a diet devoid of ultra-processed foods, a diet with >80% of ultra-processed foods caused an increase in energy intake of approximately 500kcal per day, with increased consumption of carbohydrate and fat but not protein. In 2 weeks, participants exposed to the ultra-processed diet gained approximately 1kg while participants exposed to the non-ultra-processed diet lost 1kg [32].

There is an ongoing debate about whether one macronutrient, in particular, is responsible for the increased adiposity. Some argue that if calories remain constant, the macronutrient composition of one's diet should not matter. However, a calorie is not just a calorie; its worth is dependent on the quality of the package within which it is consumed. For example, 100cal of a sugar-sweetened beverage have a drastically different nutritional profile to 100cal of salmon.

Sugar (carbohydrates)

It has long been recognized that insulin resistance is a common accompaniment of obesity. Insulin resistance is likely to be both a cause and effect of obesity. The "carbohydrate-insulin model" of obesity postulates that carbohydrates are particularly obesogenic because they elevate insulin secretion more so than protein or fat, and thereby upregulate storage of adipose tissue. Biochemical studies have shown that insulin plays key roles in regulating white adipocyte lipid accumulation, by inhibiting lipolysis as well as promoting fatty acid uptake and triglyceride synthesis (lipogenesis), and by stimulating the expression of genes involved in lipid uptake and storage [33].

There is a substantial body of evidence directly linking processed sugar intake and obesity. This association is particularly strong when the sugar is consumed in liquid form—in sugar-sweetened beverages (SSBs), colloquially known as "sodas" or "soft drinks" [34–37]. Emerging evidence demonstrates that sugar, especially in beverage form, can bypass satiety control mechanisms and directly contribute to a surplus of consumed calories [38].

A large 2013 econometric analysis of worldwide sugar availability across 175 countries revealed that for every excess 150cal of sugar (approximately one SSB per day), there was an 11-fold increase in the prevalence of type 2 diabetes, compared to an identical 150cal obtained from fat or protein. This was independent of the person's weight and physical activity level [39]. Consumption of SSBs is strongly associated with obesity in children. In a study that included over 500 ethnically diverse schoolchildren, for each additional serving of SSB consumed, BMI and frequency of obesity increased after adjustment for anthropometric, demographic, dietary, and lifestyle variables. Baseline consumption of SSB was also independently associated with a change in BMI [40] (ref). In 2007, Dubois and colleagues found that in >2000 children aged 2.5 years, regular consumers of SSBs between meals had a 2.4-fold greater odds of being overweight compared with nonconsumers, when followed up for 3 years [41].

Fats

For many years, dietary advice was based on the premise that high dietary fat causes obesity, diabetes, heart disease, and many other noncommunicable diseases. However, in recent times, there has been a resurgence in interest in lower-carbohydrate and ketogenic diets with high healthy fat content (e.g., from whole food sources such avocado, extra virgin olive oil, fatty fish).

As dietary carbohydrate is replaced by fat, postprandial spikes in the blood concentrations of glucose and insulin decrease, glucagon secretion increases, and metabolism shifts to rely more so on fat oxidation. This is associated with attenuated oxidative stress and inflammatory responses after eating [42,43], reduced hormone resistance to insulin and leptin [44,45], and improvements in features of metabolic syndrome [46].

Recent systematic reviews and meta-analyses have concluded that low-carbohydrate diets outperform low-fat diets for short- to medium-term weight loss, especially when the low-carbohydrate diet is ketogenic (approximately less than 10% of calories from carbohydrates) [47]. Although individuals with insulin sensitivity seem to respond similarly to low-fat or low-carbohydrate diets, individuals with insulin resistance, glucose intolerance, or insulin hypersecretion may lose more weight on a low-carbohydrate, high-fat diet [48,49].

One of the most dramatic changes in eating habits in the last century has been the increase in consumption of vegetable or seed oils. These polyunsaturated oils are high in Omega-6 fatty acids. As a result, the ratio of the pro-inflammatory Omega-6 fatty acids to the antiinflammatory Omega-3 fatty acids has changed from the traditional 1:1 ratio to approximately 16:1 or even higher. High dietary intake of omega-6 fatty acids leads to increases in white adipose tissue and chronic inflammation, which are associated with obesity [50].

Protein

Protein is often the forgotten macronutrient in the obesity debate, as much of the scientific community's energy is spent debating carbohydrates and fats. Dietary protein seems to be effective for body-weight management as it promotes satiety, energy expenditure with a greater thermic effect of food, and improved body composition with fat-free body mass [51,52]. Given these effects, high protein diets may be beneficial to prevent the development of a positive energy balance and subsequent weight gain. In contrast, low-protein diets may facilitate weight gain.

Is there an optimal diet to prevent obesity?

While too much of any food or food group can cause weight gain, there is consensus that a nutritious diet emphasizes *real*, whole food that is minimally processed. This can include:

- Green and nonstarchy vegetables
- Low-glycaemic fruits (e.g., berries)
- Fish, grass-fed meat, eggs
- Lentils and other legumes
- Fermented dairy
- Olive oil
- Nuts and seeds

A nutritious diet limits sugar-sweetened beverages, refined grains, seed oils, and highly processed foods.

Physical inactivity

The most recent global physical activity data for adults aged 18 years and older, from 168 countries, show that in 2016, 27.5% of adults did not meet the WHO 2010 physical activity guidelines [53]. In some countries, physical inactivity can be as high as 70%, due to changing transport and work patterns, increased use of technology, urbanization, and cultural values [54].

Mechanization of labor and transport has radically impacted daily energy expenditure, as incidental physical activity has been engineered out of daily life. Over the last 50 years in the USA, for example, daily occupation-related energy expenditure has decreased by more than 100 cal, and this reduction in energy expenditure likely accounts for a significant portion of the increase in mean body weights of women and men [55]. Massive reductions in daily energy expenditure have also occurred in countries such as China and Brazil, which have the highest absolute and relative rates of decline in total physical activity due to reductions in physical activity in the workplace [56].

Weight gain and obesity are understood to be the result of a sustained positive energy balance. It would seem logical that the rise in obesity is attributable to decreases in daily energy expenditure. If one accepts the *calories in, calories out* theory, then simply increasing the amount of physical activity should lead to weight loss. However, the body is complex and unfortunately, it is just not that simple [57]. A confluence of multiple interrelated dynamic factors influences weight loss and gain. This will be explored later in this chapter.

Sedentary behavior

Sedentary behavior has been defined as any waking behavior characterized by an energy expenditure that is less than or equal to 1.5 times the resting metabolic rate while in a sitting, reclining, or lying posture [58]. Although sedentary behavior and moderate-to-vigorous physical activity represent opposite ends of the energy expenditure continuum, they can be viewed as separate behaviors.

In children, sedentary behavior has usually been measured using sitting time or screen/television time. A 2017 Canadian systematic review concluded that higher levels of screen time/television viewing were associated with unfavorable body composition and higher cardiometabolic risk factors in children [58]. However, in 2018, The US Physical Activity Guidelines Advisory Committee found that there was only limited evidence for an association between sedentary behavior and poorer cardiometabolic health and weight status/adiposity in children; and that the evidence was somewhat stronger for screen time or television viewing than for total sedentary time [59].

According to the summary of evidence used for the 2020 WHO Guidelines on physical activity and sedentary behavior, there is low certainty evidence that sedentary behavior leads to increases in adiposity and other indicators of weight status [60].

Medical conditions

Certain medical illnesses such as hypothyroidism, polycystic ovary syndrome, and Cushing's syndrome are associated with increased rates of obesity.

Medications

Medications associated with weight gain include certain antidepressants, anticonvulsants (such as carbamazepine and valproate), some diabetes medications (insulin, sulfonylureas, and thiazolidinediones), hormones such as oral contraceptives, and most corticosteroids. Weight gain may also be seen with some antihypertensive medications and antihistamines.

Mental health and stress

For most people, emotions influence eating habits [61]. When someone experiences acute stress, their sympathetic nervous system is activated, and hormones released can result in suppressed appetite. However, when exposed to chronic stress (e.g., from daily pressures of work or study), hormonal and metabolic processes controlled by the hypothalamic–pituitary–adrenal axis can reduce lipolysis and associated sex steroids, promoting adipose tissue accumulation [ref 61]. Chronic stress may also turn people to "comfort eating"—overeating processed foods high in sugar and calories—as a result of alterations in the mesolimbic dopaminergic system that promotes feelings of reward when consuming hyper-palatable foods [62].

Poor sleep

Several studies have demonstrated increased hunger or appetite and/or increased caloric intake, especially from unhealthy food, in response to sleep restriction [63,64]. A 2016 systematic review and meta-analysis of 11 studies showed that sleep restriction (of approximately 4 h) seems to cause a person to eat an average of 385 extra calories the next day [65]. Insufficient sleep disrupts regular hormone production, causing increased levels of ghrelin and reduced leptin, leading to increased hunger—especially for calorie-dense foods with high carbohydrate content [66]. This coupled with more waking hours enabling more opportunities to eat likely results in overeating.

Socio-economic disadvantage

Although the level of overweight and obese individuals has been increasing across all socioeconomic groups, there is significant variation between groups. Obesity tends to rise with a country's economic development, as populations are able to access and consume more cheap energy-dense foods, and rely more on industrialization and mechanized transportation. In low-income countries, higher SES groups are more likely to be obese, while in high-income countries, higher SES groups are less likely to be obese–likely due to a combination of access to healthier food, time to engage in more physical activity, greater health literacy and societal ideals of thinness [67] (as well as other factors).

The obesogenic environment

The term "obesogenic environment" was first described by Boyd Swinburn, meaning that "the driver of diabetes was obesity and that obesity was just a normal physiological response to an abnormal environment" [68]. An obesogenic environment refers to "an environment that promotes gaining weight and one that is not conducive to weight loss" [69], and "the sum of influences that the surroundings, opportunities or conditions of life have on promoting obesity in individuals or populations [70].

The obvious possible drivers of the epidemic are in the food system [71]: the increased supply of cheap, palatable, energy-dense foods; improved distribution systems to make food much more accessible and convenient; and more persuasive and pervasive food marketing. Several studies have tested the hypothesis that increases in the food supply are the dominant drivers of weight gain in populations [72–74]. Results from these investigations show that the rise in food energy supply was more than sufficient to explain the rise in obesity in the US from the 1970s, and most of the weight increase in the UK since the 1980s.

A related hypothesis is that the policies put in place in the USA and other countries to increase the food supply from the 1970s led to a situation in which the abundance of food in these countries began to push up population energy intake [75]. The existing environments within a country, transport systems, active recreation opportunities, food costs and security, and culture around body size can greatly influence the effects of the population drivers on obesity [76].

Pathophysiology

At the most superficial level, the pathogenesis of obesity seems relatively straightforward. If the "calories in, calories out" theory is accepted, weight gain is the product of consuming a quantity of calories that exceeds energy expenditure. Humans can store thousands of calories of adipose tissue required for times of starvation. When humans are presented with excessive calories over an extended period of time, an innate ability to store energy results in excessive accumulation of adipose tissue and eventually obesity.

Set point theory

The pathogenesis of obesity is far more complex than a matter of balancing inputs and outputs. Body weight is the product of genetic effects, food intake, and the surrounding environment, among other factors. The central and peripheral nervous systems regulate energy balance by biological and behavioral mechanisms. For example, excessive eating is followed by increased thermogenesis, whereas increases in energy expenditure (e.g., due to vigorous exercise) may increase food intake afterward.

Bodyweight is under strong genetic and humoral control, which has come to be known universally as the "set point" theory. The theory posits that body weight depends on an internal feedback control system designed to regulate body weight to a constant "body-inherent" weight–the "set point" weight [77]. Many adults tend to remain at a stable weight for most of their lives. However, regulation of the set point may be lost, and multiple "settling points" may be found throughout life as a consequence of excessive food intake and external environmental factors. This may explain why weight lost through dieting tends to be regained over time. Growing evidence suggests that obesity is a disorder of this energy homeostasis system, rather than simply arising from the passive accumulation of excess weight [78].

Neuroendocrine dysfunction

Control of appetite, food intake, and energy storage is regulated by complex brain circuits and multiple hormones, including insulin, leptin, ghrelin, glucagon-like peptide 1 (GLP-1) peptide YY (PYY), pancreatic polypeptide (PP), and cholecystokinin (CCK). They are secreted by the gut and have a significant impact on energy balance and maintenance of homeostasis [79]. For example, leptin, produced by adipose tissue, regulates lipid metabolism by stimulating lipolysis and inhibiting lipogenesisis [80]. Leptin crosses the blood–brain barrier and acts upon receptors to downregulate appetite-stimulating neuropeptides while upregulating anorexigenic hormones [81]. Serum concentrations of leptin are elevated in obesity (as obese people have more adipose tissue), promoting leptin resistance which results in reduced satiation and appetite control—leading to overeating and weight gain.

When leptin was discovered in 1994 [82], some hypothesized that leptin could be administered as a treatment for obesity–similar to insulin being administered in diabetes. However, studies have revealed that exogenous leptin has a limited role in the treatment of common obesity, as obese people have higher levels of circulating leptin and central leptin resistance [83]. When a person loses a moderate amount of weight/adipose tissue, there is a disproportionately large decrease in circulating leptin [84]. A decrease in leptin is associated with increased sensations of hunger [85], an increase in reward-related behaviors [86], and potentially depressive symptoms [87]. Therefore, when a person loses weight, adipose tissue sends signals to the brain to try to resist any further loss of fat, triggering hunger, reward-seeking, and low mood. The pathogenesis of obesity extends far beyond the scope of this chapter. For further reading, Schwartz and colleagues' Endocrine Review [88] provides an in-depth discussion of the current understanding of obesity pathophysiology.

The effects of exercise on obesity

Governments around the world advise that overweight and obese individuals should predominantly follow two important health recommendations: eat less and move more. Exercise (a subcategory of physical activity), falls on the energy expenditure side of the equation. Each and every action produces energy expenditure, which differs depending on the type, duration, and intensity of activity. While exercise is immensely beneficial for almost all bodily processes, as explored throughout this textbook, there is conflicting evidence as to whether exercise in fact is beneficial for weight loss.

As highlighted in the summary of evidence of 2020 WHO Guidelines on physical activity and sedentary behavior [60], "The association between physical activity and adiposity in adult populations is less well established despite a large, but heterogenous, body of evidence assessing this relationship across various outcome measures (weight gain, weight change, weight control, weight stability, weight status, and weight maintenance)". In this section, exercise as a method to (i) prevent weight gain, (ii) lose weight, and (iii) maintain weight will be explored.

Prevention of weight gain

As physical activity begins to decline in adolescence, weight tends to increase [89]. For most people, excess adipose tissue tends to accumulate particularly during early and middle adulthood. In fact, the average American adult can expect to gain 0.5–1.0 kg per year from early to middle adulthood [90]. Physical activity has been found to help prevent weight gain throughout life, and help stave off the associated increased risk of type 2 diabetes, cardiovascular disease, and hypertension (among other diseases) [91].

Over a decade ago, the 2008 Physical Activity Guidelines Advisory Committee (PAGAC) Report concluded that physical activity was associated with the prevention of weight gain following weight loss [92]. The 2018 Physical Activity Guidelines Advisory Committee Scientific Report revisited the relationship between physical activity and prevention of weight gain in adults and concluded that "strong evidence demonstrates a relationship between greater amounts of physical activity and attenuated weight gain in adults, with some evidence to support that this relationship is most pronounced when physical activity exposure is above 150 min per week" [59].

The PAGAC reviewed evidence from 33 studies [59,93], with follow-up periods ranging from 1 to 20+ years. Twenty-six of 33 studies demonstrate a significant relationship between greater amounts of physical activity and attenuated weight gain in adults, however, the evidence for a specific volume threshold of physical activity that is associated with the prevention of weight gain was inconsistent. Some of the evidence reviewed supports the need to achieve at least 150 min per week of moderate-intensity physical activity [94,95] or achieve 10,000 steps per day to minimize weight gain or to prevent increases in BMI [96]. Other studies support greater amounts of physical activity to prevent or minimize weight gain, with some studies reporting this effect with greater than 150 min per week.

The 26 studies in which a significant inverse association between physical activity and weight gain was observed included adults of normal, overweight, and obese weight status. However, there was insufficient evidence to determine whether the relationship between greater amounts of physical activity and prevented weight gain in adults varies by initial weight status. Data from the China Health and Nutrition Survey [97] included in the PAGAC report suggests that the overweight and obesity risk in men with low leisure-time physical activity (LTPA) or without LTPA is approximately 2 times higher than those with high LTPA. The odds of overweight and obesity in women were increased to approximately 1.7 times in women without LTPA compared with those with high LTPA.

Similarly, an analysis of BMI and exercise trends from over 20,000 African American women aged <40 years revealed that in women with normal weight and overweight, the incidence of obesity decreased with increasing vigorous exercise (e.g., basketball, swimming, running, aerobics). When compared to less than 1 h per week, the estimated incidence rate ratios of developing obesity were significantly reduced in a graded manner: with the vigorous-intensity activity of 1–2 h per week (0.87), 3–4 h per week (0.82), 5–6 h per week (0.79), and 7 h or more per week (0.77) [98].

In the review of the studies, the PAGAC found that total leisure-time physical activity was consistently inversely associated with weight change. Studies reporting on moderate-intensity, vigorous-intensity, and moderate-to-vigorous intensity physical activity showed consistent associations with the prevention of weight gain. However, light-intensity physical activity was not associated with the prevention of weight gain.

Weight loss

For many years, people looking to lose weight have been told that they "just need to do some more exercise", with organizations and associations have promoted a controversial weight loss "rule" that originated in 1958 [99]. The rule stipulates that a pound (0.45 kg) of adipose tissue equals 3500 cal; therefore, reducing 500 cal per day, through diet or physical activity, results in about a pound of weight loss per week. Equally, adding 500 cal a day is supposed to equate to an equal weight gain. However, a growing body of evidence suggests that exercise-induced weight loss is usually small and much less than expected from an exercise-induced increase in energy expenditure. The mismatch between the amount of weight loss predicted from exercise-associated energy expenditure and actual weight loss is known as "weight compensation" [100]. Compensation could be due to multiple reasons, such as reduced metabolic rate through adaptive thermogenesis, decreased spontaneous activity, and increased appetite and food intake [101].

For example, in 1994, researchers subjected seven pairs of young adult male identical twins to a negative energy balance protocol [102]. The twins exercised on cycle ergometers twice a day, on 9 out of 10 days, over a period of 93 days, while being kept on constant daily energy and nutrient intake. The participants only lost an average of 5 kg and it was entirely accounted for by the loss of fat mass. The participants also burned 22% fewer calories through exercise than the researchers predicted. The researchers believed that either the participants' basal metabolic rates slowed down, or they were expending less energy outside of their 2-hour (almost) daily cycle. More recently, a 2009 study of sedentary, overweight postmenopausal women found that people increase their food intake after exercise—either because they thought they burned off more calories or because they were hungrier [103]. And another review of 15 different aerobic exercise studies found that people generally overestimated how much energy exercise burned and ate more when they exercised [104]. Compensatory mechanisms are powerful and are thought to be the primary reason why long-term weight loss often fails.

The highest levels of evidence have found that exercise likely has only a moderate effect on weight loss, if at all. A Cochrane review [105] that analyzed data from 43 randomized controlled clinical trials, which included nearly 3500 people, found that when compared with no treatment, exercise resulted in small weight losses across studies. Increasing exercise intensity increased the magnitude of weight loss. However, even when weight loss did not occur, the review did show that exercise still improved a range of secondary outcomes including serum lipids, blood pressure, and fasting plasma glucose. The Women's Health Study, potentially the largest study ever to investigate diet and exercise, followed 34,000 healthy American women (average age 54 years) from 1992 to 2007. Compared with women who engaged in 420 min per week (approximately 60 min per day) of moderate-intensity physical activity, those engaged in 150 to less than 420 min per week of such activity, as well as those engaged in less than 150 min per week, gained significantly more weight with no difference in weight gain between these two lesser active groups [106].

Aerobic vs resistance exercise

To date, the majority of studies that have explored the relationship between exercise and weight loss have specifically investigated the effects of aerobic or endurance activity (e.g., walking, jogging, cycling), rather than resistance exercise. Many of these studies in overweight or obese individuals undertaking supervised aerobic exercise for extended periods of time have demonstrated unsuccessful weight loss. For example, in 2007, The Dose–Response to Exercise in Women (DREW) study [107] observed no significant changes in body weight in postmenopausal women exercising at 50% and 150% of public health guidelines for 6 months. Similarly, the Inflammation and Exercise (INFLAME) study [108] found no significant change in body weight after 4 months.

Resistance training and aerobic training differ greatly in terms of the nature of the training stimulus (e.g., intermittent vs. continuous contractions, time under load, metabolic pathways). Often overlooked in a weekly exercise program, resistance exercise has been found to have a multitude of benefits [109], such as increasing muscle strength and muscle mass, maintaining bone density, decreasing cardiovascular risk profile, and improving glucose tolerance and insulin sensitivity. For these and many more benefits, the WHO recommends that all adults should also do muscle-strengthening activities at a moderate or greater intensity that involve all major muscle groups on 2 or more days a week [60].

However, research suggests that resistance training may not be beneficial for weight loss in adults. Although resistance training may contribute to the reduction of body fat, the effect on overall weight loss is minimal. In 2012, Willis et al. [110] conducted one of the largest randomized controlled trials to investigate the effects of resistance exercise vs aerobic exercise vs a combination of both on weight loss. Two-hundred and thirty-four sedentary, overweight or obese 18–70-year-olds were randomized to three groups: (1) resistance exercise only, (3 days/week, 3 sets/day, 8–12 repetitions/set); (2) aerobic exercise only, (calorically equivalent to $\sim$19 km/week at 65%–80% peak VO_2); (3) aerobic plus resistance exercise, (calorically equivalent to $\sim$19 km/week at 65%–80% peak VO_2 plus 3 days/week, 3 sets/day, 8–12 repetitions/set). The researchers found that resistance exercise alone did not reduce body mass or fat mass, that aerobic exercise was significantly better than resistance for reducing measures of body fat and body mass, and the combination of aerobic and resistance exercise did not provide an additive effect for reducing fat mass or body mass compared with aerobic exercise alone. Several other studies [111,112] investigating aerobic vs resistance exercise have shown similar results.

Interestingly, evidence suggests that aerobic and resistance exercise may be more beneficial for weight loss in overweight and obese children and adolescents (compared to adults). A 2019 network meta-analysis of randomized exercise intervention trials [113] that included nearly 3000 males and females aged 2–18 years found that aerobic exercise as well as combined aerobic and resistance exercise lead to significant reductions in BMI, fat mass, and body fat percentage. A significant reduction in body fat percentage was also found for resistance exercise alone.

Is high-intensity interval training a magic bullet for weight loss?

High-intensity interval training (HIIT) involves alternate bursts of short, intense physical activity interspersed with recovery periods. These short bursts may last from 5 s to 10 min or longer and are performed at approximately 60%–95% of a person's maximum heart rate. The recovery periods may last equally as long and are performed at 30%–50% of the maximum heart rate. For example, 30 s of hard sprinting followed by 30 s of jogging, repeated as many times as desired. There is no set formula for prescribing HIIT—the specific exercise and its intensity can be varied to meet the needs of the individual.

HIIT continues to gain popularity due to its similar health benefits to more traditional endurance training, which can be achieved in shorter periods of time. While HIIT and endurance training both lead to improved cardiovascular fitness, HIIT leads to greater improvement in VO_2 max [114]. Interestingly, HIIT is also more effective than moderate-intensity exercise at improving vascular function and markers of vascular health—perhaps due to its positive effects on cardiorespiratory fitness, cardiovascular disease risk factors, oxidative stress, inflammation, and insulin sensitivity [115].

Although some have claimed that HIIT is a "magic bullet" for weight loss, research perhaps tells a different story. A 2017 meta-analysis of 18 studies that included a total of 854 obese participants found that there was no significant difference in BMI, body weight, or waist circumference between those who engaged in HIIT compared to those who engaged in traditional (high volume continuous) exercise. However, HIIT did result in a significant reduction in body fat percentage [116]. The authors hypothesized that one of the many reasons there was no significant reduction in weight between HIIT and traditional exercise was that there was an absence of an accompanying significant dietary intervention in the included studies.

In contrast, a 2019 meta-analysis of 48 studies in overweight and obese adults [117] comparing HIIT to moderate-intensity continuous activity demonstrated that HIIT was effective in decreasing body mass, body mass, waist circumference, body fat percentage, and abdominal visceral fat area. *However,* when considering equalization between the two methods (energy expenditure or workload matched), no differences were found in any measure except body mass (for which HIIT was superior).

Weight maintenance

While exercise may not be greatly beneficial for weight loss, if someone has managed to lose weight, high volumes of exercise can help them maintain their weight (when it is accompanied by strict monitoring of calorie intake). In 2017, researchers followed 14 contestants on *The Biggest Loser* weight-loss reality television show for 6 years after their initial weight loss [118]. The contestants who managed to maintain their weight loss after 6 years engaged in 80 min of moderate-intensity exercise or 35 min of vigorous-intensity exercise per day. The direction of causality is not known, however—it may be that losing weight facilitates being physically active.

In an older study, Anderson et al. evaluated the effect of a low-fat diet (1200 kcals/day) in combination with either structured aerobic exercise or lifestyle activity (participants were advised to increase their physical activity to recommended levels) in obese women [119]. Both groups lost approximately 8 kg of weight following 16 weeks of intervention.

Weight maintenance was monitored for 1 year after the intervention, and those who were the most active lost additional weight (2 kg) whereas the group that was the least active regained a substantial amount of weight (5 kg).

Summary

On balance, evidence suggests that caloric restriction has a more significant effect on weight loss compared to exercise alone [120]. Exercising and dieting combined may have a more significant reduction in body weight and fat mass than dieting alone, however, the benefits may be marginal [121].

The "fat but fit" paradox

Conflicting evidence has previously suggested that high cardiorespiratory fitness might mitigate the detrimental effects of excess body weight on cardiometabolic health. This has come to be known as the "fat but fit" paradox [122].

While the question may be continued to be asked, a recent large-scale study has shed some light on the relationship between adiposity, physical activity, and markers of health.

In 2020, a group of researchers analyzed over 520,000 adults (average age 42 years, 32% women) in Spain [123]. Approximately 42% of participants were normal weight, 41% were overweight, and 18% were obese. The majority were inactive (63.5%), while 12.3% were insufficiently active, and 24.2% were regularly active. Approximately 30% had hypercholesterolemia, 15% had hypertension, and 3% had diabetes.

At all levels of BMI, any physical activity was linked with a lower likelihood of hypercholesterolemia, hypertension, or diabetes compared to no exercise at all. At all levels of BMI, as physical activity increased, the probability of diabetes and hypertension decreased.

Overweight and obese participants were at greater cardiovascular risk than their normal-weight peers, irrespective of physical activity levels. Compared to physically inactive normal-weight participants, physically active obese participants were twice as likely to have hypercholesterolemia, four times more likely to have diabetes, and five times more likely to have hypertension.

While the study's results suggest that one cannot be "fat and fit," there are still many questions yet to be answered—in children and adolescents, and for other health outcomes [122]. Interventions for overweight and obese populations should target both weight reduction and cardiorespiratory fitness improvement, as both are likely to improve long-term health.

Clinical practice recommendations for exercise prescribers

Exercise may not be as beneficial as once thought for weight loss, but it is still beneficial for the prevention of weight gain, and weight maintenance. Regular exercise is important for the prevention and management of noncommunicable diseases and cognitive decline, for mental health, and for general wellbeing [60]. Overweight and obese individuals who exercise regularly can reap these benefits, as well as specifically improve insulin sensitivity, lipid profiles, and overall inflammation [124].

Focus on prevention

Because overweight and obese children often become obese adults, exercise prescribers should focus on encouraging children and adolescents to be physically active during every clinical encounter. For children and adolescents, physical activity is rarely in the form of structured exercise. It can be undertaken as part of recreation and leisure (play, games, sports), physical education, transportation (wheeling, walking, and cycling), or household chores. Exercise prescribers should encourage age-appropriate activities that the child is likely to enjoy, and suggest that his or her parents offer many opportunities for the child to be active throughout the entire day.

The WHO recommends that children and adolescents should do at least an average of 60 min per day of moderate- to vigorous-intensity, mostly aerobic, physical activity, across the week. Vigorous-intensity aerobic activities, as well as those that strengthen muscle and bone, should be incorporated at least 3 days a week [60]. Exercise prescribers should emphasize that doing any level of physical activity will benefit health, even if it doesn't meet the recommended levels. Exercise prescribers should advise children and adolescents to start by doing small amounts of physical activity, and gradually increase the frequency, intensity, and duration over time. Other effective strategies to prevent childhood obesity (e.g., population-wide policies and initiatives, such as marketing restrictions on unhealthy foods and SSB to children, nutrition labeling, food taxes and subsidies, etc.) are beyond the scope of this chapter.

Increase everyday physical activity

People who are overweight or obese may not be motivated to exercise, and may actively avoid places like the gym due to fat-shaming or lack of confidence. People do not need to perform physical activity in a gym or in a structured workout to gain benefits. Many people find it easier to be active in lots of little ways throughout the day, through incidental activity. Some examples of how to reduce sedentary behavior and increase incidental activity are listed below:

- Stand up and walk when talking on the phone.
- Take the stairs instead of an escalator or elevator.
- Swap the car for the bike and cycle into work.
- Instead of trying to park the car closest to your destination, aim to park as far away as you can and walk instead.
- Clean up around the house or tend to the garden.
- Reduce all forms of screen time, including using the computer or phone, and watching television.
- When watching television, using the computer, or reading, schedule 5-min breaks every 45 min to take time to go for a quick walk.
- During television commercial breaks, perform calisthenics such as push-ups, squats, and lunges.
- Save your television watching for the gym, when using a stationary bike or treadmill. This is a form of "temptation bundling"—a way of combining "want behaviors," such as watching television, with "should behaviors," such as exercise, to ultimately encourage patients to increase should behaviors and develop healthy habits.

Prescribe combinations of exercise

All overweight and obese individuals who are planning to engage in physical activity for the first time after a period of inactivity should be encouraged to *start low, and go slow*. Exercise prescribers should recommend a gradual build-up of daily physical activity, starting with light-intensity activities, such as casual walking, which can then be increased in duration and intensity over time.

All exercise is appropriate for overweight and obese individuals, however, some types may be more appropriate than others. Obese patients often suffer from pain in their weight-bearing joints—back, hips, knees, and ankles. Exercise prescribers should alter recommendations depending on individual variation in ability. Cycling, swimming, water aerobics, weight lifting, and yoga are all good options that may put less strain on weight-bearing joints. A weekly exercise routine should comprise a combination of aerobic and muscle-strengthening activities. Importantly, overweight or obese individuals should be encouraged to continue to be active even if they do not lose weight as a result, because of physical activity's many other health benefits. Any amount of physical activity is better than none, and more is better. For health and wellbeing, WHO recommends at least 150–300 min of moderate aerobic activity per week (or the equivalent vigorous activity) for all adults [60].

Set realistic expectations

Sustained weight loss is not easy. However, even a modest weight loss of 5%–10% can result in significant health benefits. The greatest weight loss will likely take place within the first 6 months. The difficulty lies in thereafter maintaining the new lower weight, as most people are likely to regain weight due to compensatory mechanisms, as discussed above.

What actually works for weight loss?

Although the odds may be against people when it comes to losing weight, there is some helpful research that reveals the cornerstones of successful and sustained weight loss.

Started in the early 1990s, the U.S. National Weight Control Registry is a research study that includes adults who have lost at least 30 lbs. (13.6 kg) and maintained the weight loss for at least 1 year. Over 10,000 people enrolled in the study complete annual questionnaires about their current weight, diet, and exercise habits.

Analyses of the research have shown that people who have had success losing weight have a few things in common:

- Weigh themselves at least once a week [125].
- Restrict calorie intake and watch their portion sizes [126].
- Exercise regularly [127].

Nearly all adults in the registry indicated that weight loss led to improvements in their level of energy, physical mobility, general mood, self-confidence, and physical health [126].

You cannot outrun a bad diet

Consuming fewer calories, and the right calories is an essential part of a successful weight loss strategy. Countless diets have been described both for weight loss and the promotion of good health. Expert opinion remains divided on the merits of various diets, with many yet to undergo rigorous scientific investigation. Much of the research in this field is plagued by the problems of assessment of dietary intake, confounding variables, follow-up of insufficient duration, and bias due to industry funding. To compound these challenges, the mainstream media is often responsible for incomplete and incorrect reporting of the science, misinterpreting studies, and often publicizing hyperbolic and distorted content that makes for good headlines.

After the introduction of the dietary guidelines, low-fat and reduced-calorie eating was widely accepted and endorsed as the ideal dietary pattern for weight loss. However, in recent years this has been challenged, and there has been a proliferation of other diets (e.g., ketogenic, paleo, vegan) claiming to be the one true way to achieve weight loss and optimal health. There has also recently been a growing interest in "time-restricted feeding" which alternates periods of normal food intake with prolonged periods (usually 16–48 h) of restricted or no food intake. Prescribers should remind individuals that different dietary approaches can achieve a similar goal, and it is accepted that there is no one universal diet suitable for everyone. However, there is general agreement that a diet focused on real foods with minimal sugars, starches, and ultra-processed foods is most compatible with good health.

Conclusion

The prevalence of overweight and obesity is increasing at a rapid rate with dire health and economic consequences for all. Complex socioeconomic and intrapersonal factors are driving a toxic, ultra-processed food environment which is continuing to fuel the crisis. Historically, policy and implementation to tackle overweight and obesity have focused on the energy balance equation, recommending populations to exercise more and eat less. However, a growing body of evidence suggests that while helpful for preventing weight gain and for weight maintenance, exercise may not be as useful as once thought for weight loss. The foundation of healthy weight status among children, adolescents, and adults must be centered on nutrition, coupled with effective exercise prescription. Overweight and obesity will continue to be a serious threat to human health for years to come, and must be tackled using a system-based approach—both at the individual level through lifestyle interventions, as discussed throughout this chapter, and at the government level through evidence-based policy action shaping the obesogenic environment.

References

[1] S.I.G.Network. Management of obesity: a national clinical guideline. Edinburgh: Scottish Intercollegiate Guidelines Network; 2010.

[2] Keys A, et al. Indices of relative weight and obesity. Int J Epidemiol 2014;43(3):655–65.

[3] Adab P, Pallan M, Whincup PH. Is BMI the best measure of obesity? British Medical Journal Publishing Group; 2018.

[4] WHO. Obesity and overweight. 1 April 2020 21 March 2021]; Available from https://www.who.int/news-room/fact-sheets/detail/obesity-and-over weight; 2020.

[5] Swainson MG, et al. Prediction of whole-body fat percentage and visceral adipose tissue mass from five anthropometric variables. PloS one 2017;12(5), e0177175.

[6] W.H.Organization. Waist circumference and waist-hip ratio: report of a WHO expert consultation, Geneva, 8-11 December 2008; 2011.

[7] Ibrahim MM. Subcutaneous and visceral adipose tissue: structural and functional differences. Obes Rev 2010;11(1):11–8.

[8] Tanaka S, Horimai C, Katsukawa F. Ethnic differences in abdominal visceral fat accumulation between Japanese, African-Americans, and Caucasians: a meta-analysis. Acta Diabetol 2003;40(1):s302–4.

[9] Hamdy O, Porramatikul S, Al-Ozairi E. Metabolic obesity: the paradox between visceral and subcutaneous fat. Curr Diabetes Rev 2006;2(4):367–73.

[10] Wilding JP, Mooney V, Pile R. Should obesity be recognised as a disease? BMJ 2019;366:l4258. https://doi.org/10.1136/bmj.l4258.

[11] James WPT. WHO recognition of the global obesity epidemic. Int J Obes (Lond) 2008;32(7):S120–6.

[12] Bray G, et al. Obesity: a chronic relapsing progressive disease process. A position statement of the world obesity federation. Obes Rev 2017;18(7):715–23.

[13] Patankar PS. 8th Asia-Oceania conference of obesity (AOCO) 2015, Nagoya, Japan. J Obes Metab Res 2015;2(4):239.

[14] Australian Institute of Health and Welfare. Overweight & obesity. [cited 2021 21 March]; Available from https://www.aihw.gov.au/reports-data/behaviours-risk-factors/overweight-obesity/overview; 2020.

[15] OECD. Obesity update. Paris: OECD Publishing; 2017. http://www.oecd.org/els/health-systems/Obesity-Update-2017.pdf.

[16] Dobbs R, et al. How the world could better fight obesity. McKinsey Global Institute; 2016.

[17] Australian Institute of Health and Welfare. Australian Burden of Disease Study 2015: Interactive data on risk factor burden. [cited 2021 21 March]; Available from: https://www.aihw.gov.au/reports/burden-of-disease/interactive-data-risk-factor-burden/contents/overview; 2020.

[18] Colagiuri S, et al. The cost of overweight and obesity in Australia. Med J Aust 2010;192(5):260–4.

[19] Hruby A, Hu FB. The epidemiology of obesity: a big picture. Pharmacoeconomics 2015;33(7):673–89.

[20] Campbell Am L. Genetics of obesity. Aust Fam Physician 2017;46:456–9.

[21] Pigeyre M, et al. Recent progress in genetics, epigenetics and metagenomics unveils the pathophysiology of human obesity. Clin Sci 2016;130(12):943–86.

[22] Drake AJ, Reynolds RM. Focus on obesity: impact of maternal obesity on offspring obesity and cardiometabolic disease risk. Reproduction 2010;140:387–98.

[23] Northstone K, et al. Prepubertal start of father's smoking and increased body fat in his sons: further characterisation of paternal transgenerational responses. Eur J Hum Genet 2014;22(12):1382–6.

[24] de Cock M, van de Bor M. Obesogenic effects of endocrine disruptors, what do we know from animal and human studies? Environ Int 2014;70:15–24.

[25] Bammann K, et al. Early life course risk factors for childhood obesity: the IDEFICS case-control study. PLoS One 2014;9(2), e86914.

[26] Dabelea D, Harrod CS. Role of developmental overnutrition in pediatric obesity and type 2 diabetes. Nutr Rev 2013;71(Suppl. 1):S62–7.

[27] Greenway FL. Physiological adaptations to weight loss and factors favouring weight regain. Int J Obes (Lond) 2015;39(8):1188–96.

[28] Doucet E, et al. Relation between appetite ratings before and after a standard meal and estimates of daily energy intake in obese and reduced obese individuals. Appetite 2003;40(2):137–43.

[29] MacLean PS, et al. Biology's response to dieting: the impetus for weight regain. Am J Physiol Regul Integr Comp Physiol 2011;301(3):R581–600.

[30] Hall KD. Did the food environment cause the obesity epidemic? Obesity 2018;26(1):11–3.

[31] Machado PP, et al. Ultra-processed food consumption and obesity in the Australian adult population. Nutr Diabetes 2020;10(1):1–11.

[32] Hall KD, et al. Ultra-processed diets cause excess calorie intake and weight gain: an inpatient randomized controlled trial of ad libitum food intake. Cell Metab 2019;30(1):67–77.e3.

[33] Kahn BB, Flier JS. Obesity and insulin resistance. J Clin Invest 2000;106(4):473–81.

[34] Malik VS, Schulze MB, Hu FB. Intake of sugar-sweetened beverages and weight gain: a systematic review. Am J Clin Nutr 2006;84(2):274–88.

[35] Vartanian LR, Schwartz MB, Brownell KD. Effects of soft drink consumption on nutrition and health: a systematic review and meta-analysis. Am J Public Health 2007;97(4):667–75.

[36] Olsen N, Heitmann B. Intake of calorically sweetened beverages and obesity. Obes Rev 2009;10(1):68–75.

[37] Hu FB. Resolved: there is sufficient scientific evidence that decreasing sugar-sweetened beverage consumption will reduce the prevalence of obesity and obesity-related diseases. Obes Rev 2013;14(8):606–19.

[38] Caprio S. Calories from soft drinks—do they matter. N Engl J Med 2012;367(15):1462–3.

[39] Basu S, et al. The relationship of sugar to population-level diabetes prevalence: an econometric analysis of repeated cross-sectional data. PLoS One 2013;8(2), e57873.

[40] Ludwig DS, Peterson KE, Gortmaker SL. Relation between consumption of sugar-sweetened drinks and childhood obesity: a prospective, observational analysis. Lancet 2001;357(9255):505–8.

[41] Dubois L, et al. Regular sugar-sweetened beverage consumption between meals increases risk of overweight among preschool-aged children. J Am Diet Assoc 2007;107(6):924–34.

[42] Forsythe CE, et al. Comparison of low fat and low carbohydrate diets on circulating fatty acid composition and markers of inflammation. Lipids 2008;43(1):65–77.

[43] Ludwig DS. The glycemic index: physiological mechanisms relating to obesity, diabetes, and cardiovascular disease. JAMA 2002;287 (18):2414–23.

[44] Hron BM, et al. Hepatic, adipocyte, enteric and pancreatic hormones: response to dietary macronutrient composition and relationship with metabolism. Nutr Metab 2017;14(1):1–7.

[45] Ebbeling CB, et al. Effects of dietary composition on energy expenditure during weight-loss maintenance. JAMA 2012;307(24):2627–34.

[46] Mansoor N, et al. Effects of low-carbohydrate diets v. low-fat diets on body weight and cardiovascular risk factors: a meta-analysis of randomised controlled trials. Br J Nutr 2016;115(3):466–79.

[47] Bueno NB, et al. Very-low-carbohydrate ketogenic diet v. low-fat diet for long-term weight loss: a meta-analysis of randomised controlled trials. Br J Nutr 2013;110(7):1178–87.

[48] Hjorth MF, et al. Personalized dietary management of overweight and obesity based on measures of insulin and glucose. Annu Rev Nutr 2018;38:245–72.

[49] Ludwig DS, Ebbeling CB. The carbohydrate-insulin model of obesity: beyond "calories in, calories out". JAMA Intern Med 2018;178(8):1098–103.

[50] Simopoulos AP. An increase in the omega-6/omega-3 fatty acid ratio increases the risk for obesity. Nutrients 2016;8(3):128.

[51] Halton TL, Hu FB. The effects of high protein diets on thermogenesis, satiety and weight loss: a critical review. J Am Coll Nutr 2004;23(5):373–85.

[52] Westerterp-Plantenga M, et al. Dietary protein, weight loss, and weight maintenance. Annu Rev Nutr 2009;29:21–41.

[53] Guthold R, et al. Worldwide trends in insufficient physical activity from 2001 to 2016: a pooled analysis of 358 population-based surveys with 1·9 million participants. Lancet Glob Health 2018;6(10):e1077–86.

[54] W.H.Organization. Global action plan on physical activity 2018–2030: more active people for a healthier world. World Health Organization; 2019.

[55] Church TS, et al. Trends over 5 decades in US occupation-related physical activity and their associations with obesity. PLoS One 2011;6(5), e19657.

[56] Ng SW, Popkin BM. Time use and physical activity: a shift away from movement across the globe. Obes Rev 2012;13(8):659–80.

[57] Thomas DM, et al. Time to correctly predict the amount of weight loss with dieting. J Acad Nutr Diet 2014;114(6):857–61.

[58] Tremblay MS, et al. Sedentary behavior research network (SBRN)–terminology consensus project process and outcome. Int J Behav Nutr Phys Act 2017;14(1):1–17.

[59] U.D.o.Health and H. Services. Physical activity guidelines advisory committee. 2018 physical activity guidelines advisory committee scientific report; 2018.

[60] WHO. WHO guidelines on physical activity and sedentary behaviour. Geneva: World Health Organization; 2020. Licence: CC BY-NC-SA 3.0 IGO.

[61] Torres SJ, Nowson CA. Relationship between stress, eating behavior, and obesity. Nutrition 2007;23(11 – 12):887–94.

[62] Adam TC, Epel ES. Stress, eating and the reward system. Physiol Behav 2007;91(4):449–58.

[63] Reutrakul S, Van Cauter E. Sleep influences on obesity, insulin resistance, and risk of type 2 diabetes. Metabolism 2018;84:56–66.

[64] Chaput J-P, et al. The association between sleep duration and weight gain in adults: a 6-year prospective study from the Quebec family study. Sleep 2008;31(4):517–23.

[65] Al Khatib H, et al. The effects of partial sleep deprivation on energy balance: a systematic review and meta-analysis. Eur J Clin Nutr 2017;71(5):614–24.

[66] Spiegel K, et al. Brief communication: sleep curtailment in healthy young men is associated with decreased leptin levels, elevated ghrelin levels, and increased hunger and appetite. Ann Intern Med 2004;141(11):846–50.

[67] Pampel FC, Denney JT, Krueger PM. Obesity, SES, and economic development: a test of the reversal hypothesis. Soc Sci Med 2012;74(7):1073–81.

[68] Pincock S. Boyd Swinburn: combating obesity at the community level. Lancet 2011;378(9793):761.

[69] Swinburn B, Egger G, Raza F. Dissecting obesogenic environments: the development and application of a framework for identifying and prioritizing environmental interventions for obesity. Prev Med 1999;29(6):563–70.

[70] Swinburn B, Egger G. Preventive strategies against weight gain and obesity. Obes Rev 2002;3(4):289–301.

[71] Cutler DM, Glaeser EL, Shapiro JM. Why have Americans become more obese? J Econ Perspect 2003;17(3):93–118.

[72] Hall KD, et al. The progressive increase of food waste in America and its environmental impact. PLoS One 2009;4(11), e7940.

[73] Scarborough P, et al. Increased energy intake entirely accounts for increase in body weight in women but not in men in the UK between 1986 and 2000. Br J Nutr 2011;105(9):1399–404.

[74] Swinburn B, Sacks G, Ravussin E. Increased food energy supply is more than sufficient to explain the US epidemic of obesity. Am J Clin Nutr 2009;90(6):1453–6.

[75] Egger G, Swinburn B. Planet obesity: How we're eating ourselves and the planet to death. ReadHowYouWant. com; 2011.

[76] Swinburn BA, et al. The global obesity pandemic: shaped by global drivers and local environments. Lancet 2011;378(9793):804–14.

[77] Hall KD, Heymsfield SB. Models use leptin and calculus to count calories. Cell Metab 2009;9(1):3–4.

[78] Müller MJ, Bosy-Westphal A, Heymsfield SB. Is there evidence for a set point that regulates human body weight? F1000 Med Rep 2010;2:59.

[79] Mishra AK, Dubey V, Ghosh AR. Obesity: an overview of possible role (s) of gut hormones, lipid sensing and gut microbiota. Metabolism 2016;65(1):48–65.

[80] Fried SK, et al. Regulation of leptin production in humans. J Nutr 2000;130(12):3127S–31S.

[81] Arora S. Role of neuropeptides in appetite regulation and obesity—a review. Neuropeptides 2006;40(6):375–401.

[82] Singh HJ. The unfolding tale of leptin. Malays J Med Sci 2001;8(2):1.

[83] Mark AL. Selective leptin resistance revisited. Am J Physiol Regul Integr Comp Physiol 2013;305(6):R566–81.

[84] Ahima RS. Revisiting leptin's role in obesity and weight loss. J Clin Invest 2008;118(7):2380–3.

[85] Keim NL, Stern JS, Havel PJ. Relation between circulating leptin concentrations and appetite during a prolonged, moderate energy deficit in women. Am J Clin Nutr 1998;68(4):794–801.

[86] Oswal A, Yeo G. Leptin and the control of body weight: a review of its diverse central targets, signaling mechanisms, and role in the pathogenesis of obesity. Obesity 2010;18(2):221.

[87] Lu X-Y, et al. Leptin: a potential novel antidepressant. Proc Natl Acad Sci 2006;103(5):1593–8.

[88] Schwartz MW, et al. Obesity pathogenesis: an endocrine society scientific statement. Endocr Rev 2017;38(4):267–96.

[89] Kimm SY, et al. Relation between the changes in physical activity and body-mass index during adolescence: a multicentre longitudinal study. Lancet 2005;366(9482):301–7.

[90] Zheng Y, Manson JE, Yuan C, et al. Associations of weight gain from early to middle adulthood with major health outcomes later in life. JAMA 2017;318(3):255–69. https://doi.org/10.1001/jama.2017.7092.

[91] Hu F. Obesity epidemiology. Oxford University Press; 2008.

[92] U.S. Department of Health and Human Services. 2008 physical activity guidelines for Americans. Washington, DC: U.S. Department of Health and Human Services; 2008.

[93] Jakicic JM, et al. Physical activity and the prevention of weight gain in adults: a systematic review. Med Sci Sports Exerc 2019;51(6):1262.

[94] Gebel K, Ding D, Bauman AE. Volume and intensity of physical activity in a large population-based cohort of middle-aged and older Australians: prospective relationships with weight gain, and physical function. Prev Med 2014;60:131–3.

[95] Gradidge PJ-L, et al. The role of lifestyle and psycho-social factors in predicting changes in body composition in black south African women. PLoS One 2015;10(7), e0132914.

[96] Smith KJ, Gall SL, McNaughton SA, et al. Lifestyle behaviours associated with 5-year weight gain in a prospective cohort of Australian adults aged 26–36 years at baseline. BMC Public Health 2017;17(1):54. https://doi.org/10.1186/s12889-016-3931-y.

[97] Su C, et al. Longitudinal association of leisure time physical activity and sedentary behaviors with body weight among Chinese adults from China health and nutrition survey 2004–2011. Eur J Clin Nutr 2017;71(3):383–8.

[98] Rosenberg L, et al. Physical activity and the incidence of obesity in young African-American women. Am J Prev Med 2013;45(3):262–8.

[99] Wishnofsky M. Caloric equivalents of gained or lost weight. Am J Clin Nutr 1958;6:542–6.

[100] Blundell JE, et al. Cross talk between physical activity and appetite control: does physical activity stimulate appetite? Proc Nutr Soc 2003;62(3):651–61.

[101] Martin CK, et al. Effect of different doses of supervised exercise on food intake, metabolism, and non-exercise physical activity: the E-MECHANIC randomized controlled trial. Am J Clin Nutr 2019;110(3):583–92.

[102] Bouchard C, et al. The response to exercise with constant energy intake in identical twins. Obes Res 1994;2(5):400–10.

[103] Church TS, et al. Changes in weight, waist circumference and compensatory responses with different doses of exercise among sedentary, overweight postmenopausal women. PLoS One 2009;4(2), e4515.

[104] Thomas D, et al. Why do individuals not lose more weight from an exercise intervention at a defined dose? An energy balance analysis. Obes Rev 2012;13(10):835–47.

[105] Shaw KA, Gennat H, O'Rourke P, Del Mar C. Exercise for overweight or obesity. Cochrane Database Syst Rev 2006;(4):CD003817. https://doi.org/10.1002/14651858.CD003817.pub3.

[106] Lee I-M, et al. Physical activity and weight gain prevention. JAMA 2010;303(12):1173–9.

[107] Church TS, et al. Effects of different doses of physical activity on cardiorespiratory fitness among sedentary, overweight or obese postmenopausal women with elevated blood pressure: a randomized controlled trial. JAMA 2007;297(19):2081–91.

[108] Church TS, et al. Exercise without weight loss does not reduce C-reactive protein: the INFLAME study. Med Sci Sports Exerc 2010;42(4):708.

[109] Kraemer WJ, Ratamess NA, French DN. Resistance training for health and performance. Curr Sports Med Rep 2002;1(3):165–71.

[110] Willis LH, et al. Effects of aerobic and/or resistance training on body mass and fat mass in overweight or obese adults. J Appl Physiol 2012;113:1831–7.

[111] Church TS, et al. Effects of aerobic and resistance training on hemoglobin A1c levels in patients with type 2 diabetes: a randomized controlled trial. JAMA 2010;304(20):2253–62.

[112] Sigal RJ, et al. Effects of aerobic training, resistance training, or both on glycemic control in type 2 diabetes: a randomized trial. Ann Intern Med 2007;147(6):357–69.

[113] Kelley GA, Kelley KS, Pate RR. Exercise and adiposity in overweight and obese children and adolescents: a systematic review with network meta-analysis of randomised trials. BMJ Open 2019;9(11), e031220.

[114] Milanović Z, Sporiš G, Weston M. Effectiveness of high-intensity interval training (HIT) and continuous endurance training for VO 2max improvements: a systematic review and meta-analysis of controlled trials. Sports Med 2015;45(10):1469–81.

[115] Ramos JS, et al. The impact of high-intensity interval training versus moderate-intensity continuous training on vascular function: a systematic review and meta-analysis. Sports Med 2015;45(5):679–92.

[116] Türk Y, et al. High intensity training in obesity: a meta-analysis. Obes Sci Pract 2017;3(3):258–71.

[117] Andreato L, et al. The influence of high-intensity interval training on anthropometric variables of adults with overweight or obesity: a systematic review and network meta-analysis. Obes Rev 2019;20(1):142–55.

[118] Kerns JC, et al. Increased physical activity associated with less weight regain six years after "the biggest loser" competition. Obesity 2017;25(11):1838–43.

[119] Andersen RE, et al. Effects of lifestyle activity vs structured aerobic exercise in obese women: a randomized trial. JAMA 1999;281(4):335–40.

[120] Swift DL, et al. The role of exercise and physical activity in weight loss and maintenance. Prog Cardiovasc Dis 2014;56(4):441–7.

[121] Schwingshackl L, Dias S, Hoffmann G. Impact of long-term lifestyle programmes on weight loss and cardiovascular risk factors in overweight/obese participants: a systematic review and network meta-analysis. Syst Rev 2014;3(1):1–13.

[122] Ortega FB, et al. The fat but fit paradox: what we know and don't know about it. BMJ Publishing Group Ltd and British Association of Sport and Exercise Medicine; 2018.

[123] Valenzuela PL, et al. Joint association of physical activity and body mass index with cardiovascular risk: a nationwide population-based cross-sectional study. Eur J Prev Cardiol 2021. https://doi.org/10.1093/eurjpc/zwaa151, zwaa151.

[124] Dinas PC, Markati AS, Carrillo AE. Exercise-induced biological and psychological changes in overweight and obese individuals: a review of recent evidence. Int Sch Res Notices 2014;2014. https://doi.org/10.1155/2014/964627, 964627.

[125] Butryn ML, et al. Consistent self-monitoring of weight: a key component of successful weight loss maintenance. Obesity 2007;15(12):3091–6.

[126] Klem ML, et al. A descriptive study of individuals successful at long-term maintenance of substantial weight loss. Am J Clin Nutr 1997;66(2):239–46.

[127] Catenacci VA, et al. Physical activity patterns in the national weight control registry. Obesity 2008;16(1):153–61.

Chapter 9

Exercise effects in adults with asthma

Rebecca F. McLoughlin[a,b,c], Paola D. Urroz[a,b,c], Celso R.F. Carvalho[d], and Vanessa M. McDonald[a,b,c,e]

[a]National Health and Medical Research Council, Centre of Excellence in Treatable Traits, Newcastle, NSW, Australia [b]Hunter Medical Research Institute, Newcastle, NSW, Australia [c]School of Nursing and Midwifery, University of Newcastle, Newcastle, NSW, Australia [d]Department of Physical Therapy, School of Medicine, University of São Paulo, São Paulo, Brazil [e]Department of Respiratory and Sleep Medicine, John Hunter Hospital, Newcastle, NSW, Australia

Introduction

Background

Asthma is a chronic, noncommunicable disease that affects the airways and can develop at any stage in life. The airways become narrowed due to inflammation causing limited airflow. Asthma is diagnosed with lung function testing which includes the assessment of bronchodilator reversibility of airflow limitation and, airway hyperresponsiveness to provocation testing. Risk factors for asthma include genetic predisposition and environmental exposure such as allergens and air pollution.

People with asthma often report symptoms such as shortness of breath, wheezing, coughing, and chest tightness. These symptoms may present persistently and/or as an acute attack (also called a flare-up or an exacerbation). However not all asthma is the same. The severity, frequency, duration, and type of reported symptoms vary, making asthma a complex heterogeneous disease. To add to its complexity, co-morbidities often co-exist which ultimately affect asthma control and management, this is particularly concerning in more severe diseases [1]. High prevalence co-morbidities include obesity, chronic rhinosinusitis, obstructive sleep apnea, chronic obstructive pulmonary disease (COPD), anxiety, depression, allergic rhinitis, and vocal cord dysfunction [1,2]. Asthma is usually a controllable disease through addressing treatable factors with pharmacotherapy and nonpharmacotherapy interventions [3,4]. The ability to control and manage asthma is important in evaluating the severity of asthma. People with asthma who remain uncontrolled despite treatable factors having been addressed are considered to have severe asthma [5].

Epidemiology/impact

It is estimated that asthma affects over 339 million people worldwide, with an additional 100 million people projected to be diagnosed with asthma by 2025 [6,7]. The impact of asthma on mortality and morbidity is significant. In 2016, an estimated 417,918 deaths and 24.8 million disability-adjusted life years (DALY) were attributed to asthma globally [8,9]. From these statistics and given the nature of the disease, it is not surprising that the economic burden of asthma is considerable in terms of both direct medical costs (i.e., hospitalization, pharmaceutical treatment, and diagnostic tests) and indirect costs (i.e., reduced productivity and loss of income) [10]. The costs associated with asthma vary considerably from country to country [11]. In Australia alone, asthma management is estimated to cost >$770 million per year, 50% of which is attributed to prescription pharmaceuticals which are a key step in asthma management [12,13]. Patients with severe asthma, despite comprising only 3%–10% of the total asthma population [5,14], account for a substantial proportion of asthma-related morbidity, mortality [15], and healthcare costs [15,16]. In fact, over 50% of asthma-related costs are attributable to severe disease [17].

Over the last 30 years, there have been significant advances in asthma management both in terms of pharmacotherapy and self-management initiatives, resulting in improved outcomes for patients including reduced asthma mortality. However, these gains are being lost, as more recently asthma deaths have started to rise in countries such as Australia, the UK, and the USA [18]. In fact, at least one person in Australia dies every day from asthma [19]. These data are alarming

Exercise to Prevent and Manage Chronic Disease Across the Lifespan. https://doi.org/10.1016/B978-0-323-89843-0.00026-X

because asthma is a treatable disease. In response to this, there is increasing interest in nonpharmacological approaches to complement existing asthma management strategies. This includes exercise interventions to increase physical activity levels, which is the focus of this chapter.

Pathophysiology

The pathophysiology which gives rise to the symptoms of asthma involves a complex interaction between inflammation of the airways, bronchial/airway hyperresponsiveness, and airflow limitation/airway narrowing.

Airway inflammation

Inflammation of the airways in response to harmless stimuli is central to the pathophysiology of asthma [20]. This inflammation predominantly occurs in the conducting airways (bronchi and bronchioles), however, as the disease progresses, the inflammation spreads both distally to the smaller airways and proximally to the trachea [20]. Airway inflammation in asthma may be present even in the absence of symptoms, with the inflammatory response involving a complex interplay between various inflammatory cells, mediators, and the respiratory epithelium [20]. Asthma is heterogeneous and can be classified into four inflammatory phenotypes based on the predominant cell type present in the airways: neutrophilic, eosinophilic, mixed granulocytic (both increased neutrophils and eosinophils), and paucigranulocytic (normal levels of eosinophils and neutrophils) [10]. Classification is achieved using noninvasive techniques to determine the presence and/or absence of eosinophils and neutrophils in a sample of sputum [10]. Type 2 airway inflammation (immune responses in the airway involving the accumulation of eosinophils, mast cells, basophils, Th2 cells, Type 2 innate lymphoid cells (ILC2s) and IgE-producing B cells, and Type 2 mediators and cytokines (IL-4, IL-5, and IL-13) [21]) can also be assessed indirectly by measuring the concentration of nitric oxide in exhaled breath, also referred to as fractional exhaled nitric oxide (FeNO) [10].

Airway hyperresponsiveness

Airway hyperresponsiveness (AHR) is one of the hallmark characteristics of asthma [22]. It is defined as an exaggerated response to stimuli such as smoke, allergens (i.e., dust), or cold air, that would produce little to no effect in individuals without asthma [22]. This results in the narrowing and constriction of the airway smooth muscle, and subsequently airflow limitation and symptoms such as difficulty in breathing [10]. AHR can be highly variable, both among patients with asthma and within individuals themselves, with the degree of AHR generally an indication of disease severity [22]. While the mechanisms underlying AHR are only partially understood, there is increasing evidence that airway inflammation and airway remodeling (also referred to as structural changes) play important roles in its manifestation [22]. AHR can be divided into two types; chronic/persistent AHR which is suggested to be the result of airway remodeling due to the effects of chronic recurrent airway inflammation, and acute/variable AHR which is superimposed on this and reflects the acute effects of airway inflammation associated with current exposures (i.e., respiratory infections and allergens) [22]. AHR can be measured using either direct airway challenges (e.g., inhaled methacholine or histamine) or indirect airway challenges (e.g., an exercise challenge, inhaled hypertonic saline, or inhaled mannitol) [22]. Measurement of AHR can be used in the diagnosis of asthma as well as in the assessment of treatment effectiveness [10].

Airflow limitation/airway narrowing

Airway narrowing also plays an important role in the pathophysiology of asthma [10,23]. There are several factors that contribute to the development of airway narrowing and subsequently airflow limitation in asthma [10]. Contraction of airway smooth muscle is the predominant mediator of airway narrowing [10,23]. This occurs in response to numerous neurotransmitters and bronchoconstriction mediators, and as such can often be reversed by bronchodilators such as salbutamol [10]. Airway mucus hypersecretion, another pathophysiological feature of asthma, can also cause narrowing of the airways, largely due to the formation of mucus plugs [10]. Furthermore, inflammatory mediators in the airways can cause increased microvascular leakage resulting in airway edema and subsequently airway narrowing [10]. Airway thickening which results from airway remodeling is also suggested to contribute [10].

Asthma triggers

There are a number of factors that can trigger or aggravate asthma symptoms. These include stress, anxiety, fatigue, viral infections (e.g., influenza or the common cold), bacterial infections (e.g., Chlamydia pneumonia), allergens (e.g., house dust mites, pollen, and animal fur), irritants (e.g., strong smells and smoke), drugs (e.g., aspirin), occupational exposures (e.g., chemicals), air pollutants, changes in weather conditions, and dietary factors (e.g., food additives containing sulfites). Exercise can also provoke asthma-related symptoms (such as wheezing and breathlessness) [24]. In fact, exercise-induced bronchoconstriction (EIB), which is a temporary narrowing of the airways that occurs with exercise, is estimated to occur in approximately 90% of people with asthma [25]. Despite exercise being a known trigger for asthma, the benefits of exercise in this disease are well recognized. Indeed, people with asthma are recommended to engage in regular physical activity, with strategies necessary to prevent EIB [26]. These strategies include warming up prior to exercise and cooling-down postexercise, and pretreatment with a bronchodilator [26–28].

Exercise in disease

Physical inactivity in asthma

Despite the recommendation that people with asthma should engage in regular physical activity, low levels of physical activity are still reported in this population [29]. Individuals with asthma are found to spend significantly less time walking and undertaking vigorous physical activity and accumulate fewer steps/day than those without asthma [30]. In one study, individuals with asthma were reported to spend on average 60 min less per week engaged in moderate physical activity, and 67 min less per week engaged in vigorous physical activity compared to those without asthma [31]. Physical inactivity is particularly evident in the severe asthma population [32]. Individuals with severe asthma have been reported to accumulate between 5300 and 5800 steps/day [32], which is considerably less than that estimated for the asthma population in general (average of 8390 steps/day) [29]. This is consistent with a study that reported that individuals with severe asthma accumulated 21% less steps per day, and spent 17% less time undertaking moderate to vigorous physical activity (MVPA) compared with participants with less severe disease [33]. While compared with individuals without asthma, individuals with asthma have been reported to accumulate 31.4% less steps per day, and spend 47.5% less time engaging in MVPA [32].

This is important, as in addition to the negative health consequences of physical inactivity experienced by the general population, people with asthma also experience poorer respiratory functioning, increased disease severity and healthcare use, decreased physical and mental health, and decreased quality of life [34]. Conversely, higher levels of physical activity are associated with better measures of lung function, exercise capacity, asthma control, and decreased systemic inflammation [29,32,35]. A study by Dogra and colleagues found that individuals with asthma who were active (energy expenditure >3.0 kcal/kg body weight (BW) per day) reported greater perceived mental and overall health, fever mental conditions and activity limitations, and greater satisfaction compared with individuals with asthma who were moderately active (1.5 kcal/kg BW per day) or inactive (<1.5 kcal/kg BW per day) [36].

There is increasing evidence that exercise training, alongside other nonpharmacological interventions, e.g., pulmonary rehabilitation programs, has beneficial effects on exercise capacity in people with chronic lung conditions [37]. Pulmonary rehabilitation is widely recognized as a core component of the management of people with chronic lung conditions, namely COPD, bronchiectasis, interstitial lung disease, and pulmonary hypertension [37]. This multidisciplinary approach to disease management is defined as "a comprehensive intervention based on a thorough patient assessment followed by patient-tailored therapies that include, but are not limited to, exercise training, education, and behavior change, designed to improve the physical and psychological condition of people with chronic respiratory disease and to promote the long-term adherence to health-enhancing behaviors [37]." Although pulmonary rehabilitation is not routinely integrated into the management of people with asthma, there is increasing evidence that it leads to improvements in symptoms, asthma control, exercise tolerance, and quality of life in this population [38–44]. Many of the observed benefits of pulmonary rehabilitation in people with asthma have been attributed to the exercise training component of the intervention [45]. This is supported by the evidence in the literature on the effectiveness of exercise training in improving clinical outcomes in this population.

Clinical outcomes

The majority of exercise intervention studies in people with asthma conducted to date have investigated the benefits of at least 30 min of aerobic exercise (i.e., walking, swimming, cycling, calisthenics, or a combination of these), of moderate to vigorous intensity, two to three times a week for an average duration of 12 weeks (ranging from 6 to 52 weeks). A number of these are summarized in Table 1. Several physiological and psychological benefits of aerobic exercise have been reported in the literature including improved asthma control, fewer asthma symptoms and exacerbations, reduced medication and healthcare use including reduced emergency department visits, improved health-related quality of life (HRQoL), and decreased symptoms of anxiety and depression. The effect of exercise on lung function however remains unclear.

TABLE 1 Summary of clinical trials/evidence investigating the effect of exercise in asthma.

Author, year	Participants (n)[a]	Type of training	Frequency	Time	Duration	Intensity	Findings (compared to control)
Dogra, 2010 [46]	Adults with partially controlled asthma (severity NS) (n=24)	Multiple aerobic exercises (unsupervised)	5 × week	30 min	12 weeks	70%–85% HRmax	↑ Perceived asthma control, ↓ self-reported asthma symptom frequency, and severity. ↔ objective measures compared to control (ACQ score, mini-AQLQ, lung function, VO_{2max})
Dogra, 2011 [47]	Adults with partially controlled asthma (severity NS) (n=36)	Multiple aerobic exercises	3 × week	NS	12 weeks supervised, followed by 12 weeks unsupervised	70%–85% HRmax	↑ Asthma control (ACQ), ↑ perceived asthma control, ↑ mini-AQLQ, ↑ VO_{2max}, ↔ lung function parameters (FEV_1/FVC or FEV_1%pred)
Toennesen 2018 [48]	Adults with asthma (severity NS) (n=125)	High-intensity interval training + diet changes	3 × week	16–30 min	8 weeks	Up to 90% HRmax	↑ Asthma control (ACQ), ↑ AQLQ, ↑ VO_{2max}, ↔ airway (sputum cell counts, FeNO) or systemic (blood eosinophils, hs-CRP, IL-6) inflammation, ↔ lung function (FEV_1%pred, FVC%pred), ↔ AHR
Meyer, 2015 [49]	Adults with asthma (various asthma severities) (n=21)	Multiple aerobic exercises	1 × week	60 min	52 weeks	>60% HRmax	↑ SF-36, ↑ AQLQ, ↑ Cardiorespiratory fitness (↑ VO_{2max}, ↑ WR_{max})
Scott, 2013 [50]	Adults with asthma (various asthma severities) (n=38)	Aerobic and resistance training + hypocaloric diet	≥3 × week	NS	10 weeks	NS	↑ Asthma control (ACQ), ↑ AQLQ, ↑ total lung capacity, ↓ systemic inflammation (leptin, IL-6)
Emtner, 1996 [51]	Adults with mild–moderate asthma (n=22)	Indoor swimming	2 × week	45 min	10 weeks	80%–90% HRmax	↑ Lung function (FEV_1, $FEF_{25\%-75\%}$), ↓ asthma symptoms, ↑ cardiovascular functioning (↓HR)
Basaran, 2006 [52]	Children with mild–moderate asthma (n=58)	Basketball training	3 × week	60 min	8 weeks	Moderate (HRmax NS)	↑ PAQLQ, ↓ symptom score, ↑ lung function (PEF% only)
Jaakkola, 2019 [53]	Adults with mild–moderate asthma (n=105)	Individualized aerobic exercise program	3 × week	≥30 min	24 weeks	70%–80% HRmax	↑ Asthma control (ACT), ↓ shortness of breath, ↔ airflow obstruction (PEF)
Goncalves, 2008 [54]	Adults with moderate–severe asthma (n=23)	Treadmill	2 × week	30 min	12 weeks	70% HRmax	↑ HRQoL, ↑ symptom-free days, ↓ anxiety and depression scores, ↓ airway inflammation (FeNO), ↑ aerobic fitness

Study	Population	Exercise type	Frequency	Duration	Duration (weeks)	Intensity	Outcomes
Mendes, 2011 [55]	Adults with moderate–severe asthma ($n=51$)	Treadmill	2 × week	30 min	12 weeks	60%–70% HRmax	↓ ED visits, ↑ symptom-free days, ↓ airway inflammation (total and eosinophil counts, FeNO), ↑ VO$_{2max}$, ↔ lung function (FEV$_1$, FEV$_1$%pred, FVC, FVC%pred, FEV$_1$/FVC or FEF$_{25\%-75\%}$)
Mendes, 2010 [56]	Adults with moderate–severe asthma ($n=101$)	Unclear	2 × week	30 min	12 weeks	NS	↑ HRQoL, ↑symptom-free days, ↓ anxiety and depression
Franca-Pinto, 2015 [57]	Adults with moderate–severe asthma ($n=43$)	Treadmill	2 × week	35 min	12 weeks	Vigorous (HRmax NS)	↑ Symptom-free days, ↑ AQLQ, ↓ frequency of exacerbations, ↓ BHR, ↓ systemic inflammation (IL-6, MCP-1), ↔ asthma control
Freitas, 2017 [58]	Adults with moderate–severe asthma ($n=51$)	Aerobic and resistance training + hypocaloric diet	2 × week	NS	12 weeks	70%–85% HRmax	↑ Asthma control (ACQ), ↑ AQLQ, ↑ VO$_{2max}$, ↑ lung function (FEV$_1$ FVC, ERV), ↓ airway (FeNO) and systemic (CCL2, IL-4, IL-6, TNF-α, and leptin) inflammation, ↑ antiinflammatory biomarkers (IL-10, adiponectin)
Turner, 2011 [59]	Adults with moderate–severe asthma ($n=34$)	Multiple aerobic exercises	3 × week	80–90 min	6 weeks	Unclear	↑ AQLQ, ↓ asthma symptoms
Evaristo, 2020 [60]	Adults with moderate–severe asthma ($n=54$)	Treadmill	2 × week	40 min	12 weeks	60% HRmax	Compared to control, greater proportion of participants had: ↑ asthma control (ACQ), ↓ use of rescue medication

[a]Total number of participants who completed the intervention.
[b]clinically significant improvement; ↑, significant increase; ↓, significant decrease; ↔, no significant change; ACQ, asthma control questionnaire; ACT, asthma control test; BHR, bronchial hyperresponsiveness; ERV, expiratory reserve volume; FEF$_{25\%-75\%}$, forced expiratory flow between 25% and 75% of vital capacity; FeNO, fraction of exhaled nitric oxide; FEV$_1$%pred, forced expiratory volume in the first second percentage of predicted; FEV$_1$, forced expiratory volume in 1 s; FEV$_1$/FVC, ratio of the forced expiratory volume in 1 s to the forced vital capacity; FVC%pred, forced vital capacity percentage of predicted; FVC, forced vital capacity; HRQoL, health-related quality of life; NS, not specified; PEF, peak expiratory flow; SF-36, short-form 26 (general health-related quality of life questionnaire), AQLQ, Asthma Quality of Life Questionnaire; VO$_{2max}$, maximum oxygen uptake; WR$_{max}$, maximum work rate.

Improved asthma control

The main goal of asthma treatment is to achieve good asthma control [26]. Asthma control encompasses both the patients' limitations and symptoms (i.e., activity limitations, daytime and nocturnal symptoms, and use of rescue medication) and the future risk of adverse events (such as exacerbations) [26]. Several validated tools are used to assess asthma control in both the clinical and research settings. Commonly used tools are the asthma control questionnaire (ACQ) [61] and the asthma control test (ACT) [62,63]. Higher scores when using the ACQ indicate worse asthma control, with a change of ≥0.5 considered clinically important [64]. Whereas higher ACT scores indicate better asthma control, with a change of 3 points considered clinically important [63].

Exercise training has been shown to improve asthma control regardless of disease severity (mild, moderate, or severe) (Table 1) [34,46,47,50,53,58,65–67]. The exercise programs used in the majority of these studies have been supervised, with few studies examining the beneficial effect of unstructured, self-administered exercise, which may be more cost-effective and readily available to the general asthma population. In one study, while perceived asthma control was reported to significantly improve following a 12-week unsupervised aerobic exercise intervention, no significant change in measured asthma control (ACQ score) was observed [46]. In contrast, a 12-week supervised aerobic exercise intervention conducted by the same research group resulted in improvements in both perceived and measured asthma control (ACQ score), with these improvements maintained over an additional 12 weeks of self-administered, unsupervised exercise [47]. These findings suggest that supervision may be required at least initially to achieve significant clinical improvements in measured asthma control. Nonetheless, more research examining the benefits of unsupervised exercise is needed.

While supervised exercise interventions have a clear potential to improve asthma outcomes, personal preferences and barriers to exercise training may reduce patients' participation. A range of evidence-based interventions is required to address the unmet need to improve physical activity [68]. A behavior change intervention aiming to improve physical activity in adults with asthma who were physically inactive was tested [35]. The 8-week intervention comprised a weekly 40-min face-to-face counseling session, supported by a workbook and an activity monitor. The primary outcome was asthma control evaluated by the ACQ. The intervention largely improved physical activity (daily steps, time spent in moderate to vigorous physical activity, sedentary behavior) and improved asthma control [35]. These findings highlight how increased participation in daily life physical activity can lead to specific clinical benefits beyond general health.

Interestingly, evidence from several studies suggests that combined exercise and diet interventions may be more effective at improving asthma control than exercise alone [48,50]. In a multiarm study Toennesen and colleagues compared the effects of exercise (high-intensity interval training), diet (high protein/low glycaemic), and combined exercise and diet for 8 weeks in adults with asthma [48]. In addition to being high in protein with a low glycaemic index, the dietary intervention was also designed to be high in fruit, vegetables, and fish. Although improvements in asthma control from pre to postintervention were reported in all three intervention groups, clinically significant improvements were only observed in the combined exercise and diet group when compared to control. [48] Similarly, in another study asthma control improved following a combined exercise and diet intervention (10 weeks of aerobic and resistance training + hypocaloric diet) with no change observed in the exercise-only group [50]. Interestingly, participants in the combined diet and exercise group consumed significantly less total and saturated dietary fat than the exercise-only group [50]. There is increasing evidence that certain dietary components can influence clinical outcomes in asthma, predominantly due to their effect on inflammation [69]. Antioxidant-rich foods such as fruits and vegetables have been suggested to lower airway inflammation and subsequently improve asthma control [69]. Conversely, dietary components such as saturated fat have been shown to increase airway inflammation [69]. This may explain the synergistic effect of diet and exercise on asthma control observed in these studies.

Fewer asthma symptoms and exacerbations

Exercise training has also been shown to reduce the frequency and severity of asthma symptoms (i.e., shortness of breath) and increase the number of symptom-free days [46,51–55]. In one study conducted in patients with mild to moderate asthma, a 10-week supervised rehabilitation program (consisting of 2 weeks of education and daily indoor swimming exercises followed by 8 weeks of indoor swimming exercise for 45 min twice a week) resulted in fewer asthma symptoms [51]. Similarly, in patients with partially controlled asthma Dogra and colleagues reported a decrease in self-reported asthma symptom frequency and severity following a 12-week unsupervised aerobic exercise program (consisting of 30 min of aerobic exercise, strength training, and stretching). [46] While in patients with moderate to severe disease, aerobic exercise (≥30-min sessions, twice a week for 12 weeks) has been reported to result in more asthma symptom-free days [54,55,57]. Regular exercise has also been reported to reduce the risk of exacerbations [57], decrease the frequency and severity of EIB [70], and reduce both medication [60] and healthcare use [55].

Improved health-related quality of life

An important component of asthma management is the identification and treatment of impaired HRQoL, which is often assessed using questionnaires such as the asthma quality of life questionnaire (AQLQ) [26], among others. It is therefore not surprising that quality of life is one of the most commonly examined outcomes in this area of research, with the majority of studies reporting positive effects [47,49,50,54–56,71,72]. In 2020, Wu and colleagues published a systematic review that identified 10 RCTs examining the effect of regular continuous aerobic exercise (moderate-intensity, at least 20 min twice a week, with a minimum duration of 4 weeks) on HRQoL [72]. In this review, a meta-analysis of pooled data indicated that aerobic exercise significantly improved HRQoL in both adults and children with predominantly moderate to severe asthma [72]. An improvement in HRQoL was also reported in the three studies which could not be meta-analyzed [54–56], which is consistent with other research in this area [47,49,50,55,71]. While many people with asthma have some degree of impaired quality of life, compromised HRQoL has been suggested to be one of the hallmarks of severe disease [38]. Therefore, the evidence regarding the effectiveness of exercise training in improving HRQoL in the severe asthma population is particularly important.

Reduced anxiety and depression

Anxiety and depression are prevalent in the asthma population [73], particularly in those with severe asthma [74]. These co-morbidities are associated with increased symptom burden, poorer asthma control, greater frequency of exacerbations, increased healthcare utilization, poorer quality of life [74,75], and poorer treatment adherence [75]. Meta-analyses have shown that in the general population exercise training has beneficial effects on symptoms of anxiety and depression [76], with similar findings reported in the asthma population specifically [54,56,59].

Effect on lung function is inconclusive

There is inconsistent evidence in the literature regarding the benefits of exercise training on lung function in people with asthma. Findings from a number of studies indicate that exercise training has beneficial effects on lung function in this population [50–52,58,72,77–79], with two recent meta-analyses reporting significant improvements in several lung function parameters (i.e., FEV_1, PEF, FVC, FVC%pred, $FEF_{25\%–75\%}$) [72,79]. Conversely, other studies have reported no significant change in lung function following exercise, adverse or otherwise, despite reporting improvements in other clinical outcomes [46–48,55–57,59,80,81]. This suggests that the benefits of regular exercise in people with asthma are not related to the effects on lung function.

Known/potential mechanisms underlying effects

There is a paucity of evidence regarding the mechanisms responsible for the beneficial effects of exercise training in asthma. However, several potential mechanisms have been proposed.

Exercise training, in particular aerobic exercise, improves cardiorespiratory fitness which is the ability of the circulatory and respiratory systems to supply oxygen to skeletal muscles during sustained physical activity. People with asthma have been found to have lower cardiorespiratory fitness than those without asthma, which is largely due to their lower levels of physical activity [82,83]. Poor cardiorespiratory fitness has been associated with worse asthma control, whereas higher levels of cardiorespiratory fitness are associated with reduced asthma severity and increased quality of life [84]. Given this, it has been suggested that the improvements in clinical outcomes observed in people with asthma following regular exercise are related to improvements in cardiorespiratory fitness. It has been demonstrated that cardiorespiratory fitness in people with asthma can be improved with aerobic exercise training [58,85,86], indicated by an increase in maximal oxygen consumption and ventilatory threshold [87]. In one study, improved cardiorespiratory fitness following a 12-week aerobic exercise intervention was associated with improvements in asthma control [58]. Heikkinen and colleagues propose that improved oxygen uptake capacity together with an increased threshold for becoming breathless may help those with asthma to better cope with everyday life with a lower level of effort [85]. Furthermore, an increase in the ventilatory threshold has been proposed to reduce the likelihood of provoking EIB during mild to moderate exercise [82]. Indeed, regular exercise has been reported to decrease the frequency and severity of exercise-induced bronchoconstriction [70].

It has also been hypothesized that the anti-inflammatory effects of exercise training contribute to the improvements in clinical outcomes observed in people with asthma. As discussed earlier in this chapter, airway inflammation is a central feature of asthma. Regular exercise is suggested to reduce both systemic and airway inflammation, and as such a number of studies have investigated the effects of exercise training on markers of inflammation in asthma [50,55,58,88]. In adults with asthma, 12 weeks of aerobic exercise training was found to significantly reduce Type 2 airway inflammatory markers including sputum eosinophils and exhaled nitric oxide levels [55]. While Scott and colleagues also reported a significant decrease in sputum eosinophils following a 10-week aerobic exercise intervention [50], a more recent study demonstrated that regular aerobic exercise training (twice a week for 12 weeks) not only decreased biomarkers of type 2 airway (exhaled nitric oxide) and systemic inflammation (interleukin (IL)-4, IL-6, Tumor necrosis factor-alpha (TNF-α), leptin and monocyte chemoattractant-1 (MCP-1)), but also increased anti-inflammatory biomarkers (IL-10 and adiponectin) [58]. Although these studies, along with others [89], provide evidence for the anti-inflammatory benefits of exercise training in people with asthma, more research is needed in this area.

There is also evidence that exercise training decreases AHR [57,71]; another defining feature of asthma. As discussed earlier, there are two recognized components to AHR; persistent AHR which is suggested to be the result of airway remodeling due to the effects of chronic recurrent airway inflammation, and acute/variable AHR which reflects the acute effects of airway inflammation associated with current exposures [90]. In an experimental model of asthma, aerobic exercise training was shown to reverse airway inflammation and remodeling and improve respiratory mechanics [91] which may subsequently reduce AHR. While mechanistic studies in humans are lacking, in one study aerobic exercise training was found to reduce AHR and inflammatory markers (IL-6 and MCP-1) while improving quality of life and decreasing exacerbations in adults with asthma [57]. Although this study was not able to establish a direct association among the decrease in AHR, IL-6, and MCP-1, [57] there is evidence in the literature supporting the importance of these pro-inflammatory cytokines in AHR and airway inflammation in asthma [92]. Furthermore, in another study improvements in AHR following exercise training were reported to significantly contribute to improvements in quality of life [71]. However, a decrease in AHR with exercise training has not been consistently reported in the literature, with further research needed to explore this mechanism [48,89,93].

Cautions/safety

Asthma is a disease that can be mostly managed and controlled and therefore should not preclude people with asthma from engaging in physical activity including exercise training [27]. However, exercise can be a trigger for inducing asthma symptoms (i.e., EIB) resulting in safety concerns and reducing the intensity of exercise that can be achieved. Safety considerations should be taken into account when prescribing and performing exercise in this population (Fig. 1). Further, it should be noted that people with more severe diseases may not achieve satisfactory asthma control and may require additional considerations.

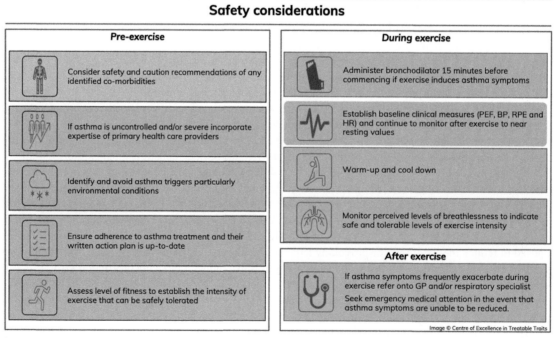

Safety considerations

Pre-exercise
- Consider safety and caution recommendations of any identified co-morbidities
- If asthma is uncontrolled and/or severe incorporate expertise of primary health care providers
- Identify and avoid asthma triggers particularly environmental conditions
- Ensure adherence to asthma treatment and their written action plan is up-to-date
- Assess level of fitness to establish the intensity of exercise that can be safely tolerated

During exercise
- Administer bronchodilator 15 minutes before commencing if exercise induces asthma symptoms
- Establish baseline clinical measures (PEF, BP, RPE and HR) and continue to monitor after exercise to near resting values
- Warm-up and cool down
- Monitor perceived levels of breathlessness to indicate safe and tolerable levels of exercise intensity

After exercise
- If asthma symptoms frequently exacerbate during exercise refer onto GP and/or respiratory specialist
- Seek emergency medical attention in the event that asthma symptoms are unable to be reduced.

Image © Centre of Excellence in Treatable Traits

FIG. 1 Summary of safety considerations for prescribing and performing exercise in adults with asthma.

Pre-exercise

Given the heterogeneous nature of asthma, a pre-exercise screening assessment should be conducted to enable tailoring of the exercise program to the individual. The Adult Pre-exercise Screening System (APSS), developed by Exercise and Sport Science Australia, Fitness Australia, and Sports Medicine Australia, is a useful screening tool in this setting. This screening system recommends that anyone who has had an asthma attack requiring immediate medical attention at any time over the previous 12 months requires guidance from an appropriate allied health professional or medical practitioner [94]. The following assessments should be considered prior to the implementation of an exercise program.

o   Presence of co-morbidities—Identify existing co-morbidities that are known to be of high prevalence in this population. Tailor the exercise program according to the safety and cautions recommended of each co-morbidity, if applicable.

o   Asthma control—Asthma is defined as uncontrolled if there are poor symptom control (Asthma control questionnaire/ Asthma control test), frequent severe attacks (exacerbations/flare-ups, hospitalization, ICU stay or mechanical ventilation), or persistent airflow limitation (reduced FEV1) [95]. Assessment of asthma control should be conducted and if asthma is uncontrolled this should be discussed with the primary care provider.

o   Severity of asthma—This is determined by assessing the level of treatment required to keep asthma controlled. If asthma is categorized as severe, the level and type of activity should be discussed with the individual's respiratory specialist and medical team [95].

o   Asthma triggers—Identify triggers that may irritate the airways causing an increase in asthma symptoms during exercise. Specifically, environmental conditions such as cold, dry air, air pollution, and pollen are common exposures during exercise. To prevent an asthma attack during exercise, it is important to avoid the identified triggers.

o   Medication—Many people with asthma are prescribed medication to manage their asthma and these medications should be identified along with their adherence to these treatments. Adherence to asthma treatment is important for managing and controlling asthma which will assist in safely maintaining an exercise program. It is important for the individual to have an up-to-date written action plan, which details appropriate short-term changes in the event of increased asthma symptoms. For those involved in competitive sport, check which medicines are permitted according to the Australian Sports Anti-Doping Authority or the World Anti-Doping Agency [27].

o   Fitness level—A baseline fitness assessment including a cardiopulmonary exercise test should be considered to establish the intensity of exercise that can be tolerated by the individual without inducing asthma symptoms [96].

During exercise

A number of measures can be implemented during an exercise session to reduce the risk of EIB and to maximize the level of exercise intensity achievable.

o   For individuals with EIB, it is recommended that a bronchodilator is administered 15 min before commencing exercise. The type of pretreatment is dependent on the individual's regular treatment regimen and may include SABA (for example salbutamol) or inhaled corticosteroid (ICS) in combination with a fast onset long-acting beta-agonist (for example budesonide—formoterol) [26–28].

o   If exercise is being prescribed in the clinical setting establish baseline clinical measures such as peak expiratory flow (PEF), blood pressure, rating of perceived exertion (Borg scale), and heart rate prior to commencing exercise. If PEF is $\leq 50\%$ prior to commencing exercise the exercise session should not take place. Continue to monitor clinical measures after the termination of exercise up to near resting values. Administer bronchodilator if PEF drops $\geq 15\%$ after exercise [54–57,60,66].

o   Individuals should warm-up prior to exercise and cool-down postexercise for 10–15 min to minimize exercise-related asthma symptoms [26–28]. Nose breathing is encouraged when possible to warm the inspired air [97,98].

o   Perceived levels of breathlessness may indicate safe and tolerable levels of exercise intensity [99]. The modified Borg dyspnea scale can be used to monitor levels of breathlessness [100].

After exercise

o   If asthma symptoms regularly increase during exercise, refer the individual onto their general practitioner and/or respiratory specialist to review their asthma management.

o   Seek emergency medical attention if acute asthma symptoms continue after implementing management strategies (e.g., Inhaled bronchodilator, written action plan).

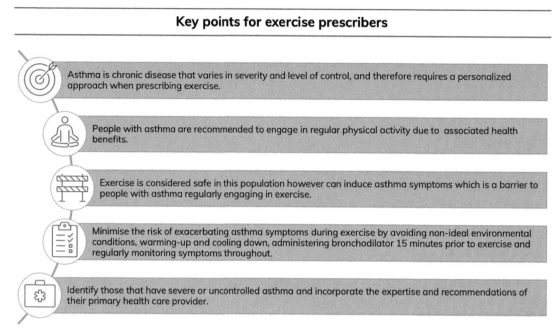

Key points for exercise prescribers

Asthma is chronic disease that varies in severity and level of control, and therefore requires a personalized approach when prescribing exercise.

People with asthma are recommended to engage in regular physical activity due to associated health benefits.

Exercise is considered safe in this population however can induce asthma symptoms which is a barrier to people with asthma regularly engaging in exercise.

Minimise the risk of exacerbating asthma symptoms during exercise by avoiding non-ideal environmental conditions, warming-up and cooling down, administering bronchodilator 15 minutes prior to exercise and regularly monitoring symptoms throughout.

Identify those that have severe or uncontrolled asthma and incorporate the expertise and recommendations of their primary health care provider.

FIG. 2 Recommendation summary for prescribing exercise and outcomes for adults with asthma.

Clinical practice recommendations for exercise prescribers

As discussed earlier, there are a number of beneficial effects of regular exercise in people with asthma including improved asthma control, fewer asthma symptoms and exacerbations, reduced medication and healthcare use including reduced emergency department visits, improved HRQoL, and decreased symptoms of anxiety and depression (Fig. 2).

The American College of Sports Medicine (ACSM) guidelines for exercise prescription in chronic conditions suggest that recommendations based on the FITT Principle (frequency, intensity, time, and type) can be applied to the vast majority of chronic conditions [101] (Fig. 2). However, heterogeneity in study design exists within the literature and specific characteristic of the FITT Principle in an asthma population remain poorly defined [102]. More research is required in applying optimal exercise prescription principles within the asthma population.

Frequency

The clinical benefits of the aerobic exercise demonstrated in asthma research studies have been achieved between 2 and 5 sessions per week [46–48,50–60]. However, the ACSM recommends aerobic exercise four to five times per week for people with chronic conditions [101]. The prescribed frequency of exercise sessions should be individualized based on the ability and tolerance of the individual.

Intensity

A consensus on the optimal intensity for an asthma population is yet to be reached. The ATS recommends that a minimum intensity of 60% (moderate-intensity) of the peak exercise capacity is required to attain greater physiologic training effects from exercise in a COPD population [103]. This may be considered in an asthma population or at limits tolerated by symptoms. This moderate level of intensity has also been demonstrated to be effective in asthma and exercise intervention research studies, with the majority of studies prescribing an intensity of 60%–70% of maximum heart rate (HRmax) [46,47,49,53–55,58,60]. High-intensity exercise has been less investigated due to concerns of provoking EIB [28]. It has been hypothesized that this is related to the increase in mouth breathing which occurs with high-intensity exercise, which increases airway exposure to pollutants and allergens by bypassing the nasal passage [104]. To our knowledge, there have been no studies conducted to date which have compared the effects of high-intensity training with low and moderate-intensity training on asthma outcomes. To establish proper exercise intensity, it is important to undertake a maximal exercise test to determine the individual's HRmax. The use of predicted HRmax in this population may not reflect the patient's exercise capacity [96].

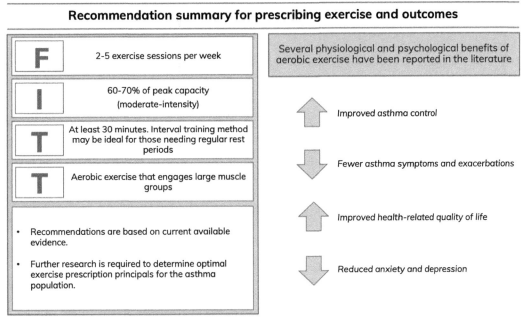

Recommendation summary for prescribing exercise and outcomes

F	2-5 exercise sessions per week
I	60-70% of peak capacity (moderate-intensity)
T	At least 30 minutes. Interval training method may be ideal for those needing regular rest periods
T	Aerobic exercise that engages large muscle groups

- Recommendations are based on current available evidence.
- Further research is required to determine optimal exercise prescription principals for the asthma population.

Several physiological and psychological benefits of aerobic exercise have been reported in the literature

Improved asthma control

Fewer asthma symptoms and exacerbations

Improved health-related quality of life

Reduced anxiety and depression

FIG. 3 Summary of key points for exercise prescribers.

Time

Ideally, an exercise session should last 30 min [101]. This duration has been tested in asthma and exercise clinical trials that report improved clinical outcomes [46,53–56]. However, some people who are symptomatic or have moderate to severe asthma may not be able to tolerate this duration. Interval training may be an alternative in order to introduce regular periods of rest in this population [103].

Type

Aerobic exercise training has been shown to be effective in an asthma population [79,86]. The ACSM recommends aerobic exercise that engages large muscle groups including walking, swimming, cycling, and light jogging [101]. It is less known about the impact of resistance training in an asthma population, however, resistance training as a component of a pulmonary rehabilitation program has been shown to be effective in an asthma cohort [105].

In addition to formal exercise prescription, it is important to recognize that engaging in overall regular physical activity should also be targeted in this population. If physical activity levels are considered below recommendations consideration and recommendations should be given to increase overall levels of physical activity including daily steps prior to engaging them in a moderate to high-intensity exercise program.

Conclusion

Physical activity levels in the asthma population are lower than that of healthy controls, particularly those with severe asthma [29,30,32,106]. Despite exercise being a trigger for inducing asthma symptoms, exercise can be performed safely in people with asthma and has been shown to improve asthma control, reduce asthma symptoms and exacerbations, improve HRQoL, and decrease symptoms of anxiety and depression. Therefore, exercise should be included as part of an asthma management plan [27] (Fig. 3). Current research suggests that improved physiological and psychological outcomes arise from engaging in moderate-intensity aerobic exercise of at least 30 min duration, two to five times per week. However, more research is required as there are no clearly defined exercise prescription guidelines for people with asthma.

References

[1] Boulet LP. Influence of comorbid conditions on asthma. Eur Respir J 2009;33(4):897.
[2] Radhakrishna N, Tay TR, Hore-Lacy F, Hoy R, et al. Profile of difficult to treat asthma patients referred for systematic assessment. Respir Med 2016;117:166–73.

[3] McDonald VM, Fingleton J, Agusti A, Hiles SA, et al. Treatable traits: a new paradigm for 21st century management of chronic airway diseases. Eur Respir J 2019;53(5):1802058. https://doi.org/10.1183/13993003.02058-2018.

[4] McDonald VM, Clark VL, Cordova-Rivera L, Wark PAB, et al. Targeting treatable traits in severe asthma: a randomised controlled trial. Eur Respir J 2020;55(3).

[5] Chung KF, Wenzel SE, Brozek JL, Bush A, et al. International ERS/ATS guidelines on definition, evaluation and treatment of severe asthma. Eur Respir J 2014;43(2):343–73.

[6] Masoli M, Fabian D, Holt S, Beasley R. The global burden of asthma: executive summary of the GINA dissemination committee report. Allergy 2004;59(5):469–78.

[7] Vos T, Abajobir AA, Abate KH, Abbafati C, et al. Global, regional, and national incidence, prevalence, and years lived with disability for 328 diseases and injuries for 195 countries, 1990–2016: a systematic analysis for the global burden of disease study 2016. Lancet 2017;390(10100):1211–59.

[8] World Health Organization. Global health estimates 2016: disease burden by cause, age, sex, by country and by region, 2000–2016; 2018 [Geneva].

[9] World Health Organization. Global health estimates 2016: deaths by cause, age, sex, by country and by region, 2000–2016; 2018 [Geneva].

[10] Global Initiative for Asthma. Online appendix. global strategy for asthma manangement and prevention, 2020, 2020. Available from: *www.ginasthma.org*.

[11] Ehteshami-Afshar S, FitzGerald JM, Doyle-Waters MM, Sadatsafavi M. The global economic burden of asthma and chronic obstructive pulmonary disease. Int J Tuberc Lung Dis 2016;20(1):11–23.

[12] Australian Institute of Health and Welfare (AIHW). Asthma, 25 August 2020. [25 January 2021] Available from: https://www.aihw.gov.au/reports/chronic-respiratory-conditions/asthma/contents/impact.

[13] Gergen PJ. Understanding the economic burden of asthma. J Allergy Clin Immunol 2001;107(5, Supplement):S445–8.

[14] Hekking PP, Wener RR, Amelink M, Zwinderman AH, et al. The prevalence of severe refractory asthma. J Allergy Clin Immunol 2015;135(4):896–902.

[15] Royal College of Physicians. London, why asthma still kills: the national review of asthma deaths (NRAD) confidential enquiry report; 2014.

[16] Sweeney J, Brightling CE, Menzies-Gow A, Niven R, et al. Clinical management and outcome of refractory asthma in the UK from the British Thoracic Society difficult asthma registry. Thorax 2012;67(8):754–6.

[17] Côté A, Godbout K, Boulet L-P. The management of severe asthma in 2020. Biochem Pharmacol 2020;179:114112.

[18] Pavord ID, Beasley R, Agusti A, Anderson GP, et al. After asthma: redefining airways diseases. Lancet 2017.

[19] ACAM. Asthma in Australia 2011. Canberra: Australian Institute of Health and Welfare; 2011.

[20] Holgate ST. Pathogenesis of asthma. Clin Exp Allergy 2008;38(6):872–97.

[21] Dunican EM, Fahy JV. The role of type 2 inflammation in the pathogenesis of asthma exacerbations. Ann Am Thorac Soc 2015;**12**(Suppl 2):S144–9.

[22] Busse WW. The relationship of airway hyperresponsiveness and airway inflammation: airway hyperresponsiveness in asthma: its measurement and clinical significance. Chest 2010;138(2 Suppl):4S–10S.

[23] An SS, Bai TR, Bates JHT, Black JL, et al. Airway smooth muscle dynamics: a common pathway of airway obstruction in asthma. Eur Respir J 2007;29(5):834.

[24] Ritz T, Rosenfield D, Steptoe A. Physical activity, lung function, and shortness of breath in the daily life of individuals with asthma. Chest 2010;138(4):913–8.

[25] Aggarwal B, Mulgirigama A, Berend N. Exercise-induced bronchoconstriction: prevalence, pathophysiology, patient impact, diagnosis and management. NPJ Prim Care Respir Med 2018;28.

[26] Global Initiative for Asthma. Global strategy for asthma manangement and prevention, 2020, 2020. Available from: *www.ginasthma.org*.

[27] National Asthma Council Australia. Australian asthma handbook, version 2.1. Melbourne: National Asthma Council Australia; 2020.

[28] Parsons JP, Hallstrand TS, Mastronarde JG, Kaminsky DA, et al. An official American Thoracic Society clinical practice guideline: exercise-induced bronchoconstriction. Am J Respir Crit Care Med 2013;187(9):1016–27.

[29] Cordova-Rivera L, Gibson PG, Gardiner PA, McDonald VM. A systematic review of associations of physical activity and sedentary time with asthma outcomes. J Allergy Clin Immunol Pract 2018;6(6):1968–1981.e2.

[30] van't Hul AJ, Frouws S, van den Akker E, van Lummel R, et al. Decreased physical activity in adults with bronchial asthma. Respir Med 2016;114:72–7.

[31] Teramoto M, Moonie S. Physical activity participation among adult Nevadans with self-reported asthma. J Asthma 2011;48(5):517–22.

[32] Cordova-Rivera L, Gibson PG, Gardiner PA, Powell H, et al. Physical activity and exercise capacity in severe asthma: key clinical associations. J Allergy Clin Immunol Pract 2018;6(3):814–22.

[33] Bahmer T, Waschki B, Schatz F, Herzmann C, et al. Physical activity, airway resistance and small airway dysfunction in severe asthma. Eur Respir J 2017;49(1):1601827.

[34] Avallone KM, McLeish AC. Asthma and aerobic exercise: a review of the empirical literature. J Asthma 2013;50(2):109–16.

[35] Freitas PD, Passos NFP, Carvalho-Pinto RM, Martins MA, et al. A behavior change intervention aimed at increasing physical activity improves clinical control in adults with asthma: a randomized controlled trial. Chest 2021;159(1):46–57.

[36] Dogra S, Baker J. Physical activity and health in Canadian asthmatics. J Asthma 2006;43(10):795–9.

[37] Spruit MA, Singh SJ, Garvey C, ZuWallack R, et al. An official American Thoracic Society/European Respiratory Society statement: key concepts and advances in pulmonary rehabilitation. Am J Respir Crit Care Med 2013;188(8):e13–64.

[38] McDonald VM, Hiles SA, Jones KA, Clark VL, et al. Health-related quality of life burden in severe asthma. Med J Aust 2018;209(S2):S28–s33.

[39] Cambach W, Chadwick-Straver RV, Wagenaar RC, van Keimpema AR, et al. The effects of a community-based pulmonary rehabilitation programme on exercise tolerance and quality of life: a randomized controlled trial. Eur Respir J 1997;10(1):104.

[40] Foglio K, Bianchi L, Ambrosino N. Is it really useful to repeat outpatient pulmonary rehabilitation programs in patients with chronic airway obstruction?: a 2-year controlled study. Chest 2001;119(6):1696–704.

[41] Manzak AS, Özyılmaz S, Atagün Güney P. Efficiency of home-based pulmonary rehabilitation in adults with asthma. Eur Respir J 2020;56 (suppl 64):5179.

[42] Majd S, Apps L, Chantrell S, Hudson N, et al. A feasibility study of a randomized controlled trial of asthma-tailored pulmonary rehabilitation compared with usual Care in Adults with severe asthma. J Allergy Clin Immunol Pract 2020;8(10):3418–27.

[43] Salandi J, Icks A, Gholami J, Hummel S, et al. Impact of pulmonary rehabilitation on patients' health care needs and asthma control: a quasi-experimental study. BMC Pulm Med 2020;20(1):267.

[44] Türk Y, Theel W, van Huisstede A, van de Geijn GM, et al. Short-term and long-term effect of a high-intensity pulmonary rehabilitation programme in obese patients with asthma: a randomised controlled trial. Eur Respir J 2020;56(1).

[45] Osadnik CR, Singh S. Pulmonary rehabilitation for obstructive lung disease. Respirology 2019;24(9):871–8.

[46] Dogra S, Jamnik V, Baker J. Self-directed exercise improves perceived measures of health in adults with partly controlled asthma. J Asthma 2010; 47(9):972–7.

[47] Dogra S, Kuk JL, Baker J, Jamnik V. Exercise is associated with improved asthma control in adults. Eur Respir J 2011;37(2):318.

[48] Toennesen LL, Meteran H, Hostrup M, Wium Geiker NR, et al. *Effects of exercise and diet in nonobese asthma patients—a randomized controlled trial*. The journal of allergy and clinical immunology. In Pract 2018;6(3):803–11.

[49] Meyer A, Günther S, Volmer T, Taube K, et al. A 12-month, moderate-intensity exercise training program improves fitness and quality of life in adults with asthma: a controlled trial. BMC Pulm Med 2015;15:56.

[50] Scott HA, Gibson PG, Garg ML, Pretto JJ, et al. Dietary restriction and exercise improve airway inflammation and clinical outcomes in overweight and obese asthma: a randomized trial. Clin Exp Allergy 2013;43(1):36–49.

[51] Emtner M, Herala M, Stålenheim G. High-intensity physical training in adults with asthma: a 10-week rehabilitation program. Chest 1996; 109(2):323–30.

[52] Basaran S, Guler-Uysal F, Ergen N, Seydaoglu G, et al. Effects of physical exercise on quality of life, exercise capacity and pulmonary function in children with asthma. J Rehabil Med 2006;38(2):130–5.

[53] Jaakkola JJK, Aalto SAM, Hernberg S, Kiihamäki SP, et al. Regular exercise improves asthma control in adults: a randomized controlled trial. Sci Rep 2019;9(1):12088.

[54] Gonçalves R, Numes M, Cukier A, Stelmach R, et al. Effects of an aerobic physical training program on psychosocial characteristics, quality of life, symptoms and exhaled nitric oxide in individuals with moderate or severe persistent asthma. Braz J Phys Ther 2008;12(2):127–35.

[55] Mendes FA, Almeida FM, Cukier A, Stelmach R, et al. Effects of aerobic training on airway inflammation in asthmatic patients. Med Sci Sports Exerc 2011;43(2):197–203.

[56] Mendes FA, Gonçalves RC, Nunes MP, Saraiva-Romanholo BM, et al. Effects of aerobic training on psychosocial morbidity and symptoms in patients with asthma: a randomized clinical trial. Chest 2010;138(2):331–7.

[57] França-Pinto A, Mendes FA, de Carvalho-Pinto RM, Agondi RC, et al. Aerobic training decreases bronchial hyperresponsiveness and systemic inflammation in patients with moderate or severe asthma: a randomised controlled trial. Thorax 2015;70(8):732–9.

[58] Freitas PD, Ferreira PG, Silva AG, Stelmach R, et al. The role of exercise in a weight-loss program on clinical control in obese adults with asthma. A randomized controlled trial. Am J Respir Crit Care Med 2017;195(1):32–42.

[59] Turner S, Eastwood P, Cook A, Jenkins S. Improvements in symptoms and quality of life following exercise training in older adults with moderate/ severe persistent asthma. Respiration 2011;81(4):302–10.

[60] Evaristo KB, Mendes FAR, Saccomani MG, Cukier A, et al. Effects of aerobic training versus breathing exercises on asthma control: a randomized trial. J Allergy Clin Immunol Pract 2020;8(9):2989–2996.e4.

[61] Juniper EF, Byrne PM, Guyatt GH, Ferrie PJ, et al. Development and validation of a questionnaire to measure asthma control. Eur Respir J 1999; 14(4):902.

[62] Nathan RA, Sorkness CA, Kosinski M, Schatz M, et al. Development of the asthma control test: a survey for assessing asthma control. J Allergy Clin Immunol 2004;113(1):59–65.

[63] Schatz M, Kosinski M, Yarlas AS, Hanlon J, et al. The minimally important difference of the asthma control test. J Allergy Clin Immunol 2009; 124(4):719–723.e1.

[64] Juniper EF, Bousquet J, Abetz L, Bateman ED. Identifying 'well-controlled' and 'not well-controlled' asthma using the asthma control question-naire. Respir Med 2006;100(4):616–21.

[65] Toennesen LL, Soerensen ED, Hostrup M, Porsbjerg C, et al. Feasibility of high-intensity training in asthma. Eur Clin Respir J 2018; 5(1):1468714.

[66] Fanelli A, Cabral ALB, Neder JA, Martins MA, et al. Exercise training on disease control and quality of life in asthmatic children. Med Sci Sports Exerc 2007;39(9).

[67] Heikkinen SAM, Mäkikyrö EMS, Hugg TT, Jaakkola MS, et al. Effects of regular exercise on asthma control in young adults. J Asthma 2018; 55(7):726–33.

[68] Holland AE, Jones AW. More movement for better control: the importance of physical activity promotion in uncontrolled asthma. Chest 2021; 159(1):1–2.

[69] Guilleminault L, Williams EJ, Scott HA, Berthon BS, et al. Diet and asthma: is it time to adapt our message? Nutrients 2017;9(11):1227.

[70] Côté A, Turmel J, Boulet L-P. Exercise and asthma. Semin Respir Crit Care Med 2018;39(01):019–28.

[71] Eichenberger PA, Diener SN, Kofmehl R, Spengler CM. Effects of exercise training on airway hyperreactivity in asthma: a systematic review and Meta-analysis. Sports Med 2013;43(11):1157–70.

[72] Wu X, Gao S, Lian Y. Effects of continuous aerobic exercise on lung function and quality of life with asthma: a systematic review and meta-analysis. J Thorac Dis 2020;12(9):4781–95.

[73] de Miguel Díez J, Hernández Barrera V, Puente Maestu L, Carrasco Garrido P, et al. Psychiatric comorbidity in asthma patients. Associated factors. J Asthma 2011;48(3):253–8.

[74] Dafauce L, Romero D, Carpio C, Barga P, et al. Psycho-demographic profile in severe asthma and effect of emotional mood disorders and hyperventilation syndrome on quality of life. BMC Psychol 2021;9(1):3.

[75] Yonas MA, Marsland AL, Emeremni CA, Moore CG, et al. Depressive symptomatology, quality of life and disease control among individuals with well-characterized severe asthma. J Asthma 2013;50(8):884–90.

[76] Wegner M, Helmich I, Machado S, Nardi AE, et al. Effects of exercise on anxiety and depression disorders: review of meta- analyses and neurobiological mechanisms. CNS Neurol Disord Drug Targets 2014;13(6):1002–14.

[77] Farid R, Azad FJ, Atri AE, Rahimi MB, et al. Effect of aerobic exercise training on pulmonary function and tolerance of activity in asthmatic patients. Iran J Allergy Asthma Immunol 2005;4(3):133–8.

[78] Refaat A, Gawish M. Effect of physical training on health-related quality of life in patients with moderate and severe asthma. Egypt J Chest Dis Tuberc 2015;64(4):761–6.

[79] Hansen ESH, Pitzner-Fabricius A, Toennesen LL, Rasmusen HK, et al. Effect of aerobic exercise training on asthma in adults: a systematic review and meta-analysis. Eur Respir J 2020;56(1).

[80] Boyd A, Yang CT, Estell K, Ms CT, et al. Feasibility of exercising adults with asthma: a randomized pilot study. Allergy Asthma Clin Immunol 2012;8(1):13.

[81] Cochrane LM, Clark CJ. Benefits and problems of a physical training programme for asthmatic patients. Thorax 1990;45(5):345.

[82] Clark CJ, Cochrane LM. Assessment of work performance in asthma for determination of cardiorespiratory fitness and training capacity. Thorax 1988;43(10):745.

[83] Garfinkel SK, Kesten S, Chapman KR, Rebuck AS. Physiologic and nonphysiologic determinants of aerobic fitness in mild to moderate asthma. Am Rev Respir Dis 1992;145(4 Pt 1):741–5.

[84] Andersen JR, Natvig GK, Aadland E, Moe VF, et al. Associations between health-related quality of life, cardiorespiratory fitness, muscle strength, physical activity and waist circumference in 10-year-old children: the ASK study. Qual Life Res 2017;26(12):3421–8.

[85] Heikkinen SAM, Quansah R, Jaakkola JJK, Jaakkola MS. Effects of regular exercise on adult asthma. Eur J Epidemiol 2012;27(6):397–407.

[86] Carson KV, Chandratilleke MG, Picot J, Brinn MP, et al. Physical training for asthma. Cochrane Database Syst Rev 2013;9:Cd001116.

[87] Ram FSF, Robinson SM, Black PN. Effects of physical training in asthma: a systematic review. Br J Sports Med 2000;34(3):162.

[88] Scott HA, Latham JR, Callister R, Pretto JJ, et al. Acute exercise is associated with reduced exhaled nitric oxide in physically inactive adults with asthma. Ann Allergy Asthma Immunol 2015;114(6):470–9.

[89] Pakhale S, Luks V, Burkett A, Turner L. Effect of physical training on airway inflammation in bronchial asthma: a systematic review. BMC Pulm Med 2013;13(1):38.

[90] Cockcroft DW, Davis BE. Mechanisms of airway hyperresponsiveness. J Allergy Clin Immunol 2006;118(3):551–9 [quiz 560-1].

[91] Silva RA, Vieira RP, Duarte ACS, Lopes FDTQS, et al. Aerobic training reverses airway inflammation and remodelling in an asthma murine model. Eur Respir J 2010;35(5):994.

[92] Barnes PJ. The cytokine network in asthma and chronic obstructive pulmonary disease. J Clin Invest 2008;118(11):3546–56.

[93] Lang JE. The impact of exercise on asthma. Curr Opin Allergy Clin Immunol 2019;19(2):118–25.

[94] Exercise & Sports Science Australia, E.i.M., Fitness Australia, and Sports Medicine Australia. Adult Pre-Exercise Screening System (APSS), 2019. [cited 2021 08 Feb]; Available from: https://www.essa.org.au/Public/ABOUT_ESSA/Adult_Pre-Screening_Tool.aspx.

[95] Centre of Excellence in Severe Asthma. Severe asthma toolkit, 2018. Available from: https://toolkit.severeasthma.org.au.

[96] Rodrigues Mendes FA, Teixeira RN, Martins MA, Cukier A, et al. The relationship between heart rate and VO(2) in moderate-to-severe asthmatics. J Asthma 2020;57(7):713–21.

[97] Thomas M, Bruton A. Breathing exercises for asthma. Breathe 2014;10(4):312.

[98] Shturman-Ellstein R, Zeballos RJ, Buckley JM, Souhrada JF. The beneficial effect of nasal breathing on exercise-induced bronchoconstriction. Am Rev Respir Dis 1978;118(1):65–73.

[99] Vermeulen F, Chirumberro A, Rummens P, Bruyneel M, et al. Relationship between the sensation of activity limitation and the results of functional assessment in asthma patients. J Asthma 2017;54(6):570–7.

[100] Borg E, Borg G, Larsson K, Letzter M, et al. An index for breathlessness and leg fatigue. Scand J Med Sci Sports 2010;20(4):644–50.

[101] Moore GE, Painter PL, Lyerly GW, Durstine JL. ACSM's exercise management for persons with chronic diseases and disabilities. 4th ed. Human Kinetics; 2016.

[102] Zampogna E, Zappa M, Spanevello A, Visca D. Pulmonary rehabilitation and asthma. Front Pharmacol 2020;11:542.

[103] Nici L, Donner C, Wouters E, Zuwallack R, et al. American thoracic society/European respiratory society statement on pulmonary rehabilitation. Am J Respir Crit Care Med 2006;173(12):1390–413.

[104] Bonini M, Silvers W. Exercise-induced bronchoconstriction: background, prevalence, and sport considerations. Immunol Allergy Clin North Am 2018;38(2):205–14.

[105] Lingner H, Ernst S, Großhennig A, Djahangiri N, et al. Asthma control and health-related quality of life one year after inpatient pulmonary rehabilitation: the ProKAR study. J Asthma 2015;52(6):614–21.

[106] Ford ES, Heath GW, Mannino DM, Redd SC. Leisure-time physical activity patterns among US adults with asthma. Chest 2003;124(2):432–7.

Chapter 10

Exercise training for people living with HIV

Hugo Ribeiro Zanetti[a,b], Camilo Luís Monteiro Lourenço[c], Mansueto Gomes-Neto[d], and Leonardo Roever[e]

[a]University Center IMEPAC, Araguari, MG, Brazil [b]Federal University of Triângulo Mineiro, Uberaba, MG, Brazil [c]Federal University of Santa Catarina, Florianópolis, SC, Brazil [d]Federal University of Bahia, Salvador, BA, Brazil [e]Federal University of Uberlândia, Uberlândia, MG, Brazil

Introduction

Infection with human immunodeficiency virus (HIV) and acquired human immunodeficiency syndrome (AIDS) are vast worldwide public health problems. Currently, it is estimated that more than 38 million people are living with HIV (PLHIV), the majority of whom are adults (36.2 million) and male (22.5 million). Furthermore, the infection remains on the rise, with an addition of 1.7 million new cases and approximately 690.000 deaths related to aids in 2019 [1].

Historically, HIV infection was recognized as a disease in the early 1980s caused by the increasing number of cases with the presence of opportunistic diseases, mainly tuberculosis, pneumonia, and rare neoplasms such as Kaposi's Sarcoma, mainly observed in men who had sex with other men [2,3]. However, after a few years, HIV was isolated and identified as a possible agent of these conditions [4,5]. Nowadays, it is known that HIV transmission occurs in circumstances where there is contact with blood or body fluids that contain the virus or with cells infected by the virus, which can often be through sexual contact, parenteral and vertical inoculation (mother to child) [6].

After exposure, HIV replication occurs through infection of the human organism's defense cells, with the integration of the provirus into the host cell genome, activation of viral replication, and production and release of new viruses. All of these processes occur with the T $CD4^+$ lymphocyte as a receptor and chemokine receptors as co-receptors. HIV infection begins as an acute clinical stage with an adaptive immune system response and progresses to chronic infection of peripheral lymphoid tissues. The natural course of the infection can be divided into three phases: (a) acute retroviral syndrome, (b) chronic middle phase, and (c) aids [7,8]. The first phase is consolidated by infection of T $CD4^+$ lymphocytes and apoptosis of infected cells [9]. After a few days, the host's humoral and cellular response is assembled, and seroconversion of the infected person occurs between 3 and 7 weeks. In the second phase, there is viral replication, mainly in the lymph nodes and spleen, and continuous cell destruction. The last phase is characterized by failure of the host's defenses, uncontrolled replication of the virus in the plasma, and severe clinical disease. The HIV infection typically progresses to AIDS after 7 to 10 years without treatment [10,11].

After the discovery of the mechanisms of action and replication of HIV in humans, there came the development of drugs to prevent its replication in the blood [12]. Highly active antiretroviral therapy (HAART) consists of a combination of drugs (fusion/entry inhibitors, reverse transcriptase inhibitors, integrase inhibitors, and protease inhibitors) with its key role being suppression of viral replication, reducing the viral load (amount of virus in the bloodstream) and opportunistic infections while preserving the T $CD4^+$ lymphocyte cell count [13]. Since then, there has been a significant decrease in the observed incidence of aids-related morbidity and mortality, as well as improvement in quality and expectancy of life [14,15].

Despite the benefits achieved by HAART, the long-term use of medications is related to a gamut of intersystem metabolic and morphological changes, which acts as a key factor for the development of chronic noncommunicable disease (CNCD), particularly cardiometabolic disease [16–18]. Moreover, due to the increased survival, there has been an increase in the incidence of CNCD, comorbidities, and/or conditions related to old age, such as functional and neurocognitive disorders, changes in the locomotor system, and social problems [19–22]. It has been shown that 28% of PLHIV have at least one comorbidity and that in 2030 this rate will be 85% [23]. Additionally, one-half of deaths in PLHIV are now related to non-HIV causes, with cardiovascular diseases leading to the rise of deaths in this population [24,25].

In addition to the problems mentioned in the paragraph above, PLHIV exhibits a high level of sedentarism/sedentary behavior compared to the general population and does not follow the available recommendation for weekly physical activity [26,27]. In addition, studies have found that the levels of physical activity in PLHIV are inversely proportional to the incidence of chronic degenerative diseases, inflammatory markers, and all cause-mortality [28–30]. Furthermore, physical inactivity favors a greater number of hospitalizations and complications from infection, and chronic use of

Exercise to Prevent and Manage Chronic Disease Across the Lifespan. https://doi.org/10.1016/B978-0-323-89843-0.00028-3

HAART. This chapter will discuss the current evidence which demonstrates that physical exercise programs lead to an increase in cardiorespiratory fitness, improvements on cardiometabolic biomarkers, muscle strength, immune function, and body composition in PLHIV, and therefore, must be considered as an integral part of care for PLHIV.

Exercise in HIV

Clinical assessment

Before participating in an exercise program PLHIV is recommended to undergo a complete cardiological examination (i.e., resting electrocardiogram and stress test with electrocardiographic monitoring), especially those that present risk factors for cardiovascular diseases [31]. The preparticipation anamnesis should check the patient's history related to immune parameters (T $CD4^+$ lymphocyte cell count and viral load), the presence of opportunistic diseases as well as other morbidities related to the use of ART and include items related to anthropometric assessment and capacities physical.

The assessment of body mass should be constantly evaluated in PLHIV due to the disease progression, chronic use of medications, and inflammation which contributes to body mass loss leading to cachexia and/or sarcopenia [32,33]. Furthermore, the evaluation of neck, chest, waist, and abdomen circumference should be included if PLHIV could present lipodystrophic syndrome [34]. Dual X-ray absorptiometry (DXA) is the gold standard to assess body composition, however, due to the high cost it becomes relatively unfeasible, hence bioimpedance or skinfold thickness (SFT) is considered useful in routine clinical practice [35]. The body fat percentage and lean body mass can be estimated using the skinfold thickness as shown in Tables 1 and 2, respectively.

Muscle strength should be assessed using tests that evaluate the capacity of a muscle group to generate strength against resistance. Thus, the tests of 1 repetition maximum (RM), 3RM, or 10RM can be used, while simultaneously respecting the initial physical capacity of the patient. In special conditions, a dynamometer can be used to assess handgrip strength [36].

Cardiorespiratory fitness, assessed through maximal oxygen consumption tests (VO2max), should be used to assess aerobic fitness, as PLHIV have a lower workload, anaerobic threshold and VO2max predicted for age. For this population, adapted or sedentary individual protocols are commonly used. Table 3 illustrates the formulas for predicted VO_2 peak and functional aerobic impairment (FAI). [37,38].

Clinical practice recommendations for exercise prescribers

Over the years, physical exercise has been seen at times as an ally, and at other times deleterious, in the control and treatment of patients with chronic diseases. This is due, in part, to the lack of clarification of some professionals, who still today do not prescribe physical exercise, for fear of worsening the condition. Currently, physical exercise is recommended for healthy people with different health conditions [39].

The current recommendations from the World Health Organization on physical activity and sedentary behavior recommend, at least, 150–300 min of aerobic physical activity with moderate weekly intensity or, at least, 75–150 min of aerobic physical activity with vigorous weekly intensity; or an equivalent combination of moderate to vigorous weekly activity for substantial health purposes [40].

Therapeutic exercise has been considered an important complementary therapy for promoting the health of PLHIV. PLHIV exercise aims to minimize the deleterious effects and complications resulting from the evolution of the disease

TABLE 1 Equation for predicted body fat percentage for PLHIV.

Men	$3.385 + 0.279 \times$ SFT midaxillary + SFT subscapular (for men)
Women	$-24.323 + 0.736 \times$ SFT suprailiac + SFT abdomen + SFT calf

TABLE 2 Equation for predicted lean body mass.

Men	$16.012 + 0.576 \times$ body mass
Women	$48,139 - 0,585 \times$ SFT triceps

TABLE 3 Equation for VO$_2$ peak and %FAI in PLHIV.

VO$_2$ peak	$[14.76 - (1379 \times T) + (0.451 \times T2) - (0.012 \times T3)]$ (men)
	$[(4.38 \times T) - 3.9]$ (women)
%FAI	$[(\text{predicted VO}_2 - \text{observed VO}_2)/\text{predicted VO}_2] \times 100$

and HAART use such as dyslipidemia, diabetes, muscle loss, while decreasing cardiovascular disease risk and promote improvement in activities of daily living (ADLs), as well as maximizing well-being and quality of life [41].

Studies have shown that PLHIV engaged in physical exercise programs experience an improvement in symptoms of anxiety, depression, and quality of life [40]. It is worth noting that all the principles of sports training must be applied to physical training programs for PLHIV.

Different modalities of exercise can be selected according to the disabilities presented by PLHIV. Recently, the combination of exercise modalities (concurrent training) has been recommended by the American College of Sports Medicine (ACSM). Specifically, these programs should include flexibility, endurance, and strength exercises [42].

Resistance exercise is well defined as the most effective method available for improving strength, endurance and muscle performance, through the principle of progressive overload. While aerobic exercise has significant effects on improving aerobic capacity as measured by maximum oxygen consumption (VO2max). Although specific guidelines for special groups form the basis for individualized prescription, the most common basic components for all exercise programs constitute the framework for prescribing the program, regardless of the population it is intended for [43].

Studies in healthy individuals demonstrate that concurrent training can reduce the specific effects of each type of exercise, but it increases the amount of improved physiological and functional outcomes overall. In patients with multiple functional impairments, the use of concurrent exercise modalities in rehabilitation programs can be a useful and complementary alternative to the use of medications, which would not be possible with any single type of exercise [44].

Endurance training

As already mentioned, PLHIV has lower VO2max compared to the general population or other vulnerable populations [45]. This situation seems to occur mainly is due to central limitations (increased heart rate response and pulmonary ventilation) to an intensity above the anaerobic threshold and/or peripheral limitations (arteriovenous oxygen difference caused by mitochondrial dysfunction and lacticaemia). Thus, PLHIV has less tolerance to effort and exercise time [46,47]. Aerobic training programs carried out continuously and/or using intervals (with active recovery) of moderate intensity, can cause physiological adaptations that increase aerobic capacity [45,48,49].

Protocols for endurance training in PLHIV have been described in the literature. O'Brien et al. suggest that a program consisting of at least 20 min of aerobic exercise per session, performed 3 ×/week with intensity between 50% and 85% of the maximum heart rate or 45 to 85% VO2max is effective in increasing VO2max, reducing the percentage of fat, without altering immune function (Fig. 1) [48]. Furthermore, Stringer et al. demonstrated that endurance exercise training performed 3 times a week for 6 weeks with high intensity (50% of the difference between VO2 peak and lactate threshold) contributed to a 13% increase in VO2peak while moderate intensity (80% of lactate threshold) did not change VO2peak values [50]. Smith et al. Also showed that 16 weeks of vigorous-intensity endurance exercise performed at 75%–90% of heart rate reserve for 40–45 min significantly increased cardiorespiratory fitness, lost weight, decreased body mass index, subcutaneous fat, abdominal girth, and fatigue [51].

Resistance training

Resistance training programs have also recently been explored within this population. It is important to remember that PLHIV may present with musculoskeletal changes, especially presarcopenia or sarcopenia, that is, a significant reduction in muscle mass [52]. Thus, resistance training is an important tool to prevent and/or reverse this condition. Research in this area has shown resistance training leads to an improvement in body mass, increase in muscle mass, the cross-sectional area of muscle, muscle strength and endurance, T CD4$^+$ lymphocyte count, reduction of body fat, immune-inflammatory markers, and improvement of metabolic profile [53–56].

Poton et al. suggest that a resistance training program should contain 5–11 exercises, with 3–4 sets of 4–15 repetitions with a load between 60% and 90% of 1RM [56]. Ghayomzadeh et al. demonstrated that 8 weeks of resistance training using

FIG. 1 Recommendation of aerobic exercise for PLHIV.

TABLE 4 Example of resistance training program for PLHIV.

Exercises	1st session	2nd session	3rd session
Squat	3 × 4–6 RM	3 × 15–20 RM	3 × 8–12 RM
Bench press	3 × 4–6 RM	3 × 15–20 RM	3 × 8–12 RM
Hamstring curls	3 × 4–6 RM	3 × 15–20 RM	3 × 8–12 RM
Frontal pull	3 × 4–6 RM	3 × 15–20 RM	3 × 8–12 RM
Seated calf raises	3 × 4–6 RM	3 × 15–20 RM	3 × 8–12 RM
Shoulder press	3 × 4–6 RM	3 × 15–20 RM	3 × 8–12 RM
RM: repetition maximum.			

elastic bands and bodyweight improves body mass, immune cells and reduces fat mass and waist circumference [57]. Moreover, our previous studies showed that 12 weeks of nonlinear periodized resistance training (Table 4) can improve body composition, inflammatory biomarkers, muscle strength, metabolic parameters, metabolic syndrome components, and immune cells [53,54,58] (Fig. 2).

Concurrent training

Concurrent training is an important strategy for PLHIV since it involves aerobic and endurance in the same session of physical exercise. Studies show that concurrent training can improve body composition (increase in muscle mass and reduction in body fat), aerobic capacity, muscle strength and endurance, metabolic and hormonal profile in addition to T CD4$^+$ lymphocyte cell count (Fig. 3) [59], and maybe more efficacious than a single-mode of exercise in isolation. Furthermore, our previous studies

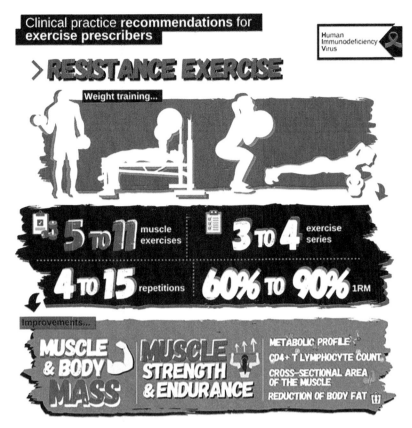

FIG. 2 Recommendations for resistance exercise for PLHIV.

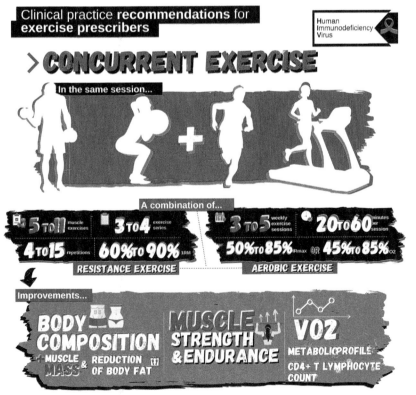

FIG. 3 Recommendations for concurrent exercise for PLHIV.

TABLE 5 Example of a concurrent training program for PLHIV.

Exercise/session	1st session	2nd session	3rd session
Squat	3 × 4–6 RM	3 × 15–20 RM	3 × 8–12 RM
Bench press	3 × 4–6 RM	3 × 15–20 RM	3 × 8–12 RM
Hamstring curls	3 × 4–6 RM	3 × 15–20 RM	3 × 8–12 RM
Frontal pull	3 × 4–6 RM	3 × 15–20 RM	3 × 8–12 RM
Seated calf raise	3 × 4–6 RM	3 × 15–20 RM	3 × 8–12 RM
Shoulder press	3 × 4–6 RM	3 × 15–20 RM	3 × 8–12 RM
Endurance	7 × 30 s: 1 min	6 × 2 min: 1 min	25 min
Intensity	85%–90% HRR	70%–50% HRR	60% HRR

HRR: heart rate reserve; *RM*: repetition maximum.

demonstrated that concurrent training in association with a statin (lower-cholesterol drug) leads to improvement in metabolic, anthropometric, hemodynamic profile as well as physical capacity [60,61] (Table 5).

Conclusion

It is well established that PLHIV has improved quality and expectancy of life through the aid of modern treatments, however, this population still presents new clinical challenges in the form of how best to reduce cardiovascular risk, HIV, and HAART-related comorbidities and mortality. The current evidence demonstrates that exercise training is an important tool to prevent and treat a range of HIV and HAART side effects. Thereby, the adherence to a physical exercise program should occur in parallel with pharmacological treatment and start as soon as possible according to the patient's history and anamnesis.

References

[1] UNAIDS. UNAIDS DATA 2018; 2018.
[2] Friedman-Kien A, et al. Kaposis sarcoma and Pneumocystis pneumonia among homosexual men—New York City and California. MMWR Morb Mortal Wkly Rep 1981;30(25):305–8.
[3] Hymes KB, et al. Kaposi's sarcoma in homosexual men-a report of eight cases. Lancet 1981;2(8247):598–600.
[4] Barré-Sinoussi F, et al. Isolation of a T-lymphotropic retrovirus from a patient at risk for acquired immune deficiency syndrome (AIDS). Science 1983;220(4599):868–71.
[5] Gallo RC, et al. Frequent detection and isolation of cytopathic retroviruses (HTLV-III) from patients with AIDS and at risk for AIDS. Science 1984;224(4648):500–3.
[6] Kariuki SM, et al. The HIV-1 transmission bottleneck. Retrovirology 2017;14(1):22.
[7] Moir S, Chun TW, Fauci AS. Pathogenic mechanisms of HIV disease. Annu Rev Pathol 2011;6:223–48.
[8] Kaplan MH. Pathogenesis of HIV. Infect Dis Clin N Am 1994;8(2):279–88.
[9] Doitsh G, et al. Cell death by pyroptosis drives CD4 T-cell depletion in HIV-1 infection. Nature 2014;505(7484):509–14.
[10] Dahabieh MS, Battivelli E, Verdin E. Understanding HIV latency: the road to an HIV cure. Annu Rev Med 2015;66:407–21.
[11] Rasmussen TA, Tolstrup M, Sogaard OS. Reversal of latency as part of a cure for HIV-1. Trends Microbiol 2016;24(2):90–7.
[12] Lu DY, et al. HAART in HIV/AIDS treatments: future trends. Infect Disord Drug Targets 2018;18(1):15–22.
[13] Pau AK, George JM. Antiretroviral therapy: current drugs. Infect Dis Clin N Am 2014;28(3):371–402.
[14] Croxford S, et al. Mortality and causes of death in people diagnosed with HIV in the era of highly active antiretroviral therapy compared with the general population: an analysis of a national observational cohort. Lancet Public Health 2017;2(1):e35–46.
[15] Poorolajal J, et al. Survival rate of AIDS disease and mortality in HIV-infected patients: a meta-analysis. Public Health 2016;139:3–12.
[16] Eyawo O, et al. Changes in mortality rates and causes of death in a population-based cohort of persons living with and without HIV from 1996 to 2012. BMC Infect Dis 2017;17(1):174.
[17] Zanetti HR, et al. Triad of the ischemic cardiovascular disease in people living with HIV? Association between risk factors, HIV infection, and use of antiretroviral therapy. Curr Atheroscler Rep 2018;20(6):30.
[18] Zanetti HR, et al. Cardiovascular complications of HIV. Int J Cardiovasc Sci 2018;31(5):538–43.

[19] Wandeler G, Johnson LF, Egger M. Trends in life expectancy of HIV-positive adults on antiretroviral therapy across the globe: comparisons with general population. Curr Opin HIV AIDS 2016;11(5):492–500.

[20] Hooshyar D, et al. Trends in perimortal conditions and mortality rates among HIV-infected patients. AIDS 2007;21(15):2093–100.

[21] Wang H, et al. Global, regional, and national life expectancy, all-cause mortality, and cause-specific mortality for 249 causes of death, 1980–2015: a systematic analysis for the global burden of disease study 2015. Lancet 2016;388(10053):1459–544.

[22] Negredo E, et al. Aging in HIV-infected subjects: a new scenario and a new view. Biomed Res Int 2017;2017:5897298.

[23] Smit M, et al. Future challenges for clinical care of an ageing population infected with HIV: a modelling study. Lancet Infect Dis 2015;15(7):810–8.

[24] Antiretroviral Therapy Cohort Collaboration. Causes of death in HIV-1-infected patients treated with antiretroviral therapy, 1996-2006: collaborative analysis of 13 HIV cohort studies. Clin Infect Dis 2010;50(10):1387–96.

[25] Feinstein MJ, et al. Patterns of cardiovascular mortality for HIV-infected adults in the United States: 1999 to 2013. Am J Cardiol 2016;117(2):214–20.

[26] Vancampfort D, et al. Global physical activity levels among people living with HIV: a systematic review and meta-analysis. Disabil Rehabil 2018;40(4):388–97.

[27] Vancampfort D, et al. Sedentary behavior in people living with HIV: a systematic review and Meta-analysis. J Phys Act Health 2017;14(7):571–7.

[28] Mustafa T, et al. Association between exercise and HIV disease progression in a cohort of homosexual men. Ann Epidemiol 1999;9(2):127–31.

[29] Young T, Busgeeth K. Home-based care for reducing morbidity and mortality in people infected with HIV/AIDS. Cochrane Database Syst Rev 2010;1, CD005417.

[30] Webel AR, et al. Cardiorespiratory fitness is associated with inflammation and physical activity in HIV+ adults. AIDS 2019;33(6):1023–30.

[31] Hsue PY, Waters DD. HIV infection and coronary heart disease: mechanisms and management. Nat Rev Cardiol 2019;16(12):745–59.

[32] Polsky B, Kotler D, Steinhart C. HIV-associated wasting in the HAART era: guidelines for assessment, diagnosis, and treatment. AIDS Patient Care STDs 2001;15(8):411–23.

[33] Kumar S, Samaras K. The impact of weight gain during HIV treatment on risk of pre-diabetes, diabetes mellitus, cardiovascular disease, and mortality. Front Endocrinol (Lausanne) 2018;9:705.

[34] Finkelstein JL, et al. HIV/AIDS and lipodystrophy: implications for clinical management in resource-limited settings. J Int AIDS Soc 2015;18:19033.

[35] Lee SY, Gallagher D. Assessment methods in human body composition. Curr Opin Clin Nutr Metab Care 2008;11(5):566–72.

[36] Thompson PD, et al. ACSM's new preparticipation health screening recommendations from ACSM's guidelines for exercise testing and prescription, ninth edition. Curr Sports Med Rep 2013;12(4):215–7.

[37] Bruce RA, Kusumi F, Hosmer D. Maximal oxygen intake and nomographic assessment of functional aerobic impairment in cardiovascular disease. Am Heart J 1973;85(4):546–62.

[38] Hand GA, et al. Impact of aerobic and resistance exercise on the health of HIV-infected persons. Am J Lifestyle Med 2009;3(6):489–99.

[39] American College of Sports Medicine Position Stand. The recommended quantity and quality of exercise for developing and maintaining cardio-respiratory and muscular fitness, and flexibility in healthy adults. Med Sci Sports Exerc 1998;30(6):975–91.

[40] Bull FC, et al. World Health Organization 2020 guidelines on physical activity and sedentary behaviour. Br J Sports Med 2020;54(24):1451–62.

[41] Dudgeon WD, et al. Physiological and psychological effects of exercise interventions in HIV disease. AIDS Patient Care STDs 2004;18(2):81–98.

[42] Garber CE, et al. American College of Sports Medicine position stand. Quantity and quality of exercise for developing and maintaining cardiorespiratory, musculoskeletal, and neuromotor fitness in apparently healthy adults: guidance for prescribing exercise. Med Sci Sports Exerc 2011;43(7):1334–59.

[43] Gomes-Neto M, et al. A systematic review of the effects of different types of therapeutic exercise on physiologic and functional measurements in patients with HIV/AIDS. Clinics (Sao Paulo) 2013;68(8):1157–67.

[44] Gomes Neto M, et al. A systematic review of effects of concurrent strength and endurance training on the health-related quality of life and cardiopulmonary status in patients with HIV/AIDS. Biomed Res Int 2013;2013, 319524.

[45] Vancampfort D, et al. Cardiorespiratory fitness levels and moderators in people with HIV: a systematic review and meta-analysis. Prev Med 2016;93:106–14.

[46] Harris M, et al. Random venous lactate levels among HIV-positive patients on antiretroviral therapy. J Acquir Immune Defic Syndr 2002;31(4):448–50.

[47] Stringer WW. Mechanisms of exercise limitation in HIV+ individuals. Med Sci Sports Exerc 2000;32(7 Suppl):S412–21.

[48] O'Brien KK, et al. Effectiveness of aerobic exercise for adults living with HIV: systematic review and meta-analysis using the cochrane collaboration protocol. BMC Infect Dis 2016;16:182.

[49] Hand GA, et al. Moderate intensity exercise training reverses functional aerobic impairment in HIV-infected individuals. AIDS Care 2008;20(9):1066–74.

[50] Stringer WW, et al. The effect of exercise training on aerobic fitness, immune indices, and quality of life in HIV+ patients. Med Sci Sports Exerc 1998;30(1):11–6.

[51] Smith BA, et al. Aerobic exercise: effects on parameters related to fatigue, dyspnea, weight and body composition in HIV-infected adults. AIDS 2001;15(6):693–701.

[52] Tehranzadeh J, Ter-Oganesyan RR, Steinbach LS. Musculoskeletal disorders associated with HIV infection and AIDS. Part I: infectious musculoskeletal conditions. Skelet Radiol 2004;33(5):249–59.

[53] Zanetti HR, et al. Does nonlinear resistance training reduce metabolic syndrome in people living with HIV? A randomized clinical trial. J Sports Med Phys Fitness 2017;57(5):678–84.

[54] Zanetti HR, et al. Nonlinear resistance training enhances the lipid profile and reduces inflammation marker in people living with HIV: a randomized clinical trial. J Phys Act Health 2016;13(7):765–70.

[55] Alves TC, et al. Resistance training with blood flow restriction: impact on the muscle strength and body composition in people living with HIV/AIDS. Eur J Sport Sci 2020;1–10.

[56] Poton R, Polito M, Farinatti P. Effects of resistance training in HIV-infected patients: a meta-analysis of randomised controlled trials. J Sports Sci 2017;35(24):2380–9.

[57] Ghayomzadeh M, et al. Effect of 8-week of hospital-based resistance training program on TCD4 + cell count and anthropometric characteristic of HIV patients in Tehran, Iran: a randomized controlled trial. J Strength Cond Res 2019;33(4):1146–55. https://doi.org/10.1519/JSC.0000000000002394.

[58] Zanetti HR, et al. Non-linear resistance training reduces inflammatory biomarkers in persons living with HIV: a randomized controlled trial. Eur J Sport Sci 2016;16(8):1232–9.

[59] Gomes Neto M, et al. Effects of combined aerobic and resistance exercise on exercise capacity, muscle strength and quality of life in HIV-infected patients: a systematic review and meta-analysis. PLoS One 2015;10(9), e0138066.

[60] Zanetti HR, et al. Effects of exercise training and statin use in people living with human immunodeficiency virus with dyslipidemia. Med Sci Sports Exerc 2020;52(1):16–24.

[61] Zanetti HR, et al. Effects of exercise training and statin on hemodynamic, biochemical, inflammatory and immune profile of people living with HIV: a randomized, double-blind, placebo-controlled trial. J Sports Med Phys Fitness 2020;60(9):1275–82.

Part IV

Middle age (35–65)

Chapter 11

Type 2 diabetes, prediabetes, and gestational diabetes mellitus

Sheri R. Colberg[a], Jenna B. Gillen[b], and Monique E. Francois[c,d]

[a]Old Dominion University, Norfolk, VA, United States [b]University of Toronto, Toronto, ON, Canada [c]University of Wollongong, Wollongong, NSW, Australia [d]Illawarra Health and Medical Research Institute, Wollongong, NSW, Australia

Introduction

Diabetes mellitus is a group of metabolic diseases characterized by an inability to produce sufficient insulin and/or to use it properly, resulting in hyperglycemia (elevations in blood glucose) [1]. Insulin, a hormone produced by the beta cells of the pancreas, is needed by the skeletal muscles, adipose tissue, and the liver to take up and use glucose, the primary simple sugar in the blood that is essential for the proper functioning of the brain and nerves [1].

Globally, as of 2021, more than 537 million people ages 20–79 years (10.5% of the world's population in this age group) were living with some type of diabetes. The International Diabetes Federation estimates that the number will rise to 11.3% (643 million) by 2030 and 12.2% (783 million) by 2045 [2,3]. Close to one in two people living with diabetes (240 million) remain undiagnosed [2]. About 90% or more of total cases are type 2 diabetes (T2D), an insulin-resistant state in which insulin production is insufficient to overcome the level of resistance to it. The resulting state of hyperglycemia (elevated blood glucose levels) is most common in older adults, but has been increasing in younger adults and youth on the heels of worldwide increases in overweight, obesity, sedentary behaviors, poor diet, other lifestyle choices, environmental factors (like pollution), and genetics [2,4].

Prediabetes (PD) is defined as a state of glucose intolerance associated with intermediate hyperglycemia, i.e., blood glucose above normal but below the level diagnostic for diabetes. PD will progress to overt T2D in approximately 25% of individuals within 3–5 years, and as many as 70% with PD will develop diabetes within their lifetimes [5]. Global cases were estimated to be 7.5% (374 million) in 2019 and projected to reach 8.0% (454 million) by 2030 and 8.6% (548 million) by 2045 [3].

Included in cases are roughly 5%–10% with type 1 diabetes (T1D). While its onset is most common in youth, it can develop at any age and most people living with the disease are adults [1,6]. It is associated with a genetic predisposition combined with an environmental trigger that causes autoimmune destruction of the insulin-producing beta cells of the pancreas and requires individuals to replace insulin through exogenous means. Physical activity and T1D are discussed fully in another chapter, whereas T2D, PD, and gestational diabetes mellitus (GDM, when hyperglycemia only presents during pregnancy) are covered in this chapter.

Lifestyle management to treat type 2 diabetes and prediabetes

Given the nature of these chronic conditions and associated comorbidities, the treatment of T2D usually consists of an individualized treatment plan that includes education, blood glucose management, cardiovascular disease risk reduction, and continued screening for microvascular changes in order to achieve and maintain optimal blood glucose, lipid, and blood pressure levels to prevent or delay health complications [4,7–11]. More optimal outcomes are typically accomplished with a program of regular physical activity, medical nutrition therapy (MNT), blood glucose monitoring, and weight loss, along with medication use and, in some cases, bariatric surgery. When medications and bariatric surgery are utilized, they should be added to lifestyle improvements rather than replace them [12]. Similarly, the goal of PD management is to prevent progression to T2D and possibly prevent, delay, or reverse other associated health complications.

Exercise to Prevent and Manage Chronic Disease Across the Lifespan. https://doi.org/10.1016/B978-0-323-89843-0.00016-7

Physical activity

Recurrent hyperglycemia (elevations in blood glucose) increases the risk for acute and chronic health issues, including cardiovascular (macrovascular) disease, microvascular diseases like retinopathy and nephropathy, and nerve damage (both autonomic and peripheral neuropathy) [7,8]. Lifestyle management, including regular physical activity, is thus a key strategy for delaying or preventing some of these potential health complications [13]. Physical activity is also a key behavior in the prevention of T2D and potentially the reversal of PD or early-stage T2D [4].

Among other studies around the globe, the U.S. Diabetes Prevention Program (DPP) demonstrated a 58% reduction in the progression of high-risk adults to T2D through a lifestyle intervention with goals of 7% weight loss and 150 min of physical activity per week [14–16]. A study in Finland reported similar risk reduction [17]. Moreover, a 10-year follow-up to the DPP (i.e., the DPP Outcomes Study) showed that the lifestyle intervention group continued to have a 34% lower risk of developing T2D [18]. DPP lifestyle modification programs have achieved clinically meaningful body weight and cardiometabolic health improvements [19] and have been implemented in various communities around the United States in an attempt to prevent T2D in at-risk adult populations [20].

Medical nutrition therapy

Often the most challenging aspect of lifestyle management, MNT is essential to the management of T2D and PD. A recent consensus report from the American Diabetes Association on this topic recommends nutrition counseling that works toward improving or maintaining blood glucose targets, weight management goals, and cardiovascular risk factors within individualized treatment goals for all adults with diabetes and PD [21]. These guidelines promote individually developed dietary plans based on metabolic, nutrition, and lifestyle requirements in place of a calculated caloric prescription because a single diet cannot adequately treat all types of diabetes or individuals.

Monitoring of glycemia

Monitoring of glucose concentrations continues to be an important part of managing T2D, particularly in insulin users but even during the first year or longer after diagnosis in noninsulin users [22]. No standard frequency for glucose monitoring has been established for individuals with T2D, but it should be performed frequently enough to help the individual meet blood glucose treatment goals. Increased frequency of monitoring is often required when individuals begin an exercise program to assess blood glucose before and after exercise and to allow safe participation and guidance on making appropriate activity, dietary, and medication adjustments. Individuals who require glucose-lowering medications, and insulin, in particular, must understand how their medications work to ensure the greatest success and safety.

Glucose monitoring around activity may also provide positive feedback regarding the regulation or progression of the exercise prescription and improve long-term exercise participation. When people with diabetes undertake an exercise program, it is helpful for them to monitor their glycemic levels closely. Monitoring and recording their levels before and after exercise (regardless of the method) is important, at least initially, because it may:

- Allow for early detection and prevention of hypoglycemia or hyperglycemia
- Determine appropriate preexercise levels to lower the risk of hypoglycemia or hyperglycemia
- Identify those who can benefit from monitoring during and after exercise
- Provide information for modifying exercise plans and recommendations
- Allow for better adjustment of diabetes regimens to manage all activities
- Motivate individuals to remain more active to better manage their diabetes

In addition to traditional fingerstick blood glucose monitoring, some individuals with T2D may use continuous or intermittently scanned glucose monitoring devices to get more frequent feedback on their glycemic levels and responses [23,24]. These devices have a small sensor placed under the skin that tracks interstitial glucose levels continuously, typically every 5 min, allowing users to see dynamic data with glucose values, trends, rate of change, time in range, and glucose levels that are too low or too high [25].

Weight loss and maintenance

Weight loss is often a therapeutic goal for those with T2D or PD because most are overweight or obese [26,27]. Moderate weight loss improves blood glucose management and decreases insulin resistance in adults with T2D [28] and PD [29].

MNT and physical activity combined are more effective than either alone in achieving and maintaining moderate weight loss [30]. Visceral, or deep abdominal, fat decreases peripheral insulin sensitivity and is a significant source of free fatty acids that may be preferentially oxidized over glucose, contributing to hyperglycemia. Exercise results in preferential mobilization of visceral body fat, likely contributing to the metabolic improvements even in the absence of significant weight loss [31,32]. Sustained exercise participation has been one of the strongest predictors of long-term maintenance of weight loss. The amount of physical activity required to maintain the weight loss is likely more than what is required to improve blood glucose management and cardiovascular health [33,34].

Being physically active with type 2 diabetes or prediabetes

Most of the benefits of physical activity for individuals with T2D come from regular, long-term participation. These benefits can include improvements in metabolic health (glucose management and insulin resistance), hypertension, lipids, body composition and weight loss or maintenance, and psychological well-being [35]. Both the frequency of aerobic training and the volume of resistance training appears to be important in lowering overall blood glucose levels in T2D [36].

Health benefits of physical activity

The health benefits associated with being regularly active are numerous. Individuals with T2D and PD can attain the same benefits and, in some cases, additional ones related to the management of blood glucose and prevention of common comorbidities. Some of these benefits are associated with changes in insulin sensitivity, blood pressure, blood lipid levels, and mental health [8,9,11,37].

Glycemia and insulin sensitivity

Like acute exercise, regular training can improve blood glucose levels. Exercise training (both aerobic and resistance) improves overall glycemic management (A1C) and glucose tolerance in adults with T2D [38,39] and PD [40,41]. In a meta-analysis, regular exercise training had a significant benefit on insulin sensitivity in adults with T2D that may persist, in some cases, even beyond 72 h after the last exercise session [42]. Following exercise training, insulin-mediated glucose disposal is improved. Insulin sensitivity of both skeletal muscle and adipose tissue can improve with or without a change in body composition [43,44]. Exercise training may improve insulin sensitivity through several mechanisms, including changes in body composition, muscle mass, fat oxidation, capillary density, and GLUT4 glucose transporters in muscle [42,45–49]. The effect of exercise on insulin action is lost within a few days, emphasizing the importance of regular participation in physical activities.

Blood pressure

Hypertension affects approximately 74% of adults with diabetes [50]. Myriad studies have shown that exercise training can lower blood pressure in adults with diabetes, and a recent meta-analysis and systematic review suggested that while both aerobic and resistance exercise appear effective in reducing systolic and diastolic blood pressure, aerobic activity may have a greater effect [51]. Reduction in blood pressure occurs more commonly in resting systolic levels; many have shown no change in diastolic blood pressure with training [52–54].

Cholesterol and lipids

The effect of exercise training on lipids is mixed and may vary with the activity and other lifestyle changes (such as MNT). In general, supervised aerobic exercise leads to a more significant lowering of total cholesterol, triglycerides, and low-density lipoprotein cholesterol than no exercise, and supervised resistance training showed more benefit than inactivity in improving total cholesterol [52]. Lipid profiles may benefit more from a combination of exercise training and weight loss. The Look AHEAD (Action for Health in Diabetes) trial found that lifestyle participants with T2D had greater decreases in triglycerides and high-density lipoprotein cholesterol than nondiabetic controls, although both groups experienced decreases in low-density lipoprotein cholesterol [43,55]. Resistance and aerobic exercise training are generally considered to have a similar impact on cardiovascular risk markers and are equally safe for adults with T2D [56]. Regular activity can also lower lipids levels in adults with PD, particularly when combined with MNT that promotes weight loss [40].

Mental health

Psychological benefits of regular exercise training have been shown for people without and with diabetes, including reduced stress, anxiety, and mild to moderate depression, and improved self-esteem [57,58]. However, those who exercise to prevent disease have significantly improved mental well-being, while it may deteriorate in adults who exercise for disease management (e.g., cardiovascular disease, end-stage renal disease, and cancer) [59]. These findings suggest that benefits may vary and that people with fewer existing health concerns may benefit the most. Mechanisms for the impact of exercise include psychological factors, such as increased self-efficacy and changes in self-concept, as well as physiological factors like increased norepinephrine transmission and brain endorphins [58].

Primary mechanisms underlying exercise improvements in blood glucose

Physical activity is a cornerstone in the prevention and treatment of T2D and PD due, in part, to the well-established effects of exercise on blood glucose management and insulin sensitivity. Individuals with T2D and PD can experience improvements in blood glucose levels following a single session of exercise or with repeated sessions (i.e., regular exercise training). These effects are largely attributable to mechanisms and alterations within skeletal muscle.

Acute exercise impact

A single session of exercise increases skeletal muscle energy demand, which can increase muscle glucose uptake 20-fold compared to rest [60]. Muscle glucose uptake during exercise occurs through insulin-independent mechanisms, whereby contraction-induced signals in muscle promote the translocation of glucose transporters (GLUT4) to the plasma membrane [61]. This translocation provides a stimulus to immediately lower blood glucose levels in individuals with PD or T2D [62], particularly if exercise is done postprandially [63]. However, blood glucose levels during exercise are also influenced by hepatic glucose production, which is stimulated by counterregulatory hormones at rest and during exercise [64]. Thus, the prevailing impact of exercise on glycemia is determined by the balance between hepatic glucose production and peripheral glucose uptake.

Acute effects following exercise

Increased insulin-independent glucose uptake generally subsides within 2 h of exercise cessation. Peripheral insulin sensitivity is enhanced via insulin-dependent mechanisms for up to 48 h following exercise [65,66]. The improvement in insulin sensitivity in the postexercise period has been attributed to increased muscle glycogen synthase activity and glycogen resynthesis and increased sensitivity of select proteins in the insulin signaling pathway [67]. Emerging evidence also suggests that acute exercise redistributes GLUT4 to a more easily recruitable site in muscle and/or alters the intrinsic capacity of GLUT4 so that greater glucose uptake occurs for a given insulin concentration [68]. Even a single session of exercise undertaken by sedentary, obese adults can acutely restore insulin sensitivity to chronic levels observed in regular exercisers [69]. As a result of the enhanced peripheral insulin sensitivity postexercise, reductions in mean blood glucose and postprandial hyperglycemia may last up to 24 h or more in individuals with T2D and PD [70–72].

Chronic exercise

Repeated sessions of exercise (exercise training) improve glycemic management and increase insulin sensitivity in adults with PD [73] and T2D [74]. A number of adaptations in skeletal muscle have been linked with improved insulin sensitivity, including increased mitochondrial content, capillary density, lipid metabolism, and GLUT4 protein content [67]. Weight loss associated with exercise training also contributes to this glycemic improvement, but weight loss is not necessary for exercise training-induced improvements in insulin sensitivity [44]. Reductions in hepatic fat content [75] and increased pancreatic beta-cell function [76] are evident in adults with T2D following exercise training and contribute to whole-body improvements in glycemia. However, training-induced improvements in insulin sensitivity can be reversed in as few as 4 days without exercise [77], highlighting that regular participation is necessary to preserve these training-induced benefits.

Impact of physical activity type on glycemic management

Most research on blood glucose improvements has focused on aerobic exercise, but resistance training and other types have been shown to similarly lower A1C after 12 weeks in adults with T2D [78]. High-intensity intervals, and even certain

TABLE 1 Benefits of types of training on blood glucose and health.

Aerobic	• Improves blood glucose management, but effects last only 2–72 h (mostly related to muscle glycogen repletion) • Improves insulin sensitivity and fat oxidation and storage in muscle • Intensity acutely affects blood glucose more than exercise volume • Prevents or delays the onset of T2D and may lead to PD reversal
High-intensity intervals	• Improves blood glucose management, but effects last similarly to continuous aerobic exercise protocols • May result in greater improvements in blood glucose levels than moderate-intensity aerobic training for some • Combined aerobic and resistance HIIE training (e.g., bodyweight squats, burpees, mountain climbers) can improve insulin sensitivity
Resistance	• Improves insulin action and fat oxidation and storage in muscle • May improve blood glucose management, primarily through increased muscle mass and prevention of muscle mass losses related to aging, disuse, and diabetes • Benefits of resistance and aerobic training on glycemic management have been demonstrated to be additive • May improve muscle quality (e.g., less marbling with fat)
Flexibility	• Little or no effect on blood glucose management (unless undertaken as part of a yoga or tai chi training program) • Flexibility training, combined with resistance training, increases the range of motion around joints • May not necessarily reduce risk of injury from activity • Yoga participation may improve blood glucose management
Balance	• Little or no effect on blood glucose management (unless balance exercises are part of a resistance training protocol) • Power training and other lower body and core resistance exercises may also improve balance • Regular balance training reduces the risk of falling

balance and flexibility protocols, may also improve insulin sensitivity and glycemic management in individuals at risk for or with T2D (as shown in Table 1).

Aerobic exercise

Moderate-intensity, continuous aerobic exercise has traditionally been recommended for individuals with T2D. Evidence from meta-analyses suggests that performing ≥150 min of continuous aerobic exercise per week leads to greater reductions in A1C than performing <150 min [79], suggesting that exercise volume is important and higher volumes may lead to greater improvements in blood glucose levels. Exercise intensity and duration should also be considered, with evidence suggesting an intensity-duration tradeoff with respect to exercise-induced improvements. Both short duration, high-intensity cycling lasting 20 min [71], and long duration, low-intensity walking lasting 60 min [80] can improve glycemic management and insulin sensitivity for at least 24 h after exercise. However, when total exercise volume is equalized during training (i.e., the same amount of calories are expended each session), performing exercise at a higher intensity leads to greater improvements in A1C [81] and oral glucose tolerance [82] than exercise performed at a lower intensity.

High-intensity interval exercise (HIIE)

Physical training that includes HIIE protocols has emerged as a popular form of exercise and is now included as an option in physical activity recommendations for adults with T2D and PD. HIIE training protocols are infinitely variable with respect to the relative intensity, duration, and number of intervals, but generally involve alternating periods of relatively intense exercise with periods of low-intensity exercise or complete rest for recovery. Many HIIE protocols are low-volume and, therefore, time-efficient (≤25 min per session), which may be appealing for individuals who cite lack of time as a barrier to regular exercise participation. One of the most well-documented low-volume HIIE protocols, which involves 10 × 1-min cycling efforts at ~90% of maximal heart rate, interspersed with 1 min of recovery (20 min total), has been shown to reduce mean blood glucose levels and postprandial excursions for 24 h following a single session [71] and improve A1C and insulin sensitivity after 8–12 weeks of training in adults with T2D [83,84].

When low-volume HIIE training is compared to higher volumes of moderate-intensity continuous aerobic exercise training, similar improvements in glycemic management and insulin sensitivity are typically observed despite large time commitment differences [85,86]. However, when exercise volume is matched between protocols, HIIE training may result in greater improvements in blood glucose levels compared to moderate-intensity continuous training [87]. However, those higher volume HIIE protocols are longer in duration and not necessarily time-efficient (e.g., 40–60 min per session). While most HIIE training protocols involve cycling, other modes of exercise are also effective. For example, interval walking involving alternating between 3 min "hard" and 3 min "easy" for 60 min lead to greater improvements in glycemic management in adults with T2D than volume-matched continuous walking after 16 weeks [88]. In addition, 8- to 20-min sessions of combined aerobic and resistance HIIE training (e.g., bodyweight squats, burpees, mountain climbers) improves insulin sensitivity in adults with T2D when performed 3 times per week for 6 weeks [89].

Resistance exercise

A single session of resistance exercise (45 min at 75% of one-repetition maximum) has been shown to reduce the prevalence of hyperglycemia for 24 h in adults with T2D [90]. When moderate- to high-intensity resistance exercise is performed two to three times weekly for 8 weeks or longer, improvements in A1C and insulin sensitivity have also been observed [91]. Skeletal muscle mechanisms contributing to resistance-training-induced improvements in insulin sensitivity appear to be similar to aerobic exercise [47]; however, there may be unique adaptations resulting from resistance exercise alone, such as increased muscle mass [92]. Indeed, the benefits of resistance training and aerobic training on glycemic management have been demonstrated to be additive [93], and both types of exercise should be undertaken by adults with T2D and PD.

Flexibility exercise

Both flexibility and balance exercises are important for health and well-being, particularly in older adults with T2D and PD. Limited joint mobility resulting in part from the formation of advanced glycation end-products, which accumulate during normal aging, are accelerated by the presence of hyperglycemia [94]. Stretching exercises increase the range of motion around joints and flexibility [95] but generally do not impact glycemia. Flexibility exercises, alone or in combination with resistance training, can increase joint range of motion in individuals with T2D and allow them to more easily engage in activities that require flexibility [95]. Moreover, flexibility programs are easy to perform and may provide the perfect introduction to a more physically active lifestyle for less fit and older adults and should be undertaken regularly, although not necessarily as a substitute for other types of training [96].

Traditional static and dynamic stretching, as well as yoga and tai chi, can provide fitness benefits, as can balance training. Inclusive of basic stretching as part of the activity, yoga improves glycemic management, lipid levels, and body composition in some adults with T2D [97–100]. Tai chi training incorporates some balance, stretching, and resistance elements and may improve glycemia, balance, neuropathic symptoms, and some aspects of quality of life in adults with T2D and neuropathy [101,102]. In those with PD, a meta-analysis of yoga intervention studies reported improvements in glycemic management, lipid profiles, and other health measures for this population as well [103].

Balance exercise

Older adults with T2D and PD should also undertake exercises that maintain or improve balance [104,105]. Many lower body and core resistance exercises can double as balance training, and even power training undertaken by adults with T2D can improve their overall body balance [106]. Many variations of balance training can reduce falls risk by improving balance and gait, even in adults with peripheral neuropathy [107,108]. At-home Wii Fit balance exercises have also been shown to reduce falls risk in the absence of significant changes in leg strength, suggesting that interventions to reduce falls risk that target intrinsic risk factors related to balance control (rather than muscle strength) may have positive benefits for older adults with T2D at risk for falling [108] (Table 1).

Impact of activity breaks and sedentary time

Engaging in prolonged periods of sedentary behavior increases risk of T2D, cardiovascular disease, and premature mortality, even when adjusted for overall physical activity levels [109,110]. In adults with T2D, prolonged periods of sedentary time are also associated with worse overall glycemic management, greater postprandial glucose spikes, and insulin resistance [111,112]. Interrupting sitting with brief, repeated activity breaks has been shown to reduce cardiometabolic disease risk factors associated with prolonged sitting. For example, performing 2–3 min of light-intensity walking or bodyweight resistance exercise every 30 min lowered postprandial blood glucose, insulin, and triglycerides in adults with T2D [113]. Walking and resistance exercise breaks also have beneficial effects on postprandial glycemic management in obese [114] and inactive [115] adults. In addition, specifically targeting the postprandial period with less frequent, but longer, activity breaks (such as 10- to 15-min walks after each meal) lowers postprandial hyperglycemia in adults with PD [116] and T2D [117,118]. Thus, in addition to recommending moderate- to vigorous-intensity physical activity, minimizing and/or breaking up prolonged periods of sitting should be encouraged in those with T2D and PD, and accumulating activity by undertaking shorter bouts of activity more frequently may work equally well in managing blood glucose levels [119].

Nutritional impact before and following exercise

The interaction between diet and exercise is important to consider when describing the acute and chronic effects of exercise on glycemic management and insulin sensitivity. Both the timing of exercise around a meal and postexercise nutrition have been shown to impact exercise responses.

Exercise timing around a meal

Performing exercise after eating a meal, as opposed to before the meal, leads to more consistent reductions in postprandial glucose concentration in those with and without T2D [63]. For example, a 20-min walk [120] or 30-min session of resistance exercise [121] performed 15–45 min following a meal lowered the meal-induced glycemic excursion to a greater extent than when the same exercise was performed before the meal. The superior benefits of postprandial exercise may be due to the additive effects of insulin-dependent and insulin-independent muscle glucose uptake and/or the insulin-induced suppression of hepatic glucose production when exercise is performed postprandially. When assessed over a 12-week aerobic exercise training program in adults with T2D, greater improvements in A1C were observed in those who performed their training sessions in the postprandial compared to the postabsorptive state [122].

Recovery (postexercise) nutrition

The time course of improvement in insulin sensitivity following a single session of exercise can be influenced by an individual's diet in the hours following exercise. Carbohydrate consumption following aerobic exercise has been shown to blunt next-day improvements in insulin sensitivity [123,124] and 24-h glycemic management [125] in healthy, recreationally active adults. However, when isoenergetic [123] or surplus quantities [126] of fat are consumed postexercise, acute improvements in insulin sensitivity are still observed. Considered together, these results suggest that acute improvements in insulin sensitivity among healthy adults are sensitive to carbohydrate intake, which is most likely a result of faster rates of muscle glycogen repletion. From a practical perspective, consuming lower-carbohydrate meals after exercise may prolong acute improvements in insulin sensitivity, but more research is needed in adults with T2D and PD.

Exercise recommendations for people with T2D and PD

In general, exercise training recommendations for adults and youth with T2D and PD differ very little from those for everyone else at the same stage of life (see recommendations summarized in Table 2). For people with these conditions, though, it is important to maintain insulin sensitivity between bouts of activity, so it is recommended that no more than two consecutive days lapse with no activity [127]. Regular moderate- to vigorous-intensity aerobic activity is recommended, along with two to three nonconsecutive days of resistance training each week. Balance training is recommended for individuals ages 40 and older or of any age with peripheral or central neuropathy. In addition, flexibility training should be included to limit the loss of range of motion around joints over time. Youth with T2D or PD can generally follow the recommendations for physical activity for youth without diabetes.

Youth

All children (5–11 years) and adolescents (12–17 years) should undertake at least 60 min of moderate- to vigorous-intensity mixed aerobic and anaerobic physical activity daily (420 min/week) [128]. Many of these youth have overweight or obesity and regular physical movement is important for weight loss and management. Physical activity patterns in youth track into adulthood, so establishing healthful ones as early as possible is important to prevent or delay cardiovascular disease, which is the major cause of early mortality and morbidity in these youth once they reach adulthood [129].

Youth can be physically active daily as part of play, games, sports, transportation, recreation, physical education, or planned exercise in the context of family, school, and community activities (e.g., volunteer, employment). This should be achieved above and beyond the incidental physical activities accumulated in the course of daily living. Reducing

TABLE 2 Recommended aerobic, resistance, flexibility, and balance training for adults with type 2 diabetes and prediabetes.

Training method	Mode	Intensity	Frequency	Duration	Progression	Important considerations
Aerobic	Walking, jogging, cycling, swimming, aquatic activities, rowing, dancing, interval training	40%–59% of VO₂R (moderate), RPE 11–12; or 60%–89% of VO₂R (vigorous), RPE 14–17	3–7 days/week, with no more than 2 consecutive days between bouts of activity	Minimum of 150–300 min/week of moderate activity or 75–150 min of vigorous activity, or a combination thereof	Rate of progression depends on many factors including baseline fitness, age, weight, health status, and individual goals; gradual progression of both intensity and volume is recommended	*Note:* Special considerations apply to all these training methods • Avoid or take precautions for exercise undertaken by insulin users • Be aware of any signs and symptoms for vascular and neurological complications
Resistance	Free weights, machines, elastic bands, or body weight as resistance; undertake 8–10 exercises involving the major muscle groups	Moderate at 50%–69% of 1-RM, or vigorous at 70%–85% of 1-RM	2–3 days/week, but never on consecutive days	10–15 repetitions per set, 1–3 sets per type of specific exercise	As tolerated Increase resistance first, followed by a greater number of sets, and then increased training frequency	• Include appropriate warm-ups and cool-downs • Use proper footwear and inspect feet daily • Avoid exercising with environmental extremes • Avoid or postpone exercise when blood glucose is not well managed • Maintain adequate hydration • Monitor glucose and follow guidelines to prevent hypoglycemia and hyperglycemia
Flexibility and balance training	Static, dynamic, or PNF stretching; balance exercises; yoga and tai chi may be appropriate for a range of motion, balance, and strength	Stretch to the point of tightness or slight discomfort	≥2–3 days/week or more; usually done with when muscles and joints are warmed up; lower body and core resistance exercises may double as balance training	10–30 s per exercise of each muscle group; 2–4 repetitions of each; balance exercises can be practiced daily or as often as possible	As tolerated; may increase range of stretch as long as not painful; balance training should be done carefully to minimize the risk of falls	

Note: *1-RM*, 1-repetition maximum; *PNF*, proprioceptive neuromotor facilitation; *RPE*, rating of perceived exertion; *VO₂R*, VO₂ reserve.

sedentary time is convincingly associated with a favorable cardiovascular profile, and several expert panels recommend limiting leisure screen time to <2 h/day [128].

Adults (18–64 years)

To achieve health benefits, all adults are recommended to undertake at least 150–300 min of moderate- or 75–150 min of vigorous-intensity aerobic physical activity per week, or an equivalent combination of both types [128,130]. Preferably, aerobic activity should be spread throughout the week, with no more than two consecutive days lapsing between bouts. Resistance training is also recommended at least two, but preferably three, days per week to promote muscle mass growth and retention (see Table 2).

Physical activity appears to lower cardiovascular mortality risk at all levels of glycemic management. For those who are initially physically inactive, doing amounts below the recommended levels can provide some health benefits and may prevent the onset of T2D in community-dwelling adults [131]. It may be appropriate to start with smaller amounts of physical activity and then gradually increase the duration, frequency, and intensity as a stepping stone to meeting these guidelines.

Older adults (≥65 years)

The goal for older adults is to maximize their functional abilities while promoting better health. It is recommended that adults aged 65 years and older should accumulate at least 150–300 min of moderate- to vigorous-intensity aerobic physical activity per week, as recommended for younger adults, if possible. It is beneficial to add muscle- and bone-strengthening activities that use the major muscle groups at least 2 days per week.

While the exercise guidelines for adults also apply to older adults, there are some additional ones that apply only to older adults (with or without diabetes or PD) [128]:

- When older adults cannot do 150 min of moderate-intensity aerobic activity a week because of chronic conditions, they should be as physically active as their abilities and conditions allow.
- Recommend exercises that maintain or improve balance, particularly if they are at risk of falling.
- Determine their level of effort for physical activity relative to their level of fitness.
- Older adults with chronic health conditions should understand whether and how their conditions affect their ability to do regular physical activity safely.

Older adults can meet these guidelines through the same means as younger adults (i.e., increased activities of daily living) to reduce the risk of comorbid disease and premature death, maintain functional independence and mobility, as well as improve fitness, body composition, bone health, cognitive function, and indicators of mental health. These guidelines may be appropriate for older adults with frailty or other comorbid conditions; however, individuals with health issues should consult a health professional to understand the types and amounts of physical activity appropriate for them based on their exercise capacity and specific health risks or limitations.

Barriers to participation and exercise maintenance

Although physical activity is a key lifestyle tool in the management of T2D and PD, many individuals with these chronic conditions fail to become or remain regularly active. In addressing this failure, it is recommended that health care providers use the broader term "physical activity" in place of the narrower term "exercise" because many types of movements have been proven to have a positive impact on both health and physical fitness and may be more accessible, even simply taking more daily steps [131]. In addition, more than one type of physical activity may be required to gain optimal glycemic and health benefits, and undertaking a variety of activities should be encouraged [130].

For individuals with T2D or PD, many of whom are physically unfit and sedentary, it may be beneficial to start by encouraging them to find ways to incorporate more unstructured activity into daily living, even if starting a more structured activity plan is implemented. While lifestyle activity does not entirely replace structured exercise training, it can help individuals increase their daily activity level and build a better fitness base. In addition, those who have successfully implemented more activity into their daily lifestyle may feel more confident, willing, and able to initiate structured forms.

Exercise barriers

Many of the barriers to exercise are similar among people with and without diabetes, but those with T2D or PD need appropriate education about exercise effects on blood glucose management and complications [132]. Some of the typical barriers include lower self-efficacy (i.e., a person's belief in his/her ability to succeed), inappropriate goal-setting, lack of access to facilities, unrealistic goals, lack of supervision or support, and inattention to cultural nuances related to physical activity. Health issues like obesity and knee and hip osteoarthritis may be barriers as they negatively impact self-efficacy. The built environment can impact a person's ability and willingness to be regularly active. The availability of facilities and pleasant places to walk is frequently an important predictor of regular physical activity. Focusing on creating more exercise-friendly environments can certainly promote greater physical activity participation.

Goal-setting and overcoming obstacles

Developing realistic goals, selecting appropriate activities, progressing slowly (to avoid injury or burnout), and getting supportive feedback can increase confidence for being active [133–135]. For example, goals that are vague, overly ambitious, or long-term may not provide enough self-motivation to maintain short-term goals. Exercise interventions should focus on self-efficacy, enjoyment, problem-solving and goal-setting, social-environmental support, and cultural nuances. Greater levels of activity are frequently related to higher levels of self-efficacy, which reflect confidence in the ability to be more active [134,136]. Counseling done by health care professionals may also be a meaningful and effective source of support [137]. Likewise, supervision of exercise sessions by qualified exercise trainers improves compliance and glycemic control [138], and individuals engaging in supervised training likely gain benefits that exceed those of exercise counseling and increased activity undertaken alone [91]. Individual cultural practices and beliefs may also influence the adoption of physical activity programs. Promoting activities that do not offend or ignore the cultural beliefs of the individual and provide culturally appropriate suggestions to help tailor a suitable exercise prescription, including yoga, tai chi, and dancing, is recommended.

Having a defined strategy is helpful for exercise maintenance. Health care providers and other professionals can assist by asking individuals to consider the following:

- How easily can they engage in their activity of choice where they live?
- How suitable is the activity in terms of their physical attributes and lifestyle?
- Have individuals identify exercise benefits they find personally motivating.
- Assist individuals in setting physical activity goals that are not too vague, ambitious, or distant; instead, set goals that are specific, measurable, attainable, realistic, and time-bound (SMART).
- Encourage individuals to establish a routine to help exercise become more habitual.
- Have individuals identify any social support systems they may have.
- Provide positive feedback and praise for meeting goals.
- Help individuals troubleshoot how to overcome any specific obstacles they have to being active related to their diabetes (such as fear of falling or hypoglycemia) or other concerns.

Technological tools to promote activity adherence

The use of technology may promote greater adherence by assisting people in overcoming some of these barriers. Use of pedometers may promote physical activity by increasing awareness of daily movement, providing motivation and visual feedback, and encouraging conversation and support among other users, family, and friends [139]. Accelerometers and other GPS-based activity trackers may also be used to detect and track activity, but are more expensive than pedometers and often require the use of sophisticated software.

Exercise safety considerations and precautions for T2D and PD

Individuals with T2D should be evaluated for potential indicators of complications, ideally at least annually. These indicators may include elevated resting pulse rate, loss of sensation or reflexes especially in the lower extremities, foot sores or ulcers that heal poorly, excessive bruising, and retinal vascular abnormalities.

Medical clearance

Clinicians should review the medical history as some individuals with T2D and PD may need medical clearance prior to starting training [140]. In making a determination regarding the necessity of such clearance, consider the following:

- Exercise participation and training status
- Body weight and body mass index
- Resting blood pressure
- Laboratory values for A1C, plasma glucose, lipids, and proteinuria
- Presence or absence of acute and chronic complications
- If chronic complications exist, the severity of those complications
- Presence of other non-diabetes-related health issues

Exercise testing

Some individuals may be recommended to undergo exercise testing before beginning a planned exercise program [140]. While most with T2D or PD can benefit from participating in regular physical activity, becoming more active is not without risk and each individual should be assessed for safety. Priority must be given to minimizing the potential adverse effects of exercise through appropriate screening, program design, monitoring, and education [140].

Exercise testing may be viewed as a barrier or unnecessary for some individuals. Discretion must be given to an individual's status when determining the need for exercise testing. For planned participation in low-intensity exercise, clinicians should use their best clinical judgment in deciding whether to recommend preexercise testing [96,127]. Conducting maximal exercise testing before starting participation in most low- or moderate-intensity activities (<60% heart rate reserve or VO$_2$ reserve) is considered unnecessary.

However, for exercise more vigorous than brisk walking (>60% heart rate reserve or VO$_2$ reserve) or exceeding the demands of everyday living, sedentary and individuals 40 years or older with diabetes will likely benefit from being assessed for conditions that might be associated with cardiovascular disease, that contraindicate certain activities, or that predispose to injuries. The assessment should include a medical evaluation and screening for blood glucose management, physical limitations, medications, and macrovascular and microvascular complications. It may also include a graded exercise test.

In general, graded stress testing with electrocardiogram (ECG) monitoring may be indicated for individuals meeting one or more of these criteria [96]:

- Age >40 years, with or without CVD risk factors other than diabetes
- Age >30 years and any of the following:
  - Type 1 or type 2 diabetes of >10 years
  - Hypertension
  - Cigarette smoking
  - Dyslipidemia
  - Proliferative or preproliferative retinopathy
  - Nephropathy including microalbuminuria
- Any of the following, regardless of age:
  - Known or suspected coronary artery disease, cerebrovascular disease, or peripheral artery disease

Medication concerns and management

In some cases, blood glucose levels should be self-monitored before an exercise session to determine whether the person can safely begin exercising, especially for anyone using insulin or glucose-lowering oral diabetes medications that cause the pancreas to release insulin (such as sulfonylureas and meglitinides), as listed in Table 3. Consideration must be given to how long and intense the exercise session will be, as well as where and when it will take place. Some individuals will need to take in supplemental carbohydrates to prevent hypoglycemia, particularly anyone using insulin [127].

Those with T2D whose glycemic levels are adequately managed by MNT and physical activity alone usually experience a reduction in blood glucose with low- to moderate-intensity exercise sessions. Timing of activities after meals can help many individuals with T2D reduce postprandial hyperglycemia [63].

If diabetes is managed by MNT or oral medications that carry little to no risk of hypoglycemia (which is most of them), the majority of individuals will not need to consume supplemental carbohydrates for exercise lasting less than 30–60 min. Extra carbohydrate is mainly required during longer duration efforts. If blood glucose levels are moderately elevated prior to exercise (up to 180 mg dL^{-1}, or 10.0 mmol L^{-1}), an activity that would normally require some carbohydrate intake may instead require less or none at all. Active individuals without diabetes benefit from consuming carbohydrates during the longer duration and higher intensity workouts, and most can use up to 70–80 g per hour during such activities [141]. In people with T2D or PD, supplemental carbohydrate or other macronutrient intake needs can be similar for performance reasons. For anyone trying to lose weight through physical activity, though, medication adjustments may be preferable to additional calorie intake if hypoglycemia is a usual response to physical activity.

Training precautions and additional considerations

When starting a regular exercise training program, most people with T2D and PD can anticipate significant and meaningful improvements, assuming the following exercise training considerations and precautions are followed:

TABLE 3 Diabetes medications with possible exercise effects.

Medication	Actions	Exercise effect
Sulfonylureas (1st generation—only generic versions available) Examples: tolbutamide, tolazamide	Increase insulin production in the pancreas	Risk of hypoglycemia
Sulfonylureas (2nd generation) Examples: glyburide, glipizide, glimepiride	Increase insulin production in the pancreas; have an insulin-sensitizing effect as well	Risk of hypoglycemia (but lower than for 1st generation sulfonylureas)
Meglitinides Examples: repaglinide, nateglinide	Increase insulin release from the pancreas	Risk of hypoglycemia during exercise that follows meals for which these are taken
Biguanides All types of metformin	Primarily decrease hepatic glucose production; may improve insulin resistance in muscles	Very low risk of hypoglycemia (possibly after prolonged or strenuous exercise)
Amylin analog Pramlintide	Decrease postmeal glucagon production; delay gastric emptying; increase satiety	Possible delayed treatment of hypoglycemia when taken near initiation of exercise
Insulin: rapid-acting, intermediate-acting, long-acting, very long-acting Examples: aspart, glulisine, lispro, regular, NPH, glargine, detemir, degludec	Lowers blood glucose levels with action dependent on the onset, peak, and duration	Risk of hypoglycemia

- Medical clearance (and exercise testing) prior to starting activities more vigorous than brisk walking may be recommended for individuals with any signs or symptoms of cardiovascular disease, longer duration of T2D, older age, or other diabetes-related complications [127].
- Individuals should not begin the exercise with blood glucose >250 mg dL^{-1} (13.9 mmol L^{-1}) if moderate or higher levels of blood or urinary ketones are present.
- Caution should be used for exercising with blood glucose >300 mg dL^{-1} (16.7 mmol L^{-1}) without excessive ketones; individuals should attempt to stay hydrated and only begin activity if feeling well [127,142].
- Both resting and exercise heart rates may be impaired in individuals with cardiac autonomic neuropathy, which can result in a higher than typical resting heart rate (usually 100 beats per minute or higher) and a blunted peak heart rate. If the heart rate is to be used to prescribe exercise intensity for such individuals, maximal heart rate should be measured rather than estimated, or ratings of perceived exertion can be used instead [143].
- Peripheral neuropathy and poor circulation may increase the risk of a foot injury and delay healing. Individuals should inspect their feet regularly for any signs of trauma and avoid weight-bearing exercise with unhealed plantar ulcers.

Exercising with health complications

Individuals with health complications arising from T2D should be made aware that certain types of activities may be contraindicated due to their condition, or that special testing/preparation may be required. Where macrovascular diseases, such as coronary artery and peripheral artery disease, are present, preexercise screening should follow the guidelines set by the American College of Sports Medicine [140] and the American Diabetes Association [96]. Adults with cardiovascular autonomic neuropathy should consult a physician and undergo an exercise stress test prior to undertaking a physical activity program, but will also generally benefit from performing moderate aerobic exercise [144], provided that it is performed with an adequate warm-up, and cool-down, and avoidance of temperature extremes.

As for microvascular complications, if an individual with T2D has severe nonproliferative or unstable proliferative retinopathy, activities that increase intraocular pressure (e.g., high-intensity activities, heavy weight lifting) or activities that involve jumping or jarring should be avoided [127]. Vigorous activities may also not be appropriate for individuals with advanced diabetic kidney disease, although moderate-intensity activities can be very beneficial [145]. Individuals with peripheral neuropathy, especially if they are lacking sensation in their legs and feet, should ensure that use proper footwear and perform regular inspections of their feet for injuries, wounds, and infections [144]. Where wounds or injuries

TABLE 4 Special precautions for diabetes-related complications.

Health complication	Precaution
Autonomic (central) neuropathy	• Be aware of an increased likelihood of hypoglycemia, abnormal BP responses, and impaired thermoregulation, as well as elevated resting and blunted maximal HR. • Use of RPE is suggested to monitor exercise intensity, and take steps to prevent dehydration and hyper/hypothermia.
Peripheral neuropathy	• Limit exercise participation that may cause foot trauma, such as prolonged hiking, jogging, or walking on uneven surfaces. • Nonweight-bearing exercises (e.g., cycling, chair exercises, swimming) may be more appropriate, although avoid aquatic exercise with unhealed ulcers. • Check feet daily for signs of trauma and redness. • Choose shoes carefully for proper fit and wear socks that keep feet dry. • Avoid activities requiring excessive balance ability.
Diabetic retinopathy	• With unstable proliferative and severe stages of retinopathy, avoid vigorous, high-intensity activities that involve breath-holding (e.g., weight lifting and isometrics) or overhead lifting. • Avoid activities that lower the head (e.g., yoga, gymnastics) or that jar the head. • In the absence of a stress test determined maximal HR, the use of RPE is recommended (10–12 on 20 scale). • If an individual has proliferative retinopathy and has recently undergone photocoagulation or surgical treatment or is not properly treated, exercise is contraindicated. • Consult an ophthalmologist for specific restrictions and limitations.
Diabetic kidney disease	• Avoid exercise that causes excessive increases in BP (e.g., weight lifting, high-intensity aerobic exercise) and refrain from breath-holding during activities. • High BP is common, and lower intensity exercise may be necessary to manage BP responses and fatigue. • Light to moderate exercise may be possible during dialysis treatments as long as electrolytes are managed.
Hypertension	• Avoid heavy weight lifting or breath-holding. • Perform dynamic exercises using large muscle groups, such as walking and cycling at a low to moderate intensity. • Follow BP guidelines for physical activity levels. • In the absence of a maximal HR determined with an exercise stress test, the use of RPE is recommended (10–12 on a 6–20 scale).
General precautions	• Maintain hydration by drinking fluids before, during, and after exercise. • Avoid exercising during the peak heat of the day or in direct sunlight. • Carry rapid-acting carbohydrate sources during all activity. • Have glucagon available to treat severe hypoglycemia.

Note: *BP*, blood pressure; *HR*, heart rate; *RPE*, rating of perceived exertion.

contraindicate weight-bearing activities, seated exercises using free weights or resistance bands should be encouraged [127]. Other considerations are listed in Table 4.

Treatment of and exercising with gestational diabetes mellitus

Hyperglycemia is one of the most common complications of pregnancy, occurring in 16.7% of pregnancies in women ages 20–48 years worldwide [2]. Of these, 80.3% were due to gestational diabetes mellitus (GDM), while 10.6% had diabetes prior to pregnancy, and 9.1% were diagnosed with diabetes (type 1 and type 2) first detected during pregnancy [2]. The prevalence of GDM remains challenging to estimate, however, as rates differ between studies due to maternal age, body mass index, and ethnicity among women [146–149]. The ongoing epidemics of obesity and diabetes have resulted in more cases of T2D in women of reproductive age, with an increase in the number of pregnant women with undiagnosed T2D in early pregnancy [1,150]. Many cases of GDM, defined as glucose intolerance first detected during pregnancy, actually arise from preexisting hyperglycemia since routine screening is not widely performed in nonpregnant women of reproductive age [1]. Even during pregnancy, screening strategies, testing methods, and diagnostic glycemic thresholds also continue to be debated [148]. Lifestyle management encompassing healthy eating and physical activity can help women with GDM manage blood glucose levels and improve health outcomes for themselves and their offspring. It is estimated that 70%–85% with GDM can meet glycemic targets with appropriate lifestyle management [151], and many women may be able to prevent the onset of GDM with lifestyle improvements [152].

Lifestyle management of GDM

Lifestyle management includes MNT, physical activity, and appropriate weight management [150]. Dietary management is critical, and pregnant women should work with a registered dietitian to develop a personalized nutrition plan to help meet their nutritional requirements (e.g., for caloric intake, micronutrients like folate, calcium, and iron). In addition, they should be educated about the amount and type of carbohydrate to consume, given that this macronutrient has the greatest immediate impact on blood glucose levels. Education on blood glucose monitoring and physical activity guidelines is also considered equally important for GDM management [150]. Similar to recommendations for all pregnancies, women with GDM are encouraged to accumulate at least 150 min of moderate-intensity physical activity each week. Ideally, physical activity should be accumulated over a minimum of 3 days per week, although the daily activity is strongly encouraged. A variety of aerobic exercise and resistance training activities are beneficial, including pelvic floor exercises [153]. As for body weight gain, current guidelines for gestational weight gain (GWG) include weekly goals based on prepregnancy body mass index [154]. For all pregnancies, GWG beyond recommended levels has been associated with an increased risk of maternal and fetal complications, such as preterm delivery, macrosomia, and Cesarean delivery [155–157].

Maternal-fetal health and outcomes in GDM

One of the major consequences of poorly managed GDM is macrosomia, or excessive fetal growth leading to large-for-gestational-age (LGA) infants. This condition increases the risk of delivery complications for mother and offspring. For example, LGA infants are more likely to be born instrumentally or via emergency C-section, require admission to the neonatal unit, and experience neonatal hypoglycemia [158]. Furthermore, the risk of hemorrhage and genital injury to the mother is up to threefold higher in women with GDM [159].

Women with GDM also have a seven-fold increased risk of developing T2D within the first 10 years following delivery [160], linearly increasing with the duration of follow-up; the estimated risks for T2DM are 19.72% at 10 years, 29.36% at 20 years, 39.00% at 30 years, 48.64% at 40 years, and 58.27% at 50 years, respectively [161]. The degree of maternal hyperglycemia is consistently associated with increased risk of future T2D, and fasting blood glucose levels during pregnancy show the strongest association, which highlights the importance of management of GDM [162,163]. Women with GDM also have a twofold increased risk for cardiovascular disease, independent of the development of T2D [164], whereas the offspring of women with GDM have an increased risk of glucose intolerance and obesity [165].

Mechanisms underlying GDM development and impact of physical activity

Hormonal changes during pregnancy result in muscular insulin resistance that ensures nutrient delivery to the growing fetus [152]. Maternal hyperglycemia results when pregnant women are unable to adequately compensate with insulin and/or regulate higher insulin resistance and hepatic gluconeogenesis during the second half of pregnancy [166,167]. To maintain euglycemia in the third trimester, pregnant women must increase their insulin secretion approximately two- to four-fold [168]. Acute and chronic prenatal exercise may reduce maternal blood glucose levels, with the impact being greater in women with diabetes [169]. This effect likely results from an increase in GLUT-4 stimulated glucose uptake into exercising skeletal muscle [170].

Studies have also shown a reduction in the number of women with GDM requiring insulin and reduced insulin requirements following exercise interventions, suggesting improvements in insulin sensitivity or insulin production [171–173]. Regular activity is particularly important for women with GDM as postprandial hyperglycemia returns quickly upon exercise cessation. Postprandial physical activity is a safe and effective strategy to help women with GDM meet postprandial glucose targets. For example, walking or cycling for 30–60 min following breakfast lowers postprandial hyperglycemia ~0.7 mmol/L (~12.6 mg/dL) [174,175], as do 30 min of low- and moderate-intensity cycling after lunch [176]. Physical activity during pregnancy may also improve maternal glucose concentrations by preventing excessive GWG [177], and women who maintain exercise during pregnancy gain less weight and body fat, have increased fitness, and lower their risk for GDM [178].

Unfortunately, many women reduce their physical activity during pregnancy due to misguided information about the risks and safety of exercise. In contrast to early views, fetal health is not compromised during exercise due to a decrease in uterine blood flow, even at higher exercise intensities; studies have shown compensatory effects to preserve fetal blood flow and oxygen delivery, such as increased hemoglobin and uterine oxygen extraction [179,180]. Prenatal exercise is neither associated with poor birth outcomes nor does it adversely affect fetal heart rate and uterine blood flow [181,182]. Furthermore, vigorous-intensity exercise in the third trimester does not compromise fetal health in low-risk pregnancies [183]. Women with GDM have the same contraindications for physical activity as pregnant women without diabetes.

Exercise recommendations and precautions for GDM

Moderate aerobic exercise is recommended to prevent excessive weight gain and maintain cardiovascular fitness during pregnancy [184], including activities like walking, swimming, and cycling. A recliner bike can make cycling more comfortable. Given that combined aerobic and resistance training activities may result in greater improvements in pregnancy outcomes than aerobic activity alone [153], both types are recommended.

Women with GDM should follow the physical activity guidelines for all pregnant women [153,185], which are similar to those for the general population:

- Accumulate at least 150 min of moderate-intensity physical activity each week spread over a minimum of 3 days per week, but ideally including activity daily.
- Incorporate a variety of aerobic exercise and resistance training activities to achieve greater benefits, along with possibly yoga and/or gentle stretching.
- In general, continue exercising during pregnancy at the same intensity as before becoming pregnant.
- Perform Kegel (pelvic floor) exercises on a daily basis to reduce urinary incontinence.

Women with GDM should also follow the same advice relating to safety precautions and contraindicated activities during pregnancy [153]:

- Pregnant women who experience light-headedness, nausea, or feel unwell when they exercise flat on their back should modify their exercise position to avoid the supine position.
- Avoid physical activity in excessive heat, especially with high humidity.
- Avoid activities that involve physical contact or the danger of falling, along with scuba diving.
- Lowlander women (i.e., living below 2500 m) should avoid physical activity at high altitudes (>2500 m).
- Women who are considering athletic competition or exercising significantly above the recommended guidelines should seek supervision from an obstetric care provider with knowledge of the impact of high-intensity physical activity on maternal and fetal outcomes.
- Maintain adequate nutrition and hydration, including drinking adequate amounts of water before, during, and after physical activity.
- Know the reasons to stop physical activity and consult a qualified healthcare provider immediately if they occur.

Conclusion

Exercise training of various types is an essential component of the treatment plan for all individuals with type 2 diabetes, prediabetes, or gestational diabetes mellitus, and the health benefits arising from regular participation are numerous. All individuals can benefit from meeting the recommended guidelines for weekly amounts and modes of physical activity for their age group, regardless of their type of diabetes. Although physical activity is a key lifestyle tool in the management of these conditions, many individuals fail to become or remain regularly active. Addressing barriers to participation is important in increasing participation. Although certain precautions may be necessary to participate, physical activity can usually improve blood glucose management, lipid levels, blood pressure, and body weight; lower mental stress, anxiety, and depression; and potentially reduce the physical and emotional burden of these metabolic diseases.

References

[1] American Diabetes Association Professional Practice Committee. 2. Classification and diagnosis of diabetes: Standards of Medical Care in Diabetes—2022. Diabetes Care 2022;45(Supplement 1):S17–38. https://doi.org/10.2337/dc22-S002.

[2] International Diabetes Federation. IDF Diabetes Atlas. 10th ed. Brussels: International Diabetes Federation; 2021.

[3] Saeedi P, et al. Global and regional diabetes prevalence estimates for 2019 and projections for 2030 and 2045: results from the international diabetes federation diabetes atlas, 9(th) edition. Diabetes Res Clin Pract 2019;157:107843.

[4] American Diabetes Association Professional Practice Committee. 3. Prevention or delay of type 2 diabetes and associated comorbidities: Standards of Medical Care in Diabetes—2022. Diabetes Care 2022;45(Supplement 1):S39–45. https://doi.org/10.2337/dc22-S003.

[5] Hostalek U. Global epidemiology of prediabetes - present and future perspectives. Clin Diabetes Endocrinol 2019;5:5.

[6] American Diabetes Association Professional Practice Committee. 14. Children and adolescents: Standards of Medical Care in Diabetes—2022. Diabetes Care 2022;45(Supplement 1):S208–31. https://doi.org/10.2337/dc22-S014.

[7] American Diabetes Association Professional Practice Committee. 12. Retinopathy, neuropathy, and foot care: Standards of Medical Care in Diabetes—2022. Diabetes Care 2022;45(Supplement 1):S185–94. https://doi.org/10.2337/dc22-S012.

[8] American Diabetes Association Professional Practice Committee. 10. Cardiovascular disease and risk management: Standards of Medical Care in Diabetes—2022. Diabetes Care 2022;45(Supplement 1):S144–74. https://doi.org/10.2337/dc22-S010.

[9] American Diabetes Association Professional Practice Committee. 8. Obesity and weight management for the prevention and treatment of type 2 diabetes: Standards of Medical Care in Diabetes—2022. Diabetes Care 2022;45(Supplement 1):S113–24. https://doi.org/10.2337/dc22-S008.

[10] American Diabetes Association Professional Practice Committee. 6. Glycemic targets: Standards of Medical Care in Diabetes—2022. Diabetes Care 2022;45(Supplement 1):S83–96. https://doi.org/10.2337/dc22-S006.

[11] American Diabetes Association Professional Practice Committee. 5. Facilitating behavior change and well-being to improve health outcomes: Standards of Medical Care in Diabetes—2022. Diabetes Care 2022;45(Supplement 1):S60–82. https://doi.org/10.2337/dc22-S005.

[12] Schauer PR, et al. Bariatric surgery versus intensive medical therapy for diabetes - 5-year outcomes. N Engl J Med 2017;376(7):641–51.

[13] Dempsey PC, et al. Managing sedentary behavior to reduce the risk of diabetes and cardiovascular disease. Curr Diab Rep 2014;14(9):522.

[14] Hamman RF, et al. Effect of weight loss with lifestyle intervention on risk of diabetes. Diabetes Care 2006;29(9):2102–7.

[15] Kriska AM, et al. Physical activity in individuals at risk for diabetes: diabetes prevention program. Med Sci Sports Exerc 2006;38(5):826–32.

[16] Wing RR, et al. Achieving weight and activity goals among diabetes prevention program lifestyle participants. Obes Res 2004;12(9):1426–34.

[17] Laaksonen DE, et al. Physical activity in the prevention of type 2 diabetes: the Finnish diabetes prevention study. Diabetes 2005;54(1):158–65.

[18] Knowler WC, et al. 10-year follow-up of diabetes incidence and weight loss in the diabetes prevention program outcomes study. Lancet 2009;374(9702):1677–86.

[19] Mudaliar U, et al. Cardiometabolic risk factor changes observed in diabetes prevention programs in US settings: a systematic review and meta-analysis. PLoS Med 2016;13(7):e1002095.

[20] Ackermann RT, et al. Translating the diabetes prevention program into the community. The DEPLOY pilot study. Am J Prev Med 2008;35 (4):357–63.

[21] Evert AB, et al. Nutrition therapy for adults with diabetes or prediabetes: a consensus Report. Diabetes Care 2019;42(5):731–54. https://doi.org/10.2337/dci19-0014 [Epub 2019 Apr 18].

[22] Malanda UL, et al. Self-monitoring of blood glucose in patients with type 2 diabetes mellitus who are not using insulin. Cochrane Database Syst Rev 2012;1:Cd005060.

[23] Battelino T, et al. Clinical Targets for continuous glucose monitoring data interpretation: recommendations from the international consensus on time in range. Diabetes Care 2019;42(8):1593–603. https://doi.org/10.2337/dci19-0028 [Epub 2019 Jun 8].

[24] Moser O, et al. Glucose management for exercise using continuous glucose monitoring (CGM) and intermittently scanned CGM (isCGM) systems in type 1 diabetes: position statement of the European Association for the Study of Diabetes (EASD) and of the International Society for Pediatric and Adolescent Diabetes (ISPAD) endorsed by JDRF and supported by the American Diabetes Association (ADA). Diabetologia 2020;2501–20 [Germany].

[25] Taylor PJ, et al. Tolerability and acceptability of real-time continuous glucose monitoring and its impact on diabetes management behaviours in individuals with type 2 diabetes - a pilot study. Diabetes Res Clin Pract 2019;155:107814.

[26] Hoskin MA, et al. Prevention of diabetes through the lifestyle intervention: lessons learned from the diabetes prevention program and outcomes study and its translation to practice. Curr Nutr Rep 2014;3(4):364–78.

[27] Gregg EW, et al. Association of an intensive lifestyle intervention with remission of type 2 diabetes. JAMA 2012;308(23):2489–96.

[28] Boulé NG, et al. Effects of exercise on glycemic control and body mass in type 2 diabetes mellitus: a meta-analysis of controlled clinical trials. JAMA 2001;286(10):1218–27.

[29] Shang Y, et al. Natural history of prediabetes in older adults from a population-based longitudinal study. J Intern Med 2019;286(3):326–40.

[30] Aguiar EJ, et al. Efficacy of interventions that include diet, aerobic and resistance training components for type 2 diabetes prevention: a systematic review with meta-analysis. Int J Behav Nutr Phys Act 2014;11:2.

[31] Johnson NA, et al. Aerobic exercise training reduces hepatic and visceral lipids in obese individuals without weight loss. Hepatology 2009;50(4):1105–12.

[32] O'Leary VB, et al. Exercise-induced reversal of insulin resistance in obese elderly is associated with reduced visceral fat. J Appl Physiol 2006;100 (5):1584–9.

[33] Donnelly JE, et al. American College of Sports Medicine position stand. Appropriate physical activity intervention strategies for weight loss and prevention of weight regain for adults. Med Sci Sports Exerc 2009;41(2):459–71.

[34] Jakicic JM, et al. Effect of a lifestyle intervention on change in cardiorespiratory fitness in adults with type 2 diabetes: results from the Look AHEAD study. Int J Obes (Lond) 2009;33(3):305–16. https://doi.org/10.1038/ijo.2008.280 [Epub 2009 Jan 20].

[35] Nomura T, et al. Regular exercise behavior is related to lower extremity muscle strength in patients with type 2 diabetes: data from the MUSCLE-std study. J Diabetes Invest 2017;14(10):12703.

[36] Umpierre D, et al. Volume of supervised exercise training impacts glycaemic control in patients with type 2 diabetes: a systematic review with meta-regression analysis. Diabetologia 2013;56(2):242–51. https://doi.org/10.1007/s00125-012-2774-z [Epub 2012 Nov 16].

[37] Anon. Facilitating behavior change and well-being to improve health outcomes: standards of medical care in diabetes-2021. Diabetes Care 2021;44 (Suppl 1):S53–s72.

[38] Winnick JJ, et al. Short-term aerobic exercise training in obese humans with type 2 diabetes mellitus improves whole-body insulin sensitivity through gains in peripheral, not hepatic insulin sensitivity. J Clin Endocrinol Metab 2008;93(3):771–8.

[39] Dunstan DW, et al. High-intensity resistance training improves glycemic control in older patients with type 2 diabetes. Diabetes Care 2002;25(10):1729–36.

[40] Tok Ö, et al. A 4-week diet with exercise intervention had a better effect on blood glucose levels compared to diet only intervention in obese individuals with insulin resistance. J Sports Med Phys Fitness 2021;61(2):287–93. https://doi.org/10.23736/S0022-4707.20.11188-5.

[41] Yan J, et al. Effect of 12-month resistance training on changes in abdominal adipose tissue and metabolic variables in patients with prediabetes: a randomized controlled trial. J Diabetes Res 2019;2019:8469739.

[42] Way KL, et al. The effect of regular exercise on insulin sensitivity in type 2 diabetes mellitus: a systematic review and Meta-analysis. Diabetes Metab J 2016;40(4):253–71.

[43] Look ARG, Wing RR. Long-term effects of a lifestyle intervention on weight and cardiovascular risk factors in individuals with type 2 diabetes mellitus: four-year results of the Look AHEAD trial. Arch Intern Med 2010;170(17):1566–75.

[44] Duncan GE, et al. Exercise training, without weight loss, increases insulin sensitivity and postheparin plasma lipase activity in previously sedentary adults. Diabetes Care 2003;26(3):557–62.

[45] Christ-Roberts CY, et al. Exercise training increases glycogen synthase activity and GLUT4 expression but not insulin signaling in overweight nondiabetic and type 2 diabetic subjects. Metabolism 2004;53(9):1233–42.

[46] Goodpaster BH, Katsiaras A, Kelley DE. Enhanced fat oxidation through physical activity is associated with improvements in insulin sensitivity in obesity. Diabetes 2003;52(9):2191–7.

[47] Holten MK, et al. Strength training increases insulin-mediated glucose uptake, GLUT4 content, and insulin signaling in skeletal muscle in patients with type 2 diabetes. Diabetes 2004;53(2):294–305.

[48] O'Gorman DJ, et al. Exercise training increases insulin-stimulated glucose disposal and GLUT4 (SLC2A4) protein content in patients with type 2 diabetes. Diabetologia 2006;49(12):2983–92.

[49] Wang Y, Simar D, Fiatarone Singh MA. Adaptations to exercise training within skeletal muscle in adults with type 2 diabetes or impaired glucose tolerance: a systematic review. Diabetes Metab Res Rev 2009;25(1):13–40.

[50] Khangura D, et al. Diabetes and hypertension: clinical update. Am J Hypertens 2018;31(5):515–21.

[51] Park S, Kim J, Lee J. Effects of exercise intervention on adults with both hypertension and type 2 diabetes mellitus: a systematic review and meta-analysis. J Cardiovasc Nurs 2020.

[52] Pan B, et al. Exercise training modalities in patients with type 2 diabetes mellitus: a systematic review and network meta-analysis. Int J Behav Nutr Phys Act 2018;15(1):72. https://doi.org/10.1186/s12966-018-0703-3.

[53] Pi-Sunyer X, et al. Reduction in weight and cardiovascular disease risk factors in individuals with type 2 diabetes: one-year results of the look AHEAD trial. Diabetes Care 2007;30(6):1374–83.

[54] Pi-Sunyer X. The Look AHEAD trial: a review and discussion of its outcomes. Curr Nutr Rep 2014;3(4):387–91.

[55] Look ARG, et al. Cardiovascular effects of intensive lifestyle intervention in type 2 diabetes. N Engl J Med 2013;369(2):145–54.

[56] Yang Z, et al. Resistance exercise versus aerobic exercise for type 2 diabetes: a systematic review and meta-analysis. Sports Med 2014; 44(4):487–99. https://doi.org/10.1007/s40279-013-0128-8.

[57] Williamson DA, et al. Impact of a weight management program on health-related quality of life in overweight adults with type 2 diabetes. Arch Intern Med 2009;169(2):163–71.

[58] Craft LL, Perna FM. The benefits of exercise for the clinically depressed. Prim Care Companion J Clin Psychiatry 2004;6(3):104–11.

[59] Gillison FB, et al. The effects of exercise interventions on quality of life in clinical and healthy populations; a meta-analysis. Soc Sci Med 2009;68 (9):1700–10.

[60] Wahren J, et al. Glucose metabolism during leg exercise in man. J Clin Invest 1971;50(12):2715–25.

[61] Kennedy JW, et al. Acute exercise induces GLUT4 translocation in skeletal muscle of normal human subjects and subjects with type 2 diabetes. Diabetes 1999;48(5):1192–7.

[62] Martin IK, Katz A, Wahren J. Splanchnic and muscle metabolism during exercise in NIDDM patients. Am J Physiol 1995;269(3 Pt 1):E583–90.

[63] Aqeel M, et al. The effect of timing of exercise and eating on postprandial response in adults: a systematic review. Nutrients 2020;12(1).

[64] Wahren J, Ekberg K. Splanchnic regulation of glucose production. Annu Rev Nutr 2007;27:329–45.

[65] Mikines KJ, et al. Effect of physical exercise on sensitivity and responsiveness to insulin in humans. Am J Physiol 1988;254(3 Pt 1):E248–59.

[66] Devlin JT, Horton ES. Effects of prior high-intensity exercise on glucose metabolism in normal and insulin-resistant men. Diabetes 1985;34(10):973–9.

[67] Wojtaszewski JF, Richter EA. Effects of acute exercise and training on insulin action and sensitivity: focus on molecular mechanisms in muscle. Essays Biochem 2006;42:31–46.

[68] Knudsen JR, et al. Prior exercise in humans redistributes intramuscular GLUT4 and enhances insulin-stimulated sarcolemmal and endosomal GLUT4 translocation. Mol Metab 2020;39:100998.

[69] Nelson RK, Horowitz JF. Acute exercise ameliorates differences in insulin resistance between physically active and sedentary overweight adults. Appl Physiol Nutr Metab 2014;39(7):811–8.

[70] Oberlin DJ, et al. One bout of exercise alters free-living postprandial glycemia in type 2 diabetes. Med Sci Sports Exerc 2014;46(2):232–8.

[71] Gillen JB, et al. Acute high-intensity interval exercise reduces the postprandial glucose response and prevalence of hyperglycaemia in patients with type 2 diabetes. Diabetes Obes Metab 2012;14(6):575–7.

[72] van Dijk JW, et al. Exercise therapy in type 2 diabetes: is daily exercise required to optimize glycemic control? Diabetes Care 2012;35(5):948–54.

[73] Dube JJ, et al. Effects of weight loss and exercise on insulin resistance, and intramyocellular triacylglycerol, diacylglycerol and ceramide. Diabetologia 2011;54(5):1147–56.

[74] Kirwan JP, et al. Effects of 7 days of exercise training on insulin sensitivity and responsiveness in type 2 diabetes mellitus. Am J Physiol Endocrinol Metab 2009;297(1):E151–6.

[75] Bacchi E, et al. Both resistance training and aerobic training reduce hepatic fat content in type 2 diabetic subjects with nonalcoholic fatty liver disease (the RAED2 randomized trial). Hepatology 2013;58(4):1287–95.

[76] Heiskanen MA, et al. Exercise training decreases pancreatic fat content and improves beta cell function regardless of baseline glucose tolerance: a randomised controlled trial. Diabetologia 2018;61(8):1817–28.

[77] Ryan BJ, et al. Moderate-intensity exercise and high-intensity interval training affect insulin sensitivity similarly in obese adults. J Clin Endocrinol Metab 2020;105(8):e2941–59.

[78] Snowling NJ, Hopkins WG. Effects of different modes of exercise training on glucose control and risk factors for complications in type 2 diabetic patients: a meta-analysis. Diabetes Care 2006;29(11):2518–27.

[79] Umpierre D, et al. Physical activity advice only or structured exercise training and association with HbA1c levels in type 2 diabetes: a systematic review and meta-analysis. JAMA 2011;305(17):1790–9.

[80] Newsom SA, et al. A single session of low-intensity exercise is sufficient to enhance insulin sensitivity into the next day in obese adults. Diabetes Care 2013;36(9):2516–22.

[81] Liubaoerjijin Y, et al. Effect of aerobic exercise intensity on glycemic control in type 2 diabetes: a meta-analysis of head-to-head randomized trials. Acta Diabetol 2016;53(5):769–81.

[82] Ross R, et al. Effects of exercise amount and intensity on abdominal obesity and glucose tolerance in obese adults: a randomized trial. Ann Intern Med 2015;162(5):325–34.

[83] Francois ME, et al. Combined interval training and post-exercise nutrition in type 2 diabetes: a randomized control trial. Front Physiol 2017;8:528.

[84] Madsen SM, et al. High intensity interval training improves Glycaemic control and pancreatic beta cell function of type 2 diabetes patients. PLoS One 2015;10(8):e0133286.

[85] Maillard F, et al. High-intensity interval training reduces abdominal fat mass in postmenopausal women with type 2 diabetes. Diabetes Metab 2016;42(6):433–41.

[86] Sjöros TJ, et al. Increased insulin-stimulated glucose uptake in both leg and arm muscles after sprint interval and moderate-intensity training in subjects with type 2 diabetes or prediabetes. Scand J Med Sci Sports 2018;28(1):77–87.

[87] Jelleyman C, et al. The effects of high-intensity interval training on glucose regulation and insulin resistance: a meta-analysis. Obes Rev 2015;16(11):942–61.

[88] Karstoft K, et al. The effects of free-living interval-walking training on glycemic control, body composition, and physical fitness in type 2 diabetic patients: a randomized, controlled trial. Diabetes Care 2013;36(2):228–36.

[89] Fealy CE, et al. Functional high-intensity exercise training ameliorates insulin resistance and cardiometabolic risk factors in type 2 diabetes. Exp Physiol 2018;103(7):985–94.

[90] van Dijk JW, et al. Both resistance- and endurance-type exercise reduce the prevalence of hyperglycaemia in individuals with impaired glucose tolerance and in insulin-treated and noninsulin-treated type 2 diabetic patients. Diabetologia 2012;55(5):1273–82.

[91] Gordon BA, et al. Resistance training improves metabolic health in type 2 diabetes: a systematic review. Diabetes Res Clin Pract 2009;83(2):157–75.

[92] Cuff DJ, et al. Effective exercise modality to reduce insulin resistance in women with type 2 diabetes. Diabetes Care 2003;26(11):2977–82.

[93] Sigal RJ, et al. Effects of aerobic training, resistance training, or both on glycemic control in type 2 diabetes: a randomized trial. Ann Intern Med 2007;147(6):357–69.

[94] Abate M, et al. Limited joint mobility in diabetes and ageing: recent advances in pathogenesis and therapy. Int J Immunopathol Pharmacol 2011;23(4):997–1003.

[95] Herriott MT, et al. Effects of 8 weeks of flexibility and resistance training in older adults with type 2 diabetes. Diabetes Care 2004;27(12):2988–9.

[96] Colberg SR, et al. Exercise and type 2 diabetes: the American College of Sports Medicine and the American Diabetes Association: joint position statement. Diabetes Care 2010;33(12):e147–67.

[97] Innes KE, Selfe TK. Yoga for adults with type 2 diabetes: a systematic review of controlled trials. J Diabetes Res 2016;2016:6979370.

[98] Jayawardena R, et al. The benefits of yoga practice compared to physical exercise in the management of type 2 diabetes mellitus: a systematic review and meta-analysis. Diabetes Metab Syndr 2018;12(5):795–805. https://doi.org/10.1016/j.dsx.2018.04.008 [Epub 2018 Apr 18].

[99] Cui J, et al. Effects of yoga in adults with type 2 diabetes mellitus: a meta-analysis. J Diabetes Invest 2017;8(2):201–9.

[100] Thind H, et al. The effects of yoga among adults with type 2 diabetes: a systematic review and meta-analysis. Prev Med 2017;105:116–26.

[101] Ahn S, Song R. Effects of tai chi exercise on glucose control, neuropathy scores, balance, and quality of life in patients with type 2 diabetes and neuropathy. J Altern Complement Med 2012;18(12):1172–8.

[102] Chao M, et al. The effects of tai chi on type 2 diabetes mellitus: a meta-analysis. J Diabetes Res 2018;2018. https://doi.org/10.1155/2018/7350567, 7350567 [eCollection 2018].

[103] Ramamoorthi R, et al. The effect of yoga practice on glycemic control and other health parameters in the prediabetic state: a systematic review and meta-analysis. PLoS One 2019;14(10):e0221067.

[104] Nelson ME, et al. Physical activity and public health in older adults: recommendation from the American College of Sports Medicine and the American Heart Association. Med Sci Sports Exerc 2007;39(8):1435–45.

[105] Physical Activity Guidelines Advisory Committee. Physical Activity Guidelines Advisory Committee Report, 2008. Washington, DC: U.S. Department of Health and Human Services; 2008. p. 683.

[106] Pfeifer LO, et al. Effects of a power training program in the functional capacity, on body balance and lower limb muscle strength of elderly with type 2 diabetes mellitus. J Sports Med Phys Fitness 2021.

[107] Morrison S, et al. Balance training reduces falls risk in older individuals with type 2 diabetes. Diabetes Care 2010;33(4):748–50.

[108] Morrison S, et al. Supervised balance training and Wii fit-based exercises lower falls risk in older adults with type 2 diabetes. J Am Med Dir Assoc 2018;19(2):185.e7–185.e13.

[109] Biswas A, et al. Sedentary time and its association with risk for disease incidence, mortality, and hospitalization in adults: a systematic review and meta-analysis. Ann Intern Med 2015;162(2):123–32.

[110] Wilmot EG, et al. Sedentary time in adults and the association with diabetes, cardiovascular disease and death: systematic review and meta-analysis. Diabetologia 2012;55(11):2895–905.

[111] Fritschi C, et al. Association between daily time spent in sedentary behavior and duration of hyperglycemia in type 2 diabetes. Biol Res Nurs 2016;18(2):160–6. https://doi.org/10.1177/1099800415600065.

[112] Cooper AR, et al. Sedentary time, breaks in sedentary time and metabolic variables in people with newly diagnosed type 2 diabetes. Diabetologia 2012;55(3):589–99.

[113] Dempsey PC, et al. Benefits for type 2 diabetes of interrupting prolonged sitting with brief bouts of light walking or simple resistance activities. Diabetes Care 2016;39(6):964–72.

[114] Larsen R, et al. Interrupting sitting time with simple resistance activities lowers postprandial Insulinemia in adults with overweight or obesity. Obesity (Silver Spring) 2019;27(9):1428–33.

[115] Gillen JB, et al. Interrupting prolonged sitting with repeated chair stands or short walks reduces postprandial insulinemia in healthy adults. J Appl Physiol (1985) 2021;130(1):104–13.

[116] DiPietro L, et al. Three 15-min bouts of moderate postmeal walking significantly improves 24-h glycemic control in older people at risk for impaired glucose tolerance. Diabetes Care 2013;36(10):3262–8.

[117] Reynolds AN, et al. Advice to walk after meals is more effective for lowering postprandial glycaemia in type 2 diabetes mellitus than advice that does not specify timing: a randomised crossover study. Diabetologia 2016;59(12):2572–8.

[118] van Dijk JW, et al. Effect of moderate-intensity exercise versus activities of daily living on 24-hour blood glucose homeostasis in male patients with type 2 diabetes. Diabetes Care 2013;36(11):3448–53.

[119] Chang CR, et al. Accumulating physical activity in short or brief bouts for glycemic control in adults with prediabetes and diabetes. Can J Diabetes 2020;44(8):759–67.

[120] Colberg SR, et al. Postprandial walking is better for lowering the glycemic effect of dinner than pre-dinner exercise in type 2 diabetic individuals. J Am Med Dir Assoc 2009;10(6):394–7. https://doi.org/10.1016/j.jamda.2009.03.015 [Epub 2009 May 21].

[121] Heden TD, et al. Postdinner resistance exercise improves postprandial risk factors more effectively than predinner resistance exercise in patients with type 2 diabetes. J Appl Physiol (1985) 2015;118(5):624–34.

[122] Verboven K, et al. Impact of exercise-nutritional state interactions in patients with type 2 diabetes. Med Sci Sports Exerc 2020;52(3):720–8.

[123] Newsom SA, et al. Energy deficit after exercise augments lipid mobilization but does not contribute to the exercise-induced increase in insulin sensitivity. J Appl Physiol (1985) 2010;108(3):554–60.

[124] Taylor HL, et al. Post-exercise carbohydrate-energy replacement attenuates insulin sensitivity and glucose tolerance the following morning in healthy adults. Nutrients 2018;10(2).

[125] Schleh MW, et al. Energy deficit required for exercise-induced improvements in Glycemia the next day. Med Sci Sports Exerc 2020;52(4):976–82.

[126] Fox AK, Kaufman AE, Horowitz JF. Adding fat calories to meals after exercise does not alter glucose tolerance. J Appl Physiol (1985) 2004;97(1):11–6.

[127] Colberg SR, et al. Physical activity/exercise and diabetes: a position statement of the American Diabetes Association. Diabetes Care 2016;39(11):2065–79.

[128] Piercy KL, et al. The physical activity guidelines for Americans. JAMA 2018;320(19):2020–8.

[129] Song SH. Early-onset type 2 diabetes: high lifetime risk for cardiovascular disease. Lancet Diabetes Endocrinol 2016;4(2):87–8.

[130] Garber CE, et al. American College of Sports Medicine position stand. Quantity and quality of exercise for developing and maintaining cardio-respiratory, musculoskeletal, and neuromotor fitness in apparently healthy adults: guidance for prescribing exercise. Med Sci Sports Exerc 2011;43(7):1334–59.

[131] Ballin M, et al. Daily step count and incident diabetes in community-dwelling 70-year-olds: a prospective cohort study. BMC Public Health 2020;20(1):1830.

[132] Lascar N, et al. Attitudes and barriers to exercise in adults with type 1 diabetes (T1DM) and how best to address them: a qualitative study. PLoS One 2014;9(9):e108019.

[133] Aljasem LI, et al. The impact of barriers and self-efficacy on self-care behaviors in type 2 diabetes. Diabetes Educ 2001;27(3):393–404.

[134] Dutton GR, et al. Relationship between self-efficacy and physical activity among patients with type 2 diabetes. J Behav Med 2009;32(3):270–7.

[135] McAuley E, Blissmer B. Self-efficacy determinants and consequences of physical activity. Exerc Sport Sci Rev 2000;28(2):85–8.

[136] Delahanty LM, Conroy MB, Nathan DM. Psychological predictors of physical activity in the diabetes prevention program. J Am Diet Assoc 2006;106(5):698–705.

[137] Armit CM, et al. Randomized trial of three strategies to promote physical activity in general practice. Prev Med 2009;48(2):156–63.

[138] Balducci S, et al. The Italian diabetes and exercise study. Diabetes 2008;57(Suppl. 1):A306–7.

[139] Lauzon N, et al. Participant experiences in a workplace pedometer-based physical activity program. J Phys Act Health 2008;5(5):675–87.

[140] Riebe D, et al. Updating ACSM's recommendations for exercise Preparticipation health screening. Med Sci Sports Exerc 2015;47(11):2473–9.

[141] Smith JW, et al. Curvilinear dose-response relationship of carbohydrate (0-120 g.h(−1)) and performance. Med Sci Sports Exerc 2013; 45(2):336–41.

[142] Riddell MC, et al. Exercise management in type 1 diabetes: a consensus statement. Lancet Diabetes Endocrinol 2017;5(5):377–90.

[143] Colberg SR, Swain DP, Vinik AI. Use of heart rate reserve and rating of perceived exertion to prescribe exercise intensity in diabetic autonomic neuropathy. Diabetes Care 2003;26(4):986–90.

[144] Colberg SR, Vinik AI. Exercising with peripheral or autonomic neuropathy: what health care providers and diabetic patients need to know. Phys Sportsmed 2014;42(1):15–23.

[145] Pongrac Barlovic D, Tikkanen-Dolenc H, Groop PH. Physical activity in the prevention of development and progression of kidney Disease in type 1 diabetes. Curr Diab Rep 2019;19(7):41.

[146] Juan J, Yang H. Prevalence, prevention, and lifestyle intervention of gestational diabetes mellitus in China. Int J Environ Res Public Health 2020;17(24).

[147] Agarwal MM. Gestational diabetes in the Arab gulf countries: sitting on a land-mine. Int J Environ Res Public Health 2020;17(24).

[148] Behboudi-Gandevani S, et al. The impact of diagnostic criteria for gestational diabetes on its prevalence: a systematic review and meta-analysis. Diabetol Metab Syndr 2019;11:11.

[149] Nguyen CL, et al. Prevalence of gestational diabetes mellitus in eastern and southeastern Asia: a systematic review and Meta-analysis. J Diabetes Res 2018;2018:6536974.

[150] Anon. Management of diabetes in pregnancy: standards of medical care in diabetes-2021. Diabetes Care 2021;44(Suppl 1):S200–s210.

[151] Mayo K, et al. The impact of adoption of the international association of diabetes in pregnancy study group criteria for the screening and diagnosis of gestational diabetes. Am J Obstet Gynecol 2015;212(2). 224.e1–9.

[152] Mottola MF, Artal R. Role of exercise in reducing gestational diabetes mellitus. Clin Obstet Gynecol 2016;59(3):620–8.

[153] Mottola MF, et al. 2019 Canadian guideline for physical activity throughout pregnancy. Br J Sports Med 2018;52(21):1339–46.

[154] Institute of Medicine and National Research Council Committee to Reexamine, I.O.M.P.W.G. In: Rasmussen KM, Yaktine AL, editors. The national academies collection: reports funded by national institutes of health, in weight gain during pregnancy: reexamining the guidelines. USA: National Academies Press; 2009 [Copyright © 2009, National Academy of Sciences.: Washington (DC)].

[155] Goldstein RF, et al. Association of gestational weight gain with maternal and infant outcomes: a systematic review and meta-analysis. JAMA 2017;317(21):2207–25.

[156] Harper LM, Tita A, Biggio JR. The institute of medicine guidelines for gestational weight gain after a diagnosis of gestational diabetes and pregnancy outcomes. Am J Perinatol 2015;32(3):239–46.

[157] Cheng YW, et al. Gestational weight gain and gestational diabetes mellitus: perinatal outcomes. Obstet Gynecol 2008;112(5):1015–22.

[158] Kc K, Shakya S, Zhang H. Gestational diabetes mellitus and macrosomia: a literature review. Ann Nutr Metab 2015;66(Suppl 2):14–20.

[159] Lazer S, et al. Complications associated with the macrosomic fetus. J Reprod Med 1986;31(6):501–5.

[160] Bellamy L, et al. Type 2 diabetes mellitus after gestational diabetes: a systematic review and meta-analysis. Lancet 2009;373(9677):1773–9.

[161] Li Z, et al. Incidence rate of type 2 diabetes mellitus after gestational diabetes mellitus: a systematic review and meta-analysis of 170,139 women. J Diabetes Res 2020;2020:3076463.

[162] Metzger BE, et al. Gestational diabetes mellitus. Correlations between the phenotypic and genotypic characteristics of the mother and abnormal glucose tolerance during the first year postpartum. Diabetes 1985;34(Suppl 2):111–5.

[163] Kim C, Newton KM, Knopp RH. Gestational diabetes and the incidence of type 2 diabetes: a systematic review. Diabetes Care 2002;25(10):1862–8.

[164] Kramer CK, Campbell S, Retnakaran R. Gestational diabetes and the risk of cardiovascular disease in women: a systematic review and meta-analysis. Diabetologia 2019;62(6):905–14.

[165] Vohr BR, Boney CM. Gestational diabetes: the forerunner for the development of maternal and childhood obesity and metabolic syndrome? J Matern Fetal Neonatal Med 2008;21(3):149–57.

[166] Catalano PM. Trying to understand gestational diabetes. Diabet Med 2014;31(3):273–81.

[167] Lain KY, Catalano PM. Metabolic changes in pregnancy. Clin Obstet Gynecol 2007;50(4):938–48.

[168] Kühl C. Etiology and pathogenesis of gestational diabetes. Diabetes Care 1998;21(Suppl 2):B19–26.

[169] Davenport MH, et al. Glucose responses to acute and chronic exercise during pregnancy: a systematic review and meta-analysis. Br J Sports Med 2018;52(21):1357–66.

[170] Richter EA, Hargreaves M. Exercise, GLUT4, and skeletal muscle glucose uptake. Physiol Rev 2013;93(3):993–1017.

[171] Jovanovic-Peterson L, Durak EP, Peterson CM. Randomized trial of diet versus diet plus cardiovascular conditioning on glucose levels in gestational diabetes. Am J Obstet Gynecol 1989;161(2):415–9.

[172] Brankston GN, et al. Resistance exercise decreases the need for insulin in overweight women with gestational diabetes mellitus. Am J Obstet Gynecol 2004;190(1):188–93.

[173] Bung P, et al. Exercise in gestational diabetes. An optional therapeutic approach? Diabetes 1991;40(Suppl 2):182–5.

[174] García-Patterson A, et al. Evaluation of light exercise in the treatment of gestational diabetes. Diabetes Care 2001;24(11):2006–7.

[175] Coe DP, et al. Postprandial walking reduces glucose levels in women with gestational diabetes mellitus. Appl Physiol Nutr Metab 2018;43(5):531–4.

[176] Avery MD, Walker AJ. Acute effect of exercise on blood glucose and insulin levels in women with gestational diabetes. J Matern Fetal Med 2001;10(1):52–8.

[177] Ruchat SM, et al. Effectiveness of exercise interventions in the prevention of excessive gestational weight gain and postpartum weight retention: a systematic review and meta-analysis. Br J Sports Med 2018;52(21):1347–56.

[178] Clapp 3rd JF. Long-term outcome after exercising throughout pregnancy: fitness and cardiovascular risk. Am J Obstet Gynecol 2008;199(5). 489.e1–6.

[179] Lotgering FK, Gilbert RD, Longo LD. Maternal and fetal responses to exercise during pregnancy. Physiol Rev 1985;65(1):1–36.

[180] Kennelly MM, et al. Exercise-related changes in umbilical and uterine artery waveforms as assessed by Doppler ultrasound scans. Am J Obstet Gynecol 2002;187(3):661–6.

[181] Davenport MH, et al. Prenatal exercise is not associated with fetal mortality: a systematic review and meta-analysis. Br J Sports Med 2019; 53(2):108–15.

[182] Skow RJ, et al. Effects of prenatal exercise on fetal heart rate, umbilical and uterine blood flow: a systematic review and meta-analysis. Br J Sports Med 2019;53(2):124–33.

[183] Beetham KS, et al. The effects of vigorous intensity exercise in the third trimester of pregnancy: a systematic review and meta-analysis. BMC Pregnancy Childbirth 2019;19(1):281.

[184] Golbidi S, Laher I. Potential mechanisms of exercise in gestational diabetes. J Nutr Metab 2013;2013:285948.

[185] Practice ACO. ACOG Committee opinion. Number 267, January 2002: exercise during pregnancy and the postpartum period. Obstet Gynecol 2002;99(1):171–3.

Chapter 12

Exercise in the management of multiple sclerosis

Narges Dargahi[a], Melina Haritopoulou-Sinanidou[b], and Vasso Apostolopoulos[a,c]

[a]Institute for Health and Sport, Victoria University, Melbourne, VIC, Australia [b]School of Biomedical Sciences, University of Queensland, Brisbane, QLD, Australia [c]The Australian Institute for Musculoskeletal Science, Western Health, Victoria University and the University of Melbourne, Melbourne, VIC, Australia

Introduction

Autoimmune disorders affect many organs and systems of the body leading to inappropriate immune responses, inflammation, and damage [1–3]. Multiple sclerosis (MS) is caused by autoimmune inflammation resulting in myelin sheath destruction in the central nervous system (CNS), a process known as demyelination. This is a result of infiltration of myelin-specific autoimmune T cells, which destroy the tissue, causing neurologic conduction deficiencies that can lead to paralysis, cognitive decline, and disability [3–7]. Typical symptoms include fatigue, spasticity, paralysis, cognitive impairment, anxiety, depression, bladder, and bowel dysfunction. Other symptoms, such as mobility, numbness, and vision impairment, may vary depending on the affected region of the CNS [6,8]. Despite decades of research, the underlying cause of MS is not clear. Nonetheless, in the late 1990s studies were published which gave clues regarding MS pathophysiology. MS is a life debilitating autoimmune disorder [4,9], that affects over 2.5 million individuals worldwide, as reported by the World Health Organization [6], which is almost 0.1% of the global population [3]. The estimated annual cost of MS is $2–3 billion, thus apart from having a substantial effect on the wellbeing of individuals, it also poses a considerable economic burden to patients with MS and the health care system [10]. Although MS does not generally reduce the life expectancy of a person, it interferes with everyday life activities, thus, treatment options become a necessity [11]. Reports into the financial impact of MS disclose the considerable costs linked with the disease and highlight the need for fundamental improvement of disability care and support for individuals with MS.

MS, in general, cannot be cured but can be controlled primarily with a number of medications aimed to reduce inflammation and pain, such as, antiinflammatory drugs, corticosteroids, immunosuppressants, and disease-modifying therapies that assist to inhibit the amassing of lesions in the CNS. However, these therapies are not only expensive, but are also associated with significant side effects, and are nontolerable by many. The use of physical therapy that boosts and improves mobility, has, therefore, become an alternative and complementary approach for patients with MS [12].

There is some uncertainty on whether exercise activities and physical therapy are beneficial for MS fatigue. Several studies primarily focused on MS fatigue as the main outcome was limited and a few studies observed nonfatigued MS patient groups, which produced heterogeneity in the data across making it hard to come to conclusions. Most studies that focused on MS fatigue demonstrate positive effects, however, it is still unclear whether some exercise interventions are greater than others, as none of the studies used comparative methods to examine different exercise regimes [13].

Physical therapy for patients with MS has been controversial for many years, due to the belief that exercise activities can pose an aggravating effect on disease symptoms including tiredness and fatigue [14]. Nonetheless, years of studies have illustrated that exercise is not only effective and safe but also it is the basis for MS rehabilitation. Exercise is thought to convey numerous benefits for MS patients and can potentially transform the entire course of medical treatments. This evidence has led to a paradigm shift in MS treatment, with new approaches using exercise activities that are tailored and prescribed based on the individual's needs. Improvement in our knowledge about the pathophysiology of inflammation and autoimmune disorders may lead to enhanced treatments for MS. Until then, more focus on studies that emphasize on unconventional treatments, such as physical therapies and activities [15,16], maybe alternative interventions to reduce the use of costly medications and adverse events associated with those medications. Such physical therapies include swimming or light aquatic exercise movements [17–19], yoga [12,20,21], Pilates [22], combined aerobic and weight training [23]; many

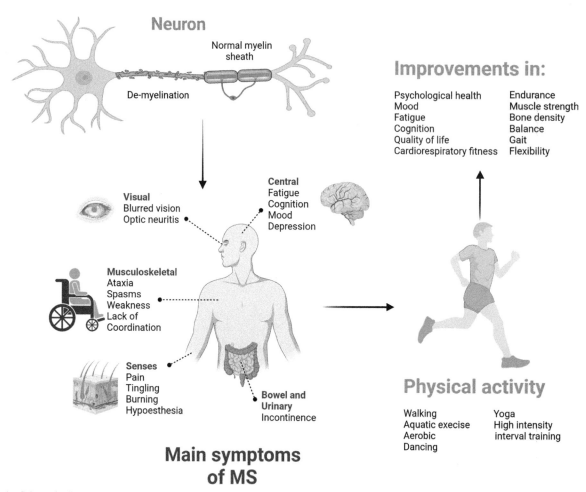

FIG. 1 Schematic diagram of MS symptoms and beneficial effects of physical activity (created using biorender.com).

of which have led to improvements in the overall quality of life of those affected by MS [15–20,22–24]. Herein, we review the studies related to the effects and impact of exercise as a primary, secondary, and tertiary treatment/prevention in MS.

Beneficial effects of physical activity in individuals with MS

MS can cause substantial cognitive and somatic impairment that leads to physical symptoms including muscle weakness, fatigue, balance issues, abnormal mobility, spasticity, as well as mental health issues such as depression. As the disease progresses, it frequently leads to a reduction in the level of physical activity performed by the individual and subsequently causes deconditioning [25]. Some clinicians use routine exercise programs that are considered to potentially have positive effects in reducing the deconditioning course and improving mobility, functioning, and other mental and physical symptoms, and are known to not worsen symptoms, causing relapse or triggering the onset of MS-related symptoms. Implementing appropriate exercise regimes has been shown to improve overall patient condition including cognition, cardiorespiratory health, fatigue, muscle strength, balance, flexibility, quality of life, and aerobic fitness [26] (Fig. 1).

Patients with MS have been shown to progressively become accustomed to resistance exercise which may lead to improvement of fatigue and ability to walk without assistance [25,27]. In addition, flexibility training reduces spasticity and inhibits agonizing contractions. Balance exercises have been shown to improve balance and decrease falls [25]. There are standard exercise protocols for MS patients, however, personalized training programs tailored to a patient's main complaints, aiming to advance strength, coordination, endurance, stability, and addressing tiredness, are more beneficial. An exercise staircase model has been proposed for exercise prescription and been shown to lead to progression in a broad spectrum of MS patients. Current data indicates that an individualized workout program that is conducted under supervision can improve functional capacity, fitness, and quality of life, as well as changeable deficiencies in patients with MS [28]. The chronic course of MS can cause permanent neurologic deficit with significant cerebral and somatic symptoms including, spasticity, balance disorder, muscle weakness, tremor, ataxia, paralysis, vision issues, vertigo, speech and swallowing impairment, cognitive impairment, decreased sensations, bowel and bladder incontinence, fatigue, pain and depression [26,29–31]. Dysfunction in the motor system in MS patients commonly occurs due to muscle weakness, fatigue, abnormal mobility mechanics, spasticity, and balance issues [25,27,32,33]. MS is known for its unpredictable, progressive nature with weaknesses and limitations of motor mechanics that may unexpectedly arise and progress faster than the predicted time [34]. It is reported that up to 50% of individuals with MS will require a walking aid 15 years after the onset of disease [35–38], and often patients decrease their activity due to their fear of symptom exacerbation [39]. Limiting activity, in turn, increases disability, fitness, walking abnormality, lack of stability, reduced muscle strength, reduced mobility, and quality of life [40,41].

Although exercise may not reverse the impairments associated with the progression of MS, it may be beneficial for the impairments resulting from deconditioning [42]. Additionally, lack of activity among MS patients increases disease-associated comorbidities such as obesity, hypertension, hypercholesterolemia, type 2 diabetes, cancer, osteoporosis, arthritis, fatigue, depression, and cardiovascular disorders [25,39]. Comorbidities can further result in increased disability, due to a decline in muscle strength and aerobic capacity, as well as increased muscular atrophy and added nervous system risks such as stroke [25]. For these reasons, although physicians have traditionally advised newly diagnosed patients with MS to avoid physical activity, it is now believed that regular exercise and training is a possible solution during the disease period, potentially limiting the deconditioning process. This helps patients reach optimal levels of activity and function and has a generally positive effect on the physical and mental symptoms of MS [25,39]. Recent data on exercise in MS, recommends activities such as aerobic, resistance, balance or combined training for MS patients, and recommends providing instructions for the efficient practice of these physical activities and outlines the effects of each exercise by summarizing the physiological view and health perspective of MS (Fig. 1).

MS patients and their physio-psychological profile

MS patients, particularly those with advanced impairment may have different musculoskeletal and cardiovascular phenotypes compared to healthy individuals [42]. Patients with MS show decreased cardiorespiratory health and aerobic capacity (expression of max-oxygen-intake/VO2peak/ VO2max of nearby 30% less) compared to healthy individuals. Dysfunctions in respiration are caused by weakened respiratory muscle and/or exterior reasons such as muscle failure and fatigue [40,43,44]. Cardiac issues including abnormal heart rate and blood pressure are caused by impairments in the autonomic regulation of cardiovascular activities [39,45]. There is also commonly an increase in muscular atrophy and a reduction in power (calculated as the isometric and isokinetic contractions of the muscles, strength and muscle mass in the entire body) [25,39,42,46]. MS patients, particularly those with spasticity also show diminished flexibility [25,42]. Additionally, up to 80% of MS patients show intolerance to high temperature which may be associated with transitory exacerbation of the clinical symptoms of MS [47,48]. This can be a concern when physical activity is applied to MS patients, because, even though exercise is advantageous, with clear health benefits, it should not result in overheating [47,48].

MS patients and cardiorespiratory fitness

The effects of aerobic training in patients with MS have been studied more extensively compared to resistance training [27]. Aerobic training, with intensities ranging from low to moderate, can improve cardiovascular fitness, quality of life, and enhance mood. Aerobic exercise is well-tolerated by those living with MS, and many demonstrate improvements in cardiorespiratory health in as little as 4 weeks [44,49]. Cardiorespiratory exercise training in MS patients improves VO2 peak (VO2 Max), breathing function, activity capability, aerobic capacity, cardiorespiratory health and decreases fatigue [44,49,50]. In fact, cardiorespiratory training has been shown to be more beneficial for the improvement of mobility and general functional capacity in MS patients compared to direct neurorehabilitation [28,51,52]. Similarly, structured cardiorespiratory exercise results in improved quality of life and mood among MS patients [53], and enhances aerobic fitness, decreases fatigue, and has beneficial effects on disability [54]. However, there is still uncertainty regarding the effects of cardiorespiratory training in MS patients with severe impairments [39].

MS patients, muscle strength and endurance

During strength training, patients use muscle contractions against a load to enhance force generation capacity, improve neuromuscular recruitment and muscular hypertrophy. A smaller number of studies have reported the benefits of strength training in MS patients [42,55]. Increased muscular strength and endurance have also been linked to other exercise interventions [44]. Generally, the strength of the upper and lower legs are often affected by MS [54,56,57], and increasing the level of strength in the lower body, could provide key benefits in these patients. In one study, the use of strength exercise in MS patients showed beneficial effects of the lower limb strength, mobility, disability, and fatigue, with improvements in plantar flexor and knee extensor strength and subsequently improved walking ability [27]. MS patients can make significant improvements in response to strength training which can improve moving capacity, tiredness [30], function, and mobility [33]. In addition, combined strength and cardiorespiratory exercises reduced fatigue in those with low disability [54]. Thus, medium intensity resistance exercise not only is safe and well-tolerated in patients with MS but in general, it can enhance function and capability of muscle in MS patients with mild impairment [27,44,50,54,58].

MS patients and the wellbeing of bones

Loss of bone density and muscle mass in MS patients can be related to inactivity and concurrent corticosteroid treatments. This can further result in osteoporosis and an increased risk of fractures. In a study of 220 females with MS, 82% were taking corticosteroids and more than 50% reported inactivity and loss of bone mass [59]. Weight-bearing activities such as low-impact aerobics, walking, and aerobic workouts can reduce the speed of bone and muscle loss in patients with MS. This makes resistance exercise a highly recommended intervention to improve and maintain bone density and muscle mass and improve overall outcomes [27,44,58].

MS patients, balance and flexibility

Due to extended inactivity and muscular spasticity, MS patients often face restricted joint motion. Although there are limited studies assessing the effectiveness of flexibility exercise regimens in patients with MS, it is a commonly recommended mode of training. Flexibility training can lead to elongation of muscles and subsequently increase range of joint motion, reduce spasticity, as well as maintenance of balance and posture [42,44]. Flexibility training has also been shown to prevent stiffness or tightness (spasticity) of muscles in the early stages of the disease, postponing the onset of spasms and contractions. These activities must be performed with the use of proprioceptive assistance to stretch muscles in the chest, hip, pelvis, and legs, while avoiding pressure on the toes to prevent aggravation of spasticity [32,39,42,44]. Another common issue among patients with MS is loss of balance and postural sway, leading to difficulty in keeping an upright posture and a slower pace of movement which may result in falls [60,61]. Balance exercises and group aquatic training have been shown to improve balance and stability in MS patients measured by the Berg Balance Scale [18,62]. Stability exercise has also been shown to reduce the rate of falls in MS patients [63,64]. However, there is not adequate evidence in balance training for nonambulatory individuals with severe MS [18,60,62].

MS patients and fatigue

Tiredness is a very common symptom in patients with MS and results in worsening neurological and other symptoms including pain, depression, anxiety, and cerebral dysfunction [44,65]. MS and its associated comorbidities commonly lead to physical inactivity and psychological disorders which also contribute to subsequent tiredness. Physical activity can result in alterations in neuroplasticity and neurological protection which decreases in long-lasting inactivity and in the regulation of the hypothalamus-pituitary–adrenal (HPA) axis which may decrease fatigue in those with MS [66]. However, the studies on the effects of exercise in managing tiredness in MS patients are scarce and inconclusive, although some studies report potential beneficial effects of exercise on fatigue [13,50,65]. The use of cooling strategies and energy storing plans combined with cardiorespiratory workouts and neural recovery rehabilitation has been shown to be effective approaches [50,52,67–72]. Aquatic training has also been shown to reduce tiredness and increase quality of life in female MS patients [18]. Establishing a safe and effective workout plan may be a key alternative to developing therapy plans for fatigue in MS and should be encouraged [65].

Quality of life in MS patients

Patients with MS have an overall reduced quality of life which is linked to worsening of disease symptoms, including deterioration of mobility, walking, and cognition [40]. Studies have shown that regular exercises can have a positive effect on the general wellbeing, energy, function, and mobility of MS patients [73], and can have long-lasting effects on improving quality of life, in terms of social and physical function [50,64,74,75].

Findings for CNS imaging and neurological morphology in MS patients

Although to date there is no solid evidence showing positive structural effects of training and exercise on brain morphology specifically in patients with MS, several studies show cardiorespiratory exercise training can affect the gray matter volume, size of the cerebral cortex, and harmony in the white matter region, along with functional coordination in the cortex and hippocampus [29,76]. In addition, relapsing–remitting MS patients who participate in the regular cardiorespiratory exercise, show reduced brain degeneration, which could be considered a protective approach. Other studies using imaging methods show some structural and functional change in the CNS of MS patients post exercise [77].

How to involve exercise programs in practice for patients with MS

It is clear that patients with MS have an overall reduced level of physical activity compared to the general population [55]. Thus, when planning functional training programs for MS patients, it is important to aim at increasing motivation, engagement, and adherence to exercise, to facilitate positive results. Importantly, there are no reports of significant adverse events associated with exercise in patients with MS [47,48,77–79]. Rehabilitation leads to increased functional independence in MS patients and is equally effective in both severely impaired and less impaired individuals. In addition, those with severe impairment show even greater benefit in regards to improvements in ataxia and cognition [80]. Overall, exercise and physical activity increase the physical function and overall wellbeing of those with less advanced levels of MS and assist in sustaining physical function in individuals with medium to severe impairments.

Exercise therapy for patients with MS

Short-term and long-term exercise therapy

Similar to the administration of immunomodulatory drug treatments, rehabilitation plans should be used on an ongoing basis, to assist in the reestablishment and maintenance of physical function and subsequent improvements in quality of life. A randomized controlled trial of 59 patients with MS studied the temporary and long-lasting effects of a blended exercise regime on fatigue, quality of life and distance walked [81]. A blended exercise program of stretching, aerobic, balancing, and resistance training for 10 weeks was implemented with a 1-year follow-up. The program led to improvements in the Berg Balance Scale, six-minute-long walk test, dependence on family support services (FSS), and quality of life but not in the expanded disability status scale. In addition, the exercise intervention significantly improved MS symptoms, although, the outcomes were not shown to be long-lasting once therapy was terminated. The study also showed that there was a similar pattern in recurrence of MS symptoms in both the exercise and control groups. Hence, continuous long-term exercises rather than short training programs, have the greatest effect on the improvement of MS symptoms [81].

Types of exercise therapies and physical interventions for MS patients

Exercise training programs ranging from walking [82], high-intensity [56,83,84], aquatic exercise [17–19], yoga [12,20,21], pilates [22], and combined training [23], have shown improvements in the overall quality of life and health of patients with MS. Among several types of exercise regimes studied in individuals with MS, aquatic exercise, yoga, stretch exercises, aerobic workouts, and balance exercises, have been the most regularly recommended. Exercise prescription and other associated aspects of rehabilitation are impacted by the degree of disability, individual preferences, time restrictions, accessibility, personal stimuli, and financial impact. Thus, training interventions should be designed for each MS individual according to their level of motivation, impairment, preferences, limitation of time, access to resources and physical therapy clinics and financial circumstances [80].

Beneficial effects of walking in MS

Weight-bearing activities such as low-impact aerobics, walking, and aerobic exercise reduce the speed of mass and bone and muscle density loss among individuals with MS. Therefore, walking and other weight-bearing aerobic activities, are highly recommended to improve and maintain bone and muscle mass [27,44,58].

Beneficial effects of aquatic exercise in MS

Therapeutic aquatic exercise interventions are recognized as important approaches for patients with MS as well as those with other disabilities. The therapists who supervise aquatic activities are thought to play an important role in the effectiveness and success of these interventions. For this reason, students are trained as therapeutic recreational trainers to develop assistance techniques that can be applied to clients. Some of these techniques include implementing aquatic treatment approaches to support individuals to attain improvements in their functional status [18].

The efficacy of an aquatic exercise training program on quality of life and fatigue in 32 women with MS was studied in a randomized controlled trial [18]. Women underwent 8 weeks of supervised aquatic exercise in a swimming pool (3 × 60-min sessions per week) [18], and quality of life questionnaires and the modified fatigue impact scale were used at baseline (time zero), week 4, and week 8. The aquatic exercise intervention led to improvements in subscales of quality of life and fatigue after 4- and 8-weeks of aquatic exercise compared to the control group, suggesting that aquatic exercise training is a beneficial addition to the management of MS symptoms [18]. A systematic review and meta-analysis study of the effects of exercise therapy on physical and mental outcomes in Iranian patients with MS [85], noted the effectiveness of exercise therapy on quality of life. However, due to the heterogeneity of the studies and other limitations specified by the authors, further studies are required to establish an effective aquatic exercise program [85]. In another study, aquatic exercise therapy interventions were assessed from a disability and recreational perspective [86], and it was again shown that aquatic exercise therapy had positive benefits. The aquatic exercise included stretching, relaxation, and aerobic workouts and lead to improvements in strength. Aquatic exercise interventions potentially deliver a broad range of beneficial effects such as increased overall quality of life and decreased burden of disability [86], making them a strong candidate for inclusion in management of MS.

Yoga practices to improve MS-related symptoms

While yoga is popular for its effectiveness in numerous neuropsychiatric disorders, it is also known as one of the most prevalent forms of unconventional and balancing exercise, which has been vastly applied to decreasing symptoms of MS [12]. Yoga has been shown to reduce fatigue in patients with MS [12]. Indeed, randomized controlled trials of yoga show significant temporary effects of yoga compared to usual care. Yoga was found to have positive effects on fatigue and mood, but not for cognitive function and muscle mobility [21]. When the risk of bias Cochrane tool was measured, the effects of yoga on mood and fatigue against bias were not robust, and there were no transient short-term or permanent/long-term effects of yoga. In addition, yoga did not show beneficial effects on serious adverse events. Despite many studies showing that yoga is an effective tool for MS, this study showed that there was no evidence to justify the recommendation of yoga as a routine intervention for patients with MS. Therefore, yoga was suggested as a treatment option for MS patients who find it hard to commit to other recommended exercise regimens [21]. In a systematic review with meta-analysis, the effect of yoga, exercise, and physiotherapy was compared in terms of the mental, physical and social quality of life of patients with MS [87]. The findings showed that yoga and a combination of exercises could not significantly influence any domains of the quality of life in patients with MS, whereas physiotherapy and aerobic exercise were shown to be more effective in improving the quality of life [87]. Although the effects of yoga interventions have been assessed on the fatigue level, MS patients showed only a slight progressive effect on fatigue [88] and there was no sufficient evidence to outline significant effects associated with yoga on the fatigue of MS patients [88]. Given that numbness and impaired mobility caused by demyelination of CNS, leads to a 56% of rate of falls among MS people, there is a need for some clinical procedures to outline an evidence-based intervention methodology that could limit falls. Studies have assessed the efficiency of various interventions designated to decrease fall rate in MS patients by comparing fall prevention in intervention groups and control groups (with no intervention) [89]. The primary outcomes were fall rates, risk of falling, number of falls per person and adverse events. The most common physical activity interventions used for MS patients, is a single exercise, as a solitary intervention, a functional electrical stimulation and exercise accompanied by education. There evidence was uncertain in regards to the effectiveness of exercise interventions on fall rate compared to the control. However, for the quantity of falls and adverse events, there was evidence that exercise interventions were beneficial in improving balance function, self-reported mobility, and objective mobility [89]. Studies that assessed treatment effects of various exercise interventions on MS fatigue showed mixed outcomes associated with intervention ability to manage this symptom. Although evidence from some studies supported the effectiveness of combined energy conservation education plus fatigue management and exercise for improvement of fatigue in MS, conclusions regarding exercise subtypes across these studies were conflicting. However, solo exercise interventions, such as yoga that combine physical activities with fatigue management (mental strategies) generate more beneficial outcomes. The evidence that supported the effectiveness of yoga and fatigue management (educational energy conservation programs) in improvement of MS fatigue was strong [90].

Similarly, controlled clinical trials examined the efficacy of exercise intervention plans and aerobic intervention programs such as, active exercise programs, yoga, and sport climbing on the cognition/ executive function of people suffering from MS [91]. Although the majority of these studies showed improvements in cognition, the total effect size was not significant regarding improvement in cognitive function. Due to the limited evidence and lack of harmony among cognition measures, exercise structures, and time, it is still uncertain whether yoga programs can positively affect the improvement of the cognitive functions of MS patients [91]. Thus, although yoga has generally been known as a harmless therapeutic approach, and has been shown to have powerful effects on decreasing many health-related symptoms in MS people, there is still a need for more studies to determine its effectiveness as a fall prevention intervention, and its role in managing other MS-related symptoms.

Aerobic exercise and MS-related symptoms

Among various forms of exercises examined by multiple studies, aerobic workouts, which consist of balance exercises and some level of socialization are highly recommended for patients with MS [80]. A meta-analysis study that examined the effect of different physical activities, including yoga, aerobic exercise, and physiotherapy on different domains of quality of life (mental, physical and social) of patients with MS, showed that physiotherapy and aerobic exercise increased the physical, mental and social satisfaction of MS patients, and could therefore be considered as a regular practice in the treatment of MS [87]. Randomized controlled studies and controlled clinical studies that recruited adults diagnosed with neurological disorders, noted that aerobic training improved cognitive function in those suffering from such disorders, including MS [92]. One study determined the effect of a cycling program on cognition and showed improvement in the choice reaction time in MS patients [92]. However, there was insufficient evidence to back up the application of aerobic training for the improvement of cognitive functions in adults with neurologic disorders, as most of the clinical trials do not include cognition as a beneficial outcome measure, and some studies do not continue the aerobics plan for a long period to ascertain its long-term effectiveness.

High-intensity interval exercise training program

High-intensity interval training (HIIT) has been around for over 100 years mostly by runners who alternate between sprints and jogging to improve endurance. However, in the last 10 years, high-intensity interval training has become popular and studies have shown that it can deliver the biggest health improvement for an exercise time. Due to its benefits to health improvement, patients with MS were included in a controlled randomized phase-2 clinical study to evaluate the effects over 24 weeks [84]. Effect of high-intensity progressive aerobic exercise (PAE) on brain magnetic resonance imaging was assessed in patients with mild to severe MS impairment, involving an exercise intervention group (instructed PAE followed by self-conducted physical training) and a control group (usual lifestyle followed by instructed PAE). The primary outcome was a change in percentage-brain-volume, measured after the 24-week intervention. Data showed no effect on percentage-brain-volume, however, there were improvements in cardiorespiratory health and the annual rate of relapse in the exercise group. Furthermore, the results neither showed evidence of PAE benefiting neuroprotection from the point of total brain degeneration in MS patients, nor did they find a significant change in the parenchymal fraction of the brain's gray matter. PAE caused decreased relapse rate and improved cardiorespiratory health, but these results were labeled level I evidence, that did not support disease-modifying effects for 24-weeks of high-intensity PAE in a percentage-brain-volume change in MS patients [84].

There is substantial data supporting the beneficial effects of long-term moderate-intensity training in clinical results in MS patients. Severely deconditioned MS patients with walking disabilities may gain benefits through HIIT, particularly if exercise is combined with the use of adaptive equipment, such as leg recumbent stepping/ recumbent arm. A study measured the practicality of a 12-week adaptive equipment and HIIT program in MS patients with walking disabilities through assessing improvements in various functions, as well as measuring clinically relevant beneficiary results following the 12-week program [56]. The clinical outcomes measured included aerobic fitness, neurological assessments, physical activity, cognitive function, fatigue, ambulation, upper arm function, and depression. Results were assessed at time zero (baseline), at 6 weeks (midpoint), and at 12 weeks (postintervention). The intervention consisted of 12-week supervised personalized HIIT sessions 2–3 times a week. Each HIIT session was 20 min long and included 10 sets of 60-s intervals (90% VO2max) followed by 60 s of active recovery. The recruitment of the study is expected to be completed by the 30th of May 2021 and data will be published later in the year [56].

In addition, a single-blind randomized controlled superiority study proposed a multimodal rehabilitation approach to successfully manage MS symptoms such as, decreased aerobic fitness and fatigue; as these are considered to be the most immobilizing symptoms that affect patient quality of life. The multidisciplinary therapy includes HIIT endurance training and educational energy conservation programs through a 3-week inpatient therapy stay, and assessed the effect of this training program on the quality of life of patients with MS following a 6-month period. Effects of combined HIIT and inpatient energy management education will be measured against moderate continuous training (MCT) and progressive muscle relaxation (PMR). Patients with MS-related fatigue (chronic progressive disease) and expanded disability status scale ≤ 6.5 will be randomized into a trial or a control group. The trial group will do HIIT three times per week and 6.5 h of inpatient energy management education twice a week for 3 weeks. HIIT sessions consist of five rounds of 1.5 min (on a cycle ergometer at 95%–100% VO2 peak) followed by breaks of 2 min unloaded pedaling to reach 60% VO2 peak, and inpatient energy management education in 6.5-h group-based sessions. The control group will do PMR twice a week and MCT three times a week for 3-weeks. PMR involves 6×1-h relaxation group sessions, while MCT includes 24 min of continuous cycling at 65% VO2peak. The primary results target quality of life outcomes and the secondary outcomes will assess cardiorespiratory fitness, inflammatory markers, fatigue, mood, self-efficacy, occupational performance, physical activity, and behavior changes. The study is aimed to provide comprehensive evidence regarding multimodal rehabilitation approaches to improve treatment for patients with MS [83]. The study is registered with ClinicalTrials.gov with reference number NCT04356248 and is currently recruiting participants.

Dancing and MS

Dance as a therapeutic intervention has been shown to produce promising outcomes among patients with neurological disorders, including those with MS [93,94]. However, in one study the effects of ballet dancing in female patients with MS noted improvements in balance and ataxia [95]. Another study of mild to moderate partnered ballroom dancing over 6 weeks showed positive effects on cognition and quality of life [96]. Salsa dancing was also investigated in a pilot study, which showed that a 4-week intervention significantly improved gait and balance [93]. A more recent study compared the effects of a 10-week choreography-based program to a control group that performed other art activities, such as word art, painting, photography or videography and music. It was shown that dance significantly influenced more parameters than other art forms. Improvements were observed in patient fatigue, physical capacity, coordination, balance and cognition [94]. Dancing is an enjoyable activity that combines motor, cognitive, limbic and sensory activity engagement [93–95]. The findings of studies focusing on dance as a possible treatment for MS patients show promising results with improvements in many areas. However, these studies are limited and further studies on this topic would be useful to establish the feasibility of using dancing as a therapeutic intervention. Future studies should focus on more dance types and investigate the effects on larger sample sizes. It would also be useful to get more follow-up measurements in order to investigate the long-term effects of these interventions.

Exercise prescription

Clinicians and patients are commonly hesitant to prescribe or engage in physical activity due to fear of injury, or disease exacerbation. However, as already shown, exercise is safe and has beneficial effects across the spectrum of MS symptoms [41]. Important considerations for prescription include the patient's physical state and capacity and risk of falling. As patients are commonly physically deconditioned, a progressive program of strength and aerobic exercise should be used to gradually build exercise tolerance. Specific exercises must be tailored to the patient's requirements, such as seated strength exercises with bands and free weights for wheelchair-bound, or unstable patients. Given the risk of falls combined with low bone density, supervised aquatic exercise is a strong candidate, however, temperature-sensitive patients may be excluded from this to prevent adverse incidents. Additionally, recumbent cycling, rowing, ski trainer, or arm cycle can be used to engage patients in aerobic exercise tailored to their physical capacity.

Conclusion

MS is an inflammatory autoimmune disorder characterized by the destruction of the myelin sheath leading to muscular spasms, inability to balance, changes in gait, fatigue and mood, depression, cognitive decline, and eventually loss of function of the legs. MS is conventionally treated with antiinflammatory drugs which work to some extent but in the long term, there are many side effects, and patients with MS stop treatment. Exercise has been studied as an alternative or adjunct treatment to improve symptoms of MS and prolong remission. Several exercise regimens have been shown to improve cardiorespiratory and aerobic fitness, endurance, muscle strength, flexibility, fatigue, mood, psychological outcomes, and overall quality of life.

References

[1] Alzabin S, Williams RO. Effector T cells in rheumatoid arthritis: lessons from animal models. FEBS Lett 2011;585(23):3649–59.

[2] Bader RA, Wagoner KL. Modulation of the response of rheumatoid arthritis synovial fibroblasts to proinflammatory stimulants with cyclic tensile strain. Cytokine 2010;51(1):35–41.

[3] Chastain EML, Miller SD. Molecular mimicry as an inducing trigger for CNS autoimmune demyelinating disease. Immunol Rev 2012;245(1): 227–38.

[4] Bar-Or A. Immunology of multiple sclerosis. Neurol Clin 2005;23(1):149–75.

[5] Steinman L. Multiple sclerosis: a coordinated immunological attack against myelin in the central nervous system. Cell 1996;85(3):299–302.

[6] Dargahi N, et al. Multiple sclerosis: immunopathology and treatment update. Brain Sci 2017;7(7).

[7] Dargahi N, Matsoukas J, Apostolopoulos V. Streptococcus thermophilus ST285 alters pro-inflammatory to antiinflammatory cytokine secretion against multiple sclerosis peptide in mice. Brain Sci 2020;10(2).

[8] Yannakakis MP, et al. Design and synthesis of non-peptide mimetics mapping the immunodominant myelin basic protein (MBP_{83-96}) epitope to function as T-cell receptor antagonists. Int J Mol Sci 2017;18(6).

[9] Compston A, Coles A. Multiple sclerosis. Lancet 2002;359(9313):1221–31.

[10] Ahmad H, et al. The increasing economic burden of multiple sclerosis by disability severity in Australia in 2017: results from updated and detailed data on types of costs. Mult Scler Relat Disord 2020;44.

[11] Rieckmann P. Improving MS patient care. J Neurol Suppl 2004;251(5).

[12] Thakur P, et al. Yoga as an intervention to manage multiple sclerosis symptoms. J Ayurveda Integr Med 2020;11(2):114–7.

[13] Andreasen AK, Stenager E, Dalgas U. The effect of exercise therapy on fatigue in multiple sclerosis. Mult Scler J 2011;17(9):1041–54.

[14] Halabchi F, et al. Exercise prescription for patients with multiple sclerosis; potential benefits and practical recommendations. BMC Neurol 2017;17(1).

[15] Mulero P, et al. Improvement of fatigue in multiple sclerosis by physical exercise is associated to modulation of systemic interferon response. J Neuroimmunol 2015;280:8–11.

[16] Ickmans K, et al. Recovery of peripheral muscle function from fatiguing exercise and daily physical activity level in patients with multiple sclerosis: a case-control study. Clin Neurol Neurosurg 2014;122:97–105.

[17] Sadeghi Bahmani D, et al. Aquatic exercising may improve sexual function in females with multiple sclerosis—an exploratory study. Mult Scler Relat Disord 2020;43, 102106.

[18] Kargarfard M, et al. Effect of aquatic exercise training on fatigue and health-related quality of life in patients with multiple sclerosis. Arch Phys Med Rehabil 2012;93(10):1701–8.

[19] Kargarfard M, et al. Randomized controlled trial to examine the impact of aquatic exercise training on functional capacity, balance, and perceptions of fatigue in female patients with multiple sclerosis. Arch Phys Med Rehabil 2018;99(2):234–41.

[20] Shohani M, et al. The effect of yoga on the quality of life and fatigue in patients with multiple sclerosis: a systematic review and meta-analysis of randomized clinical trials. Complement Ther Clin Pract 2020;39, 101087.

[21] Cramer H, et al. Yoga for multiple sclerosis: a systematic review and meta-analysis. PLoS One 2014;9(11).

[22] Marques KAP, et al. Pilates for rehabilitation in patients with multiple sclerosis: a systematic review of effects on cognition, health-related physical fitness, general symptoms and quality of life. J Bodyw Mov Ther 2020;24(2):26–36.

[23] Ozkul C, et al. Effect of combined exercise training on serum brain-derived neurotrophic factor, suppressors of cytokine signaling 1 and 3 in patients with multiple sclerosis. J Neuroimmunol 2018;316:121–9.

[24] Martin-Sanchez C, et al. Effects of 12-week inspiratory muscle training with low resistance in patients with multiple sclerosis: a non-randomised, double-blind, controlled trial. Mult Scler Relat Disord 2020;46, 102574.

[25] Hugos CL, Cameron MH, Spasticity MS. Take Control (STC) for ambulatory adults: protocol for a randomized controlled trial. BMC Neurol 2020; 20(1).

[26] Motl RW, Sandroff BM. Benefits of exercise training in multiple sclerosis. Curr Neurol Neurosci Rep 2015;15(9):1–9.

[27] White LJ, et al. Resistance training improves strength and functional capacity in persons with multiple sclerosis. Mult Scler 2004;10(6):668–74.

[28] Kastanias TV, Tokmakidis SP. The effect of exercise on functional capacity and quality of life in patients with multiple sclerosis. Arch Hell Med 2008;25(6):720–8.

[29] Motl RW, Pilutti LA. The benefits of exercise training in multiple sclerosis. Nat Rev Neurol 2012;8(9):487–97.

[30] Learmonth YC, et al. Changing behaviour towards aerobic and strength exercise (BASE): design of a randomised, phase I study determining the safety, feasibility and consumer-evaluation of a remotely-delivered exercise programme in persons with multiple sclerosis. Contemp Clin Trials 2021;102.

[31] Sá MJ. Exercise therapy and multiple sclerosis: a systematic review. J Neurol 2014;261(9):1651–61.

[32] Petajan JH, White AT. Recommendations for physical activity in patients with multiple sclerosis. Sports Med 1999;27(3):179–91.

[33] Gutierrez GM, et al. Resistance training improves gait kinematics in persons with multiple sclerosis. Arch Phys Med Rehabil 2005;86(9):1824–9.

[34] Asano M, et al. What does a structured review of the effectiveness of exercise interventions for persons with multiple sclerosis tell us about the challenges of designing trials? Mult Scler 2009;15(4):412–21.

[35] O'Connor P. Key issues in the diagnosis and treatment of multiple sclerosis: an overview. Neurology 2002;59(6 Suppl. 3):S1–S33.

[36] Goodwin E, Green C, Hawton A. What difference does it make? a comparison of health state preferences elicited from the general population and from people with multiple sclerosis. Value Health 2020;23(2):242–50.

[37] Grima DT, et al. Cost and health related quality of life consequences of multiple sclerosis. Mult Scler 2000;6(2):91–8.

[38] Grimaud J, et al. Quality of life and economic cost of multiple sclerosis. Rev Neurol 2004;160(1):23–34.

[39] Dalgas U, Stenager E, Ingemann-Hansen T. Multiple sclerosis and physical exercise: Recommendations for the application of resistance-, endurance- and combined training. Mult Scler 2008;14(1):35–53.

[40] Gallien P, et al. Physical training and multiple sclerosis. Ann Readapt Med Phys 2007;50(6):373–6.

[41] Pilutti LA, et al. The safety of exercise training in multiple sclerosis: a systematic review. J Neurol Sci 2014;343(1-2):3–7.

[42] Sandoval AEG. Exercise in multiple sclerosis. Phys Med Rehabil Clin N Am 2013;24(4):605–18.

[43] Johansson S, et al. Associations between fatigue impact and lifestyle factors in people with multiple sclerosis—the Danish MS hospitals rehabilitation study. Mult Scler Relat Disord 2021;50.

[44] White LJ, Dressendorfer RH. Exercise and multiple sclerosis. Sports Med 2004;34(15):1077–100.

[45] Huang M, Jay O, Davis SL. Autonomic dysfunction in multiple sclerosis: implications for exercise. Auton Neurosci 2015;188:82–5.

[46] Lambert CP, Lee Archer R, Evans WJ. Body composition in ambulatory women with multiple sclerosis. Arch Phys Med Rehabil 2002;83(11): 1559–61.

[47] Kaltsatou A, Flouris AD. Impact of pre-cooling therapy on the physical performance and functional capacity of multiple sclerosis patients: a systematic review. Mult Scler Relat Disord 2019;27:419–23.

[48] White AT, et al. Effect of precooling on physical performance in multiple sclerosis. Mult Scler 2000;6(3):176–80.

[49] Motl RW. Exercise and multiple sclerosis. In: Advances in experimental medicine and biology; 2020. p. 333–43.

[50] Latimer-Cheung AE, et al. Effects of exercise training on fitness, mobility, fatigue, and health-related quality of life among adults with multiple sclerosis: a systematic review to inform guideline development. Arch Phys Med Rehabil 2013;94(9):1800–1828.e3.

[51] Fry DK, et al. Randomized control trial of effects of a 10-week inspiratory muscle training program on measures of pulmonary function in persons with multiple sclerosis. J Neurol Phys Ther 2007;31(4):162–72.

[52] Rampello A, et al. Effect of aerobic training on walking capacity and maximal exercise tolerance in patients with multiple sclerosis: a randomized crossover controlled study. Phys Ther 2007;87(5):545–55.

[53] Langeskov-Christensen M, et al. Aerobic capacity in persons with multiple sclerosis: a systematic review and meta-analysis. Sports Med 2015;45 (6):905–23.

[54] DeBolt LS, McCubbin JA. The effects of home-based resistance exercise on balance, power, and mobility in adults with multiple sclerosis. Arch Phys Med Rehabil 2004;85(2):290–7.

[55] Motl RW, McAuley E, Snook EM. Physical activity and multiple sclerosis: a meta-analysis. Mult Scler 2005;11(4):459–63.

[56] Hubbard EA, Motl RW, Elmer DJ. Feasibility and initial efficacy of a high-intensity interval training program using adaptive equipment in persons with multiple sclerosis who have walking disability: study protocol for a single-group, feasibility trial. Trials 2020;21(1).

[57] Kjølhede T, Vissing K, Dalgas U. Multiple sclerosis and progressive resistance training: a systematic review. Mult Scler J 2012;18(9):1215–28.

[58] Montealegre MC, et al. Effects of a potency and hypertrophy training programs on bone mineral density and mean power in people with multiple sclerosis for 7 weeks. A preliminary study. Cult Cienc Deporte 2020;15(43):5–16.

[59] Shabas D, Weinreb H. Preventive healthcare in women with multiple sclerosis. J Womens Health Gend Based Med 2000;9(4):389–95.

[60] Cameron MH, Lord S. Postural control in multiple sclerosis: implications for fall prevention. Curr Neurol Neurosci Rep 2010;10(5):407–12.

[61] Motl RW, et al. Top 10 research questions related to physical activity and multiple sclerosis. Res Q Exerc Sport 2015;86(2):117–29.

[62] Bansi J, et al. Endurance training in MS: short-term immune responses and their relation to cardiorespiratory fitness, health-related quality of life, and fatigue. J Neurol 2013;260(12):2993–3001.

[63] Cattaneo D, et al. Effects of balance exercises on people with multiple sclerosis: a pilot study. Clin Rehabil 2007;21(9):771–81.

[64] Kasser SL, et al. Effects of balance-specific exercises on balance, physical activity and quality of life in adults with multiple sclerosis: a pilot investigation. Disabil Rehabil 2015;37(24):2238–49.

[65] Motl RW. Benefits, safety, and prescription of exercise in persons with multiple sclerosis. Expert Rev Neurother 2014;14(12):1429–36.

[66] Heine M, et al. Exercise therapy for fatigue in multiple sclerosis. Cochrane Database Syst Rev 2015;2015(9).

[67] Mostert S, Kesselring J. Effects of a short-term exercise training program on aerobic fitness, fatigue, health perception and activity level of subjects with multiple sclerosis. Mult Scler 2002;8(2):161–8.

[68] Barten LJ, et al. New approaches in the management of multiple sclerosis. Drug Des Devel Ther 2010;4:343–66.

[69] Braley TJ, Chervin RD. Fatigue in multiple sclerosis: mechanisms, evaluation, and treatment. Sleep 2010;33(8):1061–7.

[70] Barry A, et al. Impact of short-term cycle ergometer training on quality of life, cognition and depressive symptomatology in multiple sclerosis patients: a pilot study. Neurol Sci 2018;39(3):461–9.

[71] Feys P, et al. Effects of an individual 12-week community-located "start-to-run" program on physical capacity, walking, fatigue, cognitive function, brain volumes, and structures in persons with multiple sclerosis. Mult Scler J 2019;25(1):92–103.

[72] Petajan JH, et al. Impact of aerobic training on fitness and quality of life in multiple sclerosis. Ann Neurol 1996;39(4):432–41.

[73] Stuifbergen AK, et al. Exercise, functional limitations, and quality of life: a longitudinal study of persons with multiple sclerosis. Arch Phys Med Rehabil 2006;87(7):935–43.

[74] Schulz KH, et al. Impact of aerobic training on immune-endocrine parameters, neurotrophic factors, quality of life and coordinative function in multiple sclerosis. J Neurol Sci 2004;225(1-2):11–8.

[75] Dalgas U, et al. Fatigue, mood and quality of life improve in MS patients after progressive resistance training. Mult Scler 2010;16(4):480–90.

[76] Prakash RS, et al. Aerobic fitness is associated with gray matter volume and white matter integrity in multiple sclerosis. Brain Res 2010;1341:41–51.

[77] Döring A, et al. Exercise in multiple sclerosis—an integral component of disease management. EPMA J 2012;3(1).

[78] Gravesteijn AS, et al. Neuroprotective effects of exercise in people with progressive multiple sclerosis (Exercise PRO-MS): Study protocol of a phase II trial. BMC Neurol 2020;20(1).

[79] Dalgas U, Stenager E. Exercise and disease progression in multiple sclerosis: can exercise slow down the progression of multiple sclerosis? Ther Adv Neurol Disord 2012;5(2):81–95.

[80] Brown TR, Kraft GH. Exercise and rehabilitation for individuals with multiple sclerosis. Phys Med Rehabil Clin N Am 2005;16(2):513–55.

[81] Sangelaji B, et al. Effect of combination exercise therapy on walking distance, postural balance, fatigue and quality of life in multiple sclerosis patients: a clinical trial study. Iran Red Crescent Med J 2014;16(6).

[82] Pilloni G, et al. Walking in multiple sclerosis improves with tDCS: a randomized, double-blind, sham-controlled study. Ann Clin Transl Neurol 2020;7(11):2310–9.

[83] Patt N, et al. High-intensity interval training and energy management education, compared with moderate continuous training and progressive muscle relaxation, for improving health-related quality of life in persons with multiple sclerosis: study protocol of a randomized controlled superiority trial with six months' follow-up. BMC Neurol 2021;21(1).

[84] Langeskov-Christensen M, et al. Efficacy of high-intensity aerobic exercise on brain MRI measures in multiple sclerosis. Neurology 2021;96(2): e203–13.

[85] Afkar A, et al. Effect of exercise therapy on quality of life of patients with multiple sclerosis in Iran: a systematic review and meta-analysis. Neurol Sci 2017;38(11):1901–11.

[86] Scott J, et al. Aquatic therapy interventions and disability: a recreational therapy perspective. Int J Aquat Res Educ 2020;12(3).

[87] Alphonsus KB, Su Y, D'Arcy C. The effect of exercise, yoga and physiotherapy on the quality of life of people with multiple sclerosis: systematic review and meta-analysis. Complement Ther Med 2019;43:188–95.

[88] Boehm K, et al. Effects of yoga interventions on fatigue: a meta-analysis. Evid Based Complement Alternat Med 2012;2012.

[89] Hayes S, et al. Interventions for preventing falls in people with multiple sclerosis. Cochrane Database Syst Rev 2019;2019(11).

[90] Miller P, Soundy A. The pharmacological and non-pharmacological interventions for the management of fatigue related multiple sclerosis. J Neurol Sci 2017;381:41–54.

[91] Kalron A, Zeilig G. Efficacy of exercise intervention programs on cognition in people suffering from multiple sclerosis, stroke and Parkinson's disease: a systematic review and meta-analysis of current evidence. NeuroRehabilitation 2015;37(2):273–89.

[92] McDonnell MN, Smith AE, MacKintosh SF. Aerobic exercise to improve cognitive function in adults with neurological disorders: a systematic review. Arch Phys Med Rehabil 2011;92(7):1044–52.

[93] Mandelbaum R, et al. A pilot study: examining the effects and tolerability of structured dance intervention for individuals with multiple sclerosis. Disabil Rehabil 2016;38(3):218–22.

[94] Van Geel F, et al. Effects of a 10-week multimodal dance and art intervention program leading to a public performance in persons with multiple sclerosis—a controlled pilot-trial. Mult Scler Relat Disord 2020;44, 102256.

[95] Scheidler AM, et al. Targeted ballet program mitigates ataxia and improves balance in females with mild-to-moderate multiple sclerosis. PLoS One 2018;13(10), e0205382.

[96] Ng A, et al. Ballroom dance for persons with multiple sclerosis: a pilot feasibility study. Disabil Rehabil 2020;42(8):1115–21.

Chapter 13

Exercise and menopause

Serene Hilary[a], Habiba Ali[a], Leila Cheikh Ismail[b,c], Ayesha S. Al Dhaheri[a], and Lily Stojanovska[a,d]

[a]*Department of Nutrition and Health, College of Medicine and Health Sciences, United Arab Emirates University, Al-Ain, United Arab Emirates*
[b]*Department of Clinical Nutrition and Dietetics, College of Health Sciences, Research Institute of Medical and Health Sciences (RIMHS), University of Sharjah, Sharjah, United Arab Emirates* [c]*Nuffield Department of Women's & Reproductive Health, University of Oxford, Oxford, United Kingdom*
[d]*Institute for Health and Sport, Victoria University, Melbourne, VIC, Australia*

Introduction

Menopause is a natural, physiological process involving the permanent cessation of the menstrual cycle for 12 months or more, marking the end of a female's reproductive life. The menopausal transition is marked by the loss of ovarian function and subsequently a deficiency in estrogen over several years, usually between the late 40s to early 50s. The Study of Women's Health Across the Nation (SWAN) of ethnically diverse participants found that the median age for the menopausal transition was 51.4 years [1]. The SWAN study also found differences in the onset based on race where the menopausal transition in Japanese females was delayed compared to Caucasian, African-American, Hispanic, or Chinese females. Hence, the age of onset for menopausal transition is often influenced by genetic, behavioral, and environmental factors. Other factors influencing the age of onset of menopause include socioeconomic status, lower levels of education, and smoking [2]. Systematic reviews and meta-analyses have also associated weight status and physical activity levels as moderately influencing factors with the later onset of menopause [2]. Menopause may also occur due to surgical procedures such as hysterectomy with bilateral oophorectomy or the treatment of specific health conditions such as endometriosis, breast cancer, and chemotherapy medications for other types of cancers [3].

The impact of menopause

Due to the cessation of ovarian function and depleted estrogen hormone levels, the menopausal transition is marked by the onset of vasomotor, psychological, cognitive, and atrophic symptoms. These symptoms are comprehensively termed menopausal (climacteric) syndrome. An estimated 25 million females experience menopausal syndrome every year [4]. With advancements in medicine, there is an increase in life expectancy worldwide, and consequently, a projected leap of 1.2 billion postmenopausal females is likely by 2030 [5]. Eighty-five per cent of females who undergo menopausal transition report at least one associated symptom: heavy/irregular bleeding, hot flashes, palpitations, mood changes, depression, anxiety, sleep disorders, cognitive problems, and atrophic vaginitis [3]. Several national studies have reported a negative correlation between menopausal syndrome and female quality of life [6–8]. Frequent visits to physicians and the need for therapy due to menopausal syndrome contribute to a decline in quality of life. Additionally, menopausal symptoms cause a considerable economic burden due to high health expenditure. It is estimated that the average expense of hospital visits in the US for symptoms related to menopause is approximately 500 USD per case [9]. A cohort study on the health care cost related to untreated menopause symptoms found that the healthcare resource utilization was 82% higher for all-cause outpatient visits and 121% higher for outpatient visits [10]. Additionally, due to a decrease in estrogen levels, menopausal females are also at increased risk of developing chronic diseases such as type-2 diabetes [11], cancer [12], autoimmune disorders [13], cardiovascular diseases [14], and osteoporosis [15].

Stages in the menopausal transition

The changes that occur during the menopausal transition are not abrupt but occur over time. The onset of symptoms related to menopause is related to the female's reproductive age status and is classified by the Stages of Reproductive Aging Workshop (STRAW). The first STRAW classification, proposed in 2001 as chronological age, is an unreliable tool to predict the menopausal transition [16]. Earlier definitions by the World Health Organization (WHO) and the Council of Affiliated Menopause Societies (CAMS) led to terminologies such as premenopausal, perimenopause, menopausal transition, and climacteric. The STRAW system defines seven stages in the reproductive age, where −5 to −3 represents the reproductive age, −2 to −1 represents the menopausal transition, followed by the final menstrual +1 and +2, which represents the postmenopause period. The STRAW classification was later modified in 2011 and is currently the international standard for classifying reproductive aging during the menopausal transition.

Endocrine changes during menopause

Primary follicles in the ovary decrease in number with age. Estrogen, one of the predominant hormones that govern females' reproductive life, is produced by the granulosa cells of the ovary; menopause, therefore, is the inevitable consequence of aging in females. The ovarian changes during the sexual cycle in females are attributed to the two gonadotrophic hormones secreted by the anterior pituitary gland: follicle-stimulating hormone (FSH) and luteinizing hormone (LH). The target of these gonadotropins in the female body is the ovary's primary follicles. During the normal menstrual cycle in females, FSH and LH stimulate primary follicles to differentiate to granulosa and theca cells in the ovary, secreting estrogen, progesterone, and peptide hormones (inhibin A and B). The levels of inhibin B produced in the ovary are dependent on the number of differentiating follicles. Although they initially stimulate multiple ovarian follicles, FSH and LH produce only a single dominant follicle that develops and produces ovarian hormones. The estrogen produced by the ovary and inhibin B, in turn, signals negative feedback for the brain to decrease FSH production. FSH and LH's stimulatory effect is reduced due to an inadequate number of primary follicles in menopausal females, which results in decreased estrogen production and cessation of menstruation. The hypothalamic–pituitary–ovarian axis is disrupted in the absence of estrogen inhibition and inhibin hormone (Fig. 1). The LH and FSH levels are unchecked and remain high years after the onset of menopause.

Associated symptoms

The most commonly reported symptoms associated with the menopausal transition are vasomotor symptoms with hot flushes and night sweats episodes. They affect a female's quality of life through sleep disturbance, fatigue, and impacts on mental health [8]. The molecular mechanism responsible for these vasomotor symptoms is not well understood, and it is reported that estrogen is not the only factor that affects the core body temperature [3]. The symptoms of thermoregulation vary among the population, with studies pointing to differences based on genes and general health status. Hence it is

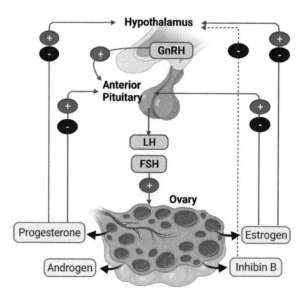

FIG. 1   The hypothalamus-pituitary-ovarian axis. Gonadotrophin-releasing hormone (GnRH) secreted by the hypothalamus stimulates the secretion of follicle-stimulating hormone (FSH) and luteinizing hormone (LH), which consequently triggers the production of estrogen, progesterone, and androgens along with peptide hormone inhibins, specifically inhibin B. Both estrogen and progesterone have positive and negative feedback on the anterior pituitary and hypothalamus, depending on the stage in the ovarian cycle. In addition, inhibin B directly inhibits gonadotropin secretion.

hypothesized that individuals who experience vasomotor symptoms have a narrower thermoregulatory zone than those without [17]. Besides vasomotor symptoms, reports suggest that about 45% of females experience some form of psychogenic symptom during the menopausal transition [18]. The symptoms that affect mental health include anger/irritability, anxiety, depression, sleep disorders, concentration loss, and loss of self-esteem or confidence. Evidence suggests that psychogenic symptoms can also be attributed to factors other than changing hormone levels, such as psychosocial or lifestyle factors. Hence, this phase of life among females has now termed a "window of vulnerability" by medical professionals [19]. Menopause is also characterized by urogenital atrophic changes such as vaginal dryness, burning, and irritation [3]. These symptoms appear over time as the menopausal transition progresses and becomes evident toward the postmenopausal stage as they are directly related to estrogen levels in the body.

Decreasing estrogen levels is also responsible for somatic symptoms such as skeletal joint and muscle problems. The broadly grouped somatic symptoms encompass symptoms related to the body, such as muscle and joint pain, numbness or tingling in the extremities, headaches, and shortness of breath [20]. The progressive loss of muscle mass and bone strength is prevalent in both sexes due to aging. However, the consequent health conditions such as sarcopenia and osteoporosis are more prevalent among older females due to low estrogen levels [21]. The menopausal transition is also associated with weight gain due to waning estrogen levels in the body [22]. In vivo studies in animals have found evidence that estrogen is an essential factor that governs weight status and fat redistribution, especially in the abdominal region [23–25]. This age group's tendency to gain weight is also compounded by a sedentary lifestyle and muscle mass loss. In addition to the symptoms described above, menopausal females have asymptomatic health considerations such as type-2 diabetes and cardiovascular diseases. The compounded effect of weight gain resulting from declining estrogen levels and decreased energy expenditure increases fat oxidation, accompanied by higher total body fat and visceral adipose tissue mass [26]. All of these health consequences predispose middle-aged females to the risk of developing type-2 diabetes and cardiovascular diseases. Evidence suggests that body weight increase is associated with hyperinsulinemia and insulin resistance in postmenopausal females [27,28], and estrogen therapy is reported to improve insulin sensitivity, decrease abdominal fatness and improve lipid profile [29]. Insomnia also increases during the menopausal transition and can become an independent risk factor for cardiovascular disease.

Treatment strategies

Various strategies are available to manage menopausal symptoms; they include nonmedical (nonprescription) interventions related to diet and physical activity and medical interventions such as hormonal replacement therapy and nonhormonal medication. Two pharmaceutical drugs are critical to nonhormonal therapeutic strategy for vasomotor symptoms; paroxetine, a selective serotonin reuptake inhibitor (FDA approved), and gabapentin (off-label) [30,31]. These drugs were first used as antidepressants, but their dose for vasomotor symptoms is lower than antidepressant dosing. However, they have potential side effects, including weight gain, loss of libido, and bloating. The most effective treatment strategy for the menopausal syndrome is hormone replacement therapy (HRT) with estrogen either alone or in combination with progesterone. HRT is primarily indicated for females with moderate to severe menopausal symptoms such as hot flushes, nights sweats, and urogenital atrophy. Other menopausal symptoms such as psychogenic symptoms, insomnia, and sexual dysfunction also improve with HRT. In addition, since estrogen is a critical factor influencing bone loss, HRT is also helpful in preventing and decreasing the risk of osteoporosis and related fractures in females.

Despite the positive health effects associated with HRT, this treatment strategy is not without controversy. Although estrogen has a cardioprotective function, the effect of HRT in preventing cardiovascular disease has provided conflicting results. The observations from the Women's Health Initiative (WHI), a study that investigated the risk and benefits of various treatment strategies such as HRT on health, is a crucial example in this regard [32]. According to the WHI, HRT effectively prevented cardiovascular complications in menopausal females without prior endothelial deterioration, while in older postmenopausal females, it posed severe thrombotic and cardiovascular risks. Currently, medical practitioners believe that HRT in the initial stages of menopause is safe for use. However, evidence from the Million Females Study by Cancer Research UK and National Health Services (NHS) suggests a direct link between HRT and breast cancer risk in females [33]. Furthermore, HRT increased the risk of breast cancer in females treated with estrogen in combination with progesterone. After analyzing the risk–benefit ratio associated with HRT from the reported evidence, in 2013, a global consensus statement was released for the appropriate use of HRT for menopause [34]. Despite these efforts, females still have growing skepticism about commencing or continuing HRT for mitigating menopausal symptoms. Consequently, there is increasing traction for alternative therapeutic strategies such as physical activity and exercise to improve quality of life during menopause and beyond.

Exercise for overall health and well-being

Physical activity was one of the driving forces in the evolution of *Homo sapiens* because performance capacity and motor skill played a crucial role in our survival, providing a clear advantage in our quest for food and defense against predators. The emergence and proliferation of our species involved conditions that required adaptation to a wide range of metabolic demands imposed by work or exercise. Lately, however, an industrialized world had eliminated much of our need for physical labor resulting in a propensity for a more sedentary lifestyle. Consequently, we also observe an alarming increase in chronic diseases across the world. Numerous studies have attributed sedentary lifestyles to the development of these chronic diseases. Although our understanding of how physical activity and exercise positively affect human health is evolving daily, we can surmise that humans are better off maintaining a physically active lifestyle.

When we engage in moderate to high-intensity physical activity, the immediate effects include improved endurance, metabolism, and energy; healthier muscles, joints, and bones; improved cognitive function; improved mental health, and healthier sleep patterns [35]. Nevertheless, incorporating a regular exercise regime is associated with a better quality of life and positive health outcomes related to health conditions such as cancers [36], cardiovascular diseases [37,38], type-2 diabetes [39], obesity [40], osteoporosis [41] and even cognitive disorders [42,43]. Substantial evidence in the literature points to the better quality of life and positive health outcomes associated with physical activity and exercise. A recently published UK Biobank cohort study that involved more than 90,000 participants observed that physical activity was not only associated with a lower risk of cardiovascular disease, it provided the best outcome for those with the highest activity levels [44]. Similarly, a retrospective cohort study in older adults with type-2 diabetes found improvements in physical quality of life, anthropometric profile, hemodynamic profile, and cardiorespiratory fitness [45]. A 12-week resistance and aerobic exercise intervention in obese middle-aged females found improvements in metabolic syndrome factors such as blood pressure, body fat, fasting glucose levels, and lipid profile [46]. The study also reported a significant reduction in visfatin, a highly expressed protein in visceral adipose tissue linked to obesity and increased health risks. Moreover, positive outcomes were reported in the HUNT cohort study, which investigated the effect of exercise on mental health [47]. The study, which involved more than 33,000 adults, found that regular leisure-time exercise of any intensity protected against future depression episodes. Evidence from intervention studies reaffirms the vital role physical activity and exercise play in improving muscle strength, improving functional capacity, modulating bone metabolism, decreasing obesity, reducing cardiovascular risk, improving mental health, and preventing type-2 diabetes. Physical activity and exercise also protect from stress-related telomere length shortening in older populations. Telomeres are the terminal ends of the chromosomes which protects the DNA throughout the cell cycle by preventing base pair loss. Telomere length decreases over time and its length governs the ability of cell to further divide, thereby triggering cell death and aging [48]. Telomere length is reported to be stable from childhood to adulthood and shows signs of decrease in the older population. Perceived stress levels in sedentary individuals is negatively associated with telomere length and among physically active individuals exercise acts as a buffer to the detrimental effects of stress on telomere length [49]. Such evidence reinforces the importance of physical activity and exercise in prolonging life among an aging population. Therefore, physical activity and exercise are inexpensive yet effective strategies to combat the symptoms associated with menopause and improve quality of life and prolong lifespan.

Physical activity and menopausal females

Physical activity and fitness are vitally important among middle-aged females to maintain a functional and independent lifestyle. Exercise and physical activity are critical in modifying menopause-related changes in health and function, which has direct implications for meeting the needs of the ever-growing population of postmenopausal females. Our earlier understanding of physical activity pertained to prescribing a particular exercise of sufficient frequency, intensity, and duration to improve physical fitness and reduce disease risk. Although the effect of physical activity on health status may be mediated primarily by the physiological changes that result from physical fitness, recent literature suggests it can also have an independent positive impact on other indicators of menopause, such as mental health [47,50,51]. It is also important to point out that the exercise prescription or plan, which generally includes determinants such as frequency, intensity, and duration, optimal in achieving one type of health outcome (e.g., weight loss), may be quite different from other health outcomes (e.g., bone mineral density or muscle strength). In general, exercise and physical activity increase the life expectancy for postmenopausal females and provide marked improvements in their quality of life [52]. Over time, randomized controlled trials have suggested the beneficial potential of exercise as an intervention strategy in controlling and preventing the symptoms and health risks associated with the menopausal transition. Exercise training in menopausal females helps reduce menopausal symptoms [35], improve muscle strength [4], increase bone metabolism [41], improve functional capacity [21], decrease obesity [4] and make positive improvements to overall mental health [47]. Exercise also effectively alleviates insulin resistance and systemic inflammation, reducing the risk of developing chronic diseases [53,54]. Physical activity in the menopausal population reduces the risk of cancer, diabetes, and heart disease. It is also reported as one of the most effective strategies to counteract sarcopenia among females [4,21]. As a complementary strategy to maintain physical and mental health during menopause, exercise benefits are well established, including its long-term effectiveness. A systematic review of randomized controlled trials in postmenopausal females between the ages of 50–65 found a positive impact of daily walking at moderate intensity combined with resistance training [55]. The study found positive health outcomes in terms of bone strength, weight and body fat, musculoskeletal fitness, motor fitness, cardiorespiratory fitness, and improvements in serum biochemical parameters. However, females of menopausal age have a tendency for sedentary behavior, which can be attributed to several social factors such as lack of facilities or adequate opportunities, high cost, safety fears, busy schedules and low accessibility to quality physical training programs [56]. Lack of motivation is also an influence that results in a sedentary lifestyle among this population.

Exercise and menopause syndrome

The menopause transition in females is characterized by the onset of associated physical and psychological symptoms that negatively impact the quality of life. These symptoms described earlier in the chapter include vasomotor, psychological, cognitive, and atrophic symptoms. Although HRT is considered the best therapeutic strategy for managing these symptoms, its adverse effects discourage females from this option. Physical activity and exercise are proposed as potential intervention strategies to mitigate symptoms during menopause. The Scientific Advisory Committee of The Royal College of Obstetricians and Gynecologists in the UK has indicated that regular aerobic physical activity can reduce menopausal symptoms [57]. Similarly, The North American Menopause Society has advised that lifestyle changes, such as engaging in regular physical activity, can help manage menopause symptoms [58]. The most predominant symptoms during menopause transition are vasomotor symptoms such as hot flushes and night sweats, with close to 85% of females reporting them in surveys [59]. The exact mechanism by which exercise affects vasomotor symptoms is not well understood. One hypothesis is that vasomotor symptoms are mitigated by exercise due to increased endorphin levels in the body [60]. During the menopausal transition, it is reported that a decrease in estrogen levels triggers a simultaneous decrease in endorphin concentrations in the hypothalamus. This endorphin decrease enhances neurotransmitter release, including norepinephrine and serotonin, which are critical activators of the thermoregulatory zones in the brain. HRT improves vasomotor symptoms by triggering an increase in hypothalamic and peripheral β-endorphin production, resulting in lower levels of norepinephrine and serotonin. It is hypothesized that exercise may have a similar effect on endorphins as HRT [61]. Indeed, studies have observed higher levels of β-endorphins in highly active individuals compared to sedentary individuals [62].

Results from large observational studies support the ability of exercise to reduce vasomotor symptoms. An Italian study reported data from over 66,000 females attending menopause clinics found that vasomotor symptoms were significantly less among active individuals than sedentary [63]. Similarly, the Melbourne Women's Midlife Health Project found that physically active females were 49% less likely to report menopause symptoms, while females who exercised less experienced more hot flushes [64]. Another large observational study involving about 12,000 multiethnic females reported higher incidents of hot flushes and night sweats among females who fall in the lower quartile of physical activity levels than physically active females of the same age [65]. Additionally, Ivarsson et al. report that only 5% of the physically active females in their study experienced vasomotor symptoms [66]. However, the data on the ability of exercise to mitigate vasomotor symptoms are conflicting. Numerous studies with smaller sample sizes report no significant positive association between exercise and vasomotor symptoms [67–69]. However, it is noteworthy that these studies showed positive associations between exercise and other menopausal symptoms, such as somatic and psychological symptoms (depression and anxiety). These observations gained traction with the publication of a Cochrane review [70]. After a systematic review and meta-analysis of randomized controlled trials, the Cochrane review concluded that there was a dearth of evidence to support the effectiveness of exercise in treating vasomotor symptoms. Following the publication of the Cochrane review, other studies have reported similar observations. For example, three months of moderate exercise intervention in one study found no significant decrease in vasomotor symptoms, but exercise improved mental health and prevented insomnia [71]. Even though the Royal College of Obstetricians and Gynecologists in the UK and the North American Menopause Society have recommended exercise as a treatment strategy for vasomotor symptoms, most randomized controlled trials have failed to find a significant positive association regardless of sample size. Consequently, there is a significant dearth of evidence for exercise recommendation for vasomotor symptoms and more research is required for evidence-based recommendations. However, the lack of evidence should not be a deterrent or barrier to physical activity during menopause because exercise is associated with positive mental health effects and reduced the risk of chronic diseases in older females.

Apart from vasomotor symptoms, menopause transition also affects mental health. Females often report symptoms of depression such as a general feeling of unhappiness, irritability, tearfulness, and lack of energy. Feelings of anxiety, mood swings, insomnia, panic attacks, cognitive problems such a forgetfulness, lack of focus, and concentration are other psychological symptoms reported during the menopausal transition. As discussed earlier, many observational studies, both large and small, among menopausal females have reported improved mental health through exercise participation. For example, a three-week stretching exercise intervention in 60 Japanese females reported significantly reduced psychological symptoms assessed through the Self-Rating Depression Scale scores [72]. Likewise, a four year observational study on the long-term effect of physical activity involving 159 menopausal females who were previously sedentary reported positive improvements in physical and mental dimensions of quality of life [73]. Another investigation in 80 community-dwelling postmenopausal females in China found that three 60 minutes sessions of walking per week for four months reported improved BMI and lower depression as assessed by the Beck Depression Inventory [74].

It is suggested that the general improvement of mental health through physical activity is the direct result of exercise serving as a distraction strategy for anxiety and depressing thoughts [60]. Group exercises also contribute by promoting more social interaction, which again benefits overall mental health. Studies indicate that exercise can increase self-esteem and self-efficacy in the population, which may also be a possible mechanism by which exercise mitigates depression [47,51,52]. Moreover, several other hypotheses based on exercise's physiological and biochemical effects could play a role in how exercise affects mental health. A predominant hypothesis is the effect of β-endorphins. As mentioned earlier in the chapter, exercise increases peripheral and hypothalamic β-endorphin levels [62]. The phenomenon termed "runner's high," which is a feeling of euphoria after an intense training session is attributed to β-endorphin, which can play a crucial role in improving symptoms of depression [50]. Similarly, the serum levels of brain-derived neurotrophic factor (BDNF), the most predominant neurotrophic factor in the brain, increase with exercise [75]. It is reported that BDNF levels are low in the older population and also among patients with depression. Low levels of BDNF can result in memory impairments, a higher risk of depression, and hippocampal degeneration. Physical activity and exercise can trigger BDNF synthesis and release by increasing blood flow to the brain [51]. Other hypotheses relate to mitochondrial dysfunction, mammalian target of rapamycin (mTOR), neurotransmitters such as serotonin, and the hypothalamic–pituitary–adrenal axis can also be attributed to the mechanistic role of improved mental health through exercise [50]. Therefore, physical activity and exercise can serve as a form of behavioral activation and effective therapeutic intervention of psychological symptoms in menopause.

Exercise for healthier bone and muscle during menopause

The higher propensity for sedentary behavior among the aging population increases health risk by loss of muscle and bone mass. The loss of bone and muscle mass in older females is directly linked to menopause due to estrogen's critical role in maintaining both these tissues. Hence apart from the symptoms related to menopause which are detrimental to female health, loss of bone and muscle mass collectively becomes a vicious cycle that makes the older females more sedentary over time. Therefore, physical activity and exercise represent an effective strategy to slow the progression of sarcopenia and delay the onset or even treat osteoporosis among older females. Evidence from literature highlights the beneficial effect of exercise on muscle strength and physical performance, which counteract the inability to perform daily life activities and musculoskeletal injuries related to sarcopenia [4,21,76]. Numerous studies also suggest the effect of exercise in delaying the onset of osteoporosis and improving muscular fitness [76–78]. Hence, physical activity and exercise are regarded as the primary nonhormonal and nonpharmacological treatments to prevent osteoporosis and related fractures, and in combination with dietary protein supplementation, it is an effective strategy to mitigate sarcopenia. Some randomized controlled trials reporting improved bone mineral density with exercise interventions are summarized in Table 1.

Both sarcopenia and osteoporosis are usually triggered in females with the onset of menopause. Hence it advised that physically inactive individuals should immediately adapt to an active lifestyle. Exercise training among osteoporotic individuals poses the risk of injuries and fractures, and therefore, females who are already osteoporotic are advised to adopt a safe and recommended exercise program to improve bone mineral density and reduce osteoporotic fractures [78]. Often a prescribed medical therapy in conjunction with physical exercise is recommended for osteoporotic females. The various exercises that can help maintain or improve bone density among older females can either be weight-bearing high/low-impact exercises or non-weight-bearing and low-impact exercises [91]. The weight-bearing high-impact exercises are not recommended for frail, osteoporotic females with low bone mass due to the risk of fractures [92]. They involve dancing, high-impact aerobics, jogging or running, gymnastics, or participating in sports. Weight-bearing low-impact exercises are suited for menopausal females who cannot participate in high-impact exercise, and they can range from walking, elliptical training, and low-impact aerobics. Strength training or resistance training exercise is ideal for menopausal females,

TABLE 1 Summary of randomized controlled trials reporting the effect of exercise on bone health in menopausal females.

Study	N	Inclusion criteria	Type and modality	Frequency, intensity, and duration	Intensity	Study outcome
Bergström et al. [79]	112	Postmenopausal Age: 45–65 years Low BMD Forearm fractures Sedentary No hormone therapy BMI: 19.9–30.9 kg/m²	Aerobic and strength training Supervised	60 min 2 days/week 12 months duration	Intensity-based on individual needs	Total hip BMD increased in the exercise group
Brentano et al. [80]	28	Postmenopausal Age: >55 years Low physical activity Low bone loss	Resistance exercise and circuit training Supervised	60 min 3 days/week 24 weeks duration	Resistance 45%–80% 1 RM with 2 min interval Circuit training 45%–60% 1 RM with no intervals	The intervention resulted in increased VO2 Max and time of exhaustion. Did not alter bone mineral density
Englund et al. [81]	48	Postmenopausal Age: >60 years Low physical activity Low bone mass	Combination of resistance, aerobic, balance, and coordination exercises Supervised	50 min 2 days/week 12 months duration	–	The intervention increased bone mineral density, improved walking speed, and isometric grip strength compared to control
Going et al. [82]	320	Postmenopausal (surgical or natural) Age: 40–65 years BMI: <33 kg/m² Nonsmoker No history of osteoporotic fractures Low physical activity	Aerobic, weight-bearing, and weight-lifting exercise, three times per week in community-based exercise facilities	30 min 3 days/week Supervised	Resistance 70%–80% 1 RM Aerobic 60% HRmax	Females who used HRT, calcium, and exercised increased femoral neck, trochanteric and lumbar spine bone mineral density. Trochanteric BMD was also significantly increased in females who exercised and used calcium without HRT compared to a negligible change in females who used HRT and did not exercise
Kemmlar et al. [83]	137	Postmenopausal Age: 48–60 years No osteoporosis or related fractures, CVD, medication affecting bones	Resistance and aerobics training Supervised/home training	60 min Supervised 20–25 min Unsupervised High frequency-1.5-2 session/week	High-intensity 70%–90% 1 RM Low intensity 50%–55% 1 RM	High-frequency exercise improved the lumbar spine and proximal femur BMD. The study concludes that an exercise frequency of at least 2 sessions/

Continued

TABLE 1 Summary of randomized controlled trials reporting the effect of exercise on bone health in menopausal females—cont'd

Study	N	Inclusion criteria	Type and modality	Frequency, intensity, and duration	Intensity	Study outcome
				Low frequency-2.5–3 sessions/week 12 years duration		week is crucial to impact bone health
Korpelainen et al. [84]	160	Postmenopausal Age: >60 years Low physical activity Low bone mass	Impact exercise training Supervised and unsupervised	45 min 7 days/week for 6 months/year supervised 20 min At home 7 days/week 30 months duration	–	Femoral neck and trochanter bone mineral density was maintained in the intervention group compared to the control. Exercise has a positive effect on bone mineral content at the trochanter
Lau et al. [85]	50	Postmenopausal Age: >60 years Low physical activity Low bone mass	Load-bearing exercises (stepping up and down a block 100 times) Supervised	15 min 4 days/week 10 months duration	Moderate to submaximal intensity	BMD at Ward's triangle and the intertrochanteric area increased significantly. Calcium supplements and exercise in combination gave significant joint effect at the femoral neck
Marques et al. [86]	71	Postmenopausal Age: >60 years Low physical activity No medication that affects bone metabolism	Resistance and aerobic exercises Supervised	Aerobic 60 min 3 days/week 8 months	Aerobic 50%–85% HRR Resistance 60%–80% 1RM	Resistance training increased BMD at trochanter and hip. Also improved body composition. Both resistance and aerobic training improved balance
Park et al. [87]	50	Postmenopausal Age: >65 years Low physical activity	Resistance, weight-bearing, balance, and posture correction training Supervised	60 min 3 days/week 12 months	Weight-bearing exercise 65%–75% HRmax	Improved gait, posture, and balance, BMD of the femoral neck and trochanter significantly increased along with improved body sway
Rhodes et al. [88]	44	Postmenopausal Age: >65 years Sedentary	Resistance training Supervised	60 min 3 days/week 52 weeks	75% 1RM	Improvement in muscle strength with the maintenance of bone mineral density
Verschueren et al. [89]	70	Postmenopausal Age: >55 years Sedentary	Whole-body vibration and resistance training Supervised	Whole-body vibration 20 min, 3d/week	Whole-body vibration Frequency 35–40 Hz	Vibration training improved isometric and dynamic muscle strength and also

TABLE 1 Summary of randomized controlled trials reporting the effect of exercise on bone health in menopausal females—cont'd

Study	N	Inclusion criteria	Type and modality	Frequency, intensity, and duration	Intensity	Study outcome
				Resistance 60 min, 3 days/week 6 months duration	Resistance first 14 weeks 2 × 15–8 RM; last 10 weeks 3 × 12 RM and 1 set × 8 RM	significantly increased BMD of the hip
Wen et al. [90]	48	Postmenopausal Age: 55–65 years Low BMD Sedentary No hormone therapy BMI: 18.5–24.9 kg/m²	Group based aerobics exercise Supervised	90 min 3 days/week 10 weeks duration	75%–85% HRR	Functional fitness components in the exercise group significantly improved by enhancement in lower and upper-limb muscular strength and cardiovascular endurance

1-RM, repetitions maximum (the greatest weight that can be moved once in good form); *BMD,* bone mineral density; *BMI,* body mass index; *CVD,* cardiovascular disease; *HR max,* heart rate max; *HR,* heart rate; *HRR,* heat rate reserve; *HRT,* hormone replacement therapy; *N,* number of randomized participants; *Supervised,* interaction with a coach.

including supervised weight training and functional movements [93]. It is essential for females who are already osteoporotic to undergo supervised exercise training, and it usually commences with non-weight-bearing and low-impact activities such as cycling, swimming, stretching, and flexibility exercises to reduce the risk of fractures. These exercises alone do not help remodel bone mineral density and are usually adopted as part of a comprehensive exercise training program. Other nonimpact exercises that are helpful for older females include exercises that help in balance and posture. Overall a menopause-friendly exercise program must include endurance (aerobic), strength, and balance exercises. It is reported that aerobic, weight-bearing, and resistance exercises effectively increase the bone mineral density of the spine in menopausal females [94]. Resistance exercise training is one of the most helpful exercise strategies to improve muscle mass and muscle strength [95]. Progressive resistance exercise training is also reported to improve endurance among older females and trigger muscle protein synthesis, thereby counteracting sarcopenia. Exercise that improves muscle mass and strength is also associated with improved bone mineral density and strength [96]. Hence exercise regimes such as resistance training can be crucial in reversing the loss of bone mass in osteoporotic individuals. Evidence from literature backs the use of resistance training for osteoporosis and sarcopenia. However, our knowledge about the effect of resistance training on these conditions is still evolving, and hence further studies are required for adopting it as a prescription strategy. A recent consensus on exercise recommendation for menopausal females with osteoporosis with or without a history of osteoporotic fractures has been cited in the literature [97–99]. They adopt the American College of Sports Medicine (ACSM) framework for exercise prescription to treat sarcopenia, specifically to promote muscle hypertrophy, strength, and power [100,101]. The recommended guidelines were recently reviewed by Agostini et al. [21]. These guidelines for exercise prescription adopt the FITT-VP principle. FITT-VP is an acronym that presents the various tenets in the exercise training program, which includes frequency (F), intensity (I), time (T), type (T) of the exercise, its volume (V) and progression (P). Table 2 is an adapted version described by Agostini et al. and provides a detailed description of the FITT-VP principle for various exercises recommended for menopause [21].

TABLE 2 Exercise recommendation for aging menopausal females.

Frequency (F)	Intensity (I)	Time (T)	Type (T)	Volume (V)	Progression (P)
Aerobic exercise recommendation					
At least 5 days a week	Moderate	30–60 min in a single session or at least 150 min per week	Weight-bearing activities include walking, jogging, dancing, or other activities where limbs support the total body weight.	≥500–1000 MET*min per week	Recommended increasing gradually any of the FITT components based on tolerance. An example is adding 5–10 min every 1–2 weeks over the first 4–6 weeks in the training program. Further upward adjustment over the next 4–8 months to meet the recommended FITT components
At least 3 days a week	Vigorous	20–60 min in a single session or at least 75 min per week			
Resistance exercise recommendations					
1–2 days a week	Beginners: ~8–12 repetitions performed near task failure, which translates to 10–14-RM or 5–8 on the 0–10 RPE scale	It depends on the exercise volume, i.e., number of sets and repetition along with the length of intervals, based on the physical activity level of the individual	Any form of movement designed to improve muscle fitness through exercises intended for a single muscle or a muscle group against external resistance. Exercise and breath techniques are crucial factors for resistance. Can include free weights, resistance machines, weight-bearing functional tasks, etc.	A single set of 8–12 repetitions per session. It is recommended not to exceed 8–10 exercises in a session	Small increments in the progress are recommended, for example, depending on muscular size and involvement and increase of 2%–10% 1-RM In case of intervals between the exercise, it is recommended to lower the resistance level by 2 weeks' worth for every week of inactivity
2–3 days a week	Intermediate and highly active individuals: from 8 to 12 repetitions performed to task failure, which translates to 8–12-RM or >8 on the 0–10 RPE scale			Two sets of 8–12 repetitions per session. It is recommended not to exceed 8–10 exercises per session	
Flexibility exercise recommendations					
≥2–3 days a week Daily exercise is recommended for better efficiency	Stretching and holding the position to the point of feeling tightness or discomfort	Holding a stretched position for at least 10–30 s is ideal, but 30–60 s gives significantly better benefit. Recommended to accumulate at least 60 s of stretched	Stretching exercise that increases the ability to move a joint through its complete ROM. It is essential to consider the individual-specific condition for flexibility	Repeat each exercise 2–4 times to attain the goal of 60 s stretch time. For example, a mix of two 30 s stretches or four 15 s stretches. Generally, a routine can be	Optimal progression is still unknown

TABLE 2 Exercise recommendation for aging menopausal females—cont'd

Frequency (F)	Intensity (I)	Time (T)	Type (T)	Volume (V)	Progression (P)
		position for each flexibility exercise. Total time depends on the number of repetitions and individual needs	exercises. Exercise includes static active flexibility, static passive flexibility, dynamic flexibility, ballistic flexibility, proprioceptive neuromuscular facilitation, etc.	completed in approximately 10 min	
Balance exercise recommendation					
Daily	Not applicable	≥15–20 min	Exercises that reduce support base in static stances such as semitandem, tandem, or one-legged stand. A dynamic or three-dimensional balance challenge such as Tai Chi, tandem walking, walking on heels or toes. Other strategies to challenge balance systems like weight shifting, reduced contact with support objects, dual-tasking, close eyes during static balance challenges, etc.	A cumulative time of 2 h per week is recommended	Progression should be from "standing still" to "dynamic" exercises. Progression of the balance challenge should also occur over time. For example, adopting a challenging exercise, removing vision or contact with support objects, dual-tasking, etc.

1-RM, one repetition maximum, the load that can be lifted only one time. *MET*min*, metabolic equivalents (MET) of energy expenditure for a physical activity performed for a given period (minutes). *Multiple RM*, the load that can be lifted no more than the specified times. *ROM*, the range of motion. *RPE*, the rate of perceived exertion on a scale of 0–10.
Adapted from Agostini D, et al. Muscle and bone health in postmenopausal females: role of protein and vitamin D supplementation combined with exercise training. Nutrients 2018;10(8).

Estrogen, cancer, and exercise

Menopause is characterized by waning estrogen levels owing to a decrease in ovarian production. However, the estrogen level in a female's body never reaches absolute zero postmenopause because, during menopause, the adrenal glands and adipose tissue produce estrogen through the aromatization of androgens [102]. Estrogen production in adipose tissue occurs by activating two enzymes, aromatase and 17β-hydroxysteriodase hydrogenase [103]. It is important to note here that menopausal transition also predisposes females to gain weight, which contributes to increased fat mass, resulting in an increased production of estrogen during menopause. While the estrogen produced by these tissues does not compensate for the ovarian deficiency, a significant level of the sex hormone does appear in circulation throughout menopause. It is also noteworthy that the estrogen produced by adipose tissue does not prevent or delay the onset of menopause in females. However, estrogen produced by excessive fat mass accumulation is reported to be detrimental to other tissue and increases breast cancer risk in older females [104,105]. The association between estrogen and breast cancer is also why HRT, which effectively mitigates menopausal syndromes, is not advised for the low symptomatic or prolonged use in females [32]. Hence obesity and breast cancer are two significant pathologies that have a solid relationship to menopause, and estrogen is proposed as the plausible mechanism connecting both conditions. Indeed, obesity is linked to increased breast cancer by 30%–50% [106]. Excessive fat accumulation is also associated with aggressive tumor development and resistance to cancer treatment [107]. Evidence suggests that exercise effectively reduces circulating estrogen levels in the body [108,109]. Although the cause of this decrease is not yet clear, overlapping lines of evidence support the working hypothesis that it is not the stress of exercise per se that causes the decrease, but rather the energy deficits caused by increased energy expenditure. Hence, physical activity-induced estrogen homeostasis can effectively combat the negative effect of circulating estrogen levels [110]. Several exercise interventions have reported improving insulin sensitivity and reduced fat mass and inflammatory biomarkers in females undergoing menopause [111,112]. Simultaneously, the lack of physical activity and sedentary behavior is often the risk factor for obesity and breast cancer in females. Overweight, obese, and sedentary postmenopausal females were reported to have elevated concentrations of circulating estrogens [113]. There is also evidence indicating low levels of sex hormone-binding receptors in these individuals, putting them at increased risk of developing cancers.

Significant evidence in the literature supports the ability of exercise in lowering circulating estrogen levels and improving overall health and well-being in females undergoing menopause. Some of the evidence from various randomized controlled trials are listed in Table 2. However, the literature also gives conflicting results regarding the impact of exercise on estrogen in postmenopausal females. An example is a study by Kemmler et al. that reports elevated estrogen levels after a single bout of exercise [114]. Similarly, another study by Ketabipoor et al. also reports a similar increase in estrogen level after the exercise program [115]. These discrepancies in observation may be the influence of body fat which affects the metabolism of estrogen. As mentioned earlier, aromatase enzyme activity in the adipose tissue can be an influencing factor, where the lower enzyme activity may effectively reduce estrogen levels [116]. The type of exercise (aerobic or resistance) may play a crucial role in modulating the hormone response. It is reported that endurance exercise increases testosterone dehydroepiandrosterone, estradiol, cortisol, and growth hormone levels, while resistance training only affects estrogen and growth hormone [117] (Table 3).

TABLE 3 Summary of randomized controlled trials reporting the effect of exercise on estrogen levels.

Study	N	Inclusion criteria	Type and modality	Frequency and duration	Intensity	Study outcome
Campbell et al. [118,119]	439	Postmenopausal Age: 50–75 years BMI: ≥25.0 kg/m² ≥23.0 kg/m² if Asian-American Sedentary No HRT	Aerobic ≥3 group supervised sessions 2 sessions at home	≥45 min 5 days/week 12 months duration	70%–85% HR max	Estrone and estrogen are reduced with diet and exercise and in combination of both. Sex hormone-binding globulin increased. Free estrogen and testosterone decreased
Copeland et al. [120]	32	Postmenopausal Age > 50 years Sedentary ± Hormonal treatment	Resistance Supervised	Duration: 2–3 sets of 10 repetitions of each of eight exercises 3 days/week. 12 weeks duration	10 RM	Increase in cortisol and DHEAS with exercise training. Increase in growth hormone
Figueroa et al. [121]	94	Postmenopausal Age: 40–65 years ± Hormonal treatment	Resistance and weight-bearing aerobic exercise Supervised	60–75 min 3 days/week. 12 months duration	70%–80% 1-RM 50%–80% HR max	Increase in total body, arm, and leg lean soft tissue mass. Decrease in leg fat mass and percentage body fat
Friedenreich et al. [122,123]	320	Postmenopausal Age: 50–74 years	Aerobic ≥3 supervised	45 min 5 days/week.	50%–60% HR max Gradual ↑	Significant reduction in estradiol and free estradiol with

TABLE 3 Summary of randomized controlled trials reporting the effect of exercise on estrogen levels—cont'd

Study	N	Inclusion criteria	Type and modality	Frequency and duration	Intensity	Study outcome
		BMI: 22–40 kg/m² Sedentary No HRT	≤2 at home	12 months duration	70%–80% HR max	an increase in sex hormone-binding globulin
Kim et al. [124]	30	Postmenopausal Obese: >32% body fat Sedentary No HRT	Aerobic (dance) Group Supervised	60 min 3 days/week 16 weeks duration	55%–65% HR max Gradual ↑ 5%/ 4 week. 70%–80% HR max	Reduced body weight, total cholesterol, glucose, insulin, HOMA-IR with intervention. Sex hormone-binding globulin levels and HDL also improved
Krishnan et al. [125]	36	Peri-menopausal Menstrual irregularities Age: 42–52 years BMI: 18,5–32,0 kg/m² Sedentary / light activity No HRT	Aerobic and resistance Supervised	60 min 6 days/week 6 months duration	50%–80% HR max	Circulating androgens Increased in both groups but independent of exercise effect. Improved insulin sensitivity in the exercise group
McTiernan et al. [126,127]	173	Postmenopausal Age: 50–75 years BMI ≥25.0 kg/m² Sedentary No HRT	Aerobic 3 supervised sessions 2 sessions at home	45 min 5 days/week. 12 months duration	40% Gradual ↑ 60%–75% HR max	A decrease in estrone, estradiol, testosterone, free testosterone, and free estradiol. Significant reduction in body fat
Monninkhof et al. [128,129]	189	Postmenopausal Age: 50–69 years BMI >22 kg/m² Sedentary No HRT	Aerobic and resistance group exercises supervised + at home	45 min 3 d /week. + 30 min/week. 12 months duration	60%–85% HR max	Significant reduction in body fat in the exercise group along with reduced levels of androgens
Oneda et al. [130]	60	Postmenopausal Hysterectomy Age: 45–60 years BMI: <30 kg/m² Sedentary No HRT	Aerobic exercise Monitored	20–50 min 3 days/week 6-month duration	Moderate	Exercise intervention reduced sympathetic nerve activity and improved muscle blood flow. It also decreases heart rate when combined with estrogen therapy
Orsatti et al. [131]	50	Postmenopausal Age: 45–70 years Sedentary No HRT	Resistance Supervised	50–60 min 3 days/week. 16 weeks duration	60%–80% 1-RM	The intervention resulted in improved muscle mass and strength. Insulin-like growth factor-1 levels were also improved significantly
Tartibian et al. [132]	79	Postmenopausal Sedentary No medications	Aerobic (walking or jogging) Monitored	40–45 min 4–6 days/ week 24 weeks duration	55%–65% HR max	Serum estrogen, osteocalcin, femoral neck bone mineral density increased. Pro-inflammatory cytokines level reduced

Continued

TABLE 3 Summary of randomized controlled trials reporting the effect of exercise on estrogen levels—cont'd

Study	N	Inclusion criteria	Type and modality	Frequency and duration	Intensity	Study outcome
Thomas et al. [133]	121	Postmenopausal Age: >50 Breast cancer survivors Least mild arthralgias No prior strength training On aromatase inhibitor prescription	Aerobic and resistance exercise Supervised	150 min/week (aerobic) 2 session/week (resistance) 12 months duration	50% Gradual ↑ 60%–85% HR max	Significant reduction in BMI and percentage body fat and increase in lean body mass
Wu et al. [134]	136	Postmenopausal Age: 45–60 years Sedentary No HRT	Walking Supervised	45 min 3 days/week. 12 months duration		Significant decrease in trunk fat mass. Increase in HDL. Improved bone mineral density
Yoo et al. [135]	28	Age >65 years No HRT	Weight-bearing walking exercises Supervised	45 min 3 days/week 3 months duration	60% HR max	Improved upper body strength, leg strength, aerobic endurance, and body composition. Trunk fat lowered. Serum estrogen levels decreased

1-RM, repetitions maximum (the greatest weight that can be moved once in good form); *BMI,* body mass index; *HR max,* heart rate max; *HR,* heart rate; *HRT,* hormone replacement therapy; *N,* number of randomized participants; ; *Supervised,* interaction with a coach.

Conclusion

Physical activity and exercise during the menopausal transition and postmenopause are vital in maintaining the quality of life in females due to lower estrogen production in the body. It benefits females by reducing menopause symptoms, preventing weight gain, improving muscle strength, increasing bone metabolism, improving functional capacity, improving mental health, improving cardiometabolic fitness, and modulating the immune system to prevent systemic inflammation. It also reduces the risk of developing osteoporosis, sarcopenia, and chronic diseases such as type-2 diabetes, cancer, and cardiovascular diseases. Exercise interventions in females of menopausal age have been crucial in demonstrating its health benefits. These studies have provided substantial evidence of reduced chronic disease risk and physical activity's ability to modulate hormone secretion in females. However, evidence-based prescription guidelines for exercise in older females are still lacking due to inconsistent observations for different studies. Another challenge to defining the benefits of exercise in menopause is the inability to draw direct associations, especially in vasomotor symptoms. Although numerous studies report the benefits of exercise on vasomotor symptoms, the data is inconclusive due to other interventions reporting no effect. These studies' limitations could be related to the sample size or the definitions used to report experiences of vasomotor symptoms in the population. Nevertheless, well-designed interventions with an appropriately sampled population are critical to elucidate the benefits of exercise during menopause.

References

[1] Gold EB, et al. Factors associated with age at natural menopause in a multiethnic sample of midlife females. Am J Epidemiol 2001;153(9):865–74.

[2] Schoenaker DA, et al. Socioeconomic position, lifestyle factors and age at natural menopause: a systematic review and meta-analyses of studies across six continents. Int J Epidemiol 2014;43(5):1542–62.

[3] O'Neill S, Eden J. The pathophysiology of menopausal symptoms. Obstet Gynaecol Reprod Med 2017;27(10):303–10.

[4] Razzak ZA, Khan AA, Farooqui SI. Effect of aerobic and anaerobic exercise on estrogen level, fat mass, and muscle mass among postmenopausal osteoporotic females. Int J Health Sci 2019;13(4):10–6.

[5] Hill K. The demography of menopause. Maturitas 1996;23(2):113–27.

[6] Rathnayake N, et al. Prevalence and severity of menopausal symptoms and the quality of life in middle-aged females: a study from Sri Lanka. Nurs Res Pract 2019;2019:2081507.

[7] Fuh JL, et al. Quality of life and menopausal transition for middle-aged females on Kinmen island. Qual Life Res 2003;12(1):53–61.

[8] Avis NE, et al. Change in health-related quality of life over the menopausal transition in a multiethnic cohort of middle-aged females: study of Females's Health Across the Nation. Menopause 2009;16(5).

[9] Kjerulff KH, et al. The cost of being a female: a national study of health care utilization and expenditures for female-specific conditions. Womens Health Issues 2007;17(1):13–21.

[10] Sarrel P, et al. Incremental direct and indirect costs of untreated vasomotor symptoms. Menopause 2015;22(3):260–6.

[11] Ren Y, et al. Association of menopause and type 2 diabetes mellitus. Menopause 2019;26(3).

[12] Dibaba DT, et al. Metabolic syndrome and risk of breast cancer mortality by menopause, obesity, and subtype. Breast Cancer Res Treat 2019;174 (1):209–18.

[13] Bansal R, Aggarwal N. Menopause in autoimmune disease and hormone replacement therapy. In: Females's health in autoimmune diseases. Springer; 2020. p. 255–67.

[14] Karvinen S, et al. Menopausal status and physical activity are independently associated with cardiovascular risk factors of healthy middle-aged females: cross-sectional and longitudinal evidence. Front Endocrinol 2019;10:589.

[15] Fistarol M, et al. Time since menopause, but not age, is associated with increased risk of osteoporosis. Climacteric 2019;22(5):523–6.

[16] Soules MR, et al. Executive summary: stages of reproductive aging workshop (STRAW) Park City, Utah, July, 2001. Menopause 2001;8 (6):402–7.

[17] Sturdee DW, et al. The menopausal hot flush: a review. Climacteric 2017;20(4):296–305.

[18] Bromberger JT, Kravitz HM. Mood and menopause: findings from the study of females's health across the nation (SWAN) over 10 years. Obstet Gynecol Clin N Am 2011;38(3):609–25.

[19] Dennerstein L, Soares CN. The unique challenges of managing depression in mid-life females. World Psychiatry 2008;7(3):137.

[20] Lu C-B, et al. Musculoskeletal pain during the menopausal transition: a systematic review and meta-analysis. Neural Plast 2020;2020:8842110.

[21] Agostini D, et al. Muscle and bone health in postmenopausal females: role of protein and vitamin D supplementation combined with exercise training. Nutrients 2018;10(8).

[22] Kapoor E, Collazo-Clavell ML, Faubion SS. Weight gain in females at midlife: a concise review of the pathophysiology and strategies for management. Mayo Clin Proc 2017;92(10):1552–8.

[23] Stubbins RE, et al. Estrogen modulates abdominal adiposity and protects female mice from obesity and impaired glucose tolerance. Eur J Nutr 2012;51(7):861–70.

[24] Xu J, et al. Estrogen improved metabolic syndrome through down-regulation of VEGF and HIF-1α to inhibit hypoxia of periaortic and intra-abdominal fat in ovariectomized female rats. Mol Biol Rep 2012;39(8):8177–85.

[25] Bryzgalova G, et al. Mechanisms of antidiabetogenic and body weight-lowering effects of estrogen in high-fat diet-fed mice. Am J Physiol Endocrinol Metab 2008;295(4):E904–12.

[26] Lovejoy JC, et al. Increased visceral fat and decreased energy expenditure during the menopausal transition. Int J Obes 2008;32(6):949–58.

[27] Howard BV, et al. Insulin resistance and weight gain in postmenopausal females of diverse ethnic groups. Int J Obes 2004;28(8):1039–47.

[28] Bonora E. Relationship between regional fat distribution and insulin resistance. Int J Obes 2000;24(2):S32–5.

[29] Salpeter SR, et al. Meta-analysis: effect of hormone-replacement therapy on components of the metabolic syndrome in postmenopausal females. Diabetes Obes Metab 2006;8(5):538–54.

[30] Butt DA, et al. Gabapentin for the treatment of menopausal hot flashes: a randomized controlled trial. Menopause 2008;15(2):310–8.

[31] Simon JA, et al. Low-dose paroxetine 7.5 mg for menopausal vasomotor symptoms: two randomized controlled trials. Menopause 2013;20 (10):1027–35.

[32] Manson JE, et al. Menopausal hormone therapy and health outcomes during the intervention and extended poststopping phases of the Females's Health Initiative randomized trials. JAMA 2013;310(13):1353–68.

[33] Anon. Breast cancer and hormone-replacement therapy in the Million Females Study. Lancet 2003;362(9382):419–27.

[34] de Villiers TJ, et al. Global consensus statement on menopausal hormone therapy. Maturitas 2013;74(4):391–2.

[35] Stojanovska L, et al. To exercise, or, not to exercise, during menopause and beyond. Maturitas 2014;77(4):318–23.

[36] Hojman P, et al. Molecular mechanisms linking exercise to cancer prevention and treatment. Cell Metab 2018;27(1):10–21.

[37] Lavie Carl J, et al. Sedentary behavior, exercise, and cardiovascular health. Circ Res 2019;124(5):799–815.

[38] Pinckard K, Baskin KK, Stanford KI. Effects of exercise to improve cardiovascular health. Front Cardiovasc Med 2019;6:69.

[39] Kirwan JP, Sacks J, Nieuwoudt S. The essential role of exercise in the management of type 2 diabetes. Cleve Clin J Med 2017;84(7 Suppl 1):S15.

[40] Philippou A, et al. The role of exercise in obesity. Diabetes Metab Syndr Clin Res Rev 2019;13(5):2861–2.

[41] Tong X, et al. The effect of exercise on the prevention of osteoporosis and bone angiogenesis. Biomed Res Int 2019;2019:8171897.

[42] Intzandt B, Beck EN, Silveira CRA. The effects of exercise on cognition and gait in Parkinson's disease: a scoping review. Neurosci Biobehav Rev 2018;95:136–69.

[43] Silveira CRA, et al. Aerobic exercise is more effective than goal-based exercise for the treatment of cognition in Parkinson's disease. Brain Cogn 2018;122:1–8.

[44] Ramakrishnan R, et al. Accelerometer measured physical activity and the incidence of cardiovascular disease: Evidence from the UK Biobank cohort study. PLoS Med 2021;18(1), e1003487.

[45] Baptista LC, et al. Effects of long-term multicomponent exercise on health-related quality of life in older adults with type 2 diabetes: evidence from a cohort study. Qual Life Res 2017;26(8):2117–27.

[46] Seo D-I, et al. Effects of 12 weeks of combined exercise training on visfatin and metabolic syndrome factors in obese middle-aged females. J Sports Sci Med 2011;10(1):222.

[47] Harvey SB, et al. Exercise and the prevention of depression: results of the HUNT cohort study. Am J Psychiatr 2017;175(1):28–36.

[48] Oeseburg H, et al. Telomere biology in healthy aging and disease. Pflugers Arch - Eur J Physiol 2010;459(2):259–68.

[49] Puterman E, et al. The power of exercise: buffering the effect of chronic stress on telomere length. PLoS One 2010;5(5), e10837.

[50] Mikkelsen K, et al. Exercise and mental health. Maturitas 2017;106:48–56.

[51] Takahashi M, et al. Effects of increased daily physical activity on mental health and depression biomarkers in postmenopausal females. J Phys Ther Sci 2019;31(4):408–13.

[52] Nguyen TM, et al. Exercise and quality of life in females with menopausal symptoms: a systematic review and meta-analysis of randomized controlled trials. Int J Environ Res Public Health 2020;17(19).

[53] Sanches IC, et al. Cardiometabolic benefits of exercise training in an experimental model of metabolic syndrome and menopause. Menopause 2012;19(5):562–8.

[54] Jeon YK, et al. Combined aerobic and resistance exercise training reduces circulating apolipoprotein J levels and improves insulin resistance in postmenopausal diabetic females. Diabetes Metab J 2020;44(1):103.

[55] Asikainen T-M, Kukkonen-Harjula K, Miilunpalo S. Exercise for health for early postmenopausal females. Sports Med 2004;34(11):753–78.

[56] Rodrigues IB, et al. Facilitators and barriers to exercise adherence in patients with osteopenia and osteoporosis: a systematic review. Osteoporos Int 2017;28(3):735–45.

[57] Royal College of Obstetricians and Gynaecologists (RCOB), Scientific Advisory Committee. Alternatives to HRT for the management of symptoms of the menopause. Scientific impact paper No. 6. [cited 2021 6 December]; Available from: https://www.rcog.org.uk/globalassets/documents/guidelines/scientific-impact-papers/sip_6.pdf; 2010.

[58] North American Menopause Society. Treatment of menopause-associated vasomotor symptoms: position statement of The North American Menopause Society. Menopause 2004;11(1):11–33.

[59] Genazzani AR, et al. The European Menopause Survey 2005: females's perceptions on the menopause and postmenopausal hormone therapy. Gynecol Endocrinol 2006;22(7):369–75.

[60] Daley AJ, Stokes-Lampard HJ, MacArthur C. Exercise to reduce vasomotor and other menopausal symptoms: a review. Maturitas 2009;63(3):176–80.

[61] Bortz 2nd WM, et al. Catecholamines, dopamine, and endorphin levels during extreme exercise. N Engl J Med 1981;305(8):466–7.

[62] Heitkamp HC, Huber W, Scheib K. β-Endorphin and adrenocorticotrophin after incremental exercise and marathon running-female responses. Eur J Appl Physiol Occup Physiol 1996;72(5):417–24.

[63] Progetto Menopausa Italia Study Group. Factors associated with climacteric symptoms in females around menopause attending menopause clinics in Italy. Maturitas 2005;52(3–4):181–9.

[64] Guthrie JR, et al. Physical activity and the menopause experience: a cross-sectional study. Maturitas 1994;20(2):71–80.

[65] Gold EB, et al. Relation of demographic and lifestyle factors to symptoms in a multi-racial/ethnic population of females 40–55 years of age. Am J Epidemiol 2000;152(5):463–73.

[66] Ivarsson T, Spetz A-C, Hammar M. Physical exercise and vasomotor symptoms in postmenopausal females. Maturitas 1998;29(2):139–46.

[67] Mirzaiinjmabadi K, Anderson D, Barnes M. The relationship between exercise, body mass index and menopausal symptoms in midlife Australian females. Int J Nurs Pract 2006;12(1):28–34.

[68] van Poppel MNM, Brown WJ. " It's my hormones, doctor"-does physical activity help with menopausal symptoms? Menopause 2008;15(1):78–85.

[69] Daley A, et al. Exercise participation, body mass index, and health-related quality of life in females of menopausal age. Br J Gen Pract 2007;57(535):130–5.

[70] Daley A, et al. Exercise for vasomotor menopausal symptoms. Cochrane Database Syst Rev 2014;11.

[71] Sternfeld B, et al. Efficacy of exercise for menopausal symptoms: a randomized controlled trial. Menopause 2014;21(4):330.

[72] Kai Y, et al. Effects of stretching on menopausal and depressive symptoms in middle-aged females: a randomized controlled trial. Menopause 2016;23(8).

[73] Mansikkamäki K, et al. Long-term effect of physical activity on health-related quality of life among menopausal females: a 4-year follow-up study to a randomised controlled trial. BMJ Open 2015;5(9), e008232.

[74] Hu L, et al. Benefits of walking on menopausal symptoms and mental health outcomes among Chinese postmenopausal females. Int J Gerontol 2017;11(3):166–70.

[75] Coelho FG, et al. Physical exercise modulates peripheral levels of brain-derived neurotrophic factor (BDNF): a systematic review of experimental studies in the elderly. Arch Gerontol Geriatr 2013;56(1):10–5.

[76] Yang P, et al. The exercise intervention may influence the dietary intake and reduce the risk of osteoporosis and sarcopenia in menopausal females. FASEB J 2017;31:967–9.

[77] Benedetti MG, et al. The effectiveness of physical exercise on bone density in osteoporotic patients. Biomed Res Int 2018;2018.

[78] Daly RM, et al. Exercise for the prevention of osteoporosis in postmenopausal females: an evidence-based guide to the optimal prescription. Braz J Phys Ther 2019;23(2):170–80.

[79] Bergstrom I, et al. Physical training preserves bone mineral density in postmenopausal females with forearm fractures and low bone mineral density. Osteoporos Int 2008;19(2):177–83.

[80] Brentano MA, et al. Physiological adaptations to strength and circuit training in postmenopausal females with bone loss. J Strength Cond Res 2008;22(6):1816–25.

[81] Englund U, et al. A 1-year combined weight-bearing training program is beneficial for bone mineral density and neuromuscular function in older females. Osteoporos Int 2005;16(9):1117–23.

[82] Going S, et al. Effects of exercise on bone mineral density in calcium-replete postmenopausal females with and without hormone replacement therapy. Osteoporos Int 2003;14(8):637–43.

[83] Kemmler W, von Stengel S. Dose–response effect of exercise frequency on bone mineral density in postmenopausal, osteopenic females. Scand J Med Sci Sports 2014;24(3):526–34.

[84] Korpelainen R, et al. Effect of impact exercise on bone mineral density in elderly females with low BMD: a population-based randomized controlled 30-month intervention. Osteoporos Int 2006;17(1):109–18.

[85] Lau EMC, et al. The effects of calcium supplementation and exercise on bone density in elderly Chinese females. Osteoporos Int 1992;2(4):168–73.

[86] Marques EA, et al. Effects of resistance and aerobic exercise on physical function, bone mineral density, OPG and RANKL in older females. Exp Gerontol 2011;46(7):524–32.

[87] Park H, et al. Effect of combined exercise training on bone, body balance, and gait ability: a randomized controlled study in community-dwelling elderly females. J Bone Miner Metab 2008;26(3):254–9.

[88] Rhodes EC, et al. Effects of one year of resistance training on the relation between muscular strength and bone density in elderly females. Br J Sports Med 2000;34(1):18.

[89] Verschueren SMP, et al. Effect of 6-month whole body vibration training on hip density, muscle strength, and postural control in postmenopausal females: a randomized controlled pilot study. J Bone Miner Res 2004;19(3):352–9.

[90] Wen HJ, et al. Effects of short-term step aerobics exercise on bone metabolism and functional fitness in postmenopausal females with low bone mass. Osteoporos Int 2017;28(2):539–47.

[91] Hoke M, et al. Impact of exercise on bone mineral density, fall prevention, and vertebral fragility fractures in postmenopausal osteoporotic females. J Clin Neurosci 2020;76:261–3.

[92] Troy KL, et al. Exercise early and often: effects of physical activity and exercise on females's bone health. Int J Environ Res Public Health 2018;15(5).

[93] Zhao R, Zhao M, Xu Z. The effects of differing resistance training modes on the preservation of bone mineral density in postmenopausal females: a meta-analysis. Osteoporos Int 2015;26(5):1605–18.

[94] Bonaiuti D, et al. Exercise for preventing and treating osteoporosis in postmenopausal females. Cochrane Database Syst Rev 2002;3, Cd000333.

[95] Cunha PM, et al. Improvement of cellular health indicators and muscle quality in older females with different resistance training volumes. J Sports Sci 2018;36(24):2843–8.

[96] Cunha PM, et al. The effects of resistance training volume on osteosarcopenic obesity in older females. J Sports Sci 2018;36(14):1564–71.

[97] Giangregorio LM, et al. Too fit to fracture: outcomes of a Delphi consensus process on physical activity and exercise recommendations for adults with osteoporosis with or without vertebral fractures. Osteoporos Int 2015;26(3):891–910.

[98] Chodzko-Zajko WJ, et al. Exercise and physical activity for older adults. Med Sci Sports Exerc 2009;41(7):1510–30.

[99] Chodzko-Zajko WJ, et al. American College of Sports Medicine position stand. Exercise and physical activity for older adults. Med Sci Sports Exerc 2009;41(7):1510–30.

[100] American College of Sports Medicine. American College of Sports Medicine position stand. Progression models in resistance training for healthy adults. Med Sci Sports Exerc 2009;41(3):687–708.

[101] Clark BC, Clark LA, Law TD. Resistance exercise to prevent and manage sarcopenia and dynapenia. Annu Rev Gerontol Geriatr 2016;36(1):205–28.

[102] Cleary MP, Grossmann ME. Obesity and breast cancer: the estrogen connection. Endocrinology 2009;150(6):2537–42.

[103] Corbould AM, Judd SJ, Rodgers RJ. Expression of types 1, 2, and 3 17β-hydroxysteroid dehydrogenase in subcutaneous abdominal and intra-abdominal adipose tissue of females. J Clin Endocrinol Metab 1998;83(1):187–94.

[104] Eliassen AH, et al. Endogenous steroid hormone concentrations and risk of breast cancer: does the association vary by a female's predicted breast cancer risk? J Clin Oncol 2006;24(12):1823–30.

[105] Hankinson SE, Eliassen AH. Endogenous estrogen, testosterone and progesterone levels in relation to breast cancer risk. J Steroid Biochem Mol Biol 2007;106(1–5):24–30.

[106] Petrelli JM, et al. Body mass index, height, and postmenopausal breast cancer mortality in a prospective cohort of US females. Cancer Causes Control 2002;13(4):325–32.

[107] Fiorio E, et al. Leptin/HER2 crosstalk in breast cancer: in vitro study and preliminary in vivo analysis. BMC Cancer 2008;8(1):1–11.

[108] McTiernan A, et al. Effect of exercise on serum estrogens in postmenopausal females: a 12-month randomized clinical trial. Cancer Res 2004;64(8):2923–8.

[109] Ennour-Idrissi K, Maunsell E, Diorio C. Effect of physical activity on sex hormones in females: a systematic review and meta-analysis of randomized controlled trials. Breast Cancer Res 2015;17(1):139.

[110] Westerlind KC, Williams NI. Effect of energy deficiency on estrogen metabolism in premenopausal females. Med Sci Sports Exerc 2007;39(7):1090–7.

[111] Mandrup CM, et al. Effects of menopause and high-intensity training on insulin sensitivity and muscle metabolism. Menopause 2018;25(2):165–75.

[112] Tartibian B, et al. A randomized controlled study examining the effect of exercise on inflammatory cytokine levels in postmenopausal females. Post Reprod Health 2015;21(1):9–15.

[113] Verkasalo PK, et al. Circulating levels of sex hormones and their relation to risk factors for breast cancer: a cross-sectional study in 1092 pre-and postmenopausal females (United Kingdom). Cancer Causes Control 2001;12(1):47–59.

[114] Kemmler W, et al. Acute hormonal responses of a high impact physical exercise session in early postmenopausal females. Eur J Appl Physiol 2003;90(1):199–209.

[115] Ketabipoor SM, Koushkie Jahromi M. Effect of aerobic exercise in water on serum estrogen and c-reactive protein and body mass index level in obese and normal weight postmenopausal females. Women's Health Bull 2015;2(3):1–6.

[116] Atkinson C, et al. Effects of a moderate intensity exercise intervention on estrogen metabolism in postmenopausal females. Cancer Epidemiol Biomark Prev 2004;13(5):868–74.

[117] Copeland JL, Consitt LA, Tremblay MS. Hormonal responses to endurance and resistance exercise in females aged 19-69 years. J Gerontol A Biol Sci Med Sci 2002;57(4):B158–65.

[118] Campbell KL, et al. Reduced-calorie dietary weight loss, exercise, and sex hormones in postmenopausal females: randomized controlled trial. J Clin Oncol 2012;30(19):2314–26.

[119] Foster-Schubert KE, et al. Effect of diet and exercise, alone or combined, on weight and body composition in overweight-to-obese postmenopausal females. Obesity (Silver Spring) 2012;20(8):1628–38.

[120] Copeland JL, Tremblay MS. Effect of HRT on hormone responses to resistance exercise in postmenopausal females. Maturitas 2004;48(4):360–71.

[121] Figueroa A, et al. Effects of exercise training and hormone replacement therapy on lean and fat mass in postmenopausal females. J Gerontol A Biol Sci Med Sci 2003;58(3):266–70.

[122] Friedenreich CM, et al. Adiposity changes after a 1-year aerobic exercise intervention among postmenopausal females: a randomized controlled trial. Int J Obes 2011;35(3):427–35.

[123] Friedenreich CM, et al. Alberta physical activity and breast cancer prevention trial: sex hormone changes in a year-long exercise intervention among postmenopausal females. J Clin Oncol 2010;28(9):1458–66.

[124] Kim JW, Kim DY. Effects of aerobic exercise training on serum sex hormone binding globulin, body fat index, and metabolic syndrome factors in obese postmenopausal females. Metab Syndr Relat Disord 2012;10(6):452–7.

[125] Krishnan S, et al. Association between circulating endogenous androgens and insulin sensitivity changes with exercise training in midlife females. Menopause 2014;21(9):967–74.

[126] McTiernan A, et al. Effect of exercise on serum estrogens in postmenopausal females: a 12-month randomized clinical trial. Cancer Res 2004;64(8):2923–8.

[127] McTiernan A, et al. Effect of exercise on serum androgens in postmenopausal females: a 12-month randomized clinical trial. Cancer Epidemiol Biomark Prev 2004;13(7):1099–105.

[128] Monninkhof EM, et al. Effect of exercise on postmenopausal sex hormone levels and role of body fat: a randomized controlled trial. J Clin Oncol 2009;27(27):4492–9.

[129] Velthuis MJ, et al. Exercise program affects body composition but not weight in postmenopausal females. Menopause 2009;16(4):777–84.

[130] Oneda B, et al. Effects of estrogen therapy and aerobic training on sympathetic activity and hemodynamics in healthy postmenopausal females: a double-blind randomized trial. Menopause 2014;21(4):369–75.

[131] Orsatti FL, et al. Plasma hormones, muscle mass and strength in resistance-trained postmenopausal females. Maturitas 2008;59(4):394–404.

[132] Tartibian B, et al. Long-term aerobic exercise and omega-3 supplementation modulate osteoporosis through inflammatory mechanisms in postmenopausal females: a randomized, repeated measures study. Nutr Metab (Lond) 2011;8:71.

[133] Thomas GA, et al. The effect of exercise on body composition and bone mineral density in breast cancer survivors taking aromatase inhibitors. Obesity 2017;25(2):346–51.

[134] Wu J, et al. Effects of isoflavone and exercise on BMD and fat mass in postmenopausal Japanese females: a 1-year randomized placebo-controlled trial. J Bone Miner Res 2006;21(5):780–9.

[135] Yoo EJ, Jun TW, Hawkins SA. The effects of a walking exercise program on fall-related fitness, bone metabolism, and fall-related psychological factors in elderly females. Res Sports Med 2010;18(4):236–50.

Chapter 14

Exercise for chronic pain

Della Buttigieg[a], Nick Efthimiou[b], and Alison Sim[c]

[a]*Melbourne, VIC, Australia* [b]*Sydney, NSW, Australia* [c]*School of Medicine, Faculty of Medicine and Health, The University of Sydney, Sydney, NSW, Australia*

The neurobiology of pain

Introduction

Pain is integral to the human experience. It is ubiquitous and omnipresent, but it has been perhaps one of the most misunderstood and misrepresented elements of the human condition. Our pervasive misunderstanding of pain was built on the Cartesian theory of pain that emerged with Rene Descartes' 1664 "Treatise of Man" (Fig. 1) [1]. This inextricably linked pain to tissue damage in our collective consciousness and it has taken us more than three centuries to begin to emerge from its' shadows and to understand our "Catastrophic misinterpretation" [2–5].

The equation of hurt and harm, or pain and damage, coupled with continuous communal reinforcement has led us to an almost society-wide "fear-avoidance" behavioral pattern. A pattern that is now postulated to underlie the chronification and persistence of pain, by driving a "vicious cycle of avoidance," disuse, deconditioning, and decreased tolerance for exercise and physical activity that eventually induces activity-dependent hyperalgesia (Fig. 2) [6–8]. No surprise then that we find ourselves seemingly on the runaway train of a chronic pain pandemic.

It's estimated that around one in five of us will suffer from persistent pain, a rate that is mirrored fairly uniformly across the globe [8–12], and pain, affecting more of us than diabetes, heart disease, and cancer combined, is the single most common reason for medical consultations. With an estimated 100 million Americans and as many as 1 billion sufferers worldwide, it is a problem that pervades all of society. It seemingly knows no bounds, but it does not affect us all equally. For example, Table 1 presents a snapshot of the incidence of chronic pain in Australia.

While this snapshot is of chronic pain in Australia it is important to note that the picture painted by this data is fairly consistent the world over, though a trend toward increased prevalence in developing countries is clearly apparent. The bottom line is that chronic pain can affect anyone, but an assortment of modifiable and non-modifiable factors contributes to increased risk [8]. Pain persistence is very strongly associated with advancing age and lower socio-economic status where poorer education, poorer health literacy, financial pressures, and a lack of early access to appropriate care, all play a role. These factors intersect sharply with sex and gender, and more sharply still with culturally and linguistically diverse populations, where care choice is often further limited by cultural and language barriers (Table 2) [8,10,11].

The impact of chronic pain

The impact of chronic pain is profound, on the individual, their social network, and the broader community. And if we are honest as practitioners, many of us would probably say that it can have a profound effect on us too. In interrogating these impacts, it is imperative that we understand and acknowledge that there are two key perspectives to be considered, the individual and the cohort. The individual stories are of the human cost to lives irrevocably changed, curtailed by distress and disability. The broader cohort and economic data are what so frequently informs policy, care allocation, and funding. These two perspectives, with much overlap, and seemingly working with the same outcomes in mind, are often juxtaposed and conflicted; frequently unable to find a middle ground. The impacts on the individual are broad-ranging and well documented.

Mood disturbance is very common in persistent pain sufferers [5,8–10,12–14]. Almost 45% experience depression and/or anxiety, which is about four times the rates seen in those without chronic pain [8,9,14]. Other psychopathology is more prevalent also. Persistent pain patients are more likely to suffer post-traumatic stress disorder (PTSD), develop substance abuse issues, experience suicidal ideation, and indeed attempt suicide [5,8,9,13,14]. In Australia alone, we lose

FIG. 1   The Cartesian theory of pain conceived of a single "wire" carrying signals from the periphery to the brain. The fundamental premise here was that if there were no fire, there could be no pain [1].

around 1000 people to prescription opioid misuse each year [9]. The WHO puts the global number at around 150,000 each year.

Sleep disorders are almost synonymous with chronic pain, affecting between 5 and 9 in every 10 patients with persisting pain [14–19]. The degree of sleep disturbance appears to correlate strongly with pain intensity and the relationship is clearly bidirectional, meaning the two problems are prone to perpetuating one another [14,15]. Managing pain early is key to minimizing sleep disturbance and managing sleep disturbance is pivotal in managing pain [5,8,14,15,17].

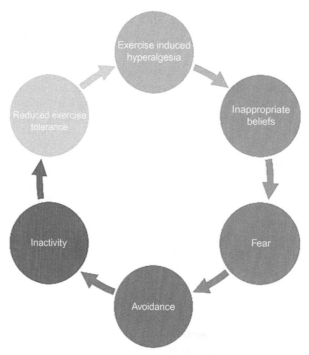

FIG. 2   Our catastrophic misunderstanding of pain and its flow-on effects. *Adapted from Meeus et al. [6].*

TABLE 1 A snapshot of chronic pain in Australia.

In 2018 3.24 million Australians lived with persistent pain

Of these 54% were women; 46% were men

Almost 70% were working age, but only ~25% maintained employment

Prevalence increased with age (1/3 of Australians aged 65+ lived with chronic pain)

Young adults were less commonly but typically more profoundly affected

25% of kids experienced chronic pain (this was moderate to severe and debilitating for ~5%)

Chronic pain is more prevalent in rural communities

Back pain is 23%–30% more likely in people living outside of major cities

Multi-disciplinary pain care is less prevalent in rural, regional, and remote areas

Adapted from Economics [9].

TABLE 2 Risk factors for chronic pain [10].

Non-modifiable	Advancing age Female sex Cultural background Socio-economic background History of trauma, injury, or interpersonal violence Heritable factors (including genetic)
Modifiable	Other pain (acute or chronic) Mental health Obesity and other co-morbidities Smoking and alcohol consumption Physical activity/exercise Sleep Nutrition Employment status and occupational factors

Cognitive decline and executive function deficits are also prevalent in chronic pain patients [8,14–16,19,20], with memory and attention deficits being estimated to affect as much as 65% of the chronic pain population [14,19]. This can be a major obstacle in conducting daily activities and is most certainly an obstacle to effective recovery.

Implications for cardiovascular health are well established. Chronic pain has been repeatedly determined as a stand-alone predictor of hypertension and a risk factor for other cardiovascular diseases and events, independent of age, sex, race, and familial history [14,21–23]. Changes to cardiovascular health and risk are thought to center around baroreflex changes gearing them toward antinociception or pain suppression. A normal adaptive change in acute pain, that when prolonged leads to baroreceptor hyposensitivity and increased risk for hypertension (HT), and increased morbidity in those with existing coronary health disease (CAD) [14,22]. Interestingly though, a patients' perceived disability has been shown to have an even greater impact on cardiovascular risk than the severity of the pain itself [14]. Perceived disability and physical activity avoidance go hand in hand, so the compounding of cardiovascular (CV) risk here is probably more obvious than may first be apparent.

Overall quality of life is significantly impacted for chronic pain patients [8,9,14,16,24–26]. Relationship decline, sexual dysfunction, reduced social interaction, increased distress, and perceived disability all play a role [14,16,20,24]. The severity of these effects largely appears to increase with pain intensity [14,20]. Where pain and disability are such that they interfere with employment continuity, further distress and financial stress typically ensue, with negative effects on relationships and social participation further compounded [14,24]. Chronic pain is the single most common reason

Australians cite for dropping out of the workforce and with almost 70% of our persistent pain population being working age and only around 25% being able to maintain employment, this is a sizable problem for not only the individual but the wider community also [9].

Physical activity is significantly reduced in those with chronic pain [5–9,11,16,20,24,25], and much like sleep disturbance, this is an element of the condition that can perpetuate pain [5–8] and indeed, requires its own management strategies and interventions [6,11,14]. Concomitant pain and physical inactivity work to not only perpetuate one another but compound the other deleterious health and quality of life impacts of chronic pain also [11,14].

The economic impact

In 2018 alone, the total financial cost of chronic pain in Australia was estimated at $73.2 billion. This included $12.2 billion in health system costs, $48.3 billion in lost productivity, and $12.7 billion in other financial costs and losses, such as informal care, and aids and modifications [9]. In America, lost productivity from pain conditions is estimated to cost $61 billion annually, in Sweden it is just shy of 80 billion SEK [27]. Quality-of-life (QOL) losses however are not just personal and come at a cost to the broader community also [9,24]. These losses can be translated into economic terms using the World Health Organization (WHO) Burden of disease methodology to account for the cost of years of life lost due to premature death (YLLs) and years of healthy life lost due to disability (YLDs). This system puts the Australian 2018 cost of quality-of-life losses at $66.1 billion. The combined financial costs and QOL costs take the Australian 2018 total cost of chronic pain to $139.3 billion. If the status quo is maintained in terms of treatments, prevalence rates, and real costs per person, this figure is expected to jump from $139.3 to $215.6 billion by 2050 [9]. While this data is again largely Australian, the trends are global. Something needs to change.

Defining pain

We have struggled to define and describe the pain and, in recent times have come to acknowledge a dearth of language to adequately capture and characterize the complexity of the pain experience. This is likely a reflection of the multifaceted nature of pain and its wide-ranging effects that commonly impact our sensory, emotional, and cognitive experiences.

From the International Association for the Study of Pain (IASP):

Pain is "An unpleasant sensory and emotional experience associated with, or resembling that associated with actual or potential tissue damage," and is expanded upon by the addition of six key notes and the etymology of the word pain for further valuable context.

- Pain is always a personal experience that is influenced to varying degrees by biological, psychological, and social factors.
- Pain and nociception are different phenomena. Pain cannot be inferred solely from activity in sensory neurons.
- Through their life experiences, individuals learn the concept of pain.
- A person's report of an experience as pain should be respected.
- Although pain usually serves an adaptive role, it may have adverse effects on function and social and psychological well-being.
- Verbal description is only one of several behaviors to express pain; inability to communicate does not negate the possibility that a human or a nonhuman animal experiences pain [28].

Chronic or persistent pain is simply pain that "persists beyond normal tissue healing time, which is assumed to be three months" [29]. This definition, scant as it is, has stood for decades unaltered. Recently however the IASP and the WHO have expanded the definition and classification of chronic pain to distinguish between primary and secondary chronic pain syndromes [28]. Critically, this update acknowledges that although chronic pain can be a symptom of another disease, it frequently occurs without clear cause and can be considered a neurological disease characterized by changes primarily to the central nervous system [14,28]. This sort of primary chronic pain syndrome accounts for 17% of chronic pain in Australia [9] and making this acknowledgment is a critical steppingstone in the development of targeted research and eventually, better treatment protocols.

Pathophysiology

Regardless of the initiating illness or injury, or indeed whether there was one at all—it is to a seemingly significant extent that the inherent plasticity and adaptive capabilities of the nervous system, encode and drive the sensitization that underpins persisting pain [5,6,8,30–32]. Understanding this unifying pathophysiology of pain persistence adds leverage to managing those with secondary pain syndromes, and crucially, adds therapeutic avenues for those with primary pain syndromes who are otherwise often left on a fruitless and often maddening search for something to treat. The pathophysiology that unifies almost all persisting pain is extremely complex, indeed entire textbooks have been devoted solely to the transition from acute to chronic pain. While pain physiology is typically considered the domain of neuroscience, the pathophysiology of chronic pain cuts across a number of complex systems and is better thought of as an integrated function of the neuroimmune-endocrine axis. There are however pragmatic ways in which we can simplify this complexity for clinical translation. The application of a top-down, bottom-up information processing approach can be readily applied with good clinical utility [33].

Bottom-up pain processing

Nociception is the fundamental bottom-up pain process, but nociception is not pain. While nociception is usually described as leading to pain perception, it is insufficient alone to explain pain [8,34] and, pain without nociception and nociception without pain are both well-substantiated phenomena [5,8,28,31,32]. Nociception is in essence a warning shot, signaling the detection of a potential threat. It is comprised of four key processes:

- Transduction
- Transmission
- Perception
- Modulation (takes place at all levels of the nervous system and influences the first three processes both directly and indirectly)

These processes are inherently adaptive and plastic to allow for amplification and attenuation of nociceptive signaling and pain perception, to balance the allocation of neurophysiological resources between the task at hand and the need for survival [8,30,34]. While we tend to talk about these modulation processes in the context of pain, illness, injury, and indeed recovery—it is important to note that sensitivity to all stimuli is continually modulated moment-to-moment by a combination of intra and exteroceptive inputs.

Key concepts in bottom-up pain processing

Bottom-up processing of nociceptive inputs is largely a function of the peripheral nervous system and spinal cord. Though both amplification and attenuation mechanisms exist at all levels of the nervous system, bottom-up processing is heavily geared toward amplification to ensure nociceptive salience and appropriate behavioral responses to threats [32,34]. To this end, there are a number of mechanisms built in that allow us to respond more readily and more intensely to potentially threatening stimuli [30–32,34,35].

Peripheral sensitization mechanisms

The first component in our protective pathways are the primary afferent nociceptors, these are our 'ground troops' in threat surveillance and detection. These nociceptors are a molecularly diverse collection of free nerve endings that respond to a broad range of stimuli, with varying degrees of sensitivity [8,13,32,34]. Each nociceptor will respond preferentially to a particular stimulus type (allowing for specificity) but the diversity of receptors within each nerve ending will also allow for generalized recruitment and contribution toward a stronger response to a non-preferred stimulus when potentially advantageous [34]. This essentially amplifies our ability to respond to intense, potentially threatening stimuli. The diversity of these nociceptive neurons allows for several modes of classification, the most useful (clinically speaking) include:

- Fiber diameter (giving fast/slow or first/second pain phenomenon).
- Transduction threshold (giving varying degrees of mechanical sensitivity and allowing for heightened mechanical sensitivity in the presence of inflammatory mediators).
- Stimulus they respond preferentially to.
- Peptidergic capacity.

These nociceptors begin to detect, transduce, and transmit threat signals in response to tissue stressors, well before tissue tolerance thresholds are reached. Meaning they build into our threat response system, a protective buffer that allows us to be predictive and protective [5,8,13,30,32,34]. Because of this, neither nociceptive signaling nor indeed pain perception are predicated on tissue damage. A clear and simple example of this can be found in our responses to potentially threatening thermal stimuli (Fig. 3).

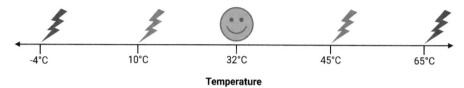

FIG. 3 The protective buffer. A simple diagram for explaining the protective buffer to patients, demonstrating clearly how nociceptive signaling and pain perception are not reliant on tissue damage.

Cutaneous heat pain thresholds are consistently lower in women than men and can vary with ambient temperature and rate of temperature change [35,36], etc., but on average it is generally accepted that humans will begin to perceive pain at around 44 or 45 °C [35]. Physiological signs of thermal injury at these temperatures, however, would take 3–6h to develop. Instantaneous or near-instantaneous cutaneous tissue injury does not develop until temperatures are >60 °C. This clearly demonstrates a 15 °C or ~30% protective buffer. A similar buffer is found at the other end of the spectrum with cold pain and injury thresholds. Cold pain is more variable than heat pain but is frequently perceived from around 6–10 °C, while the injury is not typically noted until <0 °C [35].

As nociceptor function is continually adjusted, these protective buffers are subject to continued modulation and as noted, we have numerous mechanisms for enhancing sensitivity and enhancing this buffer, particularly in the presence of inflammatory mediators. While designed to ensure protection when under a perceived threat, in the case of spontaneous chronic pain, sensitivity can blow out to such an extent that these warning signals are being fired under completely normal, innocuous stimulus conditions [8,13,30,37,38] (Fig. 5).

The peptidergic capacity of a neuron refers to its ability to generate and release its own inflammatory peptides via an axonal reflex, in response to activation. This is a key element of the initiation and maintenance of peripheral sensitization (Fig. 4) [8,13,30,37].

About 40% of our dorsal root ganglion cells are peptidergic in nature, producing neuropeptides such as substance P and Calcitonin gene-related peptide (CGRP), to initiate neurogenic inflammation [8,34]. This inflammatory response comprises vasodilation, plasma extravasation, smooth muscle contraction, and mast cell degranulation. This is typically a focal response, localized to areas of actual or potential tissue injury, that can then be augmented by central processes, in response to tissue stimulation or top-down facilitation [13,34,38].

In essence, these free nerve endings both act in response to and produce inflammation as a means of priming and enhancing the immune response to tissue injury. This drives a positive feedback loop that escalates both sensitivity and tissue repair processes [8,34,38]. G protein-coupled chemoreceptors respond to inflammatory chemicals such as histamine

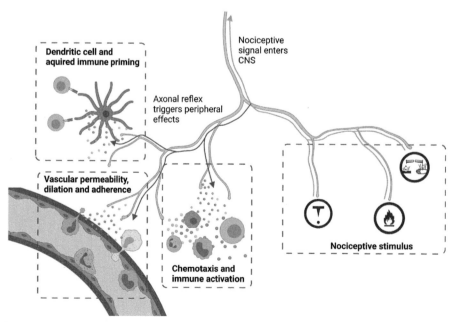

FIG. 4 The axon reflex.

prostaglandins, serotonin, bradykinins, interleukins, nerve growth factor as well as intra cellular substances such as ATP, glutamate, and adenosine [34,38]. A second messenger cascade has two-fold effects: sensitizing associated membrane channels and up-regulating receptor and neurotransmitter production signals at the neuron's nucleus [13,34,38]. The net effect of these changes can be summarized as follows:

- Existing membrane receptors become more sensitive, they open in response to lesser stimuli and stay open for longer.
- The neurons will exhibit higher frequency firing as a result.
- The neuron may also develop more receptors to the same stimulus to further boost neuronal firing and facilitate the release of enhanced neurotransmitter stores.

Clinically these changes manifest as hyperalgesia and allodynia [34,39,40]. They are a normal part of the pathophysiology of pain and tissue injury and should be thought of as a short-term enhancement of immune and recuperative behaviors. They should reverse and revert to the patient's normal baseline as the tissues heal or even earlier if the threat is assessed as being insignificant. Many of these changes can be described as semi-permanent in nature and it is important to acknowledge that when inflammatory processes persist, sensitization is also likely to persist, increasing the risk of central sensitization disorders [33,40–42]. It is also important to acknowledge however that many of these changes are inherently reversible and when pain conditions are managed well at the onset of episodes, progression to centrally sensitized states would appear to be much less likely [33,42].

Spinal sensitization mechanisms

The axonal reflex that dumps inflammatory cytokines into peripheral tissues has an orthodromic counterpart. This means that the cytokine dumping that initiates and maintains peripheral sensitization is mirrored by cytokine dumping at the dorsal horn of the spinal cord and unsurprisingly, this initiates and contributes to the maintenance of a state of spinal sensitization. Spinal sensitization has a number of unique features that are reflective of the neuroanatomical organization of the dorsal horn and cord, where primary to secondary synapses reside. The general theme of cord organization is convergence; convergence of many primary afferents onto a single secondary, the convergence of noxious and innocuous information processing, the convergence of interoception and exteroception, and importantly, parallel organization of autonomic and somatic functions (sympathetic). Finally, there is the convergence of multiple dorsal roots into the cord as stacked segments to consider [8,30,33,37,40,41].

The first and most clinically apparent difference resulting from all this convergence is that it allows for sensitivity to develop across far larger body areas. This tends to present with regional pain syndromes as opposed to the local pain seen with peripheral sensitization. It also means that spinal sensitization can give rise to pain referral within and across body systems and can also give rise to sympathetically maintained pain [8,30,33,37,40,41]. This drawing in of the autonomic nervous system and its ties to the HPA axis is pivotal to our understanding of chronic pain. All in all, spinal sensitization is a more complicated pathophysiological process than peripheral sensitization, but it is also one that reinforces peripheral processes, making the amplification cycle harder to break. This comes in the form of antidromic prostaglandin dumping from second-order relays and creates a positive feedback loop across the primary to the secondary synapse, meaning that activation of the second-order nociceptive neuron reinforces sensitization of the primary [30,32,33]. Again, this is designed to ensure nociceptive salience and will be subject to top-down dampening when deemed appropriate [13,43].

Without diving too deeply into the detailed processes here, there are some key parallels to peripheral processes that are worth covering off. Much like in peripheral sensitization processes, we see some short-term changes to neuronal behavior (activity-dependent wind-up that is inherently reversible) [13,31–33,37,43] but sustained Ca^{2+} influxes, increased intracellular Ca^{2+} mobilization and subsequent alterations in gene transcription give rise to the production of modified NMDA and AMPA receptors that effectively have longer channel opening, at lower levels of stimulus and faster associated plugging of the NMDA Mg2+ receptor blockade. When combined, these changes drive a more prolonged postsynaptic hyperexcitability in secondary afferents [13,33,37,41,43].

Activation of cord sympathetics has also been implicated in the activation and sensitization of primary afferents, and the maintenance of peripheral inflammatory processes in chronic pain patients. This is just one interaction between the sympathetic nervous system and pain, there are many more at all levels of the neuraxis but this one serves to set the scene for a key deviation from normal, acute pain physiology that we see in chronic pain. In healthy individuals, this sort of activation would commonly result in antinociception via top-down inhibitory pathways (discussed in more detail later) but in chronic pain patients where these modulatory processes get somewhat distorted—we commonly see descending inhibition replaced by descending facilitation and long-lasting nociceptive sensitization to stress hormones develop. This change is thought to be key to pain taking on a life of its own that no longer seems related or proportionate to physical stressors. Afferent hyperexcitability is also thought to be further augmented and sustained by cord glial cell activation in response to peripheral inflammatory insult and/or injury. Astrocytes and microglia are particularly involved here, orchestrating changes to neurotransmitter reuptake and extracellular ion homeostasis [13,37,44,45].

These processes are thought to be a key underpinning of the fundamentally abnormal response to sensory stimuli we see in chronic pain patients [13,40,43]. While these changes are all largely reversible it is important to acknowledge that again they are in effect, semi-permanent in nature and their self-reinforcing nature inherently reduces the plasticity of the system, particularly when sustained over longer periods. One key exception to the "sustained over long periods" notion is neuropathic pain, where dramatic structural plastic change can be noted in both neurons and glial cells, after even very brief periods. This is covered in another chapter but is an exception worthy of noting here.

Supraspinal sensitization mechanisms

These second-order nociceptive afferents are much more diverse than their primary counterparts. Somatic nociceptive relays largely arise in the dorsal horn of the cord and traverse the anterolateral system, while secondary visceral afferents largely arise in the brainstem after receiving inputs via the dorsal column medial lemniscus system [34,35,40]. Some travel ipsilaterally and some contralaterally, some encode contralateral information, and some bilateral. They terminate on tertiary neurons located in numerous parts of the brainstem and diencephalon. This diversity speaks to the diversity of function but also the imperative of survival and preservation of nociceptive salience.

In broad terms these pathways relay nociceptive information for the following purposes:

- to the thalamus and cortex for processing and perception of pain
- to the brainstem and hypothalamus for coordinating autonomic responses to pain
- to the medulla and midbrain for endogenous pain modulation
- to the limbic system for motivational-affective pain perception

Less is known about nociceptive modulation processes at the secondary to tertiary nociceptive synapses, or indeed at the third-order neurons of these pathways. But what we do know so far raises suspicions that many of the changes seen at the primary and secondary afferents to ensure nociceptive salience, may in fact be mirrored at tertiary nociceptive neurons [40]. This appears to manifest as both direct facilitation and/or disinhibition of the secondary relays [46], which are of course "top-down" processes and will be discussed further in the coming paragraphs.

Animal studies have demonstrated thalamic hyperexcitability in the presence of peripheral inflammation and injury [47] and, due to further convergence in the system, these neurons have even larger (some even have whole-body) receptive fields that may manifest clinically as chronic widespread pain (CWP) syndromes [41,48]. It is presumably also due to changes at this level that we begin to see distortions of hypothalamic and autonomic function in response to pain. These changes are characterized by sympathetic hyperarousal and to date, provide the clearest, measurable clinical indication of the acute to chronic transition.

Ostensibly, it follows that sustained changes here are a driver of the broader autonomic and endocrine dysfunction we see longer term in patients with chronic pain. This certainly appears to be the case when we look at changes to blood pressure regulation and longer term dulling of baroreceptor sensitivity [22,23]. It is probably not a far reach then to propose that these sorts of changes may underlie the myriad systemic changes and increased health risks we also see in more complex persistent pain presentations. Though largely speculative these sorts of changes are also thought to contribute to changes in the downward modulation of pain via the PAG, RVM, and amygdala. We can infer that this may manifest as impaired inhibitory modulation, the perpetuation of facilitatory modulation, altered behavior, and fear conditioning and may also be associated with anxiety states. This flow on to alterations in top-down modulation and the critical role of the amygdala will be covered in more detail in the next section.

Top-down pain processing

The top-down processing of pain-related information has two key purposes, it largely works toward the modulation of bottom-up (nociceptive) inputs and toward driving appropriate behavioral responses (both somatic and autonomic) to nociceptive stimuli. As noted previously, modulation is a constant tug of war between inhibition and facilitation to balance the allocation of resources and better the chance of survival. Top-down modulation is key to this process and is essentially tasked with determining whether nociception and pain are safe to ignore and attenuate our perception and responses to, or whether protective responses require amplification to enhance prospects for survival. This assessment appears made largely with respect to nociceptive inputs, the context of those inputs, and the previous experience and expectations of the individual [49].

The central nucleus of the amygdala (CeA) is particularly implicated here with the recent works of Wilson et al. (2019) eloquently conceptualizing the CeA as a pain rheostat that allows us to flick between the dual and opposing functions of nociceptive amplification or attenuation according to the context of the nociceptive inputs and, experience and expectations of the individual [49]. To accomplish this the CeA appears to weigh inputs from the thalamus, somatosensory cortex, and limbic cortices with nociceptive information from the cord; discharging orders to influence pain and pain-related behaviors via feedback loops to descending modulatory systems (via the substantia nigra, locus coeruleus, and Raphe nuclei of the medulla) [46,49].

These descending pain modulatory systems arise within the brainstem to send processed pain outputs to the dorsal horn (DH), where they modulate peripheral nociceptive inputs via either direct synapses or indirectly via volume transmission [13,46]. Nociceptive inhibition or antinociception is actioned and has marked evolutionary value where pain runs counter to the aspiration of survival. Nociceptive facilitation is used to prioritize protective behaviors and to enhance sensory transmissions from areas of, or adjacent to, compromised sensory surveillance. This is a system in flux, but persistent pain drives a positive feedback loop that reduces the inherent plasticity of the system and makes the return to baseline harder and harder as time goes on. Baseline should be the normal homeostatic setpoint of antinociceptive dominance when not under threat but for individuals with chronic pain, downward facilitation and nociceptive amplification commonly persist despite the absence of illness, injury, or threat.

The PAG-RVM system

The periaqueductal gray (PAG)—rostral ventromedial medulla (RVM) system is not alone in modulating nociception from above, but it is the system we currently understand the most about. The PAG plays key roles in autonomic and motivated behavior, particularly in response to threatening stimuli. It receives input from the spinomesencephalic tract and the rostral anterior cingulate cortex, hypothalamus, and amygdala to ensure moment-to-moment influence from both bottom-up and top-down processing systems [8,13,40,46]. These inputs importantly convey mechanisms by which, thoughts, emotions, stress, fear, and illness may influence nociceptive modulation within the system. The PAG projects opioid encephalin releasing neurons to the Raphe nucleus magnus in the RVM and the RVM contains a range of neurons thought to act as the final common pathway for several top-down modulatory systems [50]. These neurons then project to the superficial (and to a lesser extent deep) DH laminae where they either inhibit or facilitate A & C fiber sensory transmissions depending upon the influence of the PAG and CeA (among other inputs) [49,50].

The bulk of descending RVM neurons appear to be GABAergic and glycinergic and these are thought to mediate general anti-nociception at the DH [50]. The system's default operational mode appears to be to suppress C fiber nociception at the superficial DH. The RVM's serotonin-releasing neurons can have either inhibitory or facilitatory effects at the DH depending upon receptor subtypes they interact with. Facilitatory roles appear particularly relevant in inflammatory and neuropathic pain states. The PAG and RVM both have reciprocal connections to noradrenergic pontine nuclei, most notably the locus coeruleus. This appears to be another key component of the downward modulatory system [46,50]. Like the RVM's serotonergic neurons, noradrenergic neurons appear to exert excitatory or inhibitory effects dependent upon receptor subtypes they interact with also.

This ability to switch between downward inhibitory and facilitatory functions appears to be pivotal in our understanding of the acute to chronic transition and it would appear that "getting stuck in facilitatory mode" is a crucial step toward severe and intractable pain. While these modulatory pathways are therapeutic targets for serotonin/noradrenalin reuptake inhibitors and opioids, their influencing inputs make a clear biological connection for educational, cognitive, and talk therapies to be harnessed for analgesic effect. Their connections to cardiorespiratory centers in the brainstem that initiate and enhance antinociceptive effects in normal, healthy individuals also open avenues for exercise therapies to be used as targeted retraining in some. And indeed, there is evidence to suggest that reversal of baroreceptor desensitization with regular huff and puff exercise is coupled to improved descending inhibitory function and reductions in pain for some patients.

Diffuse noxious inhibitory control (DNIC)

DNIC refers to the curious phenomenon of pain in one part of the body inhibiting the perception of pain elsewhere [51], likely a mechanism developed for managing attentional load when under threat. The caudal medullary subnucleus reticularis dorsalis (SRD) appears to operate a parallel system, completely independent of the PAG-RVM system, to exert antinociceptive effects on the deeper wide dynamic range (WDR) neurons of the DH [51]. Under the influence of key cortical areas, the SRD conveys projections to the DH and CN V nociceptive neurons via the dorsal column medial lemniscus (DCML) system [46,50,51]. DNIC appears to be under the sway of an equally diverse range of cortical inputs that further support therapeutic pathways for non-pharmacological interventions, these include:

- Prefrontal cortex (attentional control for sensory processing, behavioral planning)
- Anterior cingulate cortex (cognition, evaluation of pain and its impact, pain avoidance)
- Midcingulate cortex (fear and unpleasantness of pain, motor responses to threat)
- Insular cortex (pain behaviors in response to PFC, ACC, amygdala, and hippocampal inputs)
- Amygdala (learned fear, fear-avoidance, anxiety, modulatory control)
- Hypothalamus (stress responses, autonomic and endocrine control)

These top-down processes are more complex again and are typically more challenging for patients to understand, but there is great value in prioritizing helping our patients to understand and indeed exploit these therapeutic avenues. There is good evidence now to suggest that impairment of endogenous inhibition and/or augmentation of endogenous facilitation are associated with chronic pain states [13,40,46,50] (particularly those where pain is spontaneous or has a life of its own because transmission is no longer predicated on presynaptic or primary afferent activity) and while these are often late changes in nociceptive pain conditions, they can occur much faster in neuropathic pain conditions and in those genetically and historically predisposed.

Selecting medicines and other therapeutic interventions that reflect the predominant pain type and sensitization mechanisms will tend to provide the patient with the best outcomes. Taking a multifaceted approach that utilizes a combination of interventions to target these mechanisms, will allow for compounding of effects and a better outcome overall. When pain is viewed as a biopsychosocial phenomenon and its potential to be or become a pathology in its own right is acknowledged, the value of exercise as both a treatment goal and therapeutic intervention becomes overwhelmingly apparent. As with all interventions, one must be thoughtful about dosing, and how and when it is best introduced.

Clinical application of exercise as a therapeutic intervention for chronic pain

Understanding the tangible changes that occur in the nervous and endocrine systems following the development of chronic pain, clinicians can be better aware of the need for a tailored approach for the use of exercise in patient with chronic pain. Overall, the sensitivity of the nervous system for the patient with persistent pain requires a much slower and considered approach to exercise.

Graded activity and graded exposure for chronic pain

Graded activity, sometimes interchangeably referred to as graded exposure, is one of the key clinical translations of the understanding of pain science and how we use this science to underpin treatment for chronic pain. Used to improve functional deficits that have come about in the presence of chronic pain, it embraces simple principles that can be adapted to most situations.

Exploring functional losses

The loss and distress that most people with pain experience is frequently a combination of the presence of the pain itself, as well as the impact that the pain has on a person's life. Functional losses seen in the chronic pain setting can be as specific as a loss of range of motion in a single joint, and as broad as the lack of stamina required to sit at the dinner table and partake in a meal with friends or family. Graded activity is a set of principles used to first explore the functional deficits and set a program of exposure-based exercises or activities designed to address the deficits.

Difference between graded exposure and graded activity

Graded activity refers to a program of activity that positively reinforces patient activity levels. Graded exposure involves a confrontation element whereby patients are exposed or challenged with concepts, situations, or movements that they are fearful of. Both approaches have been shown to be effective in improving function in the rehabilitation setting, however, graded exposure has been demonstrated to reduce catastrophizing as well as improve function [52].

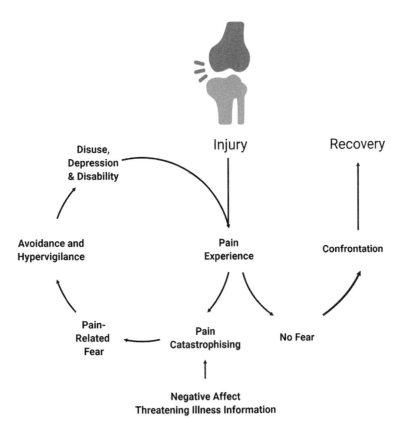

FIG. 5 Fear-avoidance cycle. *Adapted from Vlaeyen and Linton [7].*

Setting a graded exposure program

Prior to setting any form of exposure program, the practitioner needs to have spent adequate time listening to a patient's story, paying particular attention to their fears about the presence of pain and the meaning that they attach to the pain. Without having this understanding, it is possible that the best-planned exercise program may fail if patient fears have not been addressed. A common example of this is the patient who believes that the presence of pain during movement is associated with tissue damage or the potential to worsen an injury. The understandable behavioral consequences of holding these beliefs are a tendency to avoid movements or activities that cause pain. This avoidance often leads to deconditioning or further loss of function and a lowering of stamina. This cycle is referred to as the Fear-Avoidance cycle (Fig. 5) [53].

Allowing a patient to verbalize these fears can give them the sense that the practitioner has heard their concerns and can guide the practitioner in tailoring education material to provide maximum reassurance without invalidating their experience. Once these concerns have been acknowledged and validated as being very natural and common fears, the practitioner can then deliver information pertaining to chronic pain physiology. Also known as pain education or pain neurophysiology education (PNE), the aim of explaining this information to a patient is threefold:

- To separate pain from tissue damage
- To take the emphasis off pain and guide the patient to focus more on function
- To give a physiological explanation for a graded exposure approach

As covered in the earlier material regarding pain physiology, once the pain has progressed from acute to become chronic, there is less emphasis on inflammatory processes and tissue damage, and greater recognition of both central and peripheral sensitization. In de-threatening the presence of pain with pain education, practitioners can allow patients to better engage with exercise approaches knowing that they are not causing harm. As a single treatment entity for chronic pain, pain education or PNE has demonstrated small effect sizes in increasing function and reducing pain catastrophizing [54–56]. This is unlikely on its own to change the trajectory of a pain condition for an individual, however, it can be incredibly helpful to lay a foundation for a patient to engage with self-management strategies including exercise.

Graded exposure principles

During the history-taking and listening phases of the consult, a practitioner can gain a good sense of the functional losses that have occurred with the presence of pain. They can also get an understanding of what is important to the patient with regard to their rehabilitation goals. A good way to phrase the questions to explore these goals might include questions such as:

- If you had a magic wand and pain was not such a big deal for you, how would life be different?
- What things have pain taken away from you that you would like to work on getting back to doing again?

You will notice that these questions deliberately avoid discussing taking away or reducing pain and put the emphasis on function—the functional losses that are upsetting for the patient and the functional goals that they would like to work toward. This is an important distinction as a graded exposure approach does not aim to reduce pain, but rather to reduce disability and improve function. Once the patient has identified the functional goals that they would like to work on, a set of principles are applied to this task. These can include:

1. Finding a baseline of activity that does not provoke excessive pain either during the activity or in the 24 h following the activity.
2. Setting a schedule of activity or exercises that starts with this baseline and is practiced regularly.
3. Gradually increasing the parameters of this activity is a scheduled, graded fashion. These increases should be set collaboratively with the patient and practitioner and should be scaled up on a timed schedule, rather than focusing on how the patient is feeling during the exercise. The reason for this approach is that in a sensitized system in a chronic pain patient, moment-to-moment feedback from the body during exercise is often not as accurate as might be seen in a healthy person. Therefore, relying on this information to make decisions about the level of exercise can potentially either cause a patient to do too much or not enough exercise. Sticking to scheduled increases reduces this possibility. Increases of the exercise parameters may include increases to the overall time spent exercising, average heart rate, distance, weights, repetitions, or resistance. In some instances, the scheduled increases may involve different parameters at different times. For example, in a walking program, an initial grading up of the program may be to increase the time spent walking. At a later date, the grading up may be to add a slight incline on a treadmill session. The baseline activity should be sub-maximal and gradations should be small, aiming to remain sub-maximal as capacity increases. In the event that a grading up of the exercise causes significant increases in pain levels in a 24-h time period following the exercise, it is likely that the increase was either too early or too large for the patient at that time. The patient should be reassured about the temporary nature of these flare-ups and a flare-up self-management plan can be enacted to help a patient to cope with the increased pain levels until they settle. Subsequent graded increases to the program should be adjusted down with reference to this reaction. The principles of such an approach need to be explained to patients well, as many patients will either tend to err on the side of "boom and bust" attitudes (if 10 squats is good then 50 must be better) or can be too cautious and under-dose a program (it hurts so I should be careful and stop doing that exercise). A simple guide to explain this might be—on a good day do not do more, and on a bad day try doing half or a quarter of the exercises, rather than choosing to do none. This way, a level of consistency in a program can be better achieved with fewer flare-ups and gentle experiential exposure to movement in the presence of pain, which can be a powerful way to reduce a fear-avoidance cycle.

Self-efficacy

The role of self-efficacy has been identified by researchers and clinicians as an important factor in the development and maintenance of chronic pain for some time. Defined as the subjective confidence in one's own ability to achieve a task in 1977 by Bandura [57], this psychological construct has been shown to be more predictive of the development of chronic pain than any physical factors such as biomechanics, load, posture, or levels of tissue damage [7]. People who have high self-efficacy tend to have greater confidence in their ability to manage their musculoskeletal pain and are more willing to persist in the presence of setbacks. For the person with pain, this means they are more likely to attempt tasks in the presence of pain and tend to be less worried about the presence of pain. According to the research, this mental resilience seems to be a strong protective factor in the progression of acute pain to chronic pain and disability. Additionally, people with high self-efficacy do not seem to be as distressed or depressed as people with low self-efficacy and will generally have less pain (as rated by VAS scores) [58]. Understanding these concepts, as clinicians, we have an opportunity to detect the presence of low self-efficacy in our patients and direct our treatment accordingly.

Pain Self-Efficacy Questionnaire—Two-Item Short Form (PSEQ-2)

Michael K. Nicholas, PhD, Brian E. McGuire, PhD, and Ali Asghari, PhD

Please rate how **confident** you are that you can do the following things <u>at present</u>, **despite the pain**. To indicate your answer circle one of the numbers on the scale under each item, where 0 = not at all confident and 6 = completely confident.
For example:

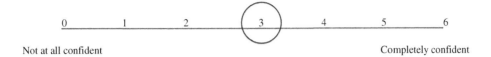

| 0 | 1 | 2 | 3 | 4 | 5 | 6 |

Not at all confident   Completely confident

Remember, this questionnaire is not asking whether or not you have been doing these things, but rather **how confident you are that you can do them at present**, <u>despite the pain</u>

1. I can do some form of work, despite the pain ("work" includes housework and paid and unpaid work).	0 Not at all confident	1	2	3	4	5	6 Completely confident
2. I can live a normal lifestyle, despite the pain.	0 Not at all confident	1	2	3	4	5	6 Completely confident

FIG. 6 The Pain Self-Efficacy Questionnaire-2 (PSEQ-2).

Measuring self-efficacy can be done using the Pain Self-Efficacy Questionnaire (PSEQ) [59] or its shorter cousin the PSEQ-2 (Fig. 6) [60], both demonstrated to be valid and reliable tools. The PSEQ-2 consists of two questions and asks the person to rank on a scale of 0–6, 0 being not confident at all, and 6 being completely confident they can complete activities and live a normal lifestyle in the presence of pain. Scores of 8 out of a total of 12 are considered to indicate a desirable level of pain self-efficacy and are likely to indicate that a patient is better placed to recover, compared to those who might score 5 or below.

Self-efficacy is an important construct to understand when setting exercise programs for patients with chronic pain. If someone has very low confidence in their body and is fearful of movements, situations, or activities, being aware of this and being able to measure it can be helpful when setting a graded exposure program. If they are starting treatment with a low PSEQ score, the things to keep in mind when setting an exercise program include:

- Reassurance in the form of pain education can reduce fear and allow patients to attempt new activities that they previously would have avoided.
- People with low self-efficacy often need higher levels of supervision when starting an exercise program. This allows for constant feedback and reassurance and for support and troubleshooting in the event of a pain flare-up.
- The baseline program should be set at a very low level, to begin with. The intentions in these early phases of an exercise program include:
  - Building a habit
  - Feeling a sense of success in completing the prescribed program
  - Building confidence in trying new movements and activities in a way that is unlikely to cause a flare-up.

Successfully starting and grading up an exercise program will naturally start to help increase a patient's self-efficacy as they are able to see the improvements and hopefully see it translate through to functional gains.

Managing pain flare-ups

Anyone trying activities that they haven't done in some time is likely to experience a level of musculoskeletal pain. Many people returning to their gym after a long break or picking up a tennis racquet for the first time in years can attest to this. Our tendency is often to expect to be able to pick up where we left off the last time, we did that activity. Unfortunately, this often leads to several days of being stiff and sore. For the chronic pain patient with a highly sensitized nervous system, new activities, or increasing the load of existing activities can easily cause a pain flare-up. Most patients would describe this experience as a temporary increase in their familiar pain. It will usually tend to occur in the hours following the exercise or the next day and can last for several days to weeks. Often these extreme pain levels are vastly disproportionate to what would be seen in a person who does not suffer chronic pain. The physiological mechanisms that are responsible for such disproportionate increases in pain levels most likely lie with both peripheral and central sensitivity changes, including reductions in descending inhibitory modulation mechanisms. A nociceptive driver may or may not be able to be identified for the individual situation, but the nervous system changes enhance and exaggerate this response [61]. This information has two important follow-ons.

Firstly, regardless of the presence or absence of inflammation or other local tissue factors that may be related to the pain experience, whatever the patient is describing their pain is what clinicians should be tuning into and believing. It is very common for chronic pain patients to have a flare-up of pain following some form of mild physical activity and it is frequent that these experiences are dismissed by health practitioners as being not real, not possible, all in the persons head or a variety of other dismissive and invalidating responses. This invalidation can be crushing for the patient in pain, who is dealing not only with the distress of high pain levels but also the sense of not being believed.

Secondly, physiology reminds us that even though the pain is real and often distressing, the heightened response is due to the nervous system changes and not due to tissue damage. Reminding patients about this in preparation for their first flare-up when starting any movement program can help to reduce the distress associated with the presence of flare-up pain.

Creating a flare-up management plan

Knowing that flare-ups are possible once we start engaging with different movements, activities, or altered loads of existing exercises, it can be helpful to have a pre-prepared flare-up management plan to help reduce the distress associated with a flare-up. In preparing such a plan collaboratively with a patient, we are setting expectations that times of increased pain are normal and somewhat expected as we try new things. Additionally, we can pre-arm them with reassurance of a couple of helpful principles for managing flare-ups. These include:

- This is temporary and will pass
- The tissues have not experienced damage
- There are some things that I can do to help myself to allow this to pass quicker and reduce the distress that I feel in the presence of the flare-up

Working with the patient, it can be helpful to create a written list to refer to in the presence of high levels of pain. The reason it is preferable to have a written list is that frequently during a flare-up, patients may also experience high levels of distress, and while in pain, their capacity to logically recall the information may be reduced. Therefore, having a "go-to" list can help patients engage with active self-management strategies and be less tempted to resort to passive approaches such as medication or healthcare seeking. When creating this list, it can be helpful to write the dot points above at the top of the list as a reminder and orientation. You can then collaboratively create a list of seven or more things that are known to the patient to reduce either their pain levels or distress levels, even if these strategies are temporary. The idea is that the patient will systematically go through the list and attempt each of the strategies. Each strategy may only reduce pain or distress by a small percentage but the act of trying serves as a great self-efficacy boosting exercise as the intention is "I am trying to help myself." Once they have completed the series of tasks, if they are still in a high level of pain and distress, they can repeat the list.

Some options that patients have found helpful include:

- Heat packs
- Hot Showers (10 min or more at a high temperature over the painful regions)
- Mindfulness or relaxation exercises
- Calling a friend for a chat
- Watching a funny TV show or YouTube comedy
- Looking at photos that make you happy
- Going for a short walk
- Making a nice cup of tea or coffee
- Listening to a favorite piece of music
- Writing/journaling

- Creative expressions or hobbies
- Sitting in the sun or nature (sunscreen on!)
- Self-massage
- Stretching exercises
- Cuddles on the couch with a partner or child
- Reading

Prompting the creation of the list might involve suggesting to the patient: what are things that make you feel relaxed, feel nice, reduce your pain a bit or lift your mood? It is important to set the expectations that these will not take the pain away completely, but should reliably reduce the pain a small amount, which is often enough to be able to better cope with the pain.

Examples of graded exposure programs

Many exercise programs fail because the patient and practitioner are not successful in identifying the baseline. Most commonly what is seen is the program being set too high with subsequent excessive pain, frustration, and abandoning of the program. For patients with significant central sensitization, the starting point of an exercise program may be very low. It is important to normalize this as part of the process as this low entry point could be viewed as being ineffective or tokenistic. Explaining to the patient that these early starting exercises are designed to build habits around exercise as well as confidence in the body to achieve these movements, is important to help with adherence to a program.

Widespread body pain

Sometimes referred to as fibromyalgia, widespread body pain presents with diffuse, chronic musculoskeletal pain, fatigue, headaches, sleep disorders, and cognitive and psychological disorders [62,63]. Various types of exercise used to treat fibromyalgia have been shown to be effective in improving overall well-being, increasing function, decreasing pain, and improving depressive symptoms. These include strength training, moderate-intensity aerobic training, hydrotherapy, and tai chi [64]. In healthy individuals, central nervous system mechanisms exist to reduce sensitivity to pain during and following exercise. These mechanisms are known as exercise-induced hypo-algesia (EIH). EIH is frequently impaired in people with chronic pain, particularly widespread body pain. This means that where healthy individuals may feel a sense of well-being and reductions in pain following exercise, many people with widespread body pain may not experience any reductions in pain sensitivity or may even experience an increase in pain sensitivity. This is thought to be related to changes in either the endogenous opioid, endocannabinoid, or serotonergic systems seen in chronic pain states [65,66]. The problem that this presents is that exercise is known to be beneficial for people with widespread body pain, however, there is a chance that pain may feel worse, not better in the short-term following exercise. This is an obvious disincentive to maintaining an exercise program designed to address widespread pain. The answer to addressing this paradox for the patient with widespread pain most likely lies in education and expectation setting around these concepts and then ensuring that the baseline starting points of the graded exposure program are low enough not to cause excessive amounts of pain following exercise. This may mean that the starting points for such a program are very low and grading them up is equally much slower than in other settings.

Headaches

One of the most important factors in directing treatment for the chronic headache patient is a diagnosis. The efficacy of treatment is dependent on tailoring the treatment to the specific type of headache presentation. To familiarize yourself with the common presenting signs and symptoms of the most common headaches, it is advisable to refer to the International Classification of Headache Disorders (ICHD) [67]. The three most common and distinct types of headaches seen in practice include migraine, tension-type headache, and cervicogenic headache. Each has quite a different etiology and as such, treatment approaches differ.

Migraine

Migraine headaches tend to respond well to exercise. The type of exercise and the appropriate dose remains to be tightly identified in the literature, however, there is moderate level evidence to suggest that aerobic exercise can reduce the number of headache days over a period of time, compared to no exercise control group [68]. It should be noted that most migraine sufferers will find that exercise during a migraine will tend to exacerbate the symptoms. Due to the lack of strong evidence pointing toward a particular type of exercise being overwhelmingly more effective than others, tailoring an exercise program for migraine patients should take into account their preferences and likelihood of continuity.

Tension-type headache

This common type of headache does not generally respond well to a single modality treatment approach. There is not a great deal of high-quality evidence to suggest that a specific type of exercise gives consistent positive outcomes. A multi-modal approach to treating this type of headache is generally considered to be the best approach. This might include medication, relaxation strategies, cognitive behavioral therapy, and exercise. It is likely that exercise in this setting would be helpful to increase overall stamina and to help facilitate descending inhibitory mechanisms [69].

Cervicogenic headache

This headache could be better conceptualized as mechanical neck pain which happens to refer to pain in the head. Sensory information from the upper three cervical spinal nerve roots is believed to interact with sensory nerve fibers in the descending tract for the trigeminal nerve in a region of the upper cervical spinal cord. It is believed that due to the close nature of the nerve pathways, a level of functional convergence may occur. This means that a bidirectional referral of painful sensations between the areas supplied by the nerves is possible. Therefore pain from the upper neck structures may be felt in the head [70]. There is a moderate level of evidence to suggest that specific neck exercises can be beneficial in the treatment of cervicogenic type headaches [70–72]. Most of these studies have focused on specific activation of deep cervical flexor musculature.

Neck and back pain

Evidence around the efficacy of the use of exercise for treating back and neck pain has been highlighted in the literature for many years. Narrow focus on the type of exercise, such as motor control exercises, pilates, resistance training, and aerobic training dominated the bulk of the research produced for many years and as such, recommendations for these types of exercises were generated in many of the guidelines for the treatment of low back pain at that time. As the research has evolved, not only have these more localized or specific types of exercises been demonstrated to be helpful in both improving function and decreasing pain, but other, simpler, and more accessible types of exercise such as walking have proven to be effective [72–74]. As we move toward better implementation of our understanding of the biopsychosocial context of common conditions like neck and back pain, so too has our application of the research regarding the efficacy of exercise treatment of these conditions. As such, many of the guidelines created for both acute and chronic low back pain place greater emphasis on exercise approaches being aimed at increasing movement levels in a way that is tailored to what the patient is likely to be able to engage with in both the short and long term [75].

Neuropathic pain

Neuropathic pain, most often described by patients as having sharp, shooting, burning, or electric shock qualities is among the most debilitating chronic pain presentations. Physiological evidence for the use of exercise in treating neuropathic pain exists, particularly the role of exercise in modulating maladaptive neuro-inflammatory changes [76–78]. Movement in various forms is frequently used in the treatment of neuropathic pain with the intention of both reducing pains and improving function. When patients are using neuropathic descriptors for their pain, it is usual for the starting point for an exercise program to be considerably lower than for more general chronic musculoskeletal conditions. In the presence of these neuropathic descriptors, as well as clinical evidence of neuropathic pain such as allodynia and hyperalgesia, it is likely that the point at which a flare-up of pain comes on may be much lower than patients without neuropathic descriptors. Therefore, when starting an exercise program for such patients it is advisable to start at a very low baseline and build up slowly according to the response. Frequent flare-ups are common in this pain population and pitching exercise at an appropriate level can be difficult to achieve. In some cases, maintaining function may be the ultimate goal whereas improving function is found to be hampered by low physical thresholds and frequent flare-ups.

Chronic pelvic pain

Chronic pelvic pain (CPP) in men and women can occur throughout various stages of life and present with varying symptoms. There is little high-quality evidence to direct exercise treatment of chronic pelvic pain [79]. There is a small amount of evidence to suggest that exercise such as yoga, stretching, dancing or aerobic training may provide good reductions in menstrual pain in females with dysmenorrhea. Participants in the studies reviewed in a recent Cochrane review exercised for 45–60 min, three times a week to achieve the clinically relevant reductions in pain [79].

Prescribing exercise for chronic pain

The role of exercise in the management of chronic pain is both direct and indirect. Direct in that it can have small to moderate positive effects on pain, and indirect, in that it can be a form of graded activity, and it can enhance general health on both physiological and psychological levels. Exercise demonstrates the potential to be beneficial in a variety of chronic pain conditions and is both low cost and accessible. Exercise instruction can be delivered effectively one to one, in small group settings, or via telehealth, making it a viable intervention across a range of clinical settings. Though it has good clinical utility, there are some barriers to uptake of exercise prescription. One such barrier is that many people report an increase in their pain intensity post-exercise, this is typically transient in the short term, but a significant barrier we must overcome. Previously in the chapter, we addressed strategies to help reframe the intention and expectation of undertaking an exercise program for sufferers of chronic pain. We will now explore some more practical aspects of exercise prescription, from establishing a baseline, to monitoring intensity and progression.

Effectiveness of exercise as an intervention for chronic pain

There is an abundance of clinical and epidemiological literature demonstrating the positive effects of exercise on numerous aspects of health, from all-cause mortality to QOL, physical function, and onset/progression of chronic disease states [80]. Health departments around the world acknowledge these benefits, with many having specific physical activity guidelines in place. These guidelines, despite small regional differences, encourage regular physical activity, daily, if possible, of varying intensities (moderate to intense) and modalities (cardiovascular and resistance training) [81]. Chronic pain patients as much as any other population need these benefits to limit and counter the effects of chronic pain but this does not mean that exercise can therefore be considered an effective intervention for chronic pain. Certainly not in isolation. Being that chronic pain is a whole-person condition, not limited to one body system or aspect of life, very few individual interventions have significant and/or lasting effects on pain. As a result, current best practice is to incorporate a multi-disciplinary management approach that layers in several strategies and allows for summation of effects [82]. To this end, exercise can be considered a positive general health intervention, which may provide improvements in chronic pain, to some degree, in some people. The critical questions are then what exercise or activity, how much, and in who?

A large Cochrane review assessed the effect of exercise on chronic non-cancer pain in adults aged >18 years [11]. It included randomized controlled trials assessing exercise/physical activity as the primary intervention and the primary outcome measured was pain. Secondary outcomes assessed included physical function, psychological function, quality of life, adherence to the prescribed intervention, healthcare use/attendance, adverse events (excluding death), and death. The review included papers looking at a variety of chronic pain presentations and included a variety of exercise interventions. The majority of participants were deemed to have mild to moderate pain at baseline, and the following outcomes were observed:

- three reviews found no statistically significant effect on pain
- three reviews found at least a 30% reduction in pain
- seven reviews found approx. 10%–20% reduction in pain
- seven reviews found statistically, but not clinically significant reductions in pain

The paper concluded that "Overall, results were inconsistent across interventions and follow-up, as exercise did not consistently bring about a change (positive or negative) in self-reported pain scores at any single point." Additionally, the authors noted the following:

- Importantly and promisingly, none of the physical and activity interventions assessed appeared to cause harm to the participants, with most adverse events being increased soreness or muscle pain, which reportedly subsided after several weeks of the intervention.
- None of the included reviews examined generalized or widespread chronic pain as a global condition, each instead examined specific conditions that included chronic pain as a symptom or result of the ongoing condition (e.g., rheumatoid arthritis, osteoarthritis, low back pain, dysmenorrhea, etc.)

- There is limited evidence of improvement in pain severity as a result of exercise. There is some evidence of improved physical function and a variable effect on both psychological function and quality of life. However, results are inconsistent, and the evidence is low quality (tier three).
- For clinicians and people with chronic pain, the evidence in this overview suggests that the broad spectrum of physical activity and exercise interventions assessed here (aerobic, strength, flexibility, core training, etc.) are potentially beneficial, though the evidence for benefit is low quality and inconsistent.
- Physical activity and exercise may improve pain severity as well as physical function and quality of life.

This review provides a good overview of the current literature on exercise for chronic pain, which suggests there is a likely small to moderate benefit on pain, independent of the type of exercise, for a variety of painful presentations. It also notes that there is potentially no effect of exercise on pain in some circumstances. What this review does not tell us, is whether certain pain conditions respond more favorably to certain types and/or dosages of exercise. Nor does it provide any insight into the individual factors that may be predictive of a positive response to exercise as an intervention for pain. The other question that remains, is whether exercise is beneficial for those who suffer from chronic primary pain—i.e., pain that is not attributable to another condition, including widespread/multi-site pain.

Potential mechanisms of exercise for pain

As discussed earlier in the chapter, chronic pain is a multi-system condition, with both bottom-up and top-down influences. Just as the entire process of pain is not fully understood, neither are the mechanisms by which exercise can influence pain. The majority of these proposed mechanisms are based on empirical observation, foundational science, and small clinical trials. As such, they are subject to change over time as new knowledge emerges.

Descending modulation

Descending or central modulation was described earlier in the chapter. Exercise shows signs of influencing modulation of nociception at both spinal and supraspinal levels, resulting in the release of inhibitory factors like endogenous opioids, GABA, and serotonin. One consideration that is relevant for chronic pain patients is that for some, descending modulation is impaired, and thus applying a systemic stressor like exercise, results in flare-ups, no matter the dose. While rare, it should be acknowledged that this condition does exist and not everyone can tolerate, let alone benefit from exercise for chronic pain [83].

Anti-inflammatory effects

Both peripheral and central sensitization of neurons occurs in association with acute and chronic inflammatory processes. Exercise is hypothesized to positively affect chronic inflammation by activation of anti-inflammatory substances, like tumor necrosis factor (TNF), interleukins, T-cells, and various other immune-modulatory cytokines [84,85].

Autonomic regulation

The autonomic nervous system works to regulate many of the automatic functions of the body. Autonomic dysregulation is associated with chronic pain, typically in the form of increased sympathetic activation (i.e., chronic stress response). Regular exercise is associated with improved autonomic regulation [86].

Mechanotransduction

Mechanotransduction defines the process by which biological cells convert mechanical forces into electrical or chemical processes. Exercise involves production and absorption of forces by the neuro-musculoskeletal system, which has widespread physiological effects. These effects may indirectly influence pain via both peripheral and central mechanisms, including but not limited to endocrine signaling and protein synthesis which may aid tissue remodeling and, influence peripheral inputs to the central nervous system [86].

Neurotrophic factors

Chronic pain is associated with cognitive decline, including the onset of dementia. This may indicate some form of chronic neurogenic inflammation, impacting neuronal cell function, growth, and repair. Exercise is known to stimulate endogenous production of brain-derived neurotrophic factor (BDNF), which helps regulate neuronal growth and repair [87,88].

Although we describe potential mechanisms by which exercise may help modulate pain, it would be remiss to not mention that for practical purposes, these things happen simultaneously. Therefore, regardless of the actual mechanisms, the benefits of exercising for chronic pain come from starting and gradually progressing in a manner that is safe and within an individual's present tolerance for adaptation. This is discussed in further detail below.

Getting started

Using exercise as an intervention for chronic pain poses a number of challenges across multiple domains. These include, but are not limited to:

- Whether somebody starts an exercise program (compliance)
- Whether somebody continues with an exercise program long enough to see positive outcomes (adherence)
- The type of exercise
- The dosage (volume/intensity/frequency)
- Initial health status (physiological and psychological)
- Socio-economic factors

To overcome these challenges, especially with a body of evidence that is inconclusive as to what is the optimal approach to using exercise as an intervention for chronic pain, a practical, patient-centered approach can be used to guide clinical decision-making.

Addressing compliance

Anecdotally, clinicians will often claim they have suggested exercise, but a patient/client has not been compliant with their recommendations. Additionally, patients themselves often cite lack of time or motivation for not beginning or adhering to an exercise program. Both of these statements and the thinking behind them are suggestive of a very unidimensional approach to behavior change that is more practitioner-centered than patient-centered, and less likely to succeed.

This has been supported in qualitative studies on the topic [89]:

> The results of this study suggest that in this sample, exercise adherence is not simply the patients lacking motivation or not having enough time in the day.
>
> Instead, the results suggest that adherence to prescribed exercise is the product of chiropractors' and patients' experiences and beliefs, the development of their clinical relationship, and the way exercise is prescribed and monitored in parallel with other treatment modalities.

While this study looked at Canadian chiropractors specifically, a similar pattern presents across multiple allied health professions and nations [89]. In the work cited above, the authors have outlined *enablers* and *barriers* to help engage people in positive behavior change, specifically as it relates to exercise prescription. These are outlined in Table 3.

When multiple variables relating to compliance are considered, we can develop the opinion that no single factor will lead to compliance for the majority of people. Thus, health professionals must understand how to implement

TABLE 3 Common enablers and barriers to exercise in people with chronic pain.

Enablers
Participants feeling involved in the process
Supervision
Pain control
Knowledge acquisition (via information sharing)
Goal sharing
Follow-up contact

Barriers
Perception of what exercise is
Lack of time
Diagnostic uncertainty
Fear of pain
Lack of fit into daily life

behavioral change strategies, in order for patients to gain the benefits of undertaking an exercise program for chronic pain [89–91].

Behavior change basics

Human behavior is complex. It is only when we view both individual and group behavior through a wider lens, we can start to make sense of it. Part of this wider lens is the effect of time on outcomes (e.g., people will smoke now because the adverse effects are delayed). Another part is the effect of meaning. Meaning becomes more important when acting in the face of both uncertainty and suffering—such as that encountered when beginning an exercise program as a treatment for chronic pain. An example of this search for meaning in the face of suffering is the increased religious beliefs/behaviors demonstrated in those who suffer from chronic musculoskeletal pain [92].

To ensure the best chance of success in both starting (and continuing) an exercise program, a person needs the proposed action to have meaning that is both related to and independent of a potential outcome. That is, if the person is seeking reduced pain and increased function, these provide a superficial source of meaning for the activity, however, empirical evidence suggests that a deeper level of meaning is more robust at sustaining behavior change. To elicit meaning, motivational interviewing techniques can be used. This form of communication ticks off a number of enablers—feeling involved, goal sharing, and knowledge acquisition. By facilitating a patient to discover their meaning, or put another way, their motivation to change their actions, the likelihood of success is increased.

Increasing adherence

While meaning and motivation can help get someone started with a new activity, for long-term benefits, this activity needs to become habit-forming. There are a variety of techniques that have been discussed to increase adherence in both the short and longer term [93]. Two examples are outlined below:

- *Shrink the change*

    By reducing the actual (or perceived) size of the change (new activity), the task becomes more achievable in the short term. With exercise for pain, a graded approach is optimal, and this fits in well with the concept of "shrink the change." In the beginning, a smaller amount of exercise is needed to elicit positive adaptations. This can gradually be built upon.
- *Anchor the activity*

    This refers to attaching the activity to something that is already habitual. For example, if someone was recommended to do some leg strengthening exercises daily, and they drank tea twice a day, they could perform a set of exercises while they waited for the tea to brew.

The most important concept in behavior change is for both parties to acknowledge that it relies on more than discipline and that setbacks are a normal experience, and should be planned for as to not create feelings of failure and despair. These strategies serve to develop habits, which are explored in goal-setting below.

Goal setting

Goal setting is a method of defining desired outcomes with a view toward orientating behaviors toward achieving the said outcome. In essence, it is a way to develop a "roadmap" to a specific achievement or state. One widely accepted way to set goals is the SMART approach. This stands for *Specific, Measurable, Attainable, Realistic, Timely*. The idea being, that when a goal has these attributes, it is more likely to be achieved. An example of a SMART goal is "I would like to have $1000 saved in 3 months." Compare this to a non-SMART goal, which is vague and has no timeline "I would like to save more money."

Applying SMART goal setting to chronic pain management

For a sufferer of chronic pain, goal setting can seem futile, especially when the desired goal might be "to have no pain". Yet this may not be realistic for many, nor does it do anything to help guide the actions one is required to take in order to achieve this. For many chronic pain sufferers, their pain has significant impacts on function, and this is where exercise can be a vital tool toward optimal pain management. Thus, making goals functional, rather than pain-focused, will typically yield better outcomes in terms of quality-of-life improvements.

An example of a SMART, functional goal for a chronic pain sufferer could be:

"I would like to increase my ability to walk for a time by 15% in 6 months".

This goal is specific—increase the ability to walk, measurable—by 15% above baseline, achievable—this rate of improvement is possible (depending on various factors), whether this is realistic depends on the baseline status, but for an able-bodied person with chronic pain this is often possible, timely—it has a defined period.

From goals to skills

The biggest limitation of setting goals, even if they are SMART, is it does not help inform you about how to achieve them. In order to achieve a certain goal, actions must be taken. These actions or behaviors should become habits. These habits should develop skills or attributes that will build toward achieving the desired goal. Continuing with the goal of increasing walking time by 15% in 6 months, we can break the goal down into certain attributes required:

- Mobility
- Balance
- Leg strength
- Cardiovascular endurance
- Psychological resilience

Based on an individual assessment (discussed later in the chapter), the areas that need the most improvement can be addressed by implementing certain behaviors. In this case, the behaviors will be in the form of exercises. Setting a SMART goal is only the first step toward actually achieving that goal.

Having a reason (meaning) for wanting to achieve that goal provides a powerful motivational tool to draw on when things are challenging—as is inevitable with any process of achievement. This can help with consistently implementing the behaviors required to develop the skills and attributes needed to achieve the goal.

Prognostic factors

When discussing exercise for pain as a management strategy, and during the goal-setting process, prognostic factors should be addressed. The following factors are associated with a higher response to exercise [94]:

- Baseline pain intensity
- Mental health status
- Age
- Prior treatment
- Drug use

In a prospective study, patients were assessed at baseline and underwent six sessions of physiotherapy exercises and education over consecutive days. The exercise programs were individualized and delivered under supervision. Follow-up assessments were conducted at discharge and 1 year. The only variable associated with a better outcome at discharge was baseline pain intensity—high pain was associated with a worse outcome. Over the long term, younger age, better baseline mental health, less prior treatments, and drug usage were associated with better outcomes at 1 year follow-up. Interestingly, the is a very low correlation between improvements in physical qualities like strength, endurance, and flexibility and improvements in pain [94]. This has been studied for a variety of clinical pain presentations.

Baseline assessment

The baseline assessment should be designed to give enough insights as to the current status of the individual's health, fitness, and psychology in order to guide the initial program design and gauge improvements in follow-up assessments. An ideal assessment is comprehensive in what information it provides, but concise and practical—that is, it does not require expensive or rare equipment to perform.

Assessments can include:

- Validated outcome measures
- Quantitative physical assessments
- Qualitative physical assessments
- Self-reported evaluation

Considering the wide variety of factors that can influence exercise prescription, along with potential and likely outcomes, it is optimal to utilize a broad range of assessments across multiple systems and domains. Specific examples of assessments from each category include, but are not limited to:

Validated outcome measures:

- *Psycho-social factors*: Identification of yellow flags, Pain Catastrophizing Scale, Fear-Avoidance Scale, K10 (mental health screening).
- Regional specific outcome measures (knee, low back, upper limb, etc.).
- Functional outcome measures (patient-specific functional scale, back pain functional scale, etc.).

Quantitative physical assessment:

- *Vital signs*: Age, height, weight, blood pressure, pulse, respiratory rate.
- Neuro-musculoskeletal testing (range of motion, force output testing, etc.).
- Validated physical tests (sit to stand, 6-min walk, etc.).
- Quantified movement screening (Y-balance test, hop test index).

Qualitative physical assessment:

- Functional tasks (ability of patient to comfortably perform a particular task)

The assessment(s) should reveal the factors you need to influence, and then be able to measure relevant change. It should also identify the current functional capacity of the patient/client, in order to guide the starting point of their exercise program. It is important to note, that it's rarely a single factor, but rather the unique interaction of all factors in that environment. The person should be the focus of the exercise program, not the results of the assessment.

Types of exercise

Before discussing which types of exercise are optimal for those suffering from chronic pain, a brief clarification of commonly used terms is needed.

Physical activity: Physical activity is defined as any bodily movement produced by skeletal muscles that require energy expenditure. By definition, all physical activity is movement, but not all physical activity is exercise.

Movement: Go in a specified direction or manner; change position, change the place, position, or state of.

Exercise: Activity requiring physical effort, carried out to sustain or improve health and fitness. An activity carried out for a specific purpose. Exercise is a sub-category of physical activity that is planned, structured, repetitive, and aims to improve or maintain one or more components of physical fitness. Exercise comes in many varieties, but for the sake of simplicity, we can broadly categorize exercise into four main groups:

1. *Cardiovascular exercise* (cyclic endurance activities such as walking, running, cycling, rowing, etc.) primarily leads to adaptations in the cardiorespiratory system
2. *Resistance training* (exercise that involves overcoming resistance, be it bodyweight, or external load, includes weightlifting, pilates) which primarily leads to adaptations in the neuromuscular and skeletal systems
3. *Flexibility training* (exercises that involve movement or positions that take joints through to their end range of motion, including stretching, yoga) which primarily leads to adaptations in the neuromuscular and skeletal systems, though these are different from those gained from resistance training
4. *Sports and skill-based activities*

Sports and skill-based activities typically involve a composite of the three generalized classes of exercise described. However, they are also typically open and chaotic versus closed or controlled, which means there is a high degree of random variability in many sports activities. While enjoyable, these do not make for an easily scalable or progressive approach, and thus are not optimal for managing chronic pain. Thus, we will exclude sports and skill-based activities from the discussion for the sake of clarity and brevity.

Is one form of exercise superior for pain relief?/What type of exercise should be performed?

Most studies comparing exercise modalities for pain management do show clear superiority/inferiority. In practical terms, an exercise program that incorporates aspects of all three exercise categories gives exposure to multiple mechanisms of action for pain management, as well as build broad physical qualities than enhance resilience and adaptability. Working toward the national guidelines for physical activity, which recommend a blend of low-moderate activity/exercise, as well as higher intensity activity/exercise and muscle-strengthening (resistance) activities/exercise is a very good target, as it promotes overall health and longevity, in addition to pain management.

While many people have a preference for one particular form of exercise, there are clear benefits to be had from engaging in these different categories/forms of exercise. So, while something is generally better than nothing, and what is preferred is more likely to get done consistently, for maximal/optimal benefit, we stand by the above approach.

Dosage and progression

Dosage and progression are important, as it relates to performing enough exercise to create a positive physiological or psychological response. Dosage also involves overall load management, with aim of minimizing the risk of significant adverse effects from exercise—pain flare-ups, soreness, fatigue, etc.

Exercise dosage can be broken down into two key components:

1. Intensity: measured as a percentage of maximum output
2. Volume: measured as total amount of work performed

These two components can be viewed from an individual session perspective, or over a longer period (weeks/months). When prescribing exercise for chronic pain, the aim is not to maximize performance, but gradually increase physical capacity, facilitate endogenous analgesic processes and provide a positive psycho-emotional experience. With this in mind, dosage does not need to be as strictly prescriptive as a typical performance-focused exercise program. In fact, there is evidence supporting self-dosage in terms of both intensity and volume results in similar physiological improvements, with a more positive affective experience for participants.

Practically, because pain fluctuates, combining self-dosage with an overall plan for each session, along with the macro progression over time allows an individual to have an internal locus of control, with enough external structure to account for these fluctuations in pain and fatigue. We will discuss how to monitor intensity/volume shortly. Applying this approach to exercise progression, while not the only way a more flexible approach tends to account for potential flare-ups, while still ensuring that over time, positive adaptation takes place. In certain circumstances, a more prescriptive approach is beneficial, however, this is dictated more by clinical experience along with trial and error, versus any definitive literature on the topic.

Managing intensity

There are two main ways to monitor exercise intensity:

1. Subjective, using a *rating of perceived exertion (RPE)* scale
2. Objective/quantified, using external sensors/monitors (heart rate monitors, GPS sensors, power meters, etc.)

RPE correlates well with measured heart rate, power output, and velocity, and requires no specialized equipment. Additionally, RPE is an established and reliable measure of exercise intensity for both aerobic exercise and resistance training [95]. While using objective data can be quite helpful in measuring exercise output and progression, when it comes to pain, improvements in physiological characteristics, like strength, endurance, and flexibility are poorly correlated to pain improvements [96]. Accordingly, this chapter will focus on how to use RPE in a graded exercise program for chronic pain management.

Rating of perceived exertion

RPE is measured according to two main scales, first described by Gunnar Borg (Table 4):

1. The 6–20 numerically rated from 6 to 20 and is designed to reflect estimated heart rate (multiply by 10)
2. The second is scaled from 1 to 10 for ease of use

Applying RPE to an exercise program for chronic pain

Exercise intensity is the key variable to manage when prescribing exercise. Intensity is described as the percentage of maximum, and as discussed, this correlates well with perceived exertion.

Intensity, by definition, will impact both volume and frequency. When initially prescribing exercise for pain, we need to account for both psychological and physiological factors and monitor for adverse effects.

TABLE 4 Borg RPE scale.

Borg RPE score	Level of exertion
6	No exertion at all
7	
7.5	Extremely light
8	
9	Very light
10	
11	Light
12	
13	Somewhat hard
14	
15	Hard (heavy)
16	
17	Very hard
18	
19	Extremely hard
20	Maximal exertion

For some, starting too slowly might be challenging, as it is perceived as too easy, or not doing anything; this can be addressed with appropriate education. For others, any form of exercise might be perceived as too intense. Education and expectation management are important factors that precede exercise prescription. A starting point of 7.5–11 on the 6–20 scale is a good general approach, as it facilitates adaptations and progressive increases in volume with minimal risks of adverse effects. There is no consensus on the optimal exercise intensity for pain management/relief, so it makes sense to start at comfortable intensity, and slowly progress, monitoring the response along the way.

How much exercise do you need?

There is much debate on how much exercise is optimal for pain management. Again, there is no clear consensus. In consideration, we refer back to the national physical activity guidelines as a goal to work toward. With that said, if someone is currently sedentary, be it due to their pain or otherwise, any amount of exercise will yield a benefit. Practically, encouraging people to start exercising, even if they do not achieve the recommended physical activity guidelines is one of the best things that can be done for both their individual health and public health at large [97]. To answer the question posed, the amount of exercise that is needed may differ from the amount that is realistically achievable, and our stance is as follows:

- Some exercise is better than no exercise
- More exercise is generally better, up to a point
- The national physical activity guidelines are a good goal to aim for in terms of total dosage

Practical aspects of dosage

Optimizing exercise dosage is a key variable to optimizing outcomes. There are numerous approaches to dosing exercise. Traditionally, for resistance exercise, this might be achieved by varying the number of exercises and prescribing a range of sets and repetitions. For endurance training, distance prescriptions (volume) are common, with pace (intensity) the variable.

These approaches undoubtedly work; however they require a lot of trial and error to optimize, and more importantly for pain management, can be easy to overdo. Another approach, that is more self-regulating, is to control the period of time prescribed and use RPE to guide intensity. For example, using endurance exercise, such as stationary cycling on an ergometer, if 15 min at an RPE of 7 (out of 20) is prescribed, then the pace/RPM and resistance is adjusted by the individual throughout to maintain that 7. Thus, the total volume is controlled. Time-based approaches work for all three easily scalable categories of exercise [97].

Making progress

The benefits of exercise come, in large part, to homeostatic adaptation following an imposed stressor. When it comes to pain management, improving physical qualities, such as strength, endurance and flexibility are poorly correlated with improvements in pain. Thus, the need to progress is less about direct pain management, and more about functional improvements, and possibly reduction of recurrence risk. One consideration that is often overlooked: demonstrated progress has a positive psychological effect. This can be helpful when changes to pain and function may be lagging indicators. Progress for the chronic pain patient should be structured, but flexible. Small increments, such as 5% in total weekly volume are supported by empirical evidence.

Section summary

Exercise is a biopsychosocial intervention for pain and should be implemented in such a manner. Considerations should be made for the unique status of each individual, including their beliefs and preferences, and advice and programming adapted accordingly. To ensure long-term success, an interactive/collaborative approach between practitioner and patient is optimal. This involves goal setting, along with mapping out the skills required to achieve the goal, and the behaviors needed to develop the skills. Once these skills and behaviors are mapped out, an exercise program can follow, which should ideally be optimized around the individual's current abilities, with consideration of their broader psycho-social environment. Additionally, appropriate dosage and progression will enhance the likelihood of success and minimize the chance of adverse effects.

Conclusion

With both the incidence and prevalence of chronic pain so high and the costs (both direct and indirect) so high, cost-effective and accessible interventions form a crucial aspect of effective long-term management. Chronic pain involves multiple body systems and has a big impact on psychological and social well-being as well. There are distinct neurobiological changes that are present in people with chronic pain, and there is sound mechanistic reasoning for the use of exercise to address some of these. We know that exercise is not a comprehensive solution for chronic pain, but it is a low-cost, highly accessible intervention that provides a multitude of positive secondary effects on health, in addition to the primary effects on chronic pain.

In a clinical setting, exercise is best delivered as part of an overall multi-modal treatment plan, with a multi-disciplinary team, within a framework of pain neurobiological education. The challenges faced by clinicians and patients alike pertaining to the use of exercise for chronic pain related to behavioral barriers to the beginning and continuing an exercise program, knowledge of each party, access to equipment and a safe and secure environment in which to exercise, a paucity of literature on what types of exercise, if any, are best for certain pain presentations or individuals, and the inherent difficulties in dosage and progression of exercise in a chronically sensitized individual who may not respond to exercise in the same manner as a healthy individual.

Future considerations for clinicians, researchers, and health policy relating to the implementation of exercise for chronic pain include accessibility and reimbursement, short-term outcome measures, long-term sustainability/adherence, optimal types and dosage of exercise for certain pain presentations or individuals, and predictive factors, for both positive and negative responses to exercise.

References

[1] Descartes R, et al. L'homme.. et un traitté de la formation du fœtus. Du mesme autheur. Charles Angot 1664.
[2] Trachsel LA, Cascella M. Pain theory. StatPearls; 2020 [Internet].
[3] Burmistr I. Theories of pain, up to Descartes and after neuromatrix: what role do they have to develop future paradigms? Pain Med 2018;3(1):6–12.
[4] Melzack R. From the gate to the neuromatrix. Pain 1999;82:S121–6.
[5] Melzack R, Wall PD. Handbook of pain management. Elsevier; 2003.
[6] Meeus M, et al. Moving on to movement in patients with chronic joint pain. Pain 2016;1(10):23–35.
[7] Vlaeyen JW, Linton SJ. Fear-avoidance and its consequences in chronic musculoskeletal pain: a state of the art. Pain 2000;85(3):317–32.
[8] McMahon SB, et al. Wall & melzack's textbook of pain e-book. Elsevier Health Sciences; 2013.

[9] Economics DA. The cost of pain in Australia. Australia: PainAustralia; 2019.

[10] Van Hecke O, Torrance N, Smith B. Chronic pain epidemiology and its clinical relevance. Br J Anaesth 2013;111(1):13–8.

[11] Geneen LJ, et al. Physical activity and exercise for chronic pain in adults: an overview of cochrane reviews. Cochrane Database Syst Rev 2017;4.

[12] Zelaya CE, et al. Chronic pain and high-impact chronic pain among US adults, 2019; 2020.

[13] Austin P. Chronic pain: a resource for effective manual therapy. Handspring Publishing Limited; 2017.

[14] Fine PG. Long-term consequences of chronic pain: mounting evidence for pain as a neurological disease and parallels with other chronic disease states. Pain Med 2011;12(7):996–1004.

[15] Husak AJ, Bair MJ. Chronic pain and sleep disturbances: a pragmatic review of their relationships, comorbidities, and treatments. Pain Med 2020;21(6):1142–52.

[16] McCracken LM, Iverson GL. Disrupted sleep patterns and daily functioning in patients with chronic pain. Pain Res Manag 2002;7(2):75–9.

[17] Marin R, Cyhan T, Miklos W. Sleep disturbance in patients with chronic low back pain. Am J Phys Med Rehabil 2006;85(5):430–5.

[18] Alsaadi SM, et al. Prevalence of sleep disturbance in patients with low back pain. Eur Spine J 2011;20(5):737–43.

[19] Whitlock EL, et al. Association between persistent pain and memory decline and dementia in a longitudinal cohort of elders. JAMA Intern Med 2017;177(8):1146–53.

[20] Breivik H, et al. Survey of chronic pain in Europe: prevalence, impact on daily life, and treatment. Eur J Pain 2006;10(4):287–333.

[21] Bruehl S, et al. Prevalence of clinical hypertension in patients with chronic pain compared to nonpain general medical patients. Clin J Pain 2005;21(2):147–53.

[22] Suarez-Roca H, et al. Contribution of baroreceptor function to pain perception and perioperative outcomes. Anesthesiology 2019;130(4):634–50.

[23] Sacco M, et al. The relationship between blood pressure and pain. J Clin Hypertens 2013;15(8):600–5.

[24] Gureje O, et al. Persistent pain and well-being: a World Health Organization study in primary care. JAMA 1998;280(2):147–51.

[25] Ataoğlu E, et al. Effects of chronic pain on quality of life and depression in patients with spinal cord injury. Spinal Cord 2013;51(1):23–6.

[26] Raja SN, et al. The revised International Association for the Study of Pain definition of pain: concepts, challenges, and compromises. Pain 2020;161(9):1976–82.

[27] Phillips CJ. The cost and burden of chronic pain. Rev Pain 2009;3(1):2–5.

[28] Treede R-D, et al. Chronic pain as a symptom or a disease: the IASP Classification of Chronic Pain for the International Classification of Diseases (ICD-11). Pain 2019;160(1):19–27.

[29] Merskey HE. Classification of chronic pain: descriptions of chronic pain syndromes and definitions of pain terms. Prepared by the International Association for the Study of Pain, Subcommittee on Taxonomy. Pain Suppl 1986;3:S1–226.

[30] Woolf CJ, Salter MW. Neuronal plasticity: increasing the gain in pain. Science 2000;288(5472):1765–8.

[31] Gold MS, Gebhart GF. Nociceptor sensitization in pain pathogenesis. Nat Med 2010;16(11):1248–57.

[32] Schaible H-G. Peripheral and central mechanisms of pain generation. In: Analgesia. Springer; 2006. p. 3–28.

[33] Harte SE, Harris RE, Clauw DJ. The neurobiology of central sensitization. J Appl Biobehav Res 2018;23(2), e12137.

[34] Bourne S, Machado AG, Nagel SJ. Basic anatomy and physiology of pain pathways. Neurosurg Clin N Am 2014;25(4):629–38.

[35] Kandel ER, et al. Principles of neural science. vol. 4. McGraw-hill New York; 2000.

[36] Bakkers M, et al. Temperature threshold testing: a systematic review. J Peripher Nerv Syst 2013;18(1):7–18.

[37] Gangadharan V, Kuner R. Pain hypersensitivity mechanisms at a glance. Dis Model Mech 2013;6(4):889–95.

[38] Hall JE. Guyton & hall physiology review e-book. Elsevier Health Sciences; 2015.

[39] Mizumura K. Peripheral mechanism of hyperalgesia-sensitization of nociceptors. Nagoya J Med Sci 1997;60:69–88.

[40] Van Griensven H, Strong J, Unruh A. Pain: a textbook for health professionals. Churchill Livingstone; 2013.

[41] Graven-Nielsen T, Arendt-Nielsen L. Peripheral and central sensitization in musculoskeletal pain disorders: an experimental approach. Curr Rheumatol Rep 2002;4(4):313–21.

[42] Woolf CJ, Chong M-S. Preemptive analgesia—treating postoperative pain by preventing the establishment of central sensitization. Anesth Analg 1993;77(2):362–79.

[43] Schaible H-G. Emerging concepts of pain therapy based on neuronal mechanisms. In: Pain control. Springer; 2015. p. 1–14.

[44] Tan AM, et al. Early microglial inhibition preemptively mitigates chronic pain development after experimental spinal cord injury. J Rehabil Res Dev 2009;46(1).

[45] Hains BC, Waxman SG. Activated microglia contribute to the maintenance of chronic pain after spinal cord injury. J Neurosci 2006;26(16):4308–17.

[46] Bolay H, Moskowitz MA. Mechanisms of pain modulation in chronic syndromes. Neurology 2002;59(5 Suppl 2):S2–7.

[47] Zhang S, et al. Central sensitization in thalamic nociceptive neurons induced by mustard oil application to rat molar tooth pulp. Neuroscience 2006;142(3):833–42.

[48] Staud R, Smitherman ML. Peripheral and central sensitization in fibromyalgia: pathogenetic role. Curr Pain Headache Rep 2002;6(4):259–66.

[49] Wilson TD, et al. Dual and opposing functions of the central amygdala in the modulation of pain. Cell Rep 2019;29(2):332–346.e5.

[50] Ossipov MH, Dussor GO, Porreca F. Central modulation of pain. J Clin Invest 2010;120(11):3779–87.

[51] Le Bars D, et al. Diffuse noxious inhibitory controls (DNIC) in animals and in man. Patol Fiziol Eksp Ter 1992;4:55–65.

[52] López-de-Uralde-Villanueva I, et al. A systematic review and meta-analysis on the effectiveness of graded activity and graded exposure for chronic nonspecific low back pain. Pain Med 2016;17(1):172–88.

[53] Crombez G, et al. Fear-avoidance model of chronic pain: the next generation. Clin J Pain 2012;28(6):475–83.

[54] Louw A, et al. The effect of neuroscience education on pain, disability, anxiety, and stress in chronic musculoskeletal pain. Arch Phys Med Rehabil 2011;92(12):2041–56.

[55] Gardner T, et al. Combined education and patient-led goal setting intervention reduced chronic low back pain disability and intensity at 12 months: a randomised controlled trial. Br J Sports Med 2019;53(22):1424–31.

[56] Moseley GL, Nicholas MK, Hodges PW. A randomized controlled trial of intensive neurophysiology education in chronic low back pain. Clin J Pain 2004;20(5):324–30.

[57] Bandura A. Self-efficacy: toward a unifying theory of behavioral change. Psychol Rev 1977;84(2):191–215.

[58] Martinez-Calderon J, et al. The role of self-efficacy on the prognosis of chronic musculoskeletal pain: a systematic review. J Pain 2018;19(1):10–34.

[59] Nicholas MK. The pain self-efficacy questionnaire: taking pain into account. Eur J Pain 2007;11(2):153–63.

[60] Nicholas MK, McGuire BE, Asghari A. A 2-item short form of the pain self-efficacy questionnaire: development and psychometric evaluation of PSEQ-2. J Pain 2015;16(2):153–63.

[61] Courtney CA, Fernández-de-Las-Peñas C, Bond S. Mechanisms of chronic pain—key considerations for appropriate physical therapy management. J Man Manip Ther 2017;25(3):118–27.

[62] Andrade A, et al. The relationship between sleep quality and fibromyalgia symptoms. J Health Psychol 2020;25(9):1176–86.

[63] Andrade A, et al. Modulation of autonomic function by physical exercise in patients with fibromyalgia syndrome: a systematic review. PM R 2019;11(10):1121–31.

[64] Busch AJ, et al. Exercise for treating fibromyalgia syndrome. Cochrane Database Syst Rev 2007;(4), CD003786.

[65] Rice D, et al. Exercise-induced hypoalgesia in pain-free and chronic pain populations: state of the art and future directions. J Pain 2019;20(11):1249–66.

[66] Polaski AM, et al. Exercise-induced hypoalgesia: a meta-analysis of exercise dosing for the treatment of chronic pain. PLoS One 2019;14(1), e0210418.

[67] Olesen J. International classification of headache disorders. Lancet Neurol 2018;17(5):396–7.

[68] Lemmens J, et al. The effect of aerobic exercise on the number of migraine days, duration and pain intensity in migraine: a systematic literature review and meta-analysis. J Headache Pain 2019;20(1):16.

[69] Jensen RH. Tension-type headache—the normal and most prevalent headache. Headache 2018;58(2):339–45.

[70] Biondi DM. Cervicogenic headache: a review of diagnostic and treatment strategies. J Am Osteopath Assoc 2005;105(4 Suppl 2):16s–22s.

[71] Fernández-de-Las-Peñas C, Cuadrado ML. Physical therapy for headaches. Cephalalgia 2016;36(12):1134–42.

[72] Gross AR, et al. Exercises for mechanical neck disorders: a cochrane review update. Man Ther 2016;24:25–45.

[73] Gordon R, Bloxham S. A systematic review of the effects of exercise and physical activity on non-specific chronic low back pain. Healthcare (Basel) 2016;4(2).

[74] Vanti C, et al. The effectiveness of walking versus exercise on pain and function in chronic low back pain: a systematic review and meta-analysis of randomized trials. Disabil Rehabil 2019;41(6):622–32.

[75] Almeida M, et al. Primary care management of non-specific low back pain: key messages from recent clinical guidelines. Med J Aust 2018;208(6):272–5.

[76] Cooper MA, Kluding PM, Wright DE. Emerging relationships between exercise, sensory nerves, and neuropathic pain. Front Neurosci 2016;10:372.

[77] Dobson JL, McMillan J, Li L. Benefits of exercise intervention in reducing neuropathic pain. Front Cell Neurosci 2014;8:102.

[78] Chhaya SJ, et al. Exercise-induced changes to the macrophage response in the dorsal root ganglia prevent neuropathic pain after spinal cord injury. J Neurotrauma 2019;36(6):877–90.

[79] Armour M, et al. Exercise for dysmenorrhoea. Cochrane Database Syst Rev 2019;9(9), Cd004142.

[80] Kelly J, Ritchie C, Sterling M. Agreement is very low between a clinical prediction rule and physiotherapist assessment for classifying the risk of poor recovery of individuals with acute whiplash injury. Musculoskelet Sci Pract 2019;39:73–9.

[81] Campbell KJ, et al. A novel, automated text-messaging system is effective in patients undergoing total joint arthroplasty. J Bone Joint Surg Am 2019;101(2):145–51.

[82] Babatunde OO, et al. Effective treatment options for musculoskeletal pain in primary care: a systematic overview of current evidence. PLoS One 2017;12(6), e0178621.

[83] Lima LV, Abner TSS, Sluka KA. Does exercise increase or decrease pain? Central mechanisms underlying these two phenomena. J Physiol 2017;595(13):4141–50.

[84] Gleeson M, et al. The anti-inflammatory effects of exercise: mechanisms and implications for the prevention and treatment of disease. Nat Rev Immunol 2011;11(9):607–15.

[85] Petersen AM, Pedersen BK. The anti-inflammatory effect of exercise. J Appl Physiol (1985) 2005;98(4):1154–62.

[86] Dunn SL, Olmedo ML. Mechanotransduction: relevance to physical therapist practice-understanding our ability to affect genetic expression through mechanical forces. Phys Ther 2016;96(5):712–21.

[87] Cao S, et al. The link between chronic pain and Alzheimer's disease. J Neuroinflammation 2019;16(1):204.

[88] Liu PZ, Nusslock R. Exercise-mediated neurogenesis in the hippocampus via BDNF. Front Neurosci 2018;12:52.

[89] Stilwell P, Harman K. 'I didn't pay her to teach me how to fix my back': a focused ethnographic study exploring chiropractors' and chiropractic patients' experiences and beliefs regarding exercise adherence. J Can Chiropr Assoc 2017;61(3):219–30.

[90] Stilwell P, Harman K. Contemporary biopsychosocial exercise prescription for chronic low back pain: questioning core stability programs and considering context. J Can Chiropr Assoc 2017;61(1):6–17.

[91] Tripodi N, et al. Osteopaths' perspectives on patient adherence to self-management strategies: a qualitative content analysis. Int J Osteopath Med 2021;41:19–26.

[92] Rippentrop EA, et al. The relationship between religion/spirituality and physical health, mental health, and pain in a chronic pain population. Pain 2005;116(3):311–21.

[93] Heath C, Heath D. Switch: how to change things when change is hard (2010). In: Made to stick: why some ideas take hold and others come unstuck. 1st ed. ARROW LTD - MASS MARKET; 2008.

[94] Cecchi F, et al. Predictors of response to exercise therapy for chronic low back pain: result of a prospective study with one year follow-up. Eur J Phys Rehabil Med 2014;50(2):143–51.

[95] Faulkner J, Eston RG. Perceived exertion research in the 21st century: developments, reflections and questions for the future. J Exerc Sci Fit 2008; 6(1):1–14.

[96] Williams N. The Borg rating of perceived exertion (RPE) scale. Occup Med 2017;67(5):404–5.

[97] Yang Z, Petrini MA. Self-selected and prescribed intensity exercise to improve physical activity among inactive retirees. West J Nurs Res 2018; 40(9):1301–18.

Chapter 15

Exercise in the management of neuropathic pain

Breanna Wright, Amy Lawton, and Douglas Wong
College of Health & Biomedicine, Victoria University, Melbourne, VIC, Australia

Introduction

Neuropathic pain is defined by the International Association for the Study of Pain (IASP) as "pain caused by a disease or lesion to the somatosensory system" [1,2]. The causes of neuropathic pain can be divided by pathological locus into central and peripheral. Central neuropathic pain is "caused by a lesion or disease of the central somatosensory system" [2], including HIV myelopathy, post-ischemic stroke, and neurodegenerative brain disorders such as Parkinson's disease or multiple sclerosis. Peripheral neuropathic pain is more common and is "caused by a lesion of the peripheral somatosensory system" [2]. This description encompasses causes such as trigeminal and post-herpetic neuralgia, painful diabetic neuropathy, painful radiculopathies, peripheral nerve entrapment, and chemotherapy-induced polyneuropathies [3,4].

Gaining an accurate measure of the prevalence and incidence of neuropathic pain in the general population is difficult due to the lack of specific diagnostic criteria [5]. However, neuropathic pain is estimated to affect 7%–10% of the global population, with epidemiological trends suggesting that this figure is steadily increasing [4–6]. The number of chronic pain patients suffering from neuropathic pain is thought to be as high as 20%–25% [7]. Females (8% vs 5.7% in men) and those over the age of 50 (8.9% vs 5.6% in those <49 years of age) are more susceptible to chronic neuropathic pain [8].

Pathophysiology

The mechanisms underlying neuropathic pain are complex, with research in this field continuing to evolve. It is beyond the scope of this chapter to discuss the pathophysiology of individual conditions or disease states that have the potential to induce neuropathic pain. Instead, an overview of the neurophysiological changes that arise secondary to nerve injury will be provided. A substantial proportion of current knowledge is drawn from animal models, and studies of peripheral, rather than central, as nerve lesions are overrepresented in existing research [9].

Neuropathic pain, along with its associated symptoms, is likely to arise from a combination of disordered neural mechanisms. Neuronal hyperexcitability, and deficiencies in pain inhibition, are among the physiological changes with the most empirical grounding. These changes occur not only in affected primary afferent fibers but also within the spinal cord and higher centers. Injured peripheral nerves appear capable of generating ectopic discharges from intact fibers—acting in effect as "irritable nociceptors" that play a key role in ongoing pain [9]. Such spontaneous discharge is attributed to ionic changes—including the increased function of sodium channels, alongside a loss of potassium channels [9–11]. At the spinal cord, transmission from injured primary afferents to second-order neurons is also enhanced, due to the upregulation of calcium channels and increased neurotransmitter release.

Hyperexcitability of the central nervous system (CNS) is mediated by the activation of N-methyl-D-aspartate (NMDA) receptors and further compounded by the inhibition of pain modulation. Decreased effectiveness of pain inhibition facilitates an increased response to noxious stimuli. Several neural pathways have been identified as part of this endogenous analgesic mechanism: each of these is unique with respect to the neurotransmitters predominantly used for neuronal communication, and each pathway contributes to a unique pain response for the individual. GABAergic, noradrenergic and serotonergic pathways have received the most attention in the study of dysfunctional pain inhibition [9].

Neuropathic pain may be associated with a diverse range of sensorimotor phenomena, as well as negative impacts upon psychosocial functioning. Ascending spinal projections are vast and include destinations responsible for the affective and

Exercise to Prevent and Manage Chronic Disease Across the Lifespan. https://doi.org/10.1016/B978-0-323-89843-0.00030-1

cognitive dimensions of a patient's pain experience. Increased neural input to the limbic system, as part of a hyperexcitable central nervous system, heightens the risk for sleep disturbances, mood disorders, and maladaptive behaviors [12].

Diagnosis and classification

As neuropathic pain is caused by a lesion or disease of the somatosensory system, it is imperative to differentiate it from other chronic pain syndromes to ensure correct treatment protocols and to improve patient outcomes [13]. Historically, neuropathic pain has been investigated and treated according to the underlying causative pathology. More recently, a mechanism-based approach to the diagnosis of neuropathic pain dependent on the individual sensory phenotypes is suggested to afford patients with a more individualized plan for treatment and management [9,14–16].

A grading system can be useful to clinicians for determining the level of confidence of which the pain presentation is neuropathic. Finnerup et al. [17] have discussed an expansion of the IASP grading system for neuropathic pain and outline the benefits of grading the "possible," "probable," and "definite" presence of neuropathic pain as useful for clinicians (Table 1).

The "possible" presence of neuropathic pain is usually uncovered throughout the history-taking process. The presence of an appropriate lesion or disease to the nervous system and a pain distribution that is neuroanatomically possible should lead the clinician to conduct further investigation into the presence of neuropathic pain [9,17,18]. For a skilled clinician, verbal sensory descriptors for neuropathic pain are readily attainable and clinically reliable–especially when reported in combination. These can include burning, crawling, prickling, itching, tingling, and shooting [14,19–21]. Sleep disturbances, anxiety, pain catastrophizing, and depression are also commonly experienced and are often more severe than in patients without neuropathic pain [9,22]. Screening tools such as the LANSS, painDETECT, Douleur Neuropathique en 4 Questions (DN4), and the Neuropathic Pain Questionnaire (NPQ) are recommended by The Assessment Committee of the Neuropathic Pain Special Interest Group (NeuPSIG) of the IASP to assist in this process [23,24]. Due to their high sensitivity and specificity, these screening tools provide good clinical utility for clinicians and utilize a range of questions related to patient symptoms and clinical examination findings [24,25]. The applicability and validity of these screening tools can be found in Table 2.

The "probable" presence of neuropathic pain utilizes the clinical examination or quantitative sensory testing (QST) to demonstrate the presence of sensory symptoms that are associated with the disease or injury to the somatosensory nervous system [17,23]. The clinical examination should include an evaluation of strength, reflexes, tone, pain, sensation, and vibration [28]. Common sensory symptoms in neuropathic pain presentations that should be evaluated include mechanical and thermal hypoesthesia, allodynia, mechanical and thermal hyperalgesia, and decreased vibration sensation [15,18,29]. These sensory changes can be found in any combination, are usually broad and widespread, can be spontaneous or evoked, and often show high levels of spatial variability between individual patients presenting with the same underlying etiology [29].

The "definite" presence of neuropathic pain relies on the confirmation of the lesion via objective diagnostic testing. As there is no gold standard investigation for neuropathic pain, this is likely to be specific to the causative lesion or disease

TABLE 1 NeuPSIG[a] grading system for determining the presence of neuropathic pain.

Level of certainty	Mode of assessment	Clinical criteria
Possible	Evaluation of patient history Use of validated screening measures	Patient report suggesting a relevant neurological lesion or disease AND a neuroanatomically plausible pain distribution
Probable	Clinical examination	Presence of negative sensory signs—partial or complete loss of one or more sensory modalities
Definite	Objective diagnostic testing	Confirmation of lesion or disease in the somatosensory nervous system

[a]IASP Special Interest Group on Neuropathic Pain.
Adapted from Finnerup NB, et al. Neuropathic pain: an updated grading system for research and clinical practice. Pain 2016;157(8):1599–606.

TABLE 2 Summary of applicability and validity of neuropathic pain screening measures.

Screening tool	Assessment	Sensitivity/ Specificity	Validated for
Leeds Assessment of Neuropathic Symptoms and Signs (LANSS)	• Five symptom items • Two clinical assessment items	• Sensitivity 82%–91% • Specificity 80%–94% [26]	Identifying dominant neuropathic pain in patients with chronic pain
painDETECT	• Seven weighted sensory items • Two items related to spatial and temporal patterns	• Sensitivity: 85% • Specificity: 80% [27]	Originally created for patients with neuropathic pain as a result of radicular lower back pain
Douleur Neuropathique 4 Questions (DN4)	• Seven symptoms items • Three clinical assessment items	• Sensitivity: 83% • Specificity: 90% [20]	Highest sensitivity for generalized polyneuropathies and central neuropathic pain Lowest for neuropathic pain as a result of trigeminal neuralgia
Neuropathic Pain Questionnaire (NPQ)	• 10 questions related to sensory responses • Two questions related to affect	Sensitivity: 66% Specificity: 74% [27]	Identifying the presence of neuropathic pain in a quick, self-administered, all YES/NO questions

Adapted from Jones RCW, Backonja MM. Review of neuropathic pain screening and assessment tools. Curr Pain Headache Rep 2013;17(9); Morgan KJ, Anghelescu DL. A review of adult and pediatric neuropathic pain assessment tools. Clin J Pain 2017;33(9):844–52; and Bouhassira D. Neuropathic pain: definition, assessment and epidemiology. Rev Neurol (Paris) 2019;175(1–2):16–25.

and may include high-resolution imaging (CT, MRI), skin biopsy, and nerve conduction studies. Establishing the definite presence of neuropathic pain can be costly and invasive and is not always necessary to achieve an optimal clinical outcome [9,17,18].

Neuropathic pain is notoriously challenging for clinicians to diagnose and manage, with many patients experiencing only partial or limited pain relief with current pharmacological protocols or interventions [14,30]. While patients with neuropathic pain will demonstrate a level of sensory change and/or sensory symptoms, these can vary greatly depending on the cause of the neuropathy and can vary between patients with the same underlying pathology [4,20,31]. Studies suggest that first-line pharmacological treatments are beneficial in less than 50% of patients [31]. The economical and personal burden of neuropathic pain is significant. As treatment protocols rarely result in meaningful change to clinical outcomes, patients are forced to endure numerous medical appointments, complex poly-pharmaceutical regimes, and other interventional procedures to maintain employment productivity, social connectivity, and quality of life [28,32,33]. Chronic neuropathic pain also increases the prevalence of sleep disturbances and symptoms of depression and anxiety [33].

Exercise as an intervention

Exercise is accepted as a primary adjunct in the management of many chronic conditions. It is a relatively safe, accessible, and cost-effective intervention. There is emerging evidence that exercise yields benefit for sufferers of neuropathic pain, through its ability to produce multiple desired clinical outcomes including improvements in pain perception, pain-related outcomes, quality of life, and mental health.

Clinical practice guidelines for the management of neuropathic pain mostly focus on pharmacological treatments and more recently, interventional procedures such as neurostimulation. As the causes of neuropathic pain are often difficult to treat, guidelines largely focus on symptom modification and pain control as the main clinical outcome [34]. Exercise has the ability to affect multiple systems in one intervention and therefore its effects are likely driven by an array of factors. This means the potential benefits of exercise in neuropathic pain lie not only in improvements in pain but also in the ability to address underlying causative factors of neuropathic pain [35].

Clinical outcomes

Clinical outcomes in neuropathic pain should take into consideration a broad range of key factors including pain, senso-rimotor deficits, and functional capacity, due to their impact on a patient's quality of life. In the context of neuropathic pain, demonstrated clinical outcomes of interventional exercise include exercise-induced hypoalgesia, decreased tactile hyper-sensitivity, decreased allodynia and hyperalgesia (including mechanical), improvements in associated sensory disturbance, increased cutaneous sensation, and improvements in balance and mobility (see Table 3) [35–37].

Patients with neuropathic pain have an increased risk of depression, anxiety, sleep disturbance, and loss of social connection which can significantly decrease their health-related quality of life [38]. For those with significant psychosocial contributors to their pain, exercise can help to address these drivers including symptoms associated with anxiety and/or depression [37]. Of particular note, is sleep disturbance, which is a recognized correlate of increased pain sensitivity. Sleep has an intimate relationship with exercise. Higher levels of exercise are known to enhance sleep duration and reciprocally, longer periods of sleep promote increased engagement in exercise [39,40]. In addition to the potential benefits for quality of life, exercise may also help counter possible side effects of medications used to treat neuropathic pain that contributes to sleep disturbance. For patients with a mixed pain presentation (elements of both nociceptive and nociplastic pain) or those with motor deficits, exercise can also help to address these contributing factors [37].

Predictors of persistent neuropathic pain that can be improved with exercise therapy include lower health status, psychological distress, and higher body mass index [41]. Other general health benefits of exercise, including improved health-related quality of life, are represented in Table 4.

TABLE 3 Summary of potential clinical outcomes and benefits of exercise in neuropathic pain management.

Clinical outcomes of exercise for neuropathic pain	Benefits of exercise as an intervention
• Exercise-induced hypoalgesia • Decreased tactile hypersensitivity • Decreased allodynia (including mechanical) • Decreased hyperalgesia (including mechanical) • Improvements in associated sensory disturbance • Increased cutaneous sensation • Improvements in balance and mobility	• High level of safety • Low economic burden for patient and public health system • Potential to address the underlying cause of neuropathic pain • Potential to address co-morbidities • Associated psychosocial benefits including mood, sleep disturbance, and quality of life

Adapted from Cooper MA, Kluding PM, Wright DE. Emerging relationships between exercise, sensory nerves, and neuropathic pain. Front Neurosci 2016;10:372; Dobson JL, McMillan J, Li L. Benefits of exercise intervention in reducing neuropathic pain. Front Cell Neurosci 2014;8; and Chimenti RL, Frey-Law LA, Sluka KA. A mechanism-based approach to physical therapist management of pain. Phys Ther 2018;98(5):302–14.

TABLE 4 Summary of the health benefits of exercise.

Health benefits of exercise—World Health Organization (2020) [42]
• Improved muscular and cardiorespiratory fitness • Improved bone health • Reduced risk of hypertension, cardiovascular disease, and some cancers • Reduced risk of depression and reduced symptoms of anxiety and depression • Reduced risk of fall and fracture • Maintenance of healthy body weight • Reduced risk of all-cause mortality • Reduced risk of type-2 diabetes and improved glycemic control in diabetic patients • Improved cognitive health • Improved sleep

Adapted from World Health Organisation. Guide to physical activity. [cited 2020 Feb 20], 2020; Available from: https://www.who.int/news-room/fact-sheets/detail/physical-activity.

Mechanisms

Exercise has the potential to induce molecular and cellular changes in the central nervous system, dorsal root ganglion, and sensory neurons. Sensory pathway remodeling and inflammatory modulation appear to be the two main mechanisms of therapeutic action. Studies show improvements in mechanical and thermal nociception, allodynia, hyperalgesia, as well as spontaneous pain (Fig. 1). Exercise increases activity in muscle-associated afferent neurons that cross other sensory pathways at the dorsal root ganglion. Such activity plays a key role in pain modulation and promotes neural plasticity. Animal studies demonstrate exercise-induced alteration of mRNA expression and upregulation of molecular mediators (neurotrophins and other growth factors). Rhythmic sensorimotor activity elicited through exercise can strengthen this neuronal circuitry assisting this neural plasticity, along with any associated peripheral or central sensitization [35,43]. Increased molecular signaling in sensory ganglia following exercise can stimulate axonal growth. Regeneration can occur as soon as 3–7 days following the commencement of exercise. This is likely due to Schwann cell proliferation which has a recognized role in phenotypic plasticity [35,44]. Animal studies demonstrate that continuous exercise can induce phenotypic change in peripheral axon terminal receptors. In hyperexcitable neurons, this has the benefit of normalizing pain thresholds and improving motor function post-exercise.

Resistance training and aerobic exercise can lead to an improvement in pain modulation. These improvements are predominantly related to metabolic and inflammatory changes that facilitate pain inhibitory pathways. One key mechanism is the nitric oxide/cyclic GMP pathway, which exerts neuro-inhibitory effects via plasma and cerebrospinal fluid. In combination with the endocannabinoid system, these play a prominent role in pain modulation and anti-nociception. This could be especially important in chronic neuropathic pain, where these systems have been shown to be dysfunctional [35,45]. In inflammatory neuropathic pain, exercise can help to reduce signaling between neurons involved in the perception of pain. Multiple molecular mechanisms have been identified, including the exercise-induced release of anti-inflammatory cytokines (such as interleukins from M2 macrophages), a reduction in pro-inflammatory cytokines from M1 macrophages, and a reduction in substance P release (pain signal propagation). These effects persist long after the cessation of exercise and are posited to be the reason for the reduction in allodynia and nociception. In animal studies, these effects have been demonstrated through swimming and treadmill running [35,36,46].

Exercise can help delay the onset of diabetic neuropathy and can play a role in the prevention of neuropathic pain following spinal cord injury. This is likely due to the preservation and promotion of sensorimotor function in peripheral nerves [35,43]. In cases of diabetes mellitus, exercise has been implicated in further benefits, including the reversal of declines in neurotrophin, the improvement of glycemic control, insulin resistance, motor nerve function, and neuropathic pain caused by peripheral neuropathy. The ability of exercise to reduce mechanical, cold, and heat hyperalgesia has been demonstrated in rodent studies. Furthermore, voluntary aerobic exercise leads to greater improvement in nociceptive symptoms [47,48]. In contrast, chemotherapy patients may experience neurotoxic side-effects such as changes in nerve fiber density and tubulin levels, causing neuropathic pain [49]. For those with established neuropathic pain, improvements in sensory function, pain, and quality of life can be seen following exercise therapy [49–51]. Current research in this field is promising, but further exploration of clinical significance is required.

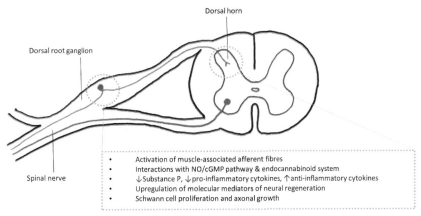

FIG. 1 Proposed impacts of exercise on neuropathic pain at the spinal cord level.

Safety

Exercise is generally well tolerated for patients with neuropathic pain. Few adverse effects have been reported in clinical trials. The majority of adverse effects reported are transient and able to be avoided through patient education, with appropriate supervision and grading [39].

Common symptoms of neuropathic pain are hyperalgesia and allodynia. Considering this, it is important to be mindful of a patient's capacity and pain thresholds in order to avoid exacerbation of these symptoms. A "start low, go slow" approach to exercise lowers the risk of transient adverse effects and patient non-compliance [39]. Low-moderate-intensity exercise has been shown to have neuroprotective effects. In contrast, high-intensity exercise may exhibit neurotoxic effects. In some patients, correct application of a Transcutaneous Electrical Nerve Stimulation (TENS) machine may assist in avoiding post-exercise hyperalgesia [52]. TENS produces analgesia through its activation of central pain modulation. It can be used both for general pain relief and to support exercise [52,53]. It is a relatively safe, non-invasive, and inexpensive adjunct that promotes patient self-efficacy [34].

In certain patient populations, exercise-related fall prevention is a top priority. A loss of sensorimotor function can accompany neuropathic pain. The risk of falls or other injuries as a result of inappropriate joint or tissue loading should always be considered when prescribing exercise. Some pharmacological agents used in the management of neuropathic pain [28] may also contribute to loss of coordination, decreased balance, and increased risk of falls. These factors combined with age-related functional decline should be a safety consideration. It is important to reassure patients that exercise can help to improve these functions and should not prevent participation [54]. Patients with specific co-morbidities may require further safety considerations. For example, for those who are undergoing other treatment such as chemotherapy, the presence of stents or cannulas should be considered in exercise prescription [50].

Clinical practice recommendations for neuropathic pain

Best practice recommendations for exercise therapy in neuropathic pain are limited, particularly when it comes to dosage and timing of programs. Despite this, exercise is included in clinical practice guidelines for neuropathic pain management. There is consensus that neuropathic pain can be managed based on sensory phenotype, rather than by the underlying etiology. Much of the research concerning exercise for neuropathic pain is grounded in animal models of diabetic neuropathy. However, these findings can be extrapolated across other painful neuropathic states [35]. In the case of radicular neuropathic pain and peripheral nerve entrapment more specific exercise guidelines can be utilized.

Goals of management should be established as a starting point [53]. Setting meaningful SMART goals (Specific, Measurable, Achievable, Relevant, and Timed) in collaboration with patients helps to improve compliance and functional outcomes [55,56]. A multidisciplinary approach including pharmacology, physical therapy, exercise, occupational therapy, and psychological intervention should be implemented. Benefits of team-based, patient-centered care include a significant reduction in pain, improvements in function, mood, catastrophizing, pain acceptance, and quality of life [26,28,34].

Currently, there is no consensus on the optimal dosage of exercise for neuropathic pain patients. In a range of other chronic health conditions, pain relief and functional improvement can be obtained with 2–3 exercises sessions of 20–30 min duration per week. For those living with chronic conditions, the World Health Organization (WHO) recommends a weekly minimum of 150–300 min of moderate-intensity aerobic physical activity [42]. This should incorporate at least 2 days of muscle-strengthening activities. It is likely that the frequency and duration of exercise needed to induce analgesic effect is lower than what is required to achieve other health benefits (Table 5). A limited amount of time should be spent engaged in sedentary activity—any amount of physical activity is better than none, with increasing cumulative effects. Any additional physical activity, even of light intensity will provide additional health benefits (Table 5).

Low- to moderate-intensity aerobic exercise should be prescribed as early as possible for maximum neuroprotective effects. Swimming, walking, and running have been researched in rodent studies with positive results. Human studies involving walking, swimming, and water exercise, strength training, balance, and tai chi all show promising results for pain management [39]. As there is no single recommended form of aerobic exercise, practicing patient-centered care and selecting a form of exercise the patient enjoys will help to improve long-term adherence.

In many cases, 2–3 sessions of exercise per week, combining progressive balance, coordination, endurance, and resistance training should yield improvements in pain and function. These programs typically range from 8 to 12 weeks, however, the few longitudinal studies available provide evidence that programs as long as 18 months can have equally positive effects [53]. For metabolically associated neuropathic pain, endurance training should also be included in a graded fashion [57]. In patients with focal neuropathic pain and sensorimotor deficit, region-specific functional programs should also be considered. For example, a hand program including scrubbing and carrying can lead to improvements in grip strength and range of motion [36].

TABLE 5 General guidelines for exercise by age and for chronic conditions.

General population by age		Chronic conditions
1–2 years	180 min, any intensity, spread throughout the day	At least 150–300 min of moderate-intensity aerobic physical activity
3–4 years	180 min spread throughout the day, with 60 min of moderate–vigorous intensity	OR r at least 75–150 min of vigorous-intensity aerobic physical activity
5–17 years	Average across a week 60 min/day moderate–vigorous intensity, mostly aerobic three sessions/week of vigorous-intensity and muscle strengthening	OR an equivalent combination of moderate- and vigorous-intensity activity throughout the week Muscle-strengthening activities at moderate or greater intensity that involve all major muscle groups on two or more days a week
18–64 years	At least 150–300 min moderate-intensity aerobic activity/week OR at least 75–100 min vigorous-intensity aerobic activity/week OR equivalent combination of moderate and vigorous-intensity activity/week Strengthening activities, moderate or higher intensity for all major muscle grounds 2+days/week Additional activity above these recommendations provides additional health benefits	Older adults should include strength and functional balance training three or more days a week to reduce risk of falls
Over 65 years	As for 18–64 years including strength and functional balance training 3 or more days a week to reduce risk of falls	

Adapted from World Health Organisation. Guide to physical activity. [cited 2020 Feb 20], 2020; Available from: https://www.who.int/news-room/fact-sheets/detail/physical-activity.

Patient education encompassing relevant pain mechanisms (nociceptive/nociplastic/neuropathic/mixed) is imperative to any exercise program. Patients should be able to distinguish between post-exercise nociceptive pain, e.g., delayed onset muscle soreness (DOMS), and a flare up of their neuropathic symptoms. It is useful to use an individual's pain response to exercise, be it positive or negative in nature, as a learning tool to manage their pain. Patients should be made aware of potential load modifications that can be made dependent on their daily pain levels. Adjustments should be made as required and ergonomic adaptations to movement should be encouraged [58]. Barriers to exercise must be addressed early in the development of a management plan. Maladaptive behaviors and beliefs such as kinesiophobia, fear-avoidance, and low self-efficacy are associated with poorer outcomes [39,59]. Principles of graded exposure can be used to decrease fear of painful movements, as well as build self-efficacy. This should be an early priority of rehabilitation. Psychological intervention such as cognitive behavioral therapy should be considered as an adjunct to exercise and medical management, to address any mental health co-morbidities and sociocultural barriers to adherence. A summary of a multidisciplinary approach to the management of neuropathic pain is outlined in Fig. 2.

Clinical exercise recommendations for radicular pain

Cervical and lumbar radicular pain can occur in isolation or may be accompanied by sensorimotor deficits as part of radiculopathy. There is evidence to support the use of exercise in both contexts with reported improvements in both pain and physical function. Further research is necessary to develop specific management protocols [60]. Screening for psychosocial risk factors including fear avoidance behaviors, especially around physical activity, is required for optimal patient management [61,62] (Fig. 3).

It is uncertain whether a structured rehabilitation program is superior to a general strengthening program or advice to stay active. There is some evidence, however, to support the use of exercises that address neural mechanosensitivity when combined with activities that promote spinal control and stability. Decisions to incorporate these exercises should be tailored to the individual presentation [50,62,63]. Patients experiencing radicular pain should be advised to remain active, particularly in the acute period, and for up to 8 weeks since symptom onset [60,61]. There are currently no clear recommendations on a single type of activity. A variety of activities should be encouraged and may include walking, swimming, endurance training, yoga, or pilates [58,64]. Aerobic conditioning should be maintained or improved, depending on the fitness of the patient, for best outcomes [57,64].

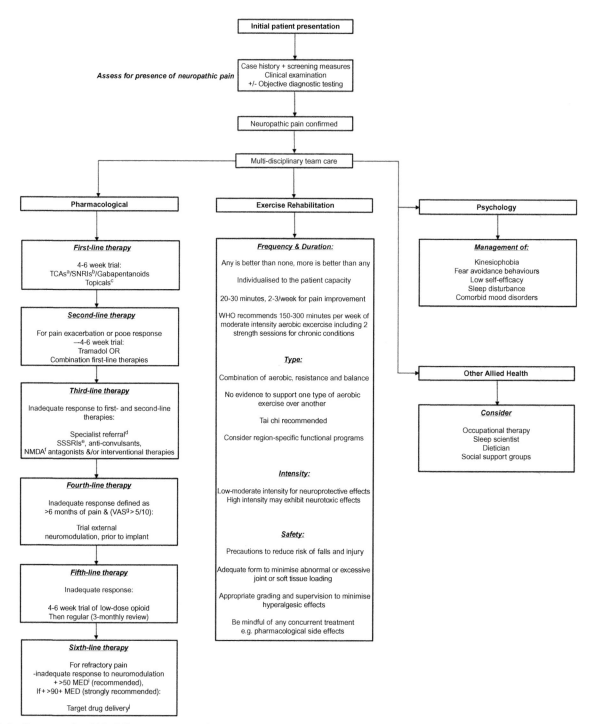

FIG. 2 Summary of multidisciplinary management for neuropathic pain.
[a]Tricyclic anti-depressants; [b]Selective noradrenaline reuptake inhibitors; [c]Topical preparations recommended for focal neuropathic pain only; [d]If not already under care of pain specialist; [e]Selective-serotonin reuptake inhibitors; [f]Glutamate receptor; [g]Visual analog scale; [h]Need >50% pain relief prior to implantation of neuromodulation device; [i]Morphine equivalent dose; intrathecal medication trial required with >50% pain relief achieved, prior to initiation of TDD. *Adapted from Bates D, et al. A comprehensive algorithm for management of neuropathic pain. Pain Med 2019;20(Suppl 1):S2–12 and Bernetti A, et al. Neuropathic pain and rehabilitation: a systematic review of international guidelines. Diagnostics (Basel) 2021;11(1).*

It is essential to address psychosocial predictors of persistent radicular neuropathic pain as a component to exercise therapy. Specific factors may include pain-related fear of movement, passive pain coping, and negative outcome expectancies [41]. By providing pain education and ongoing encouragement during an active rehabilitation program a decrease in pain and increase in patient compliance should be achieved.

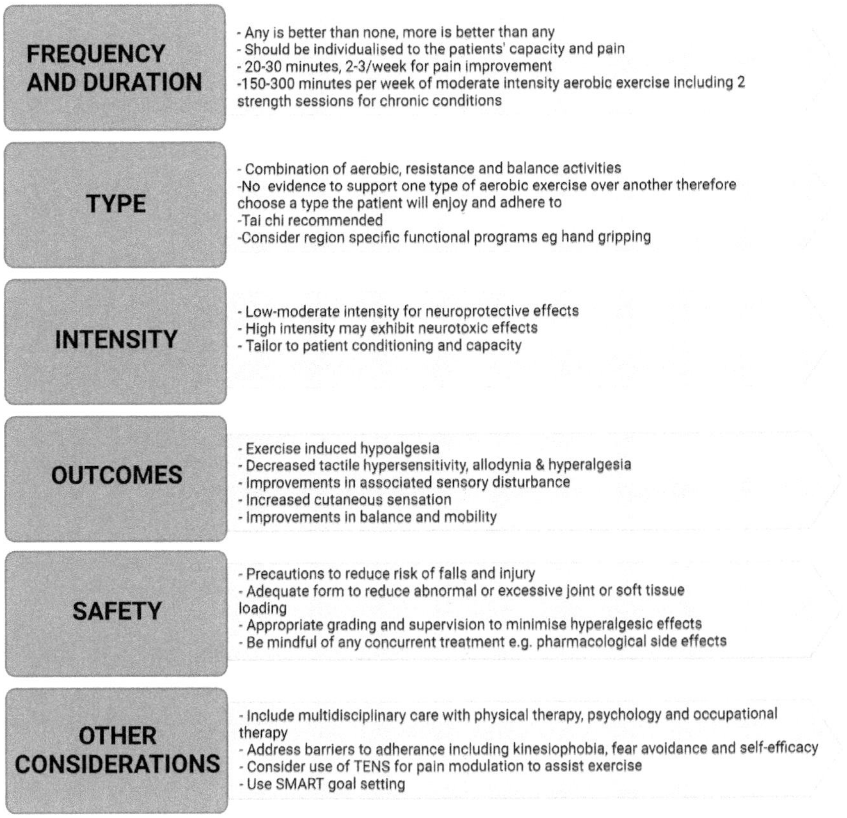

FREQUENCY AND DURATION
- Any is better than none, more is better than any
- Should be individualised to the patients' capacity and pain
- 20-30 minutes, 2-3/week for pain improvement
- 150-300 minutes per week of moderate intensity aerobic exercise including 2 strength sessions for chronic conditions

TYPE
- Combination of aerobic, resistance and balance activities
- No evidence to support one type of aerobic exercise over another therefore choose a type the patient will enjoy and adhere to
- Tai chi recommended
- Consider region specific functional programs eg hand gripping

INTENSITY
- Low-moderate intensity for neuroprotective effects
- High intensity may exhibit neurotoxic effects
- Tailor to patient conditioning and capacity

OUTCOMES
- Exercise induced hypoalgesia
- Decreased tactile hypersensitivity, allodynia & hyperalgesia
- Improvements in associated sensory disturbance
- Increased cutaneous sensation
- Improvements in balance and mobility

SAFETY
- Precautions to reduce risk of falls and injury
- Adequate form to reduce abnormal or excessive joint or soft tissue loading
- Appropriate grading and supervision to minimise hyperalgesic effects
- Be mindful of any concurrent treatment e.g. pharmacological side effects

OTHER CONSIDERATIONS
- Include multidisciplinary care with physical therapy, psychology and occupational therapy
- Address barriers to adherance including kinesiophobia, fear avoidance and self-efficacy
- Consider use of TENS for pain modulation to assist exercise
- Use SMART goal setting

FIG. 3 Summary of exercise recommendations for radicular pain. *Adapted from Berry JA, et al. A review of lumbar radiculopathy, diagnosis, and treatment. Cureus 2019;11 (10):e5934 and Blanpied PR, et al. Neck pain: revision 2017. J Orthop Sports Phys Ther 2017;47(7):A1–83.*

In some patients, a structured, region-specific rehabilitation program will be indicated. Exercises must be tailored to the individual patient presentation and should be aimed at improving mobility, motor control, and strength [58,62]. For cervical radicular pain, the cervicothoracic, shoulder girdle and upper limb regions should be addressed. For lumbar radicular pain, the lumbopelvic and lower limb regions should be addressed. Functional exercises should be included for both regions. Examples of programs can be found in Tables 6 and 7. Other protocols such as McKenzie exercises can also be utilized,

TABLE 6 Example of region-specific rehabilitation program.

Component	Exercises
Warm-up	• Awareness of the back • Pelvic tilt • Lumbar rotation • Arm movements • Whole-body movement in standing • Flexibility
Exercises	• Back rotations • Back extensor strength • Abdominal strength • Lateral gluteal strength • Posterior gluteal strength • Leg muscle strength
Dosage	• 4 progressive levels of exercises • 2–3 sets of 8–12 repetitions • Once 3×10 repetitions are reached move onto the next level

Adapted from Kjaer P, et al. GLA:D((R)) Back group-based patient education integrated with exercises to support self-management of back pain—development, theories and scientific evidence. BMC Musculoskelet Disord 2018;19(1):418.

TABLE 7 Example of region-specific rehabilitation program.

Neural tissue mobilization	• Progress from mobilizing the nerve and nerve bed to techniques which tension the nerve and nerve bed • Manual therapy and active exercise are employed to facilitate the management of sensitized nerve tissue
Range of motion exercises	• Can be performed in any plane of movement • Try to target the hypomobile segments • 3–4 times/day in the acute stage then repetitions and intensity increased as tolerance allows

Training motor function	• Exercises should be prescribed according to neuromuscular impairments revealed during the clinical assessment
	• The goal of therapeutic exercise is to restore normal motor control, function, and quality of movement, the exercises should be challenging, yet be performed with correct technique and without aggravation of symptoms
	• Exercises are prescribed in two stages:
	  • Stage 1 is training to enhance motor control
	  • Stage 2 involves resistance training

Training sensorimotor control	• Proprioception and balance exercises

Adapted from Brukner P, et al. Brukner & Khan's clinical sports medicine. vol. 1. 5th ed. McGraw Hill Australia; 2017.

especially for patients with a directional preference of movement [61]. Neurodynamic or neural mobilization techniques can be used as a component of the management plan in radicular neuropathic pain where neural mechanosensitivity exists [65,66]. For lumbar radicular pain, these can be performed in the slump or straight leg raise positions and for cervical radiculopathy, the lateral glide technique is recommended [66].

Conclusion

Exercise is an essential part of neuropathic pain management. It has the potential to improve pain, function, and quality of life for individual sufferers of neuropathic pain. Exercise represents a non-invasive, accessible, and well-tolerated component of a multidisciplinary management plan. The therapeutic benefits of exercise for neuropathic pain are only just emerging. Future research should confirm mechanisms of action in human studies, as well as produce clinical practice guidelines to provide the best recommendations on the type of exercise and dosage.

References

[1] Jensen TS, et al. A new definition of neuropathic pain. Pain 2011;152(10):2204–5.

[2] International Association for the Study of Pain. IASP terminology; 2017.

[3] Centre for Clinical Practice at NICE (UK). Neuropathic pain: the pharmacological management of neuropathic pain in adults in non-specialist settings [Internet]. London: National Institute for Health and Care Excellence (UK); 2013. PMID: 25577930.

[4] Scholz J, et al. The IASP classification of chronic pain for ICD-11: chronic neuropathic pain. Pain 2019;160(1):53–9.

[5] van Hecke O, et al. Neuropathic pain in the general population: a systematic review of epidemiological studies. Pain 2014;155(4):654–62.

[6] Cruccu G, Truini A. A review of neuropathic pain: from guidelines to clinical practice. Pain Ther 2017;6(Suppl 1):35–42.

[7] Bouhassira D, Attal N. Translational neuropathic pain research: a clinical perspective. Neuroscience 2016;338:27–35.

[8] Bouhassira D, et al. Prevalence of chronic pain with neuropathic characteristics in the general population. Pain 2008;136(3):380–7.

[9] Colloca L, et al. Neuropathic pain. Nat Rev Dis Primers 2017;3:17002.

[10] Garcia-Larrea L, Magnin M. Pathophysiology of neuropathic pain: review of experimental models and proposed mechanisms. Presse Med 2008;37(2 Pt 2):315–40.

[11] Gomez K, et al. The role of cyclin-dependent kinase 5 in neuropathic pain. Pain 2020;161(12):2674–89.

[12] Baron R, Hans G, Dickenson AH. Peripheral input and its importance for central sensitization. Ann Neurol 2013;74(5):630–6.

[13] Langley PC, et al. The burden associated with neuropathic pain in Western Europe. J Med Econ 2013;16(1):85–95.

[14] Freeman R, et al. Sensory profiles of patients with neuropathic pain based on the neuropathic pain symptoms and signs. Pain 2014;155(2):367–76.

[15] Maier C, et al. Quantitative sensory testing in the German Research Network on Neuropathic Pain (DFNS): somatosensory abnormalities in 1236 patients with different neuropathic pain syndromes. Pain 2010;150(3):439–50.

[16] Momi SK, et al. Neuropathic pain as part of chronic widespread pain: environmental and genetic influences. Pain 2015;156(10):2100–6.

[17] Finnerup NB, et al. Neuropathic pain: an updated grading system for research and clinical practice. Pain 2016;157(8):1599–606.

[18] Mistry J, et al. Clinical indicators to identify neuropathic pain in low back related leg pain: a modified Delphi study. BMC Musculoskelet Disord 2020;21(1):601.

[19] Cavalli E, et al. The neuropathic pain: an overview of the current treatment and future therapeutic approaches. Int J Immunopathol Pharmacol 2019;33. 205873841983838.

[20] Bouhassira D. Neuropathic pain: definition, assessment and epidemiology. Rev Neurol (Paris) 2019;175(1–2):16–25.

[21] Bouhassira D, Attal N. Diagnosis and assessment of neuropathic pain: the saga of clinical tools. Pain 2011;152(3 Suppl):S74–83.

[22] Forstenpointner J, et al. No pain, still gain (of function): the relation between sensory profiles and the presence or absence of self-reported pain in a large multicenter cohort of patients with neuropathy. Pain 2021;162(3):718–27.

[23] Gierthmuhlen J, et al. Can self-reported pain characteristics and bedside test be used for the assessment of pain mechanisms? An analysis of results of neuropathic pain questionnaires and quantitative sensory testing. Pain 2019;160(9):2093–104.

[24] Bennett MI, et al. Using screening tools to identify neuropathic pain. Pain 2007;127(3):199–203.

[25] Haanpaa M, et al. NeuPSIG guidelines on neuropathic pain assessment. Pain 2011;152(1):14–27.

[26] Jones RCW, Backonja MM. Review of neuropathic pain screening and assessment tools. Curr Pain Headache Rep 2013;17(9).

[27] Morgan KJ, Anghelescu DL. A review of adult and pediatric neuropathic pain assessment tools. Clin J Pain 2017;33(9):844–52.

[28] Bates D, et al. A comprehensive algorithm for management of neuropathic pain. Pain Med 2019;20(Suppl 1):S2–S12.

[29] Baron R, Förster M, Binder A. Subgrouping of patients with neuropathic pain according to pain-related sensory abnormalities: a first step to a stratified treatment approach. Lancet Neurol 2012;11(11):999–1005.

[30] Finnerup NB, et al. Pharmacotherapy for neuropathic pain in adults: a systematic review and meta-analysis. Lancet Neurol 2015;14(2):162–73.

[31] Baron R, et al. Peripheral neuropathic pain: a mechanism-related organizing principle based on sensory profiles. Pain 2017;158(2):261–72.

[32] Sadosky A, et al. Pain severity and the economic burden of neuropathic pain in the United States: BEAT neuropathic pain observational study. ClinicoEconomics Outcomes Res 2014;483.

[33] Attal N, et al. The specific disease burden of neuropathic pain: results of a French nationwide survey. Pain 2011;152(12):2836–43.

[34] Bernetti A, et al. Neuropathic pain and rehabilitation: a systematic review of international guidelines. Diagnostics (Basel) 2021;11(1).

[35] Cooper MA, Kluding PM, Wright DE. Emerging relationships between exercise, sensory nerves, and neuropathic pain. Front Neurosci 2016;10:372.

[36] Dobson JL, McMillan J, Li L. Benefits of exercise intervention in reducing neuropathic pain. Front Cell Neurosci 2014;8.

[37] Chimenti RL, Frey-Law LA, Sluka KA. A mechanism-based approach to physical therapist management of pain. Phys Ther 2018;98(5):302–14.

[38] Torta R, Ieraci V, Zizzi F. A review of the emotional aspects of neuropathic pain: from comorbidity to co-pathogenesis. Pain Ther 2017;6(Suppl 1):11–7.

[39] Ambrose KR, Golightly YM. Physical exercise as non-pharmacological treatment of chronic pain: why and when. Best Pract Res Clin Rheumatol 2015;29(1):120–30.

[40] Finan PH, Goodin BR, Smith MT. The association of sleep and pain: an update and a path forward. J Pain 2013;14(12):1539–52.

[41] Boogaard S. Predictors of persistent neuropathic pain—a systematic review. Pain Physician 2015;18(5):433–57.

[42] World Health Organisation. Guide to physical activity, 2020. [cited 2020 Feb 20]; Available from: https://www.who.int/news-room/fact-sheets/detail/physical-activity.

[43] Palandi J, et al. Neuropathic pain after spinal cord injury and physical exercise in animal models: a systematic review and meta-analysis. Neurosci Biobehav Rev 2020;108:781–95.

[44] Jessen KR, Mirsky R, Lloyd AC. Schwann cells: development and role in nerve repair. Cold Spring Harb Perspect Biol 2015;7(7):a020487.

[45] Koltyn KF, et al. Mechanisms of exercise-induced hypoalgesia. J Pain 2014;15(12):1294–304.

[46] Kami K, Tajima F, Senba E. Exercise-induced hypoalgesia: potential mechanisms in animal models of neuropathic pain. Anat Sci Int 2017;92(1):79–90.

[47] Cox ER, et al. Effect of different exercise training intensities on musculoskeletal and neuropathic pain in inactive individuals with type 2 diabetes—preliminary randomised controlled trial. Diabetes Res Clin Pract 2020;164:108168.

[48] Kluding PM, et al. The effect of exercise on neuropathic symptoms, nerve function, and cutaneous innervation in people with diabetic peripheral neuropathy. J Diabetes Complications 2012;26(5):424–9.

[49] Park JS, Kim S, Hoke A. An exercise regimen prevents development paclitaxel induced peripheral neuropathy in a mouse model. J Peripher Nerv Syst 2015;20(1):7–14.

[50] Jesson T, Runge N, Schmid AB. Physiotherapy for people with painful peripheral neuropathies: a narrative review of its efficacy and safety. Pain Rep 2020;5(5), e834.

[51] Duregon F, et al. Effects of exercise on cancer patients suffering chemotherapy-induced peripheral neuropathy undergoing treatment: a systematic review. Crit Rev Oncol Hematol 2018;121:90–100.

[52] Mokhtari T, et al. Transcutaneous electrical nerve stimulation in relieving neuropathic pain: basic mechanisms and clinical applications. Curr Pain Headache Rep 2020;24(4):14.

[53] Akyuz G, Kenis O. Physical therapy modalities and rehabilitation techniques in the management of neuropathic pain. Am J Phys Med Rehabil 2014;93(3):253–9.

[54] Schmader KE, et al. Treatment considerations for elderly and frail patients with neuropathic pain. Mayo Clin Proc 2010;85(3 Suppl):S26–32.

[55] Bovend'Eerdt TJ, Botell RE, Wade DT. Writing SMART rehabilitation goals and achieving goal attainment scaling: a practical guide. Clin Rehabil 2009;23(4):352–61.

[56] Turner-Stokes L, et al. Patient engagement and satisfaction with goal planning: impact on outcome from rehabilitation. Int J Ther Rehabil 2015;22(5):210–6.

[57] Streckmann F, et al. Exercise intervention studies in patients with peripheral neuropathy: a systematic review. Sports Med 2014;44(9):1289–304.

[58] Kjaer P, et al. GLA:D((R)) Back group-based patient education integrated with exercises to support self-management of back pain—development, theories and scientific evidence. BMC Musculoskelet Disord 2018;19(1):418.

[59] Luque-Suarez A, Martinez-Calderon J, Falla D. Role of kinesiophobia on pain, disability and quality of life in people suffering from chronic musculoskeletal pain: a systematic review. Br J Sports Med 2019;53(9):554–9.

[60] Lee JH, et al. Nonsurgical treatments for patients with radicular pain from lumbosacral disc herniation. Spine J 2019;19(9):1478–89.

[61] Berry JA, et al. A review of lumbar radiculopathy, diagnosis, and treatment. Cureus 2019;11(10), e5934.

[62] Blanpied PR, et al. Neck pain: revision 2017. J Orthop Sports Phys Ther 2017;47(7):A1–A83.

[63] Hassan F, et al. Effects of oscillatory mobilization as compared to sustained stretch mobilization in the management of cervical radiculopathy: a randomized controlled trial. J Back Musculoskelet Rehabil 2020;33(1):153–8.

[64] Van Wambeke P, et al. Low back pain and radicular pain: assessment and management. KCE report; 2017. p. 287.

[65] Brukner P, et al. Brukner & Khan's clinical sports medicine. 5th. vol. 1. McGraw Hill Australia; 2017.

[66] Basson A, et al. The effectiveness of neural mobilization for neuromusculoskeletal conditions: a systematic review and meta-analysis. J Orthop Sports Phys Ther 2017;47(9):593–615.

Chapter 16

Tendinopathy

Brett Vaughan[a,b,c], Jack Mest[d], Patrick Vallance[e], Michael Fleischmann[b,f], and Peter Malliaras[e]

[a]Department of Medical Education, Melbourne Medical School, University of Melbourne, Melbourne, VIC, Australia [b]School of Public Health, University of Technology Sydney, Sydney, NSW, Australia [c]Faculty of Health, Southern Cross University, Lismore, NSW, Australia [d]University of Canberra Health Clinics, Faculty of Health, University of Canberra, Canberra, ACT, Australia [e]Department of Physiotherapy, School of Primary and Allied Health Care, Monash University, Melbourne, VIC, Australia [f]Torrens University Australia, Melbourne, VIC, Australia

Introduction

Tendinopathy is a broad term encompassing pathological processes and pain affecting the tendon. Historically, terms such as *tendinitis* have been described in the literature. However, many of the terms used historically fail to consider the variability in tendon presentations and suggest a dominant pathophysiological process (inflammation, in the case of tendinitis), which may not be accurate for all presentations, and should be avoided. Consistent with expert consensus, this chapter will use the term *tendinopathy* throughout [1].

Broadly, tendinopathies develop secondary to repetitive load beyond its capability [2]. This overload process results in changes to the tendon microstructure and its ability to withstand the load. There are numerous theories that describe the pathogenesis of tendinopathies: mechanical (degenerative, microtrauma), altered homeostasis and inflammatory processes, and vascular and neurogenic theories [3]. However, no one theory explains the pathological and clinical presentation of the spectrum of tendinopathies and may account for the varied management approaches reported in the literature. The focus of this chapter is on the most commonly affected tendons in the upper and lower extremities [2], appreciating that tendons do exist in other areas of the body (i.e., eye). An overview of some of the terminology used in the description of exercises for tendinopathy is provided in Box 1.

Musculoskeletal anatomy of the tendon

A tendon is organized of fibrous connective tissue that attaches muscle to bone. The shape, size, and length of a tendon vary, with the most well-known and visible being the Achilles tendon at the posterior ankle and lower leg. The dominant cell type in tendons is tenocytes [4] which are embedded within a collagen and ground substance matrix. Tenocytes are mechanoresponsive, which make the tendon sensitive to changes in load and is a cell type consistent with fibroblasts [5]. Globally, the function of the tendon is to transmit muscle forces to the bone [6].

The blood supply to a tendon is limited (particularly in comparison to the related muscle) [4] and this is reflective of the function of the tendon. The tendon itself is non-contractile and requires little in the way of oxygen for function. This is an important consideration in understanding the principles of exercise rehabilitation for tendinopathies, particularly the limited and slow response of the tendon to loading interventions. The clinical considerations of tendinopathy management are illustrated in Fig. 1.

Epidemiology

The epidemiology of tendinopathies affecting the upper and lower extremities is variable. Any of the tendons in the upper or lower extremity can be affected. Tendinopathies are not restricted to individuals participating in sport or physical activity and can present in patients ranging from adolescents to older adults—albeit some tendinopathies are more prevalent in specific age groups (e.g., patellar tendinopathy in younger age groups). At present, there is limited literature on the true prevalence of tendinopathies.

Where literature does exist, prevalence data suggests that health professionals in primary care settings are likely to encounter patients experiencing tendinopathy [7–11] with Achilles tendinopathy being the most common lower limb tendinopathy. De Jonge et al. [7] evaluated the prevalence of mid-portion Achilles tendinopathy in Dutch general practice

---

Box 1 Definitions of terms used to describe exercise rehabilitation for the management of upper and lower extremity tendinopathies.

Concentric—muscle action associated with shortening of the muscle as the insertion of the muscle moves toward the origin.
  Eccentric—muscle action associated with lengthening of the muscle as the insertion of the muscle moves away from the origin.
  Isometric—muscle action associated with no change in the length of the muscle.

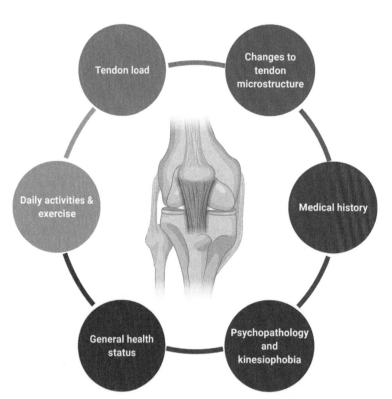

FIG. 1 Clinical considerations of tendinopathy.

presentations. These authors reported an incidence rate of 2.35 per 1000 patients for those aged between 21 and 60 years and an overall incidence of 1.85 per 1000 patients. Further work in Dutch general practice by [12] suggested that the most common lower extremity tendinopathy was that affecting the gluteal tendon (greater trochanteric pain syndrome, 4.22 per 1000 person-years) while the least common was tendinopathies affecting the thigh adductors (1.22 per 1000 person-years).

With respect to the upper extremity, the systematic review by Littlewood et al. [10] suggests the point prevalence of rotator cuff tendinopathies is between 2.0% and 14.0% and an incidence of approximately 2.9%–5.5%. Data for lateral epicondylalgia suggests that the annual incidence is approximately 2.4–4.5 cases per 1000 population [13]. The collective data described here supports the notion that tendinopathies of the extremities are relatively common [2] and health professionals in primary contact settings are likely to encounter them in their practice.

Exercise in disease

Consideration of psychosocial influences

There is an emerging literature that suggests psychosocial influences may play a role in the clinical presentation and management of patients with tendinopathies affecting the upper and lower extremities. The current literature suggests that kinesiophobia, or fear of movement, is a significant consideration with psychopathologies such as anxiety and depression being less prevalent [14–16]. Consideration should be given to screening for the presence of the aforementioned clinical concerns in order to determine the most suitable management for the patient, including consideration of psychological review.

Patient-reported outcomes

Practitioners are encouraged to utilize a range of patient-reported outcome measures (PROMs) when managing those with tendinopathy as a strategy to monitor progress [17]. Although there is no consensus as to the most appropriate PROM(s) for specific tendinopathies, practitioners may consider utilizing one or more those described in Table 1. The measures in Table 1 are not exhaustive but will be appropriate for a range of upper and lower extremity tendinopathies and are readily available.

Rotator cuff tendinopathy

Overview

The etiology of rotator cuff tendinopathies appears to be variable but is likely associated with repeated overhead movements (i.e., sport or occupation-related) or associated with systemic disease [29]. The prognosis of a rotator cuff tendinopathy is generally positive, particularly where overuse is not related to the patient's occupation [10]. However, baseline pain level, baseline disability, and past history of shoulder complaints may be associated with a poorer prognosis [10]. Additionally, psychological factors may influence complaint progression [16].

Like many musculoskeletal pathologies, diagnostic imaging findings associated with a rotator cuff tendinopathy do not appear to have prognostic value [10]. With respect to the management of a rotator cuff tendinopathy, exercise rehabilitation is considered to be one of the most effective interventions [30–33], particularly to reduce pain and improve function, including activities of daily living. Other approaches to management are highlighted in Box 2. These approaches are

TABLE 1 Patient-reported outcome measures that are suitable for use with patients experiencing tendinopathy affecting the upper or lower extremity.

General health	PROMIS Global Health 10 [18]
Mental health	Kessler 10 (K10) [19]
	Patient Health Questionnaire (PHQ9) [20] Tampa Scale of Kinesiophobia [21]
Upper extremity	Patient-Specific Functional Scale [22]
	Upper Extremity Functional Index [23]
Lower extremity	Patient-Specific Functional Scale [22]
	Lower Extremity Functional Scale [24]
	VISA-G (gluteal tendinopathy/greater trochanteric pain syndrome) [25]
	VISA-P (patellar tendinopathy) [26]
	VISA-H (hamstring tendinopathy) [27]
	VISA-A (Achilles tendinopathy) [28]

VISA—Victorian Institute of Sport Assessment.

Box 2 Non-exercise interventions for rotator cuff tendinopathy.

Manual therapy may result in improvements in pain intensity [34].

Limited evidence for the effectiveness of therapeutic ultrasound and low-level laser therapy [31,35].

Limited evidence for the effectiveness of extra-corporeal shockwave therapy [31] (may be of assistance when combined with exercise rehabilitation to address pain).

Limited evidence for the effectiveness of surgical interventions [30,36].

Oral non-steroidal anti-inflammatories may have a short term impact on pain intensity [37].

Corticosteroid injections may have a short-term effect on pain with little to no influence on prognosis or level of function [38].

Kinesiology taping may provide some benefit in improving pain-free range of motion but evidence for the use of non-elastic taping is equivocal [39].

generally considered as adjuncts to exercise or where conservative management of over 12 weeks has failed to yield significant improvements.

Exercise

With respect to exercise prescription, Littlewood et al. [32] suggest that exercises for rotator cuff tendinopathy should incorporate resistance/load, comprise at least three sets and a higher number of repetitions, and be undertaken for 12 weeks. Littlewood et al. [32] also suggest that exercises should incorporate a range of shoulder movements including elevation, external rotation, etc. Short-term changes in pain may also be achievable through open chain, closed chain, or range of motion exercises [40]. The use of load in an exercise rehabilitation program can be achieved with weights, bodyweight, or resistance bands, and one has not been shown to be superior to the other. The use of eccentric exercises over other muscle contraction exercises is equivocal [41].

Lateral elbow epicondylalgia

Overview

Lateral elbow epicondylalgia is a relatively common musculoskeletal presentation [13] that is largely self-limiting [42,43]. The etiology is thought to be a degenerative [44] or microtraumatic one [45] associated with overuse or repeated use, including in-office workers (i.e., computer/mouse use), trades (i.e., builders, carpenters), and those participating in sport (i.e., tennis, golf) [13,46,47]. The dominant arm also appears to be more common as part of this tendinopathy presentation [48] with females more likely to experience the complaint than males [49].

Exercise

Bisset et al. [50] suggest that exercise may have a positive impact on pain intensity associated with lateral epicondylalgia. Consistent with other upper extremity tendinopathies, a range of exercise approaches may be beneficial. Eccentric exercises appear to be the most common intervention for lateral epicondylalgia [43] with very limited evidence for isometric exercises [3]. The incorporation of eccentric exercises into the rehabilitation program for lateral epicondylalgia over 4–8 weeks appears to be of benefit in terms of decreased pain intensity, improved function, and increased grip strength [51–53]. Non-exercise interventions for use with lateral epicondylalgia are described in Box 3.

Gluteal tendinopathy

Overview

Gluteal tendinopathy refers to the pathology affecting the gluteus medius tendon attaching to the greater trochanter. Contemporary terminology encompassing this condition is Greater Trochanteric Pain Syndrome (GTPS). The GTPS descriptor is used as it is generally accepted that the tendon is not affected in isolation, and it is often associated with bursal irritation. Gluteal tendinopathy/GTPS is most commonly reported in those over 40 years of age and there is a distinct female dominance for presentation [57,58]. Patients presenting with GTPS will typically report pain over the region of the lateral hip,

---

Box 3 Non-exercise interventions for lateral epicondylalgia.

Manual therapy applied to the cervical spine and/or elbow may result in improvements in pain intensity [50,54] and facilitate exercise intervention [52].

Limited evidence for the effectiveness of therapeutic ultrasound and low-level laser therapy [50,54,55].

Limited evidence for the effectiveness of extra-corporeal shockwave therapy [50] but may be beneficial for recalcitrant complaints [43].

Limited evidence for the effectiveness of surgical interventions and should only be considered where condition resolution is not achieved within 6 months [13,43,48].

Oral non-steroidal anti-inflammatories may be of limited benefit [37].

Corticosteroid injections may have a short-term effect on pain with little to no influence on prognosis [42].

Platelet-rich plasma (PRP) or autologous blood injections have comparable outcomes to corticosteroid injections [42].

Outcomes of taping and bracing are equivocal [43,50,56].

and difficulties with activities of daily living including standing, walking, lying on the affected side at night, sitting, and stair use [59]. The etiology is high compressive and tensile loads within the tendon, particularly associated with excessive hip adduction angles during walking, step up, and other functional activities [60]. Consideration should also be given to the increased association between the severity of the patients' GTPS complaint and the presence (or otherwise) of psychological factors [15].

Exercise

Exercise for the management of GTPS is used frequently in clinical care [11,61] and appears to be beneficial, particularly when combined with other interventions typically resulting in greater long-term effects [62]. The use of exercise management in GTPS is superior to a "wait and see" approach in the short term, through to 12 months from onset [63]. Patient education with exercise has been shown to significantly improve the patient's overall perception of their complaint compared to corticosteroid injection [64]. Consistent with other tendinopathies, graduated load on the gluteal tendons is advocated [60]. Tendon load should be considered in exercise prescription—the stage of treatment will likely dictate the load that the tendon will be able to tolerate. In the early phases, low load exercises in the supine or prone positions may be of value in the initial phases, then progressing to higher tendon load exercises in weight-bearing and side-lying positions [65,66]. Both isometric and isotonic muscle contractions appear to be effective in GTPS exercise management [67].

Given the "tensile" and "compressive" nature of GTPS, patients should be advised to avoid crossing their legs and long periods standing with their bodyweight favoring one leg as a strategy to reduce symptoms. Further, compression of the gluteal tendon (and associated structures) occurs with the adduction movement in many stretches for the gluteal muscles. As such, these stretches are not recommended nor are other hip, lower extremity, or low back stretches that place tension on the gluteal tendons [60]. Evidence for non-exercise interventions for GTPS is presented in Box 4.

Patella tendinopathy

Overview

Patellar tendinopathy primarily presents at the tendon's insertion to the inferior pole of the patella [70]. The primary etiological factors include an increase to "store-and-release" loading within an exercise session, or inappropriate recovery between sessions [70,71]. The condition is common in people who participate in jumping sports that require repetitive store-and-release activity (e.g., basketball and volleyball), with as many as 45% of elite and 14% of non-elite jumping sports athletes suffering patellar tendinopathy [72,73]. It also impacts working capacity, with 58% of sufferers experiencing difficulty with physically demanding work [74]. Typically, the condition affects a younger population relative to other lower limb tendinopathy conditions [72,74]. Exercise rehabilitation is considered the most effective intervention for patellar tendinopathy [75,76]. Other approaches are highlighted in Box 5. These are mostly considered as adjuncts to exercise or where conservative management of over 12 weeks has failed to yield significant improvements.

Exercise

Progressive tendon loading exercise should be considered as standard initial care for patients with patellar tendinopathy [76]. This approach progresses activity from low to high tendon loading in phases, starting with isometric knee extension exercise before adding isotonic knee extension exercise (concentric-eccentric, e.g., single-leg press, knee extension). Isotonic exercises start with high repetitions and low load (4×15 repetitions) and a reduced active range of motion and are

---

Box 4 Non-exercise interventions for greater trochanteric pain syndrome (GTPS).

Limited evidence for the effectiveness of extracorporeal shockwave therapy in the short term with respect to pain but little benefit in the longer term [68].

Surgical interventions should only be considered for severe or recalcitrant complaints [63].

Corticosteroid injections may have a short-term effect on pain [62]. Further, combined exercise and corticosteroid injection are not superior to exercise alone [63].

Platelet-rich plasma (PRP) or autologous blood injections may be beneficial in low-grade complaints [63] where physical therapy interventions have failed [69].

> **Box 5 Non-exercise interventions for patellar tendinopathy.**
>
> Limited evidence for the effectiveness of manual therapy (e.g., myofascial manipulation) for reducing pain [77].
>
> No evidence for the effectiveness of bracing or taping [71].
>
> Limited evidence for the effectiveness of extracorporeal shockwave therapy for pain and disability [78], and limited evidence that there is no effect when used in-season [79].
>
> Strong evidence for the effectiveness of surgery to treat recalcitrant patellar tendinopathy in people who have already failed conservative treatment, with similar outcomes for open and arthroscopic procedures [80].
>
> Limited evidence suggests no effectiveness for cortisone injection for pain and disability [80].
>
> Very limited evidence for therapeutic ultrasound and ultrasound-guided sclerotherapy offers only limited benefits for pain and disability [80–82].
>
> Very limited evidence platelet-rich plasma injection may offer small benefits for pain and disability [80].

gradually progressed to low repetitions and high load (4×6 repetitions) and full active range of motion. When appropriate to progress further, energy storage plyometric and running exercises can be added, followed by sports-specific training [76]. These progressions can be made based on pain response, which should be within acceptable limits (approximately 3/10 or lower, where 0 = no pain and 10 = worst pain imaginable). Progressive heavy slow resistance training (concentric and eccentric) and eccentric training have also been shown to result in positive outcomes [75], although eccentric exercise can aggravate pain in the early rehabilitation phase if the tendinopathy is severe or irritable [70].

Achilles tendinopathy

Overview

The Achilles tendon is the longest tendon in the body that can withstand high loads [83]. It attaches the gastrocnemius, soleus, and plantaris muscles to the calcaneus. Achilles tendinopathy is an overuse condition that is common across the lifespan and is the most prevalent of all lower limb tendinopathies [84,85]. It is most common in distance runners, however, the condition is also common in the non-athletic population [85]. The viscoelastic properties of the Achilles tendon allow it to be used like a spring, whereby it stores and releases energy, which is required for locomotion [84]. This spring-like activity requires tensile forces to be transmitted through the tendon making it susceptible to injury particularly during running and jumping [86].

Key factors shown to be associated with the development of Achilles tendinopathy are changes in exercise behavior such as increasing propulsive activity levels in a short space of time [84,86]. This can leave the Achilles tendon insufficient time to adapt to a changing loading environment and tendinopathy can ensue. Furthermore, intrinsic factors such as hypercholesterolemia or hyperlipidemia, fluroquinolone medication, age, and metabolic diseases have been shown to be associated with the development of Achilles tendinopathy [87]. It is thought that these factors limit the Achilles tendon's ability to adapt to the changing loading environment.

Achilles tendinopathy occurs in two locations along the tendon: (a) the midportion or (b) at the insertion. Insertional Achilles tendinopathies behave slightly differently to midportion tendinopathies. The latter can be subject to compressive load during ankle dorsiflexion, as the tendon insertion (enthesis) attaches to the undersurface of the calcaneus [84]. The former location is more subject to tensile forces as it stores and releases energy during locomotion, and is typically the more common of the two anatomical locations for the tendinopathy to develop [88]. It should also be noted that pathology can also occur to the paratenon (outer tendon sheath) due to frictional forces of the tendon within the sheath secondary to repetitive ankle movements [84].

Exercise

Exercise-based interventions should be the first line of care in the management of Achilles tendinopathy [89,90], as opposed to non-exercise-based interventions (Box 6). There is a wealth of evidence investigating eccentric loading programs in Achilles tendinopathy [95]. However, there is now evidence showing that there is no added benefit in isolating the eccentric phase of contraction over other components (isometric or concentric) [95]. Currently, pain-guided, heavy-slow resistance (HSR) programs have been shown to be efficacious in the literature and associated with higher patient satisfaction than eccentric program's alone [96]. Rehabilitation should also address the stretch-shortening cycle function of

> **Box 6 Non-exercise interventions for Achilles tendinopathy.**
>
> Shockwave therapy can be of benefit when added to exercise-based interventions [78].
>
> Oral non-steroidal anti-inflammatories do not provide added benefit to rehabilitation programs [91].
>
> High volume injections with or without cortisone in the treatment of mid-portion Achilles tendinopathy when combined with an eccentric strength training program. High volume injection with corticosteroid showed better short-term improvement than without corticosteroid [92].
>
> Platelet-rich plasma (PRP) injections do not provide clear additional value in management of chronic midsubstance Achilles tendinopathy [93].
>
> Orthotic devices have limited value in addressing pain and function in Achilles Tendinopathy [94].
>
> Manual therapy may provide additional benefits as an adjunct to strength-based rehabilitation [98].

the Achilles tendon via faster plyometric exercises to help recover its spring-like properties [84,97]. Like heavy slow resistance exercises, these should be performed within specified pain parameters and progressively increased as tolerated. Rehabilitation programs should also address other identified weaknesses in the kinetic chain and biomechanical factors that may contribute to excessive load on the Achilles tendon [84].

Conclusion

Exercise should be considered as a first-line intervention in the management of tendinopathies affecting the upper and lower extremity. The diagnosis of tendinopathy is based on the clinical history and examination, incorporating tendon-specific loading tests. Clinical information gained from the history and examination can guide how best to incorporate exercise into management of a patients' tendinopathy. Diagnostic imaging is rarely indicated for tendinopathies due to its limited relation to pain and function and the diagnosis being a clinical one. The use of injection and surgical approaches should only be considered for recalcitrant tendinopathies where exercise approaches have failed. Patient education also plays a significant role in the management of tendinopathies as some of the approaches to exercise may be counterintuitive to the patient. There is also benefit in a multidisciplinary approach to tendinopathy management, particularly the involvement of a physiotherapist, osteopath, chiropractor, or exercise physiologist in program design and implementation. With exercise being a central component of a patient's management plan, it is likely that most patients will experience a positive outcome.

References

[1] Scott A, et al. ICON 2019: international scientific tendinopathy symposium consensus: clinical terminology. Br J Sports Med 2020;54(5):260.

[2] Millar NL, et al. Tendinopathy. Nat Rev Dis Primers 2021;7(1):1.

[3] Clifford C, et al. Effectiveness of isometric exercise in the management of tendinopathy: a systematic review and meta-analysis of randomised trials. BMJ Open Sport Exerc Med 2020;6(1), e000760.

[4] Benjamin M, Kaiser E, Milz S. Structure-function relationships in tendons: a review. J Anat 2008;212(3):211–28.

[5] Wang JHC, Iosifidis MI, Fu FH. Biomechanical basis for tendinopathy. Clin Orthop Relat Res 2006;443:320–32.

[6] Camargo PR, Alburquerque-Sendín F, Salvini TF. Eccentric training as a new approach for rotator cuff tendinopathy: review and perspectives. World J Orthop 2014;5(5):634–44.

[7] De Jonge S, et al. Incidence of midportion Achilles tendinopathy in the general population. Br J Sports Med 2011;45(13):1026–8.

[8] Hägglund M, Zwerver J, Ekstrand J. Epidemiology of patellar tendinopathy in elite male soccer players. Am J Sports Med 2011;39(9):1906–11.

[9] Cassel M, et al. Incidence of Achilles and patellar tendinopathy in adolescent elite athletes. Int J Sports Med 2018;39(09):726–32.

[10] Littlewood C, May S, Walters S. Epidemiology of rotator cuff tendinopathy: a systematic review. Shoulder Elbow 2013;5(4):256–65.

[11] Stephens G, et al. A survey of physiotherapy practice (2018) in the United Kingdom for patients with greater trochanteric pain syndrome. Musculoskelet Sci Pract 2019;40:10–20.

[12] Albers IS, et al. Incidence and prevalence of lower extremity tendinopathy in a Dutch general practice population: a cross sectional study. BMC Musculoskelet Disord 2016;17(1):16.

[13] Sanders Jr TL, et al. The epidemiology and health care burden of tennis elbow: a population-based study. Am J Sports Med 2015;43(5):1066–71.

[14] Mest J, et al. The prevalence of self-reported psychological characteristics of adults with lower limb tendinopathy. Muscles Ligaments Tendons J 2020;10(4):659–71.

[15] Plinsinga ML, et al. Psychological factors not strength deficits are associated with severity of gluteal tendinopathy: a cross-sectional study. Eur J Pain 2018;22(6):1124–33.

[16] Wong WK, et al. The effect of psychological factors on pain, function and quality of life in patients with rotator cuff tendinopathy: a systematic review. Musculoskelet Sci Pract 2020;47, 102173.

[17] Fleischmann M, Vaughan B. The challenges and opportunities of using patient reported outcome measures (PROMs) in clinical practice. Int J Osteopath Med 2018;28:56–61.

[18] Hays RD, et al. Development of physical and mental health summary scores from the patient-reported outcomes measurement information system (PROMIS) global items. Qual Life Res 2009;18(7):873–80.

[19] Andrews G, Slade T. Interpreting scores on the Kessler psychological distress scale (K10). Aust N Z J Public Health 2001;25(6):494–7.

[20] Kroenke K, Spitzer Robert L. The PHQ-9: a new depression diagnostic and severity measure. Psychiatr Ann 2002;32(9):509–15.

[21] Tkachuk GA, Harris CA. Psychometric properties of the tampa scale for Kinesiophobia-11 (TSK-11). J Pain 2012;13(10):970–7.

[22] Horn KK, et al. The patient-specific functional scale: psychometrics, clinimetrics, and application as a clinical outcome measure. J Orthop Sports Phys Ther 2012;42(1):30–42.

[23] Hamilton CB, Chesworth BM. A Rasch-validated version of the upper extremity functional index for interval-level measurement of upper extremity function. Phys Ther 2013;93(11):1507–19.

[24] Binkley JM, et al. The lower extremity functional scale (LEFS): scale development, measurement properties, and clinical application. Phys Ther 1999;79(4):371–83.

[25] Fearon AM, et al. Development and validation of a VISA tendinopathy questionnaire for greater trochanteric pain syndrome, the VISA-G. Man Ther 2015;20(6):805–13.

[26] Visentini PJ, et al. The VISA score: an index of severity of symptoms in patients with jumper's knee (patellar tendinosis). J Sci Med Sport 1998;1 (1):22–8.

[27] Cacchio A, De Paulis F, Maffulli N. Development and validation of a new visa questionnaire (VISA-H) for patients with proximal hamstring tendinopathy. Br J Sports Med 2014;48(6):448.

[28] Robinson JM, et al. The VISA-A questionnaire: a valid and reliable index of the clinical severity of Achilles tendinopathy. Br J Sports Med 2001;35 (5):335.

[29] Leong HT, et al. Risk factors for rotator cuff tendinopathy: a systematic review and meta-analysis. J Rehabil Med 2019;51(9):627–37.

[30] Ketola S, Lehtinen JT, Arnala I. Arthroscopic decompression not recommended in the treatment of rotator cuff tendinopathy: a final review of a randomised controlled trial at a minimum follow-up of ten years. Bone Joint J 2017;99(6):799–805.

[31] Littlewood C, May S, Walters S. A review of systematic reviews of the effectiveness of conservative interventions for rotator cuff tendinopathy. Shoulder Elbow 2013;5(3):151–67.

[32] Littlewood C, Malliaras P, Chance-Larsen K. Therapeutic exercise for rotator cuff tendinopathy: a systematic review of contextual factors and prescription parameters. Int J Rehabil Res 2015;38(2):95–106.

[33] Dominguez-Romero JG, et al. Exercise-based muscle development programmes and their effectiveness in the functional recovery of rotator cuff tendinopathy: a systematic review. Diagnostics 2021;11(3).

[34] Desjardins-Charbonneau A, et al. The efficacy of manual therapy for rotator cuff tendinopathy: a systematic review and meta-analysis. J Orthop Sports Phys Ther 2015;45(5):330–50.

[35] Desmeules F, et al. The efficacy of therapeutic ultrasound for rotator cuff tendinopathy: a systematic review and meta-analysis. Phys Ther Sport 2015;16(3):276–84.

[36] Toliopoulos P, et al. Efficacy of surgery for rotator cuff tendinopathy: a systematic review. Clin Rheumatol 2014;33(10):1373–83.

[37] Boudreault J, et al. The efficacy of oral non-steroidal anti-inflammatory drugs for rotator cuff tendinopathy: a systematic review and meta-analysis. J Rehabil Med 2014;46(4):294–306.

[38] Lin M-T, et al. Comparative effectiveness of injection therapies in rotator cuff tendinopathy: a systematic review, pairwise and network meta-analysis of randomized controlled trials. Arch Phys Med Rehabil 2019;100(2):336–349.e15.

[39] Desjardins-Charbonneau A, et al. The efficacy of taping for rotator cuff tendinopathy: a systematic review and meta-analysis. Int J Sports Phys Ther 2015;10(4):420–33.

[40] Heron SR, Woby SR, Thompson DP. Comparison of three types of exercise in the treatment of rotator cuff tendinopathy/shoulder impingement syndrome: a randomized controlled trial. Physiotherapy 2017;103(2):167–73.

[41] McCormick N. The effectiveness of eccentric loading exercises in the management of rotator cuff tendinopathy: a structured literature review. Physiotherapy 2019;105, e91.

[42] Sims SEG, et al. Non-surgical treatment of lateral epicondylitis: a aystematic review of randomized controlled trials. Hand 2014;9(4):419–46.

[43] Ma K-L, Wang H-Q. Management of lateral epicondylitis: a narrative literature review. Pain Res Manag 2020;2020:6965381.

[44] Gruchow HW, Pelletier D. An epidemiologic study of tennis elbow: incidence, recurrence, and effectiveness of prevention strategies. Am J Sports Med 1979;7(4):234–8.

[45] Winston J, Wolf JM. Tennis elbow: Definition, causes, epidemiology. In: Tennis elbow. Springer; 2015. p. 1–6.

[46] Nirschl RP. The epidemiology and health care burden of tennis elbow: a population-based study. Ann Transl Med 2015;3(10).

[47] Descatha A, et al. Lateral epicondylitis and physical exposure at work? A review of prospective studies and meta-analysis. Arthritis Care Res 2016;68 (11):1681–7.

[48] Longo UG, et al. Elbow tendinopathy. Muscles Ligaments Tendons J 2012;2(2):115.

[49] Sayampanathan AA, Basha M, Mitra AK. Risk factors of lateral epicondylitis: a meta-analysis. Surgeon 2020;18(2):122–8.

[50] Bisset L, et al. A systematic review and meta-analysis of clinical trials on physical interventions for lateral epicondylalgia. Br J Sports Med 2005;39 (7):411–22.

[51] Cullinane FL, Boocock MG, Trevelyan FC. Is eccentric exercise an effective treatment for lateral epicondylitis? A systematic review. Clin Rehabil 2013;28(1):3–19.

[52] Hoogvliet P, et al. Does effectiveness of exercise therapy and mobilisation techniques offer guidance for the treatment of lateral and medial epicondylitis? A systematic review. Br J Sports Med 2013;47(17):1112.

[53] Chen Z, Baker NA. Effectiveness of eccentric strengthening in the treatment of lateral elbow tendinopathy: a systematic review with meta-analysis. J Hand Ther 2021;34(1):18–28.

[54] Smidt N, et al. Effectiveness of physiotherapy for lateral epicondylitis: a systematic review. Ann Med 2003;35(1):51–62.

[55] Trudel D, et al. Rehabilitation for patients with lateral epicondylitis: a systematic review. J Hand Ther 2004;17(2):243–66.

[56] Eraslan L, et al. Does Kinesiotaping improve pain and functionality in patients with newly diagnosed lateral epicondylitis? Knee Surg Sports Traumatol Arthrosc 2018;26(3):938–45.

[57] Sayegh F, Potoupnis M, Kapetanos G. Greater trochanter bursitis pain syndrome in females with chronic low back pain and sciatica. Acta Orthop Belg 2004;70(5):423–8.

[58] Segal NA, et al. Greater trochanteric pain syndrome: epidemiology and associated factors. Arch Phys Med Rehabil 2007;88(8):988–92.

[59] Grimaldi A, Fearon A. Gluteal tendinopathy: integrating pathomechanics and clinical features in its management. J Orthop Sports Phys Ther 2015;45 (11):910–22.

[60] Grimaldi A, et al. Gluteal tendinopathy: a review of mechanisms, assessment and management. Sports Med 2015;45(8):1107–19.

[61] French HP, et al. Physiotherapy management of greater trochanteric pain syndrome (GTPS): an international survey of current physiotherapy practice. Physiotherapy 2020;109:111–20.

[62] Collier TS, Poole B, Bradford B. An indirect evaluation between corticosteroid injections and gluteal exercises in the management of pain in greater trochanteric pain syndrome. Phys Ther Rev 2021;26(2):139–49.

[63] Ladurner A, Fitzpatrick J, O'Donnell JM. Treatment of gluteal tendinopathy: a systematic review and stage-adjusted treatment recommendation. Orthop J Sports Med 2021;9(7). 23259671211016850.

[64] Mellor R, et al. Education plus exercise versus corticosteroid injection use versus a wait and see approach on global outcome and pain from gluteal tendinopathy: prospective, single blinded, randomised clinical trial. BMJ 2018;361, k1662.

[65] Ebert JR, et al. A systematic review of rehabilitation exercises to progressively load the gluteus medius. J Sport Rehabil 2017;26(5):418–36.

[66] McNeill W. A short consideration of exercise for gluteal tendinopathies. J Bodyw Mov Ther 2016;20(3):595–7.

[67] Clifford C, et al. Isometric versus isotonic exercise for greater trochanteric pain syndrome: a randomised controlled pilot study. BMJ Open Sport Exerc Med 2019;5(1), e000558.

[68] Seo K-H, et al. Long-term outcome of low-energy extracorporeal shockwave therapy on gluteal tendinopathy documented by magnetic resonance imaging. PLoS One 2018;13(7), e0197460.

[69] Walker-Santiago R, et al. Platelet-rich plasma versus surgery for the management of recalcitrant greater trochanteric pain syndrome: a systematic review. Arthroscopy 2020;36(3):875–88.

[70] Malliaras P, et al. Patellar tendinopathy: clinical diagnosis, load management, and advice for challenging case presentations. J Orthop Sports Phys Ther 2015;45(11):887–98.

[71] Rudavsky A, Cook J. Physiotherapy management of patellar tendinopathy (jumper's knee). J Physiother 2014;60(3):122–9.

[72] Lian ØB, Engebretsen L, Bahr R. Prevalence of jumper's knee among elite athletes from different sports: a cross-sectional study. Am J Sports Med 2005;33(4):561–7.

[73] Zwerver J, Bredeweg SW, van den Akker-Scheek I. Prevalence of Jumper's knee among nonelite athletes from different sports: a cross-sectional survey. Am J Sports Med 2011;39(9):1984–8.

[74] De Vries AJ, et al. The impact of patellar tendinopathy on sports and work performance in active athletes. Res Sports Med 2017;25(3):253–65.

[75] Kongsgaard M, et al. Corticosteroid injections, eccentric decline squat training and heavy slow resistance training in patellar tendinopathy. Scand J Med Sci Sports 2009;19(6):790–802.

[76] Breda SJ, et al. Effectiveness of progressive tendon-loading exercise therapy in patients with patellar tendinopathy: a randomised clinical trial. Br J Sports Med 2021;55:501–9. https://doi.org/10.1136/bjsports-2020-103403.

[77] Pedrelli A, Stecco C, Day JA. Treating patellar tendinopathy with fascial manipulation. J Bodyw Mov Ther 2009;13(1):73–80.

[78] Korakakis V, et al. The effectiveness of extracorporeal shockwave therapy in common lower limb conditions: a systematic review including quantification of patient-rated pain reduction. Br J Sports Med 2018;52(6):387–407.

[79] Zwerver J, et al. No effect of extracorporeal shockwave therapy on patellar tendinopathy in jumping athletes during the competitive season: a randomized clinical trial. Am J Sports Med 2011;39(6):1191–9.

[80] Everhart JS, et al. Treatment options for patellar tendinopathy: a systematic review. Arthroscopy 2017;33(4):861–72.

[81] Warden S, et al. Low-intensity pulsed ultrasound for chronic patellar tendinopathy: a randomized, double-blind, placebo-controlled trial. Rheumatology 2008;47(4):467–71.

[82] Hoksrud A, et al. Ultrasound-guided sclerosis of neovessels in patellar tendinopathy: a prospective study of 101 patients. Am J Sports Med 2012;40 (3):542–7.

[83] Komi PV, Fukashiro S, Järvinen M. Biomechanical loading of achilles tendon during normal locomotion. Clin Sports Med 1992;11(3):521–31.

[84] Cardoso TB, et al. Current trends in tendinopathy management. Best Pract Res Clin Rheumatol 2019;33(1):122–40.

[85] Scott A, Ashe MC. Common tendinopathies in the upper and lower extremities. Curr Sports Med Rep 2006;5(5):233–41.

[86] Malliaras P, O'Neill S. Potential risk factors leading to tendinopathy. Apunts Med Sport 2017;52(194):71–7.

[87] Abate M, et al. Pathogenesis of tendinopathies: inflammation or degeneration? Arthritis Res Ther 2009;11(3):235.

[88] Silbernagel KG, Hanlon S, Sprague A. Current clinical concepts: conservative management of Achilles tendinopathy. J Athl Train 2020;55(5):438–47.

[89] Malliaras P, et al. Achilles and patellar tendinopathy loading programmes. Sports Med 2013;43(4):267–86.

[90] Silbernagel KG, Brorsson A, Lundberg M. The majority of patients with Achilles tendinopathy recover fully when treated with exercise alone: a 5-year follow-up. Am J Sports Med 2011;39(3):607–13.

[91] Malmgaard-Clausen NM, et al. No additive clinical or physiological effects of short-term anti-inflammatory treatment to physical rehabilitation in the early phase of human Achilles tendinopathy: a randomized controlled trial. Am J Sports Med 2021;49(7):1711–20.

[92] Boesen AP, et al. High volume injection with and without corticosteroid in chronic midportion Achilles tendinopathy. Scand J Med Sci Sports 2019;29(8):1223–31.

[93] Nauwelaers A-K, Van Oost L, Peers K. Evidence for the use of PRP in chronic midsubstance Achilles tendinopathy: a systematic review with meta-analysis. Foot Ankle Surg 2021;27(5):486–95.

[94] Wilson F, et al. Exercise, orthoses and splinting for treating Achilles tendinopathy: a systematic review with meta-analysis. Br J Sports Med 2018;52(24):1564–74.

[95] Head J, et al. The efficacy of loading programmes for improving patient-reported outcomes in chronic midportion Achilles tendinopathy: a systematic review. Musculoskeletal Care 2019;17(4):283–99.

[96] Beyer R, et al. Heavy slow resistance versus eccentric training as treatment for Achilles tendinopathy: a randomized controlled trial. Am J Sports Med 2015;43(7):1704–11.

[97] Sancho I, et al. Education and exercise supplemented by a pain-guided hopping intervention for male recreational runners with midportion Achilles tendinopathy: a single cohort feasibility study. Phys Ther Sport 2019;40:107–16.

[98] Jayaseelan DJ, et al. Manual therapy and eccentric exercise in the management of Achilles tendinopathy. J Man Manip Ther 2017;25(2):106–14.

Chapter 17

Exercise and depression

Michael Musker[a,b,c,d]

[a]The University of South Australia, Adelaide, SA, Australia [b]South Australian Health & Medical Research Institute, Adelaide, SA, Australia [c]The University of Adelaide, Adelaide, SA, Australia [d]Flinders University, Adelaide, SA, Australia

Introduction

Depression is a high prevalence mental health disorder. According to the World Health Organization 322 million people (5.1% female vs 3.6% males) have depression or 4.4% of the world's population [1]. For most countries, this equates to approximately one in ten people having a current diagnosis of depression [2]. Human and animal research has reported that exercise ameliorates some of the pathological effects of depression, and may reverse the negative biological changes such as immuno-inflammatory and neurodegenerative processes [3,4]. The prevalence of depression increases with age, peaking at the ages of 60–65, and then only decreasing slightly with ongoing decades [1]. Depression occurs as part of several disorders described in the DSM V including major depressive disorder, bipolar disorder, cyclothymic disorder, persistent depressive disorder, and premenstrual dysphoric disorder [5]. Those who experience major depression are likely to have it for more than 6 months or for a protracted period of many decades, and approximately 75% of these people will be diagnosed with the anxious/distressed DSM specifier [2]. There is a strong relationship between depression, sleep, diet, and exercise. Regular exercise promotes healthy sleep hygiene, while diet can affect both sleep and depression symptoms [6,7]. This chapter will explore the continued debate and doubt as to whether exercise can provide sustainable improvement in the course of depression.

Following the worldwide COVID 19 pandemic, communities craved getting back to their gyms or going for long walks, emphasizing our dependency on exercise to maintain a level mood. Exercise may be a supportive intervention, but it may not be helpful for everybody, particularly those with severe psychotic depression, those with a disability or people experiencing pain. A series of meta-analyses have reported positive impacts of exercise but found the evidence to be inconsistent across studies. For example, a meta-analysis of eight studies ($n = 637$ patients) showed an absence of positive effects on global cognitive symptoms, but this evidence needs to be reviewed with caution due to the high heterogeneity of the sample population and the types of intervention within programs varied considerably in the areas of frequency, intensity, type of activity, and length of time. [8]. Another systematic review combining over 30 studies of four meta-analyses ($n = 2110$) in adolescents and children (aged 5–20 years), there is evidence to support the notion that exercise has promising treatment potential, with the reviewers pooled analysis indicating the outcome of a medium positive effect (Cohen's $d = -0.50$), with even greater improvements shown in older groups resulting in a moderate effect ($d = -0.68$) [9]. For new practitioners, it is difficult to know where to start, with over 23,288 articles about exercise and depression to choose from (PubMed platform Jan 2022). It is recommended that the novice clinician/support worker consider the holistic effects of proposed interventions and start with the latest systematic reviews. Depression interacts with many biopsychosocial and cultural components, but particularly its relationship with neural, hormonal, and nervous systems, and are controlled via the hypothalamic–pituitary axis. For some, exercise can even be a spiritual activity. Such research indicates that there are a wide variety of options for increased levels of activity to improve mood from gentle yoga to running marathons and these will be illustrated further [10].

Background

Exercise is known to produce "feel good" neurotransmitters such as beta-endorphins (a natural opioid), as well as dopamine, serotonin, and norepinephrine [11]. Combine these temporary energizing alterations in mood with the rewarding aspects of exercise, like achieving short-term goals, losing weight, and feeling physically better—exercise provides an excellent vehicle to begin engagement with clinicians to improve levels of depression. There are many features and clinical aspects of depression that should be considered when planning the introduction of exercise interventions. Challenging

Exercise to Prevent and Manage Chronic Disease Across the Lifespan. https://doi.org/10.1016/B978-0-323-89843-0.00031-3

TABLE 1 Symptoms of major depressive disorder DSM V.

1	Depressed with a low mood, feeling sad and hopeless most of the day, every day
2	Diminished interest in and withdrawal from pleasurable activity, or almost all activities
3	Either weight loss or weight gain (>5%)
4	Difficulty with sleeping, either too little sleep (agitated) or too much sleep (atypical)
5	Psychomotor agitation (movement) or being slowed down and lethargic
6	Fatigue and a low level of energy, or exhaustion from agitation
7	Feeling guilty and worthless (can be delusional and extreme)
8	Difficulty with thinking tasks and being unable to concentrate or indecisiveness
9	Thoughts of death, self-harm, or that life is not worth living

Adapted from the DSM V: American Psychiatric Association 2013 [5].

symptoms include low mood, negative self-image, poor motivation, difficulty with sleeping, anxiety, agitation, weight gain/weight loss, and ideas of self-harm and/or suicidal ideation. When depression is severe it can cause psychosis, which incorporates delusional thinking and a state of immobility known as catatonia (the person's movements are practically frozen and social withdrawal may result in complete unresponsiveness). A diagnosis of depression requires the identification of five out of nine symptoms (Table 1) which are experienced every day for at least 2 weeks as identified in the Diagnostic and Statistical Manual of Mental Disorders, Fifth Edition.

There are at least two opposing potential clinical pictures which may be present. Either the overactive, agitated individual who loses a lot of weight and sleeps poorly, which is often referred to as melancholic type depression, or the opposite picture which includes slow movement, comfort eating causing weight gain, and oversleeping. The latter symptoms are referred to as atypical type depression. When selecting a suitable exercise intervention, the type and intensity of exercise will need to be considered. The consumer's involvement in preference and choice in the type of exercise, and intensity level is likely to show a greater reduction in depression symptoms. It may be obvious that relaxing exercises should be chosen to reduce agitation and are helpful for the anxious consumer such as yoga, tai chi, Pilates, stretching, and slow mindful walking. For those with atypical depression, any type of activity that increases their movement will be helpful, for example walking, jogging, dancing, and social sports [12].

Epidemiology/impact

A risk for depression is hereditary whereby if you are genetically predisposed then you are over two and half times more likely to have an outcome of depression if one of your parents has experienced Major Depressive Disorder (an odds ratio of 2.84, 95% CI 2.31–3.49) [13]. A genome-wide association using meta-analysis found 44 significant genetic loci that may influence the phenotype (behavioral and physical outcomes) of depression [14]. We know that the cause of depression is a "gene x environment" interaction where up to 15% of the population are affected, with a lifetime prevalence of 9% for women and 5% for men [15]. Depression is known to affect other bodily systems, possibly potentiated by increased inflammation and it was noted that in an epidemiological study across 47 countries that more severe depression (including psychotic type) is often comorbid with inflammatory type disorders such as arthritis, coronary heart problems, and diabetes [16]. Previous studies have shown a bidirectional relationship of inflammatory disorders such as diabetes [17]. The body uses a messenger system called cytokines to provide feedback about levels of inflammation and to initiate a healing response, but a prolonged effect of these proinflammatory cytokines (detailed in the next section–pathophysiology) can create an autoimmune response that is known to add to the risk of depression [18]. Exercise is thought to provide a counter-response by engaging neuroprotective anti-inflammatory mechanisms in astrocytes, microglia, T cells and by reducing oxidative stress [19].

Pathophysiology

Emotions are managed by the limbic system, which is part of the hypothalamic–pituitary adrenal axis, a biofeedback loop between hormones and the nervous system to maintain equilibrium with our environment. The key brain structures include the amygdala, hypothalamus, hippocampus, and pituitary gland managing the delicate homeostasis of the human body. Serotonin, indolamine and dopamine, and catecholamine, are important neurotransmitters in depression. They are produced and depleted in the brain's nerve synapses to maintain a vigilant level of reactivity and responsiveness [20]. The well-known "serotonin hypothesis" espouses that when the neurotransmitter serotonin is low, it can lead to low mood, sluggishness, and severe depression, hence why most of the 30 available antidepressants work toward increasing the availability of serotonin or inhibit the rate at which it is broken down (metabolized to 5-hydroxyindoleacetic acid 5-HIAA), by preventing its reuptake [21]. Managing the levels of the monoamines in the brain through medication is often combined with cognitive-behavioral therapy, as well as increasing participation in physical activities such as aerobic exercise. Exercise provides beneficial changes to our neural mechanisms that moderate anhedonia such as reward sensitivity, improved sleep, positive emotional potential to stimuli, and by providing a moderated emotional state that supports the application of other interventions such as cognitive therapies [22].

The expression of our unique genetic code is affected by our environment, causing changes in the expression of messenger ribonucleic acids (mRNA) which provide the instructions to our cells to produce the proteins we need to function. Evolution through random selection results in specific coding in our genes to be slightly different but can cause similar changes to our genetic base pairs across the population, resulting in common traits. Depending on the location of a specific gene, these changes can affect protein production and the way our bodies work. These tell-tale differences are called single nucleotide polymorphisms (SNPs) and scientists are now using SNPs to examine genetic and behavior phenotypes (physical and behavioral outcomes). A polymorphism can lead to phenotypic (body shape and behavior) differences in people, for example in the fat mass and obesity (FTO) gene, which may potentially affect a person's willingness to exercise but also affect their fat mass [23]. Early trauma such as child maltreatment or other stressful life experiences may cause longitudinal epigenetic changes of a proinflammatory nature, resulting in long-term pathophysiological effects. Trauma, for example, can alter the expression rate of certain mRNA, potentially resulting in a reduction of the size of some brain structures such as hippocampal and hypothalamic volume [24,25]. This results in an impairment of cognitive functioning such as the rate at which we think, focus, or the speed at which we access memories and other functions.

Cortisol, a steroid hormone is produced in response to environmental stress and anxiety as part of the fight or flight response. Prolonged elevated levels of cortisol can lead to an alarm-type reaction which impacts the body's autoimmune system known as the inflammasome [26]. Exercise has been shown to ameliorate the neural effects of depression and chronic stress by changes to receptors in the hypothalamus reducing the potential damage of this pro-inflammatory process [27]. Therefore, exercise in the form of increased movement and targeted physical activity is an ideal way to limit the damage caused by prolonged high levels of cortisol, expending any pent-up energy and diverting the cortisol toward productive energy expenditure. Depression rarely exists without anxiety issues and both conditions are reported to be responsible for the production of cortisol and high levels of pro-inflammatory cytokines. Examples of these inflammatory markers are members of the interleukin family such as Interleukin 1 beta and Interleukin 6. These alert messengers result in a heightened activity level in our immune system [26,28]. Aerobic physical activity may counter or even reverse some of these harmful effects in many ways such as through the reduction in oxidative and nitrosative stress (O&NS), chemicals that steal important oxygen from regular cellular activity, causing damage and cell death. Exercise increases the availability of healthy anti-oxidants, influencing the production of anti-inflammatory proteins, increasing neurogenerative growth, increased vasculogenesis, and improved neurofunctionality in areas like memory [29,30].

Clinical outcomes

In a meta-analysis ($n=642$ patients) it was shown that low-intensity activities that combine cognitive treatment with physical interventions have demonstrated a significant positive effect on depression ($P=0.048$), whereas physical exercise alone may not provide global cognitive improvement (Hedges' $g=0.08$, $P=0.33$, $I^2=0\%$) [31]. Similarly, in women who experience post-natal depression, which can occur up to 18 months after childbirth, gentler low-intensity interventions such as yoga and Pilates may have a greater chance of positive outcomes such as improved mood and weight loss [32].

There are several factors to consider when encouraging someone with depression to do exercise including anhedonia (a loss of the ability to experience pleasure), volition (a lack of willpower to change), motivation (a personal reason to change), fatigue (being excessively tired), pain (arthritis or fibromyalgia), low mood (tearfulness and suicidality) and sleep difficulties (usually early morning waking, e.g., less than <4h sleep every night for at least 2 weeks). For some individuals, just getting out of bed in the morning is an effort, so it is necessary to complete a baseline assessment of the individual's capability, willingness, and feelings of self-worth. As part of a functional assessment, find out what the individual's level of activity was prior to their depression, and enquire what type of exercise they find desirable or what may genuinely personally motivate them, or at least make them willing to participate. People are motivated for their own unique reasons and this can involve their social image, or just wanting to be healthier. To positively engage consumers the simplest form of exercise that can be initiated and maintained without gym equipment is going for a walk or riding a bike but initially more fun-based activities that involve using the hands, particularly group-based activities may be helpful for people with depression. For younger people, fun team-based activities such as volleyball or badminton may be effective, but these rely on having enough participants, equipment, and supportive infrastructure, for example, the courts to play on. Some organizations may require a fully qualified sports instructor or therapist due to legal or occupational health and safety issues. When the individual has improved enough to overcome some of their symptoms of depression, possibly through a combination of cognitive-behavioral therapy, medication, and counseling; then this would be the best time to introduce longer term goals and sessional programs that manifest other motivators, reinforcers, rewards, and celebration [33]. Most exercise programs reported in the literature are planned across 8–12 weeks of intervention, offering an optimum of several sessions per week. Here is a shortlist of exercise interventions that have reportedly shown a positive reduction in depression symptoms [3]:

Walking, stretching, swimming, cycling, dancing, resistance movement, handclapping, elastic resistance movement, anti-resistance movement, yoga (Hatha), Tai Chi (Yang Taijiquan), Pilates, running, and many more. New electronic devices are being used to get people moving to improve their mental health such as virtual reality programs, often used in conjunction with a Sony PlayStation or Microsoft Xbox and others, which have the potential to provide movement feedback, and even relaxing 3D environments using devices like the Oculus Quest, but as these are emerging new technologies further research is required [34].

Known/potential mechanisms underlying effects

Animal studies have shown that exercise reduces depression through mechanisms that increase the expression of hippocampal brain-derived neurotrophic factor (BDNF) which stimulates neurogenesis, and reduces excessive apoptosis (cell death) [35]. There are many explanations for potential causes of depression and one of those is that oxidative and nitrosative stress are a result of the production of excessive reactive oxygen and nitrogen species which are harmful to cellular structures in the brain. These indicators are shown to be at higher levels in people with depression [36]. There is also evidence that pro-inflammatory biomarkers (TNFα, IL-1β, and IL-6) as well as the immunological cytokines described earlier are a symptom of depression, which like nitrosative stress, cause a toxic cellular autoimmune response [37]. One hypothesis is that heightened expression of these cytokines are thought to redirect and divert tryptophan, the precursor to serotonin, down the kynurenine pathway causing the production of kynurenic acid and quinolinic acid [38] (Fig. 1). Only a small amount of Tryptophan is metabolized into Serotonin but more than 90% is converted into Kynurenine as part of the natural cellular energy production pathway NAD+ [39]. These neurochemical changes lead to lower serotonin levels and create both neuroprotective (Kynurenic Acid) which is a N-methyl-D-aspartate antagonist (NMDA) and potentially neurotoxic metabolite (quinolinic acid) an NMDA agonist that potentiate depressive symptoms or sickness behavior [40]. Exercise can help to reduce neuroinflammation and reverse some of the potential damage, providing protection against future onset of depression and neurogenesis [39]. Regular physical activity promotes gradual epigenetic changes resulting in positive differences in gene expression that in turn cause trophic effects such as increased metabolic energy, the production of antioxidants, and induce an anti-inflammatory response [41].

Cautions/safety

Suicidal ideation and self-harm are a symptom of depression that demands caution. A systematic review elicited that there is a pooled risk ratio for suicide in consumers with a mood disorder of 12.3 (95% CI 8.9–17.1) [42]. However, exercise levels in a group of college students were associated with improved mental health and a reduced risk of suicidal ideation, indicating that increased levels of physical activity are a protective factor helping to ameliorate this risk [43]. In another study of students ($n = 167$, aged 14–25) who performed acts of non-suicidal self-harm, exercise was found to positively moderate the acts and thoughts of self-harm [44]. In an older group of veterans with depression ($n = 346$, aged 45.45 years, $SD = 14.27$) exercise was seen to reduce depressive symptoms and improve sleep, with an associated beneficial consequence of reducing suicidal ideation [45]. These research studies indicate that exercise may be a supportive intervention to reduce the symptoms of suicidal ideation and that exercise interventions are likely to have low side effects. From a biological perspective exercise offers the potential benefit of mental health improvement through neurogenerative and anti-inflammatory mechanisms, however, further research is required in this complex area [46] Fig. 2. A randomized controlled trial ($n = 54$) that compared exercise combined with cognitive-behavioral therapy (CBT) rather than just CBT alone, found that the combinatorial approach was much more successful at alleviating suicidal ideation [47]. The exercises utilized by

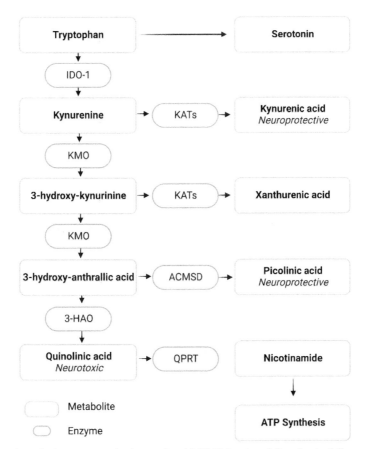

FIG. 1 Kynurenine Pathway produces both neuroprotective kynurenic acid (NMDA antagonist), and potentially neurotoxic quinolinic acid (NMDA agonist). Abbreviations: *IDO* 1, indoleamine 2,3-dioxygenase; *KMO*, kynurenine monooxygenase; *KAT*, kynurenine aminotransferase; *3HAO*, 3-hydroxyanthranillic acid oxygenase; *QPRT*, quinolinate phosphoribosyltransferase; *NAD+*, nicotinamide adenine dinucleotide; *ATP*, adenosine triphosphate.

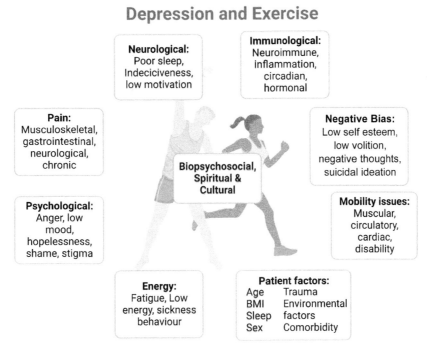

FIG. 2 Multidimensional relationship of depression and exercise.

these researchers were remarkably simple tasks like stretching, clapping hands, and walking; demonstrating that there is no need for elaborate equipment. Exercise is used as an intervention for depression across the lifespan and strong benefits have been shown in randomized controlled trials in youths, the general population, and in elderly cohorts in the form of mind–body programs, but for the older age group caution needs to be taken with the risks posed by other potential comorbid issues such as cardiovascular weakness and limited mobility [48]. The current literature presents the recurring theme that allowing self-selection in the type, frequency, intensity, and mode of exercise has a greater likelihood of success in alleviating depressive symptoms [12] (Box 1).

1. Aim for programs that last between 8 and 12 weeks, offering several sessions per week.

BOX 1 Summary of clinical practice recommendations for exercise prescribers:

1. A combinatorial approach of multiple interventions that include CBT and exercise.
2. Consider effects and side effects of medications, as well as type of depression.
3. Measure the individual's current level of motivation and activity level, do not push too hard.
4. Design new goals that gradually build up increased movement e.g., by 5% per week.
5. Consider using goals that provide clear markers of achievement (SMART).
6. Exercise should be gradually implemented as a weekly routine and lifestyle change.
7. Assist the individual to select from a diverse group of exercise styles that suit, walking, cycling, swimming, yoga, Pilates, running or team sports for example.
8. Attach new behaviors to current habits that fit well with the person's lifestyle.
9. Set milestones, record and celebrate these achievements (smart watch, or phone).
10. Use digital feedback devices to provide cardiac and pedometer feedback—e.g., Fitbit Charge.
11. Use online tracking visual aids to log ongoing improvements with apps like STRAVA.
12. Do exercise in pairs or groups where possible to provide support and encouragement.
13. Allow participants to choose their type of exercise, level, frequency, and pace.
14. Slowly work toward WHO recommendations (150–300 min of moderate activity per week). A good example is 'Couch to 5k' program described by the UK National Health Service [49] or the global phenomenon "Park Run" [50] or "Mall Walking" for the elderly [51].

Conclusion

Engaging a person in exercise can make beneficial changes to their biological, psychological, social, cultural, and spiritual well-being. Exercise is an excellent way of helping an individual feel better about themselves by experiencing the achievement of success with small goals, engaging in a community of sport, and to feel physically energized. For many people with depression, it will provide them with a goal, make them feel worthwhile, and offer opportunities for social interaction. Small walking groups are a perfect example of mixing social interaction and exercise, creating an effective community intervention to prevent or ease depression [51]. It is a restorative process in that it provides the likelihood of improved sleep and supportive biochemical changes that induce repair and increased access to energy. In this chapter, we have explored how specialist practitioners in sports and mental health, biological scientists, and clinicians are working together to develop best practice. The current available evidence supporting the benefits of exercise in depression remains uncertain, but by using emerging technologies such as wearable devices, we can elicit the objective evidence of how exercise interventions can reduce the symptoms of depression.

References

[1] WHO. Depression and other common mental disorders: global health estimates. Geneva: WHO; 2017.
[2] Hasin DS, et al. Epidemiology of adult DSM-5 major depressive disorder and its specifiers in the United States. JAMA Psychiatry 2018;75(4): 336–46. https://doi.org/10.1001/jamapsychiatry.2017.4602.
[3] Zhao JL, et al. Exercise, brain plasticity, and depression. CNS Neurosci Ther 2020;26(9):885–95.
[4] Zhuang PC, et al. Treadmill exercise reverses depression model-induced alteration of dendritic spines in the brain areas of mood circuit. Front Behav Neurosci 2019;13:93.
[5] APA. Diagnostic and statistical manual of mental disorders, fifth edition, DSM-5. Arlington, VA: American Psychiatric Association; 2013.
[6] Atkinson G, Davenne D. Relationships between sleep, physical activity and human health. Physiol Behav 2007;90(2):229–35.

[7] Lopresti AL, Hood SD, Drummond PD. A review of lifestyle factors that contribute to important pathways associated with major depression: diet, sleep and exercise. J Affect Disord 2013;148(1):12–27.

[8] Brondino N, et al. A systematic review of cognitive effects of exercise in depression. Acta Psychiatr Scand 2017;135(4):285–95.

[9] Wegner M, et al. Systematic review of meta-analyses: exercise effects on depression in children and adolescents. Front Psych 2020;11:81.

[10] Roeh A, et al. Marathon running improves mood and negative affect. J Psychiatr Res 2020;130:254–9.

[11] Bender T, et al. The effect of physical therapy on beta-endorphin levels. Eur J Appl Physiol 2007;100(4):371–82.

[12] Craft LL, Perna FM. The benefits of exercise for the clinically depressed. Prim Care Companion J Clin Psychiatry 2004;6(3):104–11.

[13] Sullivan PF, Neale MC, Kendler KS. Genetic epidemiology of major depression: review and meta-analysis. Am J Psychiatry 2000;157(10):1552–62.

[14] Wray NR, et al. Genome-wide association analyses identify 44 risk variants and refine the genetic architecture of major depression. Nat Genet 2018;50(5):668–81.

[15] Mandelli L, Serretti A. Gene environment interaction studies in depression and suicidal behavior: an update. Neurosci Biobehav Rev 2013; 37(10, Part 1):2375–97.

[16] Koyanagi A, et al. Epidemiology of depression with psychotic experiences and its association with chronic physical conditions in 47 low- and middle-income countries. Psychol Med 2017;47(3):531–42.

[17] Wu CS, Hsu LY, Wang SH. Association of depression and diabetes complications and mortality: a population-based cohort study. Epidemiol Psychiatr Sci 2020;29, e96.

[18] Musker M, Licinio J, Wong M-L. Chapter 23—inflammation genetics of depression. In: Baune BT, editor. Inflammation and immunity in depression. Academic Press; 2018. p. 411–25.

[19] Eyre H, Baune BT. Neuroimmunological effects of physical exercise in depression. Brain Behav Immun 2012;26(2):251–66.

[20] Rybakowski J. A half-century of participant observation in psychiatry. Part II: affective disorders. Psychiatr Pol 2020;54(4):641–59.

[21] Maes M, et al. The new '5-HT' hypothesis of depression: cell-mediated immune activation induces indoleamine 2,3-dioxygenase, which leads to lower plasma tryptophan and an increased synthesis of detrimental tryptophan catabolites (TRYCATs), both of which contribute to the onset of depression. Prog Neuropsychopharmacol Biol Psychiatry 2011;35(3):702–21.

[22] Brush CJ, et al. Aerobic exercise enhances positive emotional reactivity in individuals with depressive symptoms: evidence from neural responses to reward and emotional content. Ment Health Phys Act 2020;19, 100339.

[23] Antonio J, et al. A fat mass and obesity-associated gene polymorphism influences fat mass in exercise-trained individuals. J Int Soc Sports Nutr 2018;15(1):40.

[24] Gerritsen L, et al. HPA axis genes, and their interaction with childhood maltreatment, are related to cortisol levels and stress-related phenotypes. Neuropsychopharmacology 2017;42(12):2446–55.

[25] Persson N, et al. Regional brain shrinkage over two years: individual differences and effects of pro-inflammatory genetic polymorphisms. Neuroimage 2014;103.

[26] Maes M. Depression is an inflammatory disease, but cell-mediated immune activation is the key component of depression. Prog Neuropsychopharmacol Biol Psychiatry 2011;35.

[27] Wang Y, et al. Exercise amelioration of depression-like behavior in OVX mice is associated with suppression of NLRP3 inflammasome activation in hippocampus. Behav Brain Res 2016;307:18–24.

[28] Beserra AHN, et al. Can physical exercise modulate cortisol level in subjects with depression? A systematic review and meta-analysis. Trends Psychiatry Psychother 2018;40(4):360–8.

[29] Imboden C, et al. Aerobic exercise or stretching as add-on to inpatient treatment of depression: similar antidepressant effects on depressive symptoms and larger effects on working memory for aerobic exercise alone. J Affect Disord 2020;276:866–76.

[30] Moylan S, et al. Exercising the worry away: how inflammation, oxidative and nitrogen stress mediates the beneficial effect of physical activity on anxiety disorder symptoms and behaviours. Neurosci Biobehav Rev 2013;37(4):573–84.

[31] Sun M, et al. Exercise for cognitive symptoms in depression: a systematic review of interventional studies. Can J Psychiatry 2017;63(2):115–28.

[32] Saligheh M, et al. Can exercise or physical activity help improve postnatal depression and weight loss? A systematic review. Arch Womens Ment Health 2017;20(5):595–611.

[33] Bourbeau K, et al. The combined effect of exercise and behavioral therapy for depression and anxiety: systematic review and meta-analysis. Behav Sci (Basel) 2020;10(7).

[34] Cieślik B, et al. Virtual reality in psychiatric disorders: a systematic review of reviews. Complement Ther Med 2020;52, 102480.

[35] Park HS, et al. Swimming exercise ameliorates mood disorder and memory impairment by enhancing neurogenesis, serotonin expression, and inhibiting apoptosis in social isolation rats during adolescence. J Exerc Rehabil 2020;16(2):132–40.

[36] Moylan S, et al. Oxidative & nitrosative stress in depression: why so much stress? Neurosci Biobehav Rev 2014;45.

[37] D'Mello C, Swain MG. Immune-to-brain communication pathways in inflammation-associated sickness and depression. Curr Topics Behav Neurosci 2017;73–94.

[38] Leonard B, Maes M. Mechanistic explanations how cell-mediated immune activation, inflammation and oxidative and nitrosative stress pathways and their sequels and concomitants play a role in the pathophysiology of unipolar depression. Neurosci Biobehav Rev 2012;36.

[39] Savitz J. The kynurenine pathway: a finger in every pie. Mol Psychiatry 2020;25(1):131–47.

[40] Wichers MC, et al. IDO and interferon-alpha-induced depressive symptoms: a shift in hypothesis from tryptophan depletion to neurotoxicity. Mol Psychiatry 2005;10.

[41] Seo D-Y, et al. Exercise and neuroinflammation in health and disease. Int Neurourol J 2019;23(Suppl. 2):S82–92.

[42] Too LS, et al. The association between mental disorders and suicide: a systematic review and meta-analysis of record linkage studies. J Affect Disord 2019;259:302–13.

[43] Grasdalsmoen M, et al. Physical exercise, mental health problems, and suicide attempts in university students. BMC Psychiatry 2020;20(1):175.

[44] Boone SD, Brausch AM. Physical activity, exercise motivations, depression, and nonsuicidal self-injury in youth. Suicide Life Threat Behav 2016; 46(5):625–33. https://doi.org/10.1111/sltb.12240.

[45] Davidson CL, et al. The impact of exercise on suicide risk: examining pathways through depression, PTSD, and sleep in an inpatient sample of veterans. Suicide Life Threat Behav 2013;43(3):279–89.

[46] Schmitt A, Falkai P. Suicide ideation, stability of symptoms and effects of aerobic exercise in major depression. Eur Arch Psychiatry Clin Neurosci 2014;264(7):555–6.

[47] Abdollahi A, et al. Effect of exercise augmentation of cognitive behavioural therapy for the treatment of suicidal ideation and depression. J Affect Disord 2017;219:58–63.

[48] Miller KJ, et al. Comparative effectiveness of three exercise types to treat clinical depression in older adults: a systematic review and network meta-analysis of randomised controlled trials. Ageing Res Rev 2020;58, 100999.

[49] NHS. Couch to 5k: week by week. cited 22/1/21. Available from https://www.nhs.uk/live-well/exercise/couch-to-5k-week-by-week/; 2021.

[50] Sinton-Hewitt P. Park run. 21/1/21. Available from: https://www.parkrun.com/about/; 2021.

[51] Robertson R, et al. Walking for depression or depressive symptoms: a systematic review and meta-analysis. Ment Health Phys Act 2012;5(1):66–75.

Chapter 18

Yoga and mental health

Michaela C. Pascoe and Alexandra G. Parker

Institute for Health and Sport, Victoria University, Melbourne, VIC, Australia

Introduction

Yoga, a system of the ancient Vedic tradition, aims to improve health and well-being and alter states of consciousness and functioning of the mind. A variety of approaches have developed throughout different periods of history, therefore the yoga system includes a large variety of practices and techniques. Yoga as a practice has become popular throughout the world [1,2]. In Australia, yoga asana is the preferred cardio, strength, and flexibility exercise and approximately 2.18 million Australians were participating in 2017 [3]. Medical practitioners often recommend yoga to their patients, according to a national survey study from the United States [4], for a range of health and psychological concerns. Similarly, in Australia, 76.6% of general practitioners (GPs) in rural and regional New South Wales, reported referring their patients to a yoga therapist [5]. This is supported by research as a recent meta-analysis shows that yoga practice is an effective approach to reduce depressive and anxious symptoms [6].

Yoga and mental health

Yoga has become a popular way to improve psychological wellbeing [1,7,8]. Yoga-based interventions for mental health promotion and treatment have been researched in both clinical and nonclinical populations.

In clinical settings, yoga-based interventions have been shown to reduce major depressive episodes and lower the risk for dysthymia [9–12]. Yoga-based interventions have been found to decrease sadness in tsunami survivors with posttraumatic stress disorder (PTSD), and to decrease symptoms of PTSD in war-traumatized high school students [13,14]. Clinical trials and meta-analyses show that yoga-based practices decrease the severity of symptoms of depression, anxiety, pain, stress, psychological distress, and improve mental health-related quality of life, in clinical populations [6,15–20].

In nonclinical settings, systematic reviews show that yoga-based interventions decrease stress [21] and improve psychological wellbeing in healthy adults [22]. Meta-analyses have shown that yoga-based interventions decrease fatigue in both healthy people and people with chronic illnesses such as cancer, multiple sclerosis, renal disease, chronic pancreatitis, fibromyalgia, or asthma [23–25]. These findings are consistent with studies examining other physical activity interventions, such as aerobic exercise and resistance exercise training [26,27]. Yoga-based interventions may be especially suitable for those with little exercise experience, those with musculoskeletal injuries, and the elderly [28].

Known/potential mechanisms underlying effects

Mental health-related psychological processes influenced by yoga

Yoga-based interventions are shown to impact many psychological processes which impact mental health and well-being, as explored below.

Interoception

Interoception is awareness of the body or the body's internal state and is important in mental health and wellbeing [29–31]. It is an inherent component of yoga practice, as yoga involves continuous conscious monitoring or mindfulness of bodily cues and sensations [29,32]. Interoception has been associated with a range of mental health conditions and psychological constructs including empathy [33], emotional susceptibility [34], alexithymia [35], depression and anxiety [36], feelings of panic [37], generalized anxiety disorder [38], posttraumatic stress disorder (PTSD) [39], and eating disorders [40].

Exercise to Prevent and Manage Chronic Disease Across the Lifespan. https://doi.org/10.1016/B978-0-323-89843-0.00022-2

Research has found that yoga practice can impact interoceptive awareness [41,42]. One study provided preliminary evidence that in participants with PTSD, a yoga-based intervention improved interoceptive awareness and reduced symptom severity [41]. A further study found that a single yoga class improvement in interoceptive accuracy among the healthy controls, but not in patients with anorexia nervosa (AN). This study reported that patients with AN had lower interoceptive accuracy than the healthy individuals at baseline and concluded that the interoceptive awareness system appears to be impaired in AN [42].

Dispositional mindfulness

Dispositional mindfulness is a person's general level of trait mindfulness across situations and time [43,44]. Low dispositional mindfulness is associated with more depressive symptoms [45–48], posttraumatic stress disorder symptoms [49], and borderline personality disorder symptomology [50], while high dispositional mindfulness is associated with improved well-being [51], lower self-perceived stress [52] and anxiety symptoms [53].

Yoga practice can lead to an increase in dispositional mindfulness. In a cross-sectional study comprising of 455 individuals, those who rated themselves as highly involved in yoga had more dispositional and religious/spiritual well-being and lower depression-like symptoms, compared to those who rated themselves as marginally/moderately or not at all involved in yoga, indicating that there might be a dose–response relationship between the amount of yoga practice and dispositional mindfulness [54]. Six weeks of vinyasa yoga increased mindfulness and decreased anxiety and stress scores in a small pre-post study of 17 college students [55]. Another study found that college students with higher levels of dispositional mindfulness were associated with increases in self-compassion and social connectedness following a 15-week program of hatha yoga, indicating that mindfulness may be a moderator of yoga-induced changes in compassion [56]. In a randomized controlled study, undergraduate students with a diagnosis of anxiety and/or depression were randomized to 8 weeks of (1) hatha yoga-only (2) meditation and self-compassion training or (3) a no-intervention group. Both the meditation and yoga groups were seen to increase mindfulness and decrease depressive, anxiety, and stress symptoms [57].

Self-compassion

Self-compassion is the act of being compassionate to one's self [58]. Kristen Neff, a clinical research leader in this field, states that "self-compassion involves being touched by and open to one's own suffering, not avoiding or disconnecting from it, generating the desire to alleviate one's suffering and to heal oneself with kindness. Self-compassion also involves offering nonjudgmental understanding to one's pain, inadequacies, and failures, so that one's experience is seen as part of the larger human experience" [59]. Meta-analysis shows that self-compassion predicts psychological well-being [60] as higher self-compassion corresponds to lower anxiety symptoms, perceived stress, depression symptoms, and improved quality of life, happiness, and positive affect [61]. In one study, individuals experiencing depression have less self-compassion than to matched, nondepressed individuals [58]. Among adolescents and young adults with depression symptoms, self-compassion buffers against anxiety following a social stress test [62].

In a pre-post study, a 4-month residential yoga program was found to increase levels of self-compassion as well as decrease perceived stress, and the effect of the yoga program on perceived stress was found to be mediated by self-compassion [63]. In a survey of yoga teachers and students, dispositional mindfulness and self-compassion positively correlate, and both negatively correlated with depression, anxiety, and stress symptoms [64]. In college students, a 15-week program of hatha yoga increases self-compassion [56]. Increases in self-compassion have been found to mediate the mood-enhancing effects of some mindfulness-based stress reduction (MBSR) interventions, which incorporate gentle hatha yoga, after controlling for changes in mindfulness [65].

Emotion regulation

An individual's capability to regulate emotional states is an important predictor of well-being, life satisfaction, and healthy relationships while poor emotional regulation strategies, such as rumination and emotional suppression are linked to more depressive symptoms and higher intensity of negative emotions [66–68]. Better emotional regulation is associated with reduced alcohol and drug use/misuse among adolescents and young people [69]. Yoga has been shown to improve emotional regulation in adolescents, compared to physical education, and when delivered in a school setting [70]. Yoga practitioners are also seen to have reduced emotional reactivity compared to individuals who do not practice yoga, and several years of practice is associated with decreased respiratory arousal in response to negative situations [71].

Rumination

Rumination (focused attention on thoughts, past events, and symptoms associated with distress) is a maladaptive emotional regulation strategy that tends to exacerbate emotional distress [72]. Higher self-compassion is proposed to improve psychological well-being by reducing rumination and avoidance. As shown in outpatients with depression, avoidance and rumination mediate the relationship between depression symptoms and self-compassion [58]. Among undergraduate students, rumination is shown to mediate the negative association between self-compassion and depression and anxiety symptoms, respectively [73]. This suggests that the influence of self-compassion on repetitive, unproductive thinking can buffer against depression and anxiety symptoms [73].

In a randomized controlled study involving women with depression, 12 weeks of a yoga-based intervention decreased rumination, which corresponded to reductions in depression symptoms, compared to a walking intervention [74]. Similarly, in another randomized, controlled study involving women with major depression, 8 weeks of hatha yoga decreased depression and ruminations, compared to the attention-control group [75].

The stress response and mental health

The bidirectional relationship between stress and mental ill-health is well accepted [76–78]. Stressful life events are a strong predictor of depression onset [79,80]. Among young people, depression onset is often preceded by major life stressors [81].

Stressors can be both real and perceived threats to homeostasis, safety or well-being, and result in the activation of the stress response, or the "fight-flight-or-freeze" response [82]. This is characterized by an increase in blood pressure, breathing rate, heart rate among other physiological functions [82]. The stress response is controlled by the autonomic nervous system (ANS) [82], which has two components: the sympathetic nervous system (SNS) and the parasympathetic nervous system (PNS) [83]. The ANS controls involuntary body functions, such as the dilation or constriction of blood vessels, blood pressure, heartbeat, breathing, and digestion. The hypothalamic pituitary adrenal (HPA) axis, which consists of the adrenal glands, hypothalamus, and pituitary gland, is also responsible for regulating the stress response, through the modulation of stress hormones [84]. The HPA axis releases the peptide hormone, corticotropin-releasing hormone (CRH), in response to stressors, which triggers the release of the polypeptide tropic hormone, adrenocorticotropic hormone (ACTH), and the glucocorticoid steroid hormone, cortisol [84–86]. Cortisol has many metabolic and physiological functions, including the maintenance of steady supplies of blood sugar by releasing stored glucose from the liver, ensuring that sufficient energy is available to cope with ongoing or prolonged stressors [87].

Hypothalamic survival responses are especially resource-intensive, which has implications for an individual's physical and behavioral health [88]. The negative feedback loop regulating baseline HPA function and stress responses can be disrupted with too much exposure to stressful events [86,89]. The adrenal gland can increase in size and become more sensitive to ACTH, thus amplifying glucocorticoid responses to stressors [90]. Repeatedly high levels of cortisol can result in glucocorticoid receptor resistance by impairing the function of glucocorticoid receptors (such as via reduced expression, downregulation, nuclear translocation) [91–93]. This occurs in brain regions including the prefrontal cortex (and paraventricular nucleus) and hippocampus [94,95], and disrupts the glucocorticoid feedback control of the HPA axis [96–98], and baseline glucocorticoid release increases [90,99]. High ongoing cortisol can lead to damage and remodeling in the hippocampus, prefrontal cortex, and amygdala [100]. The prefrontal cortex is involved in working memory, executive function, and self-regulatory behaviors [101], the hippocampus in mood regulation [100], and the amygdala in the regulation of emotional responses (including fear, anxiety, and aggression) [102]. Individuals with mood disorders such as depression show changes in the structure and function of these regions [100] illustrating why persistent activation of amygdala-prefrontal survival responses is associated with the onset of a number of psychiatric disorders [88,103–105], which are associated with an increased inflammatory state [76,106,107].

Yoga-based interventions influence the stress response

Blood pressure

Meta-analysis of randomized controlled trials (RCTs) shows that multicomponent yoga-based interventions decreased resting diastolic blood pressure (DBP) by 3.66 mm Hg (mmHg), and systolic blood pressure (SBP) by 5.34 mmHg, compared to the time/attention control group, in 16 and 17 studies comprising of 887 and 1058 individuals respectively, from diverse populations [108]. In seven studies measuring SPB, the control group received an exercise intervention such as aerobic exercise or resistance-based exercise, and therefore the yoga decreased resting SBP more than other forms of exercise. Given this, it might be that yoga decreases resting SBP due to a combination of the mindfulness practice and cardiovascular effects of the practice [108]. Reductions in DBP and systolic blood pressure (SBP) of as little as $\geq 2\,mmHg$ have been shown to be an associate to reduce the incidence of cardiovascular disease in both hypertensive and normotensive individuals, indicating that even small reductions in blood pressure are clinically meaningful [109,110].

Heart rate

Meta-analysis of 15 RCTs ($N = 879$) shows that yoga decrease resting heart rate by 3.2 beats per minute (bpm) compared to time/attention control groups, in diverse populations [108]. An earlier meta-analysis found that aerobic exercise reduced heart rate by 5 bpm, indicating that yoga-based interventions can result in changes in heart rate not dissimilar to aerobic exercise [111].

Heart rate variability

Heart rate variability (HRV) reflects the capability of the cardiovascular system to cope with psychological and physical challenges and a higher HRV indicates better ability to switch from "fight-or-flight" to a relaxed state. [112]. A systematic review reports that yoga increased HRV and vagal dominance, which indicates increased PNS activity, from 59 studies involving 2358 participants from diverse populations [113]. This review also reported that in people who practice yoga regularly increased activity of the vagus nerve is present at rest, compared to people who do not practice yoga [113]. Another systematic review found that yoga practice improved several domains of HRV in healthy individuals or patients with any medical condition and compared to any type of control intervention [114].

Cortisol

High cortisol levels are often present in individuals with mental illness and likely reflect increased autonomic nervous system reactivity and dysregulation [115,116]. A Meta-Analysis comparing yoga to a time/attention control group reported that yoga decreased waking and evening salivary cortisol in five studies comprising 386 individuals [108].

Multicomponent yoga programs which incorporate the eight limbs of the yoga model may impact cortisol more than yoga programs focused only on postures [117]. In one RCT, university students with elevated symptoms of depression, anxiety, and stress were randomized to an integrated multicomponent yoga-based program, yoga asana as exercise, or a control group that completed surveys. Both yoga programs decreased depression symptoms and stress levels, but only the multicomponent yoga program decreased anxiety symptoms and cortisol levels [117].

Inflammatory proteins

Cytokines are signaling protein molecules that mediate and regulate immunity and inflammation [118]. Stressors, which also include psychological stressors, can induce pro-inflammatory cytokine secretion [119–121]. Stress-induced release of cytokines such as and interleukin-6 [IL-6], interleukin (IL-10), and tumor necrosis factor-alpha [TNF-α], result in HPA axis stimulation and the release of cortisol [122,123]. The increased cortisol induced a negative-feedback response, then suppress the further release of cytokines [122,124–126]. Increased inflammation is associated with adverse health outcomes, and chronic stress and inflammation can contribute to the onset and maintenance of depression and anxiety [76,127,128].

In one systematic review, 11 of 15 RCT studies reported positive effects of yoga on IL-6 ($n = 11$ studies), C-reactive protein (CRP—a pentameric protein that increases in response to inflammation) ($n = 10$ studies), and TNF-α ($n = 8$ studies), in adults with chronic inflammatory-related disorders [129]. Further, a longer duration of yoga practice was associated with greater improvements in inflammation [129]. This partially supports the findings of a meta-analysis reporting that mindfulness-based stress reduction (MBSR), which incorporates hatha yoga, decreases IL-6, but there is no impact of yoga on interleukin-8 or CRP [108]. In future research, it is worth exploring if inflammation markers are biomarkers of treatment response.

Yoga-based interventions and neurobiological changes

The etiology of depression and anxiety is intimately linked with stress and inflammation and changes in brain structure. As discussed above, depression and anxiety are charactered by hypothalamic–pituitary–adrenal (HPA) axis dysfunction, increased circulatory cortisol, increased expression of peripheral pro-inflammatory cytokines, and neurotropic factors [91,106]. In addition, meta-analyses report gross morphological cerebral changes, including smaller hippocampal and amygdala volumes and atrophy of prefrontal cortex [130].

A recent systematic review identified 11 studies assessing the effects of yoga-based practice on brain structures, function, and cerebral blood flow. Nine of these studies assessed hatha yoga and physical postures in particular, and most studies were cross-sectional in that they compared the brain of people practicing yoga to yoga naive individuals [131]. Collectively, the studies reviewed found that yoga-based practices had an effect on the function and/or structure of the prefrontal cortex, cingulate cortex, hippocampus, amygdala, and brain networks including the DMN [131].

Yoga and neurotrophic factors, neurotransmitters, and cytokines

Streeter et al. found that 12-week Iyengar yoga decreased anxiety and improved mood, compared to walking. The concentration of the neurotransmitter gamma-aminobutyric acid (GABA), positively correlates with mood scores and negatively correlates with anxiety scores [132], indicating that GABA likely plays an important role in mediating the benefits of yoga on mood. This is consistent with previous research indicating that decreased GABA levels are correlated with mood disorders, [133] and corroborates Streeter et al.'s 2007 cross-sectional study, which examined brain GABA levels in yoga practitioners, compared to a control reading group. After a 60-min yoga session, there was a 27% increase in thalamic GABA levels, [134] compared to the 13% increase observed in the current study. A possible reason could be that the 2007 study tested experienced yoga practitioners while the 2010 study recruited novice yoga practitioners.

In a small study of healthy university students, 12 weeks of yoga has been seen to reduce plasma levels of adrenalin and to increase plasma levels of serotonin, compared with a no-intervention control group [135]. In premenopausal women with chronic low back pain, 12 weeks of hatha group-based yoga was found to decrease back pain and to preserve serum levels of the monoamine neurotransmitter, serotonin (5-hydroxytryptamine [5-HT]) [136], compared to the no intervention group [137]. Furthermore, yoga was found to protect against increases in depression symptoms and to increase levels of the protein, brain-derived neurotrophic factor (BDNF) [137], which is associated with clinical changes in depression symptoms [138]. Conversely, however, in a pre-post study involving 38 healthy individuals, a 12-week hatha yoga and meditation retreat increased dispositional mindfulness, brain-derived neurotrophic factor (BDNF), the cortisol awakening response, and pro-and antiinflammatory cytokine levels [139], indicating that further research is required in order to explore differences in the impacts of yoga on neurotrophic factors, neurotransmitters, and cytokines.

Structural brain changes

Cross-sectional studies show that 13 individuals practicing yoga for more than 3 years show greater gray matter volume in the left hippocampus, compared to 13 age and sex-matched yoga-naive individuals [140], indicating that regular long-term yoga is associated with functional and functional changes in brain regions important for memory and emotional regulation [141]. A small study of seven individuals practicing multicomponent hatha yoga shows similar increases in volume in the hippocampus, parahippocampal gyrus as well as the frontal, temporal, limbic, occipital, and cerebellar regions, compared to seven yoga naïve individuals. Interestingly, hatha yoga practitioners also reported fewer cognitive failures, and the number of cognitive failures were found to be associated with having smaller gray matter volume in the limbic, frontal, occipital, and cerebellar regions, while yoga experience was positively correlated to the volume of limbic, temporal, frontal, occipital, and cerebellar regions [142].

Another study shows that 14 individuals in their 30s who practice yoga were seen to have a reduction in age-related global brain gray matter decline, compared to 14 individuals who did not practice yoga, suggesting that yoga practices may protect against age-related gray matter decline. Interestingly, the hours of weekly practice were correlated with gray matter volume in the, precuneus/posterior cingulate cortex, primary somatosensory cortex/superior parietal lobulehippocampus, and primary visual cortex [143].

Similarly, seven female adults aged 60 and above and with 8 or more years of hatha yoga experience, had increased cortical thickness in the left prefrontal cortex, including part of the middle frontal and superior frontal gyri, compared to seven yoga naive individuals. Participants between groups were matched for the amount of nonyoga physical activity that they engaged in, and therefore the differences in cortical thickness could not be attributed to overall greater physical activity among yoga-practitioners. The authors suggested that these changes indicate that yoga practice promotes neuroplastic changes in executive brain systems, which may confer therapeutic benefits [142].

In 7 healthy older adults, 6 months of yoga practice for 60 min a day, 5 times a week, resulted in an increased volume of the bilateral hippocampus [144]. Among participants with mild cognitive impairment, 12 weeks of kundalini yoga with kirtan kriya (a type of meditation from the kundalini yoga tradition) for 60 min a day once a week, did not result in changes in hippocampus volume but was associated with a trend toward the decreased volume of the dorsal anterior cingulate cortex, compared to the memory enhancement control group [145].

Only a handful of intervention studies have assessed the impact of yoga-based interventions on brain structure and function, indicating a gap in the research field. Further, the above studies were both observational or cross-sectional but also comprised of very small sample sizes, indicating a gap in this research area.

Functional brain changes

Cross-sectional research shows that in 13 individuals who have been practicing yoga for more than 3 years, that there is less brain activation in the left dorsolateral prefrontal cortex (dlPFC), which is involved in working memory, compared to 13 yoga naive individuals. This indicates that yoga-based practice may improve this component of cognitive health [140]. In elderly individuals who had practiced hatha yoga twice a week for at least 8 years, greater functional connectivity has been observed between the medial prefrontal cortex and the posterior cingulate cortex, compared to age-matched yoga-naïve individuals. The authors interpreted this to reflect a healthier cognitive aging process [146]. Finally, in four patients with mild hypertension, cerebral blood flow was seen to decrease in the right amygdala, dorsal medial cortex, and sensorimotor areas, following 12 weeks of Iyengar yoga [147].

Designing programs with participants in mind

The studies reported in the current chapter show that few have considered participant preference for the type and intensity of yoga-based intervention when designing clinical interventions or mental health promotion initiatives. Designing programs in accordance with individual preferences may facilitate engagement with, adherence to, and perhaps even impact the effectiveness of yoga programs, compared to approaches that prescribe the type and intensity of yoga. Positive emotions are an important factor in predicting adherence to physical activity and exercise generally [148,149], and autonomy is a basic psychological need fundamental to positive mental health [150,151].

Practice recommendations for clinicians

The experience of negative affect is more likely to occur when participating in higher intensity levels of exercise [148] and affect predicting adherence to exercise [148,149]. Therefore, more gentle forms of yoga may be more appropriate for individuals commencing an exercise-based program or who are starting from a low baseline of engagement in physical activity. Indeed, yoga-based interventions should be created to suit a range of individual needs and preferences, and designed to be driven by individuals, to facilitate uptake and adherence.

Conclusion

Overall, the current work demonstrates that yoga-based interventions likely improve mental health via physiological, psychological, and neurobiological pathways and that the beneficial effects of yoga-based interventions cannot be attributed to exercise-related effects of yoga asana alone. Yoga practices incorporate mindful awareness, controlled breathing, philosophical teachings, meditative techniques, as well as physical asana, all of which likely improve stress regulation and mental health outcomes. This is important to consider with regards to designing and delivering yoga-based interventions in clinical and practice settings, as better outcomes are likely to be associated with yoga-based programs that include asanas plus the other components of yoga practice.

References

[1] Ding D, Stamatakis E. Yoga practice in England 1997-2008: prevalence, temporal trends, and correlates of participation. BMC Res Notes 2014;7:172.
[2] Tindle HA, et al. Trends in use of complementary and alternative medicine by us adults: 1997-2002. Altern Ther Health Med 2005;11(1):42–9.
[3] Morgan R. Yoga participation stretches beyond pilates & aerobics. Roy Morgan; 2018.
[4] Nerurkar A, et al. When conventional medical providers recommend unconventional medicine: results of a national study. Arch Intern Med 2011;171(9):862–4.
[5] Wardle J, Adams J, Sibbritt D. Referral to yoga therapists in rural primary health care: a survey of general practitioners in rural and regional New South Wales, Australia. Int J Yoga 2014;7(1):9–16.
[6] Cramer H, et al. Yoga for depression: a systematic review and meta-analysis. Depress Anxiety 2013;30(11):1068–83.
[7] Penman S, et al. Yoga in Australia: results of a national survey. Int J Yoga 2012;5(2):92–101.
[8] Clarke TC, et al. Trends in the use of complementary health approaches among adults: United States, 2002-2012. Natl Health Stat Rep 2015;79:1–16.
[9] Banerjee B, et al. Effects of an integrated yoga program in modulating psychological stress and radiation-induced genotoxic stress in breast cancer patients undergoing radiotherapy. Integr Cancer Ther 2007;6(3):242–50.
[10] Woolery A, et al. A yoga intervention for young adults with elevated symptoms of depression. Altern Ther Health Med 2004;10(2):60–3.

[11] Sharma VK, et al. Effect of Sahaj Yoga on neuro-cognitive functions in patients suffering from major depression. Indian J Physiol Pharmacol 2006;50(4):375–83.

[12] Butler LD, et al. Meditation with yoga, group therapy with hypnosis, and psychoeducation for long-term depressed mood: a randomized pilot trial. J Clin Psychol 2008;64(7):806–20.

[13] Telles S, Singh N, Balkrishna A. Managing mental health disorders resulting from trauma through yoga: a review. Depress Res Treat 2012;2012:9.

[14] Telles S, et al. Post traumatic stress symptoms and heart rate variability in Bihar flood survivors following yoga: a randomized controlled study. BMC Psychiatry 2010;10:18.

[15] Goyal M, et al. Meditation programs for psychological stress and well-being: a systematic review and meta-analysis. JAMA Intern Med 2014;174 (3):357–68.

[16] de Manincor M, et al. Individualized yoga for reducing depression and anxiety, and improving well-being: a randomized controlled trial. Depress Anxiety 2016;33(9):816–28.

[17] Cramer H, et al. Yoga for anxiety: a systematic review and meta-analysis of randomized controlled trials. Depress Anxiety 2018;35:830–43.

[18] Cramer H, et al. A systematic review of yoga for major depressive disorder. J Affect Disord 2017;213:70–7.

[19] Cramer H, et al. Yoga for posttraumatic stress disorder—a systematic review and meta-analysis. BMC Psychiatry 2018;18(1):72.

[20] Brinsley J, et al. Effects of yoga on depressive symptoms in people with mental disorders: a systematic review and meta-analysis. Br J Sports Med 2020;55(17):992–1000. https://doi.org/10.1136/bjsports-2019-101242.

[21] Chong CS, et al. Effects of yoga on stress management in healthy adults: a systematic review. Altern Ther Health Med 2011;17(1):32–8.

[22] Hendriks T, de Jong J, Cramer H. The effects of yoga on positive mental health among healthy adults: a systematic review and meta-analysis. J Altern Complement Med 2017;23(7):505–17.

[23] Boehm K, et al. Effects of yoga interventions on fatigue: a meta-analysis. Evid Based Complement Alternat Med 2012;2012, 124703.

[24] Shohani M, et al. The effect of yoga on the quality of life and fatigue in patients with multiple sclerosis: a systematic review and meta-analysis of randomized clinical trials. Complement Ther Clin Pract 2020;39, 101087.

[25] Dong B, et al. Yoga has a solid effect on cancer-related fatigue in patients with breast cancer: a meta-analysis. Breast Cancer Res Treat 2019;177 (1):5–16.

[26] Gordon BR, et al. Association of efficacy of resistance exercise training with depressive symptoms: meta-analysis and meta-regression analysis of randomized clinical trials. JAMA Psychiat 2018;75(6):566–76.

[27] Schuch FB, et al. Exercise as a treatment for depression: a meta-analysis adjusting for publication bias. J Psychiatr Res 2016;77:42–51.

[28] Khalsa SBS. Yoga for psychiatry and mental health: an ancient practice with modern relevance. Indian J Psychiatry 2013;55(Suppl 3):S334.

[29] Gard T, et al. Potential self-regulatory mechanisms of yoga for psychological health. Front Hum Neurosci 2014;8:770.

[30] Craig AD. How do you feel? Interoception: the sense of the physiological condition of the body. Nat Rev Neurosci 2002;3(8):655–66.

[31] Farb N, Mehling WE. Editorial: Interoception, contemplative practice, and health. Front Psychol 2016;7:1898.

[32] Sullivan MB, et al. Yoga therapy and polyvagal theory: the convergence of traditional wisdom and contemporary neuroscience for self-regulation and resilience. Front Hum Neurosci 2018;12:67.

[33] Ainley V, Maister L, Tsakiris M. Heartfelt empathy? No association between interoceptive awareness, questionnaire measures of empathy, reading the mind in the eyes task or the director task. Front Psychol 2015;6:554.

[34] Calì G, et al. Investigating the relationship between interoceptive accuracy, interoceptive awareness, and emotional susceptibility. Front Psychol 2015;6:1202.

[35] Longarzo M, et al. The relationships between interoception and alexithymic trait. The self-awareness questionnaire in healthy subjects. Front Psychol 2015;6:1149.

[36] Paulus MP, Stein MB. Interoception in anxiety and depression. Brain Struct Funct 2010;214(5):451–63.

[37] Pappens M, et al. Interoceptive fear learning to mild breathlessness as a laboratory model for unexpected panic attacks. Front Psychol 2015;6:1150.

[38] Chan P-YS, et al. Respiratory sensory gating measured by respiratory-related evoked potentials in generalized anxiety disorder. Front Psychol 2015;6:957.

[39] Payne P, Levine PA, Crane-Godreau MA. Somatic experiencing: using interoception and proprioception as core elements of trauma therapy. Front Psychol 2015;6:93.

[40] Martin E, et al. Interoception and disordered eating: a systematic review. Neurosci Biobehav Rev 2019;107:166–91.

[41] Neukirch N, Reid S, Shires A. Yoga for PTSD and the role of interoceptive awareness: a preliminary mixed-methods case series study. Eur J Trauma Dissociation 2019;3(1):7–15.

[42] Demartini B, Goeta D, Marchetti M, Bertelli S, Anselmetti S, Cocchi A, et al. The effect of a single yoga class on interoceptive accuracy in patients affected by anorexia nervosa and in healthy controls: a pilot study. Eat Weight Disord 2021;26(5):1427–35. https://doi.org/10.1007/s40519-020-00950-3.

[43] Brown KW, Ryan RM, Creswell JD. Mindfulness: theoretical foundations and evidence for its salutary effects. Psychoanal Inq 2007;18(4):211–37.

[44] Kabat-Zinn J, Hanh TN. Full catastrophe living: using the wisdom of your body and mind to face stress, pain, and illness. Delta; 2009.

[45] Barnhofer T, Duggan DS, Griffith JW. Dispositional mindfulness moderates the relation between neuroticism and depressive symptoms. Personal Individ Differ 2011;51(8):958–62.

[46] Marks AD, Sobanski DJ, Hine DW. Do dispositional rumination and/or mindfulness moderate the relationship between life hassles and psychological dysfunction in adolescents? Aust N Z J Psychiatry 2010;44(9):831–8.

[47] Jimenez SS, Niles BL, Park CL. A mindfulness model of affect regulation and depressive symptoms: positive emotions, mood regulation expectancies, and self-acceptance as regulatory mechanisms. Personal Individ Differ 2010;49(6):645–50.

[48] Bränström R, Duncan LG, Moskowitz JT. The association between dispositional mindfulness, psychological well-being, and perceived health in a Swedish population-based sample. Br J Health Psychol 2011;16(2):300–16.

[49] Smith BW, et al. Mindfulness is associated with fewer PTSD symptoms, depressive symptoms, physical symptoms, and alcohol problems in urban firefighters. J Consult Clin Psychol 2011;79(5):613.

[50] Fossati A, et al. Does mindfulness mediate the association between attachment dimensions and borderline personality disorder features? A study of Italian non-clinical adolescents. Attach Hum Dev 2011;13(6):563–78.

[51] Bajaj B, Gupta R, Pande N. Self-esteem mediates the relationship between mindfulness and well-being. Personal Individ Differ 2016;94:96–100.

[52] Brown KW, Weinstein N, Creswell JD. Trait mindfulness modulates neuroendocrine and affective responses to social evaluative threat. Psychoneuroendocrinology 2012;37(12):2037–41.

[53] Hou WK, Ng SM, Wan JHY. Changes in positive affect and mindfulness predict changes in cortisol response and psychiatric symptoms: a latent change score modelling approach. Psychol Health 2015;30(5):551–67.

[54] Gaiswinkler L, Unterrainer H. The relationship between yoga involvement, mindfulness and psychological well-being. Complement Ther Med 2016;26:123–7.

[55] Lemay V, Hoolahan J, Buchanan A. Impact of a yoga and meditation intervention on students' stress and anxiety levels. Am J Pharm Educ 2019;83 (5):7001. https://doi.org/10.5688/ajpe7001.

[56] Kishida M, Molenaar PC, Elavsky S. The impact of trait mindfulness on relational outcomes in novice yoga practitioners participating in an academic yoga course. J Am Coll Health 2019;67(3):250–62.

[57] Falsafi N. A randomized controlled trial of mindfulness versus yoga: effects on depression and/or anxiety in college students. J Am Psychiatr Nurses Assoc 2016;22(6):483–97.

[58] Krieger T, et al. Self-compassion in depression: associations with depressive symptoms, rumination, and avoidance in depressed outpatients. Behav Ther 2013;44(3):501–13.

[59] Neff K. Self-compassion: an alternative conceptualization of a healthy attitude toward oneself. Self Identity 2003;2(2):85–101.

[60] MacBeth A, Gumley A. Exploring compassion: a meta-analysis of the association between self-compassion and psychopathology. Clin Psychol Rev 2012;32(6):545–52.

[61] Neff KD, Rude SS, Kirkpatrick KL. An examination of self-compassion in relation to positive psychological functioning and personality traits. J Res Pers 2007;41(4):908–16.

[62] Bluth K, et al. Does self-compassion protect adolescents from stress? J Child Fam Stud 2016;25(4):1098–109.

[63] Gard T, et al. Effects of a yoga-based intervention for young adults on quality of life and perceived stress: the potential mediating roles of mindfulness and self-compassion. J Posit Psychol 2012;7(3):165–75.

[64] Snaith N, et al. Mindfulness, self-compassion, anxiety and depression measures in South Australian yoga participants: implications for designing a yoga intervention. Complement Ther Clin Pract 2018;32:92–9.

[65] Keng SL, et al. Mechanisms of change in mindfulness-based stress reduction: self-compassion and mindfulness as mediators of intervention outcomes. J Cogn Psychother 2012;26(3):270–80.

[66] John OP, Gross JJ. Healthy and unhealthy emotion regulation: personality processes, individual differences, and life span development. J Pers 2004;72(6):1301–34.

[67] Côté S, Gyurak A, Levenson RW. The ability to regulate emotion is associated with greater well-being, income, and socioeconomic status. Emotion 2010;10(6):923.

[68] Lennarz HK, et al. Emotion regulation in action: use, selection, and success of emotion regulation in adolescents' daily lives. Int J Behav Dev 2019;43(1):1–11.

[69] Park CL, Russell BS, Fendrich M. Mind-body approaches to prevention and intervention for alcohol and other drug use/abuse in young adults. Medicines 2018;5(3):64.

[70] Daly LA, et al. Yoga and emotion regulation in high school students: a randomized controlled trial. Evid Based Complement Alternat Med 2015;2015:794928. https://doi.org/10.1155/2015/794928.

[71] Mocanu E, et al. Reasons, years and frequency of yoga practice: effect on emotion response reactivity. Front Hum Neurosci 2018;12:264.

[72] Nolen-Hoeksema S, Wisco BE, Lyubomirsky S. Rethinking rumination. Perspect Psychol Sci 2008;3(5):400–24.

[73] Raes F. Rumination and worry as mediators of the relationship between self-compassion and depression and anxiety. Personal Individ Differ 2010;48(6):757–61.

[74] Schuver KJ, Lewis BA. Mindfulness-based yoga intervention for women with depression. Complement Ther Med 2016;26:85–91.

[75] Kinser PA, et al. Feasibility, acceptability, and effects of gentle hatha yoga for women with major depression: findings from a randomized controlled mixed-methods study. Arch Psychiatr Nurs 2013;27(3):137–47.

[76] Dantzer R. Depression and inflammation: an intricate relationship. Biol Psychiatry 2012;71(1):4–5.

[77] Dantzer R, et al. Inflammation-associated depression: from serotonin to kynurenine. Psychoneuroendocrinology 2011;36(3):426–36.

[78] Maes M. The cytokine hypothesis of depression: inflammation, oxidative & nitrosative stress (IO&NS) and leaky gut as new targets for adjunctive treatments in depression. Neuroendocrinol Lett 2008;29(3):287–91.

[79] Kessler RC. The effects of stressful life events on depression. Annu Rev Psychol 1997;48:191–214.

[80] Kendler KS, et al. Stressful life events, genetic liability, and onset of an episode of major depression in women. Am J Psychiatry 1995;152 (6):833–42.

[81] Lewinsohn PM, et al. First onset versus recurrence of depression: differential processes of psychosocial risk. J Abnorm Psychol 1999;108:483–9.

[82] Charmandari E, Tsigos C, Chrousos G. Endocrinology of the stress response. Annu Rev Physiol 2005;67:259–84.

[83] Nesse RM, Bhatnagar S, Ellis B. Evolutionary origins and functions of the stress response system. In: Stress: concepts, cognition, emotion, and behavior handbook of stress series. Academic Press; 2016. p. 95–101.

[84] Ortiga-Carvalho TM, et al. Hypothalamus-pituitary-thyroid axis. Compr Physiol 2016;6(3):1387–428.

[85] Myers B, McKlveen JM, Herman JP. Neural regulation of the stress response: the many faces of feedback. Cell Mol Neurobiol 2012. https://doi.org/10.1007/s10571-012-9801-y.

[86] Herman JP, et al. Regulation of the hypothalamic-pituitary-adrenocortical stress response. Compr Physiol 2016;6(2):603–21.

[87] Hiller-Sturmhofel S, Bartke A. The endocrine system: an overview. Alcohol Health Res World 1998;22(3):153–64.

[88] Canteras NS. Hypothalamic survival circuits related to social and predatory defenses and their interactions with metabolic control, reproductive behaviors and memory systems. Curr Opin Behav Sci 2018;24:7–13.

[89] Varghese FP, Brown ES. The hypothalamic-pituitary-adrenal axis in major depressive disorder: a brief primer for primary care physicians. Prim Care Companion J Clin Psychiatry 2001;3(4):151–5.

[90] Ulrich-Lai YM, et al. Chronic stress induces adrenal hyperplasia and hypertrophy in a subregion-specific manner. Am J Physiol Endocrinol Metab 2006;291(5):E965–73.

[91] Silverman MN, Sternberg EM. Glucocorticoid regulation of inflammation and its functional correlates: from HPA axis to glucocorticoid receptor dysfunction. Ann N Y Acad Sci 2012;1261:55–63.

[92] Kunz-Ebrecht SR, et al. Differences in cortisol awakening response on work days and weekends in women and men from the Whitehall II cohort. Psychoneuroendocrinology 2004;29(4):516–28.

[93] Mackin P. The role of cortisol and depression: exploring new opportunities for treatments. Psychiatr Times 2004;21(6):92. https://www.psychiatrictimes.com/view/role-cortisol-and-depression-exploring-new-opportunities-treatments.

[94] Cohen S, et al. Chronic stress, glucocorticoid receptor resistance, inflammation, and disease risk. Proc Natl Acad Sci 2012;109(16):5995–9.

[95] Gądek-Michalska A, et al. Influence of chronic stress on brain corticosteroid receptors and HPA axis activity. Pharmacol Rep 2013;65(5):1163–75.

[96] Herman JP, Watson SJ, Spencer RL. Defense of adrenocorticosteroid receptor expression in rat hippocampus: effects of stress and strain. Endocrinology 1999;140(9):3981–91.

[97] Gómez F, et al. Hypothalamic-pituitary-adrenal response to chronic stress in five inbred rat strains: differential responses are mainly located at the adrenocortical level. Neuroendocrinology 1996;63(4):327–37.

[98] Herman JP, Adams D, Prewitt C. Regulatory changes in neuroendocrine stress-integrative circuitry produced by a variable stress paradigm. Neuroendocrinology 1995;61(2):180–90.

[99] Gray M, Bingham B, Viau V. A comparison of two repeated restraint stress paradigms on hypothalamic-pituitary-adrenal axis habituation, gonadal status and central neuropeptide expression in adult male rats. J Neuroendocrinol 2010;22(2):92–101.

[100] McEwen BS, Nasca C, Gray JD. Stress effects on neuronal structure: hippocampus, amygdala, and prefrontal cortex. Neuropsychopharmacology 2016;41(1):3.

[101] McEwen BS, Morrison JH. The brain on stress: vulnerability and plasticity of the prefrontal cortex over the life course. Neuron 2013;79(1):16–29.

[102] LeDoux J. The amygdala. Curr Biol 2007;17(20):R868–74.

[103] Iwata M, Ota KT, Duman RS. The inflammasome: pathways linking psychological stress, depression, and systemic illnesses. Brain Behav Immun 2013;31:105–14.

[104] Ventriglio A, et al. Early-life stress and psychiatric disorders: epidemiology, neurobiology and innovative pharmacological targets. Curr Pharm Des 2015;21(11):1379–87.

[105] Herbert J. Cortisol and depression: three questions for psychiatry. Psychol Med 2013;43(3):449–69.

[106] Raison CL, Capuron L, Miller AH. Cytokines sing the blues: inflammation and the pathogenesis of depression. Trends Immunol 2006;27(1):24–31.

[107] Berk M, et al. So depression is an inflammatory disease, but where does the inflammation come from? BMC Med 2013;11:200. https://doi.org/10.1186/1741-7015-11-200.

[108] Pascoe MC, Thompson DR, Ski CF. Yoga, mindfulness-based stress reduction and stress-related physiological measures: a meta-analysis. Psychoneuroendocrinology 2017;86:152–68.

[109] Turnbull F, Blood Pressure Lowering Treatment Trialists' Collaboration. Effects of different blood-pressure-lowering regimens on major cardiovascular events: results of prospectively-designed overviews of randomised trials. Lancet 2003;362(9395):1527–35. https://doi.org/10.1016/s0140-6736(03)14739-3.

[110] Wong GW, Wright JM. Blood pressure lowering efficacy of nonselective beta-blockers for primary hypertension. Cochrane Database Syst Rev 2014;(2):CD007452. https://doi.org/10.1002/14651858.CD007452.pub2.

[111] Kelley GA, Kelley KA, Tran ZV. Aerobic exercise and resting blood pressure: a meta-analytic review of randomized, controlled trials. Prev Cardiol 2001;4(2):73–80.

[112] Perna G, et al. Heart rate variability: can it serve as a marker of mental health resilience?: special section on "translational and neuroscience studies in affective disorders" section editor, Maria Nobile MD, PhD. J Affect Disord 2020;263:754–61.

[113] Tyagi A, Cohen M. Yoga and heart rate variability: a comprehensive review of the literature. Int J Yoga 2016;9(2):97.

[114] Posadzki P, et al. Yoga for heart rate variability: a systematic review and meta-analysis of randomized clinical trials. Appl Psychophysiol Biofeedback 2015;40(3):239–49.

[115] Barker ET, et al. Daily stress and cortisol patterns in parents of adult children with a serious mental illness. Health Psychol 2012;31(1):130–4.

[116] Staufenbiel SM, et al. Hair cortisol, stress exposure, and mental health in humans: a systematic review. Psychoneuroendocrinology 2013;38(8):1220–35.

[117] Smith J, et al. Is there more to yoga than exercise? Altern Ther Health Med 2011;17(3):22–9.

[118] Salim S, Chugh G, Asghar M. Inflammation in anxiety. Adv Protein Chem Struct Biol 2012;88:1–25.

[119] Ruzek MC, et al. Characterization of early cytokine responses and an interleukin (IL)-6–dependent pathway of endogenous glucocorticoid induction during murine cytomegalovirus infection. J Exp Med 1997;185(7):1185–92.

[120] Lemay LG, Vander AJ, Kluger MJ. The effects of psychological stress on plasma interleukin-6 activity in rats. Physiol Behav 1990;47(5):957–61.

[121] Dhabhar FS, et al. Stress-induced changes in blood leukocyte distribution. Role of adrenal steroid hormones. J Immunol 1996;157(4):1638–44.

[122] Turnbull AV, Rivier CL. Regulation of the hypothalamic-pituitary-adrenal axis by cytokines: actions and mechanisms of action. Physiol Rev 1999;79(1):1–71.

[123] Steensberg A, et al. IL-6 enhances plasma IL-1ra, IL-10, and cortisol in humans. Am J Physiol Endocrinol Metab 2003;285(2):E433–7.

[124] Elenkov IJ, Chrousos GP. Stress hormones, Th1/Th2 patterns, pro/antiinflammatory cytokines and susceptibility to disease. Trends Endocrinol Metab 1999;10(9):359–68.

[125] Chrousos GP. The hypothalamic–pituitary–adrenal axis and immune-mediated inflammation. N Engl J Med 1995;332(20):1351–63.

[126] Reichlin S. Neuroendocrine-immune interactions. N Engl J Med 1993;329(17):1246–53.

[127] Masi G, Brovedani P. The hippocampus, neurotrophic factors and depression: possible implications for the pharmacotherapy of depression. CNS Drugs 2011;25(11):913–31.

[128] Pascoe MC, et al. Inflammation and depression: why poststroke depression may be the norm and not the exception. Int J Stroke 2011;6(2):128–35.

[129] Djalilova DM, et al. Impact of yoga on inflammatory biomarkers: a systematic review. Biol Res Nurs 2019;21(2):198–209.

[130] Koolschijn PC, et al. Brain volume abnormalities in major depressive disorder: a meta-analysis of magnetic resonance imaging studies. Hum Brain Mapp 2009;30(11):3719–35.

[131] Gothe NP, et al. Yoga effects on brain health: a systematic review of the current literature. Brain Plast 2019;5(1):105–22. https://doi.org/10.3233/BPL-190084.

[132] Streeter CC, et al. Effects of yoga versus walking on mood, anxiety, and brain GABA levels: a randomized controlled MRS study. J Altern Complement Med 2010;16(11):1145–52.

[133] Brambilla P, et al. GABAergic dysfunction in mood disorders. Mol Psychiatry 2003;8(8):721–37.

[134] Streeter CC, et al. Yoga asana sessions increase brain GABA levels: a pilot study. J Altern Complement Med 2007;13(4):419–26.

[135] Lim S-A, Cheong K-J. Regular yoga practice improves antioxidant status, immune function, and stress hormone releases in young healthy people: a randomized, double-blind, controlled pilot study. J Altern Complement Med 2015;21(9):530–8.

[136] Cowen PJ, Browning M. What has serotonin to do with depression? World Psychiatry 2015;14(2):158.

[137] Lee M, Moon W, Kim J. Effect of yoga on pain, brain-derived neurotrophic factor, and serotonin in premenopausal women with chronic low back pain. Evid Based Complement Alternat Med 2014;2014:203173. https://doi.org/10.1155/2014/203173.

[138] Brunoni AR, Lopes M, Fregni F. A systematic review and meta-analysis of clinical studies on major depression and BDNF levels: implications for the role of neuroplasticity in depression. Int J Neuropsychopharmacol 2008;11(8):1169–80.

[139] Cahn BR, et al. Yoga, meditation and mind-body health: increased BDNF, cortisol awakening response, and altered inflammatory marker expression after a 3-month yoga and meditation retreat. Front Hum Neurosci 2017;11:315.

[140] Gothe NP, et al. Differences in brain structure and function among yoga practitioners and controls. Front Integr Neurosci 2018;12:26.

[141] Milad MR, et al. Recall of fear extinction in humans activates the ventromedial prefrontal cortex and hippocampus in concert. Biol Psychiatry 2007;62(5):446–54.

[142] Froeliger B, Garland EL, McClernon FJ. Yoga meditation practitioners exhibit greater gray matter volume and fewer reported cognitive failures: results of a preliminary voxel-based morphometric analysis. Evid Based Complement Alternat Med 2012;2012:821307. https://doi.org/10.1155/2012/821307.

[143] Villemure C, et al. Neuroprotective effects of yoga practice: age-, experience-, and frequency-dependent plasticity. Front Hum Neurosci 2015;9:281.

[144] Hariprasad V, et al. Yoga increases the volume of the hippocampus in elderly subjects. Indian J Psychiatry 2013;55(Suppl 3):S394.

[145] Yang H, et al. Neurochemical and neuroanatomical plasticity following memory training and yoga interventions in older adults with mild cognitive impairment. Front Aging Neurosci 2016;8:277.

[146] Santaella DF, et al. Greater anteroposterior default mode network functional connectivity in long-term elderly yoga practitioners. Front Aging Neurosci 2019;11:158.

[147] Cohen DL, et al. Cerebral blood flow effects of yoga training: preliminary evaluation of 4 cases. J Altern Complement Med 2009;15(1):9–14.

[148] Ekkekakis P, Parfitt G, Petruzzello SJ. The pleasure and displeasure people feel when they exercise at different intensities decennial update and progress towards a tripartite rationale for exercise intensity prescription. Sports Med 2011;41(8):641–71.

[149] Lee HH, Emerson JA, Williams DM. The exercise-affect-adherence pathway: an evolutionary perspective. Front Psychol 2016;7:1285.

[150] Craft LL, et al. Psychosocial correlates of exercise in women with self-reported depressive symptoms. J Phys Act Health 2008;5:469–80.

[151] Ryan RM, Deci EL. Self-determination theory and the facilitation of intrinsic motivation, social development, and well-being. Am Psychol 2000;55(1):68–78.

Chapter 19

Exercise for chronic heart failure

Catherine Giuliano[a], Itamar Levinger[a,b], and Mary Woessner[a]

[a]Institute for Health and Sport, Victoria University, Melbourne, VIC, Australia [b]Australian Institute for Musculoskeletal Science, University of Melbourne and Western Health, St Albans, VIC, Australia

Introduction

Definition

CHF is a clinical syndrome characterized by an inability of the heart to pump a sufficient amount of blood to meet the metabolic demands of the body [1]. CHF results in significant clinical, functional, and financial costs to individuals and the community [2,3] and is characterized by hallmark symptoms of fatigue, dyspnea, and exercise intolerance [4,5].

The prevalence of CHF is estimated between 1% and 2% of Western [6,7] and Australian populations [3,8]. In Australia, approximately 70,000 new cases of CHF are diagnosed each year [3,9] and it is one of the top 10 leading causes of death in Australian men and women, accounting for 2.8% and 1.9% of deaths each year, respectively [10]. Rates of CHF are higher among indigenous than nonindigenous Australians and in those living in rural and remote regions [8,9]. The CHF population is expected to grow, partly due to an aging population and increasing prevalence of CHF risk factors, as well as improved postmyocardial infarction survival [11].

It is also well established that older individuals are disproportionately affected by CHF, with the average age of the first presentation of CHF being approximately 76 years old [12–15]. The Framingham Heart Study remains the largest longitudinal cohort study of cardiovascular disease and found the annual prevalence of CHF in men and women to be 8 per 1000 between 50 and 59 years, which increased to 66 per 1000 between 80 and 89 years for men and 79 per 1000 for women [12].

The economic burden of CHF is considerable. Globally, an estimated $108 billion is spent on CHF-related healthcare costs each year [2], while in Australia, this amount reaches over $3.1 billion [3]. Advances in CHF treatment have seen the life expectancy of patients increasing, yet prognosis remains poor [16]: survival rates are reported between 57% and 64% at 1 year following diagnosis of CHF and between 25% and 37% at 5 years [15,17].

Etiologies

CHF can develop from multiple etiologies. In fact, all cardiovascular diseases can ultimately lead to CHF. In western, high-income regions, the leading causes of CHF are coronary artery disease (CAD), accounting for between 36% and 59% of CHF cases; and hypertension (HTN) [6]. However, CHF rarely initiates from a singular isolated event or dysfunction. Rather, the development and progression of the syndrome are more akin to a web of interrelated comorbidities and precursors that cumulatively result in the cardiac and skeletal muscle impairments common to CHF.

Diagnosis and classification

A universal diagnostic test for CHF does not exist [18,19]. A conclusive diagnosis requires a clinical judgment of signs and symptoms as well as supporting cardiac imaging. Diagnosis is challenging in practice because there is a significant overlap between the signs and symptoms of CHF and other diseases, such as chronic obstructive lung disease (COPD) [20–25]. There is also a high prevalence of comorbidities in CHF including HTN, diabetes, and obesity, which can further complicate the diagnosis [24,26]. Typical symptoms of CHF include breathlessness, fatigue, and exercise intolerance, and clinical signs of CHF include orthopnea, peripheral oedema, elevated jugular veins, and a sudden increase in weight caused by fluid retention [27].

Exercise to Prevent and Manage Chronic Disease Across the Lifespan. https://doi.org/10.1016/B978-0-323-89843-0.00014-3

Classification based on ejection fraction

Left ventricular ejection fraction (LVEF) describes the ratio of SV to left ventricular end-diastolic volume that is ejected during systole. There are three main types of CHF based on EF [27]:

- heart failure with *reduced* ejection fraction (HFrEF): EF less than or equal to 40%,
- heart failure with *mid-range* ejection fraction (HFmrEF): EF 41%–49%
- heart failure with *preserved* ejection fraction (HFpEF): EF greater than or equal to 50%.

Classifications based on symptoms

There are two common classification systems used to classify the severity of CHF: The New York Heart Association (NYHA) and the American College of Cardiology (ACC)/American Heart Association (AHA) system [28,29], as shown in Table 1. The NYHA classifies patients with CHF based on the degree of breathlessness at rest and during exercise, on a scale of I to IV, from least symptomatic (Class I) to most symptomatic (Class IV). The ACC/AHA system provides the additive benefit of identifying individuals who are at risk of developing CHF based on structural heart disease and other treatable precursors of CHF (in early-stage A and B) so that preventative treatment can be initiated in a timely manner.

Pathophysiology

Exercise intolerance (i.e., a reduction in aerobic capacity), dyspnea, and fatigue are hallmark symptoms of CHF. The reductions in aerobic capacity, as measured by oxygen consumption (VO_2), are evident both at peak (i.e., VO_{2peak}) and submaximal exercise [30]. An individual with CHF can have a VO_{2peak} that is reduced by 50% or more compared to healthy age-matched controls [31–33]. VO_{2peak} is a strong predictor of mortality in patients with CHF, with every 1 mL/kg/min reduction in VO_{2peak} resulting in an adjusted mortality hazard ratio of 1.13 over 3.5 years (95% CI 1.09–1.17) [34]. Moreover, a VO_{2peak} of less than 14 mL/kg/min is the established criterion for cardiac transplantation [35–37].

Individuals with CHF also have significant impairments during submaximal exercise as indicated by reduced ventilatory thresholds and shorter distances achieved in the six-minute walk test [38,39]. Submaximal measures of aerobic fitness have significant relevance to the performance of activities of daily living [40,41] as well as prognostic value in CHF [39,42]. In a study including 223 patients with HFrEF, a ventilatory threshold (VT) of less than 11 ml/kg/min was associated with a 5.4-fold increase in death over 6 months [42].

The "Fick Principle" [43] offers insight into the physiological mediators of aerobic capacity by identifying that VO_2 is the product of central (i.e., CO) and peripheral (i.e., arterial–venous O_2 content difference, A-VO_2) factors:

TABLE 1 A comparison of the NYHA classes and ACC/AHA stages for classification of chronic heart failure.

	ACC/AHA system		NYHA system	
	Stage	**Description**	**Class**	**Description**
Asymptomatic	Stage A	High risk for HF but without structural heart disease or symptoms of HF		–
	Stage B	Structural heart disease but without signs or symptoms of HF		–
Symptomatic	Stage C	Structural heart disease with prior or current symptoms of HF	Class I	No symptoms at rest, symptoms only at levels of exertion that would limit healthy individual
			Class II	No symptoms at rest or mild exertion, symptoms on moderate exertion
	Stage D	Refractory HF requiring specialized interventions	Class III	No symptoms at rest, symptoms at mild exertion
			Class IV	Symptoms at rest

Adapted from Hunt SA, et al. 2009 Focused update incorporated into the ACC/AHA 2005 guidelines for the diagnosis and management of heart failure in adults. Circulation 2009;119(14):e391.

$$VO_2 = CO \times A - VO_2 \text{ difference,}$$

where VO_2 is measured in mL/min and A-VO_2 difference is the difference in oxygen content of the arterial and venous system (therefore a measure of the oxygen extraction capacity of the peripheral tissues), measured in mL of O_2 per 100 mL of blood. Considering the components of CO as stroke volume (SV) and heart rate (HR) the equation can also be expressed as:

$$VO_2 = (SV \times HR) \times A - VO_2 \text{ difference}$$

A suitable increase in VO_2 peak during exercise relies on increases in both the central and peripheral components. In healthy individuals, the increase in VO_2 during maximal exercise results from an approximate two to threefold increase in HR, a 0.4-fold increase in SV, and threefold increase in A-VO_2 difference [44,45]. In CHF, however, the contributions of each component are diminished and there are both central and peripheral limitations that contribute to exercise intolerance.

Exercise intolerance in CHF: Central and peripheral causes

It was first presumed that exercise intolerance in CHF was a direct consequence of impaired CO during exercise [46], where the reduction in CO is due to reductions in both SV and HR reserves. Indeed, resting SV is up to 22% lower in patients with CHF compared to healthy controls [47] and several studies report that SV during exercise in patients with CHF rises to only 50–89 mL during exercise compared to greater than 100 mL in healthy individuals [33,47,48]. The failure to increase SV during exercise may be due to alterations in filling volume (preload), myocardial contractility, and afterload and/or failure of the Frank-Starling Mechanism [48–50]. Patients with CHF also experience chronotropic incompetence (i.e., the failure of HR to increase during exercise) [51,52], elevated resting HR (i.e., resting tachycardia), and reduced HR reserves (HRR) [48,53], which collectively also contribute to exercise intolerance [51,54].

Despite the presence of significant central limitations, these alone are not sufficient to explain exercise intolerance in CHF. For instance, several studies have failed to find direct correlations between LV function, as measured by LVEF and VO_{2peak} [55–60]. Further studies have shown that pharmacologically induced improvements in CO do not translate directly into improvements in aerobic capacity [61]. Thus, it is now accepted that peripheral maladaptations within the skeletal muscle and vasculature are the primary contributors to exercise intolerance a theory labeled the muscle hypothesis of CHF [62,63].

The muscle hypothesis is depicted in Fig. 1. According to the hypothesis, the initial reductions in LV function and CO in CHF reduce blood flow to peripheral tissues and induce a catabolic state and subsequent skeletal muscle myopathy. The resulting muscle abnormalities lead to early fatigue and dyspnea that, in turn, further contributes to physical inactivity and sets in motion a deleterious feedback loop that drives disease progression. The most recognized muscle abnormalities include ergo reflex dysfunction [64–69], mitochondrial dysfunction [70–73] and changes in muscle mass [74,75] and fiber type distribution [66–69,71,76].

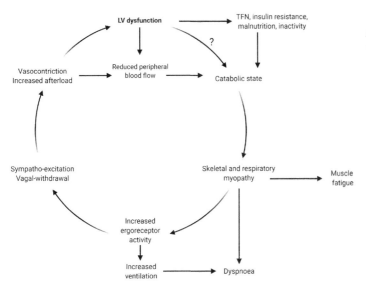

FIG. 1 The muscle hypothesis of chronic heart failure. *Adapted from Coats AJ, et al. Symptoms and quality of life in heart failure: the muscle hypothesis. Br Heart J 1994;72(2 Suppl):S36–9.*

Exercise prescription for CHF

Aerobic exercise and interval aerobic exercise

The exercise symptom burden is a major component of both HFrEF and HFpEF subtypes; however, evidence that supports exercise training is largely derived from patients with HFrEF.

The importance of regular aerobic exercise for improving aerobic capacity in patients with HFrEF is clear. A meta-regression analysis by Vromen et al. [77], including 17 studies and 2935 participants reported an increase in VO_{2peak} by 2.10 mL/kg/min following aerobic training compared to the nonexercise control group. The mean intensity of prescribed exercise of included studies ranged from 50% to 79% of VO_{2peak}, while the mean program length was 12 weeks (range 4–39) of four sessions per week (range 3–20) and of 30 min in duration (range 18–57).

An abundance of randomized controlled trials (RCTs) and meta-analyses also demonstrate the beneficial effect of aerobic exercise on quality of life and symptoms of depression [78] and on improving prognostic markers including VE/VCO_2 slope [79–81], N-terminal prohormone of brain natriuretic peptide [80,81] and vascular function as measured by flow-mediated dilation [82].

Similar benefits following aerobic exercise training are seen in patients with HFpEF, although data is limited to only a few individual studies with a small number of participants. In a meta-analysis including 10 studies and 399 participants, aerobic exercise, in comparison to nonexercise control, resulted in a group mean an increase in VO_{2peak} of 1.9 mL/kg/min (95% CI 1.3–2.5) as well as significant increases in quality of life measures [83]. An earlier meta-analysis (including eight studies and 174 participants with HFpEF) also showed significant increases in VO_{2peak} following exercise training (mean difference 2.08 mL/kg/min [95% CI 1.51–2.65]). This change equated to a 17% increase from baseline. There were also noted improvements in 6-min walking distance (32.1 m [17.2–47.1]) [84]. Taken together, aerobic exercise training appears to be an important treatment for patients with HFpEF, although further studies of greater size and higher statistical power are needed.

Interval aerobic training and high-intensity interval training

Interval training is particularly useful for individuals with limited aerobic capacity, as it allows for periods of recovery so that a greater net volume of exercise can be achieved in one exercise session. Interval training provides superior benefits to VO_{2peak} and LVEF in comparison to continuous aerobic training [85–87]. Improvements in quality of life [85] and submaximal aerobic capacity [85], however, appear similar when compared with continuous aerobic training.

High-intensity interval training (HIIT) is a form of interval training that involves high-intensity aerobic exercise (greater than 90% HRmax) for the work phase and is an effective exercise modality to increase VO_{2peak} in healthy populations [88,89]. The knowledge that greater increases in VO_{2peak} can be achieved with higher intensity exercise has generated interest for the potential utility of HIIT for patients with CHF [90]. A systematic review and meta-analysis including 13 studies and 411 patients with HFrEF (mean age range; 58–65 years) found that HIIT was superior to continuous aerobic exercise for improving VO_{2peak}. However, HIIT conferred no greater benefits than continuous aerobic exercise when examining the quality of life outcomes [91] or VE/VCO_2 slope [92].

Two primary issues have prevented the routine use of HIIT for patients with CHF. First, exercising at higher intensity exercise is associated with a greater risk of myocardial infarction or sudden death [93,94], particularly in individuals who are habitually inactive or who have cardiac risk factors [95]. Second, meta-analyses on HIIT generally recruit younger patients (mean range 57.0–65) [85,87,91] and there is insufficient data to allow for meta-analysis of HITT training in patients with HFpEF. Taken together, to date, HIIT is not routinely recommended for patients with CHF, but emerging evidence suggests it could be beneficial for individuals with CHF who are classified as a lower risk level [96].

Resistance training

Resistance training has many benefits for patients with CHF. Muscle strength is a key component of physical fitness and is important for weight management and overall cardiovascular health [97]. In older individuals, muscular strength, power, and endurance are particularly important for mobility, the performance of ADLs, quality of life, and for reducing the risk of falls and fractures [97–103]. Furthermore, it is hypothesized that resistance training can facilitate improvements in the skeletal muscle impairments that contribute to exercise tolerance in CHF.

As a standalone therapy, resistance training increases VO_{2peak}, submaximal exercise capacity, and quality of life in patients with HFrEF [104–106], while meta-analyses report superior increases in quality of life and muscular strength with combined aerobic and resistance training compared to continuous aerobic exercise [107–109]. To date, no studies have investigated resistance training as a standalone therapy in patients with HFpEF.

Combined aerobic and resistance training

The most comprehensive and recent review by the Cochrane group included 44 trials and 5783 participants (mean age range 51–81 years) with either HFpEF or HFrEF receiving exercise-based cardiac rehabilitation. All included trials involved an aerobic intervention and 14 trials included resistance training. Results from this review indicated that exercise training made little or no difference to short-term all-cause mortality at less than 1-year follow-up but may improve all-cause mortality with greater than 12-month follow-up [110]. Low-moderate quality evidence (as assessed by the GRADE method [111]) supported CR for reducing the risk of all-cause and HF-related hospital admissions and improved quality of life within 12-months of follow-up [110].

The limited benefit for mortality and hospitalizations was also demonstrated in an individual patient data meta-analysis, EXTRAMATCH II, which included 18 trials (in which all trials included aerobic exercise and six additionally included resistance training) and 3912 patients with HFrEF (mean age 61 ± 13 years, mean EF 26.7%) [112]. Results showed that CR did not have a significant effect on mortality or hospitalization (median follow-up 11.2–18.6 months) [112]. Bjarnason-Wehrens, Nebel [113] also found no difference in 6- to-12-month mortality or hospitalization in a systematic review and meta-analysis of exercise-based CR (25 studies: $n = 4481$, LVEF less than or equal to 40%) [113]. Taken together, these findings suggest that resistance training provides an additive benefit for improving quality of life and measures of muscular strength when combined with aerobic training in patients with CHF. Although patients with HFpEF are underrepresented across resistance training studies, it is likely that these patients would also likely benefit from the addition of resistance training, which is an important general recommendation for health and well-being.

Summary of recommendations

Exercise training is considered an integral part of the rehabilitation process for patients with CHF and is recommended by leading cardiac institutions around the world [27,114–117]. However, there is no universal agreement on exercise prescription (i.e., frequency, intensity, modality, duration) [118].

To date, aerobic exercise is the foundation of all exercise guidelines for patients with CHF [118]. The two most recent guidelines from Europe and Australia and New Zealand recommend moderate-intensity aerobic exercise for patients with HFrEF [27,114]. These documents cite evidence established by the 2014 Cochrane review [119] (updated in 2019) [110] and by the HF-ACTION trial [36].

Guidelines from Europe and North America support the progression of aerobic exercise to high-intensity (80%–90% of VO_2 peak) as tolerated [118,120] and suggest interval training as a useful approach for select patients, such as those who are frail [118,121,122].

The additive benefit of resistance training to general health and fitness is well recognized and an exercise program of combined aerobic and resistance training is the mainstay approach for patients with CHF [114,116,117,123]. Some institutions still take caution by recommending periods of 2–6 weeks of aerobic training before introducing resistance training [118]. Resistance training is not currently recommended *as a standalone therapy* for patients with CHF, due to it providing only a minimal benefit to aerobic capacity in comparison to other modalities.

Common among all guidelines is that syndrome severity and individual needs and risks must be taken into consideration when prescribing an exercise program. Graded exercise testing is also recommended to screen for symptomology and adverse events, as well as to determine a safe exercise intensity to prescribe initial training (Fig. 2).

Fig. 2 provides a useful decision tree for allied health professionals to consider when working with patients with CHF and was adapted from several leading society's statements and guidelines [1,106,124]. Table 2 provides a summary of exercise programming principles, adapted from Selig et al. [1].

Barriers and opportunities for exercise participation

Exercise training delivers a range of unique benefits for patients with CHF including increases in aerobic power, muscle strength, and quality of life. However, these benefits can only be actualized by long-term adherence, which remains a major concern in the CHF population [125]. The largest exercise trial in CHF to date, HF-ACTION, found that only about 40% of the participants allocated to the exercise group adhered to the suggested training volume [126]. Similarly, the coordinating study evaluating Outcomes of Advising and Counseling in HF (COACH study) found less than 40% of the study population

FIG. 2 A decision tree for exercise prescription for chronic heart failure.

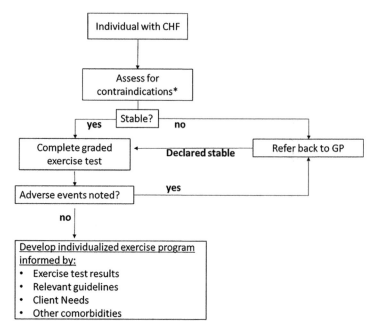

with CHF exercised regularly, despite 80% of the same population believing exercise to be an important practice for their health [127].

Attendance to formal cardiac rehabilitation programs in patients with CHF is also low, with rates of referral following hospital admission with CHF as low as 10% [128]. Barriers to exercise participation for individuals with CHF have been extensively investigated. Personal reasons for nonattendance include limited transportation assistance (ref our paper), a fear of undertaking physical activity [128] condition-specific symptoms, lack of individual motivation as well as geographical or economic barriers [129]. Service level and logistical barriers to participation also include program availability and accessibility, and the consistency and continuity in program personnel [128,130–132].

TABLE 2 A summary of exercise programming principles.

Type	Frequency	Intensity/Volume	Duration	Progression	Considerations
Aerobic training	4–7 days/week	NYHA I–II: RPE 11–14 or 40%–75% HRpeak NYHA III–IV: RPE ≤ 13 or 40%–65% of HRpeak	10–15 min at target intensity	Progress gradually until patient can complete 45–60 min	− All exercise should be below ischemic thresholds (when relevant) − Interval training can be used at a 1:1 exercise/rest ratio
Resistance training	2–3 days/week	NYHA I–II: RPE 11–15, 6–15 reps, 1 set initially NYHA III–IV: RPE: 10–13, 4–10 reps, 1 set initially	20 min at target intensity	Progress gradually until patient can complete 45–60 min. Increase sets from 1 to 2.	− Resistance training can include circuit training, Theraband, or body-weight exercise
Flexibility training	2–3 days/week	Stretch all major muscle groups	5–10 min	Progress to include more exercises as tolerated and increase range of motion of stretch	− Specific attention should be paid to stretches of the hip, knees, upper and lower spine, chest, and shoulders

Adapted from Selig SE, et al. Exercise & Sports Science Australia Position Statement on exercise training and chronic heart failure. J Sci Med Sport 2010;13 (3):288–94.

Some of these barriers can be overcome with relatively simple solutions such as patient education, adopting a motivational coaching approach, or offering reimbursements/incentives, or low-cost transport options. Practical guidelines are also available which provide recommendations for improving continuity of care within the contemporary care models of cardiac rehabilitation [130]. These recommendations include having a consistent and dedicated clinical team to provide cardiac rehabilitation across acute, subacute, and community care settings, as well as offering flexible cardiac rehabilitation models including home, telephone, and center-based programs [130].

The emergence of telehealth presents a unique opportunity to reconceptualize our traditional approaches to the delivery of exercise rehabilitation [133]. Telehealth technology can connect allied health practitioners with patients in rural and remote communities to overcome both geographical and financial barriers [134]. Recent systematic reviews and meta-analyses show that multidisciplinary telehealth exercise rehabilitation is a safe and effective alternative model of cardiac rehabilitation delivery [135–138], while targeted studies in older patients with CHF also show great promise [139]. Many patients also find telehealth-based programs to be more enjoyable than traditional program delivery [134].

Some newer technologies use mobile applications or virtual reality systems to facilitate exercise participation within a gaming platform: an approach called exergaming [133]. While there are few studies utilizing exergaming in lieu or as a component of cardiac rehabilitation, early evidence suggests these approaches are enjoyable and can improve adherence to exercise and provide significant health benefits [140,141].

Conclusion

Patients with CHF will benefit from exercise prescription involving a combination of aerobic and resistance training modalities. Exercise programming must be individualized to the patient's presenting symptoms and physical abilities, while also considering their individual motivations and resources. The emergence of telehealth technologies offers significant opportunities for improving the accessibility of exercise training for all individuals with CHF, regardless of their geographical location and other logistical barriers.

References

[1] Selig SE, et al. Exercise & Sports Science Australia Position Statement on exercise training and chronic heart failure. J Sci Med Sport 2010;13(3):288–94.

[2] Cook C, et al. The annual global economic burden of heart failure. Int J Cardiol 2014;171(3):368–76.

[3] Chen L, et al. Snapshot of heart failure in Australia. Melbourne, Australia: Mary MacKillop Institute for Health Research, Australian Catholic University; 2017.

[4] Ponikowski P, et al. 2016 ESC Guidelines for the diagnosis and treatment of acute and chronic heart failure. Eur Heart J 2016;18(8):891–975.

[5] Zambroski CH, et al. Impact of symptom prevalence and symptom burden on quality of life in patients with heart failure. Eur J Cardiovasc Nurs 2005;4(3):198–206.

[6] Mosterd A, Hoes AW. Clinical epidemiology of heart failure. Heart 2007;93(9):1137.

[7] Mosterd A, et al. Prevalence of heart failure and left ventricular dysfunction in the general population; The Rotterdam Study. Eur Heart J 1999;20(6):447–55.

[8] Sahle BW, et al. Prevalence of heart failure in Australia: a systematic review. BMC Cardiovasc Disord 2016;16:32.

[9] Clark RA, et al. Uncovering a hidden epidemic: a study of the current burden of heart failure in Australia. Heart Lung Circ 2004;13(3):266–73.

[10] Australian Institute of Health and Welfare. Australia's health 2014 in Australia's health series. Australian Institute of Health and Welfare; 2014.

[11] Australian Institute of Health and Welfare. National hospital morbidity database (NHMD). In: Trends in cardiovascular disease; 2014.

[12] Ho KK, et al. The epidemiology of heart failure: the Framingham Study. J Am Coll Cardiol 1993;22(4 Suppl A):6A–13A.

[13] Curtis LH, et al. Incidence and prevalence of heart failure in elderly persons, 1994-2003. Arch Intern Med 2008;168(4):418–24.

[14] Conrad N, et al. Temporal trends and patterns in heart failure incidence: a population-based study of 4 million individuals. Lancet 2018;391(10120):572–80.

[15] Ho KK, et al. Survival after the onset of congestive heart failure in Framingham Heart Study subjects. Circulation 1993;88(1):107–15.

[16] Jhund PS, et al. Long-term trends in first hospitalization for heart failure and subsequent survival between 1986 and 2003: a population study of 5.1 million people. Circulation 2009;119(4):515–23.

[17] Levy D, et al. Long-term trends in the incidence of and survival with heart failure. N Engl J Med 2002;347(18):1397–402.

[18] Zannad F, et al. Heart failure as an endpoint in heart failure and non-heart failure cardiovascular clinical trials: the need for a consensus definition. Eur Heart J 2008;29(3):413–21.

[19] Martindale JL, et al. Diagnosing acute heart failure in the emergency department: a systematic review and meta-analysis. Acad Emerg Med 2016;23(3):223–42.

[20] Komajda M, et al. Contemporary management of octogenarians hospitalized for heart failure in Europe: Euro Heart Failure Survey II. Eur Heart J 2009;30(4):478–86.

[21] Barsheshet A, et al. Predictors of long-term (4-year) mortality in elderly and young patients with acute heart failure. Eur J Heart Fail 2010;12 (8):833–40.

[22] Mogensen UM, et al. Clinical characteristics and major comorbidities in heart failure patients more than 85 years of age compared with younger age groups. Eur J Heart Fail 2011;13(11):1216–23.

[23] Lainscak M, et al. The burden of chronic obstructive pulmonary disease in patients hospitalized with heart failure. Wien Klin Wochenschr 2009;121 (9–10):309–13.

[24] Lien Christopher TC, et al. Heart failure in frail elderly patients: diagnostic difficulties, co-morbidities, polypharmacy and treatment dilemmas. Eur J Heart Fail 2002;4(1):91–8.

[25] Hawkins NM, et al. Heart failure and chronic obstructive pulmonary disease: diagnostic pitfalls and epidemiology. Eur J Heart Fail 2009;11 (2):130–9.

[26] Tsutsui H, Tsuchihashi-Makaya M, Kinugawa S. Clinical characteristics and outcomes of heart failure with preserved ejection fraction: lessons from epidemiological studies. J Cardiol 2010;55(1):13–22.

[27] Ponikowski P, et al. 2016 ESC Guidelines for the diagnosis and treatment of acute and chronic heart failure: the task force for the diagnosis and treatment of acute and chronic heart failure of the European Society of Cardiology (ESC). Developed with the special contribution of the Heart Failure Association (HFA) of the ESC. Eur J Heart Fail 2016;18(8):891–975.

[28] Levin R. The Criteria Committee of the New York Heart Association: Nomenclature and criteria for diagnosis of diseases of the heart and great vessels. 9th. LWW Handbooks, Boston, MA: Little, Brown and Company; 1994. p. 344.

[29] Hunt SA, et al. 2009 Focused update incorporated into the ACC/AHA 2005 guidelines for the diagnosis and management of heart failure in adults. Circulation 2009;119(14), e391.

[30] Albouaini K, et al. Cardiopulmonary exercise testing and its application. Postgrad Med J 2007;83(985):675–82.

[31] Sullivan MJ, et al. Relation between central and peripheral hemodynamics during exercise in patients with chronic heart failure. Muscle blood flow is reduced with maintenance of arterial perfusion pressure. Circulation 1989;80(4):769–81.

[32] Haykowsky MJ, et al. Determinants of exercise intolerance in elderly heart failure patients with preserved ejection fraction. J Am Coll Cardiol 2011;58(3):265–74.

[33] Dhakal BP, et al. Mechanisms of exercise intolerance in heart failure with preserved ejection fraction: the role of abnormal peripheral oxygen extraction. Circ Heart Fail 2015;8(2):286–94.

[34] O'Neill JO, et al. Peak oxygen consumption as a predictor of death in patients with heart failure receiving beta-blockers. Circulation 2005;111 (18):2313–8.

[35] Mancini DM, et al. Value of peak exercise oxygen consumption for optimal timing of cardiac transplantation in ambulatory patients with heart failure. Circulation 1991;83(3):778–86.

[36] O'Connor CM, et al. Efficacy and safety of exercise training in patients with chronic heart failure: HF-ACTION randomized controlled trial. JAMA 2009;301(14):1439–50.

[37] Peterson LR, et al. The effect of β-adrenergic blockers on the prognostic value of peak exercise oxygen uptake in patients with heart failure. J Heart Lung Transplant 2003;22(1):70–7.

[38] Zielińska D, et al. Prognostic value of the six-minute walk test in heart failure patients undergoing cardiac surgery: a literature review. Rehabil Res Pract 2013;2013:965494.

[39] Kleber FX, et al. Impairment of ventilatory efficiency in heart failure: prognostic impact. Circulation 2000;101(24):2803–9.

[40] Mezzani A, et al. Habitual activities and peak aerobic capacity in patients with asymptomatic and symptomatic left ventricular dysfunction. Chest 2000;117(5):1291–9.

[41] Spruit MA, et al. Task-related oxygen uptake and symptoms during activities of daily life in CHF patients and healthy subjects. Eur J Appl Physiol 2011;111(8):1679–86.

[42] Gitt Anselm K, et al. Exercise anaerobic threshold and ventilatory efficiency identify heart failure patients for high risk of early death. Circulation 2002;106(24):3079–84.

[43] Wasserman K, et al. In: Wasserman KH, editor. Principles of exercise testing and interpretation; including pathophysiology and clinical applications. Portland: Wolters Kluwer Health/Lippincott Williams & Wilkins; 2012.

[44] Higginbotham MB, et al. Regulation of stroke volume during submaximal and maximal upright exercise in normal man. Circ Res 1986;58(2): 281–91.

[45] Powers SK, Howley ET. Exercise physiology : theory and application to fitness and performance. NY, United States: McGraw-Hill Higher Education; 2017.

[46] Fick A. Ueber die Messung des Blutquantum in den Herzventrikeln. Sb Phys Med Ges Worzburg 1870;16–7.

[47] Fukuda T, et al. Cardiac output response to exercise in chronic cardiac failure patients. Int Heart J 2012;53(5):293–8.

[48] Piña IL, et al. Exercise and heart failure: a statement from the American Heart Association Committee on exercise, rehabilitation, and prevention. Circulation 2003;107(8):1210–25.

[49] Kemp CD, Conte JV. The pathophysiology of heart failure. Cardiovasc Pathol 2012;21(5):365–71.

[50] Sullivan MJ, Cobb FR. Central hemodynamic response to exercise in patients with chronic heart failure. Chest 1992;101(5 Suppl):340s–6s.

[51] Al-Najjar Y, Witte KK, Clark AL. Chronotropic incompetence and survival in chronic heart failure. Int J Cardiol 2012;157(1):48–52.

[52] Brubaker Peter H, Kitzman Dalane W. Chronotropic incompetence. Circulation 2011;123(9):1010–20.

[53] Orso F, Baldasseroni S, Maggioni AP. Heart rate in coronary syndromes and heart failure. Prog Cardiovasc Dis 2009;52(1):38–45.

[54] Brubaker PH, et al. Chronotropic incompetence and its contribution to exercise intolerance in older heart failure patients. J Cardiopulm Rehabil 2006;26(2):86–9.

[55] Baker BJ, et al. Relation of right ventricular ejection fraction to exercise capacity in chronic left ventricular failure. Am J Cardiol 1984;54(6):596–9.

[56] Franciosa JA, Park M, Levine TB. Lack of correlation between exercise capacity and indexes of resting left ventricular performance in heart failure. Am J Cardiol 1981;47(1):33–9.

[57] Higginbotham MB, et al. Determinants of variable exercise performance among patients with severe left ventricular dysfunction. Am J Cardiol 1983;51(1):52–60.

[58] Szlachcic J, et al. Correlates and prognostic implication of exercise capacity in chronic congestive heart failure. Am J Cardiol 1985;55(8):1037–42.

[59] Carell ES, et al. Maximal exercise tolerance in chronic congestive heart failure. Relationship to resting left ventricular function. Chest 1994;106 (6):1746–52.

[60] Cohen-Solal A. Cardiopulmonary exercise testing in chronic heart failure. In: Wasserman K, editor. Exercise gas exchange in heart disease. New York: Futura; 1996.

[61] Wilson JR, Martin JL, Ferraro N. Impaired skeletal muscle nutritive flow during exercise in patients with congestive heart failure: role of cardiac pump dysfunction as determined by the effect of dobutamine. Am J Cardiol 1984;53(9):1308–15.

[62] Coats AJ. The "muscle hypothesis" of chronic heart failure. J Mol Cell Cardiol 1996;28(11):2255–62.

[63] Coats AJ, et al. Symptoms and quality of life in heart failure: the muscle hypothesis. Br Heart J 1994;72(2 Suppl):S36–9.

[64] Belli JF, et al. Ergoreflex activity in heart failure. Arq Bras Cardiol 2011;97(2):171–8.

[65] Piepoli M, et al. Contribution of muscle afferents to the hemodynamic, autonomic, and Ventilatory responses to Exercise in patients with chronic heart failure. Circulation 1996;93(5):940–52.

[66] Sullivan MJ, et al. Altered expression of myosin heavy chain in human skeletal muscle in chronic heart failure. Med Sci Sports Exerc 1997;29 (7):860–6.

[67] Vescovo G, et al. Apoptosis in the skeletal muscle of patients with heart failure: investigation of clinical and biochemical changes. Heart 2000;84 (4):431–7.

[68] Sullivan MJ, Green HJ, Cobb FR. Skeletal muscle biochemistry and histology in ambulatory patients with long-term heart failure. Circulation 1990;81(2):518–27.

[69] Schaufelberger M, et al. Skeletal muscle fiber composition and capillarization in patients with chronic heart failure: relation to exercise capacity and central hemodynamics. J Card Fail 1995;1(4):267–72.

[70] Guzmán Mentesana G, et al. Functional and structural alterations of cardiac and skeletal muscle mitochondria in heart failure patients. Arch Med Res 2014;45(3):237–46.

[71] Drexler H, et al. Alterations of skeletal muscle in chronic heart failure. Circulation 1992;85(5):1751–9.

[72] Ning XH, et al. Signaling and expression for mitochondrial membrane proteins during left ventricular remodeling and contractile failure after myocardial infarction. J Am Coll Cardiol 2000;36(1):282–7.

[73] Schaper J, et al. Impairment of the myocardial ultrastructure and changes of the cytoskeleton in dilated cardiomyopathy. Circulation 1991;83 (2):504–14.

[74] Mancini DM, et al. Contribution of skeletal muscle atrophy to exercise intolerance and altered muscle metabolism in heart failure. Circulation 1992;85(4):1364–73.

[75] Cicoira M, et al. Skeletal muscle mass independently predicts peak oxygen consumption and ventilatory response during exercise in noncachectic patients with chronic heart failure. J Am Coll Cardiol 2001;37(8):2080–5.

[76] Middlekauff HR. Making the case for skeletal myopathy as the major limitation of exercise capacity in heart failure. Circ Heart Fail 2010;3 (4):537–46.

[77] Vromen T, et al. The influence of training characteristics on the effect of aerobic exercise training in patients with chronic heart failure: a meta-regression analysis. Int J Cardiol 2016;208:120–7.

[78] Tu RH, et al. Effects of exercise training on depression in patients with heart failure: a systematic review and meta-analysis of randomized controlled trials. Eur J Heart Fail 2014;16(7):749–57.

[79] Cipriano Jr G, et al. Aerobic exercise effect on prognostic markers for systolic heart failure patients: a systematic review and meta-analysis. Heart Fail Rev 2014;19(5):655–67.

[80] Smart NA, et al. Individual patient meta-analysis of exercise training effects on systemic brain natriuretic peptide expression in heart failure. Eur J Prev Cardiol 2012;19(3):428–35.

[81] Smart NA, Steele M. Systematic review of the effect of aerobic and resistance exercise training on systemic brain natriuretic peptide (BNP) and N-terminal BNP expression in heart failure patients. Int J Cardiol 2010;140(3):260–5.

[82] Pearson MJ, Smart NA. Aerobic training intensity for improved endothelial function in heart failure patients: a systematic review and meta-analysis. Cardiol Res Pract 2017;2017:2450202.

[83] Gomes-Neto M, et al. Effect of aerobic exercise on peak oxygen consumption, VE/VCO(2) slope, and health-related quality of life in patients with heart failure with preserved left ventricular ejection fraction: a systematic review and meta-analysis. Curr Atheroscler Rep 2019;21(11):45.

[84] Chan E, et al. Exercise training in heart failure patients with preserved ejection fraction: a systematic review and meta-analysis. Monaldi Arch Chest Dis 2016;86(1–2):759.

[85] Pattyn N, Beulque R, Cornelissen V. Aerobic interval vs. continuous training in patients with coronary artery disease or heart failure: an updated systematic review and meta-analysis with a focus on secondary outcomes. Sports Med 2018;48(5):1189–205.

[86] Smart NA, Dieberg G, Giallauria F. Intermittent versus continuous exercise training in chronic heart failure: a meta-analysis. Int J Cardiol 2013;166 (2):352–8.

[87] Haykowsky MJ, et al. Meta-analysis of aerobic interval training on exercise capacity and systolic function in patients with heart failure and reduced ejection fractions. Am J Cardiol 2013;111(10):1466–9.

[88] Milanovic Z, Sporis G, Weston M. Effectiveness of high-intensity interval training (HIT) and continuous endurance training for VO2max improvements: a systematic review and meta-analysis of controlled trials. Sports Med 2015;45(10):1469–81.

[89] Ferguson B. ACSM's guidelines for exercise testing and prescription 9th Ed. 2014. J Can Chiropr Assoc 2014;58(3):328.

[90] Ismail H, et al. Exercise training program characteristics and magnitude of change in functional capacity of heart failure patients. Int J Cardiol 2014;171(1):62–5.

[91] Gomes Neto M, et al. High intensity interval training versus moderate intensity continuous training on exercise capacity and quality of life in patients with heart failure with reduced ejection fraction: a systematic review and meta-analysis. Int J Cardiol 2018;261:134–41.

[92] Xie B, et al. Effects of high-intensity interval training on aerobic capacity in cardiac patients: a systematic review with meta-analysis. Biomed Res Int 2017;2017:5420840.

[93] Siscovick DS, et al. The incidence of primary cardiac arrest during vigorous exercise. N Engl J Med 1984;311(14):874–7.

[94] Levinger I, et al. What Doesn't kill you makes you fitter: a systematic review of high-intensity interval exercise for patients with cardiovascular and metabolic diseases. Clin Med Insights Cardiol 2015;9:53–63.

[95] Giri S, et al. Clinical and angiographic characteristics of exertion-related acute myocardial infarction. JAMA 1999;282(18):1731–6.

[96] Kuehn BM. Evidence for HIIT benefits in cardiac rehabilitation grow. Circulation 2019;140(6):514–5.

[97] Boo SH, et al. Association between skeletal muscle mass and cardiorespiratory fitness in community-dwelling elderly men. Aging Clin Exp Res 2019;31(1):49–57.

[98] Landi F, et al. Sarcopenia as a risk factor for falls in elderly individuals: results from the ilSIRENTE study. Clin Nutr 2012;31(5):652–8.

[99] Visser M, et al. Leg muscle mass and composition in relation to lower extremity performance in men and women aged 70 to 79: the health, aging and body composition study. J Am Geriatr Soc 2002;50(5):897–904.

[100] Janssen I, Heymsfield SB, Ross R. Low relative skeletal muscle mass (sarcopenia) in older persons is associated with functional impairment and physical disability. J Am Geriatr Soc 2002;50(5):889–96.

[101] Williams MA, et al. Resistance exercise in individuals with and without cardiovascular disease: 2007 update: a scientific statement from the American Heart Association Council on clinical cardiology and council on nutrition, physical activity, and metabolism. Circulation 2007;116 (5):572–84.

[102] Torres SJ, et al. Effects of progressive resistance training combined with a protein-enriched lean red meat diet on health-related quality of life in elderly women: secondary analysis of a 4-month cluster randomised controlled trial. Br J Nutr 2017;117(11):1550–9.

[103] Scott D, et al. Fall and fracture risk in sarcopenia and dynapenia with and without obesity: the role of lifestyle interventions. Curr Osteoporos Rep 2015;13(4):235–44.

[104] Giuliano C, et al. The effects of resistance training on muscle strength, quality of life and aerobic capacity in patients with chronic heart failure—a meta-analysis. Int J Cardiol 2017;227:413–23.

[105] Jewiss D, Ostman C, Smart N. The effect of resistance training on clinical outcomes in heart failure: A systematic review and meta-analysis. Int J Cardiol 2016;221:674–81.

[106] Maiorana AJ, et al. Exercise professionals with advanced clinical training should be afforded greater responsibility in pre-participation exercise screening: a new collaborative model between exercise professionals and physicians. Sports Med 2018;48(6):1293–302.

[107] Cornelis J, et al. Comparing exercise training modalities in heart failure: a systematic review and meta-analysis. Int J Cardiol 2016;221:867–76.

[108] Beckers PJ, et al. Combined endurance-resistance training vs. endurance training in patients with chronic heart failure: a prospective randomized study. Eur Heart J 2008;29(15):1858–66.

[109] Mandic S, et al. Effects of aerobic or aerobic and resistance training on cardiorespiratory and skeletal muscle function in heart failure: a randomized controlled pilot trial. Clin Rehabil 2009;23(3):207–16.

[110] Long L, et al. Exercise-based cardiac rehabilitation for adults with heart failure. Cochrane Database Syst Rev 2019;1:Cd003331.

[111] Schünemann HJ, et al. Interpreting results and drawing conclusions. In: Cochrane handbook for systematic reviews of interventions. Wiley; 2019. p. 403–31.

[112] Taylor RS, et al. Impact of exercise-based cardiac rehabilitation in patients with heart failure (ExTraMATCH II) on mortality and hospitalisation: an individual patient data meta-analysis of randomised trials. Eur J Heart Fail 2018;20(12):1735–43.

[113] Bjarnason-Wehrens B, et al. Exercise-based cardiac rehabilitation in patients with reduced left ventricular ejection fraction: the cardiac rehabilitation outcome study in heart failure (CROS-HF): a systematic review and meta-analysis. Eur J Prev Cardiol 2019;27(9):929–52. 2047487319854140.

[114] Atherton JJ, et al. National Heart Foundation of Australia and Cardiac Society of Australia and New Zealand: Guidelines for the prevention, detection, and management of heart failure in Australia 2018. Heart Lung Circ 2018;27(10):1123–208.

[115] Yancy CW, et al. 2017 ACC/AHA/HFSA focused update of the 2013 ACCF/AHA guideline for the management of heart failure: a report of the American College of Cardiology/American Heart Association task force on clinical practice guidelines and the Heart Failure Society of America. Circulation 2017;136(6):e137–61.

[116] Piepoli MF, et al. Exercise training in heart failure: from theory to practice. A consensus document of the Heart Failure Association and the European Association for Cardiovascular Prevention and Rehabilitation. Eur J Heart Fail 2011;13(4):347–57.

[117] Selig SE, et al. Exercise & Sport Science Australia Position Statement on exercise training and chronid heart failure. J Sci Med Sport 2010;13:288–94.

[118] Price KJ, et al. A review of guidelines for cardiac rehabilitation exercise programmes: is there an international consensus? Eur J Prev Cardiol 2016;23(16):1715–33.

[119] Anderson L, Taylor RS. Cardiac rehabilitation for people with heart disease: an overview of Cochrane systematic reviews. Cochrane Database Syst Rev 2014;12, Cd011273.

[120] Achttien RJ, et al. Exercise-based cardiac rehabilitation in patients with chronic heart failure: a Dutch practice guideline. Neth Hear J 2015;23(1):6–17.

[121] Pavy B, et al. French Society of Cardiology guidelines for cardiac rehabilitation in adults. Arch Cardiovasc Dis 2012;105(5):309–28.

[122] Herdy AH, et al. South American guidelines for cardiovascular disease prevention and rehabilitation. Arq Bras Cardiol 2014;103(2 Suppl 1):1–31.

[123] Balady Gary J, et al. Core components of cardiac rehabilitation/secondary prevention programs: 2007 update. Circulation 2007;115(20):2675–82.

[124] Pescatello LS, American College of Sports Medicine. ACSM's guidelines for exercise testing and prescription. 9th ed. Philadelphia: Wolters Kluwer/Lippincott Williams & Wilkins Health; 2014. p. xxiv. 456 p.

[125] Deka P, et al. Adherence to recommended exercise guidelines in patients with heart failure. Heart Fail Rev 2017;22(1):41–53.

[126] Piña IL, et al. Effects of Exercise training on outcomes in women with heart failure: analysis of HF-ACTION (heart failure—a controlled trial investigating outcomes of exercise training) by sex. JACC Heart Fail 2014;2(2):180–6.

[127] van der Wal MH, et al. Compliance in heart failure patients: the importance of knowledge and beliefs. Eur Heart J 2006;27(4):434–40.

[128] Golwala H, et al. Temporal trends and factors associated with cardiac rehabilitation referral among patients hospitalized with heart failure: findings from get with the guidelines-heart failure registry. J Am Coll Cardiol 2015;66(8):917–26.

[129] Conraads VM, et al. Adherence of heart failure patients to exercise: barriers and possible solutions: a position statement of the Study Group On Exercise Training in Heart Failure of the Heart Failure Association of the European Society of Cardiology. Eur J Heart Fail 2012;14(5):451–8.

[130] Giuliano C, et al. Cardiac rehabilitation for patients with coronary artery disease: a practical guide to enhance patient outcomes through continuity of care. Clin Med Insights Cardiol 2017;11, 1179546817710028.

[131] De Vos C, et al. Participating or not in a cardiac rehabilitation programme: factors influencing a patient's decision. Eur J Prev Cardiol 2013;20(2):341–8.

[132] McIntosh N, et al. A qualitative study of participation in cardiac rehabilitation programs in an integrated health care system. Mil Med 2017;182(9):e1757–63.

[133] Bond S, et al. Exergaming and virtual reality for health: implications for cardiac rehabilitation. Curr Probl Cardiol 2021;46(3), 100472.

[134] Kraal JJ, et al. Clinical and cost-effectiveness of home-based cardiac rehabilitation compared to conventional, Centre-based cardiac rehabilitation: results of the FIT@Home study. Eur J Prev Cardiol 2017;24(12):1260–73.

[135] Frederix I, et al. A review of telerehabilitation for cardiac patients. J Telemed Telecare 2015;21(1):45–53.

[136] Rawstorn JC, et al. Telehealth exercise-based cardiac rehabilitation: a systematic review and meta-analysis. Heart 2016;102(15):1183–92.

[137] Huang K, et al. Telehealth interventions versus center-based cardiac rehabilitation of coronary artery disease: a systematic review and meta-analysis. Eur J Prev Cardiol 2015;22(8):959–71.

[138] van Veen E, et al. E-coaching: new future for cardiac rehabilitation? A systematic review. Patient Educ Couns 2017;100(12):2218–30.

[139] Kikuchi A, et al. Feasibility of home-based cardiac rehabilitation using an integrated telerehabilitation platform in elderly patients with heart failure: a pilot study. J Cardiol 2021;78(1):66–71.

[140] Jaarsma T, et al. Increasing exercise capacity and quality of life of patients with heart failure through Wii gaming: the rationale, design and methodology of the HF-Wii study; a multicentre randomized controlled trial. Eur J Heart Fail 2015;17(7):743–8.

[141] Klompstra L, Jaarsma T, Strömberg A. Exergaming to increase the exercise capacity and daily physical activity in heart failure patients: a pilot study. BMC Geriatr 2014;14(1):1–9.

Chapter 20

Exercise interventions in women with Polycystic Ovary Syndrome

Alba Moreno-Asso[a,b], Rhiannon K. Patten[a], and Luke C. McIlvenna[a]

[a]Institute for Health and Sport, Victoria University, Melbourne, VIC, Australia [b]Australian Institute for Musculoskeletal Science, Victoria University and Western Health, Melbourne, VIC, Australia

Introduction

Background

What is PCOS?

Polycystic ovary syndrome (PCOS) is the most common endocrine condition in women of reproductive age, affecting the metabolic, reproductive, and mental health of women with the condition. PCOS is a complex and chronic condition with manifestations across the lifespan, presenting a significant health and economic burden. The primary features of PCOS include hyperandrogenism, ovulatory dysfunction, and polycystic ovaries (with a high number of follicles) [1]. Other common features of PCOS are insulin resistance (IR) and hyperinsulinemia, infertility, psychological disorders (depression and anxiety), obesity, acne, and hirsutism (excessive body hair) [2,3]. Women with PCOS are also at high-risk for developing impaired glucose tolerance (IGT), gestational diabetes mellitus (GDM), type 2 diabetes mellitus (T2DM), and cardiovascular disease (CVD) [4,5]. The etiology of PCOS is not well understood, creating a challenge for the development of effective therapies for the treatment and management of PCOS resulting in poor health and quality of life of these women [6,7].

Prevalence and diagnosis

The prevalence of PCOS is dependent upon the diagnostic criteria and population studied and ranges from 5% to 20% worldwide [8,9]. PCOS was first recognized by Stein and Leventhal in 1935 when they described a reproductive condition where they observed an association between the absence of menstruation (amenorrhoea) and polycystic ovaries [10]. However, it was not until 1990 when the first PCOS diagnostic criteria was developed by the National Institutes of Health (NIH) [11]. Currently, there are three recognized diagnostic criteria: The 1990 NIH, the American Society of Reproductive Medicine sponsored European Society of Human Reproduction and Embryology (ASRM/ESHRE) also known as the Rotterdam criteria last revised in 2003 [12] and the Androgen Excess and Polycystic Ovary Syndrome (AE-PCOS) Society criteria from 2009 [13]. As of 2012, the Rotterdam criteria became the most internationally recognized and accepted criteria for the diagnosis of PCOS.

The 1990 NIH diagnostic criteria identified only the most severe form of PCOS with chronic anovulation and hyperandrogenism required for the diagnosis. The AE-PCOS criteria require the presence of hyperandrogenism with oligo-anovulation or polycystic ovarian morphology (PCOM). Following the Rotterdam criteria, any two of the following features are required for the diagnosis:

- Ovulatory dysfunction/irregular cycles (<21 or >35 days or less than 8 cycles per year).
- Clinical or biochemical hyperandrogenism.
- Polycystic ovarian morphology on ultrasound ($\geq$12 follicles per ovary and/or an ovarian volume of >10 mL in at least one ovary).

All three diagnostic criteria require the exclusion of any other disorders that could be responsible for hyperandrogenism or any of these symptoms (e.g., hyperprolactinaemia, hyperthyroidism, and non-classic congenital adrenal hyperplasia).

Exercise to Prevent and Manage Chronic Disease Across the Lifespan. https://doi.org/10.1016/B978-0-323-89843-0.00020-9

TABLE 1 PCOS phenotypes as defined by Rotterdam Criteria (2003).

Phenotype	Hyperandrogenism	Ovulatory dysfunction (oligo- or anovulation)	Polycystic ovary morphology
A	✓	✓	✓
B	✓	✓	
C	✓		✓
D		✓	✓

Severity

Depending on the features that are present, the severity of PCOS varies widely and can be classified by different phenotypes. The Rotterdam criteria outline four phenotypes ranging from the most severe (phenotype A) to the lesser with the presence of reproductive features and absence of hyperandrogenism (phenotype D) [12] (Table 1).

Pathophysiology

Despite decades of research, the etiology of PCOS remains largely unclear due to the highly complex and multifactorial nature of the syndrome. What we know is that PCOS often clinically manifests during adolescence, and can be caused by a combination of genetic and lifestyle factors leading to an underlying hormonal imbalance, which is further exacerbated by obesity [7]. Although the heritability of PCOS has been estimated to be about 70% [14], genetic studies have not been able to fully explain this high inheritance rate. It has been proposed that PCOS may originate in the early stages of foetal development, not only from genetically-transmitted factors but also through transgenerational epigenetic alterations triggered by factors present in the uterus during pregnancy [15–17].

Hyperandrogenism and reproductive function

Hyperandrogenism is considered the main clinical hallmark of PCOS and a primary contributor to the pathophysiology of the syndrome. It is detected in approximately 80%–85% of women with PCOS [18]. Those phenotypes presenting with hyperandrogenism have been associated with a more severe form of the disease and a worse metabolic profile [19].

In the diagnosis of PCOS, biochemical hyperandrogenism is determined by elevated serum levels of androgens including; total testosterone, free testosterone, or by the free androgen index (FAI) [20]. Serum-free testosterone levels are regulated by binding to the sex-hormone-binding globulin (SHBG), produced in the liver [21]. Clinically, hyperandrogenism can manifest as hirsutism, acne, and androgenic alopecia [9,22].

In PCOS, testosterone is secreted primarily by the ovaries and, to a lesser degree, the adrenal glands. Women with PCOS have an alteration of the hypothalamic-pituitary-ovarian (HPO) axis with abnormal persistently rapid secretion of gonadotropin releasing hormone (GnRH) [23,24]. This enhanced GnRH pulsatile release favors the secretion of luteinizing hormone (LH) over the follicle-stimulating hormone (FSH) resulting in an elevated LH:FSH ratio [23]. The increased LH promotes the theca cells of the ovary to increase the synthesis of androgens, leading to hyperandrogenism [25]. This hormonal imbalance between LH and FSH, and in particular androgen excess leads to ovulatory dysfunction and infertility. Therefore, it is no surprise that approximately 70% of women with PCOS are affected by menstrual disturbances [26], while 40% suffer from infertility [27]. PCOM is also commonly diagnosed in women with PCOS with ovulatory dysfunction and hyperandrogenism. However, PCOM is not associated with the distinctive reproductive or metabolic phenotypes of PCOS [28,29].

Insulin resistance and hyperinsulinemia

Although not included in the diagnostic criteria, PCOS has a strong metabolic component with IR and compensatory hyperinsulinemia considered to be primary drivers of PCOS, affecting various tissues in the body [7,30,31]. IR is present in 38%–95% of women with PCOS [32,33], with an estimated reduction in insulin sensitivity of about 25% [34]. The prevalence of IR has been associated with the severity of the syndrome, with the most severe forms having a higher incidence of IR. Insulin sensitivity can be assessed using a range of methods, however, the gold standard is the euglycaemic-hyperinsulinaemic clamp. This method directly measures whole body glucose disposal by infusing a constant rate of insulin to raise the plasma insulin levels, and a variable rate of glucose to maintain a target level until a steady-state is achieved (glucose infusion rate equals glucose uptake) [35]. Other more common measures of insulin sensitivity include homeostatic model of insulin resistance (HOMA-IR [fasting glucose (mmol/L) × fasting insulin (mU/L)/22.5] [36]), fasting glucose, and fasting insulin.

Remarkably, women with PCOS have an intrinsic IR which is believed to be mechanistically distinct from obesity-associated IR [30,32]. However, obesity can exacerbate this PCOS-specific IR, augmenting the risk of developing IGT, GDM, T2DM, and CVD [37,38]. IR also has an adverse effect on the reproductive features of the condition. IR leads to compensatory hyperinsulinemia, which dysregulates ovarian hormone production and menstrual cyclicity, resulting in fertility issues [39,40]. Similarly, insulin directly contributes to hyperandrogenism through stimulating ovarian androgen production and increasing free androgen levels [41–45] by suppressing the hepatic production of SHBG [46,47].

Obesity

Women with PCOS have an increased prevalence of obesity with 40%–80% of women being overweight or obese [48–50]. Obesity is the main contributor to the development of PCOS and its clinical features [43,51,52], and it is suggested that the current obesity epidemic may be causing an increased prevalence and severity of the condition [53,54].

Excess body weight worsens the underlying hormonal disturbances, exacerbating both the metabolic and reproductive features of PCOS [7,55,56]. Overweight women with PCOS exhibit higher amounts of visceral fat [57], and are associated with elevated free androgens, increased FAI, and reduced SHBG compared to normal-weight women with PCOS [56]. Obesity, in particular abdominal obesity, also worsens IR and increases insulin levels [58], substantially increasing the risk of cardio-metabolic complications [4,38]. Obesity also exacerbates infertility and induces a greater risk of miscarriage and pregnancy complications [59].

Mental health

PCOS is a heterogeneous condition that is also associated with mental health consequences [22]. Fears regarding infertility, body image, and coping with the condition contribute to psychological issues, including an increased prevalence of symptoms of anxiety and depression, poor body image, and low self-esteem [60]. Women with PCOS consistently report diminished psychological wellbeing [6,61] and poorer health-related quality of life compared to women without PCOS [62,63]. The prevalence of depression rates among women with PCOS varies widely according to the population studied and the severity of depression ranging from mild to severe [3,64]. A recent meta-analysis has reported that women with PCOS are four times more likely to suffer from moderate to severe symptoms of depression and have a prevalence of depression to be 22.4% higher than non-PCOS women [3]. Similarly, the same meta-analysis reported that they are six times more likely to suffer from severe symptoms of anxiety [3]. Research indicates that BMI has only a small effect on anxiety and depression among women with PCOS, suggesting that PCOS status plays a major role [65]. Despite the evidence for an increased prevalence of poorer mental health and well-being in women with PCOS, the reason for this association is unclear and information regarding prevention and treatment strategies is limited.

Exercise in PCOS

Exercise is well established as a therapy for preventing and managing risk factors associated with chronic disease [66,67]. Whether used as a preventative measure or a treatment for chronic disease, regular exercise provides a greater quality of life [67]. These established benefits also exist for PCOS, which include improvements in a wide range of clinical outcomes. In 2011, the PCOS Australian Alliance published the first international evidence-based guidelines for the assessment and management of PCOS [60], which were later updated in 2018 [20]. These guidelines acknowledge exercise as first-line therapy, recommending a minimum of 150 min/week of moderate-intensity exercise or 75 min of vigorous-intensity exercise to maintain health or improve the clinical outcomes of PCOS. Current exercise interventions in women with PCOS have elicited a number of health benefits including improved cardio-metabolic, reproductive and mental health (Table 2) [42,68–70], which will be discussed separately throughout this chapter.

There are important considerations that need to be taken into account when designing or prescribing exercise interventions for women with PCOS, not unlike the general population. First is the type or mode of exercise. Aerobic exercise is the most commonly utilized mode of exercise in women with PCOS and is well known to improve a range of health outcomes in women with PCOS. The most consistently reported improvements are cardiorespiratory fitness, insulin sensitivity, reduced levels of androgens, and regulation of menstrual cycles [71,72]. Resistance or strength training has recently become of interest in PCOS research. Preliminary results suggest that resistance training may reduce androgen levels [73,74],

TABLE 2 Summary of beneficial effects of exercise in women with PCOS.

Health domain	Outcome measure
Cardiorespiratory fitness	↑ VO_{2peak}
Cardio-metabolic health	↑ Insulin sensitivity ↑ Healthy lipid profiles ↓ Risk of developing T2DM and CVD
Body composition	↓ Waist and hip circumferences ↓ Fat mass ↑ Lean mass
Reproductive health	↑ Menstrual regulation ↑ Ovulation ↓ Androgen levels ↑ SHBG
Mental health	↓ Symptoms of depression ↓ Symptoms of anxiety ↑ Health-related quality of life

however, further research is needed to confirm these benefits. Secondly, the intensity of exercise needs to be considered. Recently, this has become of particular interest across all population groups, including PCOS. Although moderate-intensity exercise is well established as being beneficial for reproductive and metabolic health, there is evidence to suggest that exercise of vigorous intensity may provide additional benefits [71]. To date, limited studies have been conducted comparing moderate to vigorous exercise interventions, and therefore further research is required to confirm these findings. Lastly, it is important to consider the dose of exercise which takes into consideration the duration of the intervention and the duration of the exercise sessions. A recent meta-analysis found that a minimum of 120 min per week for 12 weeks is required in order to provide positive improvements in health outcomes [71]. Adherence to exercise interventions is typically low in all populations but is especially challenging in women with PCOS. This may be due to general barriers (time commitment and low enjoyment) or PCOS-specific barriers (physical limitations, low confidence) [75]. When implementing exercise interventions for women with PCOS, it is important to consider how you might address these barriers and promote long-term adherence to exercise.

Clinical outcomes of exercise interventions in PCOS

Cardio-metabolic outcomes

Considering the beneficial effects of exercise on other insulin-resistant populations such as T2DM [76], incorporating exercise as a treatment for PCOS may be favorable. Moderate physical activity, at least three to five times per week, has consistently been shown to reduce T2DM in high-risk populations [77]. In line with this, significant improvements in insulin sensitivity are commonly reported as a result of exercise interventions in women with PCOS [78]. A previous systematic review reported improvements ranging from 9% to 30% after an exercise intervention [78]. It should be noted that the large variation in the observed improvements is likely due to the various measures used to determine insulin sensitivity. Studies that have utilized the gold standard euglycaemic-hyperinsulinaemic clamp often report large increases in insulin sensitivity [57,79,80]. However, this method is costly and difficult to undertake in a clinical setting. Studies using alternative methods (e.g., HOMA-IR) have also found significant improvements in insulin sensitivity as a result of an exercise intervention [74,81–83].

Perhaps the most commonly reported improvement is cardiorespiratory fitness as measured by VO_{2peak}. Cardiorespiratory fitness is an important indicator of health and all-cause mortality [84]. Indeed, a poor VO_{2peak} is a stronger predictor of mortality than BMI [85]. A recent meta-analysis that explored the effects of exercise in women with PCOS reported that both moderate and vigorous-intensity exercise resulted in significant increases in VO_{2peak}, with the largest improvement occurring as a result of vigorous-intensity exercise [71]. They also found that improvements in VO_{2peak} were noted in all included trials that measured VO_{2peak} with the exception of one intervention that did not adequately report exercise attendance and had an extremely high dropout rate [86].

As a large proportion of women with PCOS are overweight or obese, decreasing body weight is often the primary aim. The international evidence-based guidelines suggest a weight loss of 5%–10% in order to achieve significant clinical improvements [20]. However, inconsistent results are reported in regards to the effect of exercise for improving body composition. The impact of exercise on weight and BMI is often contradictory with one recent meta-analysis reporting a significant, although small decrease in BMI [87], while other meta-analyses report no change in BMI [71,72]. Studies reporting decreases in weight often include a dietary component within their intervention, making it difficult to distinguish the effects of exercise from the effect of diet. A meta-analysis conducted by our group reported that improvements in insulin sensitivity and VO_{2peak} are observed despite no change in BMI [71]. Meta-analyses in other chronic conditions have observed similar findings, with improvements in cardio-metabolic risk factors reported without a change in weight or BMI [88–90]. On the other hand, exercise has been found to result in improvements in other markers of body composition, including decreases in waist circumference, body fat, and increased lean muscle mass in women with PCOS [71,72,74,87]. As central or abdominal obesity is known to be associated with IR, reducing waist circumference is highly beneficial for these women [51].

Some studies have also reported an improvement in lipid profiles as a result of exercise. In particular, studies report an increase in high-density lipoprotein cholesterol (HDL-C) [74,82,86,91,92], decreased low-density lipoprotein cholesterol (LDL-C) [82,92,93], and decreased total cholesterol levels [82,92,93]. Many studies that report no change in cholesterol or triglycerides often report healthy baseline values and therefore no change is expected to occur [57]. Also, combined with the improvements in insulin sensitivity, cardiorespiratory fitness, and decreased waist circumference, exercise substantially decreases the risk of developing metabolic syndrome, T2DM, and CVD (Table 2).

Reproductive outcomes

There is also a considerable amount of research that indicates the beneficial effects of exercise for reproductive health outcomes in women with PCOS (Table 2). When considering the effects of exercise on reproductive health, there are multiple outcomes to consider. These include; changes in menstrual cycle, ovulation rates, hormonal profiles, and fertility. Regular, moderate-intensity aerobic exercise has been found to improve reproductive outcomes including menstrual cycle regulation and ovulation in young, overweight women with PCOS [78]. Overall, exercise studies have shown improvements in ovulation rates and/or menstrual cyclicity in ~50% of women with PCOS [78]. It is important to note that these improvements in menstrual cyclicity have been reported to occur both with and without changes in hormonal parameters [81,94,95]. In some studies, exercise has been found to improve FAI, total and free testosterone, LH, FSH, and SHBG, however, no changes have been reported in other studies [71]. It is also important to note that measures such as testosterone, FSH, and LH vary with the stage of the menstrual cycle, and require caution when interpreting results [96].

Multiple studies have also utilized a combined diet and exercise intervention for improving health in women with PCOS. A systematic review that compared exercise to a combined diet and exercise intervention found that exercise-alone interventions improved reproductive health outcomes in women with PCOS with limited additional benefit from diet [96]. Their analysis suggested that lifestyle interventions improve levels of FSH, SHBG, total testosterone, androstenedione, and FAI in women with PCOS. Within the existing research, there is a wide array of improvements, making it difficult to conclusively determine the effectiveness of exercise and exercise characteristics that aid in promoting benefits in reproductive health. Furthermore, at this stage, it is unknown whether a particular type or intensity of exercise is more effective for inducing improvements in reproductive function. Evaluation of exercise interventions is even more challenging due to the use of different diagnostic criteria (and phenotypes) included in research but also due to the poor reporting of the characteristics of exercise interventions (e.g., intensity and dose).

Excess body mass has been found to correlate with an increased rate of cycle disturbances [26,97], delayed time to conception, and adversely affects the response to fertility treatments [98,99]. Studies have shown that a weight loss of as little as 5% is associated with improved spontaneous ovulation rates in overweight, infertile women [100–102]. Lifestyle modifications including caloric restriction and regular exercise should be considered as the first-line treatment for overweight women with PCOS who are looking to conceive [103]. To date, no research has been conducted to determine whether exercise could achieve improved fertility rates without weight loss in PCOS. One study compared exercise to a diet group and reported significant improvements in menstrual cyclicity and fertility as a result of both groups, with a trend towards higher pregnancy rates in the exercise group [104].

Mental health outcomes

It is well reported that women with PCOS have an increased prevalence of symptoms of anxiety and depression in comparison to women without PCOS [3]. The international evidence-based guidelines for the assessment and management of

PCOS recommend that all women with PCOS are routinely screened for depression and anxiety and other aspects of emotional well-being by a suitably qualified health professional [20]. Exercise is an effective means of managing and improving mental health in healthy populations [105], in overweight women [106], and in those with chronic conditions [107]. In addition to the positive improvements in cardio-metabolic and reproductive health, there is also evidence to suggest that exercise results in positive mental health outcomes [70]. Existing exercise and lifestyle interventions have reported reduced symptoms or prevalence of anxiety and depression [108–111], improve quality of life [112,113], PCOS symptom distress (measured by the PCOSQ) [111,114], self-esteem [108], and improve body image [115]. Furthermore, physically active women with PCOS report fewer symptoms of depression in comparison to sedentary women with PCOS [75,116].

The vast majority of exercise and lifestyle intervention studies have reported both clinical benefits and significant improvements in multiple domains of health-related quality of life and mental health [70] (Table 2). However, due to the heterogeneity of exercise prescription utilized, measures used and poor reporting of exercise interventions, we are limited in regards to forming conclusions regarding the most effective exercise characteristics. Despite the benefit of exercise for improving mental health is clear, further research is required to adequately assess the impact of various exercise characteristics in order to determine the most effective strategy for improving the mental health of women with PCOS.

Molecular mechanisms of exercise-induced health benefits

Cardio-metabolic health

Peripheral tissue metabolism

The skeletal muscle is one of the first metabolic tissues to respond to exercise. In particular, it is the primary tissue for insulin- and exercise-stimulated glucose uptake through the translocation of the glucose transporter GLUT4 to the plasma membrane [117,118]. Indeed, an improvement in insulin sensitivity can occur after a single bout of exercise [119,120]. These beneficial effects result from an increase in glycogen synthesis to replenish skeletal muscle glycogen levels following exercise [121]. This process is in part regulated by AMPK, which can act as a glycogen sensor. A collection of studies has highlighted the role of AMPK and downstream signaling of TBC1D1 and TBC1D4 as regulators of post-exercise increases in glucose uptake and insulin sensitivity [122–125]. Remarkably, it has been shown that post-training improvements in skeletal muscle insulin sensitivity can be observed 24 h after the last exercise session but are absent after 4 days [126]. This underscores the importance of regular physical activity or structured exercise to induce long-term adaptations, essential for maintaining insulin sensitivity and overall metabolic health. These adaptations include, but are not limited to, increased capillarization, enhanced substrate utilization capacity, and increases in the expression of GLUT4 and mitochondrial proteins [127].

Some of these responses have been observed in women with PCOS following exercise (Fig. 1), while others have not been assessed. Following an acute bout of exercise, women with PCOS display increased skeletal muscle insulin signaling, with a reduction in the inhibitory phosphorylation of IRS-1 ser312 and increased phosphorylation of AMPK Thr172 and Akt ser473/308 [128]. It has also demonstrated that women with PCOS display an increase in skeletal muscle GLUT-4 translocation in response to a low-intensity exercise bout [128]. Furthermore, women with PCOS exhibit similar changes in insulin gene expression and metabolic signaling to those observed in healthy women in response to acute exercise [129]. Similarly, an acute bout of exercise can also influence the pathophysiology of PCOS by inducing anti-inflammatory effects through the reduction of pro-inflammatory factors; TNFα, IKKα/β, and JNK, and increasing anti-inflammatory factors; IL-6, IL-10, and IL-4 in skeletal muscle [130] (Fig. 1). Of particular interest is the upregulation of IL-6 gene expression in skeletal muscle, which can improve insulin sensitivity via the activation of AMPK [123,131]. Another positive benefit of exercise is the improved mitochondrial function and content, altering substrate utilization and whole-body metabolism. Following 12 weeks of aerobic exercise training, women with PCOS show a reduction in mitochondrial hydrogen peroxide (a marker of oxidative stress) and improvements in mitochondrial efficiency (oxidative capacity and citrate synthase activity) [132], alongside improvements in skeletal muscle insulin sensitivity.

Women with PCOS have increased adiposity, adipose tissue dysfunction, and abnormal adipose morphology [133]. Ultimately, this influences metabolic function and leads to adverse metabolic effects through the disruption of adipokine secretion and the reduction of brown adipose tissue activity [133–136]. Also, the adipose tissue of women with PCOS is insulin resistant and presents abnormal gene expression in the transforming growth factor (TGF)-beta signaling and adipogenesis pathway [137]. Many of these abnormalities or defects are responsive to exercise (Fig. 1). In women with PCOS, 16 weeks of aerobic exercise training has been shown to partially restore catecholamine and amino-acid induced adipose tissue lipolysis, independent of changes in body weight and androgen levels [138]. These improvements in lipolysis have been attributed to an increase in perilipin-3 and GTPase proteins and a reduction in lipid droplet size following exercise

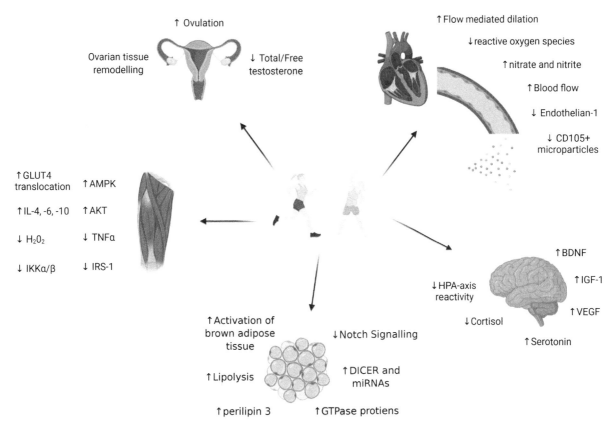

↑ Ovulation

↑ Flow mediated dilation

Ovarian tissue remodelling

↓ Total/Free testosterone

↓ reactive oxygen species

↑ nitrate and nitrite

↑ Blood flow

↓ Endothelian-1

↓ CD105+ microparticles

↑ GLUT4 translocation   ↑ AMPK

↑ IL-4, -6, -10   ↑ AKT

↓ H_2O_2 ↓ TNFα

↓ IKKα/β   ↓ IRS-1

↑ BDNF

↑ IGF-1

↓ HPA-axis reactivity

↑ VEGF

↓ Cortisol

↑ Activation of brown adipose tissue

↓ Notch Signalling

↑ Serotonin

↑ Lipolysis

↑ DICER and miRNAs

↑ perilipin 3   ↑ GTPase protiens

FIG. 1 Overview of the molecular mechanisms of exercise-induced health benefits in women with PCOS. Primary tissues and organs affected by exercise include skeletal muscle, adipose tissue, heart and endothelium, brain, and reproductive system.

training [79]. Additionally, aerobic exercise training has been shown to reduce circulating leukocytes and alter the adipokine profile in women with PCOS, switching from pro-inflammatory to anti-inflammatory profile [91]. Overall, these studies highlight how exercise training elicits beneficial effects on adipose tissue morphology, metabolism, and inflammation, leading to improvements in whole-body metabolism.

Other aspects of adipose tissue function have been explored in relation to exercise and metabolism, such as microRNAs. It has been shown that DICER, an essential factor in microRNA biogenesis, is downregulated in the adipose tissue of women with PCOS and has been related to IR [139]. A recent study has established that DICER is upregulated by skeletal muscle AMPK in response to high-intensity exercise training and can modulate metabolic responses (Fig. 1) [140]. In line with that, exercise training has been shown to regulate the release of adipokines, such as TGF-beta 2, from adipose tissue, which helps to improve insulin sensitivity and reduce inflammatory factors [141,142]. This highlights the crucial role of the exercise-inducible factors secreted by the adipose tissue that may exert beneficial metabolic effects in PCOS.

Cardiovascular system and the endothelium

Women with PCOS display endothelial dysfunction independent of obesity, and an increased risk of developing CVD due to hyperandrogenism, hyperinsulinemia, and chronic inflammation [143,144]. Endothelial function can be characterized by a reduction in endothelium dependant vasodilation [145], which is linked to a reduction in nitric oxide bioavailability and endothelial nitric oxide expression, and by an increase in reactive oxygen species production and vasoconstrictors such as endothelin-1 [146,147].

Exercise training in women with PCOS has been shown to improve endothelial function and cardiovascular health [93,148] (Fig. 1). Moderate-intensity exercise training for 16 weeks is effective for improving nitric oxide-mediated microvascular function as demonstrated by increases in cutaneous blood flow and biracial artery flow-mediated dilation, with these changes occurring in the absence of weight loss [93,148]. Also, markers of endothelial health are positively regulated in women with PCOS following a period of exercise training, resulting in a reduction in circulating CD105+ endothelial microparticles and other markers of endothelial health; vascular cell adhesion molecule-1, soluble intercellular adhesion molecule-1, and plasminogen activator inhibitor-1 [149,150]. Other potential regulators of cardiovascular health influenced by exercise (acute or training) include an increase in plasma nitrate and nitrite, both indicators of nitric oxide bioavailability [151,152]. This phenomenon tends to occur in conjunction with a reduction in arterial pressure. Indeed, it has been highlighted that these nitric oxide metabolites are decreased in women with PCOS as well as the elevated arterial pressure [153–155].

Reproductive health

Exercise training can also improve reproductive function in women with PCOS (Fig. 1) [72]. These effects are associated with improvements in body composition and reduced hormonal imbalance, which ultimately leads to regular menstrual cycles, increased ovulation rate, and an improved likelihood of pregnancy [72]. Exercise contributes to lower free androgen levels by increasing SHBG and decreasing hyperinsulinemia, which in turn causes a decrease in ovarian steroidogenesis. This reduction in androgen levels may restore the HPO axis imbalance and eventually the ovulatory function [78,156]. In this context, exercise has also been shown to normalize ovarian morphology by regulating adrenergic receptors [157]. However, the precise molecular mechanisms of exercise-induced improvements in reproductive function in women with PCOS require further investigation.

Mental health

Regular exercise has been shown to result in beneficial mental health effects in women with PCOS [70]. These effects may be attributed to the impact of exercise on the brain by increasing angiogenesis, neurogenesis, and neuroplasticity, inducing a protective effect from depression [158,159]. These changes in brain activity and structure following exercise appear to be modulated by changes in circulating neurotransmitters, neurotrophic, and inflammatory factors (Fig. 1) [160]. Physical exercise promotes the release of norepinephrine, dopamine, and serotonin, which are associated with an increased state of alertness, pleasure, and reduced level of anxiety [159]. Acute sprint interval exercise has been found to increase circulating brain-derived neurotrophic factor (BDNF), insulin-like growth factor 1 (IGF-1), and vascular endothelial growth factor (VEGF) levels immediately post-exercise, with the levels of BDNF and VEGF remaining elevated for 1 h after exercise [161]. Regular exercise and physical activity also reduce the reactivity of the hypothalamic-pituitary-adrenal axis that regulates the stress response and the sympathetic nervous system by decreasing cortisol levels both at rest and in response to a stressor [162,163]. However, the specific underlying mechanisms improving depression in women with PCOS are poorly understood, and the impact of these exercise-induced beneficial effects on mental health may depend on the physiology of each individual.

Recommendations for exercise prescription

Due to the vast number of health improvements evoked from exercise participation, it is clear why it is a first-line therapy for women with PCOS. The current international evidence-based guidelines for the assessment and management of PCOS suggest the following for prevention of weight gain and maintenance of health [20]:

- In adults from 18 to 64 years, a minimum of 150 min/week of moderate-intensity physical activity or 75 min/week of vigorous intensities or an equivalent combination of both, including muscle strengthening activities on two non-consecutive days/week.
- In adolescents, at least 60 min of moderate to vigorous-intensity physical activity/day, including those that strengthen muscle and bone at least 3 times weekly.
- Activity should be performed in at least 10-min bouts or around 1000 steps, aiming to achieve at least 30 min daily on most days.

For modest weight loss, prevention of weight gain, and greater health benefits they recommend:

- A minimum of 250 min/week of moderate-intensity activities or 150 min/week of vigorous intensity or an equivalent combination of both, and muscle strengthening activities involving major muscle groups on two non-consecutive days/week.
- Minimized sedentary, screen, or sitting time.

However, the exercise characteristics that may promote the most advantageous effects while being enjoyable and promoting adherence and compliance are yet to be determined. Therefore, exercise programs should be individualized to suit the needs of each woman.

Another relevant consideration is the high attrition rates to exercise interventions among women with PCOS [164,165]. The diminished psychological well-being in these women [75] has the potential to interfere with adherence to exercise interventions [166]. Supervised exercise studies have had much greater success in regards to attendance rates [80,94,167] and therefore are likely to contribute to greater health improvements. Fully or partially supervised programs may be more successful at enabling the promotion of self-efficacy and self-motivation and aid in long-term adherence to exercise.

Practical considerations

- Exercise testing (e.g., submaximal or maximal [if safe to do so] graded exercise test) should be conducted in order to determine baseline levels of fitness to monitor improvements and to adequately prescribe exercise intensity.
- For sedentary, untrained, or women with CVD risk factors, begin at a moderate intensity (55%–70% HR_{max}) and progress accordingly when the individual is safe to do so.
- Monitor heart rates and ratings of perceived exertion (RPE) throughout the session for safety and in order to determine when exercise progression is required.
- Conduct a thorough history (prior sports/exercise, injury/illness, barriers to exercise) to select the mode, intensity and location of exercise sessions.
- Supervised sessions may be required initially, followed by tapering of supervision, to assure long-term adherence to exercise.
- Lastly, design an exercise program that the client will enjoy in order to encourage long-term exercise participation.

Conclusion

Evidence suggests that exercise improves metabolic, reproductive, and mental health in women with PCOS, and reduces the risk factors associated with CVD and metabolic comorbidities. These health benefits are mainly mediated by improved insulin sensitivity, reduced hyperinsulinemia, and lower free androgen levels. Therefore, exercise interventions should be considered by healthcare professionals as first-line therapy for women with PCOS. The current international evidence-based guidelines for PCOS recommend a minimum of 150 min of moderate-intensity or 75 min of vigorous-intensity exercise per week. However, due to the highly complex and multifactorial nature of the syndrome, exercise programs should be individualized to suit the needs of each woman. Future research is required to elucidate the effects of exercise characteristics to determine the most effective method for improving clinical outcomes in women with PCOS.

References

[1] Ehrmann DA. Polycystic ovary syndrome. N Engl J Med 2005;352(12):1223–36.

[2] Goodarzi MO, et al. Polycystic ovary syndrome: etiology, pathogenesis and diagnosis. Nat Rev Endocrinol 2011;7(4):219–31.

[3] Cooney LG, et al. High prevalence of moderate and severe depressive and anxiety symptoms in polycystic ovary syndrome: a systematic review and meta-analysis. Hum Reprod 2017;32(5):1075–91.

[4] Moran LJ, et al. Impaired glucose tolerance, type 2 diabetes and metabolic syndrome in polycystic ovary syndrome: a systematic review and meta-analysis. Hum Reprod Update 2010;16(4):347–63.

[5] Kakoly NS, et al. Cardiometabolic risks in PCOS: a review of the current state of knowledge. Expert Rev Endocrinol Metab 2019;14(1):23–33.

[6] Deeks AA, Gibson-Helm ME, Teede HJ. Anxiety and depression in polycystic ovary syndrome: a comprehensive investigation. Fertil Steril 2010;93(7):2421–3.

[7] Teede H, Deeks A, Moran L. Polycystic ovary syndrome: a complex condition with psychological, reproductive and metabolic manifestations that impacts on health across the lifespan. BMC Med 2010;8:41.

[8] Bozdag G, et al. The prevalence and phenotypic features of polycystic ovary syndrome: a systematic review and meta-analysis. Hum Reprod 2016;31(12):2841–55.

[9] Azziz R, et al. Polycystic ovary syndrome. Nat Rev Dis Primers 2016;2:16057.

[10] Stein IF, Leventhal ML. Amenorrhea associated with bilateral polycystic ovaries. Am J Obstet Gynecol 1935;29:181–91.

[11] NIH National Institutes Of Health. Evidence-based methodology workshop on polycystic ovary syndrome, https://prevention.nih.gov/docs/programs/pcos/FinalReport.pdf; 2012.

[12] The Rotterdam ESHRE/ASRM-Sponsored PCOS Consensus Workshop Group. Revised 2003 consensus on diagnostic criteria and long-term health risks related to polycystic ovary syndrome. Fertil Steril 2004;81(1):19–25.

[13] Azziz R, et al. The androgen excess and PCOS society criteria for the polycystic ovary syndrome: the complete task force report. Fertil Steril 2009;91(2):456–88.

[14] Vink JM, et al. Heritability of polycystic ovary syndrome in a Dutch twin-family study. J Clin Endocrinol Metab 2006;91(6):2100–4.

[15] Tata B, et al. Elevated prenatal anti-Müllerian hormone reprograms the fetus and induces polycystic ovary syndrome in adulthood. Nat Med 2018;24(6):834–46.

[16] Abbott DH, Dumesic DA, Franks S. Developmental origin of polycystic ovary syndrome—a hypothesis. J Endocrinol 2002;174(1):1–5.

[17] Mimouni NEH, et al. Polycystic ovary syndrome is transmitted via a transgenerational epigenetic process. Cell Metab 2021.

[18] O'Reilly MW, et al. Hyperandrogenemia predicts metabolic phenotype in polycystic ovary syndrome: the utility of serum androstenedione. J Clin Endocrinol Metab 2014;99(3):1027–36.

[19] Dewailly D. Diagnostic criteria for PCOS: is there a need for a rethink? Best Pract Res Clin Obstet Gynaecol 2016;37:5–11.

[20] Teede HJ, et al. Recommendations from the international evidence-based guideline for the assessment and management of polycystic ovary syndrome. Fertil Steril 2018;110(3):364–79.

[21] Baptiste CG, et al. Insulin and hyperandrogenism in women with polycystic ovary syndrome. J Steroid Biochem Mol Biol 2010;122(1–3):42–52.

[22] Escobar-Morreale HF. Polycystic ovary syndrome: definition, aetiology, diagnosis and treatment. Nat Rev Endocrinol 2018;14(5):270.

[23] Blank SK, McCartney CR, Marshall JC. The origins and sequelae of abnormal neuroendocrine function in polycystic ovary syndrome. Hum Reprod Update 2006;12(4):351–61.

[24] Doi SA, et al. PCOS: an ovarian disorder that leads to dysregulation in the hypothalamic-pituitary-adrenal axis? Eur J Obstet Gynecol Reprod Biol 2005;118(1):4–16.

[25] O'Reilly MW, et al. 11-oxygenated C19 steroids are the predominant androgens in Polycystic Ovary Syndrome. J Clin Endocrinol Metabol 2016;102(3):840–8.

[26] Balen AH, et al. Andrology: Polycystic ovary syndrome: the spectrum of the disorder in 1741 patients. Hum Reprod 1995;10(8):2107–11.

[27] Sirmans SM, Pate KA. Epidemiology, diagnosis, and management of polycystic ovary syndrome. Clin Epidemiol 2013;6:1–13.

[28] Legro RS, et al. Polycystic ovaries are common in women with hyperandrogenic chronic anovulation but do not predict metabolic or reproductive phenotype. J Clin Endocrinol Metab 2005;90(5):2571–9.

[29] Murphy MK, et al. Polycystic ovarian morphology in normal women does not predict the development of polycystic ovary syndrome. J Clin Endocrinol Metab 2006;91(10):3878–84.

[30] Dunaif A, et al. Profound peripheral insulin resistance, independent of obesity, in polycystic ovary syndrome. Diabetes 1989;38(9):1165–74.

[31] Dunaif A, et al. Evidence for distinctive and intrinsic defects in insulin action in polycystic ovary syndrome. Diabetes 1992;41(10):1257–66.

[32] Stepto NK, et al. Women with polycystic ovary syndrome have intrinsic insulin resistance on euglycaemic-hyperinsulaemic clamp. Hum Reprod 2013;28(3):777–84.

[33] Moghetti P, et al. Divergences in insulin resistance between the different phenotypes of the polycystic ovary syndrome. J Clin Endocrinol Metab 2013;98(4):E628–37.

[34] Cassar S, et al. Insulin resistance in polycystic ovary syndrome: a systematic review and meta-analysis of euglycaemic-hyperinsulinaemic clamp studies. Hum Reprod 2016;31(11):2619–31.

[35] Muniyappa R, et al. Current approaches for assessing insulin sensitivity and resistance in vivo: advantages, limitations, and appropriate usage. Am J Physiol Endocrinol Metab 2008;294(1):E15–26.

[36] Matthews DR, et al. Homeostasis model assessment: insulin resistance and beta-cell function from fasting plasma glucose and insulin concentrations in man. Diabetologia 1985;28(7):412–9.

[37] Zhao L, et al. Polycystic ovary syndrome (PCOS) and the risk of coronary heart disease (CHD): a meta-analysis. Oncotarget 2016;7(23):33715–21.

[38] de Groot PC, et al. PCOS, coronary heart disease, stroke and the influence of obesity: a systematic review and meta-analysis. Hum Reprod Update 2011;17(4):495–500.

[39] Nestler JE, Clore JN, Blackard WG. The central role of obesity (hyperinsulinemia) in the pathogenesis of the polycystic ovary syndrome. Am J Obstet Gynecol 1989;161(5):1095–7.

[40] Insler V, et al. Polycystic ovaries in non-obese and obese patients: possible pathophysiological mechanism based on new interpretation of facts and findings. Hum Reprod 1993;8(3):379–84.

[41] Diamanti-Kandarakis E, Dunaif A. Insulin resistance and the polycystic ovary syndrome revisited: an update on mechanisms and implications. Endocr Rev 2012;33(6):981–1030.

[42] Moran LJ, et al. Lifestyle changes in women with polycystic ovary syndrome. Cochrane Database Syst Rev 2011;7:Cd007506.

[43] Dunaif A. Insulin resistance and the polycystic ovary syndrome: mechanism and implications for pathogenesis. Endocr Rev 1997;18(6):774–800.

[44] Nelson VL, et al. The biochemical basis for increased testosterone production in theca cells propagated from patients with polycystic ovary syndrome. J Clin Endocrinol Metab 2001;86(12):5925–33.

[45] Micic D, et al. Androgen levels during sequential insulin euglycemic clamp studies in patients with polycystic ovary disease. J Steroid Biochem 1988;31(6):995–9.

[46] Nestler JE, et al. A direct effect of hyperinsulinemia on serum sex hormone-binding globulin levels in obese women with the polycystic ovary syndrome. J Clin Endocrinol Metab 1991;72(1):83–9.

[47] Plymate SR, et al. Inhibition of sex hormone-binding globulin production in the human hepatoma (Hep G2) cell line by insulin and prolactin*. J Clin Endocrinol Metabol 1988;67(3):460–4.

[48] Azziz R, et al. The prevalence and features of the polycystic ovary syndrome in an unselected population. J Clin Endocrinol Metab 2004;89(6):2745–9.

[49] Lo JC, et al. Epidemiology and adverse cardiovascular risk profile of diagnosed polycystic ovary syndrome. J Clin Endocrinol Metab 2006;91(4):1357–63.

[50] Pasquali R, Gambineri A, Pagotto U. The impact of obesity on reproduction in women with polycystic ovary syndrome. BJOG 2006;113(10):1148–59.

[51] Barber TM, et al. Obesity and polycystic ovary syndrome. Clin Endocrinol (Oxf) 2006;65(2):137–45.

[52] Holte J, et al. Restored insulin sensitivity but persistently increased early insulin secretion after weight loss in obese women with polycystic ovary syndrome. J Clin Endocrinol Metab 1995;80(9):2586–93.

[53] Lucke JC, et al. Health across generations: findings from the Australian Longitudinal Study on Women's Health. Biol Res Nurs 2010;12(2):162–70.

[54] Tanamas SK, et al. AusDiab 2012. The Australian diabetes, obesity and lifestyle study. Baker IDI Heart and Diabetes Institute Melbourne; 2013. p. 92.

[55] Kiddy D, et al. Differences in clinical and endocrine features between obese and non-obese subjects with polycystic ovary syndrome: an analysis of 263 consecutive cases. Clin Endocrinol (Oxf) 1990;32(2):213–20.

[56] Lim SS, et al. The effect of obesity on polycystic ovary syndrome: a systematic review and meta-analysis. Obes Rev 2013;14(2):95–109.

[57] Hutchison SK, et al. Effects of exercise on insulin resistance and body composition in overweight and obese women with and without polycystic ovary syndrome. J Clin Endocrinol Metab 2011;96(1):E48–56.

[58] Carmina E, et al. Abdominal fat quantity and distribution in women with polycystic ovary syndrome and extent of its relation to insulin resistance. J Clin Endocrinol Metab 2007;92(7):2500–5.

[59] Brassard M, AinMelk Y, Baillargeon JP. Basic infertility including polycystic ovary syndrome. Med Clin North Am 2008;92(5):1163–92 [xi].

[60] Teede HJ, et al. Assessment and management of polycystic ovary syndrome: summary of an evidence-based guideline. Med J Aust 2011;195(6): S65–112.

[61] Himelein MJ, Thatcher SS. Depression and body image among women with polycystic ovary syndrome. J Health Psychol 2006;11(4):613–25.

[62] Jones GL, et al. Health-related quality of life measurement in women with polycystic ovary syndrome: a systematic review. Hum Reprod Update 2008;14(1):15–25.

[63] Coffey S, Bano G, Mason HD. Health-related quality of life in women with polycystic ovary syndrome: a comparison with the general population using the Polycystic Ovary Syndrome Questionnaire (PCOSQ) and the Short Form-36 (SF-36). Gynecol Endocrinol 2006;22(2):80–6.

[64] Moran LJ, et al. Psychological parameters in the reproductive phenotypes of polycystic ovary syndrome. Hum Reprod 2012;27(7):2082–8.

[65] Barry JA, Kuczmierczyk AR, Hardiman PJ. Anxiety and depression in polycystic ovary syndrome: a systematic review and meta-analysis. Hum Reprod 2011;26(9):2442–51.

[66] Lear SA, et al. The effect of physical activity on mortality and cardiovascular disease in 130 000 people from 17 high-income, middle-income, and low-income countries: the PURE study. Lancet 2017;390(10113):2643–54.

[67] Pedersen BK, Saltin B. Exercise as medicine—evidence for prescribing exercise as therapy in 26 different chronic diseases. Scand J Med Sci Sports 2015;25(Suppl 3):1–72.

[68] Thomson RL, Buckley JD, Brinkworth GD. Exercise for the treatment and management of overweight women with polycystic ovary syndrome: a review of the literature. Obes Rev 2011;12(5):e202–10.

[69] Cheema BS, Vizza L, Swaraj S. Progressive resistance training in polycystic ovary syndrome: can pumping iron improve clinical outcomes? Sports Med 2014;44(9):1197–207.

[70] Conte F, et al. Mental health and physical activity in women with polycystic ovary syndrome: a brief review. Sports Med 2015;45(4):497–504.

[71] Patten RK, et al. Exercise interventions in Polycystic Ovary Syndrome: a systematic review and meta-analysis. Front Physiol 2020;11(606).

[72] Benham JL, et al. Role of exercise training in polycystic ovary syndrome: a systematic review and meta-analysis. Clin Obes 2018;8(4):275–84.

[73] Miranda-Furtado CL, et al. A nonrandomized trial of progressive resistance training intervention in women with Polycystic Ovary Syndrome and its implications in telomere content. Reprod Sci 2016;23(5):644–54.

[74] Almenning I, et al. Effects of high intensity interval training and strength training on metabolic, cardiovascular and hormonal outcomes in women with Polycystic Ovary Syndrome: a pilot study. PLoS One 2015;10(9), e0138793.

[75] Banting LK, et al. Physical activity and mental health in women with polycystic ovary syndrome. BMC Womens Health 2014;14(1):51.

[76] Colberg SR, et al. Physical activity/exercise and diabetes: a position statement of the American Diabetes Association. Diabetes Care 2016;39(11):2065–79.

[77] Knowler WC, et al. Reduction in the incidence of type 2 diabetes with lifestyle intervention or metformin. N Engl J Med 2002;346(6):393–403.

[78] Harrison CL, et al. Exercise therapy in polycystic ovary syndrome: a systematic review. Hum Reprod Update 2011;17(2):171–83.

[79] Covington JD, et al. Potential effects of aerobic exercise on the expression of perilipin 3 in the adipose tissue of women with polycystic ovary syndrome: a pilot study. Eur J Endocrinol 2015;172(1):47–58.

[80] Hutchison SK, et al. Effect of exercise training on insulin sensitivity, mitochondria and computed tomography muscle attenuation in overweight women with and without polycystic ovary syndrome. Diabetologia 2012;55(5):1424–34.

[81] Thomson RL, et al. The effect of a hypocaloric diet with and without exercise training on body composition, cardiometabolic risk profile, and reproductive function in overweight and obese women with polycystic ovary syndrome. J Clin Endocrinol Metab 2008;93(9):3373–80.

[82] Orio F, et al. Oral contraceptives versus physical exercise on cardiovascular and metabolic risk factors in women with polycystic ovary syndrome: a randomized controlled trial. Clin Endocrinol (Oxf) 2016;85(5):764–71.

[83] Giallauria F, et al. Exercise training improves autonomic function and inflammatory pattern in women with polycystic ovary syndrome (PCOS). Clin Endocrinol (Oxf) 2008;69(5):792–8.

[84] Myers J, et al. Exercise capacity and mortality among men referred for exercise testing. N Engl J Med 2002;346(11):793–801.

[85] Barry VW, et al. Fitness vs. fatness on all-cause mortality: a meta-analysis. Prog Cardiovasc Dis 2014;56(4):382–90.

[86] Ladson G, et al. The effects of metformin with lifestyle therapy in polycystic ovary syndrome: a randomized double-blind study. Fertil Steril 2011;95(3), 1059-66.e1-7.

[87] Kite C, et al. Exercise, or exercise and diet for the management of polycystic ovary syndrome: a systematic review and meta-analysis. Syst Rev 2019;8(1):51.

[88] Roberts CK, Hevener AL, Barnard RJ. Metabolic syndrome and insulin resistance: underlying causes and modification by exercise training. Compr Physiol 2013;3(1):1–58.

[89] Boulé NG, et al. Effects of exercise on glycemic control and body mass in type 2 diabetes mellitus: a meta-analysis of controlled clinical trials. JAMA 2001;286(10):1218–27.

[90] Weston KS, Wisløff U, Coombes JS. High-intensity interval training in patients with lifestyle-induced cardiometabolic disease: a systematic review and meta-analysis. Br J Sports Med 2014;48(16):1227.

[91] Covington JD, et al. Higher circulating leukocytes in women with PCOS is reversed by aerobic exercise. Biochimie 2016;124:27–33.

[92] Orio F, et al. Metabolic and cardiopulmonary effects of detraining after a structured exercise training programme in young PCOS women. Clin Endocrinol (Oxf) 2008;68(6):976–81.

[93] Sprung VS, et al. Exercise training in polycystic ovarian syndrome enhances flow-mediated dilation in the absence of changes in fatness. Med Sci Sports Exerc 2013;45(12):2234–42.

[94] Vigorito C, et al. Beneficial effects of a three-month structured exercise training program on cardiopulmonary functional capacity in young women with polycystic ovary syndrome. J Clin Endocrinol Metab 2007;92(4):1379–84.

[95] Nybacka A, et al. Randomized comparison of the influence of dietary management and/or physical exercise on ovarian function and metabolic parameters in overweight women with polycystic ovary syndrome. Fertil Steril 2011;96(6):1508–13.

[96] Haqq L, et al. Effect of lifestyle intervention on the reproductive endocrine profile in women with polycystic ovarian syndrome: a systematic review and meta-analysis. Endocr Connect 2014;3(1):36–46.

[97] Norman RJ, Clark AM. Obesity and reproductive disorders: a review. Reprod Fertil Dev 1998;10(1):55–63.

[98] Balen A, et al. The influence of body weight on response to ovulation induction with gonadotrophins in 335 women with World Health Organization group II anovulatory infertility. BJOG 2006;113(10):1195–202.

[99] Palomba S, et al. Six weeks of structured exercise training and hypocaloric diet increases the probability of ovulation after clomiphene citrate in overweight and obese patients with polycystic ovary syndrome: a randomized controlled trial. Hum Reprod 2010;25(11):2783–91.

[100] Crosignani PG, et al. Overweight and obese anovulatory patients with polycystic ovaries: parallel improvements in anthropometric indices, ovarian physiology and fertility rate induced by diet. Hum Reprod 2003;18(9):1928–32.

[101] Clark AM, et al. Weight loss in obese infertile women results in improvement in reproductive outcome for all forms of fertility treatment. Hum Reprod 1998;13(6):1502–5.

[102] Kiddy DS, et al. Improvement in endocrine and ovarian function during dietary treatment of obese women with polycystic ovary syndrome. Clin Endocrinol (Oxf) 1992;36(1):105–11.

[103] Torrealday S, Patrizio P. In: Pal L, editor. Managing PCOS-related infertility: ovulation induction, in vitro fertilization, and in vitro maturation, in polycystic ovary syndrome: current and emerging concepts. New York, NY: Springer New York; 2014. p. 205–21.

[104] Palomba S, et al. Structured exercise training programme versus hypocaloric hyperproteic diet in obese polycystic ovary syndrome patients with anovulatory infertility: a 24-week pilot study. Hum Reprod 2008;23(3):642–50.

[105] Stephens T. Physical activity and mental health in the United States and Canada: evidence from four population surveys. Prev Med 1988;17(1):35–47.

[106] Rippe JM, et al. Improved psychological well-being, quality of life, and health practices in moderately overweight women participating in a 12-week structured weight loss program. Obes Res 1998;6(3):208–18.

[107] Kirk AF, et al. A randomized, controlled trial to study the effect of exercise consultation on the promotion of physical activity in people with type 2 diabetes: a pilot study. Diabet Med 2001;18(11):877–82.

[108] Galletly C, et al. Psychological benefits of a high-protein, low-carbohydrate diet in obese women with polycystic ovary syndrome—a pilot study. Appetite 2007;49(3):590–3.

[109] Lara LA, et al. Impact of physical resistance training on the sexual function of women with Polycystic Ovary Syndrome. J Sex Med 2015;12(7):1584–90.

[110] Kogure GS, et al. The effects of aerobic physical exercises on body image among women with polycystic ovary syndrome. J Affect Disord 2020;262:350–8.

[111] Thomson RL, et al. Lifestyle management improves quality of life and depression in overweight and obese women with polycystic ovary syndrome. Fertil Steril 2010;94(5):1812–6.

[112] Ribeiro VB, et al. Continuous versus intermittent aerobic exercise in the improvement of quality of life for women with polycystic ovary syndrome: a randomized controlled trial. J Health Psychol 2019;, 1359105319869806.

[113] Costa EC, et al. Aerobic training improves quality of life in women with Polycystic Ovary Syndrome. Med Sci Sports Exerc 2018;50(7):1357–66.

[114] Arentz S, et al. Combined lifestyle and herbal medicine in overweight women with Polycystic Ovary Syndrome (PCOS): a randomized controlled trial. Phytother Res 2017;31(9):1330–40.

[115] Liao LM, et al. Exercise and body image distress in overweight and obese women with polycystic ovary syndrome: a pilot investigation. Gynecol Endocrinol 2008;24(10):555–61.

[116] Lamb JD, et al. Physical activity in women with polycystic ovary syndrome: prevalence, predictors, and positive health associations. Am J Obstet Gynecol 2011;204(4), 352.e1-6.

[117] DeFronzo RA, et al. Synergistic interaction between exercise and insulin on peripheral glucose uptake. J Clin Invest 1981;68(6):1468–74.

[118] Kristiansen S, Hargreaves M, Richter EA. Exercise-induced increase in glucose transport, GLUT-4, and VAMP-2 in plasma membrane from human muscle. Am J Physiol 1996;270(1 Pt 1):E197–201.

[119] Mikines KJ, et al. Effect of physical exercise on sensitivity and responsiveness to insulin in humans. Am J Physiol 1988;254(3 Pt 1):E248–59.

[120] Richter EA, et al. Effect of exercise on insulin action in human skeletal muscle. J Appl Physiol 1985;66(2):876–85. 1989.

[121] Steenberg DE, et al. A single bout of one-legged exercise to local exhaustion decreases insulin action in nonexercised muscle leading to decreased whole-body insulin action. Diabetes 2020;69(4):578–90.

[122] Kjobsted R, et al. TBC1D4 is necessary for enhancing muscle insulin sensitivity in response to AICAR and contraction. Diabetes 2019;68(9):1756–66.

[123] Kjobsted R, et al. Enhanced muscle insulin sensitivity after contraction/exercise is mediated by AMPK. Diabetes 2017;66(3):598–612.

[124] Kjobsted R, et al. AMPK and TBC1D1 regulate muscle glucose uptake after, but not during, exercise and contraction. Diabetes 2019;68(7):1427–40.

[125] Sjoberg KA, et al. Exercise increases human skeletal muscle insulin sensitivity via coordinated increases in microvascular perfusion and molecular signaling. Diabetes 2017;66(6):1501–10.

[126] Ryan BJ, et al. Moderate-intensity exercise and high-intensity interval training affect insulin sensitivity similarly in obese adults. J Clin Endocrinol Metab 2020;105(8).

[127] Sylow L, Richter EA. Current advances in our understanding of exercise as medicine in metabolic disease. Curr Opin Physio 2019;12:12–9.

[128] Dantas WS, et al. GLUT4 translocation is not impaired after acute exercise in skeletal muscle of women with obesity and polycystic ovary syndrome. Obesity (Silver Spring) 2015;23(11):2207–15.

[129] Dantas WS, et al. Acute exercise elicits differential expression of insulin resistance genes in the skeletal muscle of patients with polycystic ovary syndrome. Clin Endocrinol (Oxf) 2017;86(5):688–97.

[130] Dantas WS, et al. Exercise-induced anti-inflammatory effects in overweight/obese women with polycystic ovary syndrome. Cytokine 2019;120:66–70.

[131] Glund S, et al. Interleukin-6 directly increases glucose metabolism in resting human skeletal muscle. Diabetes 2007;56(6):1630–7.

[132] Konopka AR, et al. Defects in mitochondrial efficiency and H2O2 emissions in obese women are restored to a lean phenotype with aerobic exercise training. Diabetes 2015;64(6):2104–15.

[133] Manneras-Holm L, et al. Adipose tissue has aberrant morphology and function in PCOS: enlarged adipocytes and low serum adiponectin, but not circulating sex steroids, are strongly associated with insulin resistance. J Clin Endocrinol Metab 2011;96(2):E304–11.

[134] Badoud F, et al. Molecular insights into the role of white adipose tissue in metabolically unhealthy normal weight and metabolically healthy obese individuals. FASEB J 2015;29(3):748–58.

[135] Shorakae S, et al. Brown adipose tissue thermogenesis in polycystic ovary syndrome. Clin Endocrinol (Oxf) 2019;90(3):425–32.

[136] Spritzer PM, et al. Adipose tissue dysfunction, adipokines, and low-grade chronic inflammation in polycystic ovary syndrome. Reproduction 2015;149(5):R219–27.

[137] Dumesic DA, et al. Adipose insulin resistance in normal-weight women with Polycystic Ovary Syndrome. J Clin Endocrinol Metab 2019;104(6):2171–83.

[138] Moro C, et al. Aerobic exercise training improves atrial natriuretic peptide and catecholamine-mediated lipolysis in obese women with polycystic ovary syndrome. J Clin Endocrinol Metab 2009;94(7):2579–86.

[139] Qin L, et al. Significant role of Dicer and miR-223 in adipose tissue of Polycystic Ovary Syndrome patients. Biomed Res Int 2019;2019:9193236.

[140] Brandao BB, et al. Dynamic changes in DICER levels in adipose tissue control metabolic adaptations to exercise. Proc Natl Acad Sci U S A 2020;117(38):23932–41.

[141] Mika A, et al. Effect of exercise on fatty acid metabolism and adipokine secretion in adipose tissue. Front Physiol 2019;10:26.

[142] Takahashi H, et al. TGF-beta2 is an exercise-induced adipokine that regulates glucose and fatty acid metabolism. Nat Metab 2019;1(2):291–303.

[143] Moreau KL. Modulatory influence of sex hormones on vascular aging. Am J Physiol Heart Circ Physiol 2019;316(3):H522–6.

[144] Usselman CW, et al. Androgens drive microvascular endothelial dysfunction in women with polycystic ovary syndrome: role of the endothelin B receptor. J Physiol 2019;597(11):2853–65.

[145] Paradisi G, et al. Polycystic ovary syndrome is associated with endothelial dysfunction. Circulation 2001;103(10):1410–5.

[146] Diamanti-Kandarakis E, et al. Inflammatory and endothelial markers in women with polycystic ovary syndrome. Eur J Clin Invest 2006;36(10):691–7.

[147] Forstermann U, Munzel T. Endothelial nitric oxide synthase in vascular disease: from marvel to menace. Circulation 2006;113(13):1708–14.

[148] Sprung VS, et al. Nitric oxide-mediated cutaneous microvascular function is impaired in polycystic ovary sydrome but can be improved by exercise training. J Physiol 2013;591(6):1475–87.

[149] Kirk RJ, et al. Circulating endothelial microparticles reduce in concentration following an exercise Programme in women with Polycystic Ovary Syndrome. Front Endocrinol (Lausanne) 2019;10:200.

[150] Thomson RL, et al. The effect of diet and exercise on markers of endothelial function in overweight and obese women with polycystic ovary syndrome. Hum Reprod 2012;27(7):2169–76.

[151] Allen JD, et al. Plasma nitrite response and arterial reactivity differentiate vascular health and performance. Nitric Oxide 2009;20(4):231–7.

[152] Rassaf T, et al. Nitric oxide synthase-derived plasma nitrite predicts exercise capacity. Br J Sports Med 2007;41(10):669–73 [discussion 673].

[153] Krishna MB, et al. Impaired arginine metabolism coupled to a defective redox conduit contributes to low plasma nitric oxide in Polycystic Ovary Syndrome. Cell Physiol Biochem 2017;43(5):1880–92.

[154] Mellembakken JR, et al. Higher blood pressure in normal weight women with PCOS compared to controls. Endocr Connect 2021;10:154–63.

[155] Meng C. Nitric oxide (NO) levels in patients with polycystic ovary syndrome (PCOS): a meta-analysis. J Int Med Res 2019;47(9):4083–94.

[156] Hakimi O, Cameron LC. Effect of exercise on ovulation: a systematic review. Sports Med 2017;47(8):1555–67.

[157] Manni L, et al. Effect of exercise on ovarian morphology and expression of nerve growth factor and alpha(1)- and beta(2)-adrenergic receptors in rats with steroid-induced polycystic ovaries. J Neuroendocrinol 2005;17(12):846–58.

[158] Cotman CW, Berchtold NC. Exercise: a behavioral intervention to enhance brain health and plasticity. Trends Neurosci 2002;25(6):295–301.

[159] Matta Mello Portugal E, et al. Neuroscience of exercise: from neurobiology mechanisms to mental health. Neuropsychobiology 2013;68(1):1–14.

[160] Cotman CW, Berchtold NC, Christie LA. Exercise builds brain health: key roles of growth factor cascades and inflammation. Trends Neurosci 2007;30(9):464–72.

[161] Kujach S, et al. Acute Sprint interval exercise increases both cognitive functions and peripheral neurotrophic factors in humans: the possible involvement of lactate. Front Neurosci 2019;13:1455.

[162] Anderson T, Berry NT, Wideman L. Exercise and the hypothalamic–pituitary–adrenal axis: a special focus on acute cortisol and growth hormone responses. Curr Opinion Endocrine Metabol Res 2019;9:74–7.

[163] Rimmele U, et al. The level of physical activity affects adrenal and cardiovascular reactivity to psychosocial stress. Psychoneuroendocrinology 2009;34(2):190–8.

[164] Brown AJ, et al. Effects of exercise on lipoprotein particles in women with polycystic ovary syndrome. Med Sci Sports Exerc 2009;41(3):497–504.

[165] Randeva HS, et al. Exercise decreases plasma total homocysteine in overweight young women with polycystic ovary syndrome. J Clin Endocrinol Metab 2002;87(10):4496–501.

[166] Huberty JL, et al. Explaining long-term exercise adherence in women who complete a structured exercise program. Res Q Exerc Sport 2008;79(3):374–84.

[167] Vizza L, et al. The feasibility of progressive resistance training in women with polycystic ovary syndrome: a pilot randomized controlled trial. BMC Sports Sci Med Rehabil 2016;8:14.

Part V

Older age

Chapter 21

Exercise for the management of osteoarthritis

Daniel Corcoran[a], Joel Hiney[a], Luke Ellis[a], Jack Feehan[a,b,c,d], and Nicholas Tripodi[a,b,c,d]

[a]Osteopathy group, College of Health and Biomedicine, Victoria University, Melbourne, VIC, Australia [b]Institute for Health and Sport, Victoria University, Melbourne, VIC, Australia [c]First Year College, Victoria University, Melbourne, VIC, Australia [d]The Australian Institute for Musculoskeletal Science, Western Health, Victoria University and the University of Melbourne, Melbourne, VIC, Australia

Osteoarthritis (OA) is a chronic and progressive joint disease that currently affects up to 7% of the global population, which approximates to 500 million individuals worldwide [1]. As of 2018, OA is a leading cause of disability and is believed to contribute to a socioeconomic cost of between 1% and 2.5% of GDP in developed countries. OA is most commonly seen in middle-aged and elderly adults. OA is more common in females compared to males [2], with females representing 42.1% of identified OA cases, compared with 32.2% in males [3]. The development of OA has been linked to a number of factors, including those which are nonmodifiable: Age, sex, genetics, race, and ethnicity; and those which are modifiable: previous joint injury, obesity, activity, occupation, diet, and muscle strength [4].

OA has a typical clinical presentation that includes the presence of one or more stiff and painful joints, which may occur insidiously, and may also be swollen. Additional features of the presentation can include local muscle weakness, reduced range of motion, and coarse crepitus [5]. OA can occur in any synovial joint in the body, however, the most commonly identified locations are the knee, hip, hand, foot, and spine [6].

Given the extensive burden of disease that OA currently poses, a number of interventions have been proposed including exercise, weight loss, and dietary interventions, rehabilitative interventions (including manual therapy, bracing, and electrical stimulation), behavioral therapies, pharmaceutical interventions, and surgical interventions [7]. Of all the currently applied interventions, exercise has been shown not only to be clinically effective but also cost-effective when compared to physician-delivered usual care [8]. Therefore, exercise is currently considered to be one of few effective primary interventions in the management of OA.

Epidemiology, societal, and global impact

OA is a common condition across the world and its burden is widespread and multifactorial. OA affects a large proportion of the global population and is the leading cause of lower-limb disability in older adults [9]. Those over the age of 50, particularly women, see an increased incidence of OA, although the difference between sexes tends to reduce at approximately 80 years of age [10]. OA commonly affects the hands, knees, hips, and spinal facet joints, however, the rates of incidence differ depending on the definition of the condition, which can be either radiographic or symptomatic. Radiographic OA is defined only by pathophysiological signs found on diagnostic imaging, whereas symptomatic disease is classified by symptoms and/or radiographic changes consistent with OA [11]. The estimates tend to vary based on the usage of these definitions, yet broadly, the trend favors higher rates of OA when diagnosed using the radiographic definition and lower rates when diagnosed using the symptomatic definition [11]. Although OA in the lower limb has been researched more than the hand, the following table offers an indication of the prevalence by region (Table 1).

The high prevalence of OA represents how widespread the condition is but it does not capture the functional impact on the affected individuals, which can be severe regardless of the location. Each joint site affected with OA results in specific functional limitations which can reduce an individual's ability to complete usual activities.

OA of the hand and fingers results in pain and stiffness with significantly reduced ability to perform tasks requiring dexterity and movement of these joints. Maximal grip strength can be reduced by as much as 10% in symptomatic sufferers of OA [12]. Tasks such as opening medication bottles, dressing, or using a telephone are negatively impacted in symptomatic OA sufferers [12]. An inability to perform these important tasks can also lead to secondary, deleterious effects

Exercise to Prevent and Manage Chronic Disease Across the Lifespan. https://doi.org/10.1016/B978-0-323-89843-0.00019-2

TABLE 1 Overall prevalence of knee, hip and hand OA (95% CIs) and heterogeneity by sex and joint site.

Joint site	OA prevalence women	OA prevalence men	OA prevalence total
Knee	27.3%[a] 95% CI [26.9–27.7] I^2=99.3%	21.0%[a] 95% CI [20.5–21.5] I^2=99.7%	23.9% 95% CI [23.6–24.2] I^2=99.8%
Hip	11.6% 95% CI [11.1–12.1] I^2=99.7%	11.5% 95% CI [11.0–12.1] I^2=99.9%	10.9% 95% CI [10.6–11.2] I^2=99.8%
Hand	43.3% 95% CI [42.6–44.0] I^2=99.1%	44.5% 95% CI [43.5–45.5] I^2=99.9%	43.3% 95% CI [42.7–42.9] I^2=100%

[a]P value< 0.01 for gender comparison using Mann–Whitney test [11].

on physical, mental, and social well-being [12]. While OA is most common in the hand, there has been significantly more research on the disease's effects on the lower limb. The global prevalence of knee OA has been estimated to be as high as 2.8% of males and 4.8% of females with a combined prevalence of 3.8% [13]. A more recent global estimate of the prevalence of knee and hip OA cases, from the global burden of OA study, was 303.1 million—a 9.3% increase in cases compared to 1990 [14]. As seen in the hand, global prevalence estimates were higher in females and increased with age reaching a peak at 95 years of age [14].

The high rates of OA globally, coupled with the impact that the condition has on those affected, leads to a significant burden of disease. In 2017, the years lived with disability (YLD) associated with OA had grown to a staggering 9.6 million, constituting a 9.6% increase from 1990 [14]. Medical costs associated with the management of OA represent a large burden, with one such example being arthroplasty, otherwise known as joint replacement. Arthroplasty is a procedure that involves removing all or part of a joint and replacing it with a prosthesis [15]. This is performed to treat the pain and disability associated with joint failure from OA. Globally, OA accounts for an estimated medical cost of 1%–2.5% of national gross domestic product [16]. When analyzing the impact of joint replacements in detail it should be noted that, due to the 10-year average lifespan of knee replacements, there is a strong likelihood that additional operations will be required; further increasing the costs and burden on both the individual and health services [17]. It should be noted that these cost estimates often only include costs directly related to healthcare, and do not include the personal costs such as days of work missed, or early retirement related to the burden of OA.

The significance of increased pain and reduced function associated with OA cannot be understated. One such relationship that is being explored is between OA and cardiovascular disease (CVD). The relationship likely stems from correlations between the two conditions as well as a mutual relationship with obesity. A working theory relates to each condition, OA and obesity, leading to a cycle of reduced activity and reduced cardiovascular health [18]. It stands to reason that cardiovascular health will be negatively impacted, and it should come as no surprise that a positive association between OA and CVD has been identified [18]. Therefore, the prevalence and resultant burden of OA are likely to increase significantly in the coming years.

Pathophysiology

The development of OA is multifactorial, consisting of both modifiable and nonmodifiable factors. Well-established nonmodifiable risk factors include age, as homeostatic mechanisms decline with time and the body becomes unable to respond to joint stressors [4]; and sex, with multiple studies noting females are more likely to develop OA than men [19,20]. Modifiable risk factors associated with the development of OA are diverse, but among the most well established are obesity and metabolic disease, sarcopenia, trauma, dietary factors, physical activity, and occupation [9].

OA is primarily identified by its typical disease pattern—degradation of articular cartilage, exposure, and sclerosis of subchondral bone, hypertrophic bony change at joint margins, and synovial inflammation [21]. This degradative process gives way to a pro-inflammatory chain reaction driven by joint biomechanics, cellular inflammatory reactions, and innate immune response within the joint. This furthers matrix degradation [22] and results in pain, swelling, reductions in a range of motion, and muscle weakness [5]. As joint physiology begins to degrade, the articular surfaces enter a pathogenic cycle; repair processes are unable to overcome the breakdown of the extracellular matrix, resulting in further inflammation and progressive joint degradation [23].

A number of cellular reactions have been implicated in the OA process, including proteases and other enzymes, proinflammatory cytokines, and growth factors [24]. Initial research into OA saw the mechanical breakdown of cartilage as the main driver of symptoms. However, more recent investigations highlight the involvement of the whole joint and the role of proinflammatory cells, which have been suggested to be a pivotal driver of the catabolic breakdown of articular tissues [25]. Specific cytokines suggested to be involved in the breakdown process include Interleukin 1β (IL-1β), Tumor necrosis factor-alpha (TNF-α), Interleukin 6 (IL-6) as well as a range of other interleukins [15,17,18,20], Leukemia inhibitory factor (LIF), and various additional chemokines [26].

As OA has primarily been associated with a pro-inflammatory state, the medical intervention has focused on addressing factors that contribute to or maintain this state. Typical pharmacological interventions include analgesics, anti-inflammatory drugs, and intra-articular therapies (including corticosteroids and other biologic agents) [27]. Nonpharmacological interventions have also been extensively investigated, with exercise being found to be effective in reducing pain symptoms and improving physical function in OA patients [28]. Additionally, a number of studies report that exercise can reduce pro-inflammatory markers in patients with chronic inflammatory diseases, noting reductions in IL-1β, IL-6, C-reactive protein, and TNFα [29,30] (Fig. 1).

While the impact of mechanical and inflammatory factors has been the predominant focus of OA research, some groups have focused on the importance of identifying and understanding peripheral and central pain processing in relation to OA [25]. Neuroplastic changes which occur as the disease progresses, have been suggested to contribute to the elevated levels of pain in patients with OA [31]. Further questions regarding the effects of sensitization have also been raised, as the correlation between structural factors, inflammatory changes, and pain appear to be poor [32]. Exercise is suggested to impact pain processing by creating endogenous analgesia, which may have a role in managing pain and sensitization in OA sufferers [33].

Exercise for osteoarthritis: A primer

As OA typically involves a cascade of pain inhibiting function, causing local muscle wasting and reduced joint movement [34], exercise is often recommended to combat these changes. In the context of OA, there has been a particular focus on three exercise modalities: aerobic exercise, resistance exercise, and aquatic exercise.

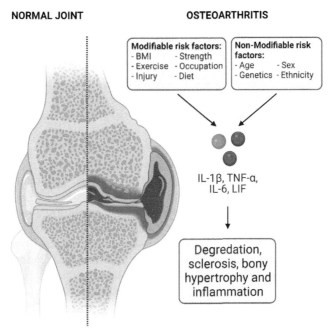

FIG. 1 Pathogenesis of osteoarthritis.

Aerobic exercise (AE) is defined as any activity using large muscle groups, maintained continuously, and is rhythmic in nature [35], with cycling, hiking, jogging, and walking being common examples [36]. Aerobic exercise has long been promoted as a modality that can offer significant changes in OA symptoms, including a reduction in pain and physical disability, as well as improvements in physical performance [37]. As obesity has been identified as a modifiable risk factor of OA, aerobic exercise combined with interventions aimed at reducing body mass index (BMI), and thereby mechanical joint forces, has been shown to effectively reduce symptoms and improve functional outcome measures in OA patients [38].

Resistance exercise (RE) is a form of exercise, whereby external weight or load is applied to provide progressive overloading of skeletal and muscular structures; with the intention to make them stronger [39]. Common examples of resistance training include the use of machines to provide external resistance, or the use of free weights. Resistance training in isolation, or in combination with aerobic exercise has been well established to have a significant impact on function and disability in OA patients [40]. Additionally, physical performance measures including walking endurance, gait speed, and balance have been noted to undergo improvement following a resistance training exercise in patients with OA [41–43].

Aquatic exercise (AqE) is defined as exercise performed in water, with the aim of improving body function relating to OA [44]. Common examples include structured hydrotherapy, aqua aerobics, swimming, or water walking. Due to the nature of water immersion and buoyancy, AqE offers a low-impact combination of resistance and aerobic activity in a noninjurious environment [42]. Various studies of AqE have noted significant improvement in pain, function, and other physical outcome measures relating to OA [45].

While all three exercise modalities appear to offer benefit, the superiority of any particular method appears unclear. Factors such as availability of equipment, facilities, and health professional supervision may play an important role in the implementation of these modalities.

The benefits of exercise for osteoarthritis

General benefits

The treatment and prevention of OA largely focus on achieving the ideal weight, muscle quality, and health of intra- and extra-articular tissues. This, in turn, reduces the global stress placed on joints while increasing their capacity to deal with the necessary load. Exercise has been shown to be at least as effective as nonsteroidal anti-inflammatory drugs (NSAIDS) and 2–3 times more effective than paracetamol [46]. High BMI is associated with an increased risk of developing OA, with impaired muscle function relating to a loss of proprioception and joint stability [46,47]. The main physiological factors affecting OA are muscle strength, cytokine release, joint proprioception, and thus joint stability [47]. Through exercise, moderate joint loading appears to prevent and also slow the development of OA. Exercise, in general, has been proven to provide analgesia through various different mechanisms and neuronal pathways. Endogenous pain modulation, in the form of exercise-induced analgesia, acts on the central, peripheral, and spinal inhibitory mechanisms, creating long-term pain modulation conditioning. Muscle contractions stimulate the Group III and IV (A-delta and C) afferents in skeletal muscle, which can activate the endogenous opioid cascade of pain modulation [48].

Aquatic exercise benefits

Aquatic exercise differs from land-based exercise as it reduces the effect of gravity and thus the axial load placed on joints. The thermal effect of exercising in warm water largely involves increased blood/fluid circulation, which helps to prevent venous stasis, poor arterial inflow, and altered cell signaling which contributes to the pathophysiology of OA [49]. Lymphatic stimulation is an added benefit of aquatic exercise, increasing the removal of toxins and inflammatory cells such as IL-1 and TNF-a, which breakdown type II collagen fibers and cause oxidative stress to the articular cartilage in the osteoarthritic joint [50,51].

Reducing compressive forces on joints promotes greater intra-articular mobility, allowing weight-bearing in such a way that may not be accessible on land. Deloading joints not only aids in increasing ROM but increases blood and lymphatic supply as well as synovial fluid viscosity [52]. Aquatic exercise lowers joint hydrostatic pressure (HP) which is suggested to be one of the primary physical stimuli that are detected by the chondrocytes situated in the cartilage matrix [53], meaning that high HP can have a damaging effect on articular cartilage by inducing apoptosis, pro-inflammatory and stress-related gene expression [54]. Exercise inducing cyclical HP has been seen to reverse the destructive effects of inflammatory markers in chondrocytes, and stimulate the expression of chondrogenic markers [53,54]. Systematic reviews conducted on the effects of aquatic exercise on osteoarthritic joints conclude that water exercises result in increased muscle strength, function, quality of life, and importantly, analgesia [55].

Resistance training

Resistance training exerts its effects by increasing muscle strength which, in turn, has an analgesic effect on joints while also providing stability to unstable joints [56]. Muscles also act to absorb shock, with stronger muscles dispersing load more evenly across the joint and reducing the potential trauma to the intra-articular tissues [57]. Resistance training promotes an increase in joint ROM and decrease in joint stiffness, while improving pain and function scores through strength exercises [58]. Other benefits include increased cartilage quality through joint loading, reduced overloading of cartilage by improved biomechanics, increased shock absorbing, and improved joint stability [59]. Reduced sensorimotor function is a consistent finding in joints affected with OA, and this leads to an inability to maintain functional joint stability. Resistance training increases the sensitivity of sensorimotor structures and aids in increasing intra-articular proprioception.

Aerobic exercise

Aerobic exercise results in increased functional capacity in muscles and tissues surrounding and acting across joints. The systemic effects of aerobic exercise, e.g., reducing sub-maximal and resting heart rate, promoting weight loss and muscular endurance all have a positive impact on the osteoarthritic joint [60].

Exercise interventions to manage osteoarthritis

With the burden of OA increasing, exercise is well-positioned to provide an effective, safe, and relatively low-cost intervention that provides symptom relief and functional improvement [61]. Clinical studies have focused on a number of measurements to ascertain the effectiveness of therapy associated with exercise in patients with OA. A number of clinical outcome measures are implemented when measuring the impact and outcomes of OA, with one systematic review finding that up to 78 patient-reported outcome measures (PROMS) were implemented among the papers they studied [62]. The 10 most frequently used PROMS identified by Lundgren-Nilsson [62] are listed in Table 2. Additional objective functional outcome measures have also been used to evaluate exercise and OA outcomes, including various walking tests, sit to stand, stair negotiation, and other testing batteries formed by multi-activity measures [63]. As measurement is primarily based on subjective outcomes (e.g., those assessed in the WOMAC), some studies have highlighted the disparity between outcome measure changes and changes in OA progression, joint ROM, disease activity, and radiographic grade [64].

Aerobic exercise and osteoarthritis

Numerous systematic reviews have been undertaken investigating the effect of aerobic exercise on the symptoms and progression of OA. The most common intervention used in aerobic programs is walking (e.g., home-, hospital-, or facility-based) [75] though other modalities of aerobic activities of lower and higher intensities, such as cycling, running, or aerobic dance have also been shown to create positive change in OA patients [76]. A systematic review carried out by Escalante et al. [77] investigated the effect of aerobic activity on OA, concluding that aerobic exercise carried out regularly provided both short- and long-term benefits on pain, joint tenderness, functional measures, and aerobic capacity. Additionally, they noted that walking may provide a "safer" form of aerobic activity as it produced fewer adverse events. These findings have been echoed in other systematic reviews such as that by Tanaka [78], which noted that AE produced significant improvements in WOMAC and muscle torque scores in OA patients. Furthermore, based on systematic review and other high-level evidence, clinical practice guidelines created by Brosseau et al. [79] advised that for knee OA; AE with or without

---

TABLE 2 10 most commonly used PROMS for OA patients.

- The Western Ontario and McMaster Universities Arthritis Index (WOMAC) [65]
- 36-Item Short Form Survey (SF-36) [66]
- The Knee Injury and Osteoarthritis Outcome Score (KOOS) [67]
- T Oxford Knee Score (OKS) [68]
- Disabilities of the Arm, Shoulder, and Hand Questionnaire (DASH) [69]
- EuroQol (EQ-5D) [70]
- 12-Item Short-Form Health Survey (SF-12) [71]
- The hip disability and osteoarthritis outcome score (HOOS) [72]
- Pain Catastrophizing Scale (PCS) [73]
- The Oxford Hip Score (OHS) [74]

resistance exercise provided significant changes in pain scores, improved function, and quality of life. This guideline was not specific to one type of aerobic activity, but rather advised AE, in general, should form a part of a more comprehensive exercise program to provide effective nonpharmaceutical analgesia to OA patients [79] (Tables 3–5).

TABLE 3 Example AE program for severe knee OA.

Environment: Nursing Home, supervised by physical therapist
Aerobic activity: Walking
Dosage: 3 × 20 minute sessions, 1 × 10 minute session over 1 week (accumulating 70 min over 1 week)
Intensity: Moderate, 3/10 rate of perceived exertion, minimum 80 steps per minute
Cautions:

1. Substantial increase in knee pain during walk >7/10
2. Substantial increase in knee pain 2 h postwalk
    a. If prewalk knee pain was 0–6/10 which had increased 3 pain points to >7/10
    b. If prewalk knee pain was 7–8/10 which had increased 2 pain points to >8–9/10
3. Patient reporting difficulty breathing which does not settle on resting, new or incessant chest pain, acute change in consciousness

Sample outcome measures: WOMAC

(Adapted from Wallis JA, Webster KE, Levinger P, Singh PJ, Fong C, Taylor NF. The maximum tolerated dose of walking for people with severe osteoarthritis of the knee: a phase I trial. Osteoarthr Cartil 2015;23(8):1285–93.)

TABLE 4 Example RE program for severe knee OA.

Environment: Physical therapy clinic or gym, supervised by physical therapist
Activity: Resistance training
Dosage: 20 min twice per week for 13 weeks
Intensity: Moderate, 60% of 1 repetition max initially. Increase 5%–10% load as able, based on patient tolerance. Rest 30–60 s between sets of exercises.
Exercises:

1. Bilateral machine leg press, 2 sets, 8–12 repetitions
2. Bilateral seated hamstring curl, 2 sets 8–12 repetitions
3. Bilateral calf raise, 2 sets, 8–12 repetitions

Sample outcome measures: Sit-to-stand, rising from the floor, stair-climbing tests and 6-min walking

(Adapted from Ciolac EG, Silva JM, Greve JM. Effects of resistance training in older women with knee osteoarthritis and total knee arthroplasty. Clinics (Sao Paulo) 2015;70(1):7–13.)

TABLE 5 Example AqE program for severe knee OA.

Environment: Heated swimming pool (1.3 m depth), supervised by physical therapist
Activity: Dance based aquatic exercise
Dosage: Three times weekly for 8 consecutive weeks
Intensity: Moderate, 4–6/10
Exercises:

1. 12 min warmup (walking forward, backward, and sideways)
2. 21 min of dance-based exercise (Alternating 5 min slow movement, 3 min fast movement for 21 min)
3. Cool down (slow walking, breathing exercises, stretching of worked muscles)

Sample outcome measures: WOMAC, treadmill 6-min walk test

(Adapted from Casilda-López J, Valenza MC, Cabrera-Martos I, Díaz-Pelegrina A, Moreno-Ramírez MP, Valenza-Demet G. Effects of a dance-based aquatic exercise program in obese postmenopausal women with knee osteoarthritis: a randomized controlled trial. Menopause 2017;24(7):768–73.)

Resistance exercise and osteoarthritis

Much like AE, RE is one of the more heavily researched interventions to improve the function of OA sufferers. As OA involves the deconditioning of musculature surrounding the affected joint, due to pain and reduced loading, clinical reasoning suggests that improving these aspects may result in an improvement in symptoms [80]. This clinical suspicion has been confirmed particularly in lower limb OA, with multiple systematic reviews finding that RE leads to significant improvements in pain and functional measures [41,81,82]. One systematic review by Turner et al. [83], covered multiple clinical trials of RE on knee OA. This review found significant improvement in WOMAC pain and function scores following structured resistance training programs of various lengths, though they noted disparity between timing and intensity of each trial, indicating uncertainty about dose–response. Another systematic review by Goh et al. [84], reviewed clinical trials comparing exercise to usual care in which they concluded that pain, function, and performance underwent a superior change in exercise over usual care. The same review noted that changes in quality of life (QoL) were not as significant when compared to usual care. For subgroups in this review, the authors made note that their analysis had reduced certainty for the efficacy of exercise on hip OA. A study by Li et al. [85], compared 17 clinical trials of various RE programs with differing parameters and outcome measures (WOMAC, VAS, OASI, and KPS) to determine their effect. Their analysis made note that both high- and low-intensity RE programs have a significant effect on pain and physical function. The authors theorized that due to the effects of OA, normalization of muscle activity and joint biomechanics may reduce pain and cartilage degradation while improving joint function.

Aquatic exercise and osteoarthritis

Exercise in an aquatic environment provides a number of benefits including improvement in muscle tone, power, and cardiovascular health [86]. Additionally, AqE, unlike many other exercise interventions, offers a reduction in weight-bearing during its performance which can be particularly beneficial in reducing pain to tolerable levels while exercising [87]. A number of systematic reviews have been undertaken to explore the effects of AqE on OA. A 2014 systematic review by Waller et al. [88] found that lower limb OA patients, who underwent AqE, had significant improvements in pain and self-reported function (WOMAC and KOOS), compared with those undertaking land-based exercise or using acetaminophen or NSAIDS. The improvement in measures of pain and function was echoed by a 2016 systematic review, which noted that AqE improved walking test times and pain levels among those undertaking the intervention [55]. The same study raised significant questions regarding AqE and muscle strength, noting that among the studies they analyzed there was limited improvement in muscle strength. One trial, in fact, reported a decrease in isokinetic strength when compared to a comparator group [89]. The same 2008 paper suggested that the use of water equipment to increase overload potential may offer a counter to the minimal resistance that AqE offers [89]. As high BMI and obesity are often barriers to exercise, AqE is an attractive intervention for this group as it offers a reduced-weight environment within which to exercise. One study by Lim et al. [42] investigated the effect of AqE in obese patients with knee OA. They found that AqE over 8 weeks offered a significant reduction in pain levels, and may aid in breaking the negative feedback cycle which occurs when other exercise is not tolerable due to pain.

Aerobic, resistance, or aquatic exercise?

As the cost of OA continues to burden healthcare systems, the consideration of efficacy, time effectiveness, and cost of interventions becomes paramount. While all three modalities may offer benefits to OA sufferers, a number of groups have investigated which of the three may be the most effective for improvements in pain and function. A systematic review comparing AE (aerobic walking) and RE analyzed 13 clinical trials involving walking or quadriceps strengthening, to determine which was superior in improving pain and reducing disability in sufferers of knee OA [90]. Their analysis concluded that both interventions were effective in improving pain and self-reported disability, but they noted that neither was significantly better than the comparator. The authors also made the specific point that due to the finding that neither was superior, patient choice and thereby patient adherence can play an important role in treatment outcome [90]. Another systematic review that investigated the same interventions, concluded that AE had the largest effect size for pain and function, followed by RE which improved multiple outcomes to varying degrees [84]. As both AE and RE class as "land-based" exercise, a number of systematic reviews have investigated land-based exercise against AqE. De Mattos et al. 2016 [55] noted in their paper that AqE and land-based exercises both provided benefits in pain and function, but the AqE effects on strength were not significant, and perhaps reduced. Similarly, a study investigating systematic reviews on OA and exercise concluded that land-based exercise, aquatic exercise, and aerobic exercise are beneficial for pain and function, yet the precise mechanisms and additional variables are not clear enough to recommend a superior intervention [87]. AqE may be superior to land-based exercise when pain levels are poorly controlled on land, thus the buoyancy provided during AqE allows exercise that otherwise might be intolerable due to pain [42].

Cautions and safety concerns regarding exercise

Exercise is considered to be low risk in OA patients, particularly when compared to typical medical interventions (NSAIDS, injection therapies, surgical intervention) [91]. The primary risk of exercise in OA relates to a transient short-term increase in pain related to participation in the exercise program [92]. Additionally, there are some adverse effects related to trauma, such as a fall or equipment failure, or misuse [93]. The long-term safety, however, appears to be excellent, with a systematic review into exercise for knee OA finding that at the 1-year mark not only were WOMAC, VAS, and KOOS scores significantly improved, but there were no adverse effects noted [91]. The summary of the current evidence base reflects that although OA sufferers may experience an increase in short-term pain due to exercise [93], the long-term outcomes reflect a significant improvement in pain and function. As a general clinical guideline, exercise of all varieties should be performed under the supervision of a licensed professional.

Challenges with exercise prescription in patient management

Difficulties providing exercise prescription-based management for OA, from the practitioner's perspective, can be related to many factors. Practitioners often face the arduous task of reframing their patients' perspectives of OA management, with current research. Past experiences often shape one's view of their body, and this can be the greatest challenge that practitioners face in obtaining the necessary "buy-in" from the patient. Dealing with unhelpful narratives is essential however, this cannot be done by dismissing a patient's understanding or attitude toward their osteoarthritic joint(s). Patients that feel listened to are generally more willing to be reeducated, and therefore open to changing their beliefs. Often, but not always, patients have been advised by previous practitioners that exercise can be beneficial for their osteoarthritic joint(s), however, pain does not permit them to perform said exercise. This is where tailored exercise programs are paramount to uncovering the beneficial effects of exercise.

Before being able to construct a patient's exercise program and manage them through it, the practitioner needs to be prepared, which includes having an understanding of an array of exercises and the patient experience. It is important for practitioners to have an appreciation for a variety of relevant exercises in order to be able to modify and adapt exercises that the patient can perform safely while also conveying the necessary benefit to the patient. Relevant exercises include strengthening exercises for the lower limb, gentle cardiovascular activity, and balancing exercises. A foundational exercise program could include three lower limb strengthening exercises, a balancing exercise, and a gentle cardiovascular exercise (see Table 6).

A key part of the practitioner's preparation is to educate the patient on what they are likely to feel during and after their exercise sessions. During exercise sessions it is important that patients exercise at an intensity that will provide benefit, the practitioner will need to outline that generally speaking, an exercise should feel arduous and potentially uncomfortable.

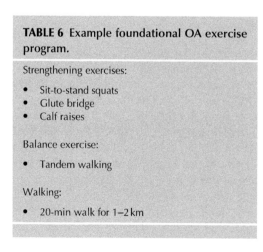

TABLE 6 Example foundational OA exercise program.

Strengthening exercises:

- Sit-to-stand squats
- Glute bridge
- Calf raises

Balance exercise:

- Tandem walking

Walking:

- 20-min walk for 1–2 km

However, the patient should not be feeling sustained levels of discomfort or pain and if so, should be encouraged to discuss this with their practitioner so that they can modify either the exercise intensity or the program itself. The eventual goal will be for the patient to be able to monitor the exercise intensity and make any necessary modifications in situ.

Another aspect of the exercise session that patients will need education for actually relates to the time after and between exercise sessions. This could be the first time that a patient has experienced delayed onset muscle soreness (DOMS) which is a common outcome of exercising. The novice exerciser will need education and reassurance that DOMS is normal and not concerning. Furthermore, the patient will need to be encouraged that despite the presence of DOMS, they should feel confident that they can still complete their exercise sessions. Lastly and most importantly, patients are going to experience flares of pain from OA on some level and this can potentially lead to regressions. Patients will see the pain as a negative outcome and is possibly due to the exercise program, leading to avoidance or cessation of the exercise program. For the practitioner, understanding that these events are likely to take place and being prepared to reassure a patient through this process is vital to the long-term success of the program.

In many ways, preparation is the key to success in preparing an exercise program and leading a patient from a novice to being able to exercise independently. The necessary preparation includes having a library of exercises from which to choose from and the confidence to be able to select, and if necessary modify, the exercises that best suit the patient. The preparation includes understanding what the novice patient is likely to experience in commencing a new exercise program for their OA, specifically managing intensity, post-exercise soreness, and pain flares. Although this can be a challenging process, prior preparation will increase the likelihood of a successful outcome for both patient and practitioner.

The deleterious effects of OA on pain and function are pronounced, leading to increasing disability in the global population. Therefore, the need to develop solutions related to the prevention and management of OA is paramount and, more importantly, these solutions need to be simple to follow for both patient and practitioner. As discussed, one such solution that is cost-effective, efficient, and efficacious is a well-rounded exercise program. Despite the aforementioned difficulties in implementing an exercise program for an OA sufferer lacking in exercise experience, with careful preparation and planning, a tailored exercise program can be designed that can produce positive effects on pain levels, physical and mental health. Therefore, when considering pathways for the patient suffering from OA, an exercise program featuring a combination of resistance and cardiovascular exercise should be given high priority.

References

[1] Anon. GBD results tool | GHDx., 2021. Internet Ghdx.healthdata.org. [cited 11 July 2021]. Available from http://ghdx.healthdata.org/gbd-results-tool.
[2] Kloppenburg M, Berenbaum F. Osteoarthritis year in review 2019: epidemiology and therapy. Osteoarthr Cartil 2020;28(3):242.
[3] Roos EM, Arden NK. Strategies for the prevention of knee osteoarthritis. Nat Rev Rheumatol 2016;12(2):92–101.
[4] Plotnikoff R, Karunamuni N, Lytvyak E, Penfold C, Schopflocher D, Imayama I, et al. Osteoarthritis prevalence and modifiable factors: a population study. BMC Public Health 2015;15:1195.
[5] Lespasio MJ, Piuzzi NS, Husni ME, Muschler GF, Guarino A, Mont MA. Knee osteoarthritis: a primer. Perm J 2017;21:16–183.
[6] O'Neill TW, McCabe PS, McBeth J. Update on the epidemiology, risk factors and disease outcomes of osteoarthritis. Best Pract Res Clin Rheumatol 2018;32(2):312–26.
[7] Murphy SL, Robinson-Lane SG, Niemiec SLS. Knee and hip osteoarthritis management: a review of current and emerging non-pharmacological approaches. Curr Treat Options Rheumatol 2016;2(4):296–311.
[8] Mazzei DR, Ademola A, Abbott JH, Sajobi T, Hildebrand K, Marshall DA. Are education, exercise and diet interventions a cost-effective treatment to manage hip and knee osteoarthritis? A systematic review. Osteoarthr Cartil 2021;29(4):456–70.
[9] Johnson VL, Hunter DJ. The epidemiology of osteoarthritis. Best Pract Res Clin Rheumatol 2014;28(1):5–15.
[10] Litwic A, Edwards MH, Dennison EM, Cooper C. Epidemiology and burden of osteoarthritis. Br Med Bull 2013;105:185–99.
[11] Pereira D, Peleteiro B, Araújo J, Branco J, Santos RA, Ramos E. The effect of osteoarthritis definition on prevalence and incidence estimates: a systematic review. Osteoarthr Cartil 2011;19(11):1270–85.
[12] Zhang Y, Niu J, Kelly-Hayes M, Chaisson CE, Aliabadi P, Felson DT. Prevalence of symptomatic hand osteoarthritis and its impact on functional status among the elderly: the Framingham study. Am J Epidemiol 2002;156(11):1021–7.
[13] Cross M, Smith E, Hoy D, Nolte S, Ackerman I, Fransen M, et al. The global burden of hip and knee osteoarthritis: estimates from the global burden of disease 2010 study. Ann Rheum Dis 2014;73(7):1323–30.
[14] Safiri S, Kolahi AA, Smith E, Hill C, Bettampadi D, Mansournia MA, et al. Global, regional and national burden of osteoarthritis 1990-2017: a systematic analysis of the Global Burden Of Disease Study 2017. Ann Rheum Dis 2020;79(6):819–28.
[15] Hart JAL. Joint replacement surgery. Med J Aust 2004;180(S5):S27–30.
[16] Allen KD, Golightly YM. State of the evidence. Curr Opin Rheumatol 2015;27(3):276–83.
[17] Felson DT, Nevitt MC, Zhang Y, Aliabadi P, Baumer B, Gale D, et al. High prevalence of lateral knee osteoarthritis in Beijing Chinese compared with Framingham Caucasian subjects. Arthritis Rheum 2002;46(5):1217–22.

[18] Kang X, Fransen M, Zhang Y, Li H, Ke Y, Lu M, et al. The high prevalence of knee osteoarthritis in a rural Chinese population: the Wuchuan osteoarthritis study. Arthritis Rheum 2009;61(5):641–7.

[19] Chaganti RK, Lane NE. Risk factors for incident osteoarthritis of the hip and knee. Curr Rev Musculoskelet Med 2011;4(3):99–104.

[20] Srikanth VK, Fryer JL, Zhai G, Winzenberg TM, Hosmer D, Jones G. A meta-analysis of sex differences prevalence, incidence and severity of osteoarthritis. Osteoarthr Cartil 2005;13(9):769–81.

[21] Ratneswaran A, Rockel JS, Kapoor M. Understanding osteoarthritis pathogenesis: a multiomics system-based approach. Curr Opin Rheumatol 2020;32(1):80–91.

[22] Woodell-May JE, Sommerfeld SD. Role of inflammation and the immune system in the progression of osteoarthritis. J Orthop Res 2020;38(2):253–7.

[23] Bijlsma JW, Berenbaum F, Lafeber FP. Osteoarthritis: an update with relevance for clinical practice. Lancet 2011;377(9783):2115–26.

[24] Iannone F, Lapadula G. The pathophysiology of osteoarthritis. Aging Clin Exp Res 2003;15(5):364–72.

[25] Glyn-Jones S, Palmer AJR, Agricola R, Price AJ, Vincent TL, Weinans H, et al. Osteoarthritis. Lancet 2015;386(9991):376–87.

[26] Kapoor M, Martel-Pelletier J, Lajeunesse D, Pelletier JP, Fahmi H. Role of proinflammatory cytokines in the pathophysiology of osteoarthritis. Nat Rev Rheumatol 2011;7(1):33–42.

[27] Walker-Bone K, Javaid K, Arden N, Cooper C. Regular review: medical management of osteoarthritis. BMJ 2000;321(7266):936–40.

[28] Iwamoto J, Sato Y, Takeda T, Matsumoto H. Effectiveness of exercise for osteoarthritis of the knee: a review of the literature. World J Orthop 2011;2 (5):37–42.

[29] Aguiar GC, Do Nascimento MR, De Miranda AS, Rocha NP, Teixeira AL, Scalzo PL. Effects of an exercise therapy protocol on inflammatory markers, perception of pain, and physical performance in individuals with knee osteoarthritis. Rheumatol Int 2015;35(3):525–31.

[30] Nader GA, Lundberg IE. Exercise as an anti-inflammatory intervention to combat inflammatory diseases of muscle. Curr Opin Rheumatol 2009;21 (6):599–603.

[31] Arendt-Nielsen L. Pain sensitisation in osteoarthritis. Clin Exp Rheumatol 2017;35 Suppl 107(5):68–74.

[32] Ohashi Y, Fukushima K, Inoue G, Uchida K, Koyama T, Tsuchiya M, et al. Central sensitization inventory scores correlate with pain at rest in patients with hip osteoarthritis: a retrospective study. BMC Musculoskelet Disord 2020;21:595.

[33] Lluch Girbés E, Nijs J, Torres-Cueco R, López CC. Pain treatment for patients with osteoarthritis and central sensitization. Phys Ther 2013;93 (6):842–51.

[34] Abhishek A, Doherty S, Maciewicz RA, Muir KR, Zhang W, Doherty M. Self-reported knee malalignment in early adult life as an independent risk for knee chondrocalcinosis. Arthritis Care Res 2011;63(11):1550–7.

[35] Wahid A, Manek N, Nichols M, Kelly P, Foster C, Webster P, et al. Quantifying the association between physical activity and cardiovascular disease and diabetes: a systematic review and meta-analysis. J Am Heart Assoc 2016;5(9), e002495.

[36] Patel H, Alkhawam H, Madanieh R, Shah N, Kosmas CE, Vittorio TJ. Aerobic vs anaerobic exercise training effects on the cardiovascular system. World J Cardiol 2017;9(2):134–8.

[37] Semanik PA, Chang RW, Dunlop DD. Aerobic activity in prevention and symptom control of osteoarthritis. PM R 2012;4(5 Suppl):S37–44.

[38] Vincent HK, Heywood K, Connelly J, Hurley RW. Obesity and weight loss in the treatment and prevention of osteoarthritis. PM R 2012;4(5 Suppl): S59–67.

[39] Phillips SM, Winett RA. Uncomplicated resistance training and health-related outcomes: evidence for a public health mandate. Curr Sports Med Rep 2010;9(4):208–13.

[40] Lange AK, Vanwanseele B, Fiatarone Singh MA. Strength training for treatment of osteoarthritis of the knee: a systematic review. Arthritis Rheum 2008;59(10):1488–94.

[41] Ettinger Jr WH, Burns R, Messier SP, Applegate W, Rejeski WJ, Morgan T, et al. A randomized trial comparing aerobic exercise and resistance exercise with a health education program in older adults with knee osteoarthritis. The Fitness Arthritis and Seniors Trial (FAST). JAMA 1997;277(1):25–31.

[42] Lim JY, Tchai E, Jang SN. Effectiveness of aquatic exercise for obese patients with knee osteoarthritis: a randomized controlled trial. PM R 2010;2 (8):723–31 [quiz 93].

[43] Messier SP, Royer TD, Craven TE, O'Toole ML, Burns R, Ettinger Jr WH. Long-term exercise and its effect on balance in older, osteoarthritic adults: results from the Fitness, Arthritis, And Seniors Trial (FAST). J Am Geriatr Soc 2000;48(2):131–8.

[44] Dong R, Wu Y, Xu S, Zhang L, Ying J, Jin H, et al. Is aquatic exercise more effective than land-based exercise for knee osteoarthritis? Medicine 2018;97(52), e13823.

[45] Lu M, Su Y, Zhang Y, Zhang Z, Wang W, He Z, et al. Effectiveness of aquatic exercise for treatment of knee osteoarthritis: systematic review and meta-analysis. Z Rheumatol 2015;74(6):543–52.

[46] Skou ST, Pedersen BK, Abbott JH, Patterson B, Barton C. Physical activity and exercise therapy benefit more than just symptoms and impairments in people with hip and knee osteoarthritis. J Orthop Sports Phys Ther 2018;48(6):439–47.

[47] Runhaar J, Luijsterburg P, Dekker J, Bierma-Zeinstra SMA. Identifying potential working mechanisms behind the positive effects of exercise therapy on pain and function in osteoarthritis; a systematic review. Osteoarthr Cartil 2015;23(7):1071–82.

[48] Koltyn KF, Brellenthin AG, Cook DB, Sehgal N, Hillard C. Mechanisms of exercise-induced hypoalgesia. J Pain 2014;15(12):1294–304.

[49] Aaron RK, Racine JR, Voisinet A, Evangelista P, Dyke JP. Subchondral bone circulation in osteoarthritis of the human knee. Osteoarthr Cartil 2018;26(7):940–4.

[50] Richardson DW, Dodge GR. Effects of interleukin-1beta and tumor necrosis factor-alpha on expression of matrix-related genes by cultured equine articular chondrocytes. Am J Vet Res 2000;61(6):624–30.

[51] Raman S, FitzGerald U, Murphy JM. Interplay of inflammatory mediators with epigenetics and cartilage modifications in osteoarthritis. Front Bioeng Biotechnol 2018;6:22.

[52] Torres-Ronda L, Del Alcázar XS. The properties of water and their applications for training. J Hum Kinet 2014;44:237–48.

[53] Montagne K, Onuma Y, Ito Y, Aiki Y, Furukawa KS, Ushida T. High hydrostatic pressure induces pro-osteoarthritic changes in cartilage precursor cells: a transcriptome analysis. PLoS One 2017;12(8), e0183226.

[54] Li Y, Zhou J, Yang X, Jiang Y, Gui J. Intermittent hydrostatic pressure maintains and enhances the chondrogenic differentiation of cartilage progenitor cells cultivated in alginate beads. Dev Growth Differ 2016;58(2):180–93.

[55] Mattos F, Leite N, Pitta A, Bento PC. Effects of aquatic exercise on muscle strength and functional performance of individuals with osteoarthritis: a systematic review. Rev Bras Reumatol 2016;56(6):530–42.

[56] Miller MS, Callahan DM, Tourville TW, Slauterbeck JR, Kaplan A, Fiske BR, et al. Moderate-intensity resistance exercise alters skeletal muscle molecular and cellular structure and function in inactive older adults with knee osteoarthritis. J Appl Physiol (1985) 2017;122(4):775–87.

[57] Magni NE, McNair PJ, Rice DA. The effects of resistance training on muscle strength, joint pain, and hand function in individuals with hand osteoarthritis: a systematic review and meta-analysis. Arthritis Res Ther 2017;19(1):131.

[58] Knoop J, Steultjens MP, Roorda LD, Lems WF, van der Esch M, Thorstensson CA, et al. Improvement in upper leg muscle strength underlies beneficial effects of exercise therapy in knee osteoarthritis: secondary analysis from a randomised controlled trial. Physiotherapy 2015;101 (2):171–7.

[59] Beckwée D, Vaes P, Cnudde M, Swinnen E, Bautmans I. Osteoarthritis of the knee: why does exercise work? A qualitative study of the literature. Ageing Res Rev 2013;12(1):226–36.

[60] Wellsandt E, Golightly Y. Exercise in the management of knee and hip osteoarthritis. Curr Opin Rheumatol 2018;30(2):151–9.

[61] van Doormaal MCM, Meerhoff GA, Vliet Vlieland TPM, Peter WF. A clinical practice guideline for physical therapy in patients with hip or knee osteoarthritis. Musculoskeletal Care 2020;18(4):575–95.

[62] Lundgren-Nilsson Å, Dencker A, Palstam A, Person G, Horton MC, Escorpizo R, et al. Patient-reported outcome measures in osteoarthritis: a systematic search and review of their use and psychometric properties. RMD Open 2018;4(2), e000715.

[63] Dobson F, Hinman RS, Hall M, Terwee CB, Roos EM, Bennell KL. Measurement properties of performance-based measures to assess physical function in hip and knee osteoarthritis: a systematic review. Osteoarthr Cartil 2012;20(12):1548–62.

[64] Johnson SR, Archibald A, Davis AM, Badley E, Wright JG, Hawker GA. Is self-reported improvement in osteoarthritis pain and disability reflected in objective measures? J Rheumatol 2007;34(1):159–64.

[65] Bellamy N, Buchanan WW, Goldsmith CH, Campbell J, Stitt LW. Validation study of WOMAC: a health status instrument for measuring clinically important patient relevant outcomes to antirheumatic drug therapy in patients with osteoarthritis of the hip or knee. J Rheumatol 1988;15(12):1833–40.

[66] Anderson C, Laubscher S, Burns R. Validation of the short form 36 (SF-36) health survey questionnaire among stroke patients. Stroke 1996;27 (10):1812–6.

[67] Roos EM, Roos HP, Ekdahl C, Lohmander LS. Knee injury and osteoarthritis outcome score (KOOS)—validation of a Swedish version. Scand J Med Sci Sports 1998;8(6):439–48.

[68] Tuğay BU, Tuğay N, Güney H, Kınıklı G, Yüksel İ, Atilla B. Oxford knee score: cross-cultural adaptation and validation of the Turkish version in patients with osteoarthritis of the knee. Acta Orthop Traumatol Turc 2016;50(2):198–206.

[69] Padua R, Padua L, Ceccarelli E, Romanini E, Zanoli G, Amadio PC, et al. Italian version of the disability of the arm, shoulder and hand (dash) questionnaire. Cross-cultural adaptation and validation. J Hand Surg Br 2003;28(2):179–86.

[70] Schweikert B, Hahmann H, Leidl R. Validation of the EuroQol questionnaire in cardiac rehabilitation. Heart 2006;92(1):62–7.

[71] Gandek B, Ware JE, Aaronson NK, Apolone G, Bjorner JB, Brazier JE, et al. Cross-validation of item selection and scoring for the SF-12 health survey in nine countries: results from the IQOLA project. International quality of life assessment. J Clin Epidemiol 1998;51(11):1171–8.

[72] Nilsdotter AK, Lohmander LS, Klässbo M, Roos EM. Hip disability and osteoarthritis outcome score (HOOS)—validity and responsiveness in total hip replacement. BMC Musculoskelet Disord 2003;4(1):10.

[73] Sullivan MJL, Bishop S, Pivik J. The pain catastrophizing scale: development and validation. Psychol Assess 1996;7:524–32.

[74] Paulsen A, Odgaard A, Overgaard S. Translation, cross-cultural adaptation and validation of the Danish version of the Oxford hip score. Bone Joint Res 2012;1(9):225–33.

[75] Loew L, Brosseau L, Wells GA, Tugwell P, Kenny GP, Reid R, et al. Ottawa panel evidence-based clinical practice guidelines for aerobic walking programs in the management of osteoarthritis. Arch Phys Med Rehabil 2012;93(7):1269–85.

[76] Regnaux JP, Lefevre-Colau MM, Trinquart L, Nguyen C, Boutron I, Brosseau L, et al. High-intensity versus low-intensity physical activity or exercise in people with hip or knee osteoarthritis. Cochrane Database Syst Rev 2015;(10).

[77] Escalante Y, Saavedra JM, Garcia-Hermoso A, Silva AJ, Barbosa TM. Physical exercise and reduction of pain in adults with lower limb osteoarthritis: a systematic review. J Back Musculoskelet Rehabil 2010;23(4):175–86.

[78] Tanaka R, Ozawa J, Kito N, Moriyama H. Efficacy of strengthening or aerobic exercise on pain relief in people with knee osteoarthritis: a systematic review and meta-analysis of randomized controlled trials. Clin Rehabil 2013;27(12):1059–71.

[79] Brosseau L, Taki J, Desjardins B, Thevenot O, Fransen M, Wells GA, et al. The Ottawa panel clinical practice guidelines for the management of knee osteoarthritis. Part three: aerobic exercise programs. Clin Rehabil 2017;31(5):612–24.

[80] Krishnasamy P, Hall M, Robbins SR. The role of skeletal muscle in the pathophysiology and management of knee osteoarthritis. Rheumatology (Oxford) 2018;57(suppl_4):iv22–33.

[81] Gür H, Cakin N, Akova B, Okay E, Küçükoğlu S. Concentric versus combined concentric-eccentric isokinetic training: effects on functional capacity and symptoms in patients with osteoarthrosis of the knee. Arch Phys Med Rehabil 2002;83(3):308–16.

[82] Schilke JM, Johnson GO, Housh TJ, O'Dell JR. Effects of muscle-strength training on the functional status of patients with osteoarthritis of the knee joint. Nurs Res 1996;45(2):68–72.

[83] Turner MN, Hernandez DO, Cade W, Emerson CP, Reynolds JM, Best TM. The role of resistance training dosing on pain and physical function in individuals with knee osteoarthritis: a systematic review. Sports Health 2020;12(2):200–6.

[84] Goh SL, Persson MSM, Stocks J, Hou Y, Welton NJ, Lin J, et al. Relative efficacy of different exercises for pain, function, performance and quality of life in knee and hip osteoarthritis: systematic review and network meta-analysis. Sports Med 2019;49(5):743–61.

[85] Li Y, Su Y, Chen S, Zhang Y, Zhang Z, Liu C, et al. The effects of resistance exercise in patients with knee osteoarthritis: a systematic review and meta-analysis. Clin Rehabil 2016;30(10):947–59.

[86] Foley A, Halbert J, Hewitt T, Crotty M. Does hydrotherapy improve strength and physical function in patients with osteoarthritis—a randomised controlled trial comparing a gym based and a hydrotherapy based strengthening programme. Ann Rheum Dis 2003;62(12):1162–7.

[87] Rahmann AE. Exercise for people with hip or knee osteoarthritis: a comparison of land-based and aquatic interventions. Open Access J Sports Med 2010;1:123–35.

[88] Waller B, Ogonowska-Slodownik A, Vitor M, Lambeck J, Daly D, Kujala UM, et al. Effect of therapeutic aquatic exercise on symptoms and function associated with lower limb osteoarthritis: systematic review with meta-analysis. Phys Ther 2014;94(10):1383–95.

[89] Lund H, Weile U, Christensen R, Rostock B, Downey A, Bartels EM, et al. A randomized controlled trial of aquatic and land-based exercise in patients with knee osteoarthritis. J Rehabil Med 2008;40(2):137–44.

[90] Roddy E, Zhang W, Doherty M. Aerobic walking or strengthening exercise for osteoarthritis of the knee? A systematic review. Ann Rheum Dis 2005;64(4):544–8.

[91] Charlesworth J, Fitzpatrick J, Perera NKP, Orchard J. Osteoarthritis—a systematic review of long-term safety implications for osteoarthritis of the knee. BMC Musculoskelet Disord 2019;20(1):151.

[92] Fransen M, McConnell S, Hernandez-Molina G, Reichenbach S. Exercise for osteoarthritis of the hip. Cochrane Database Syst Rev 2014;(4), Cd007912.

[93] Bischoff HA, Roos EM. Effectiveness and safety of strengthening, aerobic, and coordination exercises for patients with osteoarthritis. Curr Opin Rheumatol 2003;15(2):141–4.

Chapter 22

Exercise before and after orthopedic surgery

Phong Tran[a,b,c,d] and Saud Almaslmani[a,e]

[a]*Department of Orthopedic Surgery, Western Health, Melbourne, VIC, Australia* [b]*Victoria University, Footscray, VIC, Australia* [c]*Australian Institute for Musculoskeletal Science, The University of Melbourne and Western Health, St. Albans, VIC, Australia* [d]*Swinburne University, Melbourne, VIC, Australia* [e]*Department of Surgery, Faculty of Medicine in Al-Qunfudhah, Umm Al-Qura University, Makkah, Saudi Arabia*

Introduction

Orthopedic surgery is the branch of surgery concerned with the musculoskeletal system—bones, muscles, joints, and its connective tissue. Musculoskeletal conditions continue to be a common cause of pain and health dysfunctions across all age groups, affecting 30% of the population, and are the fourth-leading contributor to the burden of disease [1]. Between adolescence and 45 years of age, musculoskeletal conditions are the second leading of nonfatal disease burden cause after mental health problems, and are the leading cause for people aged 45–55 years [1,2].

Orthopedic surgeons use both surgical and nonsurgical means to treat musculoskeletal trauma and fractures, sports injuries, degenerative diseases, infections, tumors, and congenital disorders. Common elective orthopaedic surgeries include shoulder, hip, and knee procedures, and can be broadly divided into reconstructive and replacement operations. Reconstructive surgery often involves repairing soft tissue lesions, such as shoulder rotator cuff tears, stabilizing the hip joint labrum, or reconstructing the anterior cruciate ligament in the knee. It can also involve reshaping or removing bone (ostectomy), such as in the hip or shoulder, to treat impingement. Reconstructive surgery can be performed by either open incision or arthroscopic (keyhole) surgery. Replacement surgery (arthroplasty) is increasingly common for the knee, hip, and shoulder, and involves the replacement of a damaged joint with a combination of metal alloys, ceramics, and polyethylene. Joint replacements are always performed via an open incision, but incision lengths have been decreasing in size with the evolution of surgical techniques.

Improving fitness before orthopedic surgery

Activity and exercise have been demonstrated to provide important benefits across a broad spectrum of medical conditions [3,4]. In the musculoskeletal system, most conditions benefit from a period of strengthening, functional improvement, and activity modification before considering surgery. Exercises and guided training to improve muscle strength, imbalances, and endurance are recommended, and surgery should usually only be considered after nonoperative options have been exhausted after a period of 3–6 months [5]. Furthermore, when surgery is considered, patients who are fitter and healthier generally have better surgical outcomes than patients with lower fitness levels and underlying medical conditions. This has led to the concept of prehabilitation, which is the process of improving physical fitness and functional capacity before surgery to improve postoperative outcomes [6–9].

Major surgery is a stressful event for any patient, and induces pain, increases the body's catabolism and oxygen demand, and causes the patient's functional status to dip substantially [5]. The majority of patients mount an adequate stress response following surgery and regain their previous level of function. However, patients with poor preoperative physical condition might not be able to respond to the detrimental effects of surgery and subsequent hospitalization.Therefore affecting their postoperative recovery, the length of stay in hospital, and surgical outcomes [10]. A period of prehabilitation prior to surgery may be beneficial in optimizing physical and functional capacity in these patients.

Over the last two decades, the role of prehabilitation has been investigated in patients undergoing various surgical procedures ranging from cardiovascular to abdominal surgery [11–13]. There is currently high quality evidence that preoperative exercise in patients scheduled for cardiovascular surgery is well tolerated and effective. In orthopedics, there have been a number of high-quality, published systematic reviews investigating the benefits of prehabilitation, but with varying conclusions [14–16]. Prehabilitation for patients undergoing total knee and hip replacements are the most researched, with limited research for other orthopedic procedures. Earlier research into joint replacement prehabilitation demonstrated limited benefit [14,17]. However, there continues to be growing interest in prehabilitation in joint replacement because of the desire to discharge patients earlier including on the day of surgery [16,18]. More recent research suggests that prehabilitation may improve patient's quadriceps strength, pain, function, and length of stay in the hospital in knee replacements [19]. Exercise is considered the key component in prehabilitation. However, the concept of prehabilitation has evolved over the years, becoming multimodal and multidisciplinary. Modern prehabilitation includes important components such as nutrition, psychological support, and medical optimization [20,21].

Exercise after orthopedic surgery

Recovery after orthopedic surgery is optimized by a graduated strengthening and exercise program, while protecting the surgical repair to allow healing without complication and damage to the operative structures. Rehabilitation after orthopedic surgeries generally follows these principles and are best guided by a trained therapist. The phases are summarized in Fig. 1.

Depending on the particular surgery and operative findings, there may be specific restrictions on activities and exercises following the procedure in order to protect the operation site and reduce the possibility postsurgical complications. While there are numerous published protocols for rehabilitation following orthopedic surgery, the evidence supporting one particular protocol over another for a particular procedure is lacking due to the sparsity of randomized controlled trials or prospective controlled trials. Additionally, the length of immobilization following surgery, short- and long-term restrictions, and allowed sports or activities is contentious and evolving, and depends on many factors. Fortunately, many common orthopedic operations have consensus statements and systematic reviews which help guide patients and practitioners [20,21].

Peri- and postoperative recovery after surgery can pose a substantial challenge to the patient, and a thoughtful approach may reduce setbacks, limit morbidity, and help optimize functional outcomes. Rehabilitation therapists and postoperative exercise protocols should understand the underlying pathology, the findings during surgery, the structures that need to be protected following surgery, and also the individual characteristics of the patient. Creating an optimal environment for

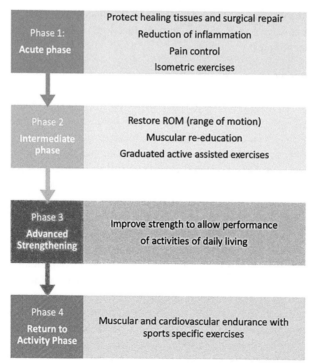

FIG. 1 Four-phase process of exercise progression after orthopedic surgery.

postoperative healing can be accomplished through a balance of protection and sufficient motion. It is important to minimize atrophy and joint stiffness while also protecting bone and soft tissue healing.

The following disease and operation guidelines are helpful guides in initiating rehabilitation. Yet, each patient's individual circumstances, procedure, and the surgeon's preference should be taken into account. Time frames associated with each postoperative phase are suggestive and not conclusive, and patient progressions should be based on the successful completion of goals of the previous phase. Each patient should be considered on a case-by-case basis with consideration of the patient's general health, psychology, and healing potential.

Patients often desire to return to sports and physical activities as soon as possible. But at this time, there is no single clinical tool available to predict and aid the timing and successful return to sport. A number of subjective and objective criteria must be considered, and ultimately, there must be agreement between the patient, treating surgeon, and supervising rehabilitation specialist. Before a return to sport can be considered, the patient should have achieved adequate strength, cardiovascular fitness, and basic athletic movements. Sports-specific high-level tasks such as running, twisting, and cutting at high-speed need to be tested in a controlled environment. During this process, it is critical that the patient's symptoms are appreciated and monitored, to allow a successful return to sport.

Unfortunately, even despite well-performed surgery, there are scenarios where a patient's return to their preferred activity or sport is not desirable or advisable. This may be due to fear of progressing a degenerative disease, uncorrectable structural abnormalities, or protecting and enhancing the longevity of joint replacements. In these cases, therapists play an important role in achieving reasonable functional goals and optimizing the patient's ability to return to modified exercise activities or alternative exercise pursuits.

Shoulder conditions

Shoulder conditions are one of the most common musculoskeletal presentations seen in general practice [22], with an annual incidence of 7% and a lifetime prevalence of 10%. 50% of patients with shoulder disorders seek medical care [23]. Common orthopedic operations of the shoulder include arthroscopic surgery for unstable shoulders after traumatic dislocations, rotator cuff repairs with treatment of associated pathology to the biceps tendon, subacromial space and acromioclavicular joint, and reverse shoulder replacements for end-stage osteoarthritis with associated rotator cuff disease.

Shoulder instability

Traumatic anterior shoulder dislocation is a common injury, especially within the second and third decades of life in young, active individuals [24], with an incidence rate in the general population between 11.2 and 26.2 per 100,000 persons yearly [25–30]. Following a first-time traumatic anterior shoulder dislocation, the risk for recurrent dislocations is high because of the pathophysiological changes in the shoulder joint [31]. The incidence of instability is greater in males, those under 30 years of age, and those participating in high demanded activities such as sport or military [30]. In individuals less than 40 years of age, the chance of having an associated labral Bankart lesion after traumatic anterior dislocation ranges from 72% to 97% [32].

Exercise before surgery

After a first-time acute dislocation, most patients will benefit from a period of shoulder strengthening and mobilization with the aid of a trained therapist. The recommended protocol displayed in Fig. 2 [33].

Exercise after surgery

If the shoulder has further instability issues despite well-guided therapy, then reconstructive surgery of the shoulder with surgical stabilization is recommended, and generally has excellent outcomes [34]. Arthroscopic repair is now the primary surgical option for the treatment of recurrent anterior shoulder instability, as studies have reported comparable results between open and arthroscopic techniques [34]. Postoperatively, the principals are to protect the repair, and then mobilize and strengthen. A sling is used for comfort for a period of 4–6 weeks. Progressive active and assisted range-of-motion exercise is initiated on the following day postsurgery, limiting external rotation to 0 degrees and limiting abduction, as well as flexion, to 45 degrees. At 4 weeks postoperatively, abduction and flexion are allowed to 90.

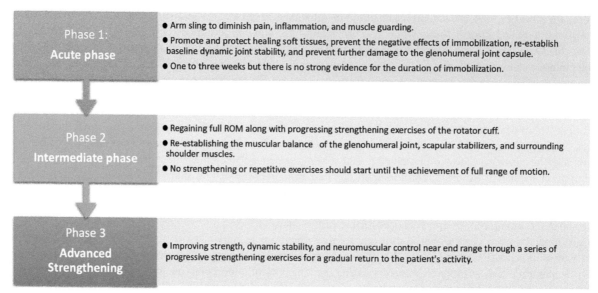

FIG. 2 Exercise therapy phases for shoulder instability before surgery.

Precautions

No shoulder external rotation with abduction for 6 weeks to protect the repaired tissues. The period of immobilization in a sling with passive movements is generally 6 weeks, however more recent literature has advocated for earlier mobilization.

Return to sport

Patients can return to sport-specific training after 3 months, and to overhead and high-contact activities after 6 months postoperatively. Studies have reported a high level of return to preinjury sports, with arthroscopic Bankart repair (71%), Latarjet procedure (73%), and open stabilization (66%) [35].

Rotator cuff disease

Approximately 1% of the adult population visit a general practitioner for shoulder pain annually [36]. The prevalence of rotator cuff related abnormalities increases with age, from 9.7% in patients aged 20 years, increasing to 54% of the patients aged over 40% and 62% in patients aged 80 years and older [37,38]. Full thickness tears of the rotator cuff are among the most frequently encountered causes of pain and dysfunction in the shoulder complex, becoming more prevalent due to an aging population and the increased functional demands in older people. In the United States, tears of the rotator cuff affect at least 10% of those over the age of 60, equating to over 5.7 million people [39], resulting in an estimated 75,000–250,000 rotator cuff surgeries performed per year [40]. In Australia, this figure is approximately 14,000 annually, with direct medical expenses for rotator cuff repair estimated at $250 million per year [41].

Exercise before surgery

As for most musculoskeletal conditions, a period of strengthening and exercises under the supervision of a therapist is recommended, which may improve the patient's symptoms to an extent where surgery is not necessary, or at least optimize the shoulder muscles and soft tissue leading up to surgery. The first line of treatment in chronic cuff tears is physical therapy with progressive rotator cuff and scapular stabilizer strengthening over 3–6 months [42]. A longer period of exercise rehabilitation should be considered in patients with rotator cuff tendinopathies, partial-thickness tears, and potentially small full-thickness tears. Numerous protocols exist and generally include daily postural exercises, active and active-assisted range of motion, active training of scapula muscles, daily anterior and posterior shoulder stretching, and exercises to strengthen the rotator cuff and periscapular muscles [43]. These exercises are commonly performed with a bar or resistance band, and the assistance of the uninvolved arm.

Strength-based exercises for rotator cuff tears should appropriately target the remaining intact cuff musculature, initiated with low load activities and progressing as patient comfort permits. Both physiotherapy and operative treatment can result in significant improvement in patient-reported outcomes for patients with symptomatic small- to medium full-thickness rotator cuff tears [44,45]. Studies have demonstrated that 75% of the patients with chronic full-thickness rotator cuff tears that undergo a 3-month supervised program of nonoperative treatment can be successfully treated without surgery [46].

Exercise after surgery

After surgery to repair the rotator cuff, commonly involving the supraspinatus tendon, immobilization in a sling for 4–6 weeks is advised to protect the tendon repair. During this time, passive movements are performed, followed by a gradual introduction of active-assisted exercises. The length of time in a sling generally depends on the tendon quality, the quality of the tendon repair and surgeon preference. The period of immobilization is contentious, with some studies advocating earlier mobilization after arthroscopic rotator cuff repair to prevent the negative effects of immobilization and to support rapid reintegration to daily living activities [47,48].

Return to sport

Patients have a high level of return to sport after arthroscopic rotator cuff surgery with 64% of athletes with a partial rotator cuff tear, returning to preinjury sports after arthroscopic subacromial decompression. 97% of elderly swimmers who underwent arthroscopic rotator cuff repair return to swimming [49] and 83% return to a satisfactory level of playing tennis after rotator cuff repair [50].

Shoulder replacement

Introduction

Osteoarthritis of the shoulder is the gradual degeneration of the articular cartilage that leads to pain and stiffness and affects approximately 32% of individuals aged over 60 [51]. As the joint surface degenerates, the subchondral bone remodels, losing its sphericity and congruity. The joint capsule also becomes thickened, leading to further loss of shoulder rotation. This painful condition is a growing problem in the aging population. Total shoulder replacement is an increasingly common operation with a 4.9% annual increase, a mean age of 73.5 years for females andreverse shoulder replacements accounting for 80.4% of total shoulder replacements. Reverse shoulder replacements are performed when the rotator cuff is pathologic, irreparable and no longer functional.

Exercise before surgery

Regular exercise is considered to be a core treatment for shoulder osteoarthritis, and it is universally recommended among treatment guidelines for all individuals with osteoarthritis, regardless of their individual presentation. Exercise has a number of potential benefits, including improving pain, physical function, and mood, as well as decreasing the risk of secondary health problems, including cardiovascular, metabolic, neurodegenerative, and bone disorders. Exercise likely reduces osteoarthritis pain by several different mechanisms, including increased central nervous system inhibition, local and systemic reductions in inflammation, psychosocial effects, and biomechanical effects at the affected joint [52]. Physical therapy is the first-line treatment and aims to reverse and prevent the stiffness and muscular deconditioning and imbalance. Multiple modalities are used, including stretching programs combined with ultrasound and moist heat. The aim should be to prepare a program that the patient can perform independently at home. Initially, physical therapy consists of a gentle stretching program and gentle isometric exercises. Progressing to active-assisted exercises once the inflammation has subsided and tolerated by the patient. The literature demonstrates that in a young, active patient with glenohumeral joint osteoarthritis, clinically meaningful short-term improvements in self-reported function and pain can be achieved with manual physiotherapy and exercise [53].

Exercise after surgery

After a shoulder replacement, the goals are a return of pain-free shoulder motion and the prevention of complications. Patients usually stay in the hospital 1–3 days for initial pain control. Rehabilitation following shoulder replacement is critical for patients to have the best possible outcomes with the least complications, but the best postoperative protocol is not clear [54].

Anatomic Total shoulder replacement

The main aim in the early recovery phase is preventing stiffness while protecting healing tissue which requires a balance of rest and exercise [55]. During an anatomic total shoulder replacement, the subscapularis is either reflected from its attachment to the proximal humerus and later repaired, or a lesser tuberosity osteotomy is performed and later fixed. Early postoperative rehabilitation precautions aim at preventing subscapularis tendon and lesser tuberosity failure [56,57]. Subscapularis failure after shoulder replacement can lead to anterior shoulder instability, pain, weakness in internal rotation as well as lower patient-reported outcome scores [58]. When comparing immediate versus 4 weeks delayed therapy, a greater subscapularis healing rate was demonstrated in the delayed group which was associated with improved patient reported outcomes and shoulder flexion [59].

Reverse total shoulder replacement

Reverse total shoulder arthroplasty (RTSA) is indicated for patients with massive rotator cuff tears [60,61]. The fixed-fulcrum kinematics of the RTSA with the glenoid as the convex articular surface allows the deltoid to be the dominant musculature for arm elevation or abduction [62,63]. Exercise therapy following reverse shoulder replacement should ensure protection from the combined movement of shoulder extension, adduction, and internal rotation (hand-behind-back posture) due to the risk of instability and to allow scar formation around the reverse articulation [64,65]. Rehabilitation of the shoulder initially involves joint protection followed by gradual tissue loading. A sling is recommended to be worn for 3–4 weeks following surgery, with early deltoid and scapular isometric exercises, and a gradual restoration of passive ROM in the first 6 weeks of recovery [65]. When passive ROM is restored, active-assisted and then active ROM progression is recommended to provide a gradual deltoid load to the acromion. Patients may perform light activities of daily living, and aerobic conditioning is also encouraged.

Return to sport

Most patients are able to return to one or more sports following shoulder arthroplasty, with anatomic total shoulder arthroplasty having the highest rate of return. Patients can return to swimming, golf, fitness noncontact activities and tennis. The overall rate of return to sport is 85.1% including 72.3% returning to an equivalent or improved level of play [66].

Hip

Hip pain is common in young to middle-aged active adults (18–50 years) and has a significant impact on physical activity and quality of life [67]. The most common cause of hip related pain in this age group is femoroacetabular impingement (FAI) syndrome and associated pathologies.

Femoroacetabular impingement and labral tears

FAI syndrome includes bony morphological changes in the hip which may cause aberrant joint forces during hip movements and possible damage to the intraarticular structures of the joint, such as the labrum, cartilage lesions, and ligamentum teres lesions [67].

Exercise before surgery

In FAI, muscle weakness of the surrounding hip muscles can lead to abnormal joint loading, causing pain and dysfunction during dynamic weight-bearing activities like walking, with strength values ranging from 11% to 28% lower than normal

[68]. A 10-week exercise program can be safely completed by adults with FAI before surgery, leading to improved strength, function, and self-reported clinical outcomes [69].

Exercise-based treatments should include hip, trunk, and functional strengthening components [70] with movement pattern retraining [71]. The most effective type of exercise, dose of exercise and progression of exercise is not yet known. It is also unclear what constitutes optimal loading and what level of pain is acceptable while exercising [72]. While ongoing physical activity and sport participation are very important for the overall health benefits, exercising with ongoing pain or performing exercises that exacerbate the pain, maybe causing further damage to the hip.

If nonoperative management of hip pain is unsuccessful, then further investigation and surgical opinion is warranted. Hip arthroscopy for the management of FAI has been demonstrated to significantly improve patient symptoms in comparison to best conservative care [73]. Prior to hip arthroscopy the maintenance of muscular strength, endurance, and function may be beneficial, however, there is a lack of published evidence evaluating the benefit of a dedicated preoperative rehabilitation program prior to hip arthroscopy.

Exercise after surgery

Rehabilitation after hip arthroscopy should consider an individualized approach that targets specific impairments [74]. Patients may have deficiencies in lower hip and trunk muscle strength, dynamic balance, single-leg squat, alterations in gait and reduced ability to jump, decelerate and perform cutting movements following hip arthroscopy [75]. There are numerous published protocols with most protocols emphasizing core strengthening and muscle stabilization exercises [76]. Aquatic exercise is a useful adjunct following hip arthroscopy with the hydrostatic forces increasing strength, reducing swelling, and improving pain modulation through muscular relaxation [77].

Postoperative phases

Phase	Time frame	Goals	Activity
Phase 1: Acute phase	First 4–6 weeks	Protecting healing tissues and minimizing pain.	• Swelling control—ice • Gluteal isometrics • Stationary bike in an upright seated position • Restrict flexion 90 degrees of flexion restriction Progression to phase 2—pain should be controlled, preferably without narcotic medication
Phase 2: Intermediate phase	Begins at 4–6 weeks	Gait normalization, restoration of ROM and reeducation of normal muscular firing patterns to provide sufficient dynamic stabilization.	• ROM should be progressed at the patient's tolerance, with consideration for provocative positions that may cause conflict within the joint • Progressive strength and endurance exercises targeting the gluteal complex, quadriceps, and hamstrings should be incorporated to restore musculature for demands of ambulation core strengthening program to increase dynamic stability—transverse abdominis and external obliques Progression to phase 3—minimal to no limitations with activities of daily living, equal or greater ROM than preoperative measures, and normal gait mechanics and should be relatively pain free

Phase 3: Advanced strengthening	Patient dependent	Improve muscular endurance, cardiovascular fitness, and dynamic stability with a gradual progression of nonimpact activities to tolerance.	• Exercises include spinning, elliptical, and aqua jogging to restore muscular and cardiovascular endurance • Weight machines and multi-joint dumbbell exercises can be utilized for global strength increases. Proprioceptive and neuromuscular training should be performed in various conditions, including single-leg tasks performed on both the surgical and nonsurgical extremities • Improved endurance will improve with ongoing low-resistance, high-repetition exercises that emphasize proper technique
Phase 4: Return to sport	Dependent on intraoperative findings, as a return to certain activities or sports may not be recommended due to the degree of hip pathology.	Return to preoperative activity without reaggravation	Currently, there is no single clinical tool available to predict a successful return to sport. Recommended criteria for return to sport include the following: • Normal and symmetrical pelvofemoral mechanics with gait, single-leg hop, double-leg drop jump, and the ability to perform straight-ahead jogging with no complaints of instability and/or pain • A minimum of 85% strength of the uninvolved leg is recommended • Testing should be adjusted as necessary to reflect the requirements of the patient's specific sport • Ability to perform sport-specific drills at a competitive intensity without pain is recommended prior to full clearance

Return to sport

Recent advances in arthroscopic technology and surgical techniques have improved clinical outcomes for athletic patients suffering from hip pathology, and current literature suggests a high rate of return to sport after surgery for FAI at 87%–93% overall [78]. The rate of return to the same level of competition following surgery for FAIS is 55%–83% in pooled studies [79]. Factors affecting return to sport include athlete demographics, duration of preoperative withdrawal from sport, the severity of pathology, type of sport, competition level, and surgical technique. The overall mean of return to sports is 7 months, with a range of 3.1–14.5 months [79].

Multiple studies have investigated return to sport following surgery for FAIS within different sports. Very high rates of return to sport (90–100%) have been reported in soccer, basketball, lacrosse, yoga, cycling, and swimming, football, baseball, running, and high-intensity interval training [78,79]. However, despite a high rate of return to sport overall, nearly all professional athletes experienced significant reductions in performance measures the first season after surgery.

Knee arthroscopy

Introduction

Meniscal tears are one of the most common knee injuries and can be secondary to trauma or degenerative changes. Acute tears are usually caused by a traumatic, twisting motion of the knee, frequently during sports [80]. Sports-related meniscal tears are often associated with anterior cruciate ligament rupture [81]. Most meniscal tears are degenerative and, in most instances, should be treated nonoperative [82].

Exercises before surgery

A home therapy program or simple rest with activity modification, ice, and nonsteroidal antiinflammatory drugs (NSAIDs) is the nonoperative management of possible meniscus tears. The therapy program goals are to minimize the effusion, normalize gait, normalize pain-free range of motion, prevent muscular atrophy, maintain proprioception, and maintain cardiovascular fitness. Choosing this course of treatment must include consideration of the patient's age, activity level, duration of symptoms, type of meniscus tear, and associated injuries such as ligamentous pathology. A trial of conservative treatment should be attempted in all but the most severe cases, such as a locked knee secondary to a displaced bucket-handle tear [83].

Exercise after surgery

The basic principle of meniscus surgery is to save as much meniscus as possible. Tears with a high probability of healing with surgical intervention are repaired. However, most tears are not reparable, and resection must be restricted to only the dysfunctional portions, preserving as much normal meniscus as possible [84]. Surgical options include partial meniscectomy or meniscus repair (and in cases of previous total or subtotal meniscectomy, meniscus transplantation). Knee arthroscopy has lower morbidity, improved visualization, faster rehabilitation, and better outcomes than open meniscal surgery, and is considered the standard of care. One study found that the arthroscopic pull-out repair of a medial meniscus root tear provided better results than partial meniscectomy [85]. Partial meniscectomy is the treatment of choice for tears in the avascular portion of the meniscus or complex tears that are not amenable to repair. The torn tissue is removed, and the remaining healthy meniscal tissue is contoured to a stable, balanced peripheral rim. Meniscus repair is recommended for any tears that occur in the vascular region (red zone or red-white zone), are longer than 1 cm, are root tears, involve greater than 50% of the meniscal thickness, or are unstable to arthroscopic probing [86–88]. A stable knee is important for successful meniscus repair and healing. Thus, associated ligamentous injuries must be addressed.

Current literature suggests that exercise therapy is as effective as arthroscopic partial meniscectomy for knee function in middle-aged patients with degenerative meniscal tears [89]. Following arthroscopic meniscectomy, patients are able to walk without support within 1–3 days [90] and can return to work within the first 2 weeks [91,92]. Patients can resume athletic training within 2–4 weeks [90–93]. When a meniscus repair is performed, the rehabilitation is typically more intensive. Many different protocols are described in the literature and with variation in the progression of knee motion, weight-bearing, and return to sports [94]. A common protocol is the avoidance of weight-bearing for 4–6 weeks, with full-motion encouraged [95].

Return to sport

Return to play after a meniscus injury is expected. The timing varies and depends on the injury, treatment, and rehabilitation protocol. In many cases, athletes can return to their sport as soon as 2–3 weeks status post arthroscopic partial meniscectomy and 6–8 weeks status post meniscal repair.

ACL reconstruction

Anterior cruciate ligament (ACL) injuries are common in the athletic population. ACL reconstruction is one of the most common procedures performed by orthopedic surgeons and the number of ACL reconstructions performed every year has steadily increased [96].

Exercises before surgery

The primary goals in the treatment of an ACL rupture are the restoration of function in the short term and the prevention of long-term pathologic changes in the knee. The preoperative and postoperative rehabilitation programs are similar with the aims of the control of swelling and the restoration of motion and strength.

Exercises after surgery

The postoperative rehabilitation program begins as soon as possible postoperatively, with quadriceps contractions for the maintenance of terminal extension. Shelbourne and Nitz published a protocol for ACL rupture which is divided into four phases (Fig. 3) [97].

Return to sport

Return to sports that involve cutting/pivoting sports is generally accepted between 6 and 9 months [98]. Clearance for return to full participation should be progressed through a structured plan involving return to practice, before progressive steps toward competition, taking into account the patient's concomitant knee injuries, specific sport and desired level of performance, and should be multidisciplinary decision involving the patient, surgeon, team physician and physical therapist. Returning to level one sports after ACL reconstruction increase in reinjury rates over 2 years by more than fourfold, and a return to sport 9 months or later after surgery and more symmetrical quadriceps strength prior to return reduce the reinjury rate [98]. The return to sport process can be considered as [99]:

1. **Returned to participation**

     The athlete is physically active, may train, but is medically, physically and/or psychologically not yet ready to return to sport.
2. **Return to sport**

     The athlete has returned to the defined sport, but the desired performance level is not yet reached.
3. **Return to performance**

     The athlete returns to the defined sport and performs at the preinjury level.

     A three-step decision-based model for a successful return to sport [100].

   - **Step 1**: deals with medical factors to evaluate the patient's health status, such as demographics, medical history, and physical and psychological examination.
   - **Step 2**: involves the sport-specific risk modifiers to evaluate participation risk, such as type of sport, competition level, limb dominance, and protective capabilities.
   - **Step 3**: deals with decision modifiers, such as timing of season, conflict of interest and internal and external pressure.

Before returning to sport, patients should pass an assessment of specific functional skills that demonstrate the appropriate quality of movement, strength, range of motion, balance and neuromuscular control of the lower extremity and body.

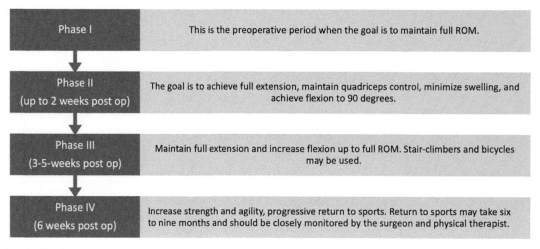

FIG. 3   ACL rehabilitation phases.

Hip and knee replacement

Introduction

Total hip and knee replacements are two of the most common procedures in orthopedic surgery, with conservative projection models predicting an increase of total hip and total knee replacements by 219% and 142% respectively between 2014 and 2046 [101]. These procedures effectively treat end-stage joint disease where nonoperative options are no longer successful. Knee and hip osteoarthritis rank highly among the global causes of disability and chronic pain, contributing significantly to impairments in function, everyday activities, sedentary behavior, and low quality of life [102,103]. Osteoarthritis of the lower limb globally affects approximately 10%–20% of people over 60 years [104]. The lifetime risk of developing symptomatic hip or knee osteoarthritis is 25.3% and 44.7%, respectively [105,106], and is the most common reason for lower limb joint replacements.

Exercises for osteoarthritis

Exercise is recognized as a core nonpharmacological therapy for the management of hip and knee osteoarthritis along with arthritis education and dietary weight management [107]. Exercise improves symptoms and the general well-being of people while being relatively safe compared to pharmacological treatments. Improved pain and functional outcomes after exercise therapy have been well demonstrated by numerous meta-analyses [108–110]. Land-based exercise has been demonstrated to reduce pain and improved physical function, and these benefits are sustained for at least a further 3–6 months after a supervised program [111]. While there are a number of published protocols, evidence for a most effective exercise regime is lacking. Most protocols involve at least six to eight individual 30-min physical therapy sessions over the course of 8 weeks or more, encompassing flexibility, strengthening, and neuromuscular coordination control exercises, such as standing weight-shifting exercises, standing balance on uneven surfaces, side-stepping, forward-backward and shuttle-walking drills, and stair walking [112]. Patients are guided on the progression of their exercises, which are tailored according to their individual physical assessments. Strengthening focuses on low-load exercises, commencing in nonweight-bearing positions and progressing to functional positions, often targeting key weak muscle groups that are deconditioned (e.g., gluteal muscle groups). In addition, aerobic exercises such as walking, cycling, or swimming for at least 30 min for 5 days a week is encouraged with the gradual progression of intensity and time of exercise, with the patient ideally incorporating the exercises into daily life [113]. Other activities that are highly recommended for the management of hip and knee arthritis are low-impact activities such as Tai Chi, Yoga, and aquatic exercises [107]. The beneficial effects of exercise seem to peak around 2 months in, and provide the best pain relief to those with knee osteoarthritis under 60 years of age who are not awaiting joint replacement surgery. However, a recent systematic review demonstrated a gradual decrease in effectiveness and exercise became no better than usual care after 9 months [114].

Prehabilitation

Patients who suffer from end-stage joint disease that has failed nonoperative management and requires joint replacement surgery often suffer from severe muscle weaknesses and functional impairments [115]. Preoperative lower limb muscle volume, muscle strength, flexibility, and functional ability has been demonstrated to be predictors of successful early recovery following joint replacements [116]. Prehabilitation exercise aims to prepare patients for surgery by improving their muscle strength, range of motion, and physical function in order to improve postoperative pain, functional recovery and quality of life.

Prehabilitation has gained significant attention in the last decade, yet studies have failed to find consistently improved postoperative outcome in hip and knee replacements following prehabilitation [117,118]. One of the likely reasons for this is the inability of patients to tolerate the recommended dosage of resistance exercise (i.e., 60%–80% of maximum effort) [103] due to pain exacerbation, therefore leading to poor compliance with the exercise program. A potential, more recent solution may be low-resistance exercises with blood-flow restriction requiring only a low load (i.e., 10%–20% of maximum muscle voluntary contraction) [119].

Exercise significantly reduces pain and improves function, performance and quality of life in people with knee and hip osteoarthritis as compared with usual care at 8 weeks. The effects are maximal around 2 months and thereafter slowly diminish, being no better than usual care at 9–18 months. Participants of a younger age, with knee osteoarthritis, and not awaiting joint replacement, may benefit more from exercise therapy [120].

Exercises after surgery

The importance of a rehabilitation program immediately following joint replacement is widely accepted, however, there is little agreement on the exercises that should comprise the regimen, the intensity of exercises, and the duration of treatment [121]. Protocols reported in literature vary in duration and involve interventions such as walking, stretching, gait retraining, aquatic therapy, quadriceps strengthening exercises, and stair climbing exercises [122]. But there are few comparative studies to demonstrate the superiority of a particular protocol or therapy. Contemporary inpatient protocols have trended toward more aggressive regimes starting on the day of surgery, to improve the length of hospital stay [123]. Clinical practice guidelines and consensus recommend that postoperative physiotherapy should include [124]:

- Supervised physical therapy with optimal settings determined by patient safety, mobility, and environmental and personal factors
- Starting within 24 h of surgery
- Cryotherapy
- Motor function training (e.g., balance, walking, movement symmetry)
- Neuromuscular electrical stimulation to improve quadriceps muscle strength, gait performance, and performance-based outcomes
- Resistance and intensity of strengthening exercise with high-intensity strength training and exercise programs during the early postacute period (i.e., within 7 days after surgery) to improve function, strength, and ROM

Return to sport

Encouraging patients to return to sport after joint replacement surgery is important but needs to be weighed against the risks of implant wear and early prosthesis failure. The goal is to balance a healthy lifestyle with preserving implanted prosthetic joints. Following joint replacement, each step can release up to 500,000 submicron-sized polyethylene particles into the patient's body [125]. These particles activate macrophages that release cytokines, which are responsible for the loosening of prosthetic implants. The volume of wear particles is related to the number of steps and joint loading [126]. Joint loading is affected by weight and type of activity. For example, during walking, peak total force is $2.3 \times$ the body weight (BW) while it is $3.0 \times$ the BW during stair climbing. Stationary cycling is $1.03 \times$ the BW, jogging is $4 \times$ the BW, and up to $22 \times$ the BW when running [127]. Current recommendations for sports and exercise after joint replacement surgery is based on expert opinion and consensus surveys rather than strong science or evidence-based medicine. Generally, there is a wide consensus on recommending low impact activities such as walking, cycling, swimming, and golf, and discouraging high-impact activities such as running and contact sports. However, with intermediate impact activities such as tennis and skiing, opinions vary [127].

Over the years, restrictions by surgeons have loosened, along with increasing patient demands and expectations, rising confidence in surgical techniques, biomaterial advancement, and innovations in joint implants [128]. Studies have not shown an increase in complication rates for patients engaging in physical activities and sports when compared to less-active control subjects. Some studies show an increase in wear rates for more active groups, but have yet to demonstrate an increase in revision rates [129].

What do patients achieve after surgery?

With the ongoing evolution of joint-replacement surgery and its increasing adaptation to a younger population, the expected goals and outcomes from these operations are no longer confined to pain management and achieving activities of daily living, with patients increasingly desiring and expecting to return to sports and higher level activities [130]. Most joint replacement patients are able to recover to an active lifestyle with a moderate level of activity. On average, joint replacements patients walk 6721 steps per day [131]. Patients are able to return to work at a rate of 91% with a mean of 6.4 weeks after a hip replacement, and 7.7 weeks after a total knee replacement. Patients are able to drive their car again at 3.7 and 4.4 weeks after hip and knee replacements, respectively [132]. The return-to-sport rate is between 54% and 98%. Higher activity levels are achieved by patients of a younger age, lower BMI, preoperative active sports practice, absence of other joint pain, or males. There is no significant difference in performance shown between hip and knee replacement patients [132].

References

[1] Australian Institute of Health Welfare. Osteoarthritis snapshot. Cat no PHE 232; 2018.

[2] Institute of health metrics and evaluation (IHME). Findings from the global burden of disease study 2017. Seattle, WA: IHME; 2018.

[3] Gehem M, Díaz PS. Shades of graying: research tackling the grand challenge of aging for Europe. The Hague Centre for Strategic Studies; 2013.

[4] Smith AK, Walter LC, Miao Y, Boscardin WJ, Covinsky KE. Disability during the last two years of life. JAMA Intern Med 2013;173(16):1506–13.

[5] Page CJ, Hinman RS, Bennell KL. Physiotherapy management of knee osteoarthritis. Int J Rheum Dis 2011;14(2):145–51.

[6] Liu Z, Qiu T, Pei L, et al. Two-week multimodal prehabilitation program improves perioperative functional capability in patients undergoing thoracoscopic lobectomy for lung cancer: a randomized controlled trial. Anesth Analg 2020;131(3):840–9.

[7] Minnella EM, Bousquet-Dion G, Awasthi R, Scheede-Bergdahl C, Carli F. Multimodal prehabilitation improves functional capacity before and after colorectal surgery for cancer: a five-year research experience. Acta Oncol 2017;56(2):295–300.

[8] Minnella EM, Awasthi R, Loiselle SE, Agnihotram RV, Ferri LE, Carli F. Effect of exercise and nutrition prehabilitation on functional capacity in esophagogastric cancer surgery: a randomized clinical trial. JAMA Surg 2018;153(12):1081–9.

[9] Gillis C, Li C, Lee L, et al. Prehabilitation versus rehabilitation: a randomized control trial in patients undergoing colorectal resection for cancer. Anesthesiology 2014;121(5):937–47.

[10] Hoogeboom TJ, Dronkers JJ, Hulzebos EH, van Meeteren NL. Merits of exercise therapy before and after major surgery. Curr Opin Anaesthesiol 2014;27(2):161.

[11] Banugo P, Amoako D. Prehabilitation. Bja Education; 2017.

[12] Howard R, Yin YS, McCandless L, Wang S, Englesbe M, Machado-Aranda D. Taking control of your surgery: impact of a prehabilitation program on major abdominal surgery. J Am Coll Surg 2019;228(1):72–80.

[13] Marmelo F, Rocha V, Moreira-Gonçalves D. The impact of prehabilitation on post-surgical complications in patients undergoing non-urgent cardiovascular surgical intervention: systematic review and meta-analysis. Eur J Prev Cardiol 2018;25(4):404–17.

[14] Cabilan CJ, Hines S, Munday J. The effectiveness of prehabilitation or preoperative exercise for surgical patients: a systematic review. JBI Database System Rev Implement Rep 2015;13(1):146–87.

[15] Almeida GJ, Khoja SS, Zelle BA. Effect of prehabilitation in older adults undergoing total joint replacement: an overview of systematic reviews. Curr Geriatr Rep 2020;9(4):280–7.

[16] Wang L, Lee M, Zhang Z, Moodie J, Cheng D, Martin J. Does preoperative rehabilitation for patients planning to undergo joint replacement surgery improve outcomes? A systematic review and meta-analysis of randomised controlled trials. BMJ Open 2016;6(2), e009857.

[17] Silver JK. Cancer rehabilitation and prehabilitation may reduce disability and early retirement. Wiley Online Library; 2014.

[18] Cross MB, Berger R. Feasibility and safety of performing outpatient unicompartmental knee arthroplasty. Int Orthop 2014;38(2):443–7.

[19] Dorr LD, Thomas DJ, Zhu J, Dastane M, Chao L, Long WT. Outpatient total hip arthroplasty. J Arthroplast 2010;25(4):501–6.

[20] Moyer R, Ikert K, Long K, Marsh J. The value of preoperative exercise and education for patients undergoing total hip and knee arthroplasty: a systematic review and meta-analysis. JBJS Rev 2017;5(12), e2.

[21] Minnella EM, Awasthi R, Gillis C, et al. Patients with poor baseline walking capacity are most likely to improve their functional status with multimodal prehabilitation. Surgery 2016;160(4):1070–9.

[22] West MA, Wischmeyer PE, Grocott MP. Prehabilitation and nutritional support to improve perioperative outcomes. Curr Anesthesiol Rep 2017;7(4):340–9.

[23] Britt H, Miller GC, Charles J, et al. General practice activity in Australia 2009–10. Gen Pract Ser 2010;27:2009–10.

[24] van der Heijden GJ. Shoulder disorders: a state-of-the-art review. Best Pract Res Clin Rheumatol 1999;13(2):287–309.

[25] Sofu H, Gürsu S, Koçkara N, Öner A, Issın A, Çamurcu Y. Recurrent anterior shoulder instability: review of the literature and current concepts. World J Clin Cases 2014;2(11):676.

[26] Krøner K, Lind T, Jensen J. The epidemiology of shoulder dislocations. Arch Orthop Trauma Surg 1989;108(5):288–90.

[27] Leroux T, Wasserstein D, Veillette C, et al. Epidemiology of primary anterior shoulder dislocation requiring closed reduction in Ontario, Canada. Am J Sports Med 2014;42(2):442–50.

[28] Liavaag S, Svenningsen S, Reikerås OA, et al. The epidemiology of shoulder dislocations in Oslo. Scand J Med Sci Sports 2011;21(6):e334–40.

[29] Romeo AA, Cohen BS, Carreira DS. Traumatic anterior shoulder instability. Orthop Clin 2001;32(3):399–409.

[30] Zacchilli MA, Owens BD. Epidemiology of shoulder dislocations presenting to emergency departments in the United States. JBJS 2010;92(3):542–9.

[31] Monk AP, Roberts PG, Logishetty K, et al. Evidence in managing traumatic anterior shoulder instability: a scoping review. Br J Sports Med 2015;49(5):307–11.

[32] Taylor DC, Arciero RA. Pathologic changes associated with shoulder dislocations. Arthroscopic and physical examination findings in first-time, traumatic anterior dislocations. Am J Sports Med 1997;25(3):306–11. https://doi.org/10.1177/036354659702500306.

[33] Wang RY, Arciero RA, Mazzocca AD. The recognition and treatment of first-time shoulder dislocation in active individuals. J Orthop Sports Phys Ther 2009;39(2):118–23.

[34] Cole BJ, L'insalata J, Irrgang J, Warner JJ. Comparison of arthroscopic and open anterior shoulder stabilization: a two to six-year follow-up study. JBJS 2000;82(8):1108.

[35] Ialenti MN, Mulvihill JD, Feinstein M, Zhang AL, Feeley BT. Return to play following shoulder stabilization: a systematic review and meta-analysis. Orthop J Sports Med 2017;5(9). 2325967117726055.

[36] Urwin M, Symmons D, Allison T, et al. Estimating the burden of musculoskeletal disorders in the community: the comparative prevalence of symptoms at different anatomical sites, and the relation to social deprivation. Ann Rheum Dis 1998;57(11):649–55.

[37] Simank H-G, Dauer G, Schneider S, Loew M. Incidence of rotator cuff tears in shoulder dislocations and results of therapy in older patients. Arch Orthop Trauma Surg 2006;126(4):235–40.

[38] Teunis T, Lubberts B, Reilly BT, Ring D. A systematic review and pooled analysis of the prevalence of rotator cuff disease with increasing age. J Shoulder Elb Surg 2014;23(12):1913–21.

[39] Werner CA. The older population, 2010. US: US Department of Commerce, Economics and Statistics Administration; 2011.

[40] Vitale MA, Vitale MG, Zivin JG, Braman JP, Bigliani LU, Flatow EL. Rotator cuff repair: an analysis of utility scores and cost-effectiveness. J Shoulder Elb Surg 2007;16(2):181–7.

[41] Chen J, Xu J, Wang A, Zheng M. Scaffolds for tendon and ligament repair: review of the efficacy of commercial products. Expert Rev Med Devices 2009;6(1):61–73.

[42] Ainsworth R, Lewis JS. Exercise therapy for the conservative management of full thickness tears of the rotator cuff: a systematic review. Br J Sports Med 2007;41(4):200–10.

[43] Kuhn JE, Dunn WR, Sanders R, et al. Effectiveness of physical therapy in treating atraumatic full-thickness rotator cuff tears: a multicenter prospective cohort study. J Shoulder Elb Surg 2013;22(10):1371–9.

[44] Kukkonen J, Joukainen A, Lehtinen J, et al. Treatment of nontraumatic rotator cuff tears: a randomized controlled trial with two years of clinical and imaging follow-up. JBJS 2015;97(21):1729–37.

[45] Moosmayer S, Lund G, Seljom US, et al. Tendon repair compared with physiotherapy in the treatment of rotator cuff tears: a randomized controlled study in 103 cases with a five-year follow-up. JBJS 2014;96(18):1504–14.

[46] Boorman RS, More KD, Hollinshead RM, et al. The rotator cuff quality-of-life index predicts the outcome of nonoperative treatment of patients with a chronic rotator cuff tear. JBJS 2014;96(22):1883–8.

[47] Arndt J, Clavert P, Mielcarek P, et al. Immediate passive motion versus immobilization after endoscopic supraspinatus tendon repair: a prospective randomized study. Orthop Traumatol Surg Res 2012;98(6):S131–8.

[48] Düzgün I, Baltacı G, Atay OA. Comparison of slow and accelerated rehabilitation protocol after arthroscopic rotator cuff repair: pain and functional activity. Acta Orthop Traumatol Turc 2011;45(1):23–33. https://doi.org/10.3944/AOTT.2011.2386. PMID: 21478659.

[49] Shimada Y, Sugaya H, Takahashi N, et al. Return to sport after arthroscopic rotator cuff repair in middle-aged and elderly swimmers. Orthop J Sports Med 2020;8(6). 2325967120922203.

[50] Bigiliani LU, Kimmel J, McCann PD, Wolfe I. Repair of rotator cuff tears in tennis players. Am J Sports Med 1992;20(2):112–7.

[51] Susa ST, Karas CS, Long NK. Treatment of glenohumeral arthritis pain utilizing spinal cord stimulation. Surg Neurol Int 2018;9.

[52] Rice D, McNair P, Huysmans E, Letzen J, Finan P. Best evidence rehabilitation for chronic pain part 5: osteoarthritis. J Clin Med 2019;8(11):1769.

[53] Crowell MS, Tragord BS. Orthopaedic manual physical therapy for shoulder pain and impaired movement in a patient with glenohumeral joint osteoarthritis: a case report. J Orthop Sports Phys Ther 2015;45(6):453–61.

[54] Blacknall J, Neumann L. Rehabilitation following reverse total shoulder replacement. Should Elb 2011;3(4):232–40.

[55] Hughes M, Neer CS. Glenohumeral joint replacement and postoperative rehabilitation. Phys Ther 1975;55(8):850–8.

[56] Caplan JL, Whitfield B, Neviaser RJ. Subscapularis function after primary tendon to tendon repair in patients after replacement arthroplasty of the shoulder. J Shoulder Elb Surg 2009;18(2):193–6.

[57] Gerber C, Pennington SD, Yian EH, Pfirrmann CA, Werner CM, Zumstein MA. Lesser tuberosity osteotomy for total shoulder arthroplasty: surgical technique. JBJS 2006;88(1_suppl_2):170–7.

[58] Bohsali KI, Wirth MA, Rockwood Jr CA. Complications of total shoulder arthroplasty. JBJS 2006;88(10):2279–92.

[59] Denard PJ, Lädermann A. Immediate versus delayed passive range of motion following total shoulder arthroplasty. J Shoulder Elb Surg 2016;25(12):1918–24.

[60] Boudreau S, Boudreau ED, Higgins LD, Wilcox 3rd RB. Rehabilitation following reverse total shoulder arthroplasty. J Orthop Sports Phys Ther 2007;37(12):734–43.

[61] Churchill JL, Garrigues GE. Current controversies in reverse total shoulder arthroplasty. JBJS Rev 2016;4(6). 01874474-201606000-00002.

[62] Berliner JL, Regalado-Magdos A, Ma CB, Feeley BT. Biomechanics of reverse total shoulder arthroplasty. J Shoulder Elb Surg 2015;24(1):150–60. https://doi.org/10.1016/j.jse.2014.08.003.

[63] Boileau P, Watkinson D, Hatzidakis AM, Hovorka I. Neer award 2005: the grammont reverse shoulder prosthesis: results in cuff tear arthritis, fracture sequelae, and revision arthroplasty. J Shoulder Elb Surg 2006;15(5):527–40. https://doi.org/10.1016/j.jse.2006.01.003.

[64] Kwaees TA, Charalambous CP. Reverse shoulder arthroplasty- -minimum age for surgery, postoperative rehabilitation and long term restrictions. A delphi consensus study. Ortop Traumatol Rehabil 2014;16(4):435–9.

[65] Romano AM, Oliva F, Nastrucci G, et al. Reverse shoulder arthroplasty patient personalized rehabilitation protocol. Preliminary results according to prognostic groups. Muscles Ligaments Tendons J 2017;7(2):263.

[66] Liu JN, Steinhaus ME, Garcia GH, et al. Return to sport after shoulder arthroplasty: a systematic review and meta-analysis. Knee Surg Sports Traumatol Arthrosc 2018;26(1):100–12.

[67] Griffin D, Dickenson E, O'donnell J, et al. The Warwick Agreement on femoroacetabular impingement syndrome (FAI syndrome): an international consensus statement. Br J Sports Med 2016;50(19):1169–76.

[68] Casartelli N, Maffiuletti N, Item-Glatthorn J, et al. Hip muscle weakness in patients with symptomatic femoroacetabular impingement. Osteoarthr Cartil 2011;19(7):816–21.

[69] Guenther JR, Cochrane CK, Crossley KM, Gilbart MK, Hunt MA. A pre-operative exercise intervention can be safely delivered to people with femoroacetabular impingement and improve clinical and biomechanical outcomes. Physiother Can 2017;69(3):204–11.

[70] Kemp JL, Mosler AB, Hart H, et al. Improving function in people with hip-related pain: a systematic review and meta-analysis of physiotherapist-led interventions for hip-related pain. Br J Sports Med 2020;54(23):1382–94.

[71] Harris-Hayes M, Czuppon S, Van Dillen LR, et al. Movement-pattern training to improve function in people with chronic hip joint pain: a feasibility randomized clinical trial. J Orthop Sports Phys Ther 2016;46(6):452–61.

[72] Kemp JL, Risberg MA, Mosler A, et al. Physiotherapist-led treatment for young to middle-aged active adults with hip-related pain: consensus recommendations from the International Hip-related pain research network, Zurich 2018. Br J Sports Med 2020;54(9):504–11.

[73] Griffin DR, Dickenson EJ, Wall PD, et al. Hip arthroscopy versus best conservative care for the treatment of femoroacetabular impingement syndrome (UK FASHIoN): a multicentre randomised controlled trial. Lancet 2018;391(10136):2225–35.

[74] Adler KL, Cook PC, Geisler PR, Yen Y-M, Giordano BD. Current concepts in hip preservation surgery: part II—rehabilitation. Sports Health 2016;8 (1):57–64.

[75] Freke M, Kemp JL, Svege I, Risberg MA, Semciw AI, Crossley KM. Physical impairments in symptomatic femoroacetabular impingement: a systematic review of the evidence. Br J Sports Med 2016;50(19):1180. https://doi.org/10.1136/bjsports-2016-096152. Erratum in: Br J Sports Med. 2019;53(20):e7. PMID: 27301577.

[76] Wall PD, Fernandez M, Griffin DR, Foster NE. Nonoperative treatment for femoroacetabular impingement: a systematic review of the literature. PM R 2013;5(5):418–26.

[77] Voight ML, Robinson K, Gill L, Griffin K. Postoperative rehabilitation guidelines for hip arthroscopy in an active population. Sports Health 2010;2 (3):222–30.

[78] Parvaresh KC, Wichman D, Rasio J, Nho SJ. Return to sport after femoroacetabular impingement surgery and sport-specific considerations: a comprehensive review. Curr Rev Musculoskelet Med 2020;13(3):213–9.

[79] Reiman MP, Peters S, Sylvain J, Hagymasi S, Mather RC, Goode AP. Femoroacetabular impingement surgery allows 74% of athletes to return to the same competitive level of sports participation but their level of performance remains unreported: a systematic review with meta-analysis. Br J Sports Med 2018;52(15):972–81.

[80] Maak TG, Fabricant PD, Wickiewicz TL. Indications for meniscus repair. Clin Sports Med 2012;31(1):1–14.

[81] Xu C, Zhao J. A meta-analysis comparing meniscal repair with meniscectomy in the treatment of meniscal tears: the more meniscus, the better outcome? Knee Surg Sports Traumatol Arthrosc 2015;23(1):164–70.

[82] Howell R, Kumar NS, Patel N, Tom J. Degenerative meniscus: pathogenesis, diagnosis, and treatment options. World J Orthop 2014;5(5):597–602. Published 2014 Nov 18 https://doi.org/10.5312/wjo.v5.i5.597.

[83] Thorlund JB, Juhl CB, Roos EM, Lohmander L. Arthroscopic surgery for degenerative knee: systematic review and meta-analysis of benefits and harms. BMJ 2015;350.

[84] Logan M, Watts M, Owen J, Myers P. Meniscal repair in the elite athlete: results of 45 repairs with a minimum 5-year follow-up. Am J Sports Med 2009;37(6):1131–4.

[85] Kim SB, Ha JK, Lee SW, et al. Medial meniscus root tear refixation: comparison of clinical, radiologic, and arthroscopic findings with medial meniscectomy. Arthroscopy 2011;27(3):346–54.

[86] Canale ST, Beaty JH. Campbell's operative orthopaedics e-book: expert consult premium edition-enhanced online features. Elsevier Health Sciences; 2012.

[87] Shybut T, Strauss EJ. Surgical management of meniscal tears. Bull NYU Hosp Jt Dis 2011;69(1):56.

[88] Yoon KH, Park KH. Meniscal repair. Knee Surg Relat Res 2014;26(2):68.

[89] Kise NJ, Risberg MA, Stensrud S, Ranstam J, Engebretsen L, Roos EM. Exercise therapy versus arthroscopic partial meniscectomy for degenerative meniscal tear in middle aged patients: randomised controlled trial with two year follow-up. BMJ 2016;354.

[90] DeHaven K. Meniscus repair in the athlete. Clin Orthop Relat Res 1985;198:31–5.

[91] Hamberg P, Gillquist J. Knee function after arthroscopic meniscectomy: a prospective study. Acta Orthop Scand 1984;55(2):172–5.

[92] Gillquist J, Oretorp N. Arthroscopic partial meniscectomy. Technique and long-term results. Clin Orthop Relat Res 1982;167:29–33.

[93] Lysholm J, Gillquist J. Arthroscopic meniscectomy in athletes. Am J Sports Med 1983;11(6):436–8.

[94] DeFroda SF, Bokshan SL, Boulos A, Owens BD. Variability of online available physical therapy protocols from academic orthopedic surgery programs for arthroscopic meniscus repair. Phys Sportsmed 2018;46(3):355–60.

[95] Lennon OM, Totlis T. Rehabilitation and return to play following meniscal repair. Oper Tech Sports Med 2017;25(3):194–207.

[96] Miyasaka K. The incidence of knee ligament injuries in the general population. Am J Knee Surg 1991;1:43–8.

[97] Shelbourne KD, Nitz P. Accelerated rehabilitation after anterior cruciate ligament reconstruction. Am J Sports Med 1990;18(3):292–9.

[98] Grindem H, Snyder-Mackler L, Moksnes H, Engebretsen L, Risberg MA. Simple decision rules can reduce reinjury risk by 84% after ACL reconstruction: the Delaware-Oslo ACL cohort study. Br J Sports Med 2016;50(13):804–8.

[99] Ardern CL, Webster KE, Taylor NF, Feller JA. Return to the preinjury level of competitive sport after anterior cruciate ligament reconstruction surgery: two-thirds of patients have not returned by 12 months after surgery. Am J Sports Med 2011;39(3):538–43.

[100] Creighton DW, Shrier I, Shultz R, Meeuwisse WH, Matheson GO. Return-to-play in sport: a decision-based model. Clin J Sport Med 2010;20 (5):379–85.

[101] Inacio MC, Graves SE, Pratt NL, Roughead EE, Nemes S. Increase in total joint arthroplasty projected from 2014 to 2046 in Australia: a conservative local model with international implications. Clin Orthop Relat Res 2017;475(8):2130–7.

[102] Cross M, Smith E, Hoy D, et al. The global burden of hip and knee osteoarthritis: estimates from the global burden of disease 2010 study. Ann Rheum Dis 2014;73(7):1323–30.

[103] Pereira D, Peleteiro B, Araujo J, Branco J, Santos R, Ramos E. The effect of osteoarthritis definition on prevalence and incidence estimates: a systematic review. Osteoarthr Cartil 2011;19(11):1270–85.

[104] Tomek IM, Goodman DC, Esty AR, Bell J-E, Fisher ES. Trends and regional variation in hip, knee and shoulder replacement. Hanover: Dartmouth College; 2010.

[105] Murphy L, Schwartz TA, Helmick CG, et al. Lifetime risk of symptomatic knee osteoarthritis. Arthritis Rheum 2008;59(9):1207–13.

[106] Murphy LB, Helmick CG, Schwartz TA, et al. One in four people may develop symptomatic hip osteoarthritis in his or her lifetime. Osteoarthr Cartil 2010;18(11):1372–9.

[107] McAlindon TE, Bannuru RR, Sullivan M, et al. OARSI guidelines for the non-surgical management of knee osteoarthritis. Osteoarthr Cartil 2014;22 (3):363–88.

[108] Nelson AE, Allen KD, Golightly YM, Goode AP, Jordan JM. A systematic review of recommendations and guidelines for the management of osteoarthritis: the chronic osteoarthritis management initiative of the US bone and joint initiative. Elsevier; 2014. p. 701–12.

[109] Uthman OA, van der Windt DA, Jordan JL, et al. Exercise for lower limb osteoarthritis: systematic review incorporating trial sequential analysis and network meta-analysis. BMJ 2013;347.

[110] Li Y, Su Y, Chen S, et al. The effects of resistance exercise in patients with knee osteoarthritis: a systematic review and meta-analysis. Clin Rehabil 2016;30(10):947–59.

[111] Fransen M, McConnell S, Hernandez-Molina G, Reichenbach S. Exercise for osteoarthritis of the hip. Cochrane Database Syst Rev 2014;4.

[112] Abbott J, Robertson M, Chapple C, et al. Manual therapy, exercise therapy, or both, in addition to usual care, for osteoarthritis of the hip or knee: a randomized controlled trial. 1: clinical effectiveness. Osteoarthr Cartil 2013;21(4):525–34.

[113] French HP, Cusack T, Brennan A, et al. Exercise and manual physiotherapy arthritis research trial (EMPART) for osteoarthritis of the hip: a multi-center randomized controlled trial. Arch Phys Med Rehabil 2013;94(2):302–14.

[114] Goh S-L, Persson MS, Stocks J, et al. Efficacy and potential determinants of exercise therapy in knee and hip osteoarthritis: a systematic review and meta-analysis. Ann Phys Rehabil Med 2019;62(5):356–65.

[115] De Groot I, Bussmann J, Stam H, Verhaar J. Actual everyday physical activity in patients with end-stage hip or knee osteoarthritis compared with healthy controls. Osteoarthr Cartil 2008;16(4):436–42.

[116] Fortin PR, Clarke AE, Joseph L, et al. Outcomes of total hip and knee replacement: preoperative functional status predicts outcomes at six months after surgery. Arthritis Rheum 1999;42(8):1722–8.

[117] Almeida GJ, Khoja SS, Zelle BA. Effect of prehabilitation in older adults undergoing total joint replacement: an overview of systematic reviews. Curr Geriatr Rep 2020;1–8.

[118] Vasta S, Papalia R, Torre G, et al. The influence of preoperative physical activity on postoperative outcomes of knee and hip arthroplasty surgery in the elderly: a systematic review. J Clin Med 2020;9(4):969.

[119] Thompson PD, Arena R, Riebe D, Pescatello LS. ACSM's new preparticipation health screening recommendations from ACSM's guidelines for exercise testing and prescription. Curr Sports Med Rep 2013;12(4):215–7.

[120] Centner C, Wiegel P, Gollhofer A, König D. Effects of blood flow restriction training on muscular strength and hypertrophy in older individuals: a systematic review and meta-analysis. Sports Med 2019;49(1):95–108.

[121] Lowe CJM, Barker KL, Dewey M, Sackley CM. Effectiveness of physiotherapy exercise after knee arthroplasty for osteoarthritis: systematic review and meta-analysis of randomised controlled trials. BMJ 2007;335(7624):812.

[122] Naylor J, Harmer A, Fransen M, Crosbie J, Innes L. Status of physiotherapy rehabilitation after total knee replacement in Australia. Physiother Res Int 2006;11(1):35–47.

[123] Larsen K, Hansen TB, Thomsen PB, Christiansen T, Søballe K. Cost-effectiveness of accelerated perioperative care and rehabilitation after total hip and knee arthroplasty. JBJS 2009;91(4):761–72.

[124] Jette DU, Hunter SJ, Jette AM. Overcoming research challenges to improve clinical practice guideline development. Oxford University Press; 2020.

[125] Callaghan J. Wear in total hip and knee replacements. J Bone Joint Surg Am 1999;81:115–36.

[126] Kuster MS, Stachowiak GW. Factors affecting polyethylene wear in total knee arthroplasty. Orthopedics 2002;25(2):S235–42.

[127] Fawaz WS, Masri BA. Allowed activities after primary total knee arthroplasty and total hip arthroplasty. Orthop Clin 2020;51(4):441–52.

[128] Thaler M, Khosravi I, Putzer D, et al. Twenty-one sports activities are recommended by the European Knee Associates (EKA) six months after total knee arthroplasty. Knee Surg Sports Traumatol Arthrosc 2021;29(3):694–709. https://doi.org/10.1007/s00167-020-06400-y.

[129] Ollivier M, Frey S, Parratte S, Flecher X, Argenson J-N. Does impact sport activity influence total hip arthroplasty durability? Clin Orthop Relat Res 2012;470(11):3060–6.

[130] Swanson EA, Schmalzried TP, Dorey FJ. Activity recommendations after total hip and knee arthroplasty: a survey of the American Association for Hip and Knee Surgeons. J Arthroplast 2009;24(6):120–6.

[131] Naal FD, Impellizzeri FM. How active are patients undergoing total joint arthroplasty?: a systematic review. Clin Orthop Relat Res 2010;468 (7):1891–904.

[132] Rondon AJ, Tan TL, Goswami K, et al. When can I drive? Predictors of returning to driving after total joint arthroplasty. J Am Acad Orthop Surg 2020;28(10):427–33.

Chapter 23

Exercise in stroke

Catherine Said[a,b,c], Kelly Bower[a], Liam Johnson[a,d,e], Erin Bicknell[b], and Natalie Fini[a]

[a]*Physiotherapy, University of Melbourne, Parkville, VIC, Australia* [b]*Physiotherapy, Western Health, St Albans, VIC, Australia* [c]*Australian Institute of Musculoskeletal Science, St Albans, VIC, Australia* [d]*School of Behavioural and Health Sciences, Australian Catholic University, Melbourne, VIC, Australia* [e]*Physiotherapy Department, Epworth HealthCare, Melbourne, VIC, Australia*

Introduction

Stroke is the second leading cause of death and disability worldwide [1]. A stroke occurs when there is an interruption of blood supply to the brain, causing cell death [2]. Approximately 85% of strokes are ischaemic, involving blood vessel blockage, with the remaining 15% caused by hemorrhages [1]. Of the almost 14 million first-ever strokes occurring each year, approximately one-third result in death, and a further third leave survivors with permanent disability [1,3]. Key risk factors for stroke include high blood pressure, smoking, diabetes, high cholesterol, obesity, and inadequate physical activity [4]. Stroke not only affects older people; approximately 60% of strokes occur in people under 65 years of age [1]. Stroke has an enormous financial cost to the individual, families, communities, and healthcare systems. A report published in 2020 estimated the direct financial cost of stroke was AUD 6.2 billion, with costs associated with premature mortality and short- and long-term disability of AUD 26 billion [3].

The outcomes of stroke are highly variable and will depend on the size and location of the lesion and an individual's personal characteristics, such as age, comorbidities, and previous level of function. Common impairments following stroke include muscle weakness, reduced motor control, loss of sensation, visual changes, spasticity, difficulty speaking or understanding speech, and changes to mood and cognition [5]. Up to 40% of stroke survivors do not achieve independent ambulation long-term [6] and around one-third are not able to walk independently outside of their homes [7]. In the early phase post-stroke, around 70% of people will have upper limb impairments [8] with only a small proportion making a complete functional recovery [8,9]. Physical activity is substantially reduced in people with stroke compared to aged-matched controls; stroke survivors average 4078 steps/day compared to an average of 8338 steps/day in controls [10].

People with stroke may also have a number of secondary complications which can impact function, participation, and quality of life. Falls are a major concern for people with stroke, with fall incidence ranging from 7% in the first week, to 73% in the first year [11]. Between 36% and 48% of individuals with chronic stroke impairments will fall at least once per year [12], which is approximately double that of healthy matched controls. Risk factors for falls following stroke include impaired balance, dependence in activities of daily living, neglect, depression, and cognitive impairment [12,13]. Falls after stroke are also associated with a high risk of injury, such as fracture [14], development of fear of falling [15], further activity restriction [16], and reduced quality of life [17]. People with stroke may also develop muscle shortening or joint contractures [18], musculoskeletal pain [19], fatigue [20], and incontinence [21]. Additionally, stroke survivors are at increased risk of future cardiovascular events, including recurrent stroke [22].

Exercise in stroke

Multiple national guidelines recognize that physical activity and exercise are safe, beneficial, and essential for all people with stroke at all stages of recovery [23–25]. Exercise plays a critical role in enhancing post-stroke recovery, ameliorating impairments and activity limitations, and reducing disability [26,27]. Exercise and physical activity may also play an important role in minimizing secondary consequences such as further medical comorbidities, recurrent stroke, deconditioning, and falls [23]. Given the heterogeneity of this population, it is important that exercise programs are developed collaboratively with appropriately trained health professionals and individually tailored to meet the stroke survivor's specific goals. Programs should take into consideration individual factors such as impairments, comorbidities, carer support, barriers and enablers to exercise, and the person's values and preferences.

Exercise to Prevent and Manage Chronic Disease Across the Lifespan. https://doi.org/10.1016/B978-0-323-89843-0.00017-9

Exercise is beneficial in all stages following stroke, however, the primary focus of an exercise program may vary depending on the stage of recovery. While there is some evidence that intensive mobilization within the first 24 h may be detrimental [28], it is well recognized that the first 3–6 months are a critical period for rehabilitation and recovery [29]. Neurophysiological changes occur rapidly and there is a period of enhanced neural plasticity, where the brain reorganizes itself and forms new synaptic connections [29]. During this stage, the focus of therapeutic exercise is often on the recovery of impairments and acquisition of normal movement patterns, and the prevention of secondary complications related to inactivity. Depending on the severity of impairments, this early stage of recovery may be undertaken within a hospital inpatient setting with a focus on preparation for safe discharge. This phase of rehabilitation generally requires the involvement of a multidisciplinary team, who work collaboratively with the stroke survivor and family members to ensure the range of post-stroke impairments are comprehensively addressed.

Exercise in the subacute (7 days to 6 months) and chronic (>6 months) stages of stroke should be aimed at improving impairments and activity limitations, with an emphasis on increasing participation in activities and engagement in long-term exercise programs. Being active long-term can help prevent physical decline and maintain independence [27]. Exercise in the sub-acute and chronic stage of recovery can occur in a range of settings, including rehabilitation centers (e.g., outpatient physiotherapy), community-based centers, such as group exercise classes or gyms, or at home (e.g., self-directed home-based training, telehealth).

Exercise and physical activity are key components in preventing post-stroke complications. Stroke survivors are at high risk of recurrent stroke and further cardiovascular disease [22]. Exercise and participation in physical activity are of paramount importance for secondary stroke prevention [30], however, evidence indicates that stroke survivors are inactive in all settings and stages of stroke recovery [10]. Exercise and physical activity participation have favorable effects on many stroke and cardiovascular disease risk factors, including blood pressure, blood lipids, blood glucose levels, and weight management [31].

Exercise may also be important in preventing falls post-stroke. Low-quality evidence supports the benefits of exercise to reduce the rate of falls after stroke (28% reduction, 8 studies, 765 participants) [11]. It is unclear what type of exercise is most beneficial for preventing falls post-stroke, but there is strong evidence in older adults that performing challenging balance-oriented and functional exercises reduces fall risk [32]. Balance outcomes are improved after stroke with approaches that are functional, repetitive, and challenge balance [33]. Exercise using devices such as force platforms, virtual reality, and robot-assisted gait training may also improve balance [34–36]. Given the high risk of fracture due to falls post-stroke, weight-bearing and strengthening exercises may also be important to improve bone health and prevent osteoporosis [37]. While more research into the effectiveness of exercise to reduce falls post-stroke is required, balance, strengthening, and weight-bearing exercises should be included in exercise programs for stroke survivors.

There are many other well-known benefits of exercise and physical activity for all adults including stroke survivors. Exercise and physical activity are effective preventative strategies for many chronic diseases [38] and have been shown to be associated with healthy aging [39]. Other benefits include positive effects on mood [40], sleep [41], and cognition [42,43]. It has also been proposed that exercise and physical activity are neuroprotective [44,45] and may reduce post-stroke fatigue [46].

Safety while exercising following stroke

Generally, it is safe for people with stroke to exercise with the benefits greatly outweighing any risks [23,24]. However, the program must be appropriately tailored to the individual's health status and current level of activity and function. Many risks associated with exercise are similar to risks for the general population, but there are some specific issues that exercise providers and stroke survivors should consider. Strategies to reduce risks may differ depending on whether the exercise is individually supervised by an exercise professional, conducted in a group setting, or completed with no formal supervision. It is recommended that people with disabilities, including stroke, seek advice from an appropriately qualified health professional to determine the most appropriate exercises and activities [24].

Prior to initiating exercise training, it is imperative that appropriate strategies are in place to reduce fall risk while exercising. The exercise area should be well lit, clear of clutter and trip hazards such as rugs and electric cords. The person exercising should wear appropriate clothing and footwear. Extra care needs to be taken when performing exercises that may challenge balance. If exercises are being completed without supervision or in a group setting, supports such as a bench, sturdy table, chair, or wall should be in close proximity so the stroke survivor can steady themselves if required. If the exercises are being completed with one-to-one supervision, the exercise professional should be positioned so they can provide assistance if required, while minimizing the risk of injury to either themselves or the stroke survivor. Care should be taken to avoid inadvertently pulling on the hemiplegic (weak) arm of the stroke survivor while providing assistance.

People who have had a stroke may be at increased risk of musculoskeletal injury. Care should be taken when prescribing shoulder exercises, particularly if the person has shoulder pain or weak shoulder muscles. People with lower limb weakness or spasticity may be at risk of knee hyperextension [47] or ankle inversion injuries [48] and they may be prescribed orthoses to assist with management. Again, care should be taken with exercise prescription to minimize knee hyperextension, ankle inversion, pain, or joint damage, and referral to a specialist physiotherapist in this area may be required.

The impact of other stroke-related impairments on the ability to safely exercise should also be considered. Difficulty with communication and cognition are common following stroke [5], therefore it is important that instructions are delivered in a clear, concise manner. Demonstration of exercises using correct technique, and resources such as written instructions, illustrations, or videos may be useful [49]. Behavioral issues, such as impulsivity, may make exercise in a group setting more challenging. Deficits in memory or poor insight may make it difficult for a person to safely and correctly perform a home exercise program. The ability to correctly set up exercise equipment may be limited by impaired upper limb function or mobility. The presence of a family member or friend to assist with setting up and monitoring safety while the person is exercising may be useful.

People who have had a stroke often have other cardiovascular risk factors such as cardiac disease, high blood pressure, and diabetes. As many people with stroke are older, they can also have other comorbidities such as osteoarthritis. Exercise prescription should take all comorbidities into account and appropriate strategies to monitor these conditions which are tailored to the individual should be in place. This could include devices to monitor heart rate or blood glucose or the use of rating scales for pain or exertion. Finally, any exercise program should include an emergency management plan. This should consider whether the person can get up from the floor in the case of a fall, how to deal with possible medical events (e.g., chest pain or low blood sugar), and who should be called in case of emergency.

What types of exercise and how much exercise should people do after stroke?

People with stroke should perform various types of exercises over the course of a week, including task-specific training, strength training, aerobic training, and balance training. More detail on each of these types of training is provided below and summarized in Table 1. Functional exercises, such as standing up from a chair or walking may provide opportunities for

TABLE 1 Exercise recommendations for people with stroke.

Exercise type	Frequency/intensity/time	Additional guidelines
Task-specific training	2 h of active task practice per day during the subacute phase[a]	Tailored, goal-oriented, repetitive practice of tasks that are challenging, progressive, and skill-based Include opportunities for practice outside scheduled rehabilitation sessions (i.e., independently or with assistance from family/carers)
Strength	2–3 days/week 1–3 sets, 10–15 repetitions, 50%–80% of 1-repetition maximum	Focus on major muscle groups Weight-bearing exercises, weights, elastic bands, pulleys
Balance	2–3 days/week Must provide a challenge to balance	Functional tasks (e.g., reaching) in sitting or standing, stepping tasks, walking tasks Minimize hand support Modify challenge by changing base of support, speed, environment, and cognitive load
Flexibility	Before or after aerobic/strength training Static stretch, hold 10 s	Focus on major muscle groups
Aerobic	*Acute stage:* as tolerated using a work/rest approach ≈10- to 20-bpm increases in resting HR; RPE ≤11 (6–20 scale; can talk full sentences but can't sing) *Subacute/chronic stage:* 3–5 days/week 40%–70% VO_2 reserve or HR reserve; 55%–80% HR_{max}; RPE 11–14 (6–20 scale) 20–60 min/session[b]; 5–10 min of warm-up and cool-down activities (e.g., easy walking or cycling)	Large-muscle activities (i.e., walking, stationary cycle ergometry, arm ergometry, arm-leg ergometry, functional activities seated exercises, if appropriate)

bpm, beats-per-minute; *HR*, heart rate; *RPE*, rating of perceived exertion; *VO_2*, volume of oxygen consumption; *HR_{max}*, heart rate maximum.
[a]Stroke Foundation [25].
[b]Total exercise time can be an accumulation of multiple short bouts of aerobic exercise pending individuals current exercise capacity.
Adapted from Billinger et al. [23].

more than one type of training. For example, standing up from a chair provides task-specific training, but can also provide strength or aerobic training. Standing up from a chair can be used for strength training by using a low chair, minimizing the use of upper limbs and few repetitions, while increasing the height of the chair to minimize strength demands and increasing the repetitions and/or speed of movement can increase aerobic demand. Exercises must be tailored to the stroke survivor's individual requirements to maximize effectiveness and can be done individually or in a group setting, depending on the person's needs and preferences. Factors that should be considered when prescribing an exercise program include salience, task specificity, challenge level, environment, progression, type of practice, instruction, demonstration, facilitation, feedback, and repetition. It is also recognized that doing some physical activity, including exercise, is better than none even if physical activity recommendations are not met [24]. For people who have not been exercising or are physically active, it is important to start slowly and gradually and independently increase frequency, intensity, and time.

Task-specific training

Following a stroke, many people have difficulty performing everyday activities such as rolling over in bed, standing up, walking, and tasks involving the upper limbs such as eating and dressing. Task-specific training of the above activities has been shown to be effective for improving function [26] and is recommended in many stroke rehabilitation guidelines [25,50,51]. Task-specific training uses principles of motor learning in the goal-directed practice of functional tasks. It relies on a thorough assessment and movement analysis. This involves observation of movement abnormalities and identification of contributing factors, such as weakness, sensory loss, or spasticity. Findings from the assessment will enable the design of an appropriate treatment plan which will address the task as well as the underlying impairments. A thorough assessment can ensure that the right treatment is delivered at the right time for each individual.

Tasks should be selected in collaboration with the stroke survivor and should be salient or meaningful and specific to their treatment goals [52]. Tasks must also be matched to the skill of the stroke survivor to provide an optimal challenge level. A task that is too difficult may cause frustration and a task that is too easy may not provide the adequate stimulus for motor skill learning [53,54]. The level of challenge needs to be continually monitored to ensure it is appropriate. Setting up a treatment environment that is contextually specific is particularly important for task-specific training. Where possible the treatment set-up should emulate the parameters of the desired task. Correct biomechanics, equipment, surfaces, variability, and distractions should be considered where possible. For example, if a person's goal is to be able to walk around their local shopping center, practicing walking in straight lines in a hospital gym environment with smooth flooring and little distraction may not effectively lead to goal attainment. While this may be an appropriate place to start, the task must be progressed to consider the environmental context [55]. Gentile's taxonomy of tasks is a useful framework for incorporating the environment and considering the progression of tasks [56].

The type of practice performed is another important consideration. Tasks can be broken down into discrete parts; for example, to retrain walking, the task can be broken down into weight shift or stepping practice. However, practice of individual parts must be translated back into practice of the whole task (e.g., walking) to allow functional task integration. Tasks can also be practiced in a blocked or variable fashion. Blocked practice occurs when one task is repeated, whereas variable practice occurs when you practice different tasks or variations of the same task in random order. Blocked practice can be more useful in earlier stages of learning and for those with reduced cognition, however, variable practice is more effective for achieving retention at later stages of learning in people with stroke [57].

Other important elements about the delivery of treatment include instruction, demonstration, facilitation/manual guidance, and feedback. Goal-oriented instructions can increase motivation and the number of repetitions performed by people with stroke [58]. Demonstration can be particularly useful for people with cognitive and language disorders. People with stroke may require hands-on facilitation or manual guidance to complete a task in the early stages of learning to assist with the completion of a task or show them how a movement will feel. Facilitation of movements can be complex and requires specific therapist training. Feedback is critical for motor skill acquisition and has an important motivational function [59]. Feedback should ideally focus only on one component of the movement to allow the person with stroke to act on the feedback. Feedback may be intrinsic (e.g., visual or proprioceptive), or extrinsic (e.g., visual or verbal feedback on performance or results). Feedback can be given immediately after each trial, delayed to provide the person time to process the movement, or summarized after a number of trials. Provision of feedback should reduce as the person progresses [60].

The final factor for consideration when prescribing task-specific exercises is repetition. Complex tasks often need to be practiced thousands of times to lead to retention and skill acquisition [61]. A recent meta-analysis demonstrated that a greater volume of rehabilitation is associated with improved outcomes following stroke. More than double the amount of additional rehabilitation is required to improve outcomes; if typical rehabilitation provides 30 min per day, 100 min per day of rehabilitation is required to improve outcomes [62]. Current Australian stroke guidelines recommend people with stroke undergoing rehabilitation complete 2 h of active practice per day [25]. It is imperative that people with stroke are provided with opportunities to practice appropriate tasks either independently or with the help of family or carers outside formal therapy sessions. Care should always be taken to ensure the task is performed as correctly as possible, to ensure the acquisition of optimal movement patterns.

Strength exercises

People with stroke experience a significant loss of muscle strength that compromises their functional capacity, mobility, and quality of life. Muscle weakness is the leading cause of disability [63] and the main impairment contributing to activity and participation limitations [64]. In a 2018 review of progressive resistance training, Dorsch et al. [64] reviewed 11 studies involving 370 people with stroke. Pooled data from six studies (163 people with stroke) demonstrated progressive resistance training significantly improves strength compared with no intervention or placebo. However, improvements in strength did not necessarily translate to improvements in activities such as walking or functional use of the upper limb. Therefore, if the goal is to improve functional activities, strength training should be combined with task-specific training. Alongside strength benefits, balance [27] and local muscle endurance [65] may be improved with resistance training after stroke. Ballistic strength training (i.e., using dynamic exercises to increase the rate of force development) shows promise as a safe and feasible alternative to traditional strength training when targeting gait speed and muscle power [66].

The review by Dorsch et al. [64] identified that progressive resistance training targeted all major lower limb muscle groups; including hip extensors, flexors, and abductors; knee extensors and flexors, and ankle plantar and dorsiflexors. Upper limb strengthening primarily targeted shoulder flexors, extensors, abductors, adductors, internal and external rotators, and elbow flexors and extensors. People with stroke should perform progressive resistance training of the major muscle groups (upper and lower extremities and trunk) and this may include the use of free weights, weight-bearing or partial weight-bearing activities, elastic bands, spring coils, or pulleys. It is recommended that strength training be performed 2–3 days/week; 1–3 sets of 10–15 repetitions of 8–10 exercises at 50%–80% of 1-repetition maximum [23].

Balance exercises

There is strong evidence that repetitive functional and challenging balance exercises improve balance outcomes for both sitting [67] and standing activities [33] in people with stroke, but there is a lack of clarity regarding the optimal type, duration, and timing of balance exercises. The effect of balance exercise on fall prevention in stroke survivors is currently unclear [11], however, there is strong evidence that challenging balance exercises to reduce falls risk in community-dwelling older adults [32]. Balance exercises may also improve functional independence and self-efficacy [68].

In general, evidence supports repetitive functional or task-specific activities that are challenging to the individual's balance [33]. This may involve activities such as practicing functional tasks or weight-shifting in a seated or standing posture, without hand support where possible (e.g., Tai Chi postures, reaching, squatting, stepping, turning, lifting, and throwing). Activities can be made more challenging in terms of speed, stability, and coordination demands (e.g., reducing the base of support, increasing object weight, size, or location). Challenges can also be moderated by modifying the environment and cognitive load. Walking activities that challenge a person's balance (e.g., obstacle courses) may also be included as well as training unexpected perturbations of varying speed, sizes, and directions. Some exercise programs which incorporate balance training that has been developed and/or implemented in stroke rehabilitation include the FAME program [69,70], the Otago Exercise Program [71], and the WEBB program [72]. A range of devices may also be considered to improve balance in people with stroke. These include force platform biofeedback devices (e.g., Balance Master) [35]; virtual reality training (e.g., Wii Balance Board) [34]; and robot-assisted gait training (e.g., Lokomat) [36]. As previously discussed, while few adverse events related to balance training post-stroke are reported in the literature, care should be taken to minimize fall risk when performing balance exercises. Although it remains unclear exactly how much balance exercise individuals with stroke should undertake to reduce the risk of falls, it is recommended that balance exercises are performed at least 2–3 times weekly [23].

Flexibility exercises

There is currently no evidence that flexibility exercises are beneficial specifically in people with stroke. A Cochrane review of 49 studies demonstrated that prolonged stretch (doses ranging from 5 min to 24 h) had little to no effect on joint mobility, pain, activity limitation, or spasticity [73], and current Australian Stroke Guidelines [25] recommend against the use of prolonged positioning or stretching to treat contractures or spasticity. The American Heart and Stroke Associations recommend stroke survivors stretch the major trunk, upper and lower limb muscles with static stretches holding each stretch for 10–30 s, 2–3 days per week, before and after aerobic and strength training [23].

Aerobic exercises

Aerobic exercise can confer multiple, meaningful benefits to people with stroke and is strongly recommended by clinical stroke management guidelines worldwide [23,25,50]. Aerobic exercise after stroke is associated with improved cardiovascular risk factors [27] and is at least as beneficial as antiplatelet and anticoagulant drugs in reducing mortality [74]. A 2020 Cochrane review examined 75 studies of physical fitness interventions in people with stroke, including aerobic (32 studies, 1631 participants), resistance (20 studies, 779 participants), and mixed training interventions (i.e., aerobic and resistance training) (23 studies, 1207 participants) [27]. Aerobic interventions, which predominantly involved walking-based activities, demonstrated improvements in fitness, walking speed and endurance, and balance. Mixed training interventions (i.e., aerobic and resistance training) can improve walking ability and balance. Along with functional benefits, aerobic exercise may indirectly positively affect post-stroke risk reduction. There are indications aerobic exercise can benefit mood, cognition, and quality of life, though further, large controlled trials are needed to consolidate the evidence. For non-ambulatory people with stroke, aerobic exercise training, including cycle ergometers and assisted walking, is safe, effective, and feasible [75].

Intensive mobilization early after stroke presents a risk of harm that may preclude the initiation of exercise training in the hyperacute stage of stroke recovery [76]. Changes to the peripheral and cerebral circulatory in the hyperacute (0–24 h) and acute (1–7 days) stages of stroke recovery and the subsequent risk of adverse circulatory effects (i.e., arterial blood pressure dysregulation), warrant a cautious approach to exercise prescription early after stroke [76]. Beyond the initial 24–48 h, however, aerobic activity should be considered an integral component of post-stroke care that can be safely implemented across the continuum of stroke recovery. Graded exercise tests prior to the initiation of an aerobic exercise program can be helpful to assess safety to participate and aid in formulating an individualized prescription [25].

Challenges to exercise following stroke

Studies have identified a number of enablers and barriers to exercise in stroke survivors. Factors that enable people with stroke to exercise vary according to personal preferences but can include external encouragement from carers and qualified personnel, and programs that provide opportunities to interact with other stroke survivors [49,77]. Having sufficient social support and a desire to be able to perform daily tasks are motivators [78]. Barriers to exercise include pain, fatigue, fear of falling, stroke-related disability, lack of social support, transport and access to equipment, and suitable exercise environments [79]. Costs, comorbid health concerns, fear of recurrent stroke, and embarrassment have also been reported [78]. Some stroke survivors believe physical activity will not improve health [79] or their condition [78] and there can be limited understanding that exercise can reduce secondary stroke risk [80]. Low self-efficacy, not feeling confident in one's body [80], not knowing where and how to exercise, and to a lesser extent, low motivation [78] have also been reported barriers, along with a lack of professional support and reported negative effects of exercise [81]. Low mood and depression may also impact physical activity. Studies have shown that older age and female sex are associated with less physical activity after stroke [82].

There are a number of strategies that may be utilized to facilitate engagement with exercise in people with stroke, including the use of behavior change techniques to optimize exercise adherence [83]. Working collaboratively with stroke survivors to identify and address modifiable barriers is likely to increase engagement with an exercise program. Programs should be tailored to consider cultural, social, and economic factors and specific impairments including mood and cognitive ability. Mobilizing social support, structuring the exercise program, utilizing and reviewing exercise diaries, and reinforcing successful performance can also increase adherence [84]. Fatigue can be managed via pacing, with the grading of session length and intensity [85] and scheduling for a time of day when the person has the most energy [25]. Education about the benefits of exercise and behavior change strategies should be customized to the individual stroke survivor's characteristics, needs, learning style, and health literacy abilities [79]. Exercise programs set away from clinical settings may also positively influence motivation to participate [86]. Extrinsic monitoring and feedback may also enhance motivation.

This can include wearable activity tracking devices to monitor steps, heart rate, and distance walked, although evidence for device effectiveness in stroke is currently limited [87]. Additionally, setting daily or session-based step targets or utilizing app- or paper-based exercise diaries to log activity and progress (e.g., number of repetitions, sets or weight) may be useful [88].

Evaluation

Regular evaluation of an exercise program is essential to ensure goals are met and allow modification of the program as the person progresses. There is a range of tools that can be used to measure the amount of exercise or physical activity a person is doing. Commercially available activity trackers can be used to monitor physical activity following stroke. Although some have been validated in the stroke population [89], the speed at which devices are developed and superseded makes it challenging to recommend using a specific device. There are many apps available to track both activities and sitting time with some giving a cue such as a sound or vibration to remind users to move after a period of time inactive. Activity diaries and questionnaires to monitor activity have also been used after stroke, although issues with recall and over-reporting have been noted with these subjective measures of physical activity [90].

There are a wide range of outcome measures that can be selected to measure the effect of exercise in people with stroke and a number of comprehensive reviews are available [91,92]. Tool selection will depend on what you want to measure, the clinometric properties, and the clinical utility of the tool [93]. Some tools, such as the Stroke Impact Scale [94] and National Institutes of Health Stroke Scale [95] have been developed specifically for people with stroke, while others such as 10-m Walk Test [96] are generic and are used for people with various health conditions. Some tools such as the Functional Independence Measure [97] require specific training. The appropriate time frame over which to evaluate the impact of any exercise program should also be considered. For example, change typically occurs more quickly in the acute phase of recovery, thus reassessment of activity following 2 weeks of intensive, task-specific training may be appropriate. In contrast, it may be unreasonable to expect meaningful changes in strength or fitness after a two-week program for someone who is in the chronic phase of recovery. A select group of outcome measures that should be considered is presented in Table 2, however, the list is not exhaustive.

TABLE 2 Selected outcome measures to evaluate the effect of exercise in people with stroke.

Assessment tool	Assessment details
Upper limb	
Dynamometry	Assesses strength, clinician administered
Fugl–Meyer Motor Assessment Upper Extremity Subscale[a,b]	Assesses sensorimotor functions, clinician administered
Action Research Arm Test (ARAT)[a,b]	Assesses upper limb activity, clinician administered
Lower limb, mobility, and balance	
Dynamometry	Assesses strength, clinician administered
Fugl–Meyer Motor Assessment Lower Extremity Subscale[a,b]	Assesses sensorimotor functions, clinician administered
10-m Walk Test[a,b]	Timed test of walking, clinician administered
Five times Sit-to-Stand[b]	Timed test of functional mobility and lower limb strength, clinician administered
Timed-Up-and-Go (TUG)[a,b]	Timed test of functional mobility and balance, clinician administered
Berg Balance Scale (BBS)[a,b]	Assesses balance and functional mobility, clinician administered
Activities-specific Balance Confidence Test[b]	Assesses balance confidence, patient reported

Continued

TABLE 2 Selected outcome measures to evaluate the effect of exercise in people with stroke—cont'd

Assessment tool	Assessment details
Cardiorespiratory fitness	
Six-Minute-Walk-Test[b,c]	Timed test of walking and aerobic capacity, clinician administered
Heart rate at a fixed submaximal workload, monitoring blood pressure[c]	Assesses aerobic capacity, clinician administered
Activities of Daily Living (ADLs)/stroke specific	
National Institutes of Health Stroke Scale[a,b]	Measures severity of symptoms and neurological deficit, clinician administered
Barthel Index[a,b]	Assesses independence with ADLs and functional mobility, clinician administered
Functional Independence Measure[a,b]	Assesses ADLs, clinician administered
Stroke Impact Scale[a,b]	Assesses health-related quality of life and participation, patient reported

[a]*Recommended by Pohl et al. [91].*
[b]*Recommended by Sullivan et al. [92].*
[c]*Recommended by MacKay-Lyons et al. [98].*

Clinical practice recommendations for exercise prescribers

This chapter provides an overview of the benefits of exercise following stroke, factors that need to be considered when developing a program, and the types of exercise that should be considered. This information is summarized for exercise prescribers in Fig. 1 and is illustrated by the accompanying case study in Box 1.

FIG. 1 Key considerations when developing an individualized exercise program with a stroke survivor.

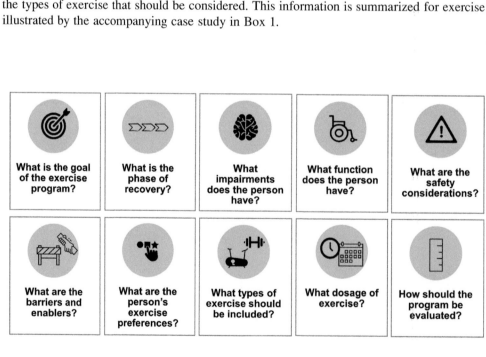

BOX 1 Case study.

Mr Jones is a 47 year old man, who had a stroke five months ago. He would like to go to the shops with his wife, but gets tired when walking and can't walk far. He is has fallen once and is nervous about falling again. He is taking medication for high blood pressure and high cholesterol and his blood sugar is 'a bit high' although he has not been diagnosed with diabetes. He did not participate in any regular exercise before his stroke. Key considerations when developing his plan are illustrated below.

 Walk with wife around local supermarket (400m): 6 weeks
Reduce falls risk/ improve balance on Berg Balance Scale by 4 points: 6 weeks
Reduce secondary stroke risk by completing 75 mins/ week moderate intensity physical activity: 6 weeks

 Subacute to chronic

 Leg weakness
Mild sensory/ proprioceptive loss
Mild expressive aphasia

 Walks independently with single point stick and ankle-foot orthosis outdoors 100m, supervision on uneven ground.
Gait speed 0.65 m/s (normative speed 1.1 m/s)
Six-minute walk test 185 m (normative distance 527m)
Ber   Balance Score 48/56

 Falls risk
Elevated blood pressure (controlled)

 Fear of falling, low self-efficacy, fatigue
Not a regular 'exerciser'
Family to assist with exercises and transport

 Health professional supervision
Individual initially, group when confident
Mix centre-based and home-based

 Aerobic, Balance
Strength
Task specific training: walking
Focus on functional exercises (e.g. sit to stand, step-ups.)

 Supervised session with health professional 1 day per week
Family to supervise home program 4 days per week.
Start with 10 min of aerobic activity 3 - 5 days per week (walking), 45% of HRR, RPE 11-14
Strength & balance training twice per week

 Six-Minute-Walk-Test   Five times Sit-to-Stand
Berg Balance Scale   Activities-specific Balance Confidence Test
Exercise diary   Blood pressure

Conclusion

Exercise is important and beneficial to all stroke survivors. Exercise and health professionals should work collaboratively with stroke survivors to develop and implement sustainable exercise programs that address meaningful goals and individual barriers to exercise, reduce disability, and minimize secondary complications of a stroke. Task-specific, aerobic, strength, and balance training should be incorporated and tailored according to the stroke survivor's goals and preferences and align with the current clinical guidelines. Measurement of stroke survivor satisfaction and progress can support motivation, adherence, and prescription of progressive exercise, as well as evaluation of program effectiveness.

References

[1] Johnson CO, et al. Global, regional, and national burden of stroke, 1990–2016: a systematic analysis for the global burden of disease study 2016. Lancet Neurol 2019;18(5):439–58.

[2] Sacco RL, et al. An updated definition of stroke for the 21st century: a statement for healthcare professionals from the American Heart Association/American Stroke Association. Stroke 2013;44(7):2064–89.

[3] Deloitte Access Economics. The economic impact of stroke in Australia, 2020; 2020.

[4] O'Donnell MJ, et al. Global and regional effects of potentially modifiable risk factors associated with acute stroke in 32 countries (INTERSTROKE): a case-control study. Lancet 2016;388(10046):761–75.

[5] Clery A, et al. Trends in prevalence of acute stroke impairments: a population-based cohort study using the South London stroke register. PLoS Med 2020;17(10).

[6] Jørgensen HS, et al. Recovery of walking function in stroke patients: the Copenhagen stroke study. Arch Phys Med Rehabil 1995;76(1):27–32.

[7] Lord SE, et al. Community ambulation after stroke: how important and obtainable is it and what measures appear predictive? Arch Phys Med Rehabil 2004;85(2):234–9.

[8] Nakayama H, et al. Recovery of upper extremity function in stroke patients: the Copenhagen stroke study. Arch Phys Med Rehabil 1994;75(4):394–8.

[9] Kwakkel G, et al. Probability of regaining dexterity in the flaccid upper limb: impact of severity of paresis and time since onset in acute stroke. Stroke 2003;34(9):2181–6.

[10] Fini NA, et al. How physically active are people following stroke? Systematic review and quantitative synthesis. Phys Ther 2017;97(7):707–17.

[11] Denissen S, et al. Interventions for preventing falls in people after stroke. Cochrane Database Syst Rev 2019;2019(10).

[12] Batchelor FA, et al. Falls after stroke. Int J Stroke 2012;7(6):482–90.

[13] Xu T, et al. Risk factors for falls in community stroke survivors: a systematic review and meta-analysis. Arch Phys Med Rehabil 2018;99(3):563–573. e5.

[14] Eng JJ, Pang MYC, Ashe MC. Balance, falls, and bone health: role of exercise in reducing fracture risk after stroke. J Rehabil Res Dev 2008;45 (2):297–314.

[15] Andersson ÅG, Kamwendo K, Appelros P. Fear of falling in stroke patients: relationship with previous falls and functional characteristics. Int J Rehabil Res 2008;31(3):261–4.

[16] Schmid AA, et al. Balance and balance self-efficacy are associated with activity and participation after stroke: a cross-sectional study in people with chronic stroke. Arch Phys Med Rehabil 2012;93(6):1101–7.

[17] Schmid AA, et al. Balance is associated with quality of life in chronic stroke. Top Stroke Rehabil 2013;20(4):340–6.

[18] Kwah LK, et al. Half of the adults who present to hospital with stroke develop at least one contracture within six months: an observational study. J Physiother 2012;58(1):41–7.

[19] Turner-Stokes L, Jackson D. Shoulder pain after stroke: a review of the evidence base to inform the development of an integrated care pathway. Clin Rehabil 2002;16(3):276–98.

[20] Cumming TB, et al. The prevalence of fatigue after stroke: a systematic review and meta-analysis. Int J Stroke 2016;11(9):968–77.

[21] Akkoç Y, et al. The course of post-stroke bladder problems and their relation with functional and mental status and quality of life: a six-month, prospective, multicenter study. Turk J Phys Med Rehabil 2019;65(4):335–42.

[22] Towfighi A, Markovic D, Ovbiagele B. Temporal trends in risk of future cardiac events among stroke survivors in the United States. Int J Stroke 2012;7:207–12.

[23] Billinger SA, et al. Physical activity and exercise recommendations for stroke survivors: a statement for healthcare professionals from the American Heart Association/American Stroke Association. Stroke 2014;45(8):2532–53.

[24] Bull FC, et al. World Health Organization 2020 guidelines on physical activity and sedentary behaviour. Br J Sports Med 2020;54(24):1451–62.

[25] Stroke Foundation. Clinical Guidelines for Stroke Management. [cited 2021 16 January]; Available from, https://informme.org.au/en/Guidelines/Clinical-Guidelines-for-Stroke-Management; 2021.

[26] Pollock A, et al. Physical rehabilitation approaches for the recovery of function and mobility following stroke. Cochrane Database Syst Rev 2014;2014(4):Cd001920.

[27] Saunders DH, et al. Physical fitness training for stroke patients. Cochrane Database Syst Rev 2020;3(3):Cd003316.

[28] Bernhardt J, et al. Efficacy and safety of very early mobilisation within 24 h of stroke onset (AVERT): a randomised controlled trial. Lancet 2015;386 (9988):46–55.

[29] Bernhardt J, et al. Agreed definitions and a shared vision for new standards in stroke recovery research: the stroke recovery and rehabilitation roundtable taskforce. Int J Stroke 2017;12(5):444–50.

[30] Turan TN, et al. Relationship between risk factor control and vascular events in the SAMMPRIS trial. Neurology 2017;88(4):379–85.

[31] D'Isabella NT, et al. Effects of exercise on cardiovascular risk factors following stroke or transient ischemic attack: a systematic review and meta-analysis. Clin Rehabil 2017;31(12):1561–72.

[32] Sherrington C, et al. Exercise for preventing falls in older people living in the community. Cochrane Database Syst Rev 2019;1.

[33] Van Duijnhoven HJR, et al. Effects of exercise therapy on balance capacity in chronic stroke: systematic review and meta-analysis. Stroke 2016;47 (10):2603–10.

[34] Corbetta D, Imeri F, Gatti R. Rehabilitation that incorporates virtual reality is more effective than standard rehabilitation for improving walking speed, balance and mobility after stroke: a systematic review. J Physiother 2015;61(3):117–24.

[35] Stanton R, et al. Biofeedback improves performance in lower limb activities more than usual therapy in people following stroke: a systematic review. J Physiother 2017;63(1):11–6.

[36] Zheng QX, et al. Robot-assisted therapy for balance function rehabilitation after stroke: a systematic review and meta-analysis. Int J Nurs Stud 2019;95:7–18.

[37] Borschmann K. Exercise protects bone after stroke, or does it? A narrative review of the evidence. Stroke Res Treat 2012;2012, 103697.

[38] Warburton DER, Bredin SSD. Health benefits of physical activity: a systematic review of current systematic reviews. Curr Opin Cardiol 2017;32 (5):541–56.

[39] Daskalopoulou C, et al. Physical activity and healthy ageing: a systematic review and meta-analysis of longitudinal cohort studies. Ageing Res Rev 2017;38:6–17.

[40] Schuch FB, et al. Physical activity and incident depression: a meta-analysis of prospective cohort studies. Am J Psychiatry 2018;175(7):631–48.

[41] 2018 Physical Activity Guidelines Advisory Committee. 2018 physical activity guidelines advisory committee scientific report. Washington, DC: U. S. Department of Health and Human Services; 2018.

[42] Johnson LG, et al. Light physical activity is positively associated with cognitive performance in older community dwelling adults. J Sci Med Sport 2016;19(11):877–82.

[43] Stillman CM, et al. Mediators of physical activity on neurocognitive function: a review at multiple levels of analysis. Front Hum Neurosci 2016;10:626.

[44] Stillman CM, Erickson KI. Physical activity as a model for health neuroscience. Ann N Y Acad Sci 2018;1428(1):103–11.

[45] Tolppanen AM, et al. Leisure-time physical activity from mid-to-late life, body mass index, and risk of dementia. Alzheimers Dement 2015;11 (4):434–443.e6.

[46] Saunders DH, Greig CA, Mead GE. Physical activity and exercise after stroke: review of multiple meaningful benefits. Stroke 2014;45(12):3742–7.

[47] Balaban B, Tok F. Gait disturbances in patients with stroke. PM&R: J Inj Funct Rehabil 2014;6(7):635–42.

[48] Deltombe T, et al. Assessment and treatment of spastic equinovarus foot after stroke: guidance from the Mont-Godinne interdisciplinary group. J Rehabil Med 2017;49(6):461–8.

[49] Banks G, et al. Exercise preferences are different after stroke. Stroke Res Treat 2012;2012:1–9.

[50] Winstein CJ, et al. Guidelines for adult stroke rehabilitation and recovery. Stroke 2016;47(6):e98–e169.

[51] Heart & Stroke Foundation: Canadian Partnership for Stroke Recovery. Evidenced-based review of stroke rehabilitation. [cited 2021 2 February]; Available from, http://www.ebrsr.com/; 2018.

[52] Bayona NA, et al. The role of task-specific training in rehabilitation therapies. Top Stroke Rehabil 2005;12(3):58–65.

[53] Guadagnoli MA, Lee TD. Challenge point: a framework for conceptualizing the effects of various practice conditions in motor learning. J Mot Behav 2004;36(2):212–24.

[54] Rosewilliam S, Roskell CA, Pandyan AD. A systematic review and synthesis of the quantitative and qualitative evidence behind patient-centred goal setting in stroke rehabilitation. Clin Rehabil 2011;25(6):501–14.

[55] Carr JH, Shepherd RT. Neurological rehabilitation: optimizing motor performance. 2nd ed. Edinburgh: Churchill Livingstone; 2010.

[56] Gentile AM, Carr JH, Shepherd RB, editors. Skill acquisition: action, movement and neuromotor processes. In: Movement science: foundations for physical therapy in rehabilitation. Rockville, MD: Aspen; 2000.

[57] Hanlon RE. Motor learning following unilateral stroke. Arch Phys Med Rehabil 1996;77(8):811–5.

[58] Hillig T, Ma H, Dorsch S. Goal-oriented instructions increase the intensity of practice in stroke rehabilitation compared with non-specific instructions: a within-participant, repeated measures experimental study. J Physiother 2019;65(2):95–8.

[59] Shumway-Cook A, Woollacott MH. Motor control: translating research into clinical practice. Lippincott Williams & Wilkins: Wolters Kluwer; 2012.

[60] Van Vliet PM, Wulf G. Extrinsic feedback for motor learning after stroke: what is the evidence? Disabil Rehabil 2006;28(13–14):831–40.

[61] Schmidt RA, Lee TD. Motor control and learning: a behavioral emphasis. 5th ed. Champaign, IL: Human Kinetics; 2011. ix, 581-ix, 581.

[62] Schneider EJ, et al. Increasing the amount of usual rehabilitation improves activity after stroke: a systematic review. J Physiother 2016;62(4):182–7.

[63] Canning CG, et al. Loss of strength contributes more to physical disability after stroke than loss of dexterity. Clin Rehabil 2004;18(3):300–8.

[64] Dorsch S, Ada L, Alloggia D. Progressive resistance training increases strength after stroke but this may not carry over to activity: a systematic review. J Physiother 2018;64(2):84–90.

[65] Ivey FM, et al. Strength training for skeletal muscle endurance after stroke. J Stroke Cerebrovasc Dis 2017;26(4):787–94.

[66] Hendrey G, et al. Feasibility of ballistic strength training in subacute stroke: a randomized, controlled, assessor-blinded pilot study. Arch Phys Med Rehabil 2018;99(12):2430–46.

[67] Veerbeek JM, et al. What is the evidence for physical therapy poststroke? A systematic review and meta-analysis. PLoS ONE 2014;9(2).

[68] Tang A, et al. The effect of interventions on balance self-efficacy in the stroke population: a systematic review and meta-analysis. Clin Rehabil 2015;29(12):1168–77.

[69] The Fitness and Mobility Exercise Program [FAME]. FAME—fitness and mobility exercise program: a group exercise program for people after stroke. [cited 2021 2 February]; Available from (n.d.) : https://fameexercise.com/.

[70] Marigold DS, et al. Exercise leads to faster postural reflexes, improved balance and mobility, and fewer falls in older persons with chronic stroke. J Am Geriatr Soc 2005;53(3):416–23.

[71] Batchelor FA, et al. Effects of a multifactorial falls prevention program for people with stroke returning home after rehabilitation: a randomized controlled trial. Arch Phys Med Rehabil 2012;93(9):1648–55.

[72] Dean C. Group task-specific circuit training for patients discharged home after stroke may be as effective as individualised physiotherapy in improving mobility. J Physiother 2012;58(4):269.

[73] Harvey LA, et al. Stretch for the treatment and prevention of contractures. Cochrane Database Syst Rev 2017;1(1):Cd007455.

[74] Naci H, Ioannidis JPA. Comparative effectiveness of exercise and drug interventions on mortality outcomes: meta-epidemiological study. Br J Sports Med 2015;49(21):1414–22.

[75] Lloyd M, et al. Physical fitness interventions for nonambulatory stroke survivors: a mixed-methods systematic review and meta-analysis. Brain Behav 2018;8(7), e01000.

[76] Marzolini S, et al. Aerobic training and mobilization early post-stroke: cautions and considerations. Front Neurol 2019;10.

[77] Robison J, et al. Resuming previously valued activities post-stroke: who or what helps? Disabil Rehabil 2009;31(19):1555–66.

[78] Nicholson S, et al. A systematic review of perceived barriers and motivators to physical activity after stroke. Int J Stroke 2013;8(5):357–64.

[79] Jackson S, Mercer C, Singer BJ. An exploration of factors influencing physical activity levels amongst a cohort of people living in the community after stroke in the south of England. Disabil Rehabil 2018;40(4):414–24.

[80] Simpson LA, et al. Exercise perceptions among people with stroke: barriers and facilitators to participation. Int J Ther Rehabil 2011;18(9):520–30.

[81] Nicholson SL, et al. A qualitative theory guided analysis of stroke survivors' perceived barriers and facilitators to physical activity. Disabil Rehabil 2014;36(22):1857–68.

[82] Thilarajah S, et al. Factors associated with post-stroke physical activity: a systematic review and meta-analysis. Arch Phys Med Rehabil 2018;99 (9):1876–89.

[83] Michie S, et al. A refined taxonomy of behaviour change techniques to help people change their physical activity and healthy eating behaviours: the CALO-RE taxonomy. Psychol Health 2011;26(11):1479–98.

[84] Prior PL, Suskin N. Exercise for stroke prevention. Stroke Vasc Neurol 2018;3(2):59–68.

[85] Zedlitz AM, et al. Cognitive and graded activity training can alleviate persistent fatigue after stroke: a randomized, controlled trial. Stroke 2012;43 (4):1046–51.

[86] Poltawski L, et al. Motivators for uptake and maintenance of exercise: perceptions of long-term stroke survivors and implications for design of exercise programmes. Disabil Rehabil 2015;37(9):795–801.

[87] Lynch EA, et al. Activity monitors for increasing physical activity in adult stroke survivors. Cochrane Database Syst Rev 2018;7:CD012543.

[88] Levy T, et al. A systematic review of measures of adherence to physical exercise recommendations in people with stroke. Clin Rehabil 2018;33 (3):535–45.

[89] Klassen TD, et al. Consumer-based physical activity monitor as a practical way to measure walking intensity during inpatient stroke rehabilitation. Stroke 2017;48:2614–7.

[90] Resnick B, et al. Inflated perceptions of physical activity after stroke: pairing self-report with physiologic measures. J Phys Act Health 2008;5 (2):308–18.

[91] Pohl J, et al. Consensus-based core set of outcome measures for clinical motor rehabilitation after stroke: a delphi study. Front Neurol 2020;11:875.

[92] Sullivan JE, et al. Outcome measures for individuals with stroke: process and recommendations from the American Physical Therapy Association neurology section task force. Phys Ther 2013;93(10):1383–96.

[93] McGinley JL, Danoudis ME. Selection of clinical outcome measures in rehabilitation of people with movement disorders: theory and practice. In: Morris RLME, editor. Rehabilitation in movement disorders. Cambridge University Press; 2013. p. 231–42.

[94] Duncan PW, et al. The stroke impact scale version 2.0. Evaluation of reliability, validity, and sensitivity to change. Stroke 1999;30(10):2131–40.

[95] Lyden PD, et al. A modified National Institutes of Health stroke scale for use in stroke clinical trials: preliminary reliability and validity. Stroke 2001;32(6):1310–7.

[96] Perera S, et al. Meaningful change and responsiveness in common physical performance measures in older adults. J Am Geriatr Soc 2006;54(5):743–9.

[97] Keith RA, et al, Eisenberg MG, editors. The functional independence measure: a new tool for rehabilitation. In: Advances in clinical Rehabilitation. New York: Springer Publishing Company, Inc; 1987.

[98] MacKay-Lyons M, et al. Aerobic exercise recommendations to optimize best practices in care after stroke: AEROBICS 2019 update. Phys Ther 2020;100(1):149–56.

Chapter 24

Chronic respiratory diseases and physical exercise

Hugo Ribeiro Zanetti[a,b], Leandro Teixeira Paranhos Lopes[c], Camilo Luís Monteiro Lourenço[d], and Leonardo Roever[e]

[a]University Center IMEPAC, Araguari, MG, Brazil [b]Federal University of Triângulo Mineiro, Uberaba, MG, Brazil [c]Brazil University, Fernandópolis, SP, Brazil [d]Federal University of Santa Catarina, Florianópolis, SC, Brazil [e]Federal University of Uberlândia, Uberlândia, MG, Brazil

Introduction

Chronic respiratory diseases (CRD) are a group of diseases affecting the airways, lungs, and associated structures. It ranks in the top three causes of death and disability worldwide [1]. The most common causative factors of CRD are exposure to air pollution, smoking, and occupational particulates [2].

Epidemiological data estimates the prevalence of CRD has increased by 39.8% compared to 1990, totaling more than 544.9 million people worldwide who are currently living with a respiratory disease [3]. Chronic obstructive pulmonary disease (COPD) remains the most prevalent disease-specific CRD, being most common among men older than 50 years [3]. Asthma is the second most prevalent CRD and is the most common chronic condition affecting children with a rising prevalence [4].

COPD is a common, preventable, and treatable disease characterized by persistent respiratory symptoms and airflow limitation due to airway and/or alveolar abnormalities. The exposure and inhalation of noxious particles or gases affect the airways, lung parenchyma, and pulmonary vasculature causing chronic inflammation that induce parenchymal tissue destruction and disruption of normal repair and defense mechanisms [5]. These physiological abnormalities lead to gas trapping and progressive airflow limitation which is characterized by reduced ventilation, forced expiratory volume in 1 s (FEV1), FEV1/forced volume capacity (FVC) ratio, hyperinflation, hypoxemia, hypercapnia, and mucus hypersecretion [6].

Asthma is a chronic inflammatory disorder of the airways in which immune cells and other cellular elements play an important role. In susceptible individuals, inflammation causes recurrent episodes of wheezing, breathlessness, chest tightness, and coughing, particularly at night or in the early morning. These episodes are usually associated with widespread but variable airflow obstruction that is typically reversible either spontaneously or with treatment [7]. The inflammation also causes an associated increase in the existing bronchial hyperresponsiveness to a variety of stimuli. Reversibility of airflow limitation may be incomplete in some patients with asthma [7].

People living with CRD present a wide range of physiological abnormalities that are associated with increased mortality and directly affect both the expectancy and quality of life. These changes go far beyond the lung, with an increase in oxidative stress, exacerbated pro-inflammatory profile, and musculoskeletal disorders such as mitochondrial dysfunction, sarcopenia, and myopathy, which all contribute to poorer prognosis [6,8–10].

Exercise in CRD

The respiratory system plays a key role during physical exercise ensuring maintenance of tissue oxygen and carbon dioxide homeostasis as well as being a critical buffer system. People with respiratory disorders have significant limitations and intolerance to physical exercise which contributes to higher levels of inactivity and sedentary behavior, as well as lower quality of life and physical capacity compared to the healthy population [11]. In this context, it is important to note that the level of daily physical activity is associated with the higher diffusing capacity of the lung for carbon dioxide, expiratory muscle strength, 6-min walking test distance, maximum oxygen uptake, and reduced pro-inflammatory circulation cytokines [12]. Therefore, it is important to be aware of and offer programs of physical exercise to people with CRD to improve disease outcomes.

Exercise to Prevent and Manage Chronic Disease Across the Lifespan. https://doi.org/10.1016/B978-0-323-89843-0.00008-8

In this scenario, pulmonary rehabilitation (PR) is one of the most important tools for people living with CRD. PR is a "comprehensive intervention based on a thorough patient assessment followed by patient tailored therapies that include, but are not limited to exercise training, education, and behavior change, designed to improve the physical and psychological condition of people with chronic respiratory disease and to promote the long-term adherence to health-enhancing behavior" [13]. PR is an essential part of the management of people with CRD, alleviating dyspnea and fatigue, improving exercise tolerance and health-related quality of life, and reducing hospital admissions, sedentary behavior, and mortality [14–16]. PR can include a wide range of exercise interventions including endurance, interval and/or resistance/strength training, and optimum benefits are achieved from programs lasting 6–8 weeks. Although this definition encourages the practice of physical exercise, it is important to emphasize that PR should be individualized considering the ideal frequency, intensity, time, and type of exercise for the patient and their capacity.

Clinical outcomes

Complete anamneses should be done considering patient signs and symptoms, evolution of the disease, and medications. Additional physical capacity evaluations in people living with CRD are required before beginning PR due to individual limitations and disease history.

Endurance capacity can be evaluated via the incremental shuttle walking test, endurance shuttle walking test, 4-m gait speed, or 6-min stepper test and sit-to-stand test. However, the methodological application, additional cost for the patient, and professional capacity are all factors that should be taken into account to facilitate the most reliable result [16]. Additionally, the 6-min walk test (6MWT) is a simple, low-cost, tolerable, and reproducible tool for people with CRD. The 6MWT should be performed in a flat place with demarcation every meter. The patient should be instructed to walk the longest distance for 6 min not being allowed to jog or run. Furthermore, the patient can interrupt the test at any time if extreme muscle discomfort or another limiting factor be reported [17].

People living with CRD commonly present peripheral weakness secondary to reduced muscle strength [18]. Thus, it is important to assess muscle strength for appropriate resistance training protocols. The one-repetition maximum (1RM) test is the gold standard for evaluating muscle strength and is defined as the maximal weight that can be lifted once using a proper technique and is a reliable and well-tolerated method for people living with CRD [19,20]. Another acceptable and low-cost method used to assess muscle strength is the handgrip dynamometry assessing grip strength [21].

Endurance exercise

Endurance exercises such as running, cycling, and swimming can improve physiological and aerobic capacity [22]. Beneficial effects are typically observed after 6–12 weeks of training and include improvements in exercise-induced hyperinflation and exertional dyspnea, heart rate recovery, and counteract muscle dysfunction [23–25]. The most common symptoms of CRD during or immediately after endurance exercise are breathlessness, leg effort, or fatigue of the quadriceps, and these symptoms should be constantly observed [26].

Current guidelines recommend a frequency of three to five times a week, an intensity >60% of maximum work rate, and a duration between 20 and 60 min per session [27,28]. However, continuous exercise modalities with low or intermittent intensity are an interesting option in people living with CRD who are unable to achieve high intensity or longer duration. In these cases, an intensity of 60%–80% of maximum work is limited by symptoms or heart rate or 4–6 on the Borg scale for least 30 min, 5 ×/week for 8 weeks [29–31] (Fig. 1).

Resistance exercise

Resistance training using free weights or machines has been proposed as an alternative and/or additional tool in PR programs that can promote adaptations that are similar to endurance training and include improvements in muscle strength, quality of life, and exercise capacity [32–36]. Resistance training through multijoint exercises promotes greater gains of muscle strength and patients should be incentivized to use it as part of their daily activities [37].

The recommended format for resistance training recommends resistance equal to 40%–50% of 1 repetition maximum (1RM) for 1–4 sets with 10–15 repetitions per set on ≥2 days per week. Some patients may be able to progress to moderate-intensity resistance training utilizing 60%–70% of 1RM. Resistance exercises should involve major muscle groups and include multijoint and single-joint exercises. Ratings of perceived exertion (RPE) of 5–6 of 10 (moderate) and 7–8 of 10 (vigorous) may be used to help guide intensity (Fig. 2).

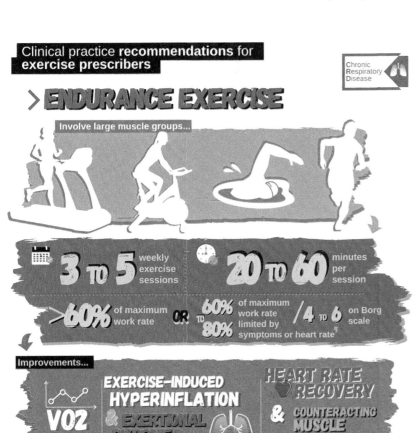

FIG. 1 Endurance exercise for chronic respiratory disease.

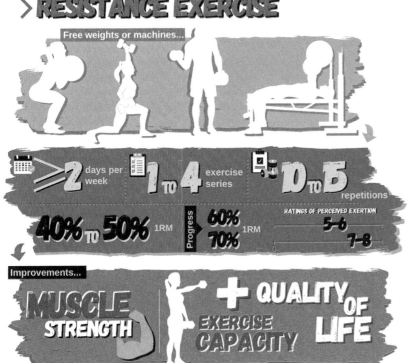

FIG. 2 Resistance exercise for chronic respiratory disease.

Respiratory muscle training

Respiratory muscle training induces improvements in inspiratory strength and endurance, functional capacity, dyspnea, and quality of life. The recommended approach is to use resistive flow breathing devices, with mechanical threshold and normocapnic hyperpnea. Most interventions have been suggested to be 30 min daily or partially, however, it has been shown that 6–10 min daily improves respiratory function [38–40]. The training intensity is recommended as 30% of maximum inspiratory power and high intensity is feasible in patients with moderate to severe COPD allowing for improved strength and endurance [41].

Conclusion

PR is an important tool that should be included in the integral care of patients with CRD, particularly those with moderate to serious complications related to peripherical muscle dysfunction that contributes to morbimortality in this population. Thus, PR using endurance training and/or resistance training is an efficacious strategy to improve cardio respiratory fitness, muscle strength, and muscle mass.

References

[1] Cruz AA. Global Surveillance, Prevention and Control of Chronic Respiratory Diseases: A Comprehensive Approach. World Health Organization; 2007.

[2] Mirza S, et al. COPD guidelines: a review of the 2018 GOLD report. Mayo Clin Proc 2018;93(10):1488–502.

[3] GBD Chronic Respiratory Disease Collaborators. Prevalence and attributable health burden of chronic respiratory diseases, 1990-2017: a systematic analysis for the global burden of disease study 2017. Lancet. Respir Med 2020;8(6):585–96.

[4] Pearce N, et al. Worldwide trends in the prevalence of asthma symptoms: phase III of the international study of asthma and allergies in childhood (ISAAC). Thorax 2007;62(9):758–66.

[5] MacNee W. Pathology, pathogenesis, and pathophysiology. BMJ 2006;332(7551):1202–4.

[6] Bourdin A, et al. Recent advances in COPD: pathophysiology, respiratory physiology and clinical aspects, including comorbidities. Eur Respir Rev 2009;18(114):198–212.

[7] Tudoric N, et al. Guidelines for diagnosis and management of asthma in adults of the Croatian respiratory society. Lijec Vjesn 2007; 129(10 − 11):315–21.

[8] Couillard A, Prefaut C. From muscle disuse to myopathy in COPD: potential contribution of oxidative stress. Eur Respir J 2005;26(4):703–19.

[9] Couillard A, Muir JF, Veale D. COPD recent findings: impact on clinical practice. COPD 2010;7(3):204–13.

[10] Rahman I, Biswas SK, Kode A. Oxidant and antioxidant balance in the airways and airway diseases. Eur J Pharmacol 2006;533(1–3):222–39.

[11] Amorim PB, et al. Barriers associated with reduced physical activity in COPD patients. J Bras Pneumol 2014;40:504–12.

[12] Garcia-Aymerich J, et al. Physical activity and clinical and functional status in COPD. Chest 2009;136(1):62–70.

[13] Spruit MA, et al. An official American Thoracic Society/European Respiratory Society statement: key concepts and advances in pulmonary rehabilitation. Am J Respir Crit Care Med 2013;188(8):e13–64.

[14] Puhan MA, et al. Pulmonary rehabilitation following exacerbations of chronic obstructive pulmonary disease. Cochrane Database Syst Rev 2016;12: CD005305.

[15] McCarthy B, et al. Pulmonary rehabilitation for chronic obstructive pulmonary disease. Cochrane Database Syst Rev 2015;2:CD003793.

[16] Mesquita R, et al. Changes in physical activity and sedentary behaviour following pulmonary rehabilitation in patients with COPD. Respir Med 2017;126:122–9.

[17] Bohannon RW, Crouch R. Minimal clinically important difference for change in 6-minute walk test distance of adults with pathology: a systematic review. J Eval Clin Pract 2017;23(2):377–81.

[18] Maltais F, et al. An official American Thoracic Society/European Respiratory Society statement: update on limb muscle dysfunction in chronic obstructive pulmonary disease. Am J Respir Crit Care Med 2014;189(9):e15–62.

[19] Levinger I, et al. The reliability of the 1RM strength test for untrained middle-aged individuals. J Sci Med Sport 2009;12(2):310–6.

[20] American Thoracic Society/European Respiratory Society. ATS/ERS statement on respiratory muscle testing. Am J Respir Crit Care Med 2002;166(4):518–624.

[21] Frontera WR, et al. A cross-sectional study of muscle strength and mass in 45- to 78-yr-old men and women. J Appl Physiol (1985) 1991; 71(2):644–50.

[22] Hottenrott K, Ludyga S, Schulze S. Effects of high intensity training and continuous endurance training on aerobic capacity and body composition in recreationally active runners. J Sports Sci Med 2012;11(3):483–8.

[23] Gimeno-Santos E, et al. Endurance exercise training improves heart rate recovery in patients with COPD. COPD 2014;11(2):190–6.

[24] Chen R, et al. Effect of endurance training on expiratory flow limitation and dynamic hyperinflation in patients with stable chronic obstructive pulmonary disease. Intern Med J 2014;44(8):791–800.

[25] Alison JA, et al. Australian and New Zealand pulmonary rehabilitation guidelines. Respirology 2017;22(4):800–19.

[26] Man WD, et al. Symptoms and quadriceps fatigability after walking and cycling in chronic obstructive pulmonary disease. Am J Respir Crit Care Med 2003;168(5):562–7.

[27] Liguori G, ACoS Medicine. ACSM's guidelines for exercise testing and prescription. Lippincott Williams & Wilkins; 2020.

[28] Rochester CL, et al. An official American Thoracic Society/European Respiratory Society policy statement: enhancing implementation, use, and delivery of pulmonary rehabilitation. Am J Respir Crit Care Med 2015;192(11):1373–86.

[29] Osadnik CR, et al. Principles of rehabilitation and reactivation. Respiration 2015;89(1):2–11.

[30] Garber CE, et al. American College of Sports Medicine position stand. Quantity and quality of exercise for developing and maintaining cardiorespiratory, musculoskeletal, and neuromotor fitness in apparently healthy adults: guidance for prescribing exercise. Med Sci Sports Exerc 2011;43(7):1334–59.

[31] Horowitz MB, Littenberg B, Mahler DA. Dyspnea ratings for prescribing exercise intensity in patients with COPD. Chest 1996;109(5):1169–75.

[32] Nyberg A, Lindstrom B, Wadell K. Assessing the effect of high-repetitive single limb exercises (HRSLE) on exercise capacity and quality of life in patients with chronic obstructive pulmonary disease (COPD): study protocol for randomized controlled trial. Trials 2012;13:114.

[33] Iepsen UW, et al. A systematic review of resistance training versus endurance training in COPD. J Cardiopulm Rehabil Prev 2015;35(3):163–72.

[34] Berry MJ, Sheilds KL, Adair NE. Comparison of effects of endurance and strength training programs in patients with COPD. COPD 2018; 15(2):192–9.

[35] Mador MJ, et al. Endurance and strength training in patients with COPD. Chest 2004;125(6):2036–45.

[36] Vonbank K, et al. Strength training increases maximum working capacity in patients with COPD—randomized clinical trial comparing three training modalities. Respir Med 2012;106(4):557–63.

[37] Paoli A, et al. Resistance training with single vs. multi-joint exercises at equal total load volume: effects on body composition, cardiorespiratory fitness, and muscle strength. Front Physiol 2017;8:1105.

[38] Zeng Y, et al. Exercise assessments and trainings of pulmonary rehabilitation in COPD: a literature review. Int J Chron Obstruct Pulmon Dis 2018;13:2013–23.

[39] Wada JT, et al. Effects of aerobic training combined with respiratory muscle stretching on the functional exercise capacity and thoracoabdominal kinematics in patients with COPD: a randomized and controlled trial. Int J Chron Obstruct Pulmon Dis 2016;11:2691–700.

[40] Langer D, et al. Inspiratory muscle training reduces diaphragm activation and dyspnea during exercise in COPD. J Appl Physiol (1985) 2018;125(2):381–92.

[41] Langer D, et al. Efficacy of a novel method for inspiratory muscle training in people with chronic obstructive pulmonary disease. Phys Ther 2015;95(9):1264–73.

Chapter 25

Exercise in cancer

Kellie Toohey[a,b,c] and Melanie Moore[a,b,c]

[a]Faculty of Health, University of Canberra, Canberra, ACT, Australia [b]Prehabilitation, Activity, Cancer, Exercise, and Survivorship (PACES) Research Group, University of Canberra, Canberra, ACT, Australia [c]Faculty of Health Clinics, University of Canberra, Canberra, ACT, Australia

Introduction to cancer

Cancer occurs when the body reproduces affected cells uncontrollably, these can stay in place without causing harm (benign) or they can break off and spread to other parts of the body (malignant). There are more than 100 types of cancers and they are usually named after the organs and the tissues where the cancer originates [1]. Cancer can occur anywhere in the human body where cell division takes place, it is a normal bodily process for cells to grow old or become unusual and die. When the abnormal cells begin to grow and multiply without dying (through a process called apoptosis), they form tumors, which essentially are lumps of tissues in the body. Cancerous cells can spread or invade other tissue nearby, they can also hitch a ride in the circulatory system to distant places (metastasis). Many cancers are a solid mass or tumor, however, cancers of the blood such as hematological cancers, generally are not.

Cancer does not discriminate with age, all ages can get cancer, however as we age the prevalence increases [2]. In 2018, there were 18.1 million new cancer cases and 9.5 million cancer-related deaths in the world. Cancer is among the leading causes of death worldwide. By 2040, the number of new cases per year is projected to increase to 29.5 million and the total number of cancer-related deaths is expected to rise to 16.4 million [3]. Commonly, cancer rates are greatest in countries whose populations have the highest life expectancy, education, and living standards (Fig. 1). But for some cancers, such as cervical cancer, the opposite is true, and the incidence rate is highest in countries in which the population ranks at a low level on these measures [3].

Cancer prevention

Cancer is the result of genetic mutations that increase cell proliferation [5,6]. The cause of this change has been attributed to both environmental factors and heredity studies [7]. Some studies have also demonstrated that certain genes may predispose people to a particular cancer diagnosis [8]. Given that a large number of cancers are caused by environmental factors that can be prevented, exercise, diet, and lifestyle changes can play an important role in reducing the number of people diagnosed.

General cancer exercise guidelines—Clinical practice recommendations for exercise prescribers

The prescription of therapeutic exercise for people diagnosed with cancer has been the center of much research over the last few decades [9–11], there are many recommendations and guidelines for clinicians to consider, however, the current evidence suggests that individually prescribed exercise is the best practice [10,11]. For the majority of people diagnosed with cancer, moderate- to high-intensity exercise would be appropriate but there is no set prescription that would be considered evidence-based for all those diagnosed [10]. The best exercise for the client would be informed by them and their goals and assessment, which would include the identification of any issues, taking into consideration their current health status and any morbidity-related risk factors.

Exercise has been shown to be beneficial before, during, and after cancer treatment, evidence suggests that it reduces treatment-related side effects such as reductions in fitness, strength, and psychological outcomes [12]. There is good evidence across several cancers on the safety and feasibility as well as the efficacy of exercise throughout the cancer continuum. This evidence has been shown across a number of cancers such as prostate, breast, colorectal, and lung cancer [13–16]. Evidence is rapidly growing, and some trials have now begun to measure the impact of exercise on many different

Males

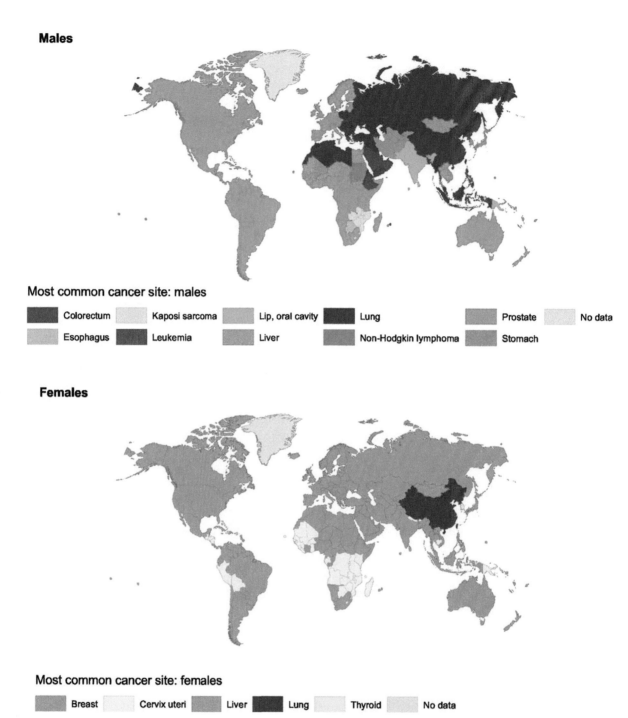

Most common cancer site: males

Colorectum	Kaposi sarcoma	Lip, oral cavity	Lung	Prostate	No data
Esophagus	Leukemia	Liver	Non-Hodgkin lymphoma	Stomach	

Females

Most common cancer site: females

Breast	Cervix uteri	Liver	Lung	Thyroid	No data

FIG. 1 Most commonly diagnosed cancers, 2012 [4].

variables caused by cancer and its treatments such as lymphoedema, bone health, neuropathies, and cachexia [17–21]. What currently seems to be missing is evidence for exercise studies outside of the general diagnoses of people who are reasonably healthy diagnosed with breast and prostate cancer. The wider cancer population includes people with co-morbidities, older people, and people at the later stages of their cancer diagnosis.

Hayes et al. (2019), designed a process for targeted exercise prescription for cancer patients which includes consideration for all cancer types, cancer-specific outcomes, and types of exercise options [10]. Fig. 2 is the process that is currently recommended, which could be used by exercise prescribers as a helpful and clinically relevant tool. It should be noted that it

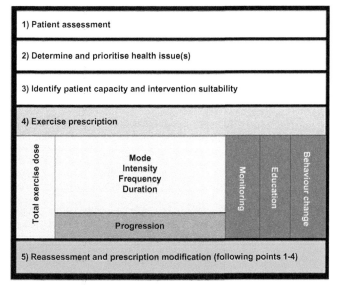

FIG. 2   Recommended process for targeted exercise prescription for cancer patients [10].

is considered best practice to engage an Accredited Exercise Physiologist/Clinical Exercise Physiologist or Physiotherapist/ Physical Therapist experienced in cancer care in the delivery of targeted exercise prescriptions for people diagnosed with cancer [10,11].

The use of screening tools such as the Exercise and Sports Science Australia (ESSA) adult pre-screening tool [22], or The American College of Sports Medicine (ACSM) guidelines for exercise testing and prescription [23] should be considered for developing a risk profile for the client and used alongside a process such as Fig. 2. However, they are general tools designed for use across populations and should not be considered the only instruments for use with people diagnosed with cancer. Additional important information about a client's biopsychosocial well-being is required to assist in making clinically relevant decisions for people diagnosed with cancer. The use of the process in Fig. 2 designed by Hayes et al. (2019) forms a more comprehensive approach to address the complex needs of a client diagnosed with cancer and offers the exercise prescriber a sound process that can be easily followed.

Hayes et al. have also supplied comprehensive exercise prescription guidelines for use with people diagnosed with cancer, which could help guide the exercise prescriber when programming for the client (Table 1).

Exercise considerations for people diagnosed with cancer have been summarized in Table 2 and adapted from Hayes et al. (2019). It is important to consider the individual client their barriers and facilitators to exercise, and what will motivate them to include movement into their lifestyle. People diagnosed with cancer come from different backgrounds and have individual needs that ought to be at the forefront of the exercise prescribers' minds when setting a program for the client.

Given the prevalence of lung cancer and the lack of good quality exercise guidelines for this particular cancer, in the next section, we will provide more detail on the topic with a comprehensive overview of what we know so far and what we aim to achieve in clinical practice when prescribing exercise for people diagnosed with lung cancer. Following this section are the current absolute and relative contraindications for exercise in general population and for those diagnosed with cancer.

Lung cancer

Bronchogenic carcinoma, commonly known as lung cancer, is heterogeneous in nature and is a highly invasive malignant neoplasm originating in the lung parenchyma or within bronchi [24]. Variance in the anatomical location and pathologic staging of the tumor explains the diversity in the presenting signs and symptoms experienced by an individual [25]. In 2020, lung cancer was the second-most commonly diagnosed cancer and was the leading cause of cancer death worldwide [26]. This accounted for approximately 2.2 million (1 in 10) new cancer diagnoses and 1.8 million (1 in 5) cancer deaths [26]. In women, lung cancer was the second most frequently diagnosed cancer (11.4%) following breast cancer (11.7%), whereas lung cancer incidence in men is most common (14.3%), followed by prostate (14.1%) and colorectal (10.6%) [26] cancer (Fig. 3).

TABLE 1 Foundation exercise prescription guidelines for cancer patients.

Weekly exercise prescription	
Mode	Mixed aerobic and resistance Priority and focus are decided based on patient goals and preferences
Intensity	At least moderate intensity
Frequency and duration	Weekly exercise volume accumulated through varying frequency of bouts per day or week, informed by patient goals, specific exercise considerations, and safety precautions
Progression, periodization, and autoregulation	• Mode, intensity, and volume can be varied with a given time period (e.g., across weeks, training cycles, or other pertinent periods), with emphasis on transition leading up to important survivorship phases (e.g., the commencement of therapy) or with new modes of exercise (e.g., impact loading) • Progressive overload can be applied as per the normal population in those with no contraindications, however may need to be slower for deconditioned individuals, or those at risk of symptom or treatment exacerbation • Prescription should allow for patient-led regulation of the program parameters depending on their specific needs
Aerobic exercise	
Mode	Ideally exercises incorporate large muscle groups, allowing for varying intensity. While commonly prescribed and recommended, but patients do not have to be restricted to walking, particularly as function and capacity improvements. Varying exercise types in line with patient capacity and function will allow for physiological and functional benefits
Intensity	Exercise of at least moderate intensity is advised, regardless of deconditioning or functionality. Low intensity should only be recommended in line with strong patient preference, or exacerbation of symptoms with higher intensity. Patient education on outcomes of exercise dose and how intensity influences this is important for compliance
Frequency and duration	For those who can achieve more than 20 min of exercises per bout, sessions should be spread across the week, avoiding consecutive days without planned exercise. Patients who through functional limitation or decondition, are unable to reach 20 min per bout, multiple sessions per day, every day are recommended
Progression	Modified through alteration to the above factors in line with patient progression
Resistance exercise	
Mode	Dynamic movements, utilizing both concentric and eccentric muscular efforts, as well as dynamic or isometric exercises, are recommended. Targeting of muscle groups negatively impacting ADLs or compromised through treatment is advised to improve outcomes. Use of machines, weights, bodyweight, or resistance bands as appropriate
Frequency, intensity, and volume	At least two sessions per week, with at least 48 h recovery before training the same muscle group again. With more regular training split programming may be required to accommodate this. Moderate to high intensity recommended, with higher volume to optimize muscle hypertrophy
Progression	Progression through modification of number, type, load, sets, or repetitions

Adapted from [10].

Lung cancer is divided into two histological subtypes: (1) small cell lung cancer (SCLC) and (2) non-small cell lung cancer (NSCLC). SCLC is an aggressive malignancy that account for approximately 15% of all lung cancer diagnoses however prognosis is extremely poor as median survival time is 15–20 month [27] despite initial staging at diagnosis and subsequent treatment [28]. Non-small cell lung cancer (NSCLC) incidence is higher, representing 85% of all new cases [29] however prognosis is still poor for individuals affected, with an overall survival based upon the Tumor, Node, and Metastasis (TNM) clinical staging at 24 and 60 months. Overall survival for Sage I ranges from 97% and 92% whereas stage IVB ranges from 10% to 0% [30]. See Fig. 4 for estimated overall survival for each pathologic lung cancer stage.

TABLE 2 Considerations for prescribing exercise for people diagnosed with cancer.

Cancer-related concerns	Potential conditions
Hematological	• Anemia • Thrombocytopenia • Neutropenia
Musculoskeletal	• Arthralgia/Aromatase inhibitor-associated (AIA) musculoskeletal syndrome • Cachexia • Sarcopenia • Bone loss • Cancerous bone tumors • Pain • Post-surgical wound healing
Systemic	• Fatigue • Fever • Infection, including cellulitis • Sleep problems • Sexual dysfunction
Cardiovascular and respiratory	• Dyspnea • Chest pain • Cardiovascular toxicity
Lymphatic	• Lymphoedema
Gastrointestinal and genitourinary	• Vomiting, nausea, loose bowl motions (fecal incontinence and diarrhea) • Urinary incontinence
Neurological	• Peripheral neuropathy • Dizziness • Cognition
Prescence of other chronic conditions	• Obesity • Type 2 diabetes mellitus • Osteoporosis • Arthritis • Depression and anxiety • Cardiovascular disease (CVD)

Adapted from acute or chronic cancer-related concerns requiring specific exercise prescription consideration table [10].

Risk factors

Lung cancer has been a stereotyped disease as "self-inflicted" [31] due to the relationship between disease development and history of tobacco smoking [32]. Although advancing age and tobacco smoking history are the predominant life-long risk factors [33], epidemiology studies have facilitated the growing recognition of disease diversity, demonstrating that 10%–20% of individuals diagnosed with lung cancer were unrelated to tobacco smoking [34].

Treatment

When diagnosed with lung cancer, standard recommended treatments depend on several factors such as the subtype (NSCLC/SCLC) and stage (early stage/advanced/recurrent) of cancer, and in more advanced stages, the patient's physical function capacity (performance status) [25].

Therapeutic modalities for lung cancer may include, surgery, targeted therapy, radiation therapy, and chemotherapy [25]. Table 3 outlines the standard treatment recommended for NSCLC based on the pathologic stage.

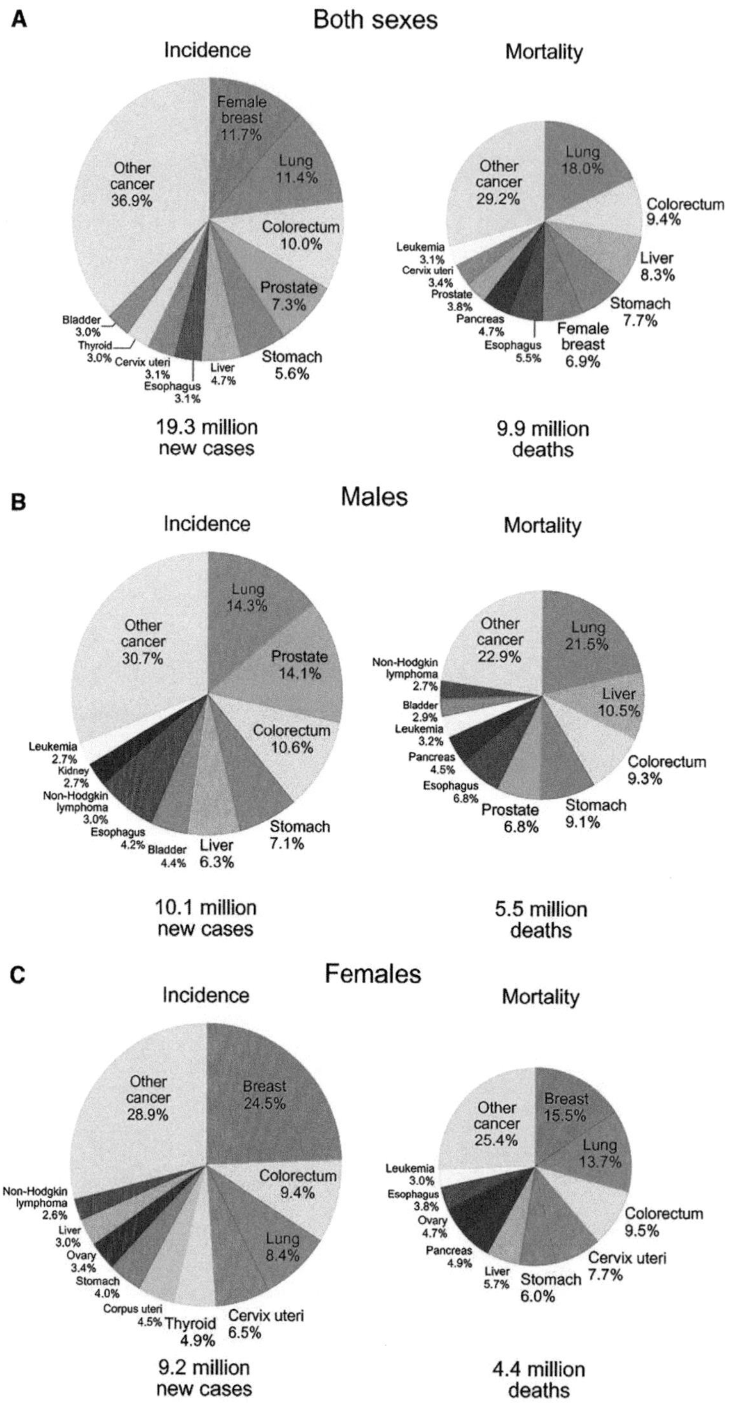

FIG. 3 Global cancer incidence and mortality in 2020. Distribution estimates of the top 10 cancers for: (A) both sexes, (B) males, and (C) females [26].

Treatment-related side effects

The extent and severity of treatment-related side effects experienced by individuals can vary with age, physical function status, comorbidities, treatment profile (type and dose). Some side effects are acute, lasting for only the duration of treatment, whereas side effect onset can be delayed or even persistent for a month or even years after treatment has finished [35]. Common physical and psychological symptoms reported by individuals with lungs cancer include anxiety, fatigue, and dyspnea. These symptoms are associated with reduced engagement in physical activity which can remain an issue throughout the entirety of treatment and beyond [36].

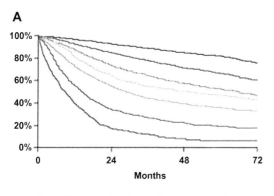
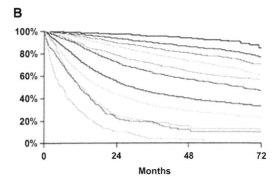

7th Ed.	Events / N	MST	24 Month	60 Month
IA	1119 / 6303	NR	93%	82%
IB	768 / 2492	NR	85%	66%
IIA	424 / 1008	66.0	74%	52%
IIB	382 / 824	49.0	64%	47%
IIIA	2139 / 3344	29.0	55%	36%
IIIB	2101 / 2624	14.1	34%	19%
IV	664 / 882	8.8	17%	6%

Proposed	Events / N	MST	24 Month	60 Month
IA1	68 / 781	NR	97%	92%
IA2	505 / 3105	NR	94%	83%
IA3	546 / 2417	NR	90%	77%
IB	560 / 1928	NR	87%	68%
IIA	215 / 585	NR	79%	60%
IIB	605 / 1453	66.0	72%	53%
IIIA	2052 / 3200	29.3	55%	36%
IIIB	1551 / 2140	19.0	44%	26%
IIIC	831 / 986	12.6	24%	13%
IVA	336 / 484	11.5	23%	10%
IVB	328 / 398	6.0	10%	0%

FIG. 4   Estimated overall survival by lung cancer stage. *MST*, median survival time; *NR*, not reported [30].

Surgical resection—Surgery is the primary treatment for early-stage NSCLC (stages I and II) with curative intent of 60%–70% [37]. However, surgery is uncommon for SCLC as staging at diagnosis is generally advanced, the current clinical oncology guidelines indicate that surgery is only considered for early-stage I [38,39]. A pneumonectomy and lobectomy is the preferred surgery procedure [40] although wedge sections tend to be suggested for older individuals with high cardio-pulmonary comorbidity burden [41]. Postoperative complications such as pain [42], fatigue [36,43], dyspnea, [44] along with impaired lung function [45] is common. Exercise capacity and level of physical activity are also typically impaired postoperatively, and decreased health-related quality of life has been reported.

Chemotherapy—94% of patients undergoing first-line chemotherapy will experience some degree of adverse effects ranging from hematologic adverse events (febrile neutropenia, infections, thrombocytopenia, and anemia), weight changes, fatigue, nausea/vomiting, diarrhea [46].

Radiotherapy—can be used as a combined treatment modality with chemotherapy or independently [47]. Radio-therapy can cause lung toxicity such as radiation pneumonitis, which is the acute phase of inflammatory 1–3 months after treatment which can cause coughing, dyspnea, and in some cases fever [48], and chronic radiation-induced injuries such as radiation fibrosis, presenting 6–24 following treatment resulting in chronic inflammation with the clinical manifestation of a dry cough, dyspnea, and insufficient respiratory exchange [48,49].

Lung cancer and exercise—Clinical practice recommendations for exercise prescribers

Many individuals, prior to a lung cancer diagnosis, have been found to engage less frequently in physical activity, leading to below-average aerobic capacity, muscular strength, and health-related quality of life compared to aged-matched healthy individuals [50]. At diagnoses, individuals will experience a high degree of disease burden, adversely impacting their physical and psychological health [50,51]. Confounding factors such as the underlying pathological processes of the disease, the cancer treatment and the patient's health profile consisting of the presence of comorbidities and a history of sedentary behavior [9,52,53] can foster poor exercise tolerance [9,52,53]. Symptoms such as dyspnea, fatigue [36], and pain can also contribute to inactivity and an ongoing cycle of exercise intolerance.

TABLE 3 Standard treatment option for NSCLC.

Stage	Treatment options
I, II-resectable	Surgery
I, II-unresectable	Radiotherapy
II-resectable	Surgery followed by chemotherapy (adjuvant)
IIIA-resectable	Surgery followed by chemotherapy
IIIA-resectable	Chemotherapy (neoadjuvant) followed by surgery
IIIA-unresectable	Sequential or combined chemoradiation
IIIA with bulky primary tumors	Radical surgery
IIIB	Chemotherapy or radiation
IIIB	Chemoradiation
IV with PS of 0 or 1	Chemotherapy (a combination of two cytotoxic drugs, platinum-based)
IV with PS of 0 or 1, with no squamous carcinoma histology, brain metastasis, significant cardiovascular disease	Bevacizumab (VEGF antibody)+first-line doublet combination chemotherapy
IV with PS of 2	Chemotherapy (single cytotoxic drug)
NSCLC with EGFR mutations	EGFR-targeted therapy (erlotinib and gefitinib)
III, IV (first-line platinum-based therapy failed with acceptable PS)	Docetaxel, pemetrexed, erlotinib, and gefitinib
Recurrent NSCLC	External palliative radiation therapy, cytotoxic chemotherapy, EGFR inhibitors (with or without EGFR mutations), EML4-ALK inhibitor (with EML-ALK translocations), surgical resection (isolated cerebral metastases), laser therapy or interstitial radiation therapy (endobronchial lesions), and stereotactic radiation surgery
Brain metastases	Whole brain radiotherapy (WBRT), surgery, stereotactic radiosurgery (SRS) with and without WBRT, systemic therapy and radiosensitisation

Adapted from [25].

Benefits of exercise

Research has shown that exercise is safe, feasible, and effective during and beyond cancer treatment [10], with many cancer survivors electing to use exercise as an effective self-management strategy [54], to reduce their physical and psychological burden of cancer [12]. The role of exercise for individuals diagnosed with lung cancer varies depending on the stage of the disease and treatment phase, however, studies have demonstrated that exercise can improve subjective and objective outcomes such as cardiorespiratory fitness, pulmonary function, muscular strength, muscle mass, fatigue, dyspnea, quality of life, psychological distress, and sleep quality [29,45,55,56].

When developing an exercise program for an individual diagnosed with lung cancer, it is important to identify their specific exercise barriers and enables. Through increasing understanding of an individual's needs, targeted strategies can be developed and implemented to promote exercise adherence [57]. Commonly reported barriers and enablers for individuals diagnosed with lung cancer can be seen in Fig. 5 [57].

Pulmonary function

Treatment for lung cancer such as surgical resection, chemotherapy, and radiation therapy along with pulmonary-specific comorbidities, can have a detrimental impact on respiratory function [58–60]. It is therefore advised that respiratory function is assessed through a spirometry test to define the therapeutic outcomes of an exercise program for a person diagnosed with lung cancer. The most utilized parameter to evaluate lung function is forced expiratory volume in 1 s (FEV_1) [29] as strong evidence has shown in chronic pulmonary disease, exercise has a small but significant increase in spirometry

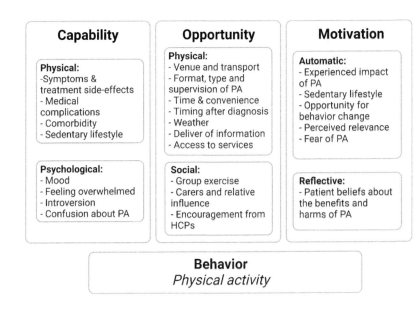

Capability	Opportunity	Motivation
Physical: -Symptoms & treatment side-effects - Medical complications - Comorbidity - Sedentary lifestyle	**Physical:** - Venue and transport - Format, type and supervision of PA - Time & convenience - Timing after diagnosis - Weather - Deliver of information - Access to services	**Automatic:** - Experienced impact of PA - Sedentary lifestyle - Opportunity for behavior change - Perceived relevance - Fear of PA
Psychological: - Mood - Feeling overwhelmed - Introversion - Confusion about PA	**Social:** - Group exercise - Carers and relative influence - Encouragement from HCPs	**Reflective:** - Patient beliefs about the benefits and harms of PA

Behavior
Physical activity

FIG. 5 Specific barriers and enablers to engaging in physical activity/exercise for people diagnosed with lung cancer using the COM-B model. *COM-B*, capability, opportunity, and motivation-behavior system; HCP, health care provider; PA, physical activity [57].

values. These improvements were seen when a multi-model exercise intervention was performed, meeting the physical activity guidelines of at least 150 min per week at a moderate intensity [61]. Preliminary evidence to support the benefits of exercise to improve pulmonary lung function in lung cancer is less robust and should be interpreted with caution.

Cardiovascular fitness

Cardiorespiratory fitness should be evaluated in individuals with lung cancer as it provides clinically significant information for surgery eligibility [52], perioperative and postoperative complication risk, and is an independent predictor of survival for individuals with lung cancer [62,63]. Individuals who present with a low $VO_{2peak} < 16$ mL/kg/min have an increased risk of postoperative complications and subsequently a reduced survival rate post curative resection in NSCLC [64]. Furthermore, a recent systematic review and meta-analysis [65] exploring the association between cardiorespiratory fitness, physical activity level, and lung cancer mortality found that there was a strong inverse correlation between cardiorespiratory fitness and level of physical activity engagement and lung cancer mortality. Individuals with a high (≥ 13.65 METS) and moderate (9.46–13.64 METS) level of cardiorespiratory fitness had reduced lung cancer mortality compared to individuals with low (< 9.45 METS) cardiorespiratory fitness. This association was also noted in individuals with a high (≥ 360 min/week), and moderate (151–359 min/week) level of physical activity, compared to inactivity or low (≤ 150 min/week) levels of physical activity engagement.

In clinical practice, the 6-min walk test is commonly used to measure functional status as it is safe and inexpensive. It can also be used as an important prognostic factor for survival in NSCLC individuals. It has been found that a 6-min walk distance of ≥ 400 m has a more positive survival rate [66] I with every 50-m increase in the 6-min walk distance from baseline, is associated with a 13% reduced risk of death [63].

It is therefore recommendations that exercise has a therapeutic role preoperatively to reduce postoperative complications and should be available for individuals preparing for surgery [57]. In a preoperative exercise program, aerobic exercise should be a priority with a combination of resistance-based exercise along with respiratory muscle training or breathing exercises [57].

Muscle strength and endurance

Individuals diagnosed with lung cancer may present with muscle dysfunction accompanied by pathological conditions, such as cachexia (a multifactorial syndrome characterized by severe muscle wasting, malnutrition, and systemic inflammation) and sarcopenia (decreased muscle mass). Many individuals diagnosed with advanced-stage lung cancer will experience unintentional weight loss >5% (69% experience cachexia) or muscle loss (47% patients experience sarcopenia) which is related to poor prognosis [67]. Chemotherapy-induced inflammatory processes, abnormal metabolism, and reduced caloric intake.

It is therefore recommended that muscular strength should be assessed in this population. A readily used assessment of strength is handgrip strength, which also has a strong correlation to all-cause mortality [68]. In individuals with lung cancer, grip strength is significantly reduced in NSCLC (stages I–IIIA) compared with healthy individuals with a mean difference of −6 kg.

Muscle strength and cardiorespiratory fitness have been demonstrated to be independent predictors of survival for people diagnosed with lung cancer, however, in clinical practice, the majority of people diagnosed present with insufficient levels of physical activity and exercise [29].

Caution and safety considerations

Pulmonary lobe resection surgery—In the first month following resection, patients have been found to have on average a 25% reduction in maximal exercise tolerance. This decrease can partly be explained by a loss in ventilatory capacity. However, Granger et al. reported that exercise tolerance declines further during a period of 6 months following lung cancer treatment, in line with observed decreases in lower limb muscle function and daily physical activity levels [57]. These changes may be associated with the observed long-term increase in symptoms of fatigue and dyspnea and impairment of health-related quality of life in these patients.

Exercise recommendations

Collaboration between treating medical oncologists, lung cancer nurses, and exercise physiologists/physiotherapists is important when developing an individualized exercise program to meet the physical and psychological needs of the individual diagnosed with lung cancer. The overall objective of the exercise program should aim to improve physical fitness and quality of life, mitigate treatment-related side effects, and promote long-term exercise adherence.

In 2010, The American College of Sports Medicine developed the first set of exercise guidelines for cancer survivors [9]. These guidelines were based on evidence drawn predominantly from breast and prostate cancer which found exercise to be safe during and after cancer treatment and was effective at improving physical fitness, function, health-related quality of life, and mitigating cancer-related fatigue (CRF). These guidelines encouraged individuals diagnosed with cancer to "avoid inactivity" and to be as physically active as possible: aim to perform at least 150 min/week of aerobic exercise, 2–3 days/week of resistance-based exercise along with daily stretching.

Since the initial ground-breaking publication, the number of rigorously designed studies globally has increased significantly which prompted the call for the 2018 American College of Sports Medicine International Multidisciplinary Roundtable on Physical Activity and Cancer Prevention and Control to present updated recommendations [69].

Due to strong and consistent evidence for the benefit of exercise on cancer-related health outcomes, exercise recommendations using the FITT (frequency, intensity, time, and type) prescription have been developed (Table 4). Although these recommendations are useful, when applying them for client care, exercise prescribers should remember that these recommendations are generalized across cancer types unless specified. Table 5 presents relative and absolute contraindicators for exercise testing and prescription and should be used as a guide by exercise prescribers to keep clients with cancer safe.

Conclusion

Increased levels of exercise and physical activity have been shown to reduce the risk of a cancer diagnosis, decrease risk of secondary cancers, and reduce mortality and morbidity. People diagnosed with cancer can benefit from exercise in many ways before, during, and after cancer treatments. Exercise has been shown to reduce the negative impact of cancer treatments across a myriad of cancers and countless research trials. The evidence within exercise and cancer is rapidly growing, recommendations are currently generalized, a client-centered approach taking into consideration the client's current health status and exercise history is best practice. There is a complexity that needs to be addressed when prescribing exercise for people diagnosed with cancer. Cancer diagnoses and treatments are highly multifaceted and impact multi-functions and organs within the human body, the most appropriate health professionals for safely prescribing exercise for clients diagnosed with cancer are Accredited Exercise Physiologists/Clinical Exercise Physiologists or Physiotherapists/Physical Therapists in the first instance, particularly during treatment. Like anyone beginning an exercise program, it is important to start low and slow and build up gradually over time, so that physical activity and exercise remains sustainable for the rest of the clients' life.

TABLE 4 Cancer-related health outcome with sufficient evidence for the development of FITT prescription.

Outcome	Exercise type	Duration and intensity	Frequency	Length of program	Cancer type
Anxiety	Aerobic	30–60 min, at 60%–80% HR$_{Max}$/VO$_{2max}$	3 sessions per week	12-weeks	Breast, colorectal, gynecological, head, and neck, hematological, and prostate
	Mixed aerobic and resistance	20–40 min, at 60%–80% HR$_{Max}$/VO$_{2max}$ 2 sets, 8–12 reps, at 65%–85% 1 RM	2–3 sessions per week	6–12 weeks	
Depression	Aerobic	30–60 min, at 60%–80% HR$_{Max}$/VO$_{2max}$	3 sessions per week	12-weeks	Breast, colorectal, prostate, hematological
	Mixed aerobic and resistance	20–40 min, at 60%–80% HR$_{Max}$/VO$_{2max}$ 2 sets, 8–12 reps, at 65%–85% 1 RM	2–3 sessions per week	6–12 weeks	
Fatigue	Aerobic	30 min, at 65% HR$_{Max}$/40% VO$_{2max}$	3 sessions per week	12-weeks	Breast, prostate, mixed
	Resistance	2 sets, 12–15 reps at 60% 1 RM	2 sessions per week	12-weeks	
	Mixed	30 min, at 65% HR$_{Max}$/40% VO$_{2max}$ 2 sets, 12–15 reps at 60% 1 RM	2–3 sessions per week	12-weeks	
Health-Related Quality of Life	Aerobic	30 min, at 60%–80% HR$_{Max}$	2–3 sessions per week	12-weeks	Breast, lung, prostate, head and neck, gynecological, bladder, mixed, hematological
	Resistance	2–3 sets, 8–15 reps at 60–75% 1 RM	2 sessions per week	12-weeks	
	Mixed	20–30 min, at 60–80% HR$_{Max}$ 2 sets, 8–15 reps at 60%–80% 1 RM	2–3 sessions per week	12-weeks	
Lymphedema	Resistance	1–3 sets, 8–15 reps at 60%–70% 1 RM	2–3 sessions per week	52-weeks	Breast
Physical function	Aerobic	30–60 min, at 60%–85% HR$_{Max}$/VO$_{2max}$	3 sessions per week	8–12 weeks	Breast, lung, colorectal, prostate, head and neck, hematological, bladder
	Resistance	2 sets, 8–12 reps at 60%–75% 1 RM	2–3 sessions per week	8–12 weeks	
	Mixed	20–40 min, at 60%–85% HR$_{Max}$/VO$_{2max}$ 2 sets, 8–12 reps at 60%–75% 1 RM	2–3 sessions per week	8–12 weeks	

Adapted from [69].

TABLE 5 Relative and absolute contraindications to exercise and testing.

CONTRAINDICATIONS TO EXERCISE & TESTING

ABSOLUTE	RELATIVE
• A RECENT SIGNIFICANT CHANGE IN RESTING ECG SUGGESTING SIGNIFICANT ISCHEMIA, RECENT MI (WITHIN 2 D), OR OTHER ACUTE CARDIAC EVENT; • UNCONTROLLED CARDIAC DYSRHYTHMIAS CAUSING SYMPTOMS OR HEMODYNAMIC COMPROMISE; • UNSTABLE ANGINA; • SYMPTOMATIC SEVERE AORTIC STENOSIS; • UNCONTROLLED SYMPTOMATIC HEART FAILURE; • ACUTE PULMONARY EMBOLUS OR PULMONARY INFARCTION; • ACUTE MYOCARDITIS OR PERICARDITIS; • SUSPECTED OR KNOWN DISSECTING ANEURYSM; • ACUTE SYSTEMIC INFECTION, ACCOMPANIED BY FEVER, BODY ACHES, OR SWOLLEN LYMPH GLANDS; *HIGH INTENSITY RT* **(80-100% 1RM) ACTIVE PROLIFERATIVE RETINOPATHY OR MODERATE OR WORSE NONPROLIFERATIVE DIABETIC RETINOPATHY.** CANCER SPECIFIC • BONE PAIN; • *DURING IV TREATMENT:* - FEBRIL ILLNESS: 38°C – EVEN IF FEELING WELL; - RESTING SBP < 85 MMHG; - HEMOGLOBIN (HB): < 8.0G/DL OR <80.0 G/L (WITHOUT CVD) < 9.0G/DL OR 90G/L (WITH CVD) - PLATELETS: < 20,000 μL - ABSOLUTE NEUTROPHILS: < 0.5 × 109/L	• Left main coronary stenosis; • Moderate stenotic valvular heart disease; • Electrolyte abnormalities (e.g., hypokalemia or hypomagnesemia); • Resting Hypertension: *Aerobic:* SBP≥ 200 mmHg and/or DBP to ≥ 110 mmHg); *Resistance:* ≥ 160/100 mmHg • Tachydysrhythmia (>100bpm) or bradydysrhythmia (<50); • Hypertrophic cardiomyopathy and other forms of outflow tract obstruction; • Neuromotor, musculoskeletal, or rheumatoid disorders that are exacerbated by exercise; • High-degree AV block; • Ventricular aneurysm; • Uncontrolled metabolic disease (e.g., diabetes, thyrotoxicosis or myxedema); • Chronic infectious disease (e.g., HIV). CANCER SPECIFIC • *During IV treatment:* - Vomiting or diarrhea within 24–36 hours; - Platelets: 20,000 – 50,000 μL (low impact, light intensity). *Relative contraindications can be suspended if exercise benefits outweigh risk. Proceed with caution.*

Adapted from the American College of Sports Medicine [23] with included considerations for those diagnosed with cancer.

References

[1] Hill CL, et al. Frequency of specific cancer types in dermatomyositis and polymyositis: a population-based study. Lancet 2001;357(9250):96–100.
[2] Kamangar F, Dores GM, Anderson WF. Patterns of cancer incidence, mortality, and prevalence across five continents: defining priorities to reduce cancer disparities in different geographic regions of the world. J Clin Oncol 2006;24(14):2137–50.
[3] World Health Organization. International agency for research on cancer, 1965-2021. [cited 2021 6.5.2021]; Available from https://www.iarc.who.int/
[4] Torre LA, et al. Global cancer incidence and mortality rates and trends—an update. Cancer Epidemiol Prev Biomark 2016;25(1):16–27.
[5] Vogelstein B, et al. Cancer genome landscapes. Science 2013;339(6127):1546–58.
[6] Garraway LA, Lander ES. Lessons from the cancer genome. Cell 2013;153(1):17–37.
[7] Mucci LA, et al. Familial risk and heritability of cancer among twins in Nordic countries. JAMA 2016;315(1):68–76.
[8] Stadler ZK, et al. Genome-wide association studies of cancer. J Clin Oncol 2010;28(27):4255.

[9] Schmitz KH, Courneya KS, Matthews C. American College of Sports Medicine roundtable on exercise guidelines for cancer survivors. J ACSM 2010;1409–26.

[10] Hayes SC, et al. The Exercise and Sports Science Australia position statement: exercise medicine in cancer management. J Sci Med Sport 2019;22(11):1175–99.

[11] Cormie P, et al. Clinical Oncology Society of Australia position statement on exercise in cancer care. Med J Aust 2018;209(4):184–7.

[12] Fuller JT, et al. Therapeutic effects of aerobic and resistance exercises for cancer survivors: a systematic review of meta-analyses of clinical trials. Br J Sports Med 2018;52(20):1311.

[13] Crawford-Williams F, et al. Interventions for prostate cancer survivorship: a systematic review of reviews. Psychooncology 2018;27(10):2339–48.

[14] Hayes SC, et al. Exercise for health: a randomized, controlled trial evaluating the impact of a pragmatic, translational exercise intervention on the quality of life, function and treatment-related side effects following breast cancer. Breast Cancer Res Treat 2013;137(1):175–86.

[15] Devin JL, et al. Cardiorespiratory fitness and body composition responses to different intensities and frequencies of exercise training in colorectal cancer survivors. Clin Colorectal Cancer 2018;17(2):e269–79.

[16] Edvardsen E, et al. High-intensity training following lung cancer surgery: a randomised controlled trial. Thorax 2015;70(3):244–50.

[17] Bloomquist K, et al. Heavy-load lifting: acute response in breast cancer survivors at risk for lymphedema. Med Sci Sports Exerc 2018;50(2):187.

[18] Cormie P, et al. Is it safe and efficacious for women with lymphedema secondary to breast cancer to lift heavy weights during exercise: a randomised controlled trial. J Cancer Surviv 2013;7(3):413–24.

[19] Winters-Stone KM, et al. Skeletal response to resistance and impact training in prostate cancer survivors. Med Sci Sports Exerc 2014;46(8):1482.

[20] Duregon F, et al. Effects of exercise on cancer patients suffering chemotherapy-induced peripheral neuropathy undergoing treatment: a systematic review. Crit Rev Oncol Hematol 2018;121:90–100.

[21] Solheim TS, et al. A randomized phase II feasibility trial of a multimodal intervention for the management of cachexia in lung and pancreatic cancer. J Cachexia Sarcopenia Muscle 2017;8(5):778–88.

[22] Norton K, Norton L. Pre-exercise screening. Exercise and Sports Science Australia and Fitness Australia and Sports Medicine Australia; 2011.

[23] American College of Sports Medicine. ACSM's guidelines for exercise testing and prescription. Lippincott Williams & Wilkins; 2013.

[24] Preusser M, et al. Recent advances in the biology and treatment of brain metastases of non-small cell lung cancer: summary of a multidisciplinary roundtable discussion. ESMO Open 2018;3(1), e000262.

[25] Lemjabbar-Alaoui H, et al. Lung cancer: biology and treatment options. Biochim Biophys Acta 2015;1856(2):189–210.

[26] Sung H, et al. Global cancer statistics 2020: GLOBOCAN estimates of incidence and mortality worldwide for 36 cancers in 185 countries. CA Cancer J Clin 2021;71:209–49.

[27] Yang S, Zhang Z, Wang Q. Emerging therapies for small cell lung cancer. J Hematol Oncol 2019;12(1):1–11.

[28] Rudin CM, et al. Treatment of small-cell lung cancer: American Society of Clinical Oncology endorsement of the American College of Chest Physicians guideline. J Clin Oncol 2015;33(34):4106.

[29] Avancini A, et al. Physical activity and exercise in lung cancer care: will promises be fulfilled? Oncologist 2020;25(3):e555–69.

[30] Goldstraw P, et al. The IASLC lung cancer staging project: proposals for revision of the TNM stage groupings in the forthcoming (eighth) edition of the TNM classification for lung cancer. J Thorac Oncol 2016;11(1):39–51.

[31] Sriram N, et al. Attitudes and stereotypes in lung cancer versus breast cancer. PLoS One 2015;10(12), e0145715.

[32] Chapple A, Ziebland S, McPherson A. Stigma, shame, and blame experienced by patients with lung cancer: qualitative study. BMJ 2004;328 (7454):1470.

[33] Villeneuve PJ, Mao Y. Lifetime probability of developing lung cancer, by smoking status, Canada. Can J Public Health 1994;**85**(6):385–8.

[34] de Groot PM, et al. The epidemiology of lung cancer. Transl Lung Cancer Res 2018;7(3):220–33.

[35] Gegechkori N, Haines L, Lin JJ. Long-term and latent side effects of specific cancer types. Med Clin North Am 2017;101(6):1053–73.

[36] Cheville AL, et al. Fatigue, dyspnea, and cough comprise a persistent symptom cluster up to five years after diagnosis with lung cancer. J Pain Symptom Manage 2011;42(2):202–12.

[37] Lang-Lazdunski L. Surgery for nonsmall cell lung cancer. Eur Respir Rev 2013;22(129):382–404.

[38] Koletsis EN, et al. Current role of surgery in small cell lung carcinoma. J Cardiothorac Surg 2009;4(1):30.

[39] Xu L, et al. Surgery for small cell lung cancer: a Surveillance, Epidemiology, and End Results (SEER) Survey from 2010 to 2015. Medicine 2019;**98** (40).

[40] Siripurapu V, et al. Pneumonectomy for cancer in the mid-atlantic US: trends and outcomes over a decade. Chest 2010;138(4):659A.

[41] Salazar MC, et al. The survival advantage of lobectomy over wedge resection lessens as health-related life expectancy decreases. JTO Clin Res Rep 2021;2(3).

[42] Wildgaard K, et al. Consequences of persistent pain after lung cancer surgery: a nationwide questionnaire study. Acta Anaesthesiol Scand 2011;55(1):60–8.

[43] Huang X, Zhou W, Zhang Y. Features of fatigue in patients with early-stage non-small cell lung cancer. J Res Med Sci 2015;20(3):1–8.

[44] Cheville AL, et al. The value of a symptom cluster of fatigue, dyspnea, and cough in predicting clinical outcomes in lung cancer survivors. J Pain Symptom Manage 2011;42(2):213–21.

[45] Cavalheri V, et al. Impairments after curative intent treatment for non-small cell lung cancer: a comparison with age and gender-matched healthy controls. Respir Med 2015;109(10):1332–9.

[46] Muthu V, et al. Adverse effects observed in lung cancer patients undergoing first-line chemotherapy and effectiveness of supportive care drugs in a resource-limited setting. Lung India 2019;36(1):32–7.

[47] Glatzer M, et al. The role of radiation therapy in the management of small cell lung cancer. Breathe 2017;13(4):e87–94.

[48] Dilalla V, et al. Radiotherapy side effects: integrating a survivorship clinical lens to better serve patients. Curr Oncol 2020;27(2):107–12.

[49] Ding N-H, Li JJ, Sun L-Q. Molecular mechanisms and treatment of radiation-induced lung fibrosis. Curr Drug Targets 2013;14(11):1347–56.

[50] Granger CL, et al. Low physical activity levels and functional decline in individuals with lung cancer. Lung Cancer 2014;83(2):292–9.

[51] Eichler M, et al. Psychological distress in lung cancer survivors at least 1 year after diagnosis—results of a German multicenter cross-sectional study. Psychooncology 2018;27(8):2002–8.

[52] Brunelli A, et al. ERS/ESTS clinical guidelines on fitness for radical therapy in lung cancer patients (surgery and chemo-radiotherapy). Eur Respir J 2009;34(1):17–41.

[53] Jones LW, et al. Exercise intolerance in cancer and the role of exercise therapy to reverse dysfunction. Lancet Oncol 2009;10(6):598–605.

[54] Shneerson C, et al. Patterns of self-management practices undertaken by cancer survivors: variations in demographic factors. Eur J Cancer Care 2015;24(5):683–94.

[55] Cavalheri V, Granger CL. Exercise training as part of lung cancer therapy. Respirology 2020;25(S2):80–7.

[56] Henshall CL, Allin L, Aveyard H. A systematic review and narrative synthesis to explore the effectiveness of exercise-based interventions in improving fatigue, dyspnea, and depression in lung cancer survivors. Cancer Nurs 2019;42(4):295–306.

[57] Granger CL, et al. Understanding factors influencing physical activity and exercise in lung cancer: a systematic review. Support Care Cancer 2017;25(3):983–99.

[58] Rivera MP, et al. Impact of preoperative chemotherapy on pulmonary function tests in resectable early-stage non-small cell lung cancer. Chest 2009;135(6):1588–95.

[59] Lopez Guerra JL, et al. Changes in pulmonary function after three-dimensional conformal radiotherapy, intensity-modulated radiotherapy, or proton beam therapy for non-small-cell lung cancer. Int J Radiat Oncol Biol Phys 2012;83(4):e537–43.

[60] Linhas A, et al. P1.04-027 changes in pulmonary function in lung cancer patients after thoracic surgery: topic: pulmonology. J Thorac Oncol 2017;12(1, Supplement):S611–2.

[61] Salcedo PA, et al. Effects of exercise training on pulmonary function in adults with chronic lung disease: a meta-analysis of randomized controlled trials. Arch Phys Med Rehabil 2018;99(12):2561–2569.e7.

[62] Licker M, et al. Preoperative evaluation of lung cancer patients. Curr Anesthesiol Rep 2014;4(2):124–34.

[63] Jones LW, et al. Prognostic significance of functional capacity and exercise behavior in patients with metastatic non-small cell lung cancer. Lung Cancer 2012;76(2):248–52.

[64] Lindenmann J, et al. Preoperative peak oxygen consumption: a predictor of survival in resected lung cancer. Cancer 2020;12(4):836.

[65] Lee J. Cardiorespiratory fitness, physical activity, walking speed, lack of participation in leisure activities, and lung cancer mortality: a systematic review and meta-analysis of prospective cohort studies. Cancer Nurs 2020;44:453–64.

[66] Kasymjanova G, et al. Prognostic value of the six-minute walk in advanced non-small cell lung cancer. J Thorac Oncol 2009;4(5):602–7.

[67] Srdic D, et al. Cancer cachexia, sarcopenia and biochemical markers in patients with advanced non-small cell lung cancer-chemotherapy toxicity and prognostic value. Support Care Cancer 2016;24(11):4495–502.

[68] Cai Y, et al. Linear association between grip strength and all-cause mortality among the elderly: results from the SHARE study. Aging Clin Exp Res 2020;33:933–41.

[69] Campbell KL, et al. Exercise guidelines for cancer survivors: consensus statement from international multidisciplinary roundtable. Med Sci Sports Exerc 2019;51(11):2375–90.

Chapter 26

Exercise in the management of motor neuron disease

Amy Lawton[a] and Maja Husaric[b]

[a]College of Health & Biomedicine, Victoria University, Melbourne, VIC, Australia [b]First Year College, Victoria University, Melbourne, VIC, Australia

Introduction

Motor neuron disease (MND) is a group of neurodegenerative disorders that includes amyotrophic lateral sclerosis (ALS), primary lateral sclerosis, progressive muscular atrophy, and progressive bulbar palsy [1]. These disorders are progressive and debilitating, with the associated neurodegeneration affecting speech, swallowing, breathing, and limb function [2]. MND, particularly its most common form ALS, is progressive and debilitating and does not currently have a cure, therefore management focuses on symptomatic relief, optimizing quality of life (QoL), and palliative care. This is best achieved through dedicated MND multidisciplinary clinics where specific, individualized care can be tailored to the complex needs present [3,4].

Epidemiology

Epidemiological data collection is challenging due to the low incidence rates of MND, however population-based registers, particularly in Europe and Australia have assisted in increasing our understanding. Collecting epidemiological data is a challenge due to the spectrum across which the different forms of MND exist. For example, there are pure lower motor neuron, mixed upper, and low motor neuron (e.g., ALS), and pure upper motor neuron variants, as well as regional variants that are restricted to arms, legs, or bulbar regions.

Economic burden

The economic impact of MND upon patients, their caregivers, and society is vast. In 2015, the total cost of MND in Australia was estimated at $2.37 billion, comprising $430.9 million in economic costs and $1.94 billion in the burden of disease costs. This equates to a $1.13 million total per person [5]. These costs can start accumulating before the diagnosis has been made and then continue to increase with advancing disability requiring medical equipment and care. Estimates of basic patient equipment costs can be greater than USD $40,000, including hospital beds, electric wheelchair, and augmentative communication equipment, while mechanical ventilation costs roughly USD $200,000 a year [6]. A large proportion of the total cost is accounted for in the final 30 days of life, particularly in rapidly progressive conditions such as MND [7].

Within the Australian healthcare system, unpaid carers, such as family and friends, often provide substantial care and the burden of additional costs can fall onto them, affecting both the patients' and their QoL. MND patients' in the community can require help 2–3 times a day or require the presence of a carer most of the time which can account for 9.5 h a day of informal care, even when there is also paid assistance [5]. In neurodegenerative conditions where the disability is less significant, these costs can make up to 43% of the total cost, therefore in MND, it is likely to be similar or higher.

Incidence and prevalence

MND is a relatively rare disease. The incidence of MND is reported to be between 0.6 and 8.4/100,000 with a higher incidence in males (60%) and Caucasian populations. Peak incidence occurs between 50 and 75 years of age and increases in incidence and mortality rates may be due to longer life expectancy [5,8,9].

Pathophysiology

Motor neurons are specialized nerve cells in the brain and spinal cord, which transmit the electrical signals to muscles to generate both conscious and automatic movements, such as swallowing and breathing. To simplify the explanation of this process we can imagine that the movement has two stages and involves two sets of neurons: the upper motor neuron located in the brain and lower motor neuron located in spine at the level where the peripheral nerve is formed.

The upper motor neuron starts in the motor cortex located in the precentral gyrus of the frontal lobe. These neurons initiate predominantly voluntary movement and their impulse travels down to the spinal cord to connect with lower motor neurons at each level of the spinal cord. The lower motor neuron cells travel out of the spinal cord and make connections with the skeletal muscle. If motor neurons of either, lower or upper motor neurons are damaged, it will affect both voluntary and automatic movement. The muscles that are supplied by the damaged nerve, may receive altered or decreased impulses which result in altered movement patterns. The muscle itself will change in morphological appearance, will become thinner (atrophic), and lose its tone [10].

There is a pattern to the change of muscle movement depending on if the damage occurs within the upper or lower motor neuron areas. Damage to lower motor neurons causes muscle weakness as the consequence of direct loss of the transmission between muscle and nerve. Although the main function of upper motor neurons is to initiate movement, these impulses from the motor cortex are influenced by other neurons that moderate muscle contraction. If the upper motor neurons are damaged, the consequence is that limbs become stiffer (spastic) as the connection between muscle and lover motor neuron remains intact however convey "moderating" impulses from the brain with decreased descending inhibition [11].

Pathological findings in MND are diverse and are associated with the clinical form of the disease. The most common form, ALS, is characterized by degeneration of the large α-motor neurons in the brainstem and spinal cord leading to progressive weakness and muscle atrophy, while loss of upper motor neurons results in spasticity and hyperreflexia. Motor neurons are the earliest and most prominently affected groups of cells, but degeneration also occurs in small interneurons in the spinal cord and cortical motor cortex cells. Small interneurons play essential roles in the organization of brain activity in particular within pyramidal systems [12,13]. The bulk loss of neurons from the multiple locations affects the motor corticospinal tracts and the sensory, spinocerebellar pathways causing neurophysiological changes in their associated tracks. As a consequence of these pathological changes, a patient with MND presents with malfunction of multiple systems, hence MND is pathologically and clinically regarded as a multisystem disease [14,15].

It is unclear why selective neurons degenerate and die; however, some current animal modeling proposes three key pathogenic processes that are associated with MND and neuron injury: (i) genetic factors, (ii) oxidative stress [16], and (iii) glutamatergic toxicity. These processes result in damage to critical target proteins such as neurofilaments and organelles such as mitochondria [17–20]. Currently, how alterations in these processes contribute to the disease remains poorly understood.

Genetic factors

In a small portion (around 10%) of patients with ALS, their genetics play a major role in pathogenesis of the disease, hence commonly denoted as the familial ALS (fALS). In this group, there is a clear link to a family history of disease. Recent research also suggests that in the remaining 90% of ALS patients, where the familial history is not apparent, genetic factors may also play significant role in disease risk and development [21,22]. The majority of ALS genes are inherited in a dominant fashion with variable and age-dependent penetrance. Significant intrafamilial and interfamilial variability in age of onset and disease progression is observed. Genes that may cause recessive disease include OPTN, SPG11, FUS, and SOD1 (specifically, the homozygous Asp90Ala mutation), among others; the UBQLN2 gene is X-linked dominant. [23]. The genes considered to be of particular interest are a group of genes that are found consistently being linked to ALS phenotypes (Table 1).

Oxidative stress

Mitochondria show early changes in ALS pathogenesis and contribute to the disease progression. Mitochondrial morphological and functional changes have been found in both human patients and SOD1 mutant mice. Mitochondrial axonal transport, fission, fusion, and mitophagy clearance may also be altered in mutant SOD1 [26]. In 20% of cases, the mutation of the SOD1 gene is associated with ALS, however, some animal modeling demonstrates coexistence of the mutation of the SOD1 gene and the irregularity of glial cells [27]. Changes in mitochondrial structure, plasticity, replication/copy number, mitochondrial DNA instability, and altered membrane potential have been demonstrated in several subsets of MNDs in both animal models and human studies. These changes are consistent with increased excitotoxicity, induction of reactive oxygen species, and activation of intrinsic apoptotic pathways [28].

TABLE 1 Genes links to ALS phenotypes.

Gene	Comments
SOD1	The first gene reported and the second most common cause of fALS SOD1-mediated toxicity involves Cu, Zn superoxide dismutase 1 (SOD1), an enzyme that converts superoxide, a toxic by-product of mitochondrial oxidative phosphorylation, to water or hydrogen peroxide [24,25] Clinically not associated with cognitive impairment, and bulbar onset is less common than in other fALS types
C9orf72	The most common cause of fALS A pathogenic hexanucleotide (G4C2) repeat expansion in the first intron of C9orf72
TARDBP	Mutations are dominantly transmitted Its protein product (TDP-43) was found to be a component of the ubiquitin-positive neuronal inclusions characteristic of both ALS and FTD
FUS	The most frequent cause of early-onset ALS (onset at age younger than 35 years)
VCP	Mutations in VCP were first reported in families with inclusion body myopathy, Paget disease, and frontotemporal dementia
MATR3	A missense mutation in MATR3, Ser85Cys, was first reported in two families with autosomal-dominant distal myopathy
CHCHD10	The Ser59Leu mutation in CHCHD10, a nuclear gene encoding a small mitochondrial protein, was first reported in a large family with late-onset motor neuron disease, cognitive decline resembling FTD, cerebellar ataxia, and myopathy

Glutamatergic toxicity

Excitotoxicity is neuronal degeneration caused by over-stimulation of the glutamate receptors. During glutamatergic neurotransmission, glutamate released from the presynaptic neuron activates ionotropic glutamate receptors present on the postsynaptic neuron. Activation of these glutamate receptors results in the influx of Na+ and Ca2+ ions into the cell, leading to depolarization and ultimately to the generation of an action potential. The astrocytic glutamate transporters glutamate–aspartate transporter (GLAST) and glutamate transporter-1 (GLT-1) and their human homologs excitatory amino acid transporter 1 (EAAT1) and 2 (EAAT2), respectively, are the major transporters that take up synaptic glutamate to maintain optimal extracellular glutamic levels, thus preventing accumulation in the synaptic cleft and ensuing excitotoxicity [27]. Experimental evidence became available that excitotoxicity could contribute to neuronal damage in stroke, neurotrauma, epilepsy, and a number of neurodegenerative disorders including amyotrophic lateral ALS but clear role between excessive levels of glutamate and neuronal death has not been established. The therapeutic drug rizoline, which affects levels [29,30] of glutamate, can improve patient condition, indicating that glutamate plays an active role in neuronal death. Recent evidence suggests that the toxicity of the SOD1 is involved to some degree in the pathogenesis of ALS [30].

MND phenotypes

The main phenotypes can be classified in four broad ways, differing based on their relative degree of UMN and LMN predominance and site of disease onset:

1. Amyotrophic lateral sclerosis (ALS) is the most common, representing approximately 70% of all MND cases [31].
2. Progressive bulbar palsy patients have swallowing, and speech dysfunction localized for more than 6 months with relative preservation of limb strength [32].
3. Progressive muscular atrophy presents with pure LMN signs [33].
4. Primary lateral sclerosis presents with pure UMN signs, exhibiting a slower disease trajectory [34].

Clinical presentation and diagnosis

Patients with motor neuron diseases may present to a primary care clinic or may be initially encountered in the inpatient setting. Timely diagnosis of these conditions is a key factor in early intervention and therapy, and accuracy of diagnosis is of extreme importance, in particular for amyotrophic lateral sclerosis with its poor prognosis.

MND is a diagnosis of exclusion. Patients may present to the general practitioner (GP) with symptoms such as asymmetric limb weakness, dysphonia, or dysphagia. The diagnosis can often be delayed, due to presumption of a peripheral nervous system disorder such as foot drop secondary to lumbar disc disease [3]. This can lead to initial referral to specialists such as a neurosurgeon or ear, nose, and throat (ENT) surgeon, preceding delayed referral to a neurologist. As a result, the mean age at which MND is diagnosed is approximately 64 years, 2 years after the mean age of symptom onset at 62 years.

Diagnosis is made on the basis of clinical findings, supplemented by targeted investigations, including electromyography and nerve conduction studies. A hallmark of ALS is the presence of both upper and lower motor neuron signs, including limb weakness, hyperreflexia, and fasciculations. Frontotemporal dementia may be present [35]. Evidence of denervation and re-innervation, involving a number of muscles, may be seen in electromyographic studies [36].

Although MND primarily affects the human motor system, cumulative evidence suggests that it also affects extramotor regions and is representative of a multi-system disease that extends beyond anterior horn cell and corticospinal tract degeneration. This includes dysfunction in cognitive, autonomic, sensory, cerebellar, and basal ganglia structures [37]. Hence, clinicians should consider a wide range of symptoms when considering MND as a diagnosis.

When approaching the patient with suspected motor neuron disease (MND), the pattern of weakness on exam helps distinguish MND from other diseases of peripheral nerves, the neuromuscular junction, or muscle. Neuropathic disorders can be broadly divided into disorders affecting the peripheral nerve processes (neuropathy) or nerve cell body (neuronopathy), can be inherited or acquired, and have different clinical courses. There are a number of key questions that can help in the clinical reasoning process [38]:

1. What parts of the nervous system are involved: motor, sensory, autonomic, or combinations of more than one system?
2. Where is the weakness (proximal, distal, or both) and is it symmetric or asymmetric?
3. If there is sensory involvement is there pain or proprioceptive loss?
4. Over what timeframe did symptoms evolve: acute (<4 weeks), subacute (4–8 weeks), or chronic?
5. Is there a family history of a similar disorder?
6. If there is motor involvement, is it upper motor neuron, lower motor neuron, or both?

The International Classification of Functioning, Health, and Disability (ICF), defines a common language for describing the impact of disease at different levels: impairment (body structure and function), limitation in activity, and participation. Within this framework, MND related impairments (weakness, spasticity), can limit "activity" or function (decreased mobility, self-care, pain) and "participation" (driving, employment, family, social reintegration) [39].

Current medical treatments

Neuroprotection and symptomatic treatments remain the cornerstone of management for MND patients (Table 2). This is typically undertaken in a multidisciplinary setting which provides a central point of services and results in improved survival when compared to standard models of outpatient care [41].

Based on the complex nature of MND, a multidisciplinary, individualized approach is recommended, based on both current and anticipated needs (Fig. 1). At diagnosis, a discussion with the patient about how much they would like to know about the condition and the involvement of family members/carers should be followed by a session of education and forward planning [43]. Multidisciplinary care should consist of medical and nursing care, as well as allied health modalities including speech therapy, physiotherapy, occupational therapy, orthotics, psychology, and nutrition, with the addition of palliative care in later stages of the disease [40–42]. This type of care may be especially important in those with bulbar disease, in order to extend survival. Multidisciplinary care will allow for ongoing cognitive assessment, as well as respiratory, medication, and liver function testing (LFT) monitoring [40,43]. Multidisciplinary care can also allow for ongoing assessment and management of the psychological impact of MND on both the patient and their caregivers [41].

Prognosis

MNDs are usually insidious in nature of onset, followed by chronic progression. There is currently no cure for these conditions. Death is usually the result of respiratory failure at an average of 2–4 years following onset. In some cases, survival can be extended to a decade or more. Biomarker models which can predict disease progression are being developed and can help to optimize the introduction of palliative care [44].

Exercise and MND

Exercise is an effective and safe intervention for many diseases, including neuromuscular disorders. Exercise has the ability to affect many organs systems, therefore its positive effects are multifaceted and widespread (Table 3) [45].

TABLE 2 Common management strategies in MND.

Common issues	Common management strategies
Medication monitoring	Regular liver function tests (LFT's)—particularly for Riluzole (disease-modifying, neuroprotective drug)
Pain	Pharmacological agents: paracetamol, NSAIDs, opioids, gabapentanoids, intra-articular steroid injections Manual therapy TENS
Spasticity	Pharmacological agents: Muscle relaxants (with LFT monitoring) Gabapentanoids Botulinum toxin injections
Respiratory dysfunction	Monitor with respiratory function tests Ventilation (non-invasive, progressing to invasive) Pharmacology: benzodiazepines, morphine
Saliva and cough effectiveness	Assisted cough techniques or devices Oral hygiene Dehumidifier, nebulizer, or suction devices Medications
Bulbar dysfunction	Speech pathology Modified diet for dysphagia and aspiration Voice amplification systems
Constipation	Maintain hydration Pharmacology- prescription of assistive agents, cessation of drying agents
Balance, strength, physical function	Exercise
Weight loss	Early involvement of a dietician Treatment of symptoms, e.g., gastroesophageal reflux, nausea Nutritional supplements
Psychological/psychiatric	Diagnostic, e.g., assessment for dementia Therapeutic, e.g., pharmacological agents and psychological intervention for mood disorders
Orthotic management	Assessment and prescription of appropriate orthoses, e.g., cervical, wrist-hand, or ankle
Community and social support	MND support groups Community groups for social interaction

Adapted from [40].

For patients with MND, it is important to maintain flexibility, strength, balance, and aerobic capacity through exercise [46]. Prevalent MND symptoms such as fatigue and weakness are associated with reduced exercise tolerance, therefore a tailored exercise rehabilitation program is necessary, based on the individuals' ability and stage of disease [47].

Currently, there is a lack of robust evidence regarding the impact of exercise on MND. Confounding factors such as small sample sizes, heterogenicity of the disease stage in participants, including control groups, and comorbidities all mean that results should be interpreted carefully [48].

Known and potential mechanisms underlying effects of exercise

MND is a difficult disease to study. This is due to its rapidly progressive nature, with no cure and mean mortality at 2.5 years following diagnosis [41]. Recruiting sufficient numbers of participants for randomized control trials is challenging, made harder by the varying symptom profile and rate of disease progression between people. Other challenging factors include a high dropout rate, due to symptom progression or comorbid health concerns, and limitations surrounding caregivers and their ability to provide support or assistance [41,48,49].

FIG. 1 MND multidisciplinary care model. *Adapted from [42].*

TABLE 3 World Health Organization health benefits of exercise.

Health benefits of exercise—World Health Organization (2020)
• Improved muscular and cardiorespiratory fitness
• Improved bone health
• Reduced risk of hypertension, cardiovascular disease, and some cancers
• Reduced risk of depression and reduced symptoms of anxiety and depression
• Reduced risk of fall and fracture
• Help maintain a healthy body weight
• Risk of all-cause mortality
• Reduced risk of type-2 diabetes and improved glycemic control in those with DM2
• Improved cognitive health
• Improved sleep

Animal studies

Mitochondria play an important role in regulatory pathways of cellular function and survival, including in MND. Studies in MND rodent models show that mitochondrial abnormalities appear prior to the clinical onset of the disease, which may suggest a causal role [28]. In a rodent model of respiratory chain complex 1 deficiency, it was shown that combining aerobic and resistance exercise activated mitochondrial biogenesis and anabolism pathways, improved respiratory chain complex activity and redox status in muscle tissue, and led to increased aerobic fitness and muscle strength [50].

Animal studies for MND, performed in SOD1 transgenic mice, have given us some insight into exercise for MND. Several studies have shown that moderate-intensity endurance exercise slowed disease progression, however high levels of endurance training may have detrimental effects [41]. Swimming-based exercise may be the most effective type of exercise and could lead to improved metabolism and survival, delay disease progression including muscular hypotrophy and have a neuroprotective response [48]. This indicates that mode, duration, and intensity and are likely important factors in developing an appropriate exercise regime for MND.

Human studies

Protein aggregation, glutamate, neuronal excitotoxicity, and oxidative stress have been implicated in the development of the symptoms of MND [40]. Regular exercise has the ability to lower oxidative stress and increase the anti-oxidative capacity of muscle fibers [48,51]. Endurance exercise increases skeletal muscle mitochondrial content and neurogenesis and has a neuroprotective effect against ischemic neuronal damage on the hippocampal region of the brain [48].

Fatigue

Fatigue constitutes two forms in MND; whole-body tiredness and reversible use-dependent muscle fatigue, both of which are partially relieved by rest [52]. Muscle fatigue results in the reduction in the force-production capacity of the muscle, and although this naturally occurs with aging, some potential mechanisms in MND and their relationship to exercise are discussed [53,54]. The perception of fatigue is the result of feedback from both the muscular system and the cardiorespiratory system regarding the presence of metabolites and substrates. In MND there are several suggested causes of muscle fatigue, including neurodegeneration, neuromuscular junction disassembly, and muscle denervation, and mitochondrial dysfunction [54].

Investigation of sensory system disturbance, rather than a motor system problem, as a cause of fatigue, is less common. Dysfunction of the proprioceptive system in mice with ALS and SMA has been demonstrated and this degeneration may occur prior to the development of neurological symptoms, which could prove clinically important as proprioceptive loss can directly affect α-motoneurons function [55,56]. Mechanoreceptors are the primary receptors of both muscle fatigue and proprioception, and the association between proprioceptive dysfunction and muscle fatigue has been demonstrated in healthy subjects [57]. The proprioceptive systems' functions include preventing excessive joint movement via reflex mechanisms, joint stabilization during static posture, and precise coordinated movement. Improving muscle spindle activity through exercise may help to normalize firing rates of group 1a muscle afferents, pre-synaptic inhibition, and the firing rates of motor neurons, leading to a decrease in fatigue [54]. More specifically, mechanoreceptor sensitivity and function can be improved through proprioceptive training and has been demonstrated in numerous neurological conditions to improve balance, pain, motor learning, and gait, although in MND the evidence is conflicting [58]. As fatigue may be related to common complaints in MND including nighttime muscle cramps, nocturia, psychological distress, social withdrawal and reduced QoL, improvements in fatigue may see concurrent improvements in these areas [53,54]. Further research is needed to investigate this.

Weakness

Progressive degeneration of motor neurons is considered to be the primary cause of muscle weakness and paralysis experienced in MND. This includes the respiratory muscles which result in decreased maximum ventilation secondary to decreased tidal volume [51]. Morphological and functional changes in skeletal muscle cells could actually precede motor neuron degeneration, suggesting that intrinsic factors may also contribute to muscle atrophy. These factors may include decreased levels of insulin-like growth factor 1 (IFG-1), with associated down-regulation of protein kinase B (Akt). Given this, exercise targeting muscular activity may prove to be beneficial through inhibition of oxidative stress [59]. Neural adaptations that occur through resistance training include cross transference, increased motor unit activation, and synchronization. Muscular adaptations that can result include muscle fiber hypertrophy, increased protein synthesis, and increased capillary density [42]. Exercise can also produce myofiber remodeling via signaling pathways that modify skeletal muscle metabolism, which may help prevent development of painful contractures [48,60]. An increase in peak ventilation seen following exercise training is likely due to an associated increase in tidal volume and respiratory frequency [51].

Clinical outcomes

Evidence of the clinical outcomes of exercise in MND is often contradictory, due to the aforementioned issues with performing robust clinical trials. Overall, exercise is suggested to result in some improvement of functional capacity, including motor control and gait, likely improvements in muscle weakness, and reduced risk of falls. It has however questionable effect on QoL and no effect on mortality.

Exercise leads to improved scores on disease-specific outcome measures, especially the functional scale of the Amyotrophic Lateral Sclerosis Functional Rating Scale (ALSFRS-R) which can be used to evaluate the functional status of ALS patients over time [41,51]. This improvement in score coincides with decreased motor deterioration, 6MWT, balance, improved aerobic fitness, and VO_2 peak, strength, and fatigue [48,58,59]. Exercise may be able to assist nociceptive pain in MND, dependent on the contributing factors. For example, shoulder pain is common due to weakness, wasting, and contractures of the shoulder girdle musculature, and this may be assisted through stretching, range of motion, and strength and conditioning exercises [40,60].

Studies show that psychological distress and QoL are not directly related to physical function or impairment. This may explain the limited impact that exercise has on psychological state and QoL ratings in MND patients. Exercise has not been found to reduce overall disease progress or impact lifespan [48,59,60]. Exercise programs can help to build capacity in all the body systems and could be especially important before the onset of significant symptoms to maintain general health and conditioning. Clinical outcomes outside of symptoms directly associated with MND can also be considered, although further research is needed to investigate these further. As reducing physical activity leads to cardiovascular deconditioning, muscular atrophy, and weakness, as well as reduced strength of connective tissue structures such as tendons, ligaments, and bone, exercise may help to prevent decline in these areas [41,51].

Looking at the effect of exercise on the broader category of neuromuscular disorders, exercise is found to be effective, leading to improvements in aerobic capacity, power, gait/ambulance, sit to stand and rise from supine, fatigue, and muscle fiber area [61]. This has been demonstrated in chronic neurological disorders including Parkinson's disease and multiple sclerosis. In Huntington's disease, it has even been demonstrated that exercise may be superior to pharmacological treatment in improving QoL [62]. New treatments for MND will hopefully provide a longer prognosis, which will increase the importance of interventions such as exercise, that have the potential to maintain functional independence and improve QoL throughout the stages of the disease [3].

Monitoring the effects of exercise in MND

Monitoring the effects of exercise on patients with MND is important. Both the efficacy and tolerance of the prescribed management plan should be assessed. Clinically, areas most relevant to exercise include; strength, functional ability, e.g., gait, bulbar functions, and psychosocial aspects [63]. Strength of respiratory muscles should be measured via spirometry readings of forced vital capacity (FVC), whereas limb strength should be measured using hand-held dynamometry (HHD). Ideally, oxidative stress and carbohydrate load should be monitored throughout the exercise program [63,64]. CPET and NIRS are valuable tools for detecting early stages of oxidative deficiency in skeletal muscles [49]. For monitoring of functional capability, the six-minute walk test has been shown to be reliable and valid in multiple neuromuscular disorders, including as a long-term measure of condition progression [65,66].

Patient-reported outcome measures

There are patient-reported outcome measures specific to MND which can be used to monitor progress of the condition and functional capacity. The revised ALS functional rating scale (ALSFRS-R) is a validated 12-item questionnaire that can be used to monitor patients with ALS [67]. It assesses functions such as gross and fine motor skills, bulbar, and breathing functions [63]. The Amyotrophic Lateral Sclerosis Assessment Questionnaire (ALSAQ) is a validated patient-reported measure, available in both a long and short form, as well as multiple languages. The long-form version ALSAQ-40 has five scales exploring physical mobility, activities of daily living and independence, eating and drinking, communication, and emotional reactions. These are rated across a five-point Likert scale from never to cannot do at all [68]. The short form ALSAQ-5 is available for when a brief assessment of ALS impact is warranted. The Fatigue Severity Scale (FSS), the Modified Fatigue Impact Scale (MFIS), and the Neurological Fatigue Index–MND (NFI-MND) can be used to monitor fatigue throughout the disease [53]. The monitoring of general health outcomes and QoL is also important to consider. The PROMIS-10 has shown to be feasible for use in ALS patients, with its score correlating with functional components of the ALSFRS-R, as well as ambulatory status [69]. There is also a range of outcome measures addressing psychological health and QoL that are specific to ALS, in an attempt to separate the physical factors associated with the disease [63]. These should be used as part of the multidisciplinary care team.

Exercise safety

In the broader category of neuromuscular disorders, exercise is considered to be safe, without major adverse effects. [61]. The safety of exercise in MND has been strongly debated due to the concern that a weak muscle will be damaged through overuse, although more recent evidence shows this not to be true [41,60,70]. It should also be considered that a cycle of musculoskeletal deconditioning and loss of the other health benefits of exercise, may compound the existing limitations of MND and be further detrimental to functional independence and QoL [41]. Concerns such as increased fatigue, pain, or muscle cramps have not been shown to occur in stretching/range of motion (ROM), resistance- or endurance-type exercise when prescribed appropriately [71]. Over-exertion should be avoided, particularly with post-exercise fatigue, muscle pain, or soreness as they may indicate over-work-induced muscle damage [70]. Although there is a growing body of evidence to show that exercise is safe in MND, the parameters need to be further studied [41,47,53]. When assessing a patients' ability to exercise, factors such as age, cardio-pulmonary health, medications, baseline degree of activity, and the rate of disease progression should all be considered. In progressive disease particularly, the risk of falls, sprains, and other complications should be strongly considered when formulating an exercise-based management plan and selecting exercise parameters [46].

Falls and transfer aids

Due to their altered gait, balance, and postural control, MND patients are at higher risk of falls and the associated increased burden of disease [64]. Patients and their carers should be educated about fall prevention strategies across the span of the disease. Orthotics and mobility aids are often needed in MND to assist mobility, support, and functional independence. Most commonly prescribed is the ankle-foot orthosis to aid with foot clearance during gait, which is impeded by ankle dorsiflexion weakness or foot drop [64]. Other initial aids may include walking sticks and crutches, progressing to pick-up frames or four-wheel walkers. These are limited by upper-limb weakness which results in the need for a wheelchair for a majority of patients. Transfer aids such as transfer belts, slide boards, and hoists can also assist with maneuvering [60]. Some patients resist the use of mobility aids, increasing their risk of falls. This is possibly due to the psychological impact of decreased independence but can also be associated with the cognitive and behavioral effects of MND [64].

Clinical practice recommendations for exercise prescribers

Due to the aforementioned limitations in performing robust clinical research into exercise for MND, there is a lack of available clinical practice recommendations. Importantly, when interpreting the literature, it should be considered that the same type of exercise might not be suitable for all MND patients, due to the heterogeneity of the disease states and stages and the majority of research being conducted in ALS. Further steps need to be taken to investigate the types and parameters of prescribed exercise to create guidelines that clinicians and patients can use [48].

Planning exercise programs

Overall, it is recommended that a low to moderate, individualized exercise program, combining stretching/range of motion, resistance, and/or endurance elements can be implemented safely for a neuroprotective response (Table 4) [48,60]. Exercises should be prescribed to the individual based on functional capacity and stage of disease [42]. Shorter, frequent

TABLE 4 Potential exercise recommendations in MND.

Type of exercise	Potential benefits	Potential parameters	Practical considerations
Aerobic	Prevention of deconditioning Improvement in mood, sleep, spasticity, and QOL Improved $VO_{2\ max}$/oxygen consumption and decreased oxidative stress	Moderate, submaximal intensity Potential exercises include walking, cycling, hydrotherapy	Consider endurance aspect The ability to talk comfortably during exercise may be used as a guide to intensity
Flexibility	Prevention and management of contractures Decrease in pain and spasticity	Target major joints, particularly shoulders, knees, and ankles	Important to both ambulatory and non-ambulatory patients due to painful nature of contractures
Resistance	Maintain muscle strength	Should be individualized, based on limitations and tolerance, and progressive Consideration of both strength and endurance should be made	Implement early, before muscle atrophy begins May be performed more than once daily

Continued

TABLE 4 Potential exercise recommendations in MND—cont'd

Type of exercise	Potential benefits	Potential parameters	Practical considerations
Proprioceptive	Improvements in balance, gait		May incorporate use of BOSU ball or other equipment
Walking	Gait speed, distance, stride length, on 6MWT ± RPE, FSS	30 min, 3 × weekly for 8 + weeks	May intersperse 5 min of walking with 5 min of rest
Hydrotherapy	Maintain ROM, muscle strength and conditioning, improve mobility, and balance	Prescribed on individual tolerance	Monitor for safety as disease progresses
Mixed–home-based	Increased lower extremity muscle strength Improved balance Reduced self-reported fatigue	Sit to stand postural alignment and core muscle activation Upright functional and endurance training Balance training and rhythmic walking Performed in modules over 10 months	
Mixed-gym-based with CPET monitoring	Improved aerobic fitness and $VO_{2\ peak}$ Improved 6MWT Maintenance of QOL	Three sessions per week for 12 weeks 15 min cycling at 80% GET 25 min strength at 60% RPM, 2 min rest between sets: – 3 × 10 dumbbell bicep curl and lateral arm raise – squat, calf raise – banded chest press and seated row 10 min proprioception on BOSU Pro balance trainer 10 min upper/lower extremity stretching on a Pancafit	

sessions can be utilized to avoid muscular fatigue from prolonged activity when indicated [72]. The role of clinician consists of education of both patient and carer, prescription of exercises with optimal technique, and safe transfers where required [64]. The program should be periodically reviewed as the disease progresses. [60]. At-home exercise has been demonstrated as safe and effective and has the advantage of assisting those with transportation challenges [59,73,74]. Tele-monitoring or the use of a digital system that can monitor vital signs and blood oxygen levels can allow for accurate monitoring, with the ability to integrate the care of community and specialist services [75].

Prediction of the functional deficits that are likely to develop, may allow these areas to be targeted in advance in order to maintain function. To provide a protective effect, exercise programs should be implemented before significant muscle atrophy occurs [41,59]. Proprioceptive training for motor control should be implemented to improve gait, balance, and pain through both peripheral and central mechanisms and to support the rest of the exercise program [54,58]. Hydrotherapy is an option that can be used to maintain ROM, muscle strength, and conditioning, improve general mobility and balance. This should be prescribed based on individual tolerance and monitored for safety as the disease progresses [46]. Engaging in a community-based exercise setting such as a hydrotherapy pool, chair yoga or tai chi can help to promote social interactions [60]. Respiratory muscle training can be used in conjunction with exercise rehabilitation to improve parameters such as cardio-respiratory function and fatigue [53]. For this training, it is recommended that a threshold-inspiratory muscle training (IMT) device, which can be set to different pressure thresholds, is used to allow for individualized programming [76].

Conclusion

MND is a progressive neurological disease, without cure. Diagnosis is often delayed resulting in mean survival rates of 2–4 years. Treatment is currently aimed at symptomatic relief, maintaining QoL, and palliative care. Much of the duty of care falls on informal caregivers such as family and friends therefore maintaining functional independence as long as possible is important for all involved.

Exercise may be an important component of a comprehensive multidisciplinary management plan in MND. It is recommended that exercise programs are individually tailored to the needs and disease stage of the patient and are started as early as possible, before significant muscular atrophy or deconditioning occurs. Further research is needed however this is impeded by the progressive nature of the disease and limited available sample sizes. In the future clinical practice guidelines that include recommendations on type and dosage of exercise will help to tailor individual programs more effectively with the aim of maintaining function and QoL.

References

[1] MND, A. What is MND? [cited 2021 05.03.2021]; Available from. https://www.mndaust.asn.au/Get-informed/What-is-MND.aspx, 2021.

[2] Huynh W, et al. Motor neuron disease. In: Hocking DR, Bradshaw JL, Fielding J, editors. Degenerative disorders of the brain. New York, NY: Routledge/Taylor & Francis Group; 2019. p. 199–228.

[3] Simon NG, et al. Motor neuron disease: current management and future prospects. Intern Med J 2015;45(10):1005–13.

[4] Turner MR, et al. Primary lateral sclerosis: consensus diagnostic criteria. J Neurol Neurosurg Psychiatry 2020;91(4):373–7.

[5] Australia M. Economic impact of MND [cited 2021 03.03.2021]; Available from. https://www.mndaust.asn.au/Influencing-policy/Economic-analysis-of-MND-(1).aspx, 2021.

[6] Hardiman O. Global burden of motor neuron diseases: mind the gaps. Lancet Neurol 2018;17(12):1030–1.

[7] Musson LS, McDermott CJ, Hobson EV. Exploring patient and public involvement in motor neuron disease research. Amyotrophic Lateral Scler Frontotemporal Degener 2019;20(7–8):511–20.

[8] Aktekin MR, Uysal H. Epidemiology of amyotrophic lateral sclerosis. Amyotrofik Lateral Skleroz Epidemiyolojisi 2020;26(3):187–96.

[9] Gundogdu B, et al. Racial differences in motor neuron disease. Amyotroph Lateral Scler Frontotemporal Degener 2014;15(1–2):114–8.

[10] Dormann D. FUScinating insights into motor neuron degeneration. EMBO J 2016;35(10):1015–7.

[11] Talbot K. Motor neuron disease. Pract Neurol 2009;9(5):303–9. BMJ Publishing Group.

[12] Rainier S, et al. Motor neuron disease due to neuropathy target esterase gene mutation: clinical features of the index families. Muscle Nerve 2011;43(1):19–25.

[13] Rosenfeld J, Strong MJ. Challenges in the understanding and treatment of amyotrophic lateral sclerosis/motor neuron disease. Neurother: J Am Soc Exp Neurother 2015;12(2):317.

[14] Ström AL, et al. Retrograde axonal transport and motor neuron disease. J Neurochem 2008;106(2):495–505.

[15] Sandercock J, et al. Riluzole for motor neurone disease. Br Med J 2001;322(7297):1305–6.

[16] Wang W, et al. The inhibition of TDP-43 mitochondrial localization blocks its neuronal toxicity. Nat Med 2016;22(8):869–78.

[17] Brenner D, et al. Heterozygous Tbk1 loss has opposing effects in early and late stages of ALS in mice. J Exp Med 2019;216(2):267–78.

[18] Heiman-Patterson TD, et al. Genetic background effects on disease onset and lifespan of the mutant dynactin p150Glued mouse model of motor neuron disease. PLoS One 2015;10(3), e0117848.

[19] Kariya S, et al. The neuroprotective factor Wld(s) fails to mitigate distal axonal and neuromuscular junction (NMJ) defects in mouse models of spinal muscular atrophy. Neurosci Lett 2009;449(3):246–51.

[20] Sun K, et al. Neuron-specific HuR-deficient mice spontaneously develop motor neuron disease. J Immunol 2018;201(1):157–66.

[21] Bäumer D. Functional genetic analysis of motor neuron disease. University of Oxford; 2010.

[22] Bäumer D, Talbot K, Turner MR. Advances in motor neurone disease. J R Soc Med 2014;107(1):14–21.

[23] Roggenbuck J, Quick A, Kolb SJ. Genetic testing and genetic counseling for amyotrophic lateral sclerosis: an update for clinicians. Genet Med 2017;19(3):267–74.

[24] Shaw PJ. Molecular and cellular pathways of neurodegeneration in motor neurone disease. J Neurol Neurosurg Psychiatry 2005;76(8):1046–57.

[25] Riva N, et al. Recent advances in amyotrophic lateral sclerosis. J Neurol 2016;263(6):1241–54.

[26] Ping S, et al. Mitochondrial dysfunction is a converging point of multiple pathological pathways in amyotrophic lateral sclerosis. J Alzheimers Dis 2010;20:311–24.

[27] Pajarillo E, et al. The role of astrocytic glutamate transporters GLT-1 and GLAST in neurological disorders: potential targets for neurotherapeutics. Neuropharmacology 2019;161.

[28] Kodavati M, Wang H, Hegde ML. Altered mitochondrial dynamics in motor neuron disease: an emerging perspective. Cell 2020;9(4):1065 [2073–4409].

[29] Le Verche V, et al. Glutamate pathway implication in amyotrophic lateral sclerosis: what is the signal in the noise? J Recept Ligand Channel Res 2011;4:1–22.

[30] Hayashi Y, Homma K, Ichijo H. SOD1 in neurotoxicity and its controversial roles in SOD1 mutation-negative ALS. Adv Biol Regul 2016;60:95–104.

[31] Vucic S, Rothstein JD, Kiernan MC. Advances in treating amyotrophic lateral sclerosis: insights from pathophysiological studies. Trends Neurosci 2014;37(8):433–42.

[32] Burrell JR, Vucic S, Kiernan MC. Isolated bulbar phenotype of amyotrophic lateral sclerosis. Amyotroph Lateral Scler 2011;12(4):283–9.

[33] Visser J, et al. Disease course and prognostic factors of progressive muscular atrophy. Arch Neurol 2007;64(4):522–8.

[34] Gordon P, et al. The natural history of primary lateral sclerosis. Neurology 2006;66(5):647–53.

[35] Neary D, Snowden J. Frontal lobe dementia, motor neuron disease, and clinical and neuropathological criteria. J Neurol Neurosurg Psychiatry 2013;84(7):713–4.

[36] Shefner JM. Strength testing in motor neuron diseases. Neurotherapeutics 2017;14(1):154–60.

[37] McCombe P, Wray N, Henderson R. Extra-motor abnormalities in amyotrophic lateral sclerosis: another layer of heterogeneity. Expert Rev Neurother 2017;17(6):561–77.

[38] Foster LA, Salajegheh MK. Motor neuron disease: pathophysiology, diagnosis, and management. Am J Med 2019;132(1):32–7.

[39] Ng L, Khan F. International classification of functioning, disability and health and motor neurone disease rehabilitation. Soc Care Neurodisability 2013;4.

[40] Lau FS, Brennan FP, Gardiner MD. Multidisciplinary management of motor neurone disease. Aust J Gen Pract 2018;47(9):593–7.

[41] Dal Bello-Haas V, Florence JM. Therapeutic exercise for people with amyotrophic lateral sclerosis or motor neuron disease. Cochrane Database Syst Rev 2013;2013(5):Cd005229.

[42] Dharmadasa T, Matamala JM, Kiernan MC. Treatment approaches in motor neurone disease. Curr Opin Neurol 2016;29(5):581–91.

[43] National Clinical Guideline Centre. National institute for health and care excellence: clinical guidelines. In: Motor neurone disease: assessment and management. UK: National Institute for Health and Care Excellence; 2016. Copyright © National Clinical Guideline Centre, 2016: London.

[44] Moura MC, Casulari LA, Novaes M. A predictive model for prognosis in motor neuron disease. J Neurol Disord 2016;4(316):2.

[45] Bull FC, et al. World Health Organization 2020 guidelines on physical activity and sedentary behaviour. Br J Sports Med 2020;54(24):1451–62.

[46] de Almeida JP, et al. Exercise and amyotrophic lateral sclerosis. Neurol Sci 2012;33(1):9–15.

[47] Lanfranconi F, Marzorati M, Tremolizzo L. Editorial: strategies to fight exercise intolerance in neuromuscular disorders. Front Physiol 2020;11:968.

[48] Tsitkanou S, et al. The role of exercise as a non-pharmacological therapeutic approach for amyotrophic lateral sclerosis: beneficial or detrimental? Front Neurol 2019;10:783.

[49] Lanfranconi F, et al. Inefficient skeletal muscle oxidative function flanks impaired motor neuron recruitment in amyotrophic lateral sclerosis during exercise. Sci Rep 2017;7(1):2951.

[50] Fiuza-Luces C, et al. Physical exercise and mitochondrial disease: insights from a mouse model. Front Neurol 2019;10:790.

[51] Ferri A, et al. Tailored exercise training counteracts muscle disuse and attenuates reductions in physical function in individuals with amyotrophic lateral sclerosis. Front Physiol 2019;10:1537.

[52] Gibbons CJ, Thornton EW, Young CA. The patient experience of fatigue in motor neurone disease. Front Psychol 2013;4:788.

[53] Gibbons C, et al. Treatment of fatigue in amyotrophic lateral sclerosis/motor neuron disease. Cochrane Database Syst Rev 2018;1:CD011005.

[54] Mohamed AA. Can proprioceptive training reduce muscle fatigue in patients with motor neuron diseases? A new direction of treatment. Front Physiol 2019;10:1243.

[55] Mentis GZ, et al. Early functional impairment of sensory-motor connectivity in a mouse model of spinal muscular atrophy. Neuron 2011;69(3):453–67.

[56] Vaughan SK, et al. Degeneration of proprioceptive sensory nerve endings in mice harboring amyotrophic lateral sclerosis–causing mutations. J Comp Neurol 2015;523(17):2477–94.

[57] Gear WS. Effect of different levels of localized muscle fatigue on knee position sense. J Sports Sci Med 2011;10(4):725.

[58] Kaya D. Exercise and proprioception. Proprioception: The Forgotten Sixth Sense; 2016.

[59] Lunetta C, et al. Strictly monitored exercise programs reduce motor deterioration in ALS: preliminary results of a randomized controlled trial. J Neurol 2016;263(1):52–60.

[60] Paganoni S, et al. Comprehensive rehabilitative care across the spectrum of amyotrophic lateral sclerosis. NeuroRehabilitation 2015;37(1):53–68.

[61] Stefanetti RJ, et al. Measuring the effects of exercise in neuromuscular disorders: a systematic review and meta-analyses. Wellcome Open Res 2020;5:84.

[62] Dauwan M, et al. Physical exercise improves quality of life, depressive symptoms, and cognition across chronic brain disorders: a transdiagnostic systematic review and meta-analysis of randomized controlled trials. J Neurol 2021;268(4):1222–46.

[63] Paganoni S, Cudkowicz M, Berry JD. Outcome measures in amyotrophic lateral sclerosis clinical trials. Clin Invest 2014;4(7):605.

[64] Dharmadasa T, et al. Motor neurone disease. Handb Clin Neurol 2018;159:345–57.

[65] Dunaway Young S, et al. Six-minute walk test is reliable and valid in spinal muscular atrophy. Muscle Nerve 2016;54(5):836–42.

[66] Eichinger K, et al. Validity of the 6 minute walk test in facioscapulohumeral muscular dystrophy. Muscle Nerve 2017;55(3):333–7.

[67] Cedarbaum JM, et al. The ALSFRS-R: a revised ALS functional rating scale that incorporates assessments of respiratory function. BDNF ALS study group (phase III). J Neurol Sci 1999;169(1–2):13–21.

[68] Jenkinson C, Fitzpatrick R. Reduced item set for the amyotrophic lateral sclerosis assessment questionnaire: development and validation of the ALSAQ-5. J Neurol Neurosurg Psychiatry 2001;70(1):70–3.

[69] De Marchi F, et al. Patient reported outcome measures (PROMs) in amyotrophic lateral sclerosis. J Neurol 2020;267(6):1754–9.

[70] Petrof BJ. The molecular basis of activity-induced muscle injury in Duchenne muscular dystrophy. Mol Cell Biochem 1998;179(1):111–24.

[71] Clawson LL, et al. A randomized controlled trial of resistance and endurance exercise in amyotrophic lateral sclerosis. Amyotrophic Lateral Scler Frontotemporal Degener 2018;19(3–4):250–8.

[72] Chen A, Montes J, Mitsumoto H. The role of exercise in amyotrophic lateral sclerosis. Phys Med Rehabil Clin N Am 2008;19(3):545–57. ix-x.

[73] Compo J, et al. Exercise intervention leads to functional improvement in a patient with spinal and bulbar muscular atrophy. J Rehabil Med Clin Commun 2020;3:1000041.

[74] Kitano K, et al. Effectiveness of home-based exercises without supervision by physical therapists for patients with early-stage amyotrophic lateral sclerosis: a pilot study. Arch Phys Med Rehabil 2018;99(10):2114–7.

[75] Braga AC, et al. Tele-monitoring of a home-based exercise program in amyotrophic lateral sclerosis: a feasibility study. Eur J Phys Rehabil Med 2018;54(3):501–3.

[76] Pinto S, Swash M, de Carvalho M. Respiratory exercise in amyotrophic lateral sclerosis. Amyotroph Lateral Scler 2012;13(1):33–43.

Chapter 27

Tai Chi exercise to improve balance and prevent falls among older people with dementia

Yolanda Barrado-Martín[a], Remco Polman[b], and Samuel R. Nyman[c]

[a]Centre for Ageing Population Studies, Research Department of Primary Care & Population Health, London, United Kingdom [b]School of Exercise & Nutrition Sciences, Queensland University of Technology, Brisbane, QLD, Australia [c]Bournemouth University Clinical Research Unit, Department of Medical Science and Public Health, Bournemouth University, Poole, United Kingdom

Introduction

Dementia is estimated to affect 46.8 million people around the world, which is set to rise to 131.5 million people by 2050 [1]. Dementia has been described as a major neurocognitive disorder in the 5th edition of the Diagnostic and Statistical Manual of Mental Disorders [2], where one or more cognitive domains are significantly impaired (i.e., attention, executive function, learning, memory, language, perceptual-motor function, and social cognition). This progressive increase in dependency to perform activities of daily living impacts not only the person with dementia but also their carer, and wider family and friends, affecting social, health, and financial circumstances. As carers need to become more active in supporting the person living with dementia, as well as keeping up with their previous responsibilities, carer burden is more likely to occur [3]. Accordingly, dementia progression has been associated with a gradual increase in care and health costs [4].

Dementia has been clinically characterized as a syndrome; an umbrella term that amalgamates different symptoms depending on the areas of the brain affected and the corresponding impact on cognitive domains [5]. The different brain regions impacted are associated with different types of dementia (listed according to their prevalence—from 63% to 2%): Alzheimer's disease, vascular dementia, mixed dementia (Alzheimer's and vascular), Lewy body dementia, frontotemporal dementia, dementia in Parkinson's disease, and other types of dementia [6]. Biologically, dementia symptoms are the manifestation of the gradual damage in the structure of the brain. The origin of such damage will depend on the type of dementia, but frequently will be due to the abnormal presence of proteins that interfere with the chemical exchanges between neurons or the lack of oxygenation of the brain [7]. Given the described impact of dementia, it has become a public health priority, with the World Health Organization advocating that countries develop a dementia policy [8].

Falls in dementia

Another challenge of our ageing populations is falls. Falls are the second most common cause of accidental death around the globe and occur more frequently among those over 65 years and who are frailer [9]. A European consensus was reached in 2005 to define a fall as "an unexpected event in which the participants come to rest on the ground, floor, or lower level" [10]. The incidence of falls per year in older people is around 30% [11,12]. Such events have a direct impact on an individual's autonomy and quality of life [13]. In fact, falls among older adults have been identified to be a strong predictor of their admission to residential care [14]. Similarly, the experience of falls could affect their willingness to partake in activities, due to a consequent fear of falls and associated loss of independence and damage to their personal identity [15]. However, as in the case of dementia, falls do not only have an impact on the individual and his or her closer social network and environment but also in the social and healthcare systems [9].

Due to the impact of falls on the individual and the cost to society of caring for those that have fallen, a variety of interventions have been developed and tested to reduce the risk factors for falls among those living in the community [11]. Nevertheless, participants' adherence to fall prevention interventions remains a challenge [16–19]. In most studies,

Exercise to Prevent and Manage Chronic Disease Across the Lifespan. https://doi.org/10.1016/B978-0-323-89843-0.00007-6

people living with dementia have been excluded even when they are between two and three times more likely to experience a fall, at higher risk of getting injured, and face a more difficult recovery including greater risk of mortality [20–22]. A recent systematic review found the risk factors for falls among people living with dementia included balance alterations, specific drugs consumption, depression, carer's distress, orthostatic hypertension, and history of falls [21]. Falls prevention interventions that work among community-dwelling individuals do not necessarily work in residential care settings [11,23]. Likewise, interventions that work in the general older adult population have not been found to work among those with dementia. In community-dwelling people living with dementia, physical activity has been attributed a protective effect against falls [20]. In this line, when fall prevention interventions targeting people living with dementia in the community have been studied, physical activity has been found to be potentially useful for this purpose [24]. Tai Chi is an exercise intervention with robust evidence for its effectiveness to prevent falls among community-dwelling older people, which also has great promise for preventing falls among people living with dementia [25].

Tai Chi

Tai Chi was originally a martial art. Nowadays it is frequently practiced as a gentle form of exercise, training aspects of physical fitness (i.e., flexibility, balance, coordination), cognitive performance (i.e., attention, memory) [26,27], as well as relaxation. For this reason, it has also been described as a mind–body exercise and there has been an interest in studying its impact on both physical and psychological outcomes (for a plain English summary, see: https://theconversation.com/tai-chi-health-benefits-what-the-research-says-130630).

Tai Chi originated from China and is based on Taoist Philosophy [28]. Several types of Tai Chi are derived from it (i.e., Chen, Yang, Sun, Wu, Hao), which are highly similar in the essential principles, though have slightly different emphases (i.e., stances and intensity) [28]. Regular participation in Tai Chi has been shown to have a variety of benefits for older adults:

- Physical improvements in:
  - Blood pressure reduction [29].
  - Grip strength (left hand) [29].
  - Walking speed [30].
  - Balance [31,32].
  - Risk of falling [31].
  - Number of falls [29,31].
- Psychological benefits in:
  - Cognitive function [30] in participants with questionable (CDR = 0.5) or no cognitive impairment (MMSE ≥24).
  - Fear of falling [29,31].

Tai chi exercise to prevent falls in dementia

Clinical outcomes

Impact of Tai Chi interventions on balance among people with dementia

Tai Chi interventions have frequently been assessed for their impact on postural balance, due to this being identified as a risk factor for falls among older adults living in the community [33]. Tai Chi interventions conducted among people living with dementia have frequently targeted an improvement in balance or the reduction of falls (62.5%). Only one of these studies did not measure balance [34] but merely the number of falls. When the balance was measured, the Berg Balance Scale (BBS) [35] was used alone in two studies [36,37]; the Unipedal Stance Test (UST) [38] and Time Up & Go (TUG) [39] were used together in one study [40]; and, in the last study, the description of the measure used was similar to the UST [41]. These measures were both used in community and long-term care settings; however, whereas BBS measures functional balance, the UST assesses static balance and the TUG dynamic balance. In this case, studies using BBS did not report improvements in balance, but maintenance [36] or a reduction similar to the control group [37]. However, those using UST (or UST and TUG) reported an improvement in participants' balance [40,41]. This suggests that not only intervention characteristics, but measures chosen could have an impact on trials' findings. It is possible that certain measures are more sensitive to balance improvements or that only specific types of balance can be enhanced by Tai Chi. In a recent RCT testing Tai Chi on people living with dementia, postural balance was the main outcome (measured with TUG), but secondary measures included functional (measured by BBS) and static balance (measured using a postural sway test) to help determine which measures were more appropriated and sensitive when used in people living with dementia [42].

According to the studies reviewed, Tai Chi had a positive impact on participants' physical performance—including balance—in most of the studies based on community-dwelling participants [40,43], even when attendance to the sessions was quite low (52.1% on average) [43]. Other interventions in this setting showed an improvement in balance performance during the first part of the intervention (1–20 weeks) and maintenance in the second half of the intervention (20–40 weeks) compared with participants in the control group (whose performance declined during the first 20 weeks [36]). Considering the degenerative nature of dementia, long-term maintenance in physical performance could be interpreted as a positive result [36]. However, in studies conducted in long-term care settings, the impact on balance has been equivocal. Choi et al. found an improvement in balance [41] whereas Saravanakumar et al. and Nowalk et al. did not [34,37]. Differences in findings could be due to variation in the level of cognitive impairment of participants, the dose provided, and use of seated Tai Chi. The dose–response relationship is supported by a systematic review that found higher dosages of exercise (≥ 3 h per week) that challenges balance had greater effect on reducing falls [44]. Nowalk et al., on the other hand, did not state the length of sessions [34], which makes it difficult to estimate the Tai Chi dosage offered, as could be particularly heterogeneous in a 24 months intervention.

Although older people living with dementia are more likely to fall than their peers without dementia [22], they have been frequently excluded from fall prevention interventions [45]. When widening the search, studies conducted among general older people at risk of cognitive decline living in the community trying to analyze the influence of exercise in balance have found positive results [26]. It has been suggested that its impact on cognition's preservation would require more long-term adherence, similarly as the improvement in balance was found to be clearer after a year of practice. This is consistent with the need of maintaining adherence to exercise interventions to reach its long-term benefits. Longer interventions that provide support to their participants, in turn, are more likely to produce a behavior change. Moreover, Lam et al's. results, pointed Tai Chi's positive impact on balance could be higher than other exercises like stretching which do not challenge balance [26]. This is in line with Nyman and Skelton's commentary which suggested that adapted versions of Tai Chi for frailer participants using more seated practice could lose their falls prevention effectiveness [46].

Impact of Tai Chi interventions on falls among people with dementia

Studies on the effectiveness of Tai Chi have frequently excluded people living with dementia [27,29,30,47–50]. Consequently, systematic reviews around this group of participants have not been found. Instead, studies including people living with dementia practicing Tai Chi, have been reported in wider reviews analyzing the effects of exercise in this condition [51,52] or the effects of different interventions seeking to prevent falls among those living in long-term care settings [51]. These reviews have found promising effects. However, systematic reviews have also reported inconsistencies [51] and large heterogeneity [52] among study results or participants' characteristics, respectively.

On the other hand, systematic reviews focusing on the impact of Tai Chi [50,53], or exercise interventions including Tai Chi [54], have been conducted on older adults' cognition. Reviews that have included a number of studies that included people living with dementia (5 out of 20) have reported the positive impact of Tai Chi in global cognitive function (but not on specific domains) [53]. Tai Chi studies conducted among people living with dementia and in the general population have reported both cognitive and physical benefits [36,40]. Most of these studies including people living with dementia have been conducted in long-term care setting environments [34,37,41,55,56] and it is unclear whether similar benefits are achieved for community-dwelling older adults [36,40,43,57]. In particular, it has been suggested that differences in the effects of Tai Chi could be due to the characteristics of participants living in both settings, the environment where these interventions are delivered, and the characteristics of the intervention itself [58,59].

Most studies included complementary interventions in parallel (i.e., other types of exercise, a cognitive-behavioral intervention, and support group) [36,43] which cloud inference of Tai Chi's effectiveness. When two or more different interventions are delivered together, it is not possible to determine which is causing an impact on study results, it might be that only one of the various interventions is (in)effective, or that the conjunction of various interventions explains its (in) effectiveness. In long-term care settings, however, this has been less common, as only one of the papers included a therapeutic intervention delivered at the same time [34]. This could be due to community interventions aiming to address various elements (i.e., behavioral, physical, and cognitive functioning) in the person living with dementia to, in turn, help reduce carer burden and indirectly delay the person living with dementia's move to long-term care settings.

Community studies have aimed to improve both physical and cognitive functioning [36,43], whereas Tai Chi interventions conducted in long-term care settings seek to improve physical aspects [34,37,41], particularly determinants related to falls. This could be explained by the amount of people falling in care settings, which is three times higher than those living in the community [60]. Similarly, there are some differences between the aims of studies and primary and secondary outcomes. Some are targeting the prevention of falls itself [34,37,40,41] and focus on incidence of falls and balance outcomes, whereas others target the reduction of pain (i.e., in participants' knee) among participants with dementia [56]. Probably because Tai Chi has been described as a mind–body exercise, some authors have focused on its potential to improve cognitive functioning [55]. In this case, the common standing position to challenge balance was replaced by a seated version of Tai Chi.

In studies where Tai Chi did not significantly reduce the incidence of falls [34,57], low intensity and low adherence rates [57], together with the need for tailoring the intervention to participant's needs [34] have been suggested as possible explanations. In neighboring literature, where a Taoist form of Tai Chi was examined in people with Mild Cognitive Impairment (MCI), findings showed no impact on cognition or mobility. The type of Tai Chi used, the level of frailty of participants, dosage (including intensity and frequency), participants' skeptical attitudes, and low adherence, have been pointed as possible explanations of their results and differences with previous research [48].

Delivery of Tai Chi exercise

Tai Chi prescription in terms of length of the sessions, frequency, and duration of the intervention has been mostly described across papers. Only one of the studies identified in our review failed to provide the duration of each session [34]. This is in line with Wu, MacDonald, and Pescatello's systematic review around the ways of reporting dosage and session content in Tai Chi interventions designed to improve balance among older people in general [61]. Also consistent with the results of this systematic review [61], a number of papers failed to provide information on their instructional methods (i.e., style of Tai Chi) [34,43], others describe the style as "traditional" [36] or "modified" [37] which according to the authors of a systematic review [61] should be equally considered as absent. These papers did not point to a dose–response of the Tai Chi intervention nor reflected (when specified) on the most suitable style of Tai Chi, which is again in line with Wu et al's. review. Moreover, a lack of guidelines to adapt these interventions to preserve their effectiveness was identified. However, a meta-analysis around exercise interventions to prevent falls in older adults recommended a dosage of 3 h or more a week of exercise that challenged balance (such as Tai Chi) [44].

Only one study reported having used a dyadic approach in a community [40], where the person living with dementia takes part together in the intervention with a friend or family member. This approach, combined with the use of behavioral techniques to facilitate motivation, was found beneficial to enhance participants' motivation, attention, and adherence to the intervention [40]. No challenges related to this approach were reported, although authors admit the high level of motivation of carers taking part in the intervention could have had a positive effect on their results [40]. For this reason, they recommended further research into the different "types of caregiver interaction" established among different types of carers (i.e., relatives, friends, neighbors, volunteers) [40]. Another study did report recruitment in dyads, but carers were not actually involved in supporting people living with dementia during the class [43]. Most studies did not report a dyadic approach at any stage [34,41,55,56], or provided insufficient information to assess if support was provided one-to-one by volunteers [37].

Tai Chi-based trials reporting qualitative findings are limited and support previous findings on the underuse of qualitative methodologies alongside RCTs in healthcare interventions [62,63]. Only one study [43] reported a change in their study protocol to implement a qualitative data analysis. This paper did also include some integration of qualitative and quantitative results in its discussion section [64]. By doing this, the authors were able to explain the importance of making exercise movements relevant to participants to facilitate their engagement and enhance intervention outcomes. These techniques could have been relevant to overcome some participants' initial skepticism toward the intervention or exercise in general, as some of them thought "movements seemed unusual or silly" [64], but this ceased during the intervention when a participant admitted, "maybe it's not worthless" [64]. Another two papers unexpectedly reported an exit interview in their discussion [36,40] to highlight participants' enjoyment of the intervention or carers' suggestion to have more group-based sessions. However, they did not provide any methodological information nor refer to further publications.

Acceptability of and adherence to Tai Chi interventions

Only one RCT study delivering only Tai Chi and including people living with dementia has reported the use of qualitative methods [65]. This study, which was delivered in a combination of class and home-based practice, confirmed Tai Chi's safety for people living with dementia in both settings. Likewise, an improvement in the quality of life and a reduction of the number of falls over the 6-months study period of those living with dementia in the intervention group was also reported [42]. The qualitative study of acceptability and adherence, during the pilot and RCT phases of this clinical trial, found

Tai Chi to be an enjoyable activity for both people living with dementia and their family carers [66]. Adherence to the intervention was facilitated by its socializing component, participants' willingness to benefit from their practice, and their ability to learn the movements with repetition [67,68]. Whereas additional commitments, health issues, or difficulties following the booklet provided to support home practice were the main barriers to adherence [67,68].

For the other studies reviewed, acceptability was reported by authors following their perceptions about participants' satisfaction with the intervention. Some authors, despite not having reported the use of qualitative methods in their studies, however, have pointed the need for listening to participants' opinions and perceptions regarding their involvement in Tai Chi interventions. For example: "The high level of interest for yoga and tai chi amongst the older participants needs to be explored. What were the perceptions and experiences of the older participants?" [37].

When barriers and facilitators for participant adherence have been reported, these came mainly from lessons learned by authors or their recommendations for future research. These included the use of dyadic approaches and behavior change techniques to improve participants' adherence [40] and the creation of a motivational and friendly environment that allows social interaction [43] in interventions conducted in the community. If these were based in long-term care settings, then the decision of carrying out the whole intervention in the same setting [34] was also perceived as a facilitator. On the other hand, care routines in long-term care settings, together with professional carers not facilitating participants' attendance to the sessions were perceived as barriers [37].

Occasionally, carers or people living with dementia's opinions have been reported, but only as anecdotal comments [36,40] gathered in interventions conducted in the community. On these occasions, a demand for more group-based sessions by carers has been interpreted as a possible facilitator of adherence [40]. Also, people living with dementia could adhere more to the intervention when feeling their improvement or enjoying the social interactions in group-based interventions [36].

Exercise interventions

Besides Tai Chi, several exercise interventions have been studied in people living with dementia with the potential to "improve health, mood, and quality of life" [69]. A systematic review including 30 trials, involving a total of 2020 older participants experiencing cognitive impairment and dementia, supported the positive effects of exercise on health-related physical fitness and cognitive functioning [70]. A potential to delay this functional decline was also supported in a recent systematic review of systematic reviews [71].

Although some of the interventions designed for people living with cognitive impairment in the community have reported only the use of exercise [72], frequently they have also applied behavior change techniques with the aim of enhancing adherence. For instance, exercise interventions have commonly been accompanied by: (a) Problem-solving interventions [69,73,74]; (b) Goal-setting techniques [74,75]; and (c) Self-monitoring (i.e., using pedometers), feedback, and reinforcement procedures [74]. When goal-setting techniques have been used with older people (including those living with dementia), several difficulties have been identified. Elements like participants' ability to establish health behavior-related goals and the use of inadequate tools to assess their achievement or procedures to review such goals have been found to be detrimental [75]. In these multicomponent interventions, improvements in physical and psychological functioning have been shown. In addition, these interventions improved quality of life in people living with dementia and their carers [76].

Existing reports of exercise research involving people living with dementia have aimed to study its impact on cognitive and physical function or on their mood and any behavioral challenges associated (i.e., agitation, aggression, sundowning, restlessness, shouting and screaming, and losing inhibitions) [77,78]. Lowery et al. reported no effects on mood in a walking intervention conducted in a clinical trial context, although carers' burden was reduced [79]. This negative result could have been influenced by low intervention adherence and dosage, as other research has found a positive impact on behavioral outcomes. Nascimento, Teixeira, Gobbi, Gobbi, and Stella, for instance, who only analyzed data coming from high adherent participants, with a minimum attendance to 75% of the exercise sessions [80], found a reduction in most of the scores on the Neuropsychiatric Inventory (NPI) [81]. Similar results were found in Stella et al., although authors warn that such positive results would need to be taken with caution as researchers assessing the study were not blinded to participant's allocation [82]. Other exercise interventions have included social interaction as one of the key elements, together with memory training and body awareness [64]. In their intervention, Wu et al. included 11 people living with mild-to-moderate Alzheimer's disease attending a day center, who were asked to take part in an integrative intervention consisting of conventional and complementary types of exercise (such as Tai Chi) [64]. Feedback from instructors, carers, and the study team revealed positive changes in the functional (i.e., "during a class activity involving patting the legs, Doris looked at her legs and said 'you're waking up!'") emotional ("She often smiled, breathed deeply and commented about feeling 'more peaceful'"), and social ("'Well, I never thought so when younger, so I didn't speak up, but now, I've started to speak up more here, and I plan to more") spheres; which were observable in terms of reduced anxiety and friendship development during the sessions [64].

The variability in exercise frequency, duration, and intensity [59], as well as the additional component delivered (i.e., psychoeducational) during the intervention period, makes it difficult to establish which is the effective component in such interventions. Similarly, in dual-task interventions, it is not feasible to distinguish which is the effective task in the intervention, if both are having an impact on the research outcomes, or whether the combination of both tasks is what makes these interventions effective. Andrade et al., for instance, explored the positive impact of a multimodal exercise in postural control and frontal lobe cognitive functions; however, they used a dual-task intervention to test their hypothesis which involved cognitive tasks [83].

Exercise interventions designed for people living with dementia have also employed different approaches (i.e., dyadic and nondyadic). Frequently, quantitative studies have included exercise interventions targeting the person living with dementia [84] or the carer [73,85] separately. Interventions for people living with dementia tend to deliver only the exercise component, aim to improve gait performance and use a group setting. In contrast, exercise programs targeting carers are frequently multicomponent (including different types of intervention such as support and increase of exercise practice) and delivered individually through the telephone.

Acceptability of and adherence to exercise interventions

Exercise has shown potential benefits for people living with dementia; however, as for the older population in general [26,58,86,87], people living with dementia's adherence to these interventions is a challenge [40]. A systematic review around exercise interventions targeting people living with dementia to prevent falls in the community [24] pointed that attrition rates varied from 4.5 to 50% depending on the study and length of follow-up. This highlights that not only adherence to the intervention is a challenge, but also participants' retention in the research process. The most common reasons for dropping out among these community-dwelling participants were: Admission to care settings, poor health, and not willing to complete the program, and in some cases death. Reasons for missing sessions were: The lack of an exercise habit before starting the intervention, need of support while practising, temporary illnesses, or holiday periods [88].

There are contrasting opinions regarding the suitability of home-based or group-based interventions when involving people living with dementia across the literature. Papers reporting home-based interventions [73] argue and refer to aligned literature where participants prefer individually tailored interventions delivered in their own homes. In contrast, group-based interventions highlight the role of the social component as a facilitator for participants' engagement in the intervention [89]. However, as reported for Tai Chi interventions, not only people should be offered different types of cost-effective interventions to fit their varied needs [46], but these will need to be offered in different formats (i.e., home- and class-based) which meet their preferences.

An exercise intervention conducted among people living with dementia showed that 92.9% of participants allocated to the home-based exercise arm participated in 50% or more sessions, compared to 78.6% of participants in the group-based intervention over 12 months [90]. However, the duration, intensity, and modality of exercise were not necessarily the same for both groups, as the home-based intervention was tailored to everyone's needs. On the other hand, two systematic reviews of studies conducted among general older people, have provided different measures of adherence such as average of classes attended or average of full attending participants. Whereas a systematic review around home-based exercise interventions reported an average full attendance of 21% [91]; another systematic review of group-based exercise interventions reported an average attendance to the total amount of sessions of 74% [92]. An additional systematic review of interventions to prevent falls among older adults living in the community [19] found a higher median adherence to class-based (73.2%) than to home-based (52%) interventions over 12 months.

A number of lessons can be learnt from the previous studies reviewed:

(a) The need for tailoring home-based exercises: Tailoring exercises to participants' needs is not only important to facilitate their adherence to the intervention [93,94], but also to ensure safe practice.
(b) The inclusion of diverse informal carers: The inclusion of other carers different from the primary carer, who could prefer to take the intervention time as respite or be less motivated to the exercise practice than other carers, has been suggested as a gap in research [95].
(c) The use of behavior change techniques [96]: Goal-setting techniques [85], in form of action plans [97], have been frequently used to encourage participants adherence to the intervention.

(d) A continuous telephone follow-up during the intervention period: A continuous contact with participants during the intervention has been used in previous studies [85,97] to encourage participants' engagement in trials and to monitor intervention's safety [65,93].

Summary of lessons learnt from the exercise literature in general for the application of Tai Chi with older people with dementia

The positive impact of exercise on people living with dementia's physical and mental health has been well recognized. The use of behavior change techniques, as well as the socializing component and perceived benefits of the interventions, have helped enhance their adherence to interventions. Exercise interventions in the context of dementia have tended to target people living with dementia and their carers separately. However, when a dyadic approach has been used, qualitative methods have been underused and the role of the carers has been poorly described. Future research needs to address this to further understand the impact of their joint participation in adherence and, in turn, in research outcomes in both members. Similarly, there is a lack of process evaluations conducted along RCTs testing exercise interventions in people living with dementia. There is a need to understand the views of participants involved in exercise interventions and explore their views in complex interventions.

Clinical practice recommendations for exercise prescribers

- If possible, build up to at least 3h per week of exercise practice.
- Encourage dyadic participation, where the person living with dementia and family carer/friend take part together, where possible.
- Explore acceptability and adherence qualitatively to better adjust exercise to the needs of people living with dementia and their carers if invited to take part together.
- Use behavior change techniques and explore its impact on adherence.
- Create a motivational and friendly environment.
- Where possible, offer a combination of class and home-based practice.
- Highlight and explain the benefits attributed to each exercise.
- When using visual materials, consider using a DVD if part of the intervention will take place at home.
- Keep in touch with participants regularly, to resolve questions and encourage adherence to the intervention.
- Develop programs specifically designed for people living with dementia.
- Pilot: Test programs with people living with dementia and allow time to introduce changes and adapt the intervention to their needs, before running a full program.
- Provide opportunities for socializing as part of the intervention to enhance enjoyment and adherence to the exercise program.

Conclusion

Exercise interventions, in general, and Tai Chi in particular, have shown a positive impact both on the physical and psychological outcomes of older adults. Particularly, its impact on postural balance and the reduction of falls events in people living with dementia has been promising. Often the effect of these interventions has been reduced by low adherence and high attrition rates. The tendency to use quantitative methods to assess the effectiveness of these interventions has possibly limited our understanding of participants' experiences. This, in turn, has left a gap in research to identify ways of improving sustained adherence to exercise interventions in older people living with dementia. Future research needs to study these outcomes in high-quality studies where participants are given the opportunity to express their views on these interventions. Exploring participants' characteristics together with their views will facilitate making recommendations of which interventions are more likely to have a significant impact on participants.

References

[1] Alzheimer's Disease International. Dementia Statistics, 2015. [cited 2016 15.02.2016]; Available from: http://www.alz.co.uk/research/statistics.

[2] American Psychiatric Association. Diagnostic and statistical manual of mental disorders (DSM-5®). 5th ed. Arlington, VA, USA: American Psychiatric Publishing; 2013.

[3] Canonici AP, et al. Functional dependence and caregiver burden in Alzheimer's disease: a controlled trial on the benefits of motor intervention. Psychogeriatrics 2012;12(3):186–92.

[4] Joling KJ, et al. Predictors of societal costs in dementia patients and their informal caregivers: a two-year prospective cohort study. Am J Geriatr Psychiatry: Off J Am Assoc Geriatr Psychiatry 2015;23(11):1193–203.

[5] Bundy R, Minihane AM. Diet, exercise and dementia: the potential impact of a mediterranean diet pattern and physical activity on cognitive health in a UK population. Nutr Bull 2018;43(3):284–9.

[6] Alzheimer's Society. Demographics. [30/08/2016]; Available from., 2014, https://www.alzheimers.org.uk/site/scripts/documents_info.php?documentID=412.

[7] Alzheimer's Society. Factsheet 400LP: what is dementia?; 2017.

[8] WHO. Toward a dementia plan: A WHO guide. World Health Organization; 2018.

[9] WHO. WHO global report on falls prevention in older age; 2007.

[10] Lamb SE, et al. Development of a common outcome data set for fall injury prevention trials: the prevention of falls network Europe consensus. J Am Geriatr Soc 2005;53(9):1618–22.

[11] Gillespie LD, et al. Interventions for preventing falls in older people living in the community. Cochrane Database Syst Rev 2012;2012(9), CD007146.

[12] Peel NM. Epidemiology of falls in older age. Can J Aging 2011;30(1):7–19.

[13] National Institute for Health and Care Excellence. Falls in older people: Assessing risk and prevention. NICE; 2013.

[14] Tinetti ME, Williams CS. Falls, injuries due to falls, and the risk of admission to a nursing home. N Engl J Med 1997;337(18):1279–84.

[15] Yardley L, Smith H. A prospective study of the relationship between feared consequences of falling and avoidance of activity in community-living older people. Gerontologist 2002;42(1):17–23.

[16] Osho O, Owoeye O, Armijo-Olivo S. Adherence and attrition in fall prevention exercise programs for community-dwelling older adults: a systematic review and meta-analysis. J Aging Phys Act 2018;26(2):304–26.

[17] Nyman S. Psychosocial issues in engaging older people with physical activity interventions for the prevention of falls. Can J Aging 2011;30(1):45–55.

[18] Nyman SR, Victor CR. Older people's recruitment, sustained participation, and adherence to falls prevention interventions in institutional settings: a supplement to the Cochrane systematic review. Age Ageing 2011;40(4):430–6.

[19] Nyman S, Victor C. Older people's participation in and engagement with falls prevention interventions in community settings: an augment to the cochrane systematic review. Age Ageing 2012;41(1):16–23.

[20] Allan LM, et al. Incidence and prediction of falls in dementia: a prospective study in older people. PLoS ONE 2009;4(5):e5521.

[21] Fernando E, et al. Risk factors associated with falls in older adults with dementia: a systematic review. Physiother Can 2017;69(2):161–70.

[22] Shaw FE. Falls in older people with dementia. Geriatr Aging 2003;6:37–40.

[23] Cameron ID, et al. Interventions for preventing falls in older people in care facilities and hospitals. Cochrane Database Syst Rev 2018;9(9).

[24] Burton E, et al. Effectiveness of exercise programs to reduce falls in older people with dementia living in the community: a systematic review and meta-analysis. Clin Interv Aging 2015;10:421–34.

[25] Nyman SR. Tai Chi for the prevention of falls among older adults: a critical analysis of the evidence. J Aging Phys Act 2020;1–10.

[26] Lam LCW, et al. A 1-year randomized controlled trial comparing mind body exercise (Tai Chi) with stretching and toning exercise on cognitive function in older Chinese adults at risk of cognitive decline. J Am Med Dir Assoc 2012;13(6). p. 568.e15-20.

[27] Wayne PM, et al. Tai Chi training may reduce dual task gait variability, a potential mediator of fall risk, in healthy older adults: cross-sectional and randomized trial studies. Front Hum Neurosci 2015;9.

[28] Lam P. History of Tai Chi. [01/09/2016]; Available from., 2007, http://taichiforhealthinstitute.org/history-of-tai-chi-2/.

[29] Wolf SL, et al. Reducing frailty and falls in older persons: an investigation of Tai Chi and computerized balance training. Atlanta FICSIT group. Frailty and injuries: cooperative studies of intervention techniques. J Am Geriatr Soc 1996;44(5):489–97.

[30] Sun J, et al. Tai chi improves cognitive and physical function in the elderly: a randomized controlled trial. J Phys Ther Sci 2015;27(5):1467–71.

[31] Li F, et al. Tai Chi and fall reductions in older adults: a randomized controlled trial. J Gerontol A Biol Sci Med Sci 2005;60(2):187–94.

[32] Voukelatos A, et al. A randomized, controlled trial of tai chi for the prevention of falls: the Central Sydney tai chi trial. J Am Geriatr Soc 2007;55(8):1185–91.

[33] Vieira ER, Palmer RC, Chaves PHM. Prevention of falls in older people living in the community. BMJ 2016;353.

[34] Nowalk MP, et al. A randomized trial of exercise programs among older individuals living in two long-term care facilities: the FallsFREE program. J Am Geriatr Soc 2001;49(7):859–65.

[35] Berg KO, et al. Measuring balance in the elderly: validation of an instrument. Can J Publ Health 1992;83(Suppl 2):S7–S11.

[36] Burgener SC, et al. The effects of a multimodal intervention on outcomes of persons with early-stage dementia. Am J Alzheimers Dis Other Demen 2008;23(4):382–94.

[37] Saravanakumar P, et al. The influence of Tai Chi and yoga on balance and falls in a residential care setting: a randomised controlled trial. Contemp Nurse 2014;48(1):76–87.

[38] Hurvitz EA, et al. Unipedal stance testing as an indicator of fall risk among older outpatients. Arch Phys Med Rehabil 2000;81(5):587–91.

[39] Podsiadlo D, Richardson S. The timed "up & go": a test of basic functional mobility for frail elderly persons. J Am Geriatr Soc 1991;39(2):142–8.

[40] Yao L, et al. Fall risk-relevant functional mobility outcomes in dementia following dyadic Tai Chi exercise. West J Nurs Res 2012;35(3):281–96.

[41] Choi JH, Moon JS, Song R. Effects of Sun-style Tai Chi exercise on physical fitness and fall prevention in fall-prone older adults. J Adv Nurs 2005;51(2):150–7.

[42] Nyman SR, et al. Randomised controlled trial of the effect of Tai Chi on postural balance of people with dementia. Clin Interv Aging 2019;14:2017–29.

[43] Barnes DE, et al. Preventing loss of independence through exercise (PLIÉ): a pilot clinical trial in older adults with dementia. PLoS ONE 2015;10(2): e0113367.

[44] Sherrington C, et al. Exercise to prevent falls in older adults: an updated systematic review and meta-analysis. Br J Sports Med 2017;57:1750–8.

[45] Hasselmann V, et al. Are exergames promoting mobility an attractive alternative to conventional self-regulated exercises for elderly people in a rehabilitation setting? Study protocol of a randomized controlled trial. BMC Geriatr 2015;15.

[46] Nyman SR, Skelton D. The case for Tai Chi in the repertoire of strategies to prevent falls among older people. Perspect Public Health 2017;132 (2):85–6.

[47] Chen KM, et al. The effects of a simplified Tai-Chi exercise program (STEP) on the physical health of older adults living in long-term care facilities: a single group design with multiple time points. Int J Nurs Stud 2008;45:501–7. https://doi.org/10.1016/j.ijnurstu.2006.11.008.

[48] Fogarty JN, et al. Taoist Tai Chi(R) and memory intervention for individuals with mild cognitive impairment. J Aging Phys Act 2016;24(2):169–80.

[49] Walsh JN, et al. Impact of short- and long-term Tai Chi mind-body exercise training on cognitive function in healthy adults: results from a hybrid observational study and randomized trial. Global Adv Health Med 2015;4(4):38–48.

[50] Zheng G, et al. Tai Chi and the protection of cognitive ability: a systematic review of prospective studies in healthy adults. Am J Prev Med 2015;49(1):89–97.

[51] Cameron ID, et al. Interventions for preventing falls in older people in care facilities and hospitals. Cochrane Database Syst Rev 2012;12:1–184.

[52] Forbes D, et al. Exercise programs for people with dementia. Cochrane Database Syst Rev 2015;4:CD006489. https://doi.org/10.1002/14651858. CD006489.pub4.

[53] Wayne PM, et al. Effect of Tai Chi on cognitive performance in older adults: systematic review and Meta-analysis. J Am Geriatr Soc 2014;62(1):25–39.

[54] Kelly ME, et al. The impact of exercise on the cognitive functioning of healthy older adults: a systematic review and meta-analysis [abstract]. Ageing Res Rev 2014;16:12–31.

[55] Cheng S-T, et al. Mental and physical activities delay cognitive decline in older persons with dementia. Am J Geriatr Psychiatry 2014;22(1):63–74.

[56] Tsai P-F, et al. A supplemental report to a randomized cluster trial of a 20-week Sun-style Tai Chi for osteoarthritic knee pain in elders with cognitive impairment. Complement Ther Med 2015;23(4):570–6.

[57] Day L, et al. Impact of Tai-Chi on falls among Preclinically disabled older people. A randomized controlled trial. J Am Med Dir Assoc 2015;16:420–6. https://doi.org/10.1016/j.jamda.2015.01.089.

[58] Hill KD, et al. Individualized home-based exercise programs for older people to reduce falls and improve physical performance: a systematic review and meta-analysis. Maturitas 2015;82(1):72–84.

[59] McCurry SM, et al. Predictors of short- and long-term adherence to a daily walking program in persons with Alzheimer's disease. Am J Alzheimers Dis Other Dementias 2010;25(6):505–12.

[60] Care Inspectorate and NHS Scotland. Managing falls and fractures in care homes for older people – good Practise resource. Revised edition; 2016.

[61] Wu Y, MacDonald HV, Pescatello LS. Evaluating exercise prescription and instructional methods used in Tai Chi studies aimed at improving balance in older adults: a systematic review. J Am Geriatr Soc 2016.

[62] Gibson G, et al. The scope for qualitative methods in research and clinical trials in dementia. Age Ageing 2004;33(4):422–6.

[63] Lewin S, Glenton C, Oxman A. Use of qualitative methods alongside randomised controlled trials of complex healthcare interventions: methodological study. BMJ 2009;339.

[64] Wu E, et al. Preventing loss of Independence through exercise (PLIE): qualitative analysis of a clinical trial in older adults with dementia. Aging Ment Health 2015;19(4):353–62.

[65] Nyman SR, et al. A randomised controlled trial comparing the effectiveness of Tai Chi alongside usual care with usual care alone on the postural balance of community-dwelling people with dementia: protocol for the TACIT trial (TAi ChI for people with dementia). BMC Geriatr 2018;18.

[66] Barrado-Martín Y, et al. Acceptability of a dyadic Tai Chi intervention for older people living with dementia and their informal Carers. J Aging Phys Act 2019;27(2):166–83.

[67] Barrado-Martín Y, et al. People living with dementia and their family carers' adherence to home-based Tai Chi practice. Dementia 2021;20 (5):1586–603.

[68] Barrado-Martín Y, et al. Adherence to the class-based component of a Tai Chi exercise intervention for people living with dementia and their informal Carers. J Aging Phys Act 2021;29(5):721–34.

[69] Logsdon RG, McCurry SM, Teri L. A home health care approach to exercise for persons with Alzheimer's disease. Care Manag J 2005;6:90–7.

[70] Heyn P, Abreu BC, Ottenbacher KJ. The effects of exercise training on elderly persons with cognitive impairment and dementia: a meta-analysis. Arch Phys Med Rehabil 2004;85(10):1694–704.

[71] Laver K, et al. Interventions to delay functional decline in people with dementia: a systematic review of systematic reviews. BMJ Open 2016;6(4):1.

[72] Frederiksen KS, et al. Moderate-to-high intensity aerobic exercise in patients with mild to moderate Alzheimers disease: a pilot study. Int J Geriatr Psychiatry 2014;29(12):1242–8.

[73] Farran CJ, et al. A lifestyle physical activity intervention for caregivers of persons with Alzheimer's disease. Am J Alzheimers Dis Other Demen 2008;23(2):132–42.

[74] Logsdon RG, et al. Making physical activity accessible to older adults with memory loss: a feasibility study. Gerontologist 2009;49(Suppl 1):S94–9.

[75] Haas R, Mason W, Haines TP. Difficulties experienced in setting and achieving goals by participants of a falls prevention programme: a mixed-methods evaluation. Physiother Can 2014;66(4):413–22.

[76] Logsdon RG, McCurry SM, Teri L. Evidence-based interventions to improve quality of life for individuals with dementia. Alzheimers Care Today 2007;8(4):309–18.

[77] Alzheimer's Society. Factsheet 509LP: Dementia and aggressive behaviour; 2013.

[78] Alzheimer's Society. Factsheet 525LP: Changes in behaviour; 2015.

[79] Lowery D, et al. The effect of exercise on behavioural and psychological symptoms of dementia: the EVIDEM-E randomised controlled clinical trial. Int J Geriatr Psychiatry 2014;29(8):819–27.

[80] Nascimento CMC, et al. A controlled clinical trial on the effects of exercise on neuropsychiatric disorders and instrumental activities in women with Alzheimers disease. Revista Brasileira De Fisioterapia (São Carlos (São Paulo, Brazil)) 2012;**16**(3):197–204.

[81] Cummings JL. The neuropsychiatric inventory: assessing psychopathology in dementia patients. Neurology 1997;48(5 Suppl 6):S10–6.

[82] Stella F, et al. Attenuation of neuropsychiatric symptoms and caregiver burden in Alzheimer's disease by motor intervention: a controlled trial. Clinics 2011;66(8):1353–60.

[83] Andrade LP, et al. Benefits of multimodal exercise intervention for postural control and frontal cognitive functions in individuals with Alzheimer's disease: a controlled trial. J Am Geriatr Soc 2013;61(11):1919–26.

[84] Schwenk M, et al. Improvements in gait characteristics after intensive resistance and functional training in people with dementia: a randomised controlled trial. BMC Geriatr 2014;14.

[85] Connell CM, Janevic MR. Effects of a telephone-based exercise intervention for dementia caregiving wives: a randomized controlled trial. J Appl Gerontol 2009;28(2):171–94.

[86] Hawley-Hague H, et al. Older adults' uptake and adherence to exercise classes: Instructors' perspectives. J Aging Phys Act 2016;24(1):119–28.

[87] Suttanon P, et al. Factors influencing commencement and adherence to a home-based balance exercise program for reducing risk of falls: perceptions of people with Alzheimer's disease and their caregivers. Int Psychogeriatr 2012;24(7):1172–82.

[88] Wesson J, et al. A feasibility study and pilot randomised trial of a tailored prevention program to reduce falls in older people with mild dementia. BMC Geriatr 2013;13:89.

[89] Dal Bello-Haas VPM, et al. Lessons learned: feasibility and acceptability of a telehealth-delivered exercise intervention for rural- dwelling individuals with dementia and their caregivers. Rural Remote Health 2014;14:2715.

[90] Pitkälä KH, et al. Effects of the Finnish Alzheimer disease exercise trial (FINALEX): a randomized controlled trial. JAMA Intern Med 2013;173 (10):894–901.

[91] Simek EM, McPhate L, Haines TP. Adherence to and efficacy of home exercise programs to prevent falls: a systematic review and meta-analysis of the impact of exercise program characteristics. Prev Med 2012;55(4):262–75.

[92] McPhate L, Simek EM, Haines TP. Program-related factors are associated with adherence to group exercise interventions for the prevention of falls: a systematic review. J Physiother 2013;59(2):81–92.

[93] Close JCT, et al. Can a tailored exercise and home hazard reduction program reduce the rate of falls in community dwelling older people with cognitive impairment: protocol paper for the i-FOCIS randomised controlled trial. BMC Geriatr 2014;14:89. https://doi.org/10.1186/1471-2318-14-89.

[94] Teri L, et al. Exercise and activity level in Alzheimer's disease: a potential treatment focus. J Rehabil Res Dev 1998;35:411–9.

[95] McCurry SM, et al. Increasing walking and bright light exposure to improve sleep in community-dwelling persons with Alzheimer's disease: results of a randomized, controlled trial. J Am Geriatr Soc 2011;59(8):1393–402.

[96] Nyman S, Adamczewska N, Howlett N. Systematic review of behaviour change techniques to promote participation in physical activity among people with dementia. Br J Health Psychol 2018;23(1):148–70.

[97] Brown D, et al. Development of an exercise intervention to improve cognition in people with mild to moderate dementia: dementia and physical activity (DAPA) trial, registration ISRCTN32612072. Physiotherapy 2015;101:126–34. https://doi.org/10.1016/j.physio.2015.01.002.

Chapter 28

Osteosarcopenia and exercise

Troy Walker[a], Jordan Dixon[b], Ian Haryono[b], and Jesse Zanker[b,c,d]

[a]*Global Obesity Centre (GLOBE), Institute for Health Transformation, Deakin University, Geelong, VIC, Australia* [b]*Australian Institute for Musculoskeletal Science, The University of Melbourne and Western Health, St. Albans, VIC, Australia* [c]*Department of Medicine—Western Health, The University of Melbourne, St. Albans, VIC, Australia* [d]*Institute for Health Transformation, Deakin University, Geelong, VIC, Australia*

Background

Globally, advances in medicine have paved the way for people to live longer [1] . While increased life expectancy is a feat of these developments in medicine and economics, with age comes greater likelihood of comorbidity, disability, and dependency. Musculoskeletal diseases represent a significant burden in older persons and a major cost to health systems worldwide [1]. Diseases of importance in this demographic include conditions of accelerated musculoskeletal loss such as sarcopenia (loss of muscle mass, strength, and/or function) and osteoporosis (reduced bone density with increased fracture risk) [2].

The relationship between reduced volume and quality of muscle and bone was described as a "hazardous duet" [3] and coined as "sarco-osteopenia" by Binkley and Buehring in 2009 [4]. Over time, the evidence examining this combined syndrome has grown substantially, giving rise to a novel and now established geriatric syndrome—"osteosarcopenia" [5].

The global burden of osteosarcopenia is set to rise. Functional disability and limitations (e.g., the difficulty persons may experience with tasks known as activities of daily living (ADLs)), has a significant, and negative impact on the lives of older people. Without timely diagnosis and intervention for disorders of muscle and bone, older persons face threats to their mobility and independence, both of which are key in achieving quality of life in older persons [6].

Currently, clinical evidence regarding nonpharmacological interventions for osteosarcopenia remains in its infancy [7] but is growing. Further, there is increasing interest in the shared pathophysiological mechanisms behind dysfunction of muscle and bone [8]. Research in this domain may offer unique targets for pharmacological therapy for osteosarcopenia in the future [8]. This chapter aims to inform health professionals, students, and persons interested in osteosarcopenia with useful information on the pathophysiology, epidemiology, clinical assessment, and management of osteosarcopenia with a particular focus on practical interventions. The most important intervention for osteosarcopenia is exercise and therefore this chapter deeply examines the benefits, risks, and practical applications of exercise in the prevention and management of osteosarcopenia.

Epidemiology

In contrast to osteoporosis and sarcopenia considered individually, there is less research on the epidemiology of osteosarcopenia [2]. This is in part because osteosarcopenia is a more recently established syndrome than its component parts—osteoporosis/-penia and sarcopenia. Determining the prevalence of osteosarcopenia depends upon the definition of osteosarcopenia applied. This complexity is further increased when considering population differences for what is considered "normal" or "abnormal."

In a population of community-dwelling older adults, it was confirmed that the prevalence of osteosarcopenia increases with age. In men aged 60–64 years old, the prevalence was 14.3%, increasing to 59.4% at ≥75 years. In women aged 60–64 years, the prevalence is 20.3% but increasing more than twofold to 48.5% in those aged ≥75 years [9].

A recent systematic review and meta-analysis that encompassed five definitions of osteosarcopenia across 17 studies found a wide range of prevalence's across different populations, suggesting the prevalence of osteosarcopenia was 5%–37% across inpatient and community settings [10]. The highest prevalence and highest risk of osteosarcopenia were in the population that has experienced a fall or fracture [10].

The high prevalence of osteosarcopenia particularly in persons aged over 75 years and in those who have experienced a fall or fracture suggests that clinicians should be suspicious of and actively consider osteosarcopenia in persons at risk.

Exercise to Prevent and Manage Chronic Disease Across the Lifespan. https://doi.org/10.1016/B978-0-323-89843-0.00027-1

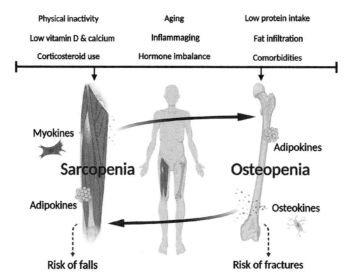

FIG. 1 Risk factors, muscle-bone crosstalk, and the pathophysiology of osteosarcopenia. *From Kirk B, Zanker J, Duque G. Osteosarcopenia: epidemiology, diagnosis, and treatment-facts and numbers. J Cachexia Sarcopenia Muscle 2020;11(3):609–18; under the Creative Commons license.*

Pathophysiology

There are numerous mechanisms that individually, or in combination, may explain the development of osteosarcopenia in older persons. The pathophysiology of muscle and bone loss may be broadly considered in terms of mechanical, molecular, cellular, genetic, and nutritional factors. These factors are summarized in Fig. 1, from Kirk et al. [11].

Mechanical loading

Muscle contraction imposes mechanical forces on bone. If the force of muscle contraction exceeds a certain threshold, bone resorption shifts in favor of bone formation [12]. Mechanical loading is vital for the maintenance of the bone-muscle unit. The reduction in physical activity often observed with aging can reduce useful mechanical stressors on bone with the result being net loss of muscle and bone [12].

Molecular and cellular factors

The communication between muscle and bone, also known as muscle–bone crosstalk (Fig. 1), is mediated by a number of hormones and molecules. The dysregulation of this communication is key in the development of osteosarcopenia. Hormonal factors such as gonadal sex hormones (estrogen in women and testosterone in men), growth hormone and insulin-like growth factor 1 (GH/IGF1) play a role in the development of muscle and bone in early life, maintenance in midlife, and the preservation of these tissues in later life [13]. Both GH and IGF1 exert an anabolic influence on osteoblasts (bone-building cells) and muscle [14]. The expression of estrogen receptors on bone and muscle as well as exposure to adequate levels of estrogen is vital to the preservation of bone and muscle mass with age. Lower levels of estrogen in postmenopausal women without treatment with exogenous estrogen is a strong risk factor for osteoporosis and minimal trauma fracture [14]. Estrogen derived from metabolism of androgens (male sex hormones, e.g., testosterone) plays a role in preserving bone and muscle mass in males. Low testosterone levels in older men result in reduced protein synthesis, subsequent loss of muscle mass, and reduced bone density [13].

Muscle–bone crosstalk is also orchestrated by a family of cytokines (cell-signaling molecules) [15]. These include osteokines (released from bone), myokines (released from muscle), and adipokines (released from adipose tissue or fat). Osteokines such as osteocalcin and sclerostin modulate bone turnover and bone mineral density and induce muscle turnover [15]. Myostatin, a cytokine released from muscle, plays an important role in the proliferative capacity of muscle and bone cells [16]. Myostatin induces protein degradation in muscle and inhibits osteoblastic differentiation (i.e., bone growth cell maturation) in bone, resulting in a net loss of muscle and bone mass [17]. Several other myokines such as IL-6, irisin, IGF-1, BDNF and FGF2 exert effects on bone and the myokines with lipolytic effects (e.g., the breaking down of fat) such as IL-6, irisin, and LIF (released during exercise) stimulate thermogenesis by effects on adipocytes [15].

Increasing fat accumulation with aging creates a more "lipotoxic" environment. Adipokines modulate bone and muscle metabolism through leptin, resistin, adiponectin, and TNFα [15]. There is also increasing evidence of the importance dysregulation of mesenchymal stem cells, which reside in muscle bone and fat, in the development of osteosarcopenia [18].

Genetic factors

Genetic polymorphisms, which describe the difference in DNA sequences and traits among individuals and populations, play a role in the development of muscle and bone pathology in adults. Implicated genes include glycine-N-acyltransferase (GLYAT), methyltransferase like 21C (METTL21C), and peroxisome proliferator-activated receptor gamma coactivator 1-alpha (PGC-1 α) [11]. Polymorphisms of these genes have been associated with bone loss and muscle atrophy [5].

Nutritional factors

The production of type 1 collagen and synthesis of actin and myosin in skeletal muscle is dependent on dietary protein and its constituent amino acids [14]. Dietary protein stimulates the release of anabolic hormones such as IGF-1 and other growth factors involved in the regeneration of skeletal muscle and bone. Further, calcium and vitamin D work in concert to support bone structure. Vitamin D regulates hundreds of genes involved in calcium and phosphate homeostasis in the skeleton. Vitamin D and its receptor are also important components in the development of muscle known as myogenesis [19].

Risk factors

Age is the major risk factor for the development of osteosarcopenia. Other, reversible or modifiable risk factors include low levels of physical activity in both early and later life, immobilization, low intake of essential nutrients, malnutrition, and obesity [20]. These are summarized in Fig. 1 above.

Aging

Stem cell senescence (the loss of a cell's ability to divide and grow with age) is a major contributor to the development of osteosarcopenia [21]. Possible mechanisms behind stem cell senescence include increased reactive oxygen species (that can cause cellular damage), reduced transcriptional control, and telomere shortening (which describe changes and problems at the DNA-level) [3].

The composition of the human body changes with advancing age. On average, there is an increase in body fat and a decrease in muscle mass both in absolute terms and proportionally, which often occurs despite overall body weight remaining stable. This excess fat (or adipose tissue) in combination with low muscle mass, when reaching critical levels has been termed "sarcopenic obesity." Sarcopenic obesity and is associated with increased risk of falls and fractures in older people [22].

Nutrition and malnutrition

The utilization of essential nutrients such as protein, vitamin D, and calcium impact the structure and function of muscle and bone [14]. The availability and utilization of these nutrients to regulate bodily tissues are dependent on multiple cellular pathways. By ensuring adequate energy, protein, and micronutrient intake, and correcting vitamin D, calcium, and protein deficiencies, bone density can be increased and fracture risk reduced [16].

Inflammation

States of increased inflammation, termed "proinflammatory states," are risk factors for osteosarcopenia and its complications. Where osteosarcopenia is characterized by a disproportionate loss of muscle and bone compared with fat, cachexia, which can occur in cancer and other proinflammatory states, describes a loss of body weight (with loss of both muscle and fat mass), and carries related but distinct risks to osteosarcopenia [8]. Other comorbidities associated with cachexia such as chronic obstructive pulmonary disease (COPD), diabetes, congestive heart disease, through pro-inflammatory mechanisms, expedite the process of bone and muscle loss [5]. Chronic energy surplus (increases in food intake or reductions in physical activity) associated with aging can also accelerate the decline in muscle and bone mass through increased fatty infiltration, the development of a proinflammatory state, and lipotoxic effects on surrounding cells [15]. At the population level, epidemiological studies have shown associations between both osteoporosis and sarcopenia and elevated C-reactive protein (CRP), which is a marker of active inflammation [23].

Smoking and alcohol

Both smoking and alcohol intake are well-established risk factors for fracture due to multifactorial effects on bone and muscle, through the process of inflammation and generation of reactive oxygen species [24]. Heavy intake of alcohol (>210 g/week) is associated with low muscle mass due to direct toxic effect on muscle fibers [25].

Clinical outcomes

There is some inconsistency among the literature as to whether osteosarcopenia is associated with worse clinical outcomes than its component parts, osteoporosis and sarcopenia [8]. Part of this inconsistency in findings relates to differences in the definitions applied coupled with variable study populations. However, several papers have identified worse outcomes and associations, including depression, malnutrition, falls, fractures, and mobility limitations [26].

Using various definitions of sarcopenia, it was found that older persons with osteosarcopenia showed an increased rate of falls (odds ratio 2.83–3.63; $P < 0.05$) and fractures (odds ratio from 3.86–4.38; $P < 0.05$) compared to non-osteosarcopenic individuals [26]. Fractures are a very important clinical outcome associated with significant morbidity and mortality. Following a hip fracture approximately half of previously ambulatory individuals are unable to mobilize independently and approximately 55% of individuals >90 years old are no longer able to live independently after fracture [27].

In an observational study of Korean patients with hip fractures, it was determined that the presence of osteosarcopenia was associated with a one-year mortality rate of 15.1%, compared to osteoporosis (5.1%) and sarcopenia alone (10.3%) [28]. Studies reporting no associations with worse outcomes in persons with osteosarcopenia compared to those with osteoporosis and sarcopenia alone include two longitudinal studies, which did not show greater risk of falls, fractures [29], or mortality [30] in those with osteosarcopenia.

Definition and clinical assessment

While osteosarcopenia is well established as the presence of sarcopenia and osteoporosis or osteopenia, there is no universal definition of sarcopenia. An exploration of the nuances of controversy regarding the definition of sarcopenia is beyond the scope of this chapter and therefore this section on assessment will focus on that which is practically relevant to clinicians.

Typically, the term "older adults" includes all people over the age of 65 years. In the Australian context, it includes Aboriginal and Torres Strait Islander peoples over the age of 50 or 55 years and non-Indigenous people over the age of 65 years [31].

Osteoporosis and sarcopenia and under-detected and undertreated in older adults [11]. Should either condition be suspected or confirmed, clinicians should be suspicious of osteosarcopenia given the high rate of co-occurrence between conditions. Those at risk of osteosarcopenia may present with a fall, fracture, describe feeling "weak" or with reduced ability to undertake ADLs, or be without any symptoms. A screening test, the SARC-F, has been included in one recent definition of sarcopenia [32,33].

A comprehensive assessment for osteosarcopenia comprises a history (e.g., medical comorbidities, social, medications, falls, and fractures) risk factor identification (see "Risk Factors" above), physical assessments and targeted investigations [20]. Fig. 2, from Kirk et al [20], illustrates a possible diagnostic and management approach to osteosarcopenia. For the isolated management of either sarcopenia or osteoporosis individually, we point readers to recent guidelines [2,34].

Assessment of bone

The presence of a minimal trauma fracture is diagnostic for osteoporosis, regardless of bone mineral density. Clinicians should consider BMD testing via Dual Xray Absorptiometry (DXA) in all adults over 50 or at risk of or with a history of fracture, postmenopausal women, men over the age of 70, or adults with a condition (e.g., rheumatoid arthritis) or medication (e.g., steroids such as prednisolone) that contribute to bone loss [2]. Bone mineral density is best described by the "T-score," which compares the individual's BMD with that of a sex-matched healthy young adult. The World Health Organization (WHO) has defined osteopenia as a T-score between −1 and −2.5, and osteoporosis as a T-score less than −2.5 or the presence of a minimal trauma fracture [35]. In some settings, resource limitations may prevent the assessment of BMD

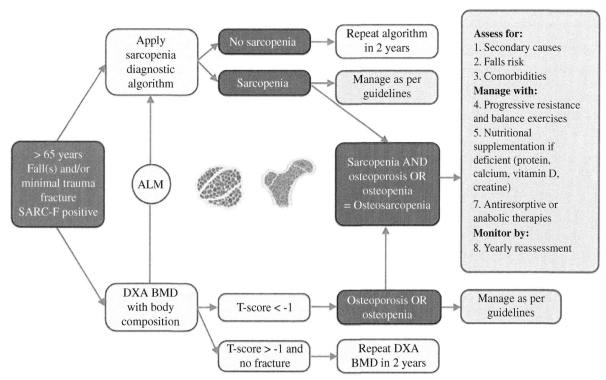

FIG. 2 Diagnosis and management algorithm for osteosarcopenia. *BMD* =bone mineral density. *DXA* =dual-X-ray absorptiometry. *SARC-F* =Screening tool for sarcopenia. *From Kirk B, Miller S, Zanker J, Duque G. A clinical guide to the pathophysiology, diagnosis and treatment of osteosarcopenia. Maturitas. 2020;140:27–33 reproduced with permission.*

with DXA. The FRAX is a validated tool used for fracture risk estimation and can be applied in the absence of BMD assessment [36]. FRAX is a free, online resource and has been validated in over 80% of global populations [2].

Assessment of muscle

A physical examination, exploring for comorbidities, pain, and contributory factors to musculoskeletal health and ill-health is a key component of an assessment for osteosarcopenia. However, there are some specific physical assessments to be included in an assessment of sarcopenia. Clinicians should familiarize themselves with the various definitions of sarcopenia (Table 1) and the preferred definition and cut-points within their institution or region. Using the same definition within and between individuals is key for establishing consistent management approaches to osteosarcopenia.

First, muscle characteristics may be described in terms of strength, physical performance, and quality or quantity. While previous definitions of sarcopenia required measures of all three, the most recent definition of sarcopenia by the Sarcopenia Definition and Outcomes Consortium (SDOC) only requires a measure of muscle strength (grip strength by handheld dynamometer) and physical performance (normal walking speed) [40]. The SDOC algorithm supports diagnosis of sarcopenia in settings of fewer resources as imaging technologies such as DXA are not required for diagnosis, and sarcopenia diagnosed by the SDOC has been associated with important negative outcomes in high-risk older adults with a history fall or fracture [41]. Previous, yet well-established, definitions of sarcopenia give clinicians the option to choose measures of muscle strength (grip strength; or 5-times chair sit to stand test) and physical performance (normal walking speed; 400m walk test; Timed up and Go Test; or Short Physical Performance Battery (SPPB)) [32] for assessment. While this creates flexibility in approaches, the broad application of different measures of muscle creates wide variability in prevalence estimates and diagnosis of sarcopenia [42]. The authors of this chapter do not suggest one definition of sarcopenia is superior to another and support clinician discretion in their application of a diagnostic algorithm for sarcopenia.

Muscle quality or quantity can be estimated using a range of imaging modalities. It is important for clinicians not to confuse the term "muscle mass" with the surrogate measures of muscle quantity or quality that currently available imaging techniques produce. DXA estimates "lean mass" (which is a surrogate and over-estimate of muscle mass, and includes connective tissues and is influenced by hydration status). DXA-derived lean mass is widely used in past sarcopenia

TABLE 1 Recent operational definitions of sarcopenia.

Component	Cut-points
Initial European Working Group on Sarcopenia in Older People (EWGSOP) [37] **adopted by the Australian and New Zealand Society for Sarcopenia and Frailty Research (ANZSSFR) in 2018** [38] Sarcopenia = Low lean mass and low muscle strength and low physical performance	
Low muscle (lean) quantity	ALM using whole-body DXA: *Adjusted for height (m²)* Men: $< 7.26\,kg/m^2$ Women: $< 5.50\,kg/m^2$
Low muscle strength	Hand grip strength using dynamometer: Men: $<30\,kg$ Women: $<20\,kg$
Low physical performance	Men and women over 4 m course: Gait speed: $\leq 0.8\,m/s$
Revised European Working Group on Sarcopenia in Older People [32] Probable sarcopenia = Low muscle strength Sarcopenia = Low muscle strength and low muscle quality or quantity Severe sarcopenia = Low muscle strength and low muscle quality or quantity and low physical performance	
Low muscle quantity	ASM adjusted for height (m²) using whole-body DXA: Men: $<7.0\,kg/m^2$ Women: $<5.5\,kg/m^2$ ASM adjusted for height (m²) using DXA: Men: $<20.0\,kg/m^2$ Women: $<15.0\,kg/m^2$
Low muscle strength	Hand grip strength using dynamometer: Men: $<27\,kg$ Women: $< 16\,kg$ Chair stand: $>15\,s$ for 5 rises
Low physical performance	Gait speed: $\leq 0.8\,m/s$ SPPB: ≤ 8 point score TUG: $\geq 20\,s$ 400 m walk test: non-completion of $\geq 6\,min$
Asian Working Group for Sarcopenia: 2019 consensus update [39] Sarcopenia: Low ASM and low muscle strength OR low physical performance Severe sarcopenia: Low ASM and low muscle strength and low physical performance	
Low muscle quantity	ASM adjusted for height (m²) using whole-body DXA: Men: $<7.0\,kg/m^2$ Women: $<5.54\,g/m^2$ SSMI adjusted for height (m²) using BIA: Men: $<7.0\,kg/m^2$ Women: $<5.7\,kg/m^2$
Low muscle strength	Hand grip strength using dynamometer: Men: $<28\,kg$ Women: $< 18\,kg$
Low physical performance	Gait speed: $<1.0\,m/s$ Chair stand: $>12\,s$ for 5 rises SPPB: ≤ 9 point score

TABLE 1 Recent operational definitions of sarcopenia—cont'd

Component	Cut-points
Sarcopenia Definition and Outcomes Consortium [40] Sarcopenia=Low muscle strength (grip strength) and low physical performance (walking speed)	
Low muscle strength	Hand grip strength using dynamometer: Men: <35.5 kg Women: <20 kg
Low physical performance	Gait speed: <0.8 m/s

ALM =appendicular lean mass. *ASM* =appendicular skeletal muscle mass. *BIA* =bioelectrical impedance analysis. *BMI* =body mass index. *DXA* =dual energy X-ray absorptiometry. *SPPB* =short physical performance battery.

definitions, however, DXA-derived low lean mass has shown a lack of consistent association with negative outcomes in older adults and was not selected as a key variable in the SDOC's data-driven definition of sarcopenia [40]. However, with respect to osteosarcopenia, there is dual benefit in assessment of BMD with DXA as clinicians may request assessment of appendicular lean mass (ALM) which in sarcopenia definitions is typically adjusted for height^2 [32].

Less commonly used imaging modalities include bioimpedance analysis (BIA) which measures "fat-free mass;" computerized tomography (CT) which measures "cross-sectional area" of muscle, or CSA; and Magnetic Resonance Imaging, or MRI, which measures "muscle volume." An exciting, novel technique, known as the "d3-creatine dilution technique" is a direct measure of muscle mass and has shown associations with negative outcomes in older men, and found to be a superior predictor of mobility disability than DXA ALM in this group [43]. However, this technique remains in the validation phase and further study is required prior to adoption into clinical settings.

Other investigations

Additional investigations are recommended to assist clinicians in addressing modifiable risk factors for osteosarcopenia. A base set of investigations that will support clinicians to identify most secondary causes of osteosarcopenia (beyond those revealed in the clinical history) include serum vitamin D (25OH), calcium, parathyroid hormone, and serum testosterone in men [44].

Management of osteosarcopenia: A focus on exercise and nutrition

In this section, we outline practical considerations regarding exercise and nutrition prescriptions for older adults with osteosarcopenia. While there are numerous pharmacological options for osteoporosis, there are no approved pharmacological treatments specifically for osteosarcopenia. Therefore, in line with the scope of this book, we have focused management principles on exercise and nutrition. The pharmacologic interventions for osteoporosis, along with future possibilities for pharmacological therapy for sarcopenia and osteosarcopenia are well-described elsewhere [2,45].

Overview of exercise in musculoskeletal syndromes and disease

Any formal exercise independent of the mode (e.g., cardiorespiratory-dominant or musculoskeletal-dominant) has varying yet favorable effects on both bone and muscle tissue. For exercise to be most effective in osteosarcopenia, the primary goal should be consistency regardless of the mode applied [46]. Although some studies suggest that cardiorespiratory-dominant exercises do not meaningfully stimulate bone tissue, this is not necessarily the case for skeletal muscle tissue and generally, any exercise has wide-ranging health benefits compared with no exercise [47,48].

The ideal "dose" of exercise (i.e., how often and how intensely) depends on the guidelines being followed and the evidence informing them. Australian recommendations for older adults suggest aiming for around 30 min of moderate physical activity on most days of the week [49], however, these general recommendations do not target those with specific health conditions such as osteosarcopenia. Much of the literature that focuses on specific exercise programs recommend two to three times per week, with each session lasting between 30 and 60 min [50].

The optimal exercise program for older adults living with osteosarcopenia includes progressive resistance training (PRT) and balance training. These two modes of exercise appear to be the most beneficial in preventing or managing osteosarcopenia [51–55].

The rationale behind both regular resistance training and balance training is their combined effects on bone and skeletal muscle loading as well as proprioceptive awareness—all of which converge with the goal of improving strength, physical functioning (walking and climbing stairs), mobility, and position sense while simultaneously decreasing falls and fracture risk.

Mechanism of effect—Progressive resistance training (PRT)

During aging, there is a gradual net decline of bone and muscle tissue starting from as early as the third decade of life. The mechanical loading forces associated with regular PRT stimulate both osteoblastogenesis and myofibrillar hypertrophy through muscle protein synthesis [56,57].

Importantly, PRT also increases total daily energy expenditure and this produces a higher rate of relative endogenous fatty acid oxidation, particularly within skeletal muscle tissue (including intramuscular adipose tissue) [58–60]. This may well be a mechanism by which the effects of osteosarcopenic obesity can be mitigated and reversed. Beyond osteosarcopenic obesity, increased fatty acid oxidation via PRT may reduce the negative effects that excessive adipose tissue can have on the musculoskeletal system.

PRT may also improve balance, but the literature is mixed. One study in older adults concluded that PRT alone does not improve some aspects of balance whereas another more recent study concluded that it did [61,62]. Interestingly, both of these studies only used expensive machine-based resistance equipment. Less expensive and more accessible free weight equipment (e.g., dumbbells, barbells, kettlebells, and jugs filled with water) may serve to improve balance in addition to strength through the postural stability needed to hold the weight in different positions when performing resistance exercise. This is an area in need of further study.

Mechanism of effect—Balance training

The three major components of the nervous system responsible for position sense include the somatosensory, vestibular, and visual pathways [63,64]. All of these diminish with aging and can impair both balance and postural stability.

The neural pathways that are responsible for balance respond favorably to focused and repeated postural challenges often involving isolated or combined movements of the lower limbs, upper limbs, neck, trunk and eyes (Fig. 3). Training of the somatosensory, vestibular, and visual systems using balance training variations can augment motor learning patterns which support the musculoskeletal system to move more efficiently (or faster and with less error) (Fig. 4) [65,66].

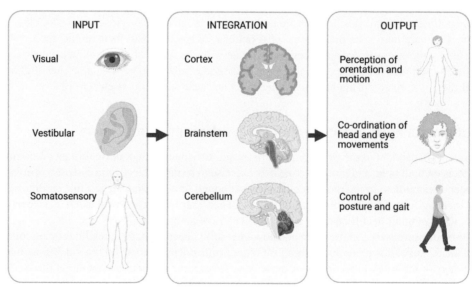

FIG. 3 The pathways of sensory inputs, integration, and output for balance in humans.

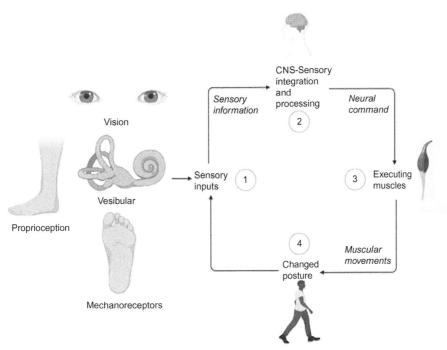

FIG. 4   Human postural control and orientation. *Adapted from Modig F. Effects of acute alcohol intoxication on human sensory orientation and postural control. Faculty of Medicine, Lund University; 2013.*

Despite the importance of consistent balance training on both falls and fracture risk, most of the existing evidence supports the use of multimodal exercise, although some isolated balance-training programs have been effective [67]. This suggests that an exercise program combining PRT and balance training together may be optimal in older persons with osteosarcopenia, but requires further research.

Safety, precautions and adverse effects

The safety of exercise for individuals and groups needs to be considered, and individuals thoroughly assessed for safety risks, prior to older adults with osteosarcopenia commencing exercise programs.

While the literature suggests that focused exercise prescriptions carry minimal risk of adverse effects, it is nonetheless important to consider precautions and safety and modify recommendations according to risk [68].

Potential adverse effects of exercise

Some of the more frequent, albeit uncommon, complications associated with a wide range of exercise-related activities in older people include [68]:

- Spinal (or vertebral) compression fractures. These fractures have an association with forward bending or spinal flexion. Spinal compression fractures are more commonly reported in postmenopausal females with osteoporosis. Pathological fractures may also be a sequalae of other primary bone disease and metastatic disease affecting bone.
- Back pain and joint pain. This may be a result of pain arising in the innervated joints within the spine itself or the appendicular joints of the arms and legs. Joint pain can arise through either overuse or acute injury [69]. Clinicians should consider that the manifestation of pain, however, is not always nor strictly a direct result of musculoskeletal damage [70,71].
- Generalized muscle soreness experienced after certain modes of exercise—also called delayed onset muscle soreness (DOMS). DOMS is often considered a natural response to increased muscular stress following exercise in the initial phases. DOMS may be decreased by reducing eccentric load early in the initiation of a new exercise program. DOMS could be considered adverse if it leads to individual's ceasing the exercise program and experiencing the indirect and long-term complications of cessation [72].

- Exacerbation of chronic non-musculoskeletal conditions and comorbidities that can also impact the ability to safely perform very high-intensity exercise, such as uncontrolled hypertension, coronary artery disease, and cardiomyopathies [73].

Potential adverse effects of exercise in older persons with osteosarcopenia

The literature examining the potential adverse effects of exercise on those living with osteosarcopenia is scarce. One systematic review found no major adverse effects of exercise, although DOMS was reported in some in the beginning phases of the exercise program and both knee and shoulder pain were reported by some individuals [7]. Notably, these adverse effects were specific to resistance training and are likely more relevant for older adults with osteosarcopenia compared with the adverse effects associated with general exercise.

Those engaging in exercise programs, particularly older adults who may be inexperienced with formal exercise, require supervision and training to minimize risk of adverse effects. Improper lifting and movement techniques are associated with low skill acquisition or excessive loading, and may increase the risk of adverse effects and injury [69,74,75].

An additional consideration is that osteosarcopenia is characterized by both decreased muscle strength and physical functioning, which independently increases the risk of fall and fracture. These risks may be most relevant in the balance training settings, where the focus is on challenges to the somatosensory, vestibular, and visual systems which are upheld by muscle strength and function.

Clinical practice recommendations for exercise prescribers

The prescription of exercise for older adults with osteosarcopenia should be undertaken with the same level of care applied to the prescription of any therapeutic intervention. This includes consultation with the older adult on exercise options, goals of therapy, explanation of benefits and risks, "dose" (timing and frequency), and timeframes for reassessment.

The primary exercise prescription in managing osteosarcopenia is PRT. PRT that is individualized, performed consistently and with elements of progression has been shown to be the most effective mode of exercise in managing osteosarcopenia [7,51–53,73]. Beyond the specifics of an exercise prescription, health professionals should also focus on how to best incorporate and encourage exercise adherence and the associated goals and motivations of the individual [76]. Motivational interviewing—a form of conversing with older adults in a clinical context which explores affirmations, encourages reflections and positively promotes self-analysis by empowering the older adult—is a tool the practitioner may be able to utilize to inspire exercise adherence [77].

The most recent systematic review of PRT for osteosarcopenia included two randomized control trials (RCTs) comprised of 106 older adults with osteosarcopenia [7]. The exercise protocols studied are represented in Tables 2 and 3. In these RCTs, there was moderate-quality evidence that PRT increased muscle mass, strength, and quality [51,55]. Changes in strength and quality tended to precede changes in muscle (or lean) mass. PRT increased lumbar spine BMD and maintained hip BMD over 12 and 18 months, with low-quality evidence. PRT showed no effect on physical performance, such as normal walking speed [7].

In addition to resistance exercise, prescribers may also wish to consider the role of general exercise and its potential benefits for chronic pain, given exercise may in general reduce the experience of certain types of pain, increase physical and physiological functioning and importantly—increase the quality of life [78].

Individual exercise preferences of older adults should be built upon, rather than discouraged, even if there is an absence of evidence demonstrating benefit of the individual's exercise of choice on osteosarcopenia. Individuals may prefer exercise types such as yoga, water aerobics, endurance training, or competitive sport. It is our opinion that clinicians should view this as an opportunity to promote, explain, and demonstrate how to add optimal exercise for osteosarcopenia to the individual's existing exercise program.

Tables 2 and 3 are presented for readers to consider and clinicians to adapt, modify, and implement according to the client's values and preferences.

We present Table 4 to provide an example of how the programs listed in Tables 2 and 3 could be modified to match the goals and abilities of the older adult engaging in a program. The authors recommend that only certified clinicians trained in developing and implementing exercise programs for older adults perform such adaptations. Exercise prescribers may wish to utilize a mix, combination, or variation of the options listed below. The authors note that these descriptions draw on the available evidence to provide practical options for exercise prescribers, but have not been established in a clinical trial on

TABLE 2 The use of elastic bands resistance training.

- Demographic: 63 community-dwelling females with osteosarcopenia aged 65–80 years, living in Shahrekord, Iran.
- Frequency: Three times per week.
- Duration: 60 min per session (plus a 10-min mobility warm-up and five-minute cooldown) for 12 weeks total.
- Volume/Intensity: Both progressively increased using inexpensive TheraBand equipment for the 12 weeks of the study starting with one set of 10–12 repetitions for the first 4 weeks and the addition of a second set for the remaining 8 weeks. TheraBand tensile strength increased from easier (yellow) to harder (to red, then to green) based on individual reports of difficulty. Either one or two exercises (an upper quarter, a lower quarter, or both) were performed in any one session.
- Muscle groups trained: Legs, back, abdomen, chest, shoulders and arms.
- Results: The 30-s chair-stand test (a measure of lower limb strength and endurance) as well as handgrip strength (a measure of upper limb strength) both increased significantly in the exercise group compared with the control group.
- Advantages: TheraBand equipment is relatively inexpensive at around <$20USD per set, they are generally safe and can complement exercise in any location making them user-friendly and convenient.
- Disadvantages: Theraband equipment may not confer the same progression of resistance as free weights and may not maximize hypertrophic effects and concomitant strength gains. The use of a Theraband can create difficulty in quantifying resistance load (in kilograms) given each band has a set resistance without a qualified weight.

From Banitalebi E, Faramarzi M, Ghahfarokhi MM, SavariNikoo F, Soltani N, Bahramzadeh A. Osteosarcopenic obesity markers following elastic band resistance training: A randomized controlled trial. Exp Gerontol. 2020;135:110884.

TABLE 3 The use of machine-based resistance training.

- Demographic: 43 community-dwelling males with osteosarcopenia aged 73–91 years, living in Erlangen, Germany.
- Frequency: Two times per week.
- Duration: Around 45 min per session on average with follow-ups for up to 78 weeks.
- Volume/Intensity: A periodized high-intensity protocol with one set in a repetition range of 5–10. The intensity of effort was defined as non-repetition maximum with some remaining repetitions in reserve.
- Muscle groups trained: All major and minor groups via leg press, leg extension, leg curl, leg abduction and adduction, lateral pulldowns, rows, back extensions, inverse fly, bench press, military press, lateral raises, and crunches.
- Results: BMD in the lumbar spine increased in the resistance training group compared with the control group. Skeletal muscle mass index increased significantly in the resistance training group and decreased significantly in the control group. Maximal hip and leg extensor strength increased significantly in the training group compared with the control group.
- Advantages: Machines are easy to use and do not require as much technical competence compared with free weights. They have set weight progressions which support the tracking of strength progression throughout any training period.
- Disadvantages: Machines require significant floor space and can be inconvenient from a storage perspective. To maximize benefit a gym membership is usually required to use all the relevant machines which can be expensive. While simple for introducing people to resistance training, people can be caught in a potentially aberrant movement pattern if not properly coached or supervised.

From Lichtenberg T, von Stengel S, Sieber C, Kemmler W. The Favorable Effects of a High-Intensity Resistance Training on Sarcopenia in Older Community-Dwelling Men with Osteosarcopenia: The Randomized Controlled FrOST Study. Clin Interv Aging. 2019;14:2173–86.

participants with osteosarcopenia. Drawing on the available evidence, we posit that the optimal non-pharmacological approach to osteosarcopenia would include PRT or multimodal training (e.g., balance training) and nutritional considerations, which are outlined below. These are summarized in Fig. 5, from Kirk et al [11].

1. Progressive resistance training

The key variable for increasing both muscle mass and strength is training volume [79]. Any resistance training program will need to ensure that adequate volume is one of the foundations. Drawing from the previous RCTs on resistance training in osteosarcopenia, the optimal training volume is likely to be comprised of at least two, 45–60 min sessions, per week.

Intensity can be measured in many different ways and one of the more accepted methods measuring the intensity of effort is based on the rate of perceived exertion (RPE) scale, also called the Borg scale (Table 5) [80]. The RPE scale asks the participant to self-rate how difficult their exercise felt from 1 to 10, with 1 being the easiest and 10 being the most difficult. Another measure of intensity is the "intensity of load" which is often described as the amount of weight used in a set, often expressed as a percentage of one-repetition maximum (%1RM) [81].

TABLE 4 Upper and lower body split routine.

Frequency: Performed twice per week with at least 36–48h between each workout.

Duration: Between 30[a]-60min per session.

Intensity: Aiming for a %1RM between 60% and 80% or a repetition range between 10 and 20. Using the RPE scale, the rating to aim for would likely range between 6 and 8 (hard to very hard).

Video examples of the exercises below are available from a range of online resources but individual descriptions are beyond the scope of this chapter.

[Day One—Monday] Upper Body:

- Barbell Bench Press [1 set of 10–15]
- Seated Cable Row [1 set of 10–15]
- Chest Press Machine [1 set of 10–15]
- Lateral Pulldown [1 set of 10–15]
- Dumbbell Hammer Curl [1 set of 10–15 each arm]
- Cable Triceps Pushdown [1 set of 10–15]
- Cable Face Pull [1 set of 10–15]
- Shoulder Press Machine [1 set of 10–15]
- Standing Cable Front Raise [1 set of 10–15]
- Pallof Press [1 set of 10–15]
- Wall or Kneeling Push-Ups [1–3 sets of 10–15][#]
- Kneeling Plank to Plank progression [1–3 sets of 20–60s][#]
- Kettlebell/Dumbbell/Water Bottle Farmers Walks [1–3 sets of 10-m paces][#]
- Dumbbell/Water Bottle One-Arm Bent Over Row [1–3 sets of 10–15][#]

[Day Two—Thursday] Lower Body:

- Leg Press [1 set of 10–15]
- Calf Press [1 set of 15–20]
- Seated Leg Extension Machine [1 set of 10–15]
- Goblet/Water Bottle Box Squat on adjustable box [1 set of 10–15][b]
- Lateral Band Walk with band around midfoot [1–3 sets of 10–15][b]
- Dumbbell/Water Bottle Romanian Deadlift [1–3 sets of 10–15]
- Lying/Seated Leg Curl Machine [1 set of 10–15]
- Stability Ball Wall Sit [1–3 sets of 10–15][b]
- Supported Standing Leg Kickbacks with band [1–3 sets of 10–15 each leg][b]
- Supported/Seated Calf Raise [1–3 sets of 15–20][b]

[a]Note that these programs are best supervised, especially when commencing a new resistance-training protocol. Also, strength gains will achieve diminishing returns as each new set is introduced therefore only one set may be necessary as is shown below (compared to two or three sets). This may help to reduce boredom and also potentially minimize injury risk from same muscle group overuse. However, adding more sets may also add marginally better strength gains but is more time-consuming [37]. Repetition ranges below are not definitive and all exercises prescribed would be gauged on clinical assessment and management considerations of each individual based on any possible co-morbidities, access to equipment, mental health status, cultural, gender, and language considerations.

[b]Increased sets are added to both upper and lower exercises that can be done at home to increase exercise training volume and allow for affordable access to some older adults who may be in a socioeconomic disadvantaged position.

Further, there appears to be some advantage in the use of free weights as compared with machine-based resistance training when measured by strength gains but also subjective perception of strength improvements in tasks of daily life [82]. Exercise prescribers could consider developing a training program with free weights for those individuals seeking potentially greater challenge and improvement.

2. Multimodal training

This format of training combines multiple methods aiming to induce musculoskeletal gains. Multimodal training should always include PRT as the foundation but can incorporate other modes of training that may further enhance the effect on the musculoskeletal system.

Beyond strength training, other common modes include balance, flexibility, and endurance training. All forms of training have their respective effects on fitness and the addition of balance training may have the most optimal effect

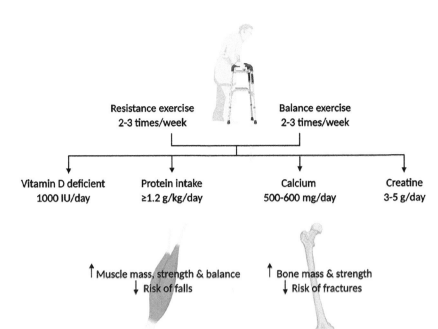

FIG. 5 Lifestyle treatments for osteosarcopenia. *From Kirk B, Zanker J, Duque G. Osteosarcopenia: epidemiology, diagnosis, and treatment-facts and numbers. J Cachexia Sarcopenia Muscle 2020;11(3):609–18; under the Creative Commons license.*

TABLE 5 The Rate of perceived exertion (RPE) scale.

Rating	Descriptor
0	Rest
1	Very, very easy
2	Easy
3	Moderate
4	Somewhat hard
5	Hard
6	Between hard and very hard
7	Very hard
8	Between very hard and maximal
9	Between very hard and maximal
10	Maximal

Adapted from Egan A, Winchester J, Foster C, McGuigan M. Using Session RPE to Monitor Different Methods of Resistance Exercise. J Sports Sci Med. 2006;5(2):289–95.

on osteosarcopenia (Table 6) [63,85]. Some balance exercises also contain components of bodyweight resistance training by default, thus potentially having a synergistic effect on strength and balance [86].

There is also evidence that impact training, such as jumps, hops, bounds, and similar movements can stimulate BMD in older people [87,88]. However, older people with existing fractures of the vertebrae may require individualization of more moderate compared with vigorous activity intensity [89]. Resistance training programs can be easily modified if there is concern for high-impact training in some individuals.

TABLE 6 Basic static and dynamic balance protocol.

Protocol: Performed twice per week and can be done at home or in a fitness center on consecutive days, nonconsecutive days, or interspersed with a resistance training program.
Duration: Around 20–30 min per session depending on starting age and physical fitness.
[Day One—Tuesday] Static[a] Balance:

- Tandem Stance [30 s each leg in front, to increasing time up to 60 s with progression and using cushion]
- Standing—Feet Together [60 s with eyes open on firm surface, progressing to longer stance phases with cushion, progressing further to eyes closed]
- Single-Leg Stance [30 s each leg with supported chair progressions into unsupported with progression]
- Supported Single Leg Stance with plantarflexion and dorsiflexion [1 set of 10–15 s in isometric plantarflexion on the toes followed by 1 set of 10–15 s in isometric dorsiflexion on the heels with progression to longer holds as strength and balance increase][b]
- Sit-to-Stands [1 set of 10 repetitions progressing to increasing repetitions or sets and narrowing foot positions][b]

[Day Two—Wednesday or Friday] Dynamic[a] Balance:

- Tandem Gait [2–3 sets of 10 paces up and back using a balance beam or piece of tape with supports or using the wall in a narrow hallway with progression to unsupported]
- Chair Transfers [1–2 sets of 10 from chair to chair with progressions being increasing set number or repetitions]
- Modified Clock Reach [1–2 sets of 3–4 excursions of each arm supported with chair and a small reach at each part of the clockface]
- Stair Step-Ups [1 set of 6–10 alternating step-ups onto a single stair supported with progressions to unsupported]
- Lateral (Side) Walking [1–2 sets of 10 paces up and back leading with dominant then switching to nondominant leg with progression to longer distances covered or more sets added]
- Modified Star Excursions [1–2 sets of toe touches in front, on the side and on a backward oblique angle with each leg supported with a chair on the side of the stable leg]

[a]Static balance keeps the body fixed in one place and stable over its center of mass. Static balance training uses a chair in most instances. Even if progressions take place without a chair, it is still a good idea to keep the chair within reaching distance for support if needed. Wobble boards, Bosu balance balls, and rocker boards are seldom needed and a standard lounge chair cushion often will suffice for adding a compliant surface (compared to a firm surface like the ground). Dynamic balance is a more complex task than static balance that uses more complex multi-planar movements and how those movements occur around the base of support based on the vertical axis of the body [83,84].
[b]Both Single-Leg Stance and Sit-to-stands are also a form of bodyweight resistance training. Sit-to-stand may initially need to be done supported when arising from a chair for those with poor lower leg strength.
Adapted from Alfieri FM, et al. Functional mobility and balance in community-dwelling elderly submitted to multisensory versus strength exercises. Clin Interv Aging; 2010;5:181–5.

While some evidence also exists for combined endurance training (e.g., continuous cycling, swimming, jogging, etc.) with strength training as effective in muscle and bone wasting conditions, it has been argued that it may simply be that the exercise combination is more effective for older adults because of a total increase in workout volume [86].

3. Nutritional considerations

There is a paucity of literature on optimal nutritional management for older persons with osteosarcopenia [90,91]. However, dietary changes should only be considered as supplemental to PRT or multimodal training. The major nutrients involved in muscle and bone metabolism include dietary protein, vitamin D, calcium, long-chain omega-3 polyunsaturated fatty acids, and supplemental creatine monohydrate.

Protein

Protein sources containing all of the essential amino acids that typically have a higher concentration of the branched-chain amino acid leucine are most beneficial in stimulating muscle protein synthesis [92]. Whey protein found in dairy contains the most absorbable leucine content. Leucine supplementation, when added at a range between 3 and 5 g to any given meal in older men and women, showed significant augmentation of muscle protein synthesis in both lower (0.8 g/kg/d) and higher (1.2 g/kg/d) daily protein intakes. Similar augmentation of muscle protein synthesis occurred when 3 g of leucine was provided by supplementation with 10 g of whey protein, compared with a 25 g serving of whey protein which naturally contained 3 g of leucine [93,94]. This equivalent effect may help mitigate issues of reduced appetite which can occur as part of aging, particularly as it relates to protein intake [90].

Vitamin D

Deficiency of vitamin D is linked to increased disability and diminishing physical performance [95,96]. Sunlight provides the greatest source for vitamin D by stimulating the conversion to the active form, and nutritional sources include oily fish, egg yolks, and some fortified foods. Therefore, supplementation is recommended for older people who are more at risk of deficiency or proven deficient with 1000 IU/day per day [53].

Calcium

Relatively higher intakes of dietary calcium have been shown to increase bone mineral density by decreasing the amount of bone calcium leeched to maintain adequate blood calcium levels [45]. Therefore, sufficient intake may offset a decline in bone density and potential bone pathologies such as osteoporosis and osteomalacia. Dairy foods such as milk and yoghurt are the greatest source of calcium and older adults are advised to consume 1000–1300 mg/day, with supplementation only provided if dietary intake is suboptimal and there are no contraindications [45].

Long-chain omega-3 polyunsaturated fatty acids (LCn-3PUFAs)

Eicosapentaenoic acid (EPA) and docosahexaenoic acid (DHA) are found in fatty fish [97]. The proposed mechanism for LCn-3PUFAs is both its anti-inflammatory effects on reducing circulating cytokines (which act to increase skeletal muscle fatty infiltration) and they may also promote muscle protein synthesis via the mTOR signaling pathway serving to promote muscle cell growth [92,98,99]. If fatty fish cannot be consumed about twice per week, supplementation with fish oil is recommended. There is no international consensus on dose however the emphasis should be placed on recommending amounts at or higher than the recommendations made by expert groups of 250–500 mg/day or EPA and DHA [100]. However, the effect of LCn-3PUFAs specifically on older adults with osteosarcopenia and its associated outcomes has not been specifically studied and therefore cannot be recommended for the indication of osteosarcopenia alone.

Creatine

Creatine is found in very small amounts in animal flesh like red meat, poultry, and fish. However, creatine monohydrate is best as a supplement given that the dose in foods may not be sufficient to aid in muscle strength and size. Creatine is often taken at an absolute dose of around 20 g/day in an initial dose over 1–2 weeks before being tapered to lower doses of 3–5 g/day in longer term use [101]. Literature strongly suggests that use of creatine monohydrate in both shorter and longer term periods increases muscle size, strength, and physical performance in older adults [94,102,103].

Future considerations and conclusion

In this chapter, we have highlighted the clinical importance of a recently established geriatric syndrome, osteosarcopenia. While osteosarcopenia carries significant risks for older adults, we have presented recent evidence and practical options for clinicians and those living with osteosarcopenia. Through application of evidence-based exercise programs, namely progressive resistance training, coupled with optimal nutrition and protein consumption, the negative effect of osteosarcopenia on the lives of older adults may be mitigated. There is a great opportunity in the knowledge that simple, cost-effective interventions (e.g., exercise) can be provided to treat osteosarcopenia. In the future, further research is required to solidify the role of exercise in the treatment of osteosarcopenia, coupled with possible pharmacologic interventions simultaneously targeting and improving muscle and bone health.

Glossary of exercise terminology

Mode refers to the type of exercise performed: Resistance training (also called weight training or strength training), aerobic training (also called endurance training) and balance training (also called proprioceptive training) are all different modes of exercises.

Volume refers to the total amount of work being done (often over a certain period of time). Volume is arguably the most important variable to increase muscular hypertrophy but also helps increase muscle strength.

Intensity refers to the level of effort the person puts into their exercise. The measure for intensity in the case of resistance training is usually in the form of a percentage of one-repetition maximum (%1RM) or a rating of perceived exertion (RPE also called the Borg scale). Intensity of load (expressed as %1RM) is a key in mediating muscle strength.

Frequency refers to the number of times training or exercise is performed usually expressed as "per-week." E.g., Three training sessions per week indicate the volume would be three.

Duration refers to the time or how long the session lasts. Usually defined in minutes when it comes to resistance exercise or balance training.

Moderate intensity physical activity refers to physical activity where there is an increased heart rate and some shortness of breath but that a conversation of talking can still be sustained.

Vigorous intensity physical activity refers to physical activity where the heart rate increases a lot more and there is shortness of breath to the point of no longer being able to talk or hold a conversation.

References

[1] World Health Organization. World report on ageing and health. Geneva: WHO; 2015.

[2] Zanker J, Duque G. Osteoporosis in older persons: old and new players. J Am Geriatr Soc 2019;67(4):831–40.

[3] Crepaldi G, Maggi S. Sarcopenia and osteoporosis: a hazardous duet. J Endocrinol Invest 2005;28(10 Suppl):66–8.

[4] Binkley N, Buehring B. Beyond FRAX: it's time to consider "sarco-osteopenia". J Clin Densitom 2009;12(4):413–6.

[5] Hirschfeld HP, Kinsella R, Duque G. Osteosarcopenia: where bone, muscle, and fat collide. Osteoporos Int 2017;28(10):2781–90.

[6] Boonen S, et al. Functional outcome and quality of life following hip fracture in elderly women: a prospective controlled study. Osteoporos Int 2004;15(2):87–94.

[7] Atlihan R, Kirk B, Duque G. Non-pharmacological interventions in osteosarcopenia: a systematic review. J Nutr Health Aging 2021;25(1):25–32.

[8] Zanker J, Duque G. Osteosarcopenia: the path beyond controversy. Curr Osteoporos Rep 2020;18(2):81–4.

[9] Fahimfar N, et al. Prevalence of osteosarcopenia and its association with cardiovascular risk factors in Iranian older people: Bushehr elderly health (BEH) program. Calcif Tissue Int 2020;106(4):364–70.

[10] Nielsen BR, et al. Sarcopenia and osteoporosis in older people: a systematic review and meta-analysis. Eur Geriatr Med 2018;9(4):419–34.

[11] Kirk B, Zanker J, Duque G. Osteosarcopenia: epidemiology, diagnosis, and treatment-facts and numbers. J Cachexia Sarcopenia Muscle 2020; 11(3):609–18.

[12] Daly RM, et al. Effects of a 12-month supervised, community-based, multimodal exercise program followed by a 6-month research-to-practice transition on bone mineral density, trabecular microarchitecture, and physical function in older adults: a randomized controlled trial. J Bone Miner Res 2020;35(3):419–29.

[13] Bikle DD, et al. Role of IGF-I signaling in muscle bone interactions. Bone 2015;80:79–88.

[14] Dolan E, Sale C. Protein and bone health across the lifespan. Proc Nutr Soc 2019;78(1):45–55.

[15] Kirk B, et al. Muscle, bone, and fat crosstalk: the biological role of myokines, osteokines, and adipokines. Curr Osteoporos Rep 2020; 18(4):388–400.

[16] Kirk B, Al Saedi A, Duque G. Osteosarcopenia: a case of geroscience. Aging Med (Milton) 2019;2(3):147–56.

[17] Al Saedi A, et al. Rapamycin affects palmitate-induced lipotoxicity in osteoblasts by modulating apoptosis and autophagy. J Gerontol A Biol Sci Med Sci 2020;75(1):58–63.

[18] Feehan J, et al. Bone from blood: characteristics and clinical implications of circulating osteogenic progenitor (COP) cells. J Bone Miner Res 2021;36(1):12–23.

[19] Hayes A, et al. The effect of yearly-dose vitamin D supplementation on muscle function in mice. Nutrients 2019;11(5):1097.

[20] Kirk B, et al. A clinical guide to the pathophysiology, diagnosis and treatment of osteosarcopenia. Maturitas 2020;140:27–33.

[21] Laurent MR, et al. Age-related bone loss and sarcopenia in men. Maturitas 2019;122:51–6.

[22] Scott D, et al. Sarcopenic obesity and its temporal associations with changes in bone mineral density, incident falls, and fractures in older men: the concord health and ageing in men project. J Bone Miner Res 2017;32(3):575–83.

[23] Bano G, et al. Inflammation and sarcopenia: a systematic review and meta-analysis. Maturitas 2017;96:10–5.

[24] Degens H, Gayan-Ramirez G, van Hees HW. Smoking-induced skeletal muscle dysfunction: from evidence to mechanisms. Am J Respir Crit Care Med 2015;191(6):620–5.

[25] Shenkman BS, et al. Effect of chronic alcohol abuse on anabolic and catabolic signaling pathways in human skeletal muscle. Alcohol Clin Exp Res 2018;42(1):41–52.

[26] Sepúlveda-Loyola W, et al. The joint occurrence of osteoporosis and sarcopenia (osteosarcopenia): definitions and characteristics. J Am Med Dir Assoc 2020;21(2):220–5.

[27] Grigoryan KV, Javedan H, Rudolph JL. Orthogeriatric care models and outcomes in hip fracture patients: a systematic review and meta-analysis. J Orthop Trauma 2014;28(3):e49–55.

[28] Yoo JI, et al. Osteosarcopenia in patients with hip fracture is related with high mortality. J Korean Med Sci 2018;33(4), e27.

[29] Scott D, et al. Does combined osteopenia/osteoporosis and sarcopenia confer greater risk of falls and fracture than either condition alone in older men? The concord health and ageing in men project. J Gerontol A Biol Sci Med Sci 2019;74(6):827–34.

[30] Balogun S, et al. Prospective associations of osteosarcopenia and osteodynapenia with incident fracture and mortality over 10 years in community-dwelling older adults. Arch Gerontol Geriatr 2019;82:67–73.

[31] Royal Australian College of General Practitioners. Standards for general practices. Sydney: Royal Australian College of General Practitioners; 2010.

[32] Cruz-Jentoft AJ, et al. Sarcopenia: revised European consensus on definition and diagnosis. Age Ageing 2019;48(1):16–31.

[33] Malmstrom TK, et al. SARC-F: a symptom score to predict persons with sarcopenia at risk for poor functional outcomes. J Cachexia Sarcopenia Muscle 2016;7(1):28–36.

[34] Dent E, et al. International clinical practice guidelines for sarcopenia (ICFSR): screening, diagnosis and management. J Nutr Health Aging 2018; 22(10):1148–61.

[35] Kanis JA, et al. The diagnosis of osteoporosis. J Bone Miner Res 1994;9(8):1137–41.

[36] Kanis JA, et al. FRAX and the assessment of fracture probability in men and women from the UK. Osteoporos Int 2008;19(4):385–97.

[37] Cruz-Jentoft AJ, et al. Sarcopenia: European consensus on definition and diagnosis: report of the European working group on sarcopenia in older people. Age Ageing 2010;39(4):412–23.

[38] Zanker J, et al. Establishing an operational definition of sarcopenia in Australia and New Zealand: Delphi method based consensus statement. J Nutr Health Aging 2019;23(1):105–10.

[39] Chen LK, et al. Asian working group for Sarcopenia: 2019 consensus update on sarcopenia diagnosis and treatment. J Am Med Dir Assoc 2020;21 (3):300–307.e2.

[40] Bhasin S, et al. Sarcopenia definition: the position statements of the sarcopenia definition and outcomes consortium. J Am Geriatr Soc 2020;68 (7):1410–8.

[41] Kirk B, et al. Sarcopenia definitions and outcomes consortium (SDOC) criteria are strongly associated with malnutrition, depression, falls, and fractures in high-risk older persons. J Am Med Dir Assoc 2021;22(4):741–5. https://doi.org/10.1016/j.jamda.2020.06.050.

[42] Phu S, et al. Agreement between initial and revised European working group on sarcopenia in older people definitions. J Am Med Dir Assoc 2019; 20(3):382–383.e1.

[43] Zanker J, et al. Walking speed and muscle mass estimated by the D(3)-creatine dilution method are important components of sarcopenia associated with incident mobility disability in older men: a classification and regression tree analysis. J Am Med Dir Assoc 2020;21(12):1997–2002.e1.

[44] Johnson K, et al. Yield and cost-effectiveness of laboratory testing to identify metabolic contributors to falls and fractures in older persons. Arch Osteoporos 2015;10:226.

[45] Fatima M, Brennan-Olsen SL, Duque G. Therapeutic approaches to osteosarcopenia: insights for the clinician. Ther Adv Musculoskelet Dis 2019;11, 1759720x19867009.

[46] Aartolahti E, et al. Long-term strength and balance training in prevention of decline in muscle strength and mobility in older adults. Aging Clin Exp Res 2020;32(1):59–66.

[47] Beck BR, et al. Exercise and sports science Australia (ESSA) position statement on exercise prescription for the prevention and management of osteoporosis. J Sci Med Sport 2017;20(5):438–45.

[48] Yoo SZ, et al. Role of exercise in age-related sarcopenia. J Exerc Rehabil 2018;14(4):551–8.

[49] Sims J, et al. National physical activity recommendations for older Australians: Discussion document. Canberra: Australian Government Department of Health and Ageing; 2006.

[50] Mayer F, et al. The intensity and effects of strength training in the elderly. Dtsch Arztebl Int 2011;108(21):359–64.

[51] Lichtenberg T, et al. The favorable effects of a high-intensity resistance training on sarcopenia in older community-dwelling men with osteosar-copenia: the randomized controlled FrOST study. Clin Interv Aging 2019;14:2173–86.

[52] Kemmler W, et al. High intensity resistance exercise training to improve body composition and strength in older men with osteosarcopenia. Results of the randomized controlled Franconian osteopenia and sarcopenia trial (FrOST). Front Sports Act Living 2020;2:4.

[53] Kemmler W, et al. Effects of high-intensity resistance training on osteopenia and sarcopenia parameters in older men with osteosarcopenia-one-year results of the randomized controlled Franconian osteopenia and sarcopenia trial (FrOST). J Bone Miner Res 2020;35(9):1634–44.

[54] Kemmler W, et al. Effects of high intensity dynamic resistance exercise and whey protein supplements on osteosarcopenia in older men with low bone and muscle mass. Final results of the randomized controlled FrOST study. Nutrients 2020;12(8):2341.

[55] Banitalebi E, et al. Osteosarcopenic obesity markers following elastic band resistance training: a randomized controlled trial. Exp Gerontol 2020;135, 110884.

[56] Bethel M, et al. The changing balance between osteoblastogenesis and adipogenesis in aging and its impact on hematopoiesis. Curr Osteoporos Rep 2013;11(2):99–106.

[57] Powers SK, Kavazis AN, McClung JM. Oxidative stress and disuse muscle atrophy. J Appl Physiol (1985) 2007;102(6):2389–97.

[58] Ghasemikaram M, et al. Effects of 16 months of high intensity resistance training on thigh muscle fat infiltration in elderly men with osteosarco-penia. Geroscience 2021;43(2):607–17.

[59] Toth MJ, Tchernof A. Lipid metabolism in the elderly. Eur J Clin Nutr 2000;54(Suppl 3):S121–5.

[60] Hunter GR, McCarthy JP, Bamman MM. Effects of resistance training on older adults. Sports Med 2004;34(5):329–48.

[61] Schlicht J, Camaione DN, Owen SV. Effect of intense strength training on standing balance, walking speed, and sit-to-stand performance in older adults. J Gerontol A Biol Sci Med Sci 2001;56(5):M281–6.

[62] Lee IH, Park SY. Balance improvement by strength training for the elderly. J Phys Ther Sci 2013;25(12):1591–3.

[63] Alfieri FM, et al. Functional mobility and balance in community-dwelling elderly submitted to multisensory versus strength exercises. Clin Interv Aging 2010;5:181–5.

[64] Weeks HM, Therrien AS, Bastian AJ. The cerebellum contributes to proprioception during motion. J Neurophysiol 2017;118(2):693–702.

[65] Wong JD, et al. Can proprioceptive training improve motor learning? J Neurophysiol 2012;108(12):3313–21.

[66] Wong JD, Wilson ET, Gribble PL. Spatially selective enhancement of proprioceptive acuity following motor learning. J Neurophysiol 2011; 105(5):2512–21.

[67] Miko I, et al. Effect of a balance-training programme on postural balance, aerobic capacity and frequency of falls in women with osteoporosis: a randomized controlled trial. J Rehabil Med 2018;50(6):542–7.

[68] Kunutsor SK, et al. Adverse events and safety issues associated with physical activity and exercise for adults with osteoporosis and osteopenia: a systematic review of observational studies and an updated review of interventional studies. J Frailty Sarcopenia Falls 2018;3(4):155–78.

[69] Bahr R, et al. The IOC manual of sports injuries: an illustrated guide to the management of injuries in physical activity. John Wiley & Sons; 2012.

[70] O'Sullivan PB, et al. Cognitive functional therapy: an integrated behavioral approach for the targeted management of disabling low Back Pain. Phys Ther 2018;98(5):408–23.

[71] Rabey M, et al. Multidimensional pain profiles in four cases of chronic non-specific axial low back pain: an examination of the limitations of contemporary classification systems. Man Ther 2015;20(1):138–47.

[72] Hotfiel T, et al. Advances in delayed-onset muscle soreness (DOMS): part I: pathogenesis and diagnostics. Sportverletz Sportschaden 2018; 32(4):243–50.

[73] Fragala MS, et al. Resistance training for older adults: position statement from the national strength and conditioning association. J Strength Cond Res 2019;33(8):2019–52.

[74] Hedrick A, Wada H. Weightlifting movements: do the benefits outweigh the risks? Strength Cond J 2008;30(6):26–35.

[75] Khodaee M, Waterbrook AL, Gammons M. Sports-related fractures, dislocations and trauma: advanced on- and off-field management. Springer International Publishing; 2020.

[76] Kraemer WJ, Ratamess NA. Fundamentals of resistance training: progression and exercise prescription. Med Sci Sports Exerc 2004;36(4):674–88.

[77] Sullivan G, Promidor A. Exercise for aging adults. In: A Guide for Practitioners. Switzerland: Springer International Publishing; 2015. p. 155.

[78] Geneen LJ, et al. Physical activity and exercise for chronic pain in adults: an overview of Cochrane reviews. Cochrane Database Syst Rev 2017; 4(4):Cd011279.

[79] Csapo R, Alegre LM. Effects of resistance training with moderate vs heavy loads on muscle mass and strength in the elderly: a meta-analysis. Scand J Med Sci Sports 2016;26(9):995–1006.

[80] Egan A, et al. Using session RPE to monitor different methods of resistance exercise. J Sports Sci Med 2006;5(2):289–95.

[81] Thompson SW, et al. The effectiveness of two methods of prescribing load on maximal strength development: a systematic review. Sports Med 2020;50(5):919–38.

[82] Schott N, Johnen B, Holfelder B. Effects of free weights and machine training on muscular strength in high-functioning older adults. Exp Gerontol 2019;122:15–24.

[83] Neptune R, Vistamehr A. Dynamic balance during human movement: measurement and control mechanisms. J Biomech Eng 2018; 141(7):0708011–07080110.

[84] Zhuang J, et al. The effectiveness of a combined exercise intervention on physical fitness factors related to falls in community-dwelling older adults. Clin Interv Aging 2014;9:131–40.

[85] Otero M, et al. The effectiveness of a basic exercise intervention to improve strength and balance in women with osteoporosis. Clin Interv Aging 2017;12:505–13.

[86] McLeod JC, Stokes T, Phillips SM. Resistance exercise training as a primary countermeasure to age-related chronic disease. Front Physiol 2019;10:645.

[87] Watson S, et al. High-intensity resistance and impact training improves bone mineral density and physical function in postmenopausal women with osteopenia and osteoporosis: the LIFTMOR randomized controlled trial. J Bone Miner Res 2019;34(3):572.

[88] Hong AR, Kim SW. Effects of resistance exercise on bone health. Endocrinol Metab 2018;33(4):435–44.

[89] Giangregorio LM, et al. Too fit to fracture: outcomes of a Delphi consensus process on physical activity and exercise recommendations for adults with osteoporosis with or without vertebral fractures. Osteoporos Int 2015;26(3):891–910.

[90] De Rui M, et al. Dietary strategies for mitigating osteosarcopenia in older adults: a narrative review. Aging Clin Exp Res 2019;31(7):897–903.

[91] Chew J, et al. Nutrition mediates the relationship between osteosarcopenia and frailty: a pathway analysis. Nutrients 2020;12(10):2957.

[92] Tessier AJ, Chevalier S. An update on protein, leucine, Omega-3 fatty acids, and vitamin D in the prevention and treatment of sarcopenia and functional decline. Nutrients 2018;10(8):1099.

[93] Murphy CH, et al. Leucine supplementation enhances integrative myofibrillar protein synthesis in free-living older men consuming lower- and higher-protein diets: a parallel-group crossover study. Am J Clin Nutr 2016;104(6):1594–606.

[94] Devries MC, et al. Leucine, not total protein, content of a supplement is the primary determinant of muscle protein anabolic responses in healthy older women. J Nutr 2018;148(7):1088–95.

[95] Toffanello ED, et al. Vitamin D and physical performance in elderly subjects: the Pro.V.A study. PLoS One 2012;7(4), e34950.

[96] Sohl E, et al. Vitamin D status is associated with physical performance: the results of three independent cohorts. Osteoporos Int 2013;24(1):187–96.

[97] Abedi E, Sahari MA. Long-chain polyunsaturated fatty acid sources and evaluation of their nutritional and functional properties. Food Sci Nutr 2014;2(5):443–63.

[98] Laplante M, Sabatini DM. mTOR signaling at a glance. J Cell Sci 2009;122(Pt 20):3589–94.

[99] Gray SR, Mittendorfer B. Fish oil-derived n-3 polyunsaturated fatty acids for the prevention and treatment of sarcopenia. Curr Opin Clin Nutr Metab Care 2018;21(2):104–9.

[100] Vannice G, Rasmussen H. Position of the academy of nutrition and dietetics: dietary fatty acids for healthy adults. J Acad Nutr Diet 2014; 114(1):136–53.

[101] Candow DG, et al. Effectiveness of creatine supplementation on aging muscle and bone: focus on falls prevention and inflammation. J Clin Med 2019;8(4):488.

[102] Chilibeck PD, et al. Effect of creatine supplementation during resistance training on lean tissue mass and muscular strength in older adults: a meta-analysis. Open Access J Sports Med 2017;8:213–26.

[103] Candow DG, Chilibeck PD, Forbes SC. Creatine supplementation and aging musculoskeletal health. Endocrine 2014;45(3):354–61.

Chapter 29

Exercise and older adults receiving home care services

Elissa Burton[a,b] and Anne-Marie Hill[c]

[a]Curtin School of Allied Health, Curtin University, Perth, WA, Australia [b]enAble Institute, Curtin University, Perth, WA, Australia [c]School of Allied Health, WA Centre for Health & Ageing, University of Western Australia, Perth, WA, Australia

Introduction

The number of older adults in Australia is increasing, with 15.9% of the total population aged 65 years and over, and this is predicted to continue rising over the coming decades [1]. During the latter years of life, there is an increased incidence of long-term or chronic health conditions, such as arthritis, cancer, and dementia leading to older adults experiencing decline in physical, psychological, or cognitive health [2]. These older adults are frequently supported by family and friends (i.e., informal caregivers) [3]. For example, in Australia, over 2.5 million people provide informal care for older adults, 32% as primary caregivers [3]. However formal support is also often required to enable these older adults to continue living independently in their home. In Australia there are two main types of formal government-subsidized home care services; (1) Entry-level care called the Commonwealth Home Support Program, and (2) more complex care called Home Care Packages [4].

The Commonwealth Home Support Program (CHSP) can either provide short term services called reablement/restorative care or more ongoing services. Reablement services are usually short-term in nature, delivered by an interdisciplinary team (e.g., physiotherapist, occupational therapist, nurse and/or social worker) and focus on the older adult setting and achieving their own goals, with the aim of enabling them to return to living independently without the need for ongoing services. If an older adult is finding everyday activities such as cleaning floors and bathrooms or getting assistance with meals, personal care or home maintenance are becoming more difficult, receiving ongoing services from the CHSP may assist them to stay living independently for longer. People aged 65 years and older (50 years or older for Aboriginal or Torres Strait Islander people) or 50 years or older (45 years of older for Aboriginal or Torres Strait Islander people) and on low income, homeless, or at risk of being homeless are eligible to apply if they are having difficulty undertaking their daily activities [5]. These services are usually available within days of receiving an assessment. It must be noted that ongoing home care services, once initiated, remain, or increase over time and often only cease when the older adult moves into a residential aged care facility or dies.

Home care packages are designed for older people who have complex or intensive needs [6]. There are currently graduated levels of packages in Australia ranging from a basic care needs package (Level 1) of $9000 per year through to a Level 4 high care needs package worth approximately $52,000 [6]. Older adults who need coordinated services to remain living independently or for younger people with a disability, dementia, or other care needs that are not met through other services, are eligible for these packages [6]. However, there is a waiting time for people to receive these packages that ranges from 3 to 6 months for Level 1, with Level 2–4 packages taking more than a year, although once the services are commenced they are ongoing until no longer required [6].

In Australia in 2019–20 over 1450 aged care organizations were funded to deliver CHSP services and 920 organizations provided home care packages [7,8]. Almost two-thirds of older Australians receiving home care services received home support ($3.3 billion), 21% residential care ($13.4 billion), 13% home care ($3.4 billion) and 2% restorative care (e.g., reablement) [7]. This equated in 2019–2020 to 839,373 CHSP clients, 142,436 home care consumers, 24,775 flexible care consumers (e.g., reablement and transition care) and over 366,000 people employed in the Australian aged care workforce [7].

Other countries throughout the world, particularly first world, also offer home care services for older adults. Reablement services are delivered in countries such as the United Kingdom, New Zealand, and Scandinavian countries. The United

Exercise to Prevent and Manage Chronic Disease Across the Lifespan. https://doi.org/10.1016/B978-0-323-89843-0.00018-0

States of America, Canada, and many European countries also offer ongoing home care services, and they are often delivered by local or state government organizations. This model of care differs from Australia where service coordination and funding has recently moved under federal government jurisdictions. Due to policy and funding restrictions, home care services are delivered differently across the globe [9].

Adults in Australia using home care services tend to be younger (60% under 84 years) [10] than those living in residential aged care (around 72% are over 80 years) [11], with more women than men using the services [12]. Older adults in the metropolitan areas use home care services more than people in rural areas [13]. Compared to 10 years ago the number of older adults receiving home care services in Australia has increased from 44,100 in 2009 to over 106,000 in 2019, an increase of 142% [13]. Community-dwelling older adults receiving home care services also fall 50% more often than those not receiving services, are frailer and have poorer functional ability [14]. Many people living in the community with dementia also receive home care packages, as their care needs are rated as high and are usually complex in nature [15]. These packages are essential to assist them to continue living independently in the community and provide much needed support for their partners and other caregivers [16].

Unfortunately for those receiving home care services, the promotion of physical activity or exercise and reducing sedentary behavior is limited or nonexistent in many services. There are other services available that may include physiotherapy, occupational therapy, day centers, and respite care [6]. However, apart from day centers (which are predominantly run by non-health professionals), these are usually time limited and not freely accessible, and usually occur due to a sudden injury or illness (e.g., physiotherapy—exercises to assist specific injury) or a change in functional status requiring home modifications (e.g., rail in the toilet). This is a serious gap in providing preventative care for older adults receiving home care because there is strong evidence that older adults who engage in regular physical activity have lower rates of all-cause mortality, coronary heart disease, high blood pressure, stroke, type 2 diabetes, colon cancer, and breast cancer, a higher level of cardiorespiratory and muscular fitness, healthier body mass and composition. In addition, older adults who exercise demonstrate higher levels of functional health, a lower risk of falling, and better cognitive function; have reduced risk of moderate and severe functional limitations and role limitations [16–18].

Physiotherapy services provided for older adults who have specific rehabilitation requirements will often encourage physical activity, however, these are time-limited and often focus on a particular injury, perhaps after a fall, rather than improving an older person's physical function or mood through physical activity or exercise. Some reablement services in Australia have also included physical activity or exercise programs but often it is not an essential component of the service but is included only if the staff member delivering the service determines it would benefit the older adult. A survey of older clients revealed that only 13.7% engaged in strength training [19]. Since older adults around the world receive home care services to enable them to continue living independently, it is critical that these services incorporate exercise and physical activity guidelines with the aim of maintaining and improving physical function, reducing falls, and enhancing participation in daily life. It is particularly important that older adults receiving ongoing packages and services are encouraged to undertake and maintain a physical activity program, as home care assistance usually results in these older adults doing less housework, gardening, and other physical activities. The following sections will synthesize the recent evidence for prescribing exercise and physical activity for older adults who receive home care services. Advice for health practitioners who work with these older adults will be summarized and information for caregivers and family is also presented.

Exercise and physical activity for older adults receiving home care services

Exercise and physical activity guidelines for older adults

It is important to raise awareness of service providers, health professionals, and older adults that key national guidelines now strongly recommend that ALL older adults regardless of age or chronic health conditions engage in a physical activity program. Therefore, home care services should incorporate the following guidelines into their training, policies, and practice. In 2020, the World Health Organization (WHO) published new guidelines for physical activity and sedentary behavior [20]. The new guidelines include recommendations for older adults, including those living with chronic conditions or disabilities. For those older adults who do no physical activity, the guidelines recommend that engaging in any amount of physical activity is better than nothing [20]. In general, the WHO suggests a medical certificate is not required to begin being active, but that people should start slowly and build up to meeting the recommended guidelines (see Table 1) [20]. It is also stated that there are no major risks to participating in physical activity for people living with a disability and that the health and functional benefits far outweigh any potential risks [20].

TABLE 1 Physical activity guidelines for older adults.

Place of origin	Guidelines
WHO [20]	• Be physically active on a regular basis, something is better than nothing • Do at least **150–300** min of moderate-intensity *aerobic* physical activity, or at least **75–150** min vigorous-intensity *aerobic* physical activity, or a combination each week • Do varied multicomponent physical activity that emphasizes *functional balance and strength training* at moderate or greater intensity on *3 or more days a week*, to enhance functional capacity and to prevent falls
Australia [21]	• Do some form of physical activity, no matter age, weight, health problems or abilities • Be active every day in as many ways as possible, doing a range of physical activities that incorporate *fitness, strength, balance, and flexibility* • Accumulate at least **30** min moderate-intensity physical activity, preferably on all days • For those not active, start at a level that is manageable and build up over time the amount, type, and frequency of activity • Those who have always been vigorously active should continue to participate at their capability, provided safety procedures and guidelines are adhered to
United States [22]	• Move more, sit less. Some physical activity is better than none. Adults who sit less and do any moderate-to-vigorous physical activity gain some health benefits. • For substantial health benefits: do at least **150–300** min *(2.5–5* h) per week moderate-intensity or **75–150** min *(1.25–2.5* h) vigorous-intensity *aerobic* physical activity each week or a combination. Ideally, this is spread across the week • For additional health benefits: participate in **>300** min *(>5* h) moderate-intensity physical activity each week • Do *muscle-strengthening activities* of moderate or greater intensity that involve major muscle groups *twice a week* • Specific to older adults • Do multicomponent physical activity that includes *balance* as well as *aerobic and muscle-strengthening activities* each week • Determine level of effort for physical activity relative to current level of fitness • Those with chronic conditions should understand whether and how conditions affect their ability to do regular physical activity safely • Those who cannot meet the recommended minutes above due to chronic conditions should be as physically active as their abilities and conditions allow.
United Kingdom [23]	• Participate in daily physical activity to gain health benefits (i.e., for physical, mental, well-being and social functioning benefits). Some physical activity is better than none, more daily physical activity provides greater health benefits • Aim to maintain or improve physical functioning by participating in activities that improve or maintain *muscle strength, balance, and flexibility* at least **2 days a week**. These can be combined with a session of moderate aerobic activity or could be additional session especially aimed at these components of fitness • Each week accumulate at least **150** min moderate-intensity *aerobic* activity, building up gradually if current levels are below this. Older adults can also participate in **75** min vigorous-intensity activity each week or a combination or vigorous and moderate. *Weight-bearing activities* which create impact through the body help to maintain bone health • Break up prolonged periods of being sedentary with light activity where possible, or at least with standing, as this also has health benefits

The Australian physical activity guidelines for older Australians differ slightly from the latest WHO guidelines (see Table 1) because they are less prescriptive, providing a selection of activities that should be included daily and allowing older adults to determine which type of activity they will choose to participate in [21]. The Australian physical activity guidelines for older adults are now over a decade old and due to the research evidence identified since then, it is recommended they be updated in the near future.

The United States (US) physical activity guidelines were published in 2018 and the chapter on older adults provides a positive start, being named Active Older Adults [22]. These guidelines discuss the importance of maintaining physical function and mobility into the latter years and promote being active over sedentary behavior, "move more—sit less" [22]. The first four recommendations are the same as for adults (18–64 years) and the last four are specific to older adults (i.e., 65+ years). Like the Australian and WHO guidelines they recommend those who are not currently active or have chronic conditions begin at a suitable level and progress over time.

The United Kingdom physical activity guidelines were published in 2019 [23]. Like the US guidelines, they are promoting older adults to move more, even if it is light intensity. They also recommend strength, balance, and flexibility activities be undertaken at least twice a week, as well as aerobic activity which is prescriptively consistent across all of the included guidelines [23]. The UK guidelines also included weight-bearing activity to maintain bone health, which is particularly important as people get older and their risk of falling increases.

Motivators and barriers to older adults receiving home care services being physically active

National surveys reveal that many older adults are inactive or have very low levels of activity. In the US, a large review examined older adults' levels of physical activity and found that there is a clear age-related decline in physical activity. This review noted that only 41.7% of adults 65–74 years old meet physical activity guidelines and only 18.4% of those over 85 years [24]. The Australian national survey also identified that levels of physical activity declined in older age and of older adults aged 65 and over, 69% of men and 75% of women were insufficiently active [25]. Therefore, it is important that older adults receiving home care services and the health care professionals who deliver services to this population understand what barriers and enablers are relevant to engaging in physical activity in this population.

Very few studies have explored why older adults receiving home and community care services (i.e., not residential aged care) are not physically active or conversely what would motivate them to being physically active. A mixed-method study by Burton and colleagues [26] surveyed almost 1500 older adults receiving home care services (either short-term restorative services or ongoing home care services). The average age of this group was 82.2 (±5.9) years, 79.3% were female and two-thirds lived alone [26].

The main reasons provided for being physically active were to maintain or improve well-being and for health or fitness [26]. A 102-year old participant explained that she felt much better by staying active than when sitting around and that for the past 40 years she had completed a 15-min set of exercises prior to getting out of bed each morning [26]. Other participants talked about not wanting to sit for too long as they would then feel stiff. Some described setting challenges for themselves each day to keep moving, such as walking up and down the driveway 20–30 times [26], or when it was raining outside, doing laps of the house intermittently across the day.

Other motivators for undertaking physical activity were for enjoyment and social engagement, including with family, was viewed as important. Spending time with grandchildren who are active and wanting to throw or kick balls around with their grandparents was also identified as an incentive for staying physically active [26]. Some participants were motivated to be active because it was their form of transport [26]. Many enjoyed walking, particularly because they could get out and about and see things, and gardening was also identified as a source of physical activity [26].

The key factor older adults receiving home care services identified as a barrier to being active was ongoing injury or illness [26]. Many discussed how tired they got during or after doing anything active. Unfortunately, there were also some misconceptions held about participating in physical activity when living with certain health conditions, for example, asthma or arthritis. A number of older adults receiving home care services stated these two conditions precluded them from physical activity. These findings are similar to other studies in older populations, which also identified that illness or injury were major barriers to being active [27–29]. However, these perceptions are not supported by research, as there is strong evidence that shows that strength training in particular can assist older adults living with arthritis [30] and that it can also be beneficial for people with asthma [31]. Healthcare professionals, as well as care workers who are working with older adults receiving home care on a regular basis need to be aware of these misconceptions and provide accurate information and referrals to appropriate health professionals (e.g., accredited exercise physiologist, physiotherapist) to guide them in starting to be more physically active.

Feeling "too old" or age was ranked the second-highest barrier to being physically active [26]. The Australian physical activity guidelines explicitly recommend being active regardless of age, whereas the other recommended guidelines outlined above, highlighted the presence of chronic health conditions, more than age. Feeling too old or perceiving older age as a barrier to being physically active, is a belief or perception not only among adults receiving home care services but expressed by many community-dwelling older people not receiving services also [19,32].

Losing the ability to drive, public transport being difficult to access and getting assistance to move around was also viewed as a barrier to being active [26]. Some older adults reported difficulty getting onto buses and that there were financial constraints to using taxis [26]. Others reported examples such as seeking the opportunity to go to hydrotherapy, but the home care service transport assistance was not offered at the time of the hydrotherapy session. Cost, not being interested and not knowing how to be physically active were also lesser ranked barriers for this group [26].

A review of reviews exploring the enablers and barriers to older adults participating in strength and balance activities identified many other factors that could either motivate or prevent an older adult from undertaking this type of activity [33]. Preventing deterioration (i.e., disability), reducing the risk of falling, feeling more alert, and having better concentration were some of the 92 motivators found in one of the included reviews [34]. Reducing pain and improving mental health benefits were also identified in a number of reviews [33]. Barriers to participating in strength and balance training were similar to those found in the home care population, for example, poor health, pain, and lack of social support or access to facilities were also identified [34]. Again, there were some barriers identified that have no evidence supporting them, such as a fear of looking too muscular and risk of having a heart attack, stroke, or death when participating in strength training [34]. This reinforces the need for older adults to access accurate information when considering becoming physically active, and maybe a reason why the WHO physical activity guidelines included statements that physical activity is safe (i.e., low risk compared to possible health benefits gained) and the majority of older adults do not require a medical clearance before commencing, as long as they start slowly and progress over time [20]. This was acknowledged by all of the guidelines included in this chapter (Table 1).

Table 2 provides a summary of intrinsic individual factors and external social and environmental motivators that should be considered when encouraging an older adult to commence and sustain a program of physical activity. Caregivers, family, and service providers should address these factors to increase the number of people engaging in physical activity.

Physical activity levels of those receiving home care services

One study compared the physical activity levels (using the Physical Activity Scale for the Elderly, PASE) of older adults receiving reablement to ongoing home care services. The authors found that those receiving short-term reablement were significantly more physically active than those receiving ongoing services [35]. When compared to studies of older adults living in the community who are not receiving services the activity levels of the reablement group were comparable to those with a functional impairment [35]. However, the ongoing home care group had considerably lower physical activity levels [35]. People receiving either home care service (i.e., reablement or ongoing) who were younger, in better physical condition, had good mobility and had no diagnosis of depression were more likely to be physically active [35].

TABLE 2 Older adults' motivations and external factors that encourage engagement in physical activity.

Motivators—Intrinsic	
Psychological benefits	**Physical benefits**
• Mental function (e.g., improved alertness, concentration, stimulate mind) • Relieve stress/relaxing • Mental health benefits (e.g., better mood, positive outlook, confidence) • Exercise self-efficacy • Improved wellbeing • Enjoy exercising	• Health (e.g., improved energy, better sleep) • Physical fitness (e.g., strength, endurance, flexibility, balance, coordination) • Physical functioning (e.g., improved walking) • Good health/had a health scare • Reduce pain/injury/illness including chronic conditions • Appearance (e.g., manage weight, muscle tone)
Motivators—External	
Social Support	**Environmental considerations**
• Encouragement from peers and staff, spouse, family, friends, health professional (doctor) • Increased social activity, sense of belonging • Observing others being active (e.g., family/friends participating in strength training)	• Access to facility/equipment, convenient location • Access to organized programs • Individualized program (e.g., level of exercising difficulty being addressed, exercising at own pace, gym, or facility atmosphere) • Professional staff (e.g., access, knowledge, interaction, competence of staff all essential)

Note: from Burton E, et al. Identifying motivators and barriers to older community-dwelling people participating in resistance training—a cross sectional study. J Sports Sci 2017;35(15):1523–1532.

Physical activity preferences of those receiving home care services

As part of identifying motivators and barriers to older adults' engagement in physical activity, it is important to understand what types of physical activity this group of older adults would prefer. Health professionals and service providers should seek to discuss each clients' preferences in order to encourage older adults receiving home care services to commence participation or engage in more physical activity. Burton, et al. completed a qualitative study using semi-structured interviews to identify what types of physical activity older adults receiving home care preferred [36]. In this study, physical activity was purposely not defined, because the authors wanted to explore what this population viewed as physical activity, as well as what they liked to engage in [36]. Walking, as with older adults not receiving services within the home was the activity of choice. Many participants (average age 84.2 (±7.18) years) had spent their childhoods and early adult life walking for transport for example, walking to school, the shops, and to the bus to go to and from work [36]. Those with a little more money had ridden bikes in their childhood, but this was more sporadic than walking. In their later years (time of the interviews) interviewees spoke of consciously getting up from their chair and walking around the house to increase their movement or walking to the shops to get groceries on most days [36]. Some had a specific journey mapped out for their walk each day, that included beautiful scenery and wildlife. This cohort also saw gardening (including sweeping and racking outside) and housework as physical activity, although many did not enjoy cleaning the house. They often spoke of these chores being included in their daily routines [36]. One centenarian perceived housework as the best activity and stated "you can have all your yoga and what have you [sic] there's more exercise in house work than anything" [36, p.174]. The home care participants also described doing activities to stay mentally active and these included crossword puzzles, craft, and reading [36]. All participants, except one, knew being physically active was important and some commented that it often stopped them from thinking about negative thoughts, many believed being mentally active was as important as being physically active in order to lead a happy and healthy life [36].

These findings demonstrate the importance of discussing the mental benefits of physical activity with this group of older adults. Five participants noted that a health professional had discussed with them how important being physically active was, using quotes such as "if you don't use it you lose it" [36]. Forms of media, such as television and the newspaper were also important forms of educating them on the importance of being physically active [36]. A small number also had family members encouraging them to be more active [36]. The type of activities identified, however, were unstructured and not exercise- or sport-related. They described incidental, lifestyle activity being more enjoyable than the traditional structured sessions. This needs to be considered when prescribing physical activity to this population because they may be unlikely to commence and continue structured exercise or physical activity programs.

Effectiveness of physical activity programs for older adults receiving home care services

Eighteen studies, including 10 randomized controlled trials (RCTs) were included in a recent systematic review investigating the effectiveness of physical activity programs for older adults receiving home care services in the community [37]. Many of the studies used multifactorial interventions, with physical activity or exercise not the primary outcome for the study. The physical activity or exercise programs that were included involved lifestyle activity (i.e., incorporating exercises into daily activities), strength and/or balance training, repetitive functional activity, and multicomponent activities [37–39].

An average of 13.5% of participants withdrew from the studies showing over 85% enjoyed participating in these types of activities. The participants engaged in the exercise programs on approximately 4 days per week, however, the longer interventions saw a drop in adherence over time [37]. Mobility, strength, and balance did not improve significantly for the intervention groups and further research is required to gain a better understanding of how to improve physical and cognitive health through regular exercise or physical activity in this population.

More broadly many other trials have included older frailer people who share similar chronic health conditions. Multicomponent exercise interventions, such as aerobic and resistance training, as well as balance interventions have reported positive effects on aspects such as mobility, balance performance and muscle strength in older frail adults [40]. However, again similar to those receiving home care services the optimal exercise program characteristics (i.e., type, intensity of activity) remain unclear [40].

In summary, more trials that evaluate single physical activity interventions and also motivational techniques for delivering them, including telehealth, coaching, and care worker support are necessary in order to better understand whether physical activity interventions delivered to home care clients can improve their physical function and ability to live independently for longer. However, overall, health professionals and service care providers working with older people receiving home care services can follow guideline recommendations to safely prescribe physical activity for this population.

Effectiveness of physical activity programs for older adults receiving reablement services in the home

A similar review was recently published to the above described, however, this systematic scoping review only included reablement studies and not long-term, ongoing home care services [41]. Fifty-one reablement studies were included, with most studies mentioning physical activity or exercise interventions but very few providing detailed information, particularly on intensity [41], as mentioned above [37]. Very few interventions were aimed at increasing physical activity levels of the participants or attempted to reduce sedentary behavior [41]. None of the studies explored the physical activity experiences in a reablement setting with older adults, their health care providers or family members [41]. Outcomes of physical fitness such as mobility, strength, and balance were reported but it was not possible to conduct a meta-analysis due to insufficient evidence across the studies [41]. Interestingly, the review found that physical activity guidelines such as those mentioned earlier were rarely discussed and that inactivity and sedentary behavior among older adults receiving reablement services was given little consideration [41]. This is problematic as guidelines confirm that older adults with chronic health conditions and of any age should be engaging in physical activity as per guideline recommendations [20,21,23].

Suggestions for future research from this review included interventions being described more explicitly, including all exercises, loads (e.g., kgs, lbs. lifted) or intensities (e.g., sets, reps) used and progressions across the intervention need to be specifically detailed [41]. Older people, health professionals, and family perspectives are also needed to better understand how uptake and adherence can be maximized for the older person, to assist them to live independently for as long as they choose [41]. Meanwhile, older adults, their caregivers, and family (supported where necessary by care service providers) should priorities encouraging these older adults to increase and maintain regular levels of physical activity. An overall summary of motivators that should be addressed are presented in Table 2 earlier.

Falls and older adults receiving home care services

There is established strong evidence from over 100 randomized trials that exercise reduces falls and injurious falls, even in older adults who have the chronic disease [42]. Exercise that includes strength, balance, and functional components is particularly effective in reducing falls [43]. Additionally, multifactorial programs that include fall risk assessment and treatment for identified risk factors (such as vision correction and home modifications) are effective [42]. Therefore, all older adults receiving home care should be encouraged to take up this evidence-based intervention, since as described earlier in the chapter, exercise is also highly beneficial for improving cognitive and physical functional ability [16–18]. In general, 30% of people aged 65 years and over living in the community, without home care services, will fall each year [44]. Two studies, conducted 10 years apart, explored the prevalence rates of falls across 10 home care organizations and thousands of older home care clients, and found the rates of falling did not differ significantly with a slight increase from 45% [45] to 48% [14]. However, these rates are 50% higher than those found for community-dwelling older adults not receiving home care services [14]. Having balance issues and perceiving oneself as being at high risk of falling were associated with a higher likelihood of falling in the future [14]. Older adults receiving home care services who had been referred to a falls prevention service or program had a reduced likelihood of falling in the future [14]. This survey also noted that only 27% of survey recipients who had fallen in the past year had been referred to a falls prevention service or program [14], strongly suggesting that older adults receiving home care require more active management of their fall risk by service providers.

Cautions/safety

Older adults receiving home care services are a vulnerable population, with an increased incidence of multiple comorbidities, including heart disease and diabetes, that result in physical or mental or psychosocial disability [46]. Hence, it is important that these older clients commence physical activity programs slowly and progress through at a pace that is safe and permissible to them as recommended by a number of the physical activity guidelines. While medical assessment is not compulsory prior to commencing physical activity, older adults receiving services who may be frail and have multiple medical conditions can benefit from a thorough medical review by their doctor. In Australia, these reviews can allow the doctor to refer the older adult to programs that are specifically funded by the government, whereby they can receive services from an accredited health professional [47]. Previous research has also identified that older adults are more likely to engage in exercise if encouraged by their doctor or other health professional and if they receive professional support and advice [28,34,48].

When an older adult is undertaking exercise or physical activity in or around their home, it is important that they are advised about safety precautions. These include wearing the correct footwear (i.e., closed in shoes, not slippers), making sure the area is clear where the activities will be undertaken (i.e., nothing on the floor they could trip over), when

FIG. 1 Aspects of safety to consider when exercising.

completing balance activities start near a firm benchtop, if practicing sit-to-stands starting on a chair that is firm and of reasonable height and as they progress moving to a lower, softer chair needs to be considered (Fig. 1). If practicing tasks such as getting up off the ground from a lying (supine) position the older person should make sure that someone is present who has the ability to help them up, if they have difficulty completing this task. A systematic review identified that there is limited evidence about what programs improve older home care clients' ability to get up from the floor after a fall, although interventions including resistance training may be beneficial [49]. Therefore, being able to call for help using a security alarm, mobile phone or similar device may be beneficial for some clients to assist them to more confidently undertake physical activity and exercise [43]. Older adults who receive home care services should be given a written checklist to ensure they understand how to engage in such programs, and it is helpful if they can receive care worker or caregiver support at commencement. Social support of this nature is important as it is known to increase levels of adherence and maintenance of the activity [26,34,45]. Safety is important, but as described in the recommended physical activity guidelines above, should not preclude an older person receiving home care services from being more physically active, even if it is simply encouraging them to move more, and more often each day.

Advice for health professionals

Health professionals working with older adults receiving home care services should be aware of the general principles of exercise and physical activity prescription. Exercise prescription, either for short-term reablement programs or ongoing development of a physical activity program, for older adults should always be undertaken using a comprehensive functional approach that includes assessment of physical, cognitive, mental function, and identifies social support structures [50]. Additionally, since these frailer, older adults identify barriers to engaging in physical activity, it is important to understand the principles of behavior change and how to effectively address these barriers with the older person. Extensive research regarding health behavior change indicates that adults must have good levels of motivation, awareness and knowledge, and social and environmental supports if they are to engage in suitable health actions, such as doing exercise [51]. Providing feedback, monitoring, knowledge about how to engage in physical activity and developing motivational aspects to a prescribed program will increase the older clients' adherence and make it more likely that they will experience subsequent benefits of the program [52].

Health professionals delivering these projects should be aware of the many safety aspects that need to be considered. Providing initial instruction and printed or online resources can assist in maintaining safety, as well as increasing adherence to the program. Additionally, ongoing programs for adults receiving home care services will usually be supervised by care workers rather than health professionals. Therefore, care worker training should include mandatory sections on encouraging their clients to keep active and engage in suitable programs.

Clinical practice recommendations

For exercise prescribers/clinicians
- Encourage older adults receiving home care services to move more (as a minimum) and sit less.
- Try and dispel inaccuracies about exercise and chronic conditions.
- Encourage participation in lifestyle activity that they enjoy that builds strength and balance.
- If an older adult receiving home care falls, refer them to a falls prevention program or service and follow up that they have been.
- Provide high quality resources with links to online programs if the client has internet capability.

For older adults receiving home care and their families/friends
- Maintaining or building fitness, strength and balance is essential for living independently.
- Find activities you enjoy and do them regularly.
- It is important to move more and sit less, even if it is around the house or in the garden.
- If you are able to get up off the floor independently (make sure someone is around when you try) it is a good skill to practice on a regular basis.
- Remember—Age is only a number, not a barrier to doing things!
- Ask your care service provider for a health professional referral if you have medical conditions that make it difficult to do exercise. Your therapist can assist you to develop a suitable activity plan that will maintain your mobility.

References

[1] Australian Bureau of Statistics. 3101.0—Australian demographic statistics, Jun 2019, 2019. [cited 2021 16 January]; Available from: https://www.abs.gov.au/ausstats/abs@.nsf/0/1CD2B1952AFC5E7ACA257298000F2E76#:~:text=Over%20the%2020%20years%20between,1946%20and%201964)%20turn%2065.

[2] NCOA Healthy Aging Team. Top 10 chronic conditions in adults 65+ and what you can do to prevent or manage them, 2017. [cited 2021 2 February]; Available from: https://www.ncoa.org/blog/10-common-chronic-diseases-prevention-tips/.

[3] Australian Bureau of Statistics. Disability, ageing and carers. Australia: summary of findings, 2018, 2018. [cited 2020 23 August]; Available from: https://www.abs.gov.au/ausstats/abs@.nsf/mf/4430.0.

[4] Commonwealth of Australia: My Aged Care. Help at home, 2021. [cited 2021 16 January]; Available from: https://www.myagedcare.gov.au/help-at-home.

[5] Commonwealth of Australia: My Aged Care. Commonwealth home support programme, 2021. [cited 2021 16 January]; Available from: https://www.myagedcare.gov.au/help-at-home/commonwealth-home-support-programme.

[6] Commonwealth of Australia: My Aged Care. Home care packages, 2021. [cited 2021 16 January 2021]; Available from: https://www.myagedcare.gov.au/help-at-home/home-care-packages.

[7] Australian Institute of Health and Welfare. ROACA summary (report on the operation of the aged care act), 2020. [cited 2021 16 January]; Available from: https://www.gen-agedcaredata.gov.au/www_aihwgen/media/ROACA/ROACA-Summary-2020.pdf.

[8] Commonwealth of Australia: Department of Health. Summary of the 2019-20 report on the operation of the Aged Care Act 1997, 2020. [cited 2021 16 January]; Available from: https://www.health.gov.au/resources/publications/summary-of-the-2019-20-report-on-the-operation-of-the-aged-care-act-1997.

[9] World Health Organization. In: Tarricone R, Tsouros A, editors. Home care in Europe. Italy: Universita Commerciale Luigi Bocconi; 2008.

[10] Australian Institute of Health and Welfare. Older Australia at a glance, 2018. [cited 2021 1 February]; Available from: https://www.aihw.gov.au/reports/older-people/older-australia-at-a-glance/contents/service-use/aged-care.

[11] Australian Bureau of Statistics. 4430.0—disability, ageing and carers, Australia: summary of findings, 2015. 2015 [cited 2021 1 February]; Available from: https://www.abs.gov.au/ausstats/abs@.nsf/Lookup/4430.0main+features302015.

[12] Australian Institute of Health and Welfare. People using aged care, 2020. [cited 2021 January 16]; Available from: https://www.gen-agedcaredata.gov.au/Topics/People-using-aged-care.

[13] Australian Institute of Health and Welfare. Explore people using aged care, 2020. [cited 2021 January 16]; Available from: https://www.gen-agedcaredata.gov.au/Topics/People-using-aged-care/Explore-people-using-aged-care.

[14] Burton E, et al. Falls prevention in community care: 10 years on. Clin Interv Aging 2018;13:261–9.

[15] Australian Institute of Health and Welfare. Explore people's care needs in aged care; 2020 [cited 2021 January 16].

[16] Vogel T, et al. Health benefits of physical activity in older patients: a review. Int J Clin Pract 2009;63(2):303–20.

[17] Chodzko-Zajko W, et al. American college of sports medicine position stand. Exercise and physical activity for older adults. Med Sci Sports Exerc 2009;41(7):1510–30.

[18] Fern A. Benefits of physical activity in older adults: programming modifications to enhance exercise experience. ACSMs Health Fit J 2009;13(5):12–6.

[19] Burton E, et al. Identifying motivators and barriers to older community-dwelling people participating in resistance training—a cross sectional study. J Sports Sci 2017;35(15):1523–32.

[20] Bull F, et al. World Health Organization 2020 guidelines on physical activity and sedentary behaviour. BMJ 2020;54:1451–62.

[21] Sims J, et al. Physical activity recommendations for older Australians. Australas J Ageing 2010;29(2):81–7.

[22] U.S. Department of Health and Human Services. Physical activity guidelines for Americans. 2nd ed. Washington, DC: U.S. Department of Health and Human Services; 2018.

[23] Department of Health & Social Care. UK chief medical officers' physical activity guidelines, 2019. Available from: https://assets.publishing.service.gov.uk/government/uploads/system/uploads/attachment_ data/file/832868/uk-chief-medical-officersphysical-activity-guidelines.pdf.

[24] Keadle S, et al. Prevalence and trends in physical activity among older adults in the United States: a comparison across three national surveys. Prev Med 2016;89:37–43.

[25] Australian Institute of Health and Welfare. Insufficient physical activity, 2020. [cited 2021 1 February]; Available from: https://www.aihw.gov.au/reports/risk-factors/insufficient-physical-activity/contents/physical-inactivity.

[26] Burton E, Lewin G, Boldy D. Barriers and motivators to being physically active for older home care clients. Phys Occup Ther Geriatr 2013;31(1):21–36.

[27] Burton E, et al. Why do seniors leave resistance training programs? Clin Interv Aging 2017;12:585–92.

[28] Hill A-M, et al. Factors associated with older patients' engagement in exercise after hospital discharge. Arch Phys Med Rehabil 2011;92(9):1395–403.

[29] Pettigrew S, et al. A typology of factors influencing seniors' participation in strength training in gyms and fitness centres. J Aging Phys Act 2018;26(3):492–8.

[30] Latham N, Liu C-J. Strength training in older adults: the benefits for osteoarthritis. Clin Geriatr Med 2010;26(3):445–59.

[31] Panagiotou M, Koulouris N, Rovina N. Physical activity: a missing link in asthma care. J Clin Med 2020;9(3):706.

[32] Baert V, et al. Motivators and barriers for physical activity in older adults with osteoporosis. J Geriatr Phys Ther 2015;38(3):105–14.

[33] Cavill N, Foster C. Enablers and barriers to older people's participation in strength and balance activities: a review of reviews journal of frailty. J Frailty Sarcopenia Falls 2018;3(2):105–13.

[34] Burton E, et al. Motivators and barriers for older people participating in resistance training: a systematic review. J Aging Phys Act 2017;25(2):311–24.

[35] Burton E, Lewin G, Boldy D. Physical activity levels of older adults receiving a home care service. J Aging Phys Act 2013;21(2):140–54.

[36] Burton E, Lewin G, Boldy D. Physical activity preferences of older home care clients. Int J Older People Nurs 2015;10(3):170–8.

[37] Burton E, et al. Physical activity programs for older people in the community receiving home care services: systematic review and meta-analysis. Clin Interv Aging 2019;14:1045–64.

[38] King A, et al. Assessing the impact of a restorative home care service in New Zealand: a cluster randomised controlled trial. Health Soc Care Community 2012;4:336–74.

[39] Parsons J, et al. Randomized controlled trial to determine the effect of a model of restorative home care on physical function and social support among older people. Arch Phys Med Rehabil 2013;94:1015–22.

[40] de Labra C, et al. Effects of physical exercise interventions in frail older adults: a systematic review of randomized controlled trials. BMC Geriatr 2015;15:154.

[41] Mjøsund H, et al. Integration of physical activity in reablement for community dwelling older adults: a systematic scoping review. J Multidiscip Healthc 2020;13:1291–315.

[42] Tricco A, et al. Comparisons of interventions for preventing falls in older adults: a systematic review and meta-analysis. JAMA 2017;318(17):1687–99.

[43] McKenna A, et al. Purchasing and using personal emergency response systems (PERS): how decisions are made by community-dwelling seniors in Canada. BMC Geriatr 2015;15:81.

[44] World Health Organisation. WHO global report on falls prevention in older age. Geneva, Switzerland: World Health Organisation; 2007.

[45] Smith J, Lewin G. Home care clients' participation in fall prevention activities. Australas J Ageing 2008;27(1):38–42.

[46] Mitzner T, et al. Older adults' needs for home health care and the potential for human factors interventions. Proc Hum Factors Ergon Soc Annu Meet 2009;53(1):718–22.

[47] Australian Government: The Department of Health. Chronic disease management—individual allied health services under medicare—provider information, 2014. [cited 2021 1 February]; Available from: https://www1.health.gov.au/internet/main/publishing.nsf/Content/health-medicare-health_pro-gp-pdf-allied-cnt.htm.

[48] Taylor D. Physical activity is medicine for older adults. Postgrad Med J 2014;90(1059):26–32.

[49] Burton E, et al. Are interventions effective in improving the ability of older adults to rise from the floor independently? A mixed method systematic review. Disabil Rehabil 2020;42(6):743–53.

[50] Lewis C, Bottomley J. Geriatric physical therapy: a clinical approach. Pearson: Connecticut (USA); 2003.

[51] Michie S, van Stralen M, West R. The behaviour change wheel: a new method for characterising and designing behaviour change interventions. Implement Sci 2011;6:42.

[52] Michie S, et al. The behavior change technique taxonomy (v1) of 93 hierarchically clustered techniques: building an international consensus for the reporting of behavior change interventions. Ann Behav Med 2013;46(1):81–95.

Chapter 30

Exercise in Parkinson's disease

Eleanor M. Taylor[a,*], Dylan Curtin[a,*], Joshua J. Hendrikse[a], Claire J. Cadwallader[a], Julie C. Stout[a], Trevor T-J. Chong[a,b,c], and James P. Coxon[a]

[a]*The Turner Institute for Brain and Mental Health, School of Psychological Sciences, Monash University, Melbourne, VIC, Australia* [b]*Department of Neurology, Alfred Health, Melbourne, VIC, Australia* [c]*Department of Clinical Neurosciences, St Vincent's Hospital, Melbourne, VIC, Australia*

Introduction

Parkinson's disease (PD) is the fastest growing neurological disorder globally—affecting 6.1 million people worldwide in 2016, with expectations for this number to surpass 13 million by 2040 [1,2]. The defining neuropathological hallmark of the disease is the progressive loss of dopaminergic neurons within specific areas of the basal ganglia—a key area of the brain responsible for motor control [3]. The resulting deficiency of dopamine within the basal ganglia leads to a group of characteristic motor symptoms, including the hallmark feature of bradykinesia (a slowness of movement), as well as muscular rigidity and/or rest tremor [4]. As the disease progresses, a range of additional motor features can also emerge, including postural reflex disturbances (e.g., flexed postures of the trunk and limbs), postural instability, speech and swallowing problems, as well as "freezing of gait" (a temporary inability to move the feet forward) and falls [5].

Although once viewed solely as a movement disorder, it is now clear that PD also involves a range of nonmotor symptoms [6]. These symptoms are thought to relate to neurodegeneration outside the motor system, including the pathological spread of Lewy Bodies (abnormal aggregations of α-synuclein and other proteins) to extrastriatal regions as well as dysfunction within additional neurotransmitter systems (e.g., serotonergic, noradrenergic, and cholinergic; [3]). Common nonmotor symptoms of PD include neuropsychiatric symptoms (e.g., depression, apathy), fatigue, sleep disturbance, autonomic dysfunction (e.g., orthostatic hypotension, constipation), and sensory symptoms (e.g., pain, hyposmia) [7]. Deficits in cognitive functions such as higher-order executive skills (e.g., cognitive flexibility, abstract thinking, and cognitive inhibition), attention, and working memory are also common [8]. The impact of nonmotor symptoms is substantial and of great concern to patients and their families due to implications for quality of life, caregiver burden, hospital admissions, and health-related costs [9–14].

Clinical diagnosis and progression

In the absence of any definitive diagnostic tests, a clinical diagnosis of PD is based on the presence of the cardinal motor features (bradykinesia and either rigidity and/or rest tremor) in addition to further supporting and exclusionary criteria. The average age of onset is in the late 50s, although this can vary from <40 years to >80 years [15]. Progression of the disease is characterized by worsening of both motor and nonmotor features. Motor symptoms typically emerge on one side of the body, and while this asymmetry persists throughout the disease course, these symptoms also begin to affect both sides of the body in later stages. The Movement Disorder Society (MDS) Unified Parkinson's Disease Rating Scale (UPDRS) is a standardized measure to assess the various motor symptoms and signs of PD (Fig. 1; [16]). The Hoehn and Yahr stage is a complementary measure that outlines the milestones in disease progression from mild unilateral symptoms (Stage 1) through the end-stage non-ambulatory state (Stage 5; [17]).

Although motor symptoms remain the core diagnostic feature, nonmotor symptoms are often the first symptoms to emerge in the disease progression, and they become increasingly prevalent over the course of the illness [7]. For example, dementia—a key concern for many patients—is present in 70%–80% of those patients surviving for >20 years after their initial diagnosis [18,19]. In combination, the motor and nonmotor symptoms of late-stage PD contribute substantially to disability and mortality.

---

* Joint first authors: These authors contributed equally to this work.

Exercise to Prevent and Manage Chronic Disease Across the Lifespan. https://doi.org/10.1016/B978-0-323-89843-0.00023-4

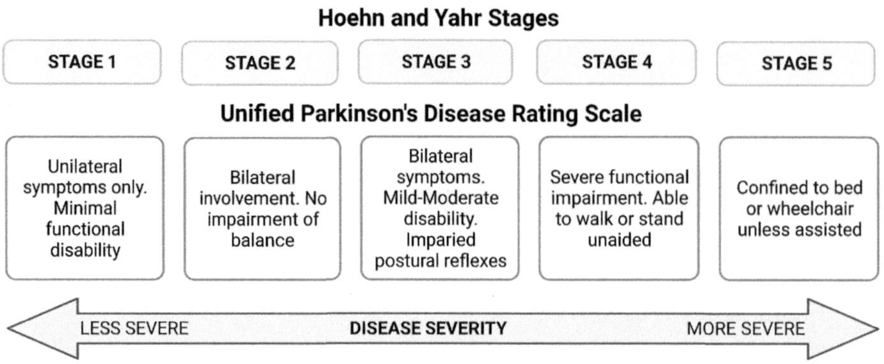

Hoehn and Yahr Stages

STAGE 1	STAGE 2	STAGE 3	STAGE 4	STAGE 5

Unified Parkinson's Disease Rating Scale

Unilateral symptoms only. Minimal functional disability	Bilateral involvement. No impairment of balance	Bilateral symptoms. Mild-Moderate disability. Imparied postural reflexes	Severe functional impairment. Able to walk or stand unaided	Confined to bed or wheelchair unless assisted

← LESS SEVERE   **DISEASE SEVERITY**   MORE SEVERE →

FIG. 1 The Hoehn and Yahr, and Movement Disorder Society Unified Parkinson's Disease Rating Scales.

Treatments

There are currently no treatments capable of slowing or forestalling the progression of PD; however, there are several established treatments to alleviate the symptoms. Medication management forms a mainstay of such treatment, primarily involving drugs that act on dopaminergic pathways (e.g., levodopa). Such medications are effective in alleviating motor symptoms and improving quality of life, but are associated with complications, particularly as the disease progresses, and either medication dose is increased or new pharmacological agents are introduced. Common complications include wearing-off effects (as motor symptoms reemerge between doses of medications), dyskinesias, and behavioral side effects (e.g., increased risk of impulsive-compulsive behaviors). Moreover, these pharmacological treatments offer limited benefit for nonmotor symptoms [20,21], particularly those that are driven by neurodegeneration outside the dopaminergic pathways [22]. Deep brain stimulation is a surgical approach that can significantly reduce motor fluctuations in later disease stages [23], although this intervention has strict safety criteria and is inherently invasive. Nonpharmacological interventions can play an important role in complementing treatments, both in further reducing motor symptoms as well as managing nonmotor symptoms that are otherwise difficult to treat [24]. Accumulating evidence indicates that exercise is a safe and feasible intervention to alleviate both motor and nonmotor symptoms of PD [25].

Exercise in disease

Exercise has long been proposed as an adjunct treatment for people with PD, both for overall brain health and disease-related symptoms [26]. Large observational studies in PD patients demonstrate a positive association between regular physical exercise and better quality of life, overall physical function, and lower caregiver burden [27–29]. In recent years, the findings of several controlled trials have also shown direct benefits of exercise on both motor and nonmotor symptoms [30]. The majority of research examining the benefits of exercise in PD focuses on early disease stages. This is when balance and gait impairment are less likely to interfere with adherence, and when exercise is thought to offer the greatest benefits to patients. Much of the research discussed below focused on the early stages of the disease.

Exercise and motor symptoms

Several recent high-quality studies have shown that regular cardiorespiratory exercise can attenuate the progression of motor symptoms relative to a nonexercise control group [31–33]. These studies typically measure motor symptoms using the MDS-UPDRS motor scale, with higher scores indicating more severe motor symptoms. For example, a study by van der Kolk et al. [33] compared the effects of a 6-month program of high-intensity cardiorespiratory exercise (30–45 min of cycling, three times weekly) relative to a nonexercise control group. Results showed that participants who completed the exercise program scored, on average, 1.3 points higher on the MDS-UPDRS motor scale at the end of 6 months, compared to an average of 5.6 points higher for those who did not exercise. This indicates a less rapid progression of motor symptoms in those who completed the exercise program in comparison to the control group. Encouragingly, similar benefits have been demonstrated across trials conducted both in controlled experimental environments and in remote home settings [32,33]. In investigating the benefits of exercise on PD motor symptoms, researchers are showing increasing interest in activities that include both physical and cognitive demands (Box 1).

Box 1 Cognitively demanding exercise

Activities such as dance, boxing, Tai Chi or yoga may be particularly beneficial for PD patients, as they integrate physical activity with coordination, balance and learning. In a study by Duncan and Earhart [9], participants with PD were assigned to a 12-month tango dance program or a control group. Participants who completed the dance program demonstrated significant improvements in overall motor symptoms measured by the MDS-UPDRS at 3 months, with even greater improvements observed at 6- and 12-month time-points. This result has been supported by a number of similar studies [36], though it is unclear whether these improvements are greater than those observed following aerobic or resistance training [37]. Similarly, structured Tai Chi programs have been shown to improve motor performance in PD. Specifically research shows it may improve balance, reducing falls, and enhance performance on functional tasks involving reaching and rising from a chair [38], though findings in this area are not consistent [39]. In sum, the addition of cognitive demands to physical activity may represent a promising new way to engage PD patients in exercise, though research in this area is ongoing.

Resistance training has also shown benefits for motor symptoms in PD [34]. One study by Corcos et al. (2013) [35] assessed the relative efficacy of progressive resistance training on motor symptoms for PD patients over 2 years. They found significant improvements in MDS-UPDRS scores at 6, 12, and 18 months. The greatest improvement was observed following 24 months of upper- and lower-body strength training, as participants scored, on average, 7.4 points lower on the UPDRS scale. This indicates the ongoing benefits of progressive resistance training over time. Taken together, there is now good evidence that both cardiorespiratory exercise and resistance training are feasible and clinically relevant interventions for improving overall motor symptom presentation in PD.

Due to the highly variable presentations of PD, clinicians may wish to tailor exercise programs to the specific needs of the individual. For example, treadmill training may result in improvements in gait speed [40], and reduce gait disturbance (e.g., freezing; [41]). Resistance training, on the other hand, appears to have less impact on gait speed and stamina, but improves muscle strength, balance, and overall mobility [34,42]. Improvements in muscle strength can reduce risk of falls [43], in addition to improving the performance of daily tasks that people which people with PD often find difficult, such as rising from a chair [44]. These distinct outcomes allow clinicians to tailor exercise programs depending on the specific needs of the individual.

Exercise and nonmotor symptoms

Exercise and cognition

Cognitive changes have been highlighted as a key area of concern for PD patients [45], however, evidence suggests that exercise may improve some of the cognitive side effects of the disease. In healthy adults, cardiorespiratory exercise is known to improve performance across various cognitive domains, including executive functioning [46–48]. Accumulating evidence supports the use of exercise to improve cognitive functioning in PD patients who do not suffer from dementia. This is of great interest to clinicians, as the mild to moderate reductions in cognition that occur in PD can reduce patients' capacity to live independently, and thereby represent a significant contributor to disability and reduced quality of life [49]. Greater cognitive decline is associated with increased caregiver burden, and greater likelihood of institutionalization [50].

A key area of cognitive decline in PD is in executive function skills, such as cognitive flexibility, abstract thinking, and cognitive inhibition [51]. Improvements in these areas have been found following basic treadmill training [52], resistance training [53], cycling [54], and combined cardiorespiratory and resistance exercise programs [55]. Benefits are also seen in other key cognitive domains. In one promising study, David et al. [56] compared outcomes on measures of attention and working memory for PD patients across two forms of resistance exercise training over 24 months (consistent weight training vs progressive weight training). Results showed improvements in the cognitive domains of attention and working memory at both 12 and 24 months, with no significant difference in cognitive improvement between the two programs. Taken together, these studies indicate that cognitive decline across several key cognitive domains associated with PD can be ameliorated or slowed by a number of different exercise programs.

Exercise and sleep

Another nonmotor symptom that exercise may address is sleep disturbance. Between 74% and 98% of PD patients report some form of sleep disturbance [57–59]. The type of sleep disturbance experienced by PD patients is varied, and may include waking during the night, rapid eye movement (REM) sleep disorders, daytime sleepiness, and insomnia [60].

In healthy elderly people, a range of different exercises (e.g., Tai Chi, resistance training) can improve self-perceived sleep quality [61,62], however, emerging evidence suggests that these exercise-related benefits may also apply to PD patients. A 6-month multimodal exercise program involving stretching, resistance training, and balance exercises was found to improve self-reported sleep quality for PD patients [63]. Similarly, a study by Silva-Batista et al. [64] assessed the impact of a 12-week progressive resistance training program. Following exercise, patients reported significant improvements in self-reported sleep quality, reduced sleep disturbance, and reduced daytime sleepiness compared to those who did not exercise [64].

In addition to the self-reported improvements outlined above, the benefits of exercise on sleep in PD are further supported by more objective measures. A study by Amara et al. [65] used polysomnography, a laboratory-based sleep assessment, to assess the sleep quality of PD patients who completed a 16-week program involving high-intensity resistance training and functional mobility training. Consistent with findings from earlier self-report sleep studies, participants in the exercise group demonstrated improvements in overall sleep efficiency compared to a control group, quantified by reduced waking after sleep onset and increased overall time spent asleep [65]. While research in this area is limited, positive outcomes have been found across several exercise modalities and intensities, suggesting a generalized benefit of exercise on sleep quality in PD. Taken together these studies suggest structured exercise programs can improve both subjective and objective measures of sleep quality in people with PD.

Exercise and mood

There is accumulating evidence showing that exercise can improve symptoms of depression in PD patients [31,66–68]. In two notable controlled trials, regular exercise comprising strengthening exercise [69] or a combination of cardiorespiratory and strengthening exercise [70] was found to improve affect and reduce depressive symptoms. However, other studies have not observed improvements in depressive symptoms [33,44,53], though this may reflect mild symptom presentations pre-intervention. Exercise is a well-established treatment for older adults with depression [71], particularly when combined with other established therapies (e.g., cognitive behavioral therapy, antidepressant medications). More research is required to understand how it may benefit PD populations. There is also a need for further research investigating the effects of exercise on other mood-related symptoms (e.g., apathy, anxiety).

Autonomic dysfunction

Studies investigating the effects of exercise on autonomic dysfunction in PD are limited, though this represents an area of emerging interest [72]. One randomized controlled trial showed that a 12-week program of resistance training improved cardiac sympathetic modulation, quantified by decreased heart rate variability and blood pressure response, as compared to a no-exercise control group [73]. If supported by further trials, improvements in sympathetic modulation may help to alleviate symptoms and rates of cardiovascular morbidity and mortality in PD. Further studies addressing the effects of exercise on cardiovascular dysfunction as well as other autonomic features (e.g., constipation, urinary dysfunction) are needed in the future.

Summary

Regular exercise shows benefits for various PD symptoms, though there are varying levels of support for the effects of exercise on different symptoms (Fig. 2). There is strong evidence showing improvements, or slowing of progression, in overall motor symptoms following cardiorespiratory or resistance exercise. Similarly, there is consistent evidence showing improvements in cognition, particularly executive function, following exercise. While there is some evidence for the use of exercise to address additional nonmotor features such as mood and autonomic dysfunction more research is required to draw definitive conclusions. The potential for exercise to benefit nonmotor symptoms is promising, particularly as these are not improved by common PD medication regimes, and warrants further research. Another area with emerging support is research demonstrating a link between exercise and specific motor symptom improvements. This would allow for greater tailoring of exercise for individual patient needs.

Known and potential mechanisms underlying exercise benefits

Although there is much to be learnt about the mechanisms of exercise in PD, there is accumulating evidence from animal models and humans that exercise exerts direct effects on the brain in PD. We focus here only on cardiorespiratory exercise, as research investigating the mechanisms of other forms of exercise (i.e., resistance training) remains preliminary. A central theme to research exploring the mechanism of exercise in PD is *neuroplasticity*, which refers to the ongoing capacity of the

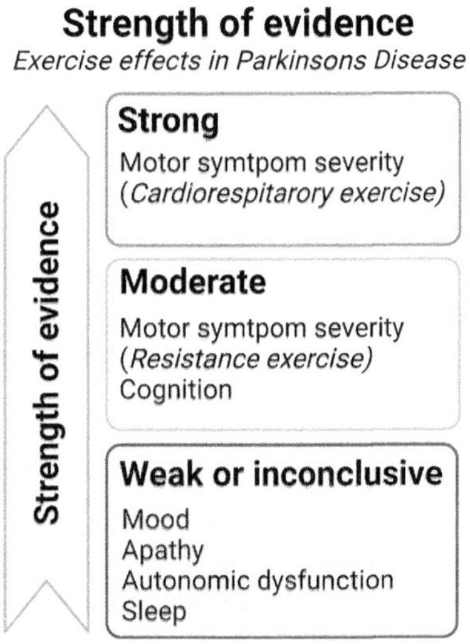

Strength of evidence
Exercise effects in Parkinsons Disease

Strength of evidence

Strong
Motor symtpom severity
(*Cardiorespitarory exercise*)

Moderate
Motor symtpom severity
(*Resistance exercise*)
Cognition

Weak or inconclusive
Mood
Apathy
Autonomic dysfunction
Sleep

FIG. 2 Strength of evidence for motor and nonmotor symptoms of Parkinson's disease.

brain to respond to intrinsic and extrinsic stimuli by reorganizing its connections, structure, and function [74]. Neuroplastic changes occur throughout the lifespan and can be both adaptive (i.e., when associated with an improvement in motor and cognitive functions) or maladaptive (i.e., associated with a loss of function). In PD, it is hypothesized that exercise can promote a range of adaptive neuroplastic changes such as improvements in dopamine transmission, particularly when exercise is undertaken early in the disease course. These changes may counter the underlying neurodegenerative processes and thereby ameliorate symptoms and even slow the disease course [26,75].

Insights into the effects of exercise on the brain in PD have tended to come from rodent models, owing to the challenges of directly studying the brain in humans. In these studies, rodents are administered with neurotoxins that selectively target the dopaminergic nigrostriatal system and cause *parkinsonian* motor deficits, mimicking those observed in humans with PD. Although not unequivocal, there is evidence that regular cardiovascular exercise can reverse striatal dopaminergic cell loss in these models [76–78]. There is also evidence that exercise can decrease markers of brain damage [79] and enhance angiogenesis (i.e., the generation of new blood vessels that branch from preexisting vessels) [80].

Though limited, there are some studies in humans that support findings from animal models. For example, several cohort studies have shown increased dopamine release [81] and dopamine receptor D2 binding potential [82] following regular cardiorespiratory exercise. Taken together, these results indicate regular cardiorespiratory exercise can exert direct benefits for the dopaminergic system and overall brain health in PD.

Improvements in brain structure and function may be supported by cellular and molecular factors implicated in plasticity [83]. One way that exercise may exert neuroprotective effects in PD is via the upregulation of neurotrophic factors, which are known to support the growth and survival of neurons [70]. In particular, parkinsonian mouse models have shown increases in levels of two key neurotrophic factors—brain-derived neurotrophic factor (BDNF) and glial cell line-derived neurotrophic factor (GDNF)—within the striatum following daily exercise [84,85]. Complementary evidence in humans has also shown that serum BDNF levels are significantly increased after 1 month of treadmill exercise in a small group of patients with PD, as compared to a nonexercise control group with PD [86]. Further research is needed to establish a link between increased neurotrophic levels and improved symptomatology in people with PD.

Safety considerations

Generally, exercise studies in PD report low rates of exercise-related adverse events, including those involving moderate and high intensities [32]. However, there are some risks, especially in people with advanced disease stages (i.e., Hoen and Yahr stages >3), where balance and gait impairment become more apparent. Particular risks for people with PD include fall-related injuries and cardiovascular complications [87]. As for any person with a preexisting medical condition, it is recommended that PD patients are first screened for contraindications to exercise (e.g., cardiac complications, poorly managed diabetes) and receive appropriate supervision throughout their exercise program.

Falls

Balance issues, reduced muscle strength, and low confidence, in combination with lower bone density can increase both the risk and severity of fall-related injuries in PD (e.g., fractures) [88,89]. A key strategy for minimizing these risks is having patients exercise during the medication "ON" state (i.e., a time of the day when their symptoms are least severe). Furthermore, focus can be placed on types of exercise that match a patient's capabilities and interests, while also providing sufficient challenge and incorporating input from both the patient and exercise professional. For example, people who experience freezing of gait could be encouraged to exercise on a stationary ergometer rather than a treadmill. Water-based exercises are also another alternative, which have similar benefits as other forms of cardiovascular exercise for motor symptoms [90], and can also help to increase confidence with exercise [67].

Cardiac events

There is a risk that people with PD, especially those who are elderly or sedentary, who start with a new exercise program may experience cardiovascular complications. Although PD studies report low rates of cardiac events during clinical trials, the potential impact of a cardiac event is substantial (e.g., myocardial infarction). In addition to the general recommendations above, special consideration should be given to cardiovascular conditions common in PD such as chronotropic incompetence (the inability of the heart to increase its rate in line with increased activity or demand), which can make it difficult for patients to reach standardized age-based heart rate zones [91,92]. Using subjective ratings of exertion (e.g., the BORG scale) in addition to heart rate monitoring and other objective measures (e.g., blood lactate measurement, ventilation rate) is one way to gain more reliable feedback about exercise intensity and reduce the risk of over-exertion. Exercise-induced hypotension [93] is also another adverse effect to consider, particularly where this could cause additional injuries (e.g., fainting at the top of a set of stairs). Patient and support staff should be made aware of this risk so that it can be appropriately managed by supervision.

Clinical practice recommendations for exercise prescribers

As there are gaps in the scientific literature regarding the optimal exercise parameters for people with PD, the current guidelines for exercise in PD are based on healthy adults, with provisions to accommodate for the challenges and goals relevant to PD [94]. These guidelines recommend a range of exercise types including aerobic, resistance, flexibility, and balance exercises across the week. Research outlined in this chapter has highlighted improvements in both motor and nonmotor symptoms following exercise at low, moderate, and high intensities and across various modalities. There is likely no blanket optimal exercise program for all PD patients, therefore the prescription of exercise duration, intensity, and frequency can be guided by individual factors such as goals of the exercise program (e.g., improving balance vs cardiorespiratory fitness), interests, and personal barriers.

Adherence and barriers

Exercise has well-established benefits in PD; however, patients do not always include exercise as part of their treatment program. Although attendance is generally strong within clinical trials [33,66,95] there are additional barriers that patients face in everyday life. These may include reduced self-confidence, mobility issues (e.g., postural instability, gait impairment), and difficulties related to nonmotor symptoms such as depression, anxiety, apathy, and fatigue. In contrast, factors that can motivate patients include social support by family, friends, and peers as well as education about the benefits of exercise [30,87]. Individual and/or group supervision can also play a key role for PD patients when exercising, both for increasing safety as well as enhancing engagement with exercise. Ideally this supervision is provided in-person; however, it is also possible to use other aids (e.g., mobile apps) to provide remote supervision [33], particularly where in-person supervision is difficult.

Conclusion

Exercise is a feasible and clinically meaningful intervention in PD. There is strong evidence supporting both cardiovascular and resistance exercise in managing overall motor symptomatology. Combining exercise with cognitive demands involved in activities such as dance and Tai Chi may be particularly beneficial for PD patients, as they integrate physical activity with coordination, balance, and concentration. Accumulating evidence supports the use of exercise to improve a range of nonmotor symptoms, including delaying global cognitive decline as well as improving performance within specific cognitive domains such as executive functions, attention, and working memory. Sleep and mood disturbances (e.g., depression) can also be alleviated with exercise. From a mechanistic perspective, there is evidence from human and animal studies showing that exercise can promote a range of adaptive neuroplastic changes in PD (e.g., promote the release of key neurotrophins and dopamine, as well as reverse dopaminergic cell loss within the striatum). Key safety considerations for people with PD include fall-related injuries and cardiovascular complications. Supervision and psychoeducation about these factors can help to significantly reduce their risk as well as increase engagement in exercise. Guidelines for the prescription of exercise in PD recommend a range of exercise types including aerobic, resistance, flexibility, and balance exercises across the week. There is likely no blanket optimal exercise program for all PD patients, therefore the prescription of exercise duration, intensity, and frequency can be guided by individual factors such as goals of cardiorespiratory fitness, interests, personal barriers and the goals of the exercise program (e.g., improving balance vs cardiovascular fitness).

Acknowledgments

J.S. is supported by an Australian National Health and Medical Research Council (NHMRC) Investigator Grant 1173472.

T.C. is supported by the Australian Research Council (DP 180102383, DE 180100389). J.C. is supported by an Australian Research Council Discovery project (DP 200100234).

Glossary

Bradykinesia	Slowness of voluntary movement with progressive reduction in speed and amplitude of repetitive actions.
Dopamine	Neurotransmitter produced by the substantia nigra in the basal ganglia. Responsible for transmission of signals between nerve cells that control movement.
Dyskinesia	Involuntary movements (e.g., nodding, jerking, twisting) that occur most frequently when levodopa concentrations are at their maximum.
Levodopa	A chemical precursor of dopamine which can be taken orally. It is converted to dopamine and crosses the blood brain barrier.
Hoehn & Yahr scale	A 5-point system that outlines the milestones in progression of the illness from mild unilateral symptoms through the end stage non-ambulatory state.
Motor fluctuations ("ON" and "OFF" periods)	Alterations between periods of good motor symptom control ("ON" time) and periods of reduced motor symptom control ('OFF' time).
Neurotrophic Factors	Proteins that are thought to be a key mechanism underlying the benefits of exercise for the brain in health and Parkinson's disease.
Rest tremor	A tremor which occurs when the affected limb or body part is at rest. Typically oscillates at a frequency of 4–5 hertz (Hz) per second.
Rigidity	Stiffness and increased tone of muscles.
Unified Parkinson's Disease Rating Scale (UPDRS)	The most widely used measure to assess the various motor symptoms and signs of Parkinson's disease.

References

[1] Dorsey ER, Bloem BR. The Parkinson pandemic—a call to action. JAMA Neurol 2018;75(1):9–10.

[2] Dorsey ER, et al. Global, regional, and national burden of Parkinson's disease, 1990–2016: a systematic analysis for the global burden of disease study 2016. Lancet Neurol 2018;17(11):939–53.

[3] Kalia LV, Lang AE. Parkinson's disease [in eng]. Lancet 2015;386(9996):896–912. https://doi.org/10.1016/s0140-6736(14)61393-3.

[4] Tolosa E, Wenning G, Poewe W. The diagnosis of Parkinson's disease. Lancet Neurol 2006;5(1):75–86.

[5] Jankovic J. Parkinson's disease: clinical features and diagnosis [in eng]. J Neurol Neurosurg Psychiatry 2008;79(4):368–76. https://doi.org/10.1136/jnnp.2007.131045.

[6] Obeso J, et al. Past, present, and future of Parkinson's disease: a special essay on the 200th anniversary of the Shaking Palsy. Mov Disord 2017;32(9):1264–310.

[7] Schapira AH, Chaudhuri KR, Jenner P. Non-motor features of Parkinson disease. Nat Rev Neurosci 2017;18(7):435–50.

[8] Svenningsson P, Westman E, Ballard C, Aarsland D. Cognitive impairment in patients with Parkinson's disease: diagnosis, biomarkers, and treatment. Lancet Neurol 2012;11(8):697–707.

[9] Duncan RP, Earhart GM. Randomized controlled trial of community-based dancing to modify disease progression in Parkinson disease. Neurorehabil Neural Repair 2012;26(2):132–43.

[10] Boller F, Caputi N. History of neuropsychology in Parkinson's disease. Rev Neurol 2017;173(10):607.

[11] Duncan GW, et al. Health-related quality of life in early Parkinson's disease: the impact of nonmotor symptoms. Mov Disord 2014;29(2):195–202.

[12] Fletcher P, Leake A, Marion MH. Patients with Parkinson's disease dementia stay in the hospital twice as long as those without dementia. Mov Disord 2011;26(5):919.

[13] Kudlicka A, Clare L, Hindle JV. Quality of life, health status and caregiver burden in Parkinson's disease: relationship to executive functioning. Int J Geriatr Psychiatry 2014;29(1):68–76.

[14] Vossius C, Larsen JP, Janvin C, Aarsland D. The economic impact of cognitive impairment in Parkinson's disease. Mov Disord 2011;26(8):1541–4.

[15] Poewe W, et al. Parkinson disease [in eng]. Nat Rev Dis Primers 2017;3:17013. https://doi.org/10.1038/nrdp.2017.13.

[16] Goetz CG, et al. Movement disorder society-sponsored revision of the unified Parkinson's Disease Rating Scale (MDS-UPDRS): scale presentation and clinimetric testing results. Mov Disord 2008;23(15):2129–70.

[17] Goetz CG, et al. Movement Disorder Society task force report on the Hoehn and Yahr staging scale: status and recommendations the movement disorder society task force on rating scales for Parkinson's disease. Mov Disord 2004;19(9):1020–8.

[18] Aarsland D, Kurz MW. The epidemiology of dementia associated with Parkinson disease. J Neurol Sci 2010;289(1–2):18–22.

[19] Hely MA, Reid WG, Adena MA, Halliday GM, Morris JG. The Sydney multicenter study of Parkinson's disease: the inevitability of dementia at 20 years [in eng]. Mov Disord 2008;23(6):837–44. https://doi.org/10.1002/mds.21956.

[20] Connolly BS, Lang AE. Pharmacological treatment of Parkinson disease: a review [in eng]. JAMA 2014;311(16):1670–83. https://doi.org/10.1001/jama.2014.3654.

[21] Meissner WG, et al. Priorities in Parkinson's disease research. Nat Rev Drug Discov 2011;10(5):377–93.

[22] Nonnekes J, Timmer MH, de Vries NM, Rascol O, Helmich RC, Bloem BR. Unmasking levodopa resistance in Parkinson's disease. Mov Disord 2016;31(11):1602–9.

[23] Bronstein JM, et al. Deep brain stimulation for Parkinson disease: an expert consensus and review of key issues. Arch Neurol 2011;68(2):165.

[24] Bloem BR, de Vries NM, Ebersbach G. Nonpharmacological treatments for patients with Parkinson's disease. Mov Disord 2015;30(11):1504–20.

[25] Bouça-Machado R, et al. Physical activity, exercise, and physiotherapy in Parkinson's disease: defining the concepts. Mov Disord Clin Pract 2020;7(1):7–15.

[26] Ahlskog JE. Aerobic exercise: evidence for a direct brain effect to slow Parkinson disease progression. Mayo Clin Proc 2018;93(3):360–72. Elsevier.

[27] Baatile J, Langbein W, Weaver F, Maloney C, Jost M. Effect of exercise on perceived quality of life of individuals with Parkinson's disease. J Rehabil Res Dev 2000;37(5):529–34.

[28] Fang X, et al. Association of levels of physical activity with risk of Parkinson disease: a systematic review and meta-analysis. JAMA Netw Open 2018;1(5):e182421.

[29] Oguh O, Eisenstein A, Kwasny M, Simuni T. Back to the basics: regular exercise matters in Parkinson's disease: results from the National Parkinson Foundation QII registry study. Parkinsonism Relat Disord 2014;20(11):1221–5.

[30] Schootemeijer S, van der Kolk NM, Bloem BR, de Vries NM. Current perspectives on aerobic exercise in people with Parkinson's disease. Neurotherapeutics 2020;17:1418–33. https://doi.org/10.1007/s13311-020-00904-8.

[31] Uc EY, et al. Phase I/II randomized trial of aerobic exercise in Parkinson disease in a community setting [in eng]. Neurology 2014;83(5):413–25. https://doi.org/10.1212/wnl.0000000000000644.

[32] Schenkman M, et al. Effect of high-intensity treadmill exercise on motor symptoms in patients with de novo Parkinson disease: a phase 2 randomized clinical trial. JAMA Neurol 2018;75(2):219–26.

[33] van der Kolk NM, et al. Effectiveness of home-based and remotely supervised aerobic exercise in Parkinson's disease: a double-blind, randomised controlled trial. Lancet Neurol 2019;18(11):998–1008.

[34] Chung CLH, Thilarajah S, Tan D. Effectiveness of resistance training on muscle strength and physical function in people with Parkinson's disease: a systematic review and meta-analysis. Clin Rehabil 2016;30(1):11–23.

[35] Corcos DM, et al. A two-year randomized controlled trial of progressive resistance exercise for Parkinson's disease. Mov Disord 2013;28(9):1230–40.

[36] Lötzke D, Ostermann T, Büssing A. Argentine tango in Parkinson disease–a systematic review and meta-analysis. BMC Neurol 2015;15(1):1–18.

[37] Rawson KS, McNeely ME, Duncan RP, Pickett KA, Perlmutter JS, Earhart GM. Exercise and parkinson disease: comparing tango, treadmill and stretching. J Neurol Phys Ther 2019;43(1):26.

[38] Liu H-H, Yeh N-C, Wu Y-F, Yang Y-R, Wang R-Y, Cheng F-Y. Effects of Tai Chi exercise on reducing falls and improving balance performance in Parkinson's disease: a meta-analysis. Parkinsons Dis 2019;2019, 9626934.

[39] Amano S, et al. The effect of Tai Chi exercise on gait initiation and gait performance in persons with Parkinson's disease. Parkinsonism Relat Disord 2013;19(11):955–60.

[40] Ni M, Hazzard JB, Signorile JF, Luca C. Exercise guidelines for gait function in Parkinson's disease: a systematic review and meta-analysis. Neurorehabil Neural Repair 2018;32(10):872–86.

[41] Bello O, Marquez G, Camblor M, Fernandez-Del-Olmo M. Mechanisms involved in treadmill walking improvements in Parkinson's disease. Gait Posture 2010;32(1):118–23.

[42] Cruickshank TM, et al. The effect of multidisciplinary rehabilitation on brain structure and cognition in Huntington's disease: an exploratory study. Brain Behav 2015;5(2), e00312.

[43] Allen NE, et al. The effects of an exercise program on fall risk factors in people with Parkinson's disease: a randomized controlled trial. Mov Disord 2010;25(9):1217–25.

[44] Shulman LM, et al. Randomized clinical trial of 3 types of physical exercise for patients with Parkinson disease [in eng]. JAMA Neurol 2013;70(2):183–90. https://doi.org/10.1001/jamaneurol.2013.646.

[45] Goldman JG, et al. Cognitive impairment in Parkinson's disease: a report from a multidisciplinary symposium on unmet needs and future directions to maintain cognitive health. npj Parkinson's Dis 2018;4(1):1–11.

[46] Erickson KI, et al. Physical activity, cognition, and brain outcomes: a review of the 2018 physical activity guidelines. Med Sci Sports Exerc 2019;51 (6):1242.

[47] Prakash RS, Voss MW, Erickson KI, Kramer AF. Physical activity and cognitive vitality. Annu Rev Psychol 2015;66:769–97.

[48] Smith PJ, et al. Aerobic exercise and neurocognitive performance: a meta-analytic review of randomized controlled trials. Psychosom Med 2010;72 (3):239.

[49] Dirnberger G, Jahanshahi M. Executive dysfunction in Parkinson's disease: a review. J Neuropsychol 2013;7(2):193–224.

[50] Aarsland D, Larsen JP, Tandberg E, Laake K. Predictors of nursing home placement in Parkinson's disease: a population-based, prospective study. J Am Geriatr Soc 2000;48(8):938–42.

[51] Kudlicka A, Clare L, Hindle JV. Executive functions in Parkinson's disease: systematic review and meta-analysis. Mov Disord 2011;26 (13):2305–15.

[52] Picelli A, et al. Effects of treadmill training on cognitive and motor features of patients with mild to moderate Parkinson's disease: a pilot, single-blind, randomized controlled trial. Funct Neurol 2016;31(1):25.

[53] Cruise K, Bucks R, Loftus A, Newton R, Pegoraro R, Thomas M. Exercise and Parkinson's: benefits for cognition and quality of life. Acta Neurol Scand 2011;123(1):13–9.

[54] Nadeau A, et al. A 12-week cycling training regimen improves gait and executive functions concomitantly in people with Parkinson's disease. Front Hum Neurosci 2017;10:690.

[55] Tanaka K, de Quadros Jr AC, Santos RF, Stella F, Gobbi LTB, Gobbi S. Benefits of physical exercise on executive functions in older people with Parkinson's disease. Brain Cogn 2009;69(2):435–41.

[56] David FJ, et al. Exercise improves cognition in Parkinson's disease: the PRET-PD randomized, clinical trial. Mov Disord 2015;30(12):1657–63.

[57] Factor SA, McAlarney T, Sanchez-Ramos JR, Weiner WJ. Sleep disorders and sleep effect in Parkinson's disease. Mov Disord 1990;5(4):280–5.

[58] Lees AJ, Blackburn NA, Campbell VL. The nighttime problems of Parkinson's disease. Clin Neuropharmacol 1988;11(6):512–9.

[59] Nausieda PA, Weiner WJ, Kaplan LR, Weber S, Klawans HL. Sleep disruption in the course of chronic levodopa therapy: an early feature of the levodopa psychosis. Clin Neuropharmacol 1982;5(2):183–94.

[60] Chahine LM, Amara AW, Videnovic A. A systematic review of the literature on disorders of sleep and wakefulness in Parkinson's disease from 2005 to 2015. Sleep Med Rev 2017;35:33–50.

[61] Li F, Fisher KJ, Harmer P, Irbe D, Tearse RG, Weimer C. Tai Chi and self-rated quality of sleep and daytime sleepiness in older adults: a randomized controlled trial. J Am Geriatr Soc 2004;52(6):892–900.

[62] Ferris LT, Williams JS, Shen CL, O'Keefe KA, Hale KB. Resistance training improves sleep quality in older adults a pilot study. J Sports Sci Med 2005;4(3):354.

[63] Nascimento CMC, Ayan C, Cancela JM, Gobbi LTB, Gobbi S, Stella F. Effect of a multimodal exercise program on sleep disturbances and instrumental activities of daily living performance on P arkinson's and Alzheimer's disease patients. Geriatr Gerontol Int 2014;14(2):259–66.

[64] Silva-Batista C, et al. Resistance training improves sleep quality in subjects with moderate Parkinson's disease. J Strength Cond Res 2017;31 (8):2270–7.

[65] Amara AW, et al. Randomized, controlled trial of exercise on objective and subjective sleep in Parkinson's disease. Mov Disord 2020;35(6):947–58.

[66] Altmann LJ, et al. Aerobic exercise improves mood, cognition, and language function in Parkinson's disease: results of a controlled study. J Int Neuropsychol Soc 2016;22(9):878–89.

[67] Cugusi L, et al. Effects of a Nordic walking program on motor and non-motor symptoms, functional performance and body composition in patients with Parkinson's disease. NeuroRehabilitation 2015;37(2):245–54.

[68] Tollár J, Nagy F, Hortobágyi T. Vastly different exercise programs similarly improve parkinsonian symptoms: a randomized clinical trial. Gerontology 2019;65(2):120–7.

[69] Canning CG, et al. Exercise for falls prevention in Parkinson disease: a randomized controlled trial. Neurology 2015;84(3):304–12.

[70] Park H, Poo M-M. Neurotrophin regulation of neural circuit development and function. Nat Rev Neurosci 2013;14(1):7–23.

[71] Schuch FB, Vancampfort D, Rosenbaum S, Richards J, Ward PB, Stubbs B. Exercise improves physical and psychological quality of life in people with depression: a meta-analysis including the evaluation of control group response. Psychiatry Res 2016;241:47–54.

[72] Amara AW, Memon AA. Effects of exercise on non-motor symptoms in Parkinson's disease. Clin Ther 2018;40(1):8–15.

[73] Kanegusuku H, et al. Effects of progressive resistance training on cardiovascular autonomic regulation in patients with Parkinson disease: a randomized controlled trial [in eng]. Arch Phys Med Rehabil 2017;98(11):2134–41. https://doi.org/10.1016/j.apmr.2017.06.009.

[74] Cramer SC, et al. Harnessing neuroplasticity for clinical applications. Brain 2011;134(6):1591–609. https://doi.org/10.1093/brain/awr039.

[75] Petzinger GM, Fisher BE, McEwen S, Beeler JA, Walsh JP, Jakowec MW. Exercise-enhanced neuroplasticity targeting motor and cognitive circuitry in Parkinson's disease [in eng]. Lancet Neurol 2013;12(7):716–26. https://doi.org/10.1016/s1474-4422(13)70123-6.

[76] Fisher BE, et al. Exercise-induced behavioral recovery and neuroplasticity in the 1-methyl-4-phenyl-1, 2, 3, 6-tetrahydropyridine-lesioned mouse basal ganglia. J Neurosci Res 2004;77(3):378–90.

[77] Vučcković MG, et al. Exercise elevates dopamine D2 receptor in a mouse model of Parkinson's disease: in vivo imaging with [18F] fallypride. Mov Disord 2010;25(16):2777–84.

[78] Yoon M-C, et al. Treadmill exercise suppresses nigrostriatal dopaminergic neuronal loss in 6-hydroxydopamine-induced Parkinson's rats. Neurosci Lett 2007;423(1):12–7.

[79] Al-Jarrah MD, Jamous M. Effect of endurance exercise training on the expression of GFAP, S100B, and NSE in the striatum of chronic/progressive mouse model of Parkinson's disease. NeuroRehabilitation 2011;28(4):359–63.

[80] Al-Jarrah M, Jamous M, Al Zailaey K, Bweir SO. Endurance exercise training promotes angiogenesis in the brain of chronic/progressive mouse model of Parkinson's disease. NeuroRehabilitation 2010;26(4):369–73.

[81] Sacheli MA, et al. Exercise increases caudate dopamine release and ventral striatal activation in Parkinson's disease. Mov Disord 2019;34 (12):1891–900.

[82] Fisher BE, et al. Treadmill exercise elevates striatal dopamine D2 receptor binding potential in patients with early Parkinson's disease. Neuroreport 2013;24(10):509–14.

[83] Voss MW, Vivar C, Kramer AF, van Praag H. Bridging animal and human models of exercise-induced brain plasticity [in eng]. Trends Cogn Sci 2013;17(10):525–44. https://doi.org/10.1016/j.tics.2013.08.001.

[84] Tajiri N, et al. Exercise exerts neuroprotective effects on Parkinson's disease model of rats. Brain Res 2010;1310:200–7.

[85] Real C, Ferreira A, Chaves-Kirsten G, Torrao A, Pires R, Britto L. BDNF receptor blockade hinders the beneficial effects of exercise in a rat model of Parkinson's disease. Neuroscience 2013;237:118–29.

[86] Frazzitta G, et al. Intensive rehabilitation increases BDNF serum levels in parkinsonian patients: a randomized study. Neurorehabil Neural Repair 2014;28(2):163–8.

[87] Schootemeijer S, et al. Barriers and motivators to engage in exercise for persons with Parkinson's disease. J Parkinsons Dis 2020;10(4):1293–9. https://doi.org/10.3233/JPD-202247.

[88] Fink HA, Kuskowski MA, Orwoll ES, Cauley JA, Ensrud KE, O. F. I. M. S. Group. Association between Parkinson's disease and low bone density and falls in older men: the osteoporotic fractures in men study. J Am Geriatr Soc 2005;53(9):1559–64.

[89] Van Den Bos F, Speelman AD, Samson M, Munneke M, Bloem BR, Verhaar HJ. Parkinson's disease and osteoporosis. Age Ageing 2013;42(2): 156–62.

[90] Carroll LM, Volpe D, Morris ME, Saunders J, Clifford AM. Aquatic exercise therapy for people with Parkinson disease: a randomized controlled trial. Arch Phys Med Rehabil 2017;98(4):631–8.

[91] Merola A, et al. Autonomic dysfunction in Parkinson's disease: a prospective cohort study. Mov Disord 2018;33(3):391–7.

[92] Speelman AD, et al. Cardiovascular responses during a submaximal exercise test in patients with Parkinson's disease. J Parkinsons Dis 2012;2 (3):241–7.

[93] Low DA, Vichayanrat E, Iodice V, Mathias CJ. Exercise hemodynamics in Parkinson's disease and autonomic dysfunction. Parkinsonism Relat Disord 2014;20(5):549–53.

[94] Gallo PM, Garber CE. Parkinson's disease: a comprehensive approach to exercise prescription for the health fitness professional. ACSMs Health Fit J 2011;15(4):8–17.

[95] Fisher BE, et al. The effect of exercise training in improving motor performance and corticomotor excitability in people with early Parkinson's disease. Arch Phys Med Rehabil 2008;89(7):1221–9.

Chapter 31

Exercise and Alzheimer's disease

Susan Irvine and Kathy Tangalakis

First Year College, Victoria University, Melbourne, VIC, Australia

Introduction

Alzheimer's disease (AD) is an incurable and degenerative neurological disease and is the most common cause of dementia, estimated to contribute to about 60%–75% of cases [1]. In 2006, the global prevalence of AD was 27 million people, with the number projected to quadruple to 106 million by 2050 [2]. With an aging population, there is no doubt that AD is fast becoming one of the biggest global public health problems with immense social and economic costs for individuals and society. While there is currently no known cure for AD, it is estimated that if interventions could delay both disease onset and progression by 1 year, there would be nearly 9.2 million fewer cases of the disease in 2050 [2].

According to the World Health Organization, AD and other forms of dementia ranked as the 7th leading cause of death worldwide and 2nd in middle-high income countries [3]. Women are affected disproportionally to men, with a threefold increase in women's mortality rates attributed to AD and other dementias over the last 20 years, making this the 5th leading cause of death for women globally [3]. The disease is classified according to age of symptom onset. Early Onset Alzheimer's Disease (EOAD), usually presents between ages 40 and 65, and occurs in less than 1% of AD. Late-Onset Alzheimer's Disease (LOAD), typically presents after the age of 65 and is associated with complex interactions between genetics, age and lifestyle risk factors [1].

Pathophysiology

One hundred and fifteen years ago, Alois Alzheimer first described the neuropathology of AD in an atrophied brain from a 56-year-old woman who presented with dementia symptoms [4]. Since then an extensive amount of research has been conducted on the extracellular amyloid plaques and intracellular neurofibrillary tangles associated with AD and the complex interplay of factors and pathways [5,6].

The plaque-forming Aβ peptides originate from abnormal proteolysis of the transmembrane protein amyloid precursor protein (APP), by the sequential enzymatic actions of β-secretase and γ-secretase to produce toxic amyloidogenic Aβ42 peptide fragments, which accumulate and self-assemble into plaques and oligomers, causing synaptic damage and neuronal death [5,6]. Subsequent neuroinflammation and immune cell activation play crucial roles in AD's pathogenesis and development [7].

Abnormal hyperphosphorylation of the tau protein causes the formation of neurofibrillary tangles inside neurons, destabilizing the microtubules and impairing the transport of essential nutrients and molecules, causing neuronal death and brain atrophy [8]. While the complete interplay between amyloid plaques and tau pathology remains elusive, it is the number of neurofibrillary tangles in the neocortex and not Aβ-plaque deposition, which is strongly correlated with AD's severity [9,10]. The abnormal tau accumulation is highly predictable, first appearing in the transentorhinal cortex [9]. Cognitive impairment becomes evident when abnormal tau deposits reach the hippocampus (responsible for learning and memory), resulting in the symptoms of the disease, which become worse and more widespread as the tau deposits appear in the neocortex [11]. It is currently postulated that these abnormal species of tau spread to different regions of the brain within exosomes associated with cell-to-cell communication [12] or by prion-like seeding [13].

More recently, mechanisms including impaired brain metabolism, blood–brain barrier (BBB) disruption, calcium homeostasis disturbance, mitochondrial dysfunction, and increased oxidative stress, have been implicated in the pathology of the disease [14]. These mechanisms provide insight into the disease process, but also offer potential targets for treatment.

Exercise to Prevent and Manage Chronic Disease Across the Lifespan. https://doi.org/10.1016/B978-0-323-89843-0.00009-X

Clinical manifestations

Clinical manifestations of AD are seen several decades after the onset of the pathophysiological changes in the brain. In the preclinical stage, biomarkers can be measured in the cerebrospinal fluid, the brain and blood, but symptoms of cognitive decline are absent [15]. It has recently been reported that the presence of preclinical disease (amyloid positive without other biomarkers) does not necessarily signal a high likelihood of developing AD [16]. In the mild cognitive impairment (MCI) stage, in addition to the presence of biomarkers, mild symptoms of cognitive decline present above those expected for that age, usually relating to memory, but do not interfere with everyday activities [17]. In the third stage, the neuronal damage to different parts of the brain is reflected in the mild–moderate–severe noticeable cognitive decline over many years. In addition to memory loss, there are personality and behavioral changes, and problems with language, attention, orientation, visuospatial and executive functioning (problem-solving, dealing with household affairs) [18]. As the disease progresses, the person cannot take care of themselves or undertake everyday tasks such as walking and eating, requiring full-time care and assistance and ultimately resulting in death.

Exercise

Exercise is defined as a physical activity that is "planned, structured and repetitive" instead of physical activity that involves "the movement of skeletal muscles resulting in energy expenditure exceeding the resting state" [19] (p. 127). The terms activity and exercise are often used interchangeably in the literature [19] however, for clarity, consistency, and interpretation of studies, this definition of exercise will be used to explore the impact on individuals with AD in this chapter.

Regular exercise throughout the lifespan is associated with considerable health benefits, including benefits to brain health [20], with demonstrated improvements observed among persons with mild cognitive decline [20]. AD's progressive neurological nature is mainly characterized by a decline in cognitive function ultimately impacting quality of life [21].

The next section will explore the clinical outcomes of exercise on cognition and physical functioning including preventing falls and activities of daily living; behavioral and psychological symptoms, quality of life, and safety in individuals with AD. Also, the future direction for research and practice will be discussed.

Exercise and cognitive function

Cognitive functions are the mental abilities involved in reasoning and decision-making, thinking, learning, memory, and attention [22]. According to Barnes et al. [23], exercise may positively affect three areas of the brain affected by vascular physiology, neurogenesis, and hippocampal volumes, impacting memory and executive function.

A Cochrane review of 11 randomized, controlled trials (RCTs) performed in older adults without cognitive impairment implied that aerobic exercise positively affects motor function, auditory awareness, cognitive rapidity, and visual attention [24]. Although these studies focused on preventing dementia, including AD, results from experimental studies report inconsistent results on the impact of exercise on the symptoms of persons with AD.

Yu et al. undertook a single-group repeated-measures design to deliver a 6-month cycling intervention to community-dwelling older adults with mild-to-moderate AD. The Assessment Scale–Cognitive (ADAS-Cog) test was used to measure cognition at three-time points: baseline, 3 and 6 months. The ADAS-Cog scores remained unchanged over the 6 months [25]. Concurring with these findings, Hoffman and colleagues reported results from a RCT of 200 persons with mild AD allocated to a supervised exercise group (60-min sessions three times a week for 16 weeks) and a control group allocated to routine health care. It concluded that cognition was preserved in people in the supervised exercise group with high attendance and intensity [26]. Other studies report cognitive improvement using multicomponent approaches to exercise interventions.

Sampaio et al. conducted a quasi-experimental, non-randomized study using multicomponent exercise intervention comprising aerobic, muscle strengthening, flexibility, balance, and postural exercises. The Mini-Mental State Examination (MMSE) was used to measure the cognitive state of people with AD in residential care homes. Measures were taken at three data points: baseline, 3 and 6 months ($n = 15$ experimental and $n = 15$ control group completed the study). Cognitive ability was significantly better in the experimental group from the one, three-month period, while the control group decreased over time [27]. In contrast, Toots et al. reported no significant impact on global cognition or executive function compared to a control group from a RCT examining the effects of exercise on cognition of nursing home residents, using the MMSE and the AD Assessment Scale [28].

Heterogeneity in intervention methods and measures may be a reason for the inconsistent results as demonstrated in a systematic review of 869 patients diagnosed with AD from 13 RCTs conducted [21]. The intervention groups received an exercise intervention. Eight studies demonstrated that exercise improves cognitive function for individuals with AD, while the remaining five reported no positive benefits [21].

Due to the lack of methodological rigor identified in studies, results must be treated with caution [21,29–32]. Hernandez et al. reported results of a systematic review of 10 years of studies on benefits of exercise and AD, including longitudinal, randomized, and non-randomized studies. The results in 13 of the 14 studies revealed a positive effect of exercise routines in individuals with AD. However, 12 of the 14 studies had small sample sizes ranging from 4 to 17 people, and 12 studies did not have a control group [29]. Another systematic review of RCTs reported little support for positive effects of exercise on cognition in individuals with AD, mainly due to studies reporting only positive results using global cognition as a measure and not accurate neurophysiological assessments of each cognitive domain [30]. Similar results were reported by Song and colleagues [31], who concluded that, although physical exercise, particularly aerobic exercise, benefits global cognition, evidence for the impact of exercise on domain-specific cognition remains ambiguous, requiring future experimental studies with rigorous design [31]. The lack of information on the recommended or prescribed exercise for testing cognition in AD is another contributor to the inconsistent results reported in the literature [32,33]. Future studies require standardized protocols and large-scale longitudinal studies to determine the exercise best suited to improving people's cognition with AD.

Vascular function

The association between the effects of exercise on improving peripheral vascular function (PVF) and the associated impact on cerebral vascular and cognitive function in persons with AD requires further research. Pedrinolla et al. reported an exercise-induced improvement in peripheral vascular function in a RCT involving 72 individuals with AD, using high-intensity aerobic and strength training exercises. Before and after the 6-month exercise regime, vascular function (measured by a passive-leg movement test), flow-mediated dilation and blood sampling for vascular endothelial growth factor, arterial blood flow and shear rate were measured. This study indicated that exercise training improves PVF, the researchers asserting the improved PVF may also positively impact cerebral vascular function [34]. However, to confirm this, RCT studies examining the direct effect of exercise on cerebral blood flow are required.

Physical functioning

Cognitive decline in elderly patients with AD is associated with gait and balance disorders [35]. There is evidence of accelerated deterioration in physical performance, especially balance and mobility, in people with dementia, including people with AD [36], contributing to the high incidents of falls and associated fractures, disability, and mortality [37,38].

Falls prevention

Falls are common in older people, with people aged 65 and over hospitalized due to falls in 2016–17 [39]. A third of people over the age of 65 suffer a fall, which is exacerbated in older people with dementia, with up to 80% falling each year [39]. The accelerated decline in physical functioning in individuals with AD can significantly impact balance and mobility, resulting in serious consequences such as fractures and subsequent effects on morbidity and mortality.

Visuospatial impairment and executive and attention dysfunction might contribute to wandering behaviors (roaming and becoming lost or confused about location) when a person is trying to navigate environments [40]. There are many safety considerations involved with wandering away from familiar surroundings. The effect on depth perception also makes it difficult for people with AD to navigate going down stairs, increasing the risk of falls [41].

For over a decade, evidence has indicated that exercise in older people with cognitive impairment also has a positive impact on falls prevention [42,43], including reports of a significant relationship between physical exercise and the decline in the prevalence of falls in people with AD [36,44–46]. Programs involving a high challenge to balance, performed at least three times a week, have shown the most significant impact on preventing falls [42].

Balance plays an important role in falls prevention [36,47]. Ahn and Kim examined the effect of a RCT involving resistance exercise on muscular function among mild AD patients. The experimental group performed upper and lower extremity exercises three times a week for 5 months. Static balance improved, as did muscle strength and gait, indicating resistance training could be an effective exercise program to improve physical functioning in persons with mild AD [47]. Similarly, Suttanon et al. reported outcomes of a RCT involving people with mild to moderate AD, participating in a home-based exercise program. Significantly more improvement was seen in falls risk, with improved balance and mobility in the exercise group compared to the control group [36]. Hill et al. reported similar results from meta-analyses of studies using RCTs and quasi-experimental designs in community settings, with improvements in physical activity and balance, mobility, and muscle strength while five studies reported no significant effect from the intervention [48]. Toots et al. also reported no significant impact of exercise on reducing the incidents of falls, in a cluster control trial using a high-intensity functional exercise program, on individuals with dementia in residential care [45]. Contributing factors to the contrasting results in these studies may be due to the different approaches to the methodology including the prescribed exercise programs.

Activities of daily living

Disease progression contributes to decreased functional ability and loss of independence to perform activities of daily living (ADL). ADL includes basic activities such as dressing, hygiene, continence and eating, to more complex tasks including meal preparation, housework, finances, and leisure activities [49]. Impaired brain function, involving the frontal lobes and subcortical circuitry of the brain [50] contributes to the reduced ability to perform ADL.

Performing ADL involves executive functioning or high-level thought processing, including the ability to set goals, plan and execute the plan in a sequential manner, self-monitoring, and control of behaviors and activities [50]. In AD, executive functioning and the multiple processes involved in activities, such as cooking and driving are significantly impacted as the disease progresses [51], resulting in significant risk to safety [32]. Other features of impaired executive functioning in dementia include poor judgment, disorganization, socially inappropriate behavior, difficulty making plans and the lack of insight into how their behaviors and decision-making impact others.

The efficacy of exercise in maintaining or enhancing functional ability and the ability to perform ADL has been reported in several studies in aged care residential [46,52–54] and community settings [55–57]. Yu et al. examined the impact of a 6-month cycling intervention, single-group repeated-measures design, on community-dwelling older adults with mild-to-moderate AD. Functional ability was measured using the Disability Assessment for Dementia (DAD). Results indicated functional ability was unchanged over the 6 months [25].

Rolland et al. conducted a RCT incorporating five residential care facilities to assess the relationship of exercise on ADL performance. Participants were allocated to an experimental group (1 hour, twice-weekly of walk, strength, balance, and flexibility training) or routine medical care. Measurement of the ability to perform ADL was assessed using the Katz Index of ADL. Physical performance was evaluated using 6-meter walking speed, the get-up-and-go test, and the one-leg-balance test. ADL mean change from the experimental group's baseline score showed a slower decline than the control group [46].

Maintaining and improving physical functional ability through exercise programs for individuals with AD is associated with decreased carer burden [58,59]. Carers play a significant role in caring for people with AD in the community. According to Kim et al., [58] a predictor of caregiver burden in a sample of 302 caregivers, found the greatest amount of variance (16%) arose from problems with activities of daily living ADLs. Reports have highlighted that the success of exercise programs for individuals with cognitive impairment rely on caregivers' support [36], and for this reason supervised programs should be a routine part of the management of AD.

Henskens et al. conducted a six-month double parallel randomized controlled trial evaluating the effects of ADL training, a multicomponent aerobic and strength exercise training, on residents with dementia in residential care facilities. No significant effects were found on ADL performance following the combined ADL and exercise intervention (participants $n = 16$–22, varying in the experimental and control groups) [54]. According to the authors, it remains unclear which type of exercise is best for improving ADL performance. Contrary to these findings, an earlier RCT provided evidence that a one-year exercise program in persons in residential care, can significantly slow the decline in individuals' functional ability with AD, allowing them to perform ADL independently [53]. Further longitudinal studies are required to test the effect of combined ADL training with exercise on ADL.

A decline in functional ability to independently feed is a component of the progressive nature of AD. It contributes to weight loss, which is reported in up to 40% of people with AD [60]. Takada et al. suggest that associated impaired cognitive function and behavioral and psychological symptoms in AD contribute to weight loss in this population [61]. Chen and colleagues conducted a 6-month intervention in 60 patients with AD residing in a nursing home [62]. They were divided into hand exercise and control groups. Patients in the control group maintained their daily routine. The improvement in the Edinburgh Feeding Evaluation in Dementia scale in the hand exercise group was more significant than in the control group. The ability to self-feed was greater in the experimental group, as was the accuracy of eating action and coordination of eating action. The authors concluded hand exercise as a safe and effective intervention to improve the feeding and eating habits of people with AD [62]. Further RCTs are required to provide conclusive evidence for this type of focused physical exercise improving self-feeding, food intake and weight.

Behavioral and psychological symptoms

Behavioral and psychological symptoms of dementia (BPSD) are characteristic features of AD and are associated with rapid cognitive decline and a high rate of mortality [63]. A high percentage, 80% of people present with at least one symptom [64]. Associated symptoms of BPSD included verbal aggressiveness, emotional disinhibition, apathy and memory impairment. Verbal aggressiveness is generally associated with moderate to severe dementia [65], and apathy and memory impairment with mild dementia [64]. Spaccavento et al. reported that neuropsychiatric symptoms, for example aberrant behaviors such as wandering and purposeless activities, apathy and hallucination, were more severe among AD individuals with malnutrition [66].

Aberrant motor behaviors in patients with dementia can be extremely challenging for carers and often associated with unsafe behaviors exhibited by the person with AD [67]. Aberrant motor behaviors in patients with neurodegenerative disease are associated with morphometric changes to the right dorsal anterior cingulate cortex and left premotor cortex [68]. However, these behaviors may also be associated with a decline in cerebral blood flow in the left parietal–temporal lobe [46]. Harwood et al. reported that behavioral problems in people with AD were significantly correlated with a decline in the MMSE score and a positive correlation between ADL and severity of aberrant motor behaviors [69].

Moderate and/or severe psychopathological symptoms may require pharmacological interventions [70]. However, non-pharmacological interventions such as behavioral therapy should be the first treatment line because pharmacological treatment for behavioral therapy is not always effective [70]. However, few studies report on the impact of exercise on psychopathological symptoms associated with AD [71]. The severity of neurological decline associated with AD and the inability to engage with an exercise program may account for the limited studies related to exercise.

One study conducted by Stella et al. evaluated the impact of a 60-min exercise program on individuals with mild to moderate AD, in which 16 people with AD were allocated to the intervention group and 16 to a control group [71]. The program incorporated aerobic exercises, flexibility, strength, agility, and balance exercises, conducted three times per week over 6 months. Individuals in the intervention group had a significant reduction in neuropsychiatric conditions than the control group [71]. A limitation of the study was the sample size and exclusion of individuals with severe disease. In addition, the study had methodological limitations that may have influenced the results. The participants were not blinded, and the carers assessment may have presented more positive symptomology. Large, rigorous RCTs with long–term follow-up is required to validate the results found in this study.

Several studies report exercise as having a positive impact on reducing depression in individuals with AD [57]. An RCT of 153 individuals with AD was conducted to determine whether a home-based exercise program, combined with caregiver training in behavioral management techniques, would reduce depression in individuals over 3 months. Individuals in the intervention group had improved Cornell Depression Scale in Dementia scores, while the individuals in the routine medical care group had worse scores. For individuals with higher depression scores at baseline, the intervention group improved significantly more at 3 months and maintained that improvement at 24 months [57]. The study conducted by Yu et al. [72] also reported that individualized moderate-intensity cycling three times a week for 6 months reduced depression in people with AD. However, positive effects of exercise are not consistent, for example, Kurz et al. using motor exercises 3 hours a week for 4 weeks, reported no significant reduction in depression in people with AD [73].

Similarly, Rolland et al. carried out a study where exercise was performed by 132 patients with mild-to-severe AD for 1 hour, twice weekly and showed no significant effects on depression, although this could be a result of the low level of exercise used [46]. Studies on other psychological symptoms and behaviors, other than depression in AD, are scarce [74]. Lowery et al. conducted an RCT study to evaluate an exercise program's effectiveness on the behavioral and psychological symptoms of dementia. The exercise program involved a prescribed walking regimen designed to become progressively intensive and last between 20 and 30 min, at least five times per week. One hundred and thirty-one individuals with AD, living in the community with significant behavioral and psychological symptoms associated with dementia, were involved in the study. There was no significant difference in behavioral and psychological symptoms as measured by the Neuropsychiatric Inventory at week 12, between the group receiving the exercise regimen and those that did not [74].

Exercise and quality of life

Cognitive impairment is often the main presenting feature of an individual presenting with AD and is associated with poor quality of life (QOL) [72]. Exercise is known to improve the QOL in older adults; therefore, the assumption exists that these results should be similar in people with various types of dementia. However, the ability to measure QOL in people with dementia is challenging and a difficult concept to measure [75]. Self-reports in individuals with no cognitive impairment are known to be subjective [76] and therefore, could be more problematic in individuals with varying levels of dementia. An alternative measure of QOL using a proxy rater is also subjective and unreliable and could account for inconsistent study outcomes [75].

In the few studies examining the impact of exercise on the QOL of people with AD, study outcomes are inconsistent [75,77]. A meta-analysis of 13 interventional studies reported a small or non-significant improvement in QOL post-intervention [75]. Dauwan et al. reported inconclusive results from a meta-analysis of exercise and QOL in people with chronic brain disorders, including AD [77]. Hoffmann et al. [26] conducted a RCT to examine the effect of exercise on QOL in people with AD, using the European Quality of Life–5 Dimensions (EQ5D) scales. The results showed no significant difference from baseline measures. Suppose exercise improves ADL, with an associated level of independence and social interactions with others in supervised group exercise programs or 1:1 at home. In that case, the outcome may positively impact the individual's QOL [78]; therefore supervised exercise programs should be part of the health intervention strategy to manage individuals with AD.

Caregivers experience high levels of stress and emotional burden and are associated with compassion fatigue [79]. Exercise programs for people with AD decrease carer burden by 40% due to the associated respite provided by supervised community exercise programs [25]. A patient's degree of independence in performing most ADLs has been found to be closely related to caregiver burden [80]. Hence, an exercise that positively improves a person's physical and cognitive ability may impact QOL for both the person with AD and the carer [80].

Safety and exercise

Given the advanced age of participants and associated cognitive decline, safety during exercise intervention is paramount. Safety, including the prevention of falls, is an essential consideration in interventional studies involving exercise in individuals with AD. According to the literature, minimal adverse events occur during exercise interventions in this population. According to a meta-analysis of 16 studies, 10 studies explored adverse events and of those 3 found minor adverse events, possibly related to the intervention [59]. Other studies [26,36] reported no adverse effects of exercise during the study period, with only one study directly reporting minor adverse events (five falls) in the exercise group [46].

In another meta-analysis [77] of 65 studies reporting on the safety of exercise, 45 of which found no adverse events related to exercise and 18 studies reported only minor injuries such as soft tissue injuries and incidents of falls with complete recovery, with the researchers concluding that supervised exercise programs can be safely prescribed [77]. Although many studies report safety and risk associated with an exercise intervention, all studies should be bound ethically to report adverse events. Given exercise may reduce the incidents of falls, it is imperative that supervised exercise programs be offered to manage this chronic condition. This is even more relevant for the one-third of people with dementia living alone in their own homes [81].

Summary of the benefits of exercise

Although AD is a degenerative condition with no known cure, exercise programs appear to be a potential non-pharmacological intervention to improve AD symptoms, including some components of cognition, physical functioning relating to activities associated with daily living, the risk of falls, QOL and depression (Fig. 1).

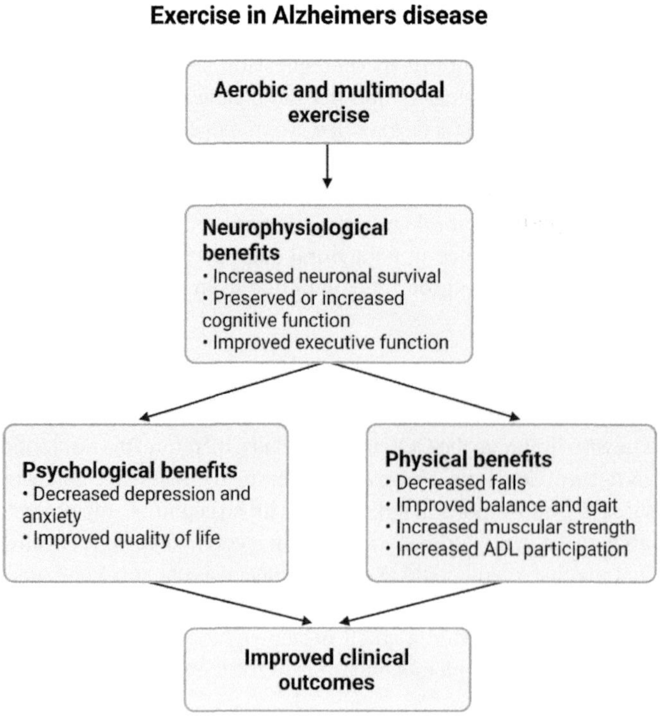

FIG. 1 Summary of the benefits of exercise.

Conclusion

Evidence from various forms of exercise programs on AD symptoms varies, with some studies providing evidence of improvements in symptoms of AD and other studies reporting little or no improvement. Lack of information on the recommended or prescribed exercise for testing cognition in AD may contribute to the inconsistent results reported in the literature. Further, the lack of methodological rigor and consistency requires standardized protocols, large, rigorous RCTs with long-term follow-up to provide accurate insights into the effects of exercise for individuals with AD [29].

References

[1] Alzheimer's Association. Alzheimer's disease facts and figures. Alzheimers Dement 2019;15(3):321–87.

[2] Brookmeyer R, et al. Forecasting the global burden of Alzheimer's disease. Alzheimers Dement 2007;3(3):186–91.

[3] WHO. The top 10 causes of death., 2020, https://www.who.int/news-room/fact-sheets/detail/the-top-10-causes-of-death.

[4] Stelzmann RA, Norman Schnitzlein H, Reed Murtagh F. An English translation of Alzheimer's 1907 Paper,"über eine eigenartige erkankung der hirnrinde". Clin Anat 1995;8(6):429–31.

[5] Walker LC. Aβ Plaques. Free Neuropathol 2020;1(31):1–42.

[6] Long JM, Holtzman DM. Alzheimer disease: an update on pathobiology and treatment strategies. Cell 2019;179(2):312–39.

[7] Heneka MT, et al. Neuroinflammation in Alzheimer's disease. Lancet Neurol 2015;14(4):388–405.

[8] Serrano-Pozo A, et al. Neuropathological alterations in Alzheimer disease. Cold Spring Harb Perspect Med 2011;1(1):a006189.

[9] Braak H, Braak E. Neuropathological stageing of Alzheimer-related changes. Acta Neuropathol 1991;82(4):239–59.

[10] Arriagada PV, et al. Neurofibrillary tangles but not senile plaques parallel duration and severity of Alzheimer's disease. Neurology 1992;42(3 Pt 1):631–9.

[11] Braak H, Braak E. Staging of Alzheimer's disease-related neurofibrillary changes. Neurobiol Aging 1995;16(3):271–8 [discussion 278-84].

[12] Saman S, et al. Exosome-associated tau is secreted in tauopathy models and is selectively phosphorylated in cerebrospinal fluid in early Alzheimer disease. J Biol Chem 2012;287(6):3842–9.

[13] Furman JL, et al. Widespread tau seeding activity at early Braak stages. Acta Neuropathol 2017;133(1):91–100.

[14] Song K, et al. Oxidative stress-mediated blood-brain barrier (BBB) disruption in neurological diseases. Oxid Med Cell Longev 2020;2020:4356386.

[15] Sperling RA, et al. Toward defining the preclinical stages of Alzheimer's disease: recommendations from the National Institute on Aging-Alzheimer's Association workgroups on diagnostic guidelines for Alzheimer's disease. Alzheimers Dement 2011;7(3):280–92.

[16] Brookmeyer R, Abdalla N. Estimation of lifetime risks of Alzheimer's disease dementia using biomarkers for preclinical disease. Alzheimers Dement 2018;14(8):981–8.

[17] Albert MS, et al. The diagnosis of mild cognitive impairment due to Alzheimer's disease: recommendations from the National Institute on Aging-Alzheimer's Association workgroups on diagnostic guidelines for Alzheimer's disease. Alzheimers Dement 2011;7(3):270–9.

[18] McKhann GM, et al. The diagnosis of dementia due to Alzheimer's disease: recommendations from the National Institute on Aging-Alzheimer's Association workgroups on diagnostic guidelines for Alzheimer's disease. Alzheimers Dement 2011;7(3):263–9.

[19] Caspersen Carl J, Powell Kenneth E, Christenson Gregory M. Physical activity, exercise, and physical fitness: definitions and distinctions for health-related research. Public Health Rep (1974-) 1985;100(2):126–31.

[20] WHO. Adopting a healthy lifestyle helps reduce the risk of dementia: New WHO Guidelines recommend specific interventions for reducing the risk of cognitive decline and dementia., 2019, https://www.who.int/news/item/14-05-2019-adopting-a-healthy-lifestyle-helps-reduce-the-risk-of-dementia.

[21] Du Z, et al. Physical activity can improve cognition in patients with Alzheimer's disease: a systematic review and meta-analysis of randomized controlled trials. Clin Interv Aging 2018;13:1593–603.

[22] Fischer A, et al. Modelling the impact of functionality, cognition, and mood state on awareness in people with Alzheimer's disease. Int Psychogeriatr 2019;1–11.

[23] Barnes DE, et al. Preventing loss of independence through exercise (PLIÉ): a pilot clinical trial in older adults with dementia. PLoS One 2015;10(2), e0113367.

[24] Angevaren M, et al. Physical activity and enhanced fitness to improve cognitive function in older people without known cognitive impairment. Cochrane Database Syst Rev 2008;(3):CD005381.

[25] Yu F, et al. Impact of 6-month aerobic exercise on Alzheimer's symptoms. J Appl Gerontol 2015;34(4):484–500.

[26] Hoffmann K, et al. Moderate-to-high intensity physical exercise in patients with Alzheimer's disease: a randomized controlled trial. J Alzheimers Dis 2016;50(2):443–53.

[27] Sampaio A, et al. Effects of a multicomponent exercise program in institutionalized elders with Alzheimer's disease. Dementia 2019;18(2):417–31.

[28] Toots A, et al. Effects of exercise on cognitive function in older people with dementia: a randomized controlled trial. J Alzheimers Dis 2017;60(1):323–32.

[29] Hernández SS, et al. What are the benefits of exercise for Alzheimer's disease? A systematic review of the past 10 years. J Aging Phys Act 2015;23(4):659–68.

[30] Cammisuli DM, et al. Aerobic exercise effects upon cognition in Alzheimer's disease: a systematic review of randomized controlled trials. Arch Ital Biol 2018;156(1–2):54–63.

[31] Song D, et al. The effectiveness of physical exercise on cognitive and psychological outcomes in individuals with mild cognitive impairment: a systematic review and meta-analysis. Int J Nurs Stud 2018;79:155–64.

[32] Farina N, Rusted J, Tabet N. The effect of exercise interventions on cognitive outcome in Alzheimer's disease: a systematic review. Int Psychogeriatr 2014;26(1):9–18.

[33] Jia R-X, et al. Effects of physical activity and exercise on the cognitive function of patients with Alzheimer disease: a meta-analysis. BMC Geriatr 2019;19(1):181.

[34] Pedrinolla A, et al. Exercise training improves vascular function in patients with Alzheimer's disease. Eur J Appl Physiol 2020;120(10):2233.

[35] Castrillo A, et al. Gait disorder in a cohort of patients with mild and moderate Alzheimer's disease. Am J Alzheimers Dis Other Demen 2016;31(3):257–62.

[36] Suttanon P, et al. Feasibility, safety and preliminary evidence of the effectiveness of a home-based exercise programme for older people with Alzheimer's disease: a pilot randomized controlled trial. Clin Rehabil 2013;27(5):427–38.

[37] Robertson DA, Savva GM, Coen RF. Cognitive function in the prefrailty and frailty syndrome. J Am Geriatr Soc 2014;62:2118–24.

[38] Casas-Herrero A, Anton-Rodrigo I, Zambom-Ferraresi F. Effect of a multicomponent exercise programme (VIVIFRAIL) on functional capacity in frail community elders with cognitive decline: study protocol for a randomized multicentre control trial. Trials 2019;20:362.

[39] AIHW. Trends in hospitalised injury due to falls in older people 2007–08 to 2016–17, 2019. Available from: https://www.aihw.gov.au/reports/injury/trends-in-hospitalised-injury-due-to-falls.

[40] O'Brien L, et al. Visual mechanisms of spatial disorientation in Alzheimer's disease. Cereb Cortex 2001;11(11):1083–92.

[41] Heerema E, Chaves C. Dementia effects on activities of daily living., 2020, https://www.verywellhealth.com/dementia-daily-living-adls-97635.

[42] Sherrington C, et al. Exercise to prevent falls in older adults: an updated systematic review and meta-analysis. Br J Sports Med 2017;51(24):1750–8.

[43] Chan WC, et al. Efficacy of physical exercise in preventing falls in older adults with cognitive impairment: a systematic review and meta-analysis. J Am Med Dir Assoc 2015;16(2):149–54.

[44] Dodd K, et al. Feasibility, safety and preliminary evidence of the effectiveness of a home-based exercise programme for older people with Alzheimer's disease: A pilot randomized controlled trial. SAGE Publications; 2013.

[45] Toots A, et al. The effects of exercise on falls in older people with dementia living in nursing homes: a randomized controlled trial. J Am Med Dir Assoc 2019;20(7):835–842.e1.

[46] Rolland Y, et al. Exercise program for nursing home residents with Alzheimer's disease: a 1-year randomized, controlled trial. J Am Geriatr Soc 2007;55(2):158–65.

[47] Ahn N, Kim K. Effects of an elastic band resistance exercise program on lower extremity muscle strength and gait ability in patients with Alzheimer's disease. J Phys Ther Sci 2015;27(6):1953–5.

[48] Hill KD, et al. Individualized home-based exercise programs for older people to reduce falls and improve physical performance: a systematic review and meta-analysis. Maturitas 2015;82(1):72–84.

[49] Rao AK, et al. Systematic review of the effects of exercise on activities of daily living in people with Alzheimer's disease. Am J Occup Ther 2014;68(1):50–6.

[50] Yu F, Vock DM, Barclay TR. Executive function: responses to aerobic exercise in Alzheimer's disease. Geriatr Nurs 2018;39(2):219–24.

[51] McGuinness B, et al. Executive functioning in Alzheimer's disease and vascular dementia. Int J Geriatr Psychiatry 2010;25(6):562–8.

[52] Roach KE, et al. A randomized controlled trial of an activity specific exercise program for individuals with alzheimer disease in long-term care settings. J Geriatr Phys Ther 2011;34(2):50–6.

[53] Buchner DM. Exercise slows functional decline in nursing home residents with Alzheimer's disease. Aust J Physiother 2007;53(3):204.

[54] Henskens M, et al. Effects of physical activity in nursing home residents with dementia: a randomized controlled trial. Dement Geriatr Cogn Disord 2018;46(1–2):60–80.

[55] Venturelli M, Scarsini R, Schena F. Six-month walking program changes cognitive and ADL performance in patients with Alzheimer. Am J Alzheimers Dis Other Demen 2011;26(5):381–8.

[56] Vreugdenhil A, et al. A community-based exercise programme to improve functional ability in people with Alzheimer's disease: a randomized controlled trial. Scand J Caring Sci 2012;26(1):12–9.

[57] Teri L, et al. Exercise plus behavioral management in patients with Alzheimer disease: a randomized controlled trial. JAMA 2003;290(15):2015–22.

[58] Kim S-Y, et al. A systematic review of the effects of occupational therapy for persons with dementia: a meta-analysis of randomized controlled trials. NeuroRehabilitation 2012;31(2):107–15.

[59] Almeida S, Gomes da Silva M, Marques A. Home-based physical activity programs for people with dementia: systematic review and meta-analysis. Gerontologist 2019;60(8):600–8.

[60] Aziz NA, et al. Weight loss in neurodegenerative disorders. J Neurol 2008;255(12):1872–80.

[61] Takada K, et al. Grouped factors of the 'SSADE: signs and symptoms accompanying dementia while eating' and nutritional status—an analysis of older people receiving nutritional care in long-term care facilities in Japan. Int J Older People Nurs 2017;12(3).

[62] Chen L-L, et al. Effects of hand exercise on eating action in patients with Alzheimer's disease. Am J Alzheimers Dis Other Demen 2019;34(1):57–62.

[63] Kimura A, et al. Malnutrition is associated with behavioral and psychiatric symptoms of dementia in older women with mild cognitive impairment and early-stage Alzheimer's disease. Nutrients 2019;11(8):1951.

[64] Lyketsos CG, et al. Prevalence of neuropsychiatric symptoms in dementia and mild cognitive impairment: results from the cardiovascular health study. JAMA 2002;288:1475–83.

[65] Gonfrier S, et al. Course of neuropsychiatric symptoms during a 4-year follow up in the REAL-FR cohort. J Nutr Health Aging 2012;16:134–7.

[66] Spaccavento S, et al. Influence of nutritional status on cognitive, functional and neuropsychiatric deficits in Alzheimer's disease. Arch Gerontol Geriatr 2009;48(3):356–60.

[67] Shinoda-Tagawa T, et al. Resident-to-resident violent incidents in nursing homes. JAMA 2004;291(5):591–8.

[68] Rosen HJ, et al. Neuroanatomical correlated of behavioral disorders in dementia. Brain 2005;128:2612–25.

[69] Harwood DG, et al. Relationship of behavioral and psychological symptoms to cognitive impairments and functional status in Alzheimer's disease. Int J Geriatr Psychiatry 2000;15:393–400.

[70] Tan E, et al. Current approaches to the pharmacological treatment of Alzheimer's disease. vol. 47. Royal Australian College of General Practitioners; 2018. p. 586–92.

[71] Stella F, et al. Attenuation of neuropsychiatric symptoms and caregiver burden in Alzheimer's disease by motor intervention: a controlled trial. Clinics (Sao Paulo) 2011;66(8):1353–60.

[72] Yu F, et al. Affecting cognition and quality of life via aerobic exercise in Alzheimer's disease. West J Nurs Res 2013;35(1):24–38.

[73] Kurz A, et al. Cognitive rehabilitation in patients with mild cognitive impairment. Int J Geriatr Psychiatry 2009;24(2):163–8.

[74] Lowery D, et al. The effect of exercise on behavioural and psychological symptoms of dementia: the EVIDEM-E randomised controlled clinical trial. Int J Geriatr Psychiatry 2014;29(8):819–27.

[75] Ojagbemi A, Akin-Ojagbemi N. Exercise and quality of life in dementia: a systematic review and meta-analysis of randomized controlled trials. J Appl Gerontol 2019;38(1):27–48.

[76] Dinsmore D, Alexander P. A critical discussion of deep and surface processing: what it means, how it is measured, the role of context, and model specification. Educ Psychol Rev 2012;24(4):499–567.

[77] Dauwan M, et al. Physical exercise improves quality of life, depressive symptoms, and cognition across chronic brain disorders: a transdiagnostic systematic review and meta-analysis of randomized controlled trials. J Neurol 2019;1.

[78] Dan M, Boca I-C, Sere CR. Enhancing quality of life in Alzheimer's dementia by adapted physical activities. Sport & Society/Sport si Societate 2012;12(1):12–9.

[79] Lynch SH, Lobo ML. Compassion fatigue in family caregivers: a Wilsonian concept analysis. J Adv Nurs 2012;68(9):2125–34.

[80] Hall D, et al. Variables associated with high caregiver stress in patients with mild cognitive impairment or Alzheimer's disease: implications for providers in a co-located memory assessment clinic. J Ment Health Couns 2014;36(2):145–59.

[81] Alzheimer's Association. Facts and Figues., 2012, https://www.alz.org/alzheimers-dementia/facts-figures.

Chapter 32

Exercise and the elderly: Gait and balance

Hanatsu Nagano, William Anthony Sparrow, and Rezaul Begg

Institute for Health and Sport, Victoria University, Melbourne, VIC, Australia

Introduction

The negative effects on health and well-being due to aging are widely recognized and advances in the health and exercise sciences have strongly encouraged senior individuals to maintain adequate exercise for healthy aging. There are a range of recommended exercise interventions for senior adults but in a broader context, the goal is to promote health by slowing or possibly reversing age-associated declines in cognitive and motor function. Health can be divided into several essential domains including physiological, psychological, and social factors. Importantly, these multiple factors interact strongly and, therefore, a decline in one can negatively affect the others, exacerbating general health problems. In contrast, prevention measures addressing primary aging-related physiological changes, such as reduced muscle strength, may have positive flow-on effects on overall health. Due to the strong correlations between aging processes, it is important to identify the initial causes and design exercise interventions to prevent a health-related domino effect. As such, the focus of this chapter is falls prevention for senior adults because falls-related injuries are one of the primary causes of lost independence and medical costs. In all demographically aging societies, such measures are critical to maintaining sustainable national social security systems due to the considerable flow-on effects of falls-related injuries to a range of mental and physical health problems.

The epidemiology of aging-related falls

Worldwide, about one in three adults over 65 years fall every year and 9%–20% of falls cause severe injuries, such as hip fractures [1]. Falls are the leading cause of nonfatal injury (27%) and also the third leading direct cause of injury-related death [2]. The annual medical costs of falls are estimated at close to $3 billion in Australia and on an individual basis, the average cost per falls-related hospitalization is reported to exceed $20,000 [3–6]. In Australia, medical costs are predicted to triple in the next 30 years due to the increasing proportion of older adults relative to the population as a whole from 16% now to 22% by 2050 [7]. In addition to costs, falls have a negative impact on an individual's quality of life. It is estimated that about half of the hip fractures due to falling result in permanent loss of independent lifestyles and 20%–30% of those lead to death [8–10]. To meet these challenges falls prevention should be a priority for national and state Governments and society as a whole.

There is an ongoing worldwide effort to develop interventions to reduce injury-related falls. Current interventions for falls prevention include strength and balance exercises, medication management, and the installation of home safety devices. According to previous studies [10–13], about half of falls in older people occur while they are walking outdoors. Despite the risk of falling, maintaining mobility by walking is one of the highly recommended exercise interventions for older adults. Further understanding of age-related effects on gait and balance is, therefore, essential for the assessment of aging-related falls risk and developing effective prevention strategies.

Changes to gait with aging

Locomotion is fundamental to our everyday lives and is important in maintaining independent lifestyles and quality of life (QOL). It is, however, necessary to understand how gait function declines with age due to a variety of interacting factors such as falls and consequent injuries, which presents critical health risks to older people. Aging-related changes to gait and balance can be attributed to deterioration in sensorimotor function [14,15]. These functional declines are considered the primary intrinsic cause of falls in older adults [16]. Aging induces a systematic and progressive decline in foot sensitivity [17], lower limb muscle strength and power [18], lower limb joint flexibility [19], and muscle activation [20].

Exercise to Prevent and Manage Chronic Disease Across the Lifespan. https://doi.org/10.1016/B978-0-323-89843-0.00005-2

Aging is associated with reduced strength in the quadriceps, hamstrings, dorsiflexors, and plantar flexors in addition to restricted joint range of motion (ROM) due to decreased maximum flexion and extension [21].

Motor control also declines with aging and negatively affects reaction speed, the precision of end-point control (i.e., the foot) and movement variability [22]. Muscle activation during walking is modified in older adults, as reflected in decreased activation of some muscles and increased co-activation of others [23]. For example, deficits in ankle dorsiflexor muscle activation can change the foot's swing phase trajectory, reduce foot-ground clearance and increase the risk of tripping [24]. With aging, afferent sensory signals from the feet are attenuated due to reduced plantar fat pad thickness, flattening of the longitudinal arch, and the development of claw and hammer toes [25]. Reduced foot sensitivity has, furthermore, been associated with an increased risk of falling due to balance loss [26]. The essential problem, therefore, is that sensorimotor declines with aging, such as those identified above, influence the mechanics of walking, i.e., the individual's gait biomechanics, impairing their capacity to maintain balance.

Reduced walking speed is the most well-documented aging effect on gait, due to shorter step length and prolonged double support time, the interval when both feet are in contact with the ground [27–30]. A further adaptation to spatiotemporal gait parameters as we get older is wider steps, which help preserve balance but with additional energy costs [31–33]. Despite increased step width, falls are, however, more frequent in "older" and frailer individuals, suggesting that age-related balance impairment cannot be fully compensated by walking with wider steps [34,35].

Inconsistent gait patterns over multiple-step cycles described as "gait variability" indicate disturbance to rhythmic stepping, possibly due to declines in neurological function [22]. Gait asymmetry, representing the difference in gait pattern variables between the two lower limbs, has also been used to evaluate walking health in terms of interlimb coordination ability [36]. Between-limb variation in gait mechanics and variability across multiple steps reflect aging-related gait impairments and, as discussed below, aging-related neurological conditions, such as hemiplegia due to stroke, are also associated with, sometimes considerable, deficiencies in one limb [37]. Similarly, Parkinson's disease is characterized by increased variability in step-to-step timing, making the patient more prone to falls [38]. Unlike general exercise interventions, strategies to remediate fine lower limb control are more difficult to design, requiring specific, targeted, gait-training procedures. For example, technology-based gait assessment, utilizing 3D motion capture systems and wearable motion sensors, can provide the patient with detailed quantitative information of their walking by targeting specific gait variables.

Fear of falling and cautious gait patterns

Not only do physiological and neurological factors affect gait and balance by reducing the ability to process sensory inputs but all older adults' motor skills appear to be influenced by the intrinsic psychological phenomenon of becoming increasingly cautious or fearful [39,40]. The term often applied to aging effects on gait due to psychological preparatory responses to potential hazards is "fear of falling" [27,32,41,42].

Fear of falling has been identified as a primary factor in cautious gait adaptations, such as shorter step length and associated reduction in walking speed [14]. Multiple factors, including self-awareness of decline in physical capacities, such as lower limb strength, slower reaction speed, visual or auditory deterioration, and poor balance, in addition to a history of falls or activity restriction, may trigger caution-based gait adaptations [43,44]. The onset of fear of falling tends to initiate a cycle of negative "chain reactions", for example, regulation of outdoor activities leading to loss of sociality, depression, accelerating physical decline and further escalation of fear of falling. Through exercise and other interventions, senior individuals may be helped in dealing with the fear of falling, encouraging active healthy aging.

Greater caution has also been linked to wider steps, a response common to many gait disorders, which has the effect of increasing the medio-lateral base of support to protect against sideways balance loss [45]. Cautious gait control may also be observed in relatively mobile and confident individuals, who do not show signs of impaired walking on unobstructed, firm, level surfaces [40,42,46]. Caution emerges in response to more challenging or unfamiliar gait tasks, such as destabilizing treadmill walking, stepping on or off raised surfaces, and high attention-demanding environments. When assessing relatively healthy, active older people's walking it may, therefore, be useful to observe their responses to more challenging gait tasks that may reveal weaknesses not seen in everyday walking [47].

Aging and associated pathology

Natural aging processes accompany gradual deterioration of cellular functions, such as respiration and division, negatively affecting overall physical health [48,49]. In addition to intrinsic, progressive aging effects on sensorimotor functions, aging-related pathologies also increase with age, further compromising balance and increasing falls risk. Pathological gaits have a variety of distinctive features but can be more generally classified based on direct causes, including (i) loss of strength, (ii) neurological disorders, (iii) pain, (iv) medication effects, and (v) other conditions such as visual and auditory impairments.

1. Loss of strength

    Muscle atrophy and sarcopenia become more prevalent with age as the size and number of muscle cells reduce [50,51]. Overall gait function and balance recovery are impaired due to reduced muscle strength [52] but conditions related to aging processes also accelerate strength loss. For example, diabetes, osteoporosis, reduced hormonal secretion (e.g., growth hormone, testosterone) and chronic diseases (e.g., cancer, heart failure, chronic obstructive pulmonary disease) are more common with aging, directly contributing to reduced muscle strength and also indirectly due to reduced everyday physical activity and less engagement in structured exercise, such as walking for recreation [53–55]. There are also rare but seriously disabling conditions, such as amyotrophic lateral sclerosis, that cause progressive, severe strength loss [56].

2. Neurological disorders

    Aging is associated with neurological disorders that impair gait and balance and increase falls risk [57]. These neurological disorders affect balance due to slowed reactions, impaired spatial perception, loss of fine-motor skills and deteriorated executive functions [58,59]. Dementia increases falls risk due to a range of interacting effects, such as the presence of Lewy bodies and reduced psychosocial health [60,61]. Parkinson's disease is also a critical risk factor for falls often accompanying "freezing" of gait, which occurs more frequently during gait initiation, turning behavior or obstacle negotiation [62]. Multiple sclerosis is a progressive disease that seriously impairs neuromuscular function and, as a consequence, increases fall risk [63]. Similarly, the risk of falling is 150% higher in people with chronic stroke (1 year poststroke) compared with age- and gender-matched controls [64].

3. Pain

    Pain due to osteoarthritis becomes more prevalent with aging. Osteoarthritis reduces soft tissue regeneration, with diminished articular cartilage compromising joint shock absorption, leading to microfractures. One primary symptom of osteoarthritis is joint pain, in which the knee is highly affected in approximately 10% of seniors with considerable associated effects on gait control [65,66]. With aging and related conditions, pain due to foot deformities also causes gait impairments. Diabetic foot, for example, causes walking impairments due to neuropathy [67] and falls risk is also increased.

4. Medication

    Older adults are prescribed medications to treat a range of health conditions, but antihypertensive drugs, sedatives, tranquilizers, neuroleptics, antidepressants, benzodiazepines, and antiinflammatories increase falls risk by more than 20% [68]. In addition, polypharmacy (more than five medicines daily) also increases falls risk [69]. Medications are essential for many aging-related health conditions, but they may exacerbate gait and balance problems. Alternative approaches, including exercise interventions, should be considered to reduce the dependence on medications for conditions affecting lower limb muscle and joint function.

5. Other factors

    There are many other factors that increase falls risk among senior adults. For example, visual and hearing impairments that disturb spatial perception, dietary considerations associated with overall health and falls risk and risk factors in the home, such as tripping hazards, can be critical external factors [70–72].

Aging effects on balance and tripping

Aging effects on balance

Balance can be maintained under static conditions but any movement is likely to pose some risk of stability loss. Balance is defined by the relationship between the whole-body center of mass (COM) and the base of support (BOS), the area defined by the stance foot, shaded in Fig. 1. In the static condition (i.e., standing), the COM is easily maintained within the BOS, and balance is secured, as illustrated in Fig. 1C. Walking, however, can be described as a continuum of ongoing balance loss and balance recovery. Progression requires displacing the COM forward, potentially causing loss of balance but the following recovery step maintains stability by establishing a new, more anterior, base of support, preserving the COM within the BOS [73–75]. The location and speed of COM movement is, therefore, an important determinant of balance control [76], reflecting the ability to control the center of mass and establish an adequate BOS by moving the foot quickly and accurately, the primary requirements for stable gait.

FIG. 1   Illustration of the center of mass (COM) and base of support (BOS) relationship (A) during walking; (B) taking a step and (C) static standing.

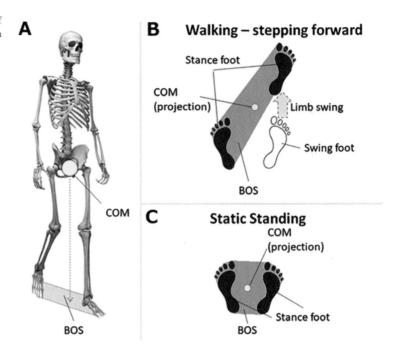

The above balance responses tend to decline with age, reflected in reduced COM control and/or failure to maintain an adequate BOS to prevent the COM from moving beyond the BOS boundary. When the latter occurs balance cannot be restored and the individual will fall. When only 1 ft is on the walking surface, in the single-limb stance phase, balance is ensured by taking a step forward with the opposite limb, containing the COM securely within the BOS and avoiding balance loss. Exercises for falls prevention should be designed to address this function by increasing the muscle strength and reaction speed required to take long, fast, balance recovery steps [77].

Aging effects on tripping

Although balance recovery is important for preventing a fall, stabilizing movements are required only when walking is perturbed. It is, therefore, important to consider how we can mitigate the likelihood of perturbations to walking. Tripping is the leading cause of balance loss when walking, accounting for up to 53% of falls in older individuals [11,39,78]. Slipping is the second leading cause of falls but its occurrence depends primarily on the interfacial friction between the walking surface and the footwear. Such considerations are beyond the scope of this chapter but they involve surface friction due to construction materials (wood, tiles, etc.) and contaminants such as water and other liquids, and slip-resistant footwear designs.

Tripping is caused by unexpected swing foot contact with an obstacle or the walking surface, with sufficient force to destabilize the walker. Tripping hazards typically include everyday objects on the floor and raised or uneven surfaces, such as footpaths and carpet edges. Tripping occurs in the gait cycle swing phase, when the foot is off the ground and moving forward. The most important biomechanical variable for understanding tripping is, therefore, the swing foot's ground clearance (see Fig. 2).

Foot-ground (or obstacle) clearance is determined by the magnitude of the vertical component of the swing trajectory and when the foot moves below obstacle height there is obstacle contact and increased risk of stability loss, sometimes leading to a fall. The risk of tripping is greatest at the swing phase event Minimum Foot Clearance (MFC), approximately mid-swing at the local minimum vertical displacement of the swing foot (Fig. 2). MFC is hazardous because in addition to low clearance, it occurs at maximum horizontal velocity when obstacle contact will be more forceful. Furthermore, MFC occurs in single limb support when only the opposite stance foot is on the ground, further increasing the risk of gait instability and reducing the base of support (BOS). Maintaining sufficient swing foot margin above the walking surface at MFC across multiple steps is, therefore, fundamental to reducing tripping probability [79].

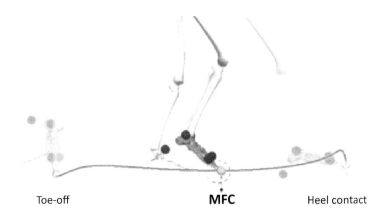

FIG. 2 (Left) Swing foot clearance and minimum foot clearance—MFC; (Right) mean and variability of MFC.

Swing Foot Clearance

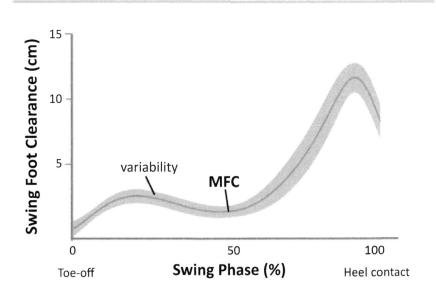

In summary, intervention strategies to reduce tripping risk should aim to achieve safer MFC, characterized as high, consistent, and symmetrical control of the most distal segment of the foot (i.e., toe). The next section summarizes general principles of exercise interventions to (1) improve dynamic balance and (2) reduce tripping risk.

Exercise interventions for falls prevention

There are specific considerations when older adults engage in physical activity but despite age-appropriate modifications, general principles for effective exercise training are applicable. Increased muscular strength can be gained by the "overload" of targeted muscles and supplementation of muscle-building proteins [80]. Vigorous cardiovascular training requires extra care, to ensure that heart rate is maintained within a safe range. It is also important to provide a secure environment to minimize external risks, such as slipping and tripping hazards and poor illumination. Other considerations include adequate nutrition via a balanced diet and sufficient exposure to sunlight to synthesize Vitamin D for calcium absorption [81]. Locomotor function is fundamental to maintaining everyday activities; exercise interventions to promote safe walking are, therefore, particularly important, while being cautious in prescribing walking activities to minimize falls risk.

Balance exercise interventions

Balance is maintained by taking an effective recovery step to reestablish the BOS and in well-controlled gait mechanics the COM path should be symmetrical between right-leading and left-leading steps and consistent across step cycles. Energy efficiency is a priority for functionally optimized movements and optimum COM control is considered to reflect energy-efficient gait mechanics [82].

Aging can, however, impair functionally efficient COM control due to declines in physical capacity and/or as a consequence of prioritizing safety over energy cost. Accordingly, exercise interventions for senior individuals should focus on functionally efficient gait patterns, for example using strength training to sustain vigorous gait mechanics. To encourage more symmetrical interlimb coordination, the weaker side should be compensated and rhythmic exercises can be used to reduce step-to-step variability and encourage more consistent step patterns. COM control is dependent on the efficient coordination of all limb segments but trunk stabilizing muscle groups are particularly important, giving rise to the now popular fitness industry concept of "core stability" [83]. One such exercise tradition emphasizing core stability is Pilates, originating as a veterans rehabilitation program. Pilates can be conducted without specialized equipment (mat-Pilates) but apparatus such as Cadillac, Wundachair, and Reformer are also used (Fig. 3). As Pilates was originally developed for seriously injured individuals, the design of these devices was modeled on hospital beds and wheelchairs. Pilates exercises at moderate intensity are suitable for developing postural stabilizers and often involve finely coordinated motions. Pilates or similar core stability exercises may, therefore, help develop COM control.

As described earlier, maintaining the COM-BOS relationship determines balance, and effective recovery steps are required to reestablish the BOS as we move forward. Attempts to enlarge the BOS may not, however, be most adaptive. Taking wider steps, for example, maybe useful in increasing the medio-lateral BOS but this adaptation is often seen in gait-impaired individuals. An awareness of balance problems may, therefore, encourage this gait adaptation to compensate for declining balance control. Strength training may also be useful to assist longer steps (i.e., increasing step length) to extend the BOS boundary forward. Reestablishing the newly secured BOS is time-critical because the falling COM's momentum must be arrested before it accelerates beyond the capacity of counteracting recovery steps. Strength of the recovery step, therefore, determines whether falling momentum can be successfully opposed, and exercises to increase the eccentric work output of primary lower limb muscles (i.e., quadriceps, tibialis anterior, soleus, and gastrocnemius) may help in maintaining dynamic stability.

In summary, exercise interventions for balance can be designed to train trunk stabilizers for COM control, increase overall lower limb strength to sustain energy-efficient "vigorous" walking and improve reaction speeds to assist more effective recovery steps. These interventions are expected to increase the probability of balance recovery but exercise programs can also target measures to reduce the risk of balance perturbation. In this chapter, we will address exercise interventions to help prevent tripping, one of the most frequent causes of falls-related instability.

Exercises to reduce tripping risk

The ability to maintain balance is fundamental to preventing a fall but it is also important to avoid balance perturbations. As outlined earlier, maintaining sufficient swing foot height is the essential condition for avoiding balance threats due to obstacles. Of all lower limb joint motions (hip, knee, and ankle) ankle dorsiflexion is most efficient in increasing ground clearance and reducing tripping risk [84]. The Tibialis Anterior (TA) muscle is the primary ankle dorsiflexor and exercises to develop TA strength should be incorporated into exercise programs for older adults. In addition to promoting safer ground clearance, ankle dorsiflexion at heel contact is also important in providing a foot-ground contact angle that can distribute impact forces and reduce the stress on foot tissues [85]. As described in the following section, in addition to

FIG. 3 Pilates exercises for seniors (left) WundaChair, (right) Reformer.

WundaChair   Reformer

strengthening exercises, improved motor control using biofeedback training, electrical muscle stimulation (EMS) of the TA, and footwear interventions can also be considered for tripping prevention.

Technology-based exercise interventions

General exercise guidelines for seniors are widely available [86] but more specific approaches may be required for symptom-specific gait disorders, especially those designed to remediate gait problems due to neurological conditions. It is unknown whether general training methods are effective for neural aspects of gait control but more specialized interventions have been designed for neurologically impaired gait functions. Exercise intervention for senior adults' gait training can, therefore, incorporate specialized training procedures for neurological deficits and recent gait-biomechanics-based technological advances now provide viable rehabilitation options.

Treadmill walking

Compared to overground walking, older adults' gait declines in treadmill walking [40]. As seen in aging-related declines in overground locomotion, slower walking due to shorter step length, longer double support, and increased step width are observed when older adults walk on a motor-driven treadmill [42]. These gait changes are typical of fearful walking but if safety can be ensured treadmill-based exercise training can promote more consistent steady-state walking patterns [87]. Due to the constant walking speed and invariant surface, the treadmill serves as a pacemaker to reduce gait variability. Seniors with gait disorders due to neurological impairments, such as poststroke and Parkinson's patients, have been shown to benefit from treadmill training [88]. Older adults tend to initially perceive treadmill walking as challenging but safety procedures, such as a harness and stability assistance using the handrails, may overcome the fear of falling.

Biofeedback gait training

Biofeedback gait training utilizing a 3D motion capture system can intervene in microlevel gait adjustments. Loss of fine control of the distal toe segment is often the consequence of a stroke, increasing tripping risk in poststroke individuals. Biofeedback training can increase MFC while reducing variability, the optimum adaptation to prevent tripping falls. Ankle dorsiflexors (e.g., tibialis anterior) are the primary muscles in elevating the swing foot but strength training alone cannot reduce step-to-step variability [84]. Tripping is a low probability event, such that it is important to consistently maintain MFC height because only the infrequent low clearance steps increase tripping risk.

In response to this problem, Begg et al. [89] devised a biofeedback gait training procedure to assist poststroke individuals in safer swing foot control using real-time monitoring of foot-ground clearance. Their technology is based on the precision of motion capture systems capable of monitoring a foot-ground clearance of less than 2 cm. An infrared light-emitting marker attached to the front part of the shoe is tracked in real-time, with swing foot motion presented on a screen during treadmill walking (Fig. 4). Prior to undertaking a program of training trials with biofeedback, a patient-specific target, increased MFC range was determined in a baseline walking condition. Participants were then requested to control their MFC within the safe range between the upper and lower boundary shown in Fig. 4 [90].

The application of this technology is unlimited, such that most gait parameters can be presented visually in real-time and walkers can be trained to match the presented criterion. Teran-Yengle et al. [91] applied the same concept to knee sagittal joint kinematics to prevent knee hyperextension. Thus, 3D motion capture systems can be used to assist the patient to visualize their locomotion pattern in real-time and correct their walking accordingly. Electromyography (EMG) has also been applied in biofeedback training, helping patients to monitor and voluntarily control ankle dorsiflexors [92]. To combat tripping, the leading cause of falls among senior individuals, biofeedback training could be utilized extensively in neurological rehabilitation programs [93].

Wearable gait assisting technologies

While biofeedback training based on data from 3D motion capture systems can provide accurate position-time data in clinic or laboratory settings, wearable systems can be utilized to assist gait control in the everyday environment. Inertial measurement units (IMUs) provide a practical wearable technology that can obtain acceleration, joint angles, and angular kinematics [94]. While there are challenges in obtaining accurate spatial coordinates, due to the considerable advantages of portability and cost, wearable technology is a promising direction for practical gait monitoring and rehabilitation [95].

FIG. 4 Real-time biofeedback training to improve swing foot control at MFC. *Figure taken from Nagano H, Said CM, James L, Begg RK. Feasibility of using foot–ground clearance biofeedback training in treadmill walking for poststroke gait rehabilitation. Brain Sci 2020; 10 (12): 978.*

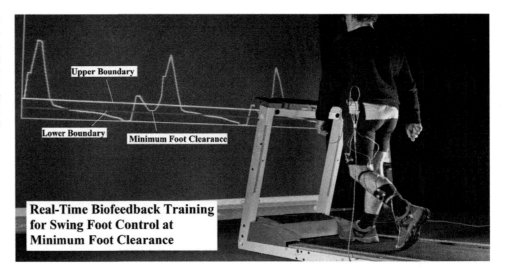

Smart shoes are one such application that can monitor foot motion continuously and, when connected with other applications (e.g., a mobile phone), walking performance can be recorded and training functions may be incorporated [96,97]. For practical biofeedback training, smart footwear development is expected to be a promising direction for assisting senior adults to maintain their locomotor abilities. Through an online data cloud, clinicians would be able to access gait data for locomotion and general health monitoring and use this information to prescribe highly individualized training programs.

Electrical muscle stimulation (EMS)

EMS is relatively a new approach to movement treatment and training and by activating muscles via electrodes on the skin the rhythm and intensity of contraction can be controlled to match a target pattern. People with mobility problems can benefit from this technology by maintaining muscle strength and EMS applications designed for more frail seniors with pathological conditions are becoming available [98]. While physical training, such as outdoor walking or group exercise classes, have advantages, EMS can provide positive effects for seniors with more limited mobility. No apparent side effects have been reported but safety precautions require further investigation such as contraindications for cardiac pacemaker users.

Application of gait assisting device

Older adults are often constrained by related health problems and, as discussed throughout the chapter, many of these, directly and indirectly, influence locomotive functions. Exercise can slow the progression of age-associated gait deterioration and possibly reverse some gait impairments. For the frailer aging population, however, general exercise intervention may not be applicable due to limited physical ability. For this reason, technology-based assistive devices can be utilized to assist everyday locomotion or, in more advanced applications, enhance neuro-rehabilitation using powered exoskeletons that actively move the limbs. Wearable assistive devices are primarily of two types, passive and active; the latter employ an external power source (e.g., an electric motor) to activate the joints while passive devices do not.

Passive devices are developed using structural ergonomic designs that optimize movements to prevent injury and enhance performance without external power assistance. One approach to understanding the mechanics of passive exoskeletons is to focus on the mechanical energy demands of movements [99]. In walking, foot contact impact is the main source of external energy input and walking mechanics should be constrained by the device to maximize mechanical energy efficiency [100,101]. Shoe-insole design can, for example, incorporate structures that assist foot motion to minimize mechanical energy loss. Similarly, improved control of the foot center of pressure and ankle joint motion can potentially reduce various gait-associated injury risks, such as balance loss, tripping, and knee osteoarthritis [97,102,103]. Passive mechanics can be also utilized in designing ankle orthoses that can absorb foot-ground contact impact via "energy harvesting," to be used later in the gait cycle to assist the propulsive forces required when the foot pushes off the ground. Inefficient mechanical energy usage can also impose a burden on lower limb tissues, leading to longer-term effects such as foot deformity and overuse injuries.

Active exoskeletons are operated fully or in part by a power source, assisting motor function by actively controlling the body's mechanical energy and associated movements [104]. Active exoskeletons can be externally driven to control the intensity and frequency of movements and cutting-edge technology is now focused on joint actuation based on individual-specific inputs. These more advanced devices can support the user's intended movements rather than simply perform automated operations defined by the external system. Inputs can be based on kinematic and kinetic information (e.g., velocity, acceleration joint angle, force, and pressure) or bio-signals (e.g., muscle contraction, body temperature, heart rate), described as the "triggers" for active exoskeleton actuation [105]. When seniors with heavily impaired muscular strength or motor control attempt to move, for example standing or walking, active exoskeletons can provide sufficient support to achieve the intended motion.

One example of an active exoskeleton is the Hybrid Assistive Limb (HAL), described as the "first wearable cyborg" and it is being used to help people with movement impairments, such as wheelchair users and poststroke individuals, to regain their mobility [106]. HAL's innovation is in bio-signal sensing technology capable of detecting microsignals, even without actual movements (Fig. 5). HAL can, therefore, sense the wearer's intention and provide joint actuation to reproduce the required movement, which has been reported to be effective in rebuilding damaged neural networks [107].

Summary and recommendations

Exercise plays a crucial role for senior adults in maintaining their locomotor function, which is necessary for healthy aging. Falls risk minimization can be an important goal of exercise intervention. For example, it can be focused on enhancing whole body balance and strengthening specific muscles that have links to falling, such as the ankle dorsiflexor muscle to assist with foot clearance during mid-swing. Technology-based interventions offer more specialized interventions for neurologically impaired gait functions. Exercise intervention for older adults' gait training can, therefore, incorporate more specialized training procedures, with recent advances in gait-biomechanics and associated technologies, outlined in Table 1, now providing viable rehabilitation options.

One advantage of technology-based interventions is their effects on neurological functions, which are otherwise difficult to treat using conventional exercise methods. In regard to tripping prevention, particularly among stroke patients, biofeedback training has effectively improved swing foot clearance. Biofeedback approaches can be also applied to healthy seniors to help them acquire fine-motor control skills for injury prevention. EMS and other assistive devices can be helpful

 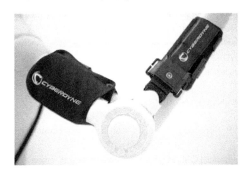

FIG. 5 Hybrid assistive limb (HAL)—active exoskeleton (left) ankle, (right) single-joint (Prof. Sankai, University of Tsukuba, Cyberdyne Inc.)

TABLE 1 Technological devices to support exercise effects for seniors.

	Strength	Neural function	Prevention	Rehabilitation	Mobility aid
Treadmill	△	○	△	○	×
Biofeedback	△	◎	◎	◎	×
EMS	◎	○	△	○	×
Passive assistive device	○※	×	◎	△	○
Active exoskeleton	△	◎	○	◎	◎

◎ = highly effective, ○ = effective, △ = can be effective, × = not effective, ※ possible to incorporate training effects.

for older adults with related health conditions. In addition to the care required when older adults participate in any physical activity, exercise prescribers should be familiar with the options presented by technology-based interventions. As technology advances, these systems will be used increasingly as a considerably more effective intervention for senior adults to maintain their walking abilities, one of the most fundamental physical capacities for healthy aging.

References

[1] Watson WL, Ozanne-Smith J. Injury surveillance in Victoria, Australia: developing comprehensive injury incidence estimates. Accid Anal Prev 2000;32:277–86.

[2] Stevens JA, Corso PS, Finkelstein EA, Miller TR. The costs of fatal and nonfatal falls among older adults. Inj Prev 2006;12:290–5.

[3] Australian Government: Department of Health and Ageing. Home & community care: slips and falls learning resource: falls risk and falls prevention. In: National Slips and Falls Prevention Project; 2007. p. 1–24.

[4] Hill K, Schwarz J, Flicker L, Carroll S. Falls among healthy, community-dwelling, older women: a prospective study of frequency, circumstances, consequences and prediction accuracy. Aust N Z J Public Health 1999;23:41–8.

[5] Stalenhoef PA, Diederiks JPM, Knottnerus JA, Kester ADM, Crebolder HFJM. A risk model for the prediction of recurrent falls in community-dwelling elderly: a prospective cohort study. J Clin Epidemiol 2002;55(11):1088–94.

[6] National Injury Prevention Advisory Council Australia. Directions in injury prevention a report from the National Injury Prevention Advisory Council. Australia: Safety Education; 1999.

[7] Australian Government. Australia to 2050: future challenges. Commonwealth of Australia; 2010.

[8] Cassell E, Clapperton A. Preventing injury in Victorian seniors aged 65 years and older. Hazard VISU 2008;67:1–24.

[9] Hung WW, Egol KA, Zuckerman JD, Siu AL. Hip fracture management tailoring care for the older patient. Am Med Assoc 2012;307(20):2185–96.

[10] Sherrington C, Menz HB. An evaluation of footwear worn at the time of fall related hip fracture. Age Ageing 2003;32:310–4.

[11] Berg WR, Alessio HM, Mills EM, Tong C. Circumstances and consequences of falls in independent community dwelling older adults. Age Ageing 1997;26:261–8.

[12] Li W, Keegan THM, Stemfeld B, Sidney S, Quesenberry CP, Kelsey JL. Outdoor falls among middle-aged and older adults: a neglected public health problem. Am J Public Health 2006;96(7):1192–200.

[13] Curry LC, Hogstel MO, Davis GC. Functional status in older women following hip fracture. J Adv Nurs 2003;42(4):347–54.

[14] Menz HB, Lord SR, Fitzpatrick RC. A structural equation model relating impaired sensorimotor function, fear of falling and gait patterns in older people. Gait Posture 2007;25(2):243–9.

[15] Perry SD, Radtke A, Mcllroym WE, Fernie GR, Maki BE. Efficacy and effectiveness of a balance-enhancing insole. J Gerontol 2008;63A(6): 595–602.

[16] Richardson JP, Knight AL. Common problems of the elderly. Fundamentals of family medicine. In: Taylor RB, David AK, Fields SA, Phillips DM, Scherger JE, editors. Fundamentals of family medicine. New York, NY: Springer; 2003. https://doi.org/10.1007/978-0-387-21745-1_5.

[17] Perry SD. Evaluation of age-related plantar-surface insensitivity and onset age of advanced insensitivity in older adults using vibratory and touch sensation tests. Neurosci Lett 2006;392(1–2):62–7.

[18] Puthoff ML, Nielsen DH. Relationships among impairments in lower-extremity strength and power, functional limitations, and disability in older adults. Phys Ther 2007;87(10):1334–47.

[19] Kang HG, Dingwell JB. Effects of walking speed, strength and range of motion on gait stability in healthy older adults. J Biomech 2008; 41(14):2899–905.

[20] Solnik S, Rider P, Steinweg K, DeVita P, Hortobágyi T. Teager–Kaiser energy operator signal conditioning improves EMG onset detection. Eur J Appl Physiol 2010;110:489–98.

[21] Perry MC, Carville SF, Smith ICH, Rugherford OM, Newham D. Strength, power output and symmetry of leg muscles: effect of age and history of falling. Eur J Appl Physiol 2007;100:553–61.

[22] Hausdorff JM, Rios DA, Edelberg HK. Gait variability and fall risk in community-living older adults: a 1-year prospective study. Arch Phys Med Rehabil 2001;82(8):1050–6.

[23] Hortobagyi T, Solnik S, Gruber A, Rider P, Steinweg K, Helseth J, DeVita P. Interaction between age and gait velocity in the amplitude and timing of antagonist muscle coactivation. Gait Posture 2009;29:558–64.

[24] Morse CI, Thom JM, Davis MG, Fox KR, Birch KM, Narici MV. Reduced plantarflexor specific torque in the elderly is associated with a lower activation capacity. Eur J Appl Physiol 2004;92(1–2):219–26.

[25] Burnfield JM, Mohamed O. The influence of walking speed and footwear on plantar pressures in older adult. Clin Biomech 2004;19(1):78–84.

[26] Zehr EP, Stein RB. What functions do reflexes serve during human locomotion? Prog Neurobiol 1999;58:185–205.

[27] Hollman JH, McDade EM, Petersen RC. Normative spatiotemporal gait parameters in older adults. Gait Posture 2011;34(1):111–8.

[28] Macellari V, Giacomozzi C, Saggini R. Spatial-temporal parameters of gait: reference data and a statistical method for normality assessment. Gait Posture 1999;10:171–81.

[29] Winter DA. The biomechanics and motor control of human gait: Normal, elderly and pathological. 2nd ed. Waterloo Biomechanics; 1991.

[30] Whittle M. Gait analysis: An introduction. 4th ed. Butterworth-Heinemann Elsevier; 2007.

[31] Hurt CP, Rosenblatt N, Crenshaw JR, Grabiner MD. Variation in trunk kinematics influences variation in step width during treadmill walking by older and younger adults. Gait Posture 2010;31:461–4.

[32] Ko SU, Gunter KB, Costello M, Aum H, MacDonald S, White K, Snow CM, Hayes WC. Stride width discriminates gait of side-fallers compared to otherdirected fallers during overground walking. J Aging Health 2007;19(2):200–12.

[33] Pandy MG, Lin YC, Kim HJ. Muscle coordination of mediolateral balance in normal walking. J Biomech 2010;43:2055–64.

[34] Åberg AC, Frykberg GE, Halvorsen K. Medio-lateral stability of sit-to-walk performance in older individuals with and without fear of falling. Gait Posture 2010;31:438–43.

[35] Wezenberg D, de Haan A, van Bennekom CAM, Houdijk H. Mind your step: metabolic energy cost while walking an enforced gait pattern. Gait Posture 2011;33:544–9.

[36] LaRoche DP, Cook SB, Mackala K. Strength asymmetry increases gait asymmetry and variability in older women. Med Sci Sports Exerc 2012;44 (11):2172–81.

[37] Patterson KK, Parafianowicz I, Danells CJ, Closson V, Verrier MC, Staines WR, Black SE, Mcllroy WE. Gait asymmetry in community-ambulating stroke survivors. Arch Phys Med Rehabil 2008;89(2):304–10.

[38] Hausdorff JM. Gait dynamics in Parkinson's disease: common and distinct behavior among stride length, gait variability, and fractal-like scaling. Chaos 2009;19(2), 026113.

[39] Prince F, Corriveau H, Hebert R, Winter DA. Gait in the elderly. Gait Posture 1997;5:128–35.

[40] Wass E, Taylor N, Matsas A. Familiarisation to treadmill walking in unimpaired older people. Gait Posture 2005;21:72–9.

[41] Bock O, Beurskens R. Changes of locomotion in old age depend on task setting. Gait Posture 2010;32:645–9.

[42] Nagano H, Begg RK, Sparrow WA, Taylor S. A comparison of treadmill and overground walking effects on step cycle asymmetry in young and older individuals. J Appl Biomech 2013;29(2):188–93.

[43] Vellas BJ, Wayne SJ, Romero LJ, Baumgartner RN, Garry PJ. Fear of falling and restriction of mobility in elderly fallers. Age Ageing 1997;26 (3):189–93.

[44] Zjistra GAR, van Haastregt JCM, van Eijk JTM, van Rossum E, Stalenhoef PA. Prevalence and correlates of fear of falling, and associated avoidance of activity in the general population of community-living older people. Age Ageing 2007;36(3):304–9.

[45] Osoba MY, Rao AK, Agrawal SK, Lalwani AK. Balance and gait in the elderly: a contemporary review. Laryngoscope Investig Otolaryngol 2019;4 (1):143–53.

[46] Schniepp R, Kugler G, Wuehr M, Eckl M, Huppert D, Huth S, Pradhan C, Jahn K, Brandt T. Quantification of gait changes in subjects with visual height intolerance when exposed to heights. Front Hum Neurosci 2014;8:963.

[47] Thaler-Kall K, Peters A, Thorand B, Grill E, Autenrieth CS, Horsch A, Meisinger C. Description of spatio-temporal gait parameters in elderly people and their association with history of falls: results of the population-based cross-sectional KORA-age study. BMC Geriatr 2015;15:32.

[48] MedlinePlus. Aging changes in organs, tissues, and cells, https://medlineplus.gov; 2021.

[49] Mays Hoopes LL. Aging and cell division. Nat Educ 2010;3(9):55.

[50] Siparsky PN, Kirkendall DT, Garrett WE. Muscle changes in aging: understanding sarcopenia. Sports Health 2014;6(1):36–40.

[51] Wilkinson DJ, Piasecki M, Atherton J. The age-related loss of skeletal muscle mass and function: measurement and physiology of muscle fibre atrophy and muscle fibre loss in humans. Ageing Res Rev 2018;47:123–32.

[52] Gadelha AB, Ricci Neri SG, de Oliveira RJ, Bottaro M, de David AC, Vainshelboim B, Lima RM. Severity of sarcopenia is associated with postural balance and risk of falls in community-dwelling older women. Exp Aging Res 2018;44(3):258–69.

[53] Kalyani RR, Corriere M, Ferrucci L. Age-related and disease-related muscle loss: the effect of diabetes, obesity, and other diseases. Lancet Diabetes Endocrinol 2014;2(10):819–29.

[54] Krantz E, Trimpou P, Landin-Wilhelmsen K. Effect of growth hormone treatment on fractures and quality of life in postmenopausal osteoporosis: a 10-year follow-up study. J Clin Endocrinol Metabol 2015;200(9):3251–9.

[55] Kraemer WJ, Ratamess NA, Hymer WC, Nindl BC, Fragala MS. Growth hormone(s), testosterone, insulin-like growth factors, and cortisol: roles and integration for cellular development and growth with exercise. Front Endocrinol 2020;11:33.

[56] Longinetti E, Fang F. Epidemiology of amyotrophic lateral sclerosis: an update of recent literature. Curr Opin Neurol 2019;32(5):771–6.

[57] Pirker W, Katzenschlager R. Gait disorders in adults and the elderly. Wien Klin Wochenschr 2017;123:81–95.

[58] Possin KL. Visual spatial cognition in neurodegenerative disease. Neurocase 2010;16(6):466–87.

[59] Rabinovici GD, Stephens ML, Possin KL. Executive dysfunction. Continuum (Minneap Minn) 2015;21(3):646–59.

[60] Fernando E, Fraser M, Hendriksen J, Kim CH, Muir-Hunter SW. Risk factors associated with falls in older adults with dementia: a systematic review. Physiother Can 2017;69(2):161–70.

[61] Joza S, Camicioli R, Ba F. Falls in Parkinson's disease and Lewy body dementia. In: Montero-Odasso M, Camicioli R, editors. Falls and cognition in older persons. Cham: Springer; 2019. p. 191–210.

[62] Paulo HS, Pelicioni PHS, Menant JC, Latt MD, Lord SR. Falls in Parkinson's disease subtypes: risk factors, locations and circumstances. Int J Environ Res Public Health 2019;16(12):2216.

[63] Carling A, Forsberg A, Nilsagård Y. Falls in people with multiple sclerosis: experiences of 115 fall situations. Clin Rehabil 2018;32(4):526–35.

[64] Batchelor FA, Mackintosh SM, Said CM, Hill KD. Falls after stroke. Int J Stroke 2012;7:482–90.

[65] Peat G, McCarney R, Croft P. Knee pain and osteoarthritis in older adults: a review of community burden and current use of primary health care. Ann Rheum Dis 2001;60(2):91–7.

[66] Anderson AS, Loeser RF. Why is osteoarthritis an age-related disease? Best Pract Res Clin Rheumatol 2011;24(1):15.

[67] Vinik AI, Strotmeyer ES, Nakave AA, Patel CV. Diabetic neuropathy in older adults. Clin Geriatr Med 2008;24(3):407–v.

[68] de Jong MR, Van der Elst M, Hartholt KA. Drug-related falls in older patients: implicated drugs, consequences, and possible prevention strategies. Ther Adv Drug Saf 2013;4(4):147–54.

[69] Dhalwani NN, Fahami R, Sathanapally H, Seidu S, Davies MJ, Khunti K. Association between polypharmacy and falls in older adults: a longitudinal study from England. BMJ Open 2017;7, e016358.

[70] Gopinath B, McMahon C, Burlutsky G, Mitchell P. Hearing and vision impairment and the 5-year incidence of falls in older adults. Age Ageing 2016;45:409–14.

[71] Esquivel MK. Nutritional assessment and intervention to prevent and treat malnutrition for fall risk reduction in elderly populations. Am J Lifestyle Med 2017;12(2):107–12.

[72] Pynoos J, Steinman BA, Nguyen AQD. Environmental assessment and modification as fall-prevention strategies for older adults. Clin Geriatr Med 2010;26(4):633–44.

[73] Drury CG, Wooley SM. Visually-controlled leg movements embedded in a walking task. Ergonomics 1995;38:714–22.

[74] Brooks V. The neural basis of motor control. New York, NY: Oxford University Press; 1986.

[75] Soderberg GL. Kinesiology: Application to pathological motion. Altimore, MD: Williams & Wilkins; 1986.

[76] Hof AL, Gazendam MGJ, Sinke WE. The condition for dynamic stability. J Biomech 2005;38:1–8.

[77] Nagano H, Levinger P, Downie C, Hayes A, Begg R. Contribution of lower limb eccentric work and different step responses to balance recovery among older adults. Gait Posture 2015;42(3):257–62.

[78] Blake AJ, Morgan K, Bendall MJ, Dallosso H, Ebrahim SBJ, Arie THD, Fentem PH, Bassey EJ. Falls by elderly people at home prevalence and associated factors. Age Ageing 1988;17:365–72.

[79] Begg R, Best R, Dell'Oro L, Taylor S. Minimum foot clearance during walking: strategies for the minimisation of trip-related falls. Gait Posture 2007;25:191–8.

[80] Powers SK, Howley ET. Exercise physiology: theory and application to fitness and performance. 5th ed. New York: McGrawHill; 2004.

[81] Nair R, Maseeh A. Vitamin D: the "sunshine" vitamin. J Pharmacol Pharmacother 2012;3(2):118–26.

[82] Patla AE, Sparrow WA. Factors that have shaped human locomotor structure and behavior: the 'joules' in the crown. In: Sparrow WA, editor. The energetics of human activity. Champaign, Illinois: Human Kinetics Publishers; 2000. p. 43–65.

[83] Bilven KCH, Anderson BE. Core stability training for injury prevention. Sports Health 2013;5(6):514–22.

[84] Moosabhoy MA, Gard SA. Methodology for determining the sensitivity of swing leg toe clearance and leg length to swing leg joint angles during gait. Gait Posture 2006;24(4):493–501.

[85] Chan CW, Rudins A. Foot biomechanics during walking and running. Mayo Clin Proc 1994;69:448–61.

[86] Australian Government Department of Health. Recommendations on physical activity for health for older Australians. Australian Government Department of Health; 2013. viewed 16, March, 2021 https://www1.health.gov.au/internet/main/publishing.nsf/Content/phd-physical-rec-older-guidelines.

[87] Park J, Park S, Kim Y, Woo Y. Comparison between treadmill training with rhythmic auditory stimulation and ground walking with rhythmic auditory stimulation on gait ability in chronic stroke patients: a pilot study. NeuroRehabilitation 2015;37(2):193–202.

[88] Earhart GM, Williams AJ. Treadmill training for individuals with Parkinson disease. Phys Ther 2012;92(7):893–7.

[89] Begg RK, Tirosh O, Said CM, Sparrow WA, Steinberg N, Levinger P, Galea MP. Gait training with real-time augmented toe-ground clearance information decreases tripping risk in older adults and a person with chronic stroke. Front Hum Neurosci 2014;8:243.

[90] Begg R, Galea MP, James L, Sparrow WAT, Levinger P, Khan F, Said CM. Real-time foot clearance biofeedback to assist gait rehabilitation following stroke: a randomized controlled trial protocol. Trials 2019;20:317.

[91] Teran-Yengle P, Birkhofer R, Weber MA, Patton K, Thatcher E, Yack HJ. Efficacy of gait training with real-time biofeedback in correcting knee hyperextension patterns in young women. J Orthop Sports Phys Ther 2011;41(12):948–52.

[92] Moreland JD, Thomson MA, Fuoco AR. Electromyographic biofeedback to improve lower extremity function after stroke: a meta-analysis. Arch Phys Med Rehabil 1998;79:134–40.

[93] Nagano H, Said CM, James L, Begg RK. Feasibility of using foot–ground clearance biofeedback training in treadmill walking for poststroke gait rehabilitation. Brain Sci 2020;10(12):978.

[94] Patel M, Pavic A, Goodwin VA. Wearable inertial sensors to measure gait and posture characteristic differences in older adult fallers and non-fallers: a scoping review. Gait Posture 2020;76:110–21.

[95] Qiu S, Liu L, Zhao H, Wang Z, Jiang Y. MEMS inertial sensors based gait analysis for rehabilitation assessment via multi-sensor fusion. Micromachines 2018;9:442.

[96] Carbonaro N, Lorussi F, Tognetti A. Assessment of a smart sensing shoe for gait phase detection in level walking. Electronics 2016;5:78.

[97] Nagano H, Begg RK. Shoe-insole technology for injury prevention in walking. Sensors 2018;18(5):1468.

[98] Mosole S, Zampieri S, Furlan S, Carraro U, Löefler S, Kern H, Volpe P, Nori A. Effects of electrical stimulation on skeletal muscle of old sedentary people. Gerontol Geriatr Med 2018;4:1–11.

[99] Panizzolo FA, Bolgiani C, Di Liddo L, Annese E, Marcolin G. Reducing the energy cost of walking in older adults using a passive hip flexion device. J Neuroeng Rehabil 2019;16:117.

[100] Cavagna GA, Zamboni A. The sources of external work in level walking and running. J Physiol 1976;262:639–57.

[101] Collett J, Dawes H, Howells K, Elswowrth C, Izadi H, Sackley C. Anomalous centre of mass energy fluctuations during treadmill walking in healthy individuals. Gait Posture 2007;26:400–6.

[102] Nagano H, Begg R. Ageing-related gait adaptations to knee joint kinetics: implications for the development of knee osteoarthritis. Appl Sci 2020;10:8881.

[103] Nagano H, Begg R. A shoe-insole to improve ankle joint mechanics for injury prevention among older adults. Ergonomics 2021. https://doi.org/10.1080/00140139.2021.1918351.

[104] Sawicki GS, Beck ON, Kang I, Young AJ. The exoskeleton expansion: improving walking and running economy. J Neuroeng Rehabil 2020;17:25.

[105] Chen B, Ma H, Qin L, Gao F, Chan K, Law S, Qin L, Liaob W. Recent developments and challenges of lower extremity exoskeletons. J Orthop Translat 2016;5:26–37.

[106] Nilsson A, Vreede KS, Häglund V, Kawamoto H, Sankai Y, Borg J. Gait training early after stroke with a new exoskeleton—the hybrid assistive limb: a study of safety and feasibility. J Neuroeng Rehabil 2014;11:92.

[107] Wall A, Borg J, Palmcrantz S. Clinical application of the hybrid assistive limb (HAL) for gait training - a systematic review. Front Syst Neurosci 2015;9:48.

Chapter 33

Exercise and cognition in aging

Claire J. Cadwallader[a], Eleanor M. Taylor[a], Trevor T-J. Chong[a,b,c], Dylan Curtin[a], Joshua J. Hendrikse[a], Julie C. Stout[a], and James P. Coxon[a]

[a]The Turner Institute for Brain and Mental Health, School of Psychological Sciences, Monash University, Melbourne, VIC, Australia [b]Department of Neurology, Alfred Health, Melbourne, VIC, Australia [c]Department of Clinical Neurosciences, St Vincent's Hospital, Melbourne, VIC, Australia

Introduction

The global population is aging. According to the United Nations, in 2019, 9.1% of the global population was comprised of people aged 65 and over. Projecting 30 years into the future (2050), this proportion is expected to more than double [1]. The impacts of the population age distribution shift will be felt across almost all sectors of society, from increased demand for health services to reductions in the growth of the labor force and changes in financial expenditure. As rates of disability and activity limitations increase within the aging population [2], so too will the pressure placed on residential care, community-based care, and family carers [3]. Hence, there is a need to consider methods of reducing the age-related decline in function, maximizing quality of life for the aging population, and enabling people to live independently for longer. Prescribing exercise to support brain health and cognition in older adults is one way this may be achieved.

This chapter focuses on "healthy aging," defined here as the changes in brain structure, chemistry, and function that occur with age, in the absence of disease pathology. Therefore, discussions of pathological brain changes commonly seen in aging, such as those associated with Alzheimer's Disease are addressed in other chapters of this book. Here we discuss current evidence surrounding the benefits of exercise for cognition in aging, along with proposed underlying neural mechanisms driving these effects. We also provide recommendations for the safe prescription of exercise to support cognition in older adults.

Cognitive decline associated with healthy aging is a significant predictor of functional decline (i.e., difficulty undertaking daily tasks and personal care) in the older population [4]. However, the relationship between cognition and aging is complex. A popular conceptual framework developed by Horn and Cattell [5] distinguishes between two types of cognitive abilities, referred to as fluid and crystallized skills. Crystalized skills refer to those that are accumulated across the lifespan with experience, including acquired knowledge and language abilities such as listening and communication skills. These crystallized skills appear to be largely unaffected by the process of aging, often showing ongoing improvements up to very advanced age [6,7]. In contrast, fluid skills are a broad group of cognitive abilities (e.g., information processing speed, attention, memory, visuospatial skills, executive function skills) that allow us to process information and solve problems in new and novel situations [8]. It is these fluid skills that appear to be more susceptible to age-related decline, with mild progressive reductions often seen from mid-adulthood [9].

Regarding the processes underlying these cognitive changes, neuroimaging studies have established that the brain progressively atrophies from mid-adulthood, starting as early as an individual's 20s, and advancing at a rate of approximately 5% per decade after age 40 [10]. This rate of decline may increase further as an individual reaches their 70s [11]. Age-related brain atrophy affects both gray matter (neuronal cell bodies, astrocytes) and white matter (myelinated axons), and occurs at disproportionately higher rates in some brain areas compared to others. The most significant gray matter volume losses are seen in the prefrontal cortex and caudate nucleus (structures important for executive function skills and goal-directed behavior), and medial temporal lobe structures including the hippocampus (areas crucial for learning and memory) [12–15]. Age-related gray matter atrophy has been associated with declines in attention, verbal abilities [16], memory, executive functioning [17], and visuospatial skills [18]. White matter volume loss occurs most prominently within frontal white matter tracts, and to a lesser extent in white matter tracts located in the temporal and parietal lobes [19–21]. This white matter change has been linked with reductions in information processing speed, attention, memory, visuospatial skills, and executive functions, as well as an increase in the rate of cognitive decline [21,22,23]. Finally, functional brain imaging methods have shown age-related reductions in the modulation of brain activity in response to task demands and reduced connectivity and coherence within functional networks that support higher level cognitive functions [24–26].

At a cellular level, synaptic neuroplasticity and neurogenesis are crucial for many aspects of fluid cognitive skills. Synaptic neuroplasticity refers to the brain's ability to reorganize its neuronal connections in response to changing environmental demands [27]. This function is particularly important for learning and memory. As we age, the capacity of our brain to undergo a type of synaptic neuroplasticity known as long-term potentiation (LTP; an increased strength of the connection between neurons) is thought to decline [28–31]. Neurogenesis, which refers to the generation of new neurons and occurs in specific areas of the brain such as the hippocampus [32,33], also declines with age [34,35]. There are many hypotheses as to why these changes occur. These include an age-related disruption in the production and activity of neurochemicals that support synaptic neuroplasticity and neurogenesis, e.g., brain-derived neurotrophic factor (BDNF) [36], insulin-like growth factor, (IGF-1) [37–39], dopamine transmission [40,41] and insulin sensitivity [42–44]. Other potential mechanisms include age-related disruption to important neurotransmitter systems [45–47], changes in the balance of cellular signaling ions [48], and/or changes in the expression of genes that contribute to cell function [49]. Both large-scale changes, such as atrophy and disruption of brain networks, and smaller-scale changes, e.g., disruption of cellular and molecular processes supporting synaptic neuroplasticity and neurogenesis, likely underlie the cognitive decline associated with aging.

Importantly, there is a high level of variability in the degree of cognitive decline between individuals in aging [50,51], as well as in the rate of underlying brain changes [12]. This has led to the concepts of (i) "successful aging," referring to those individuals who experience less age-related cognitive decline and age-related brain change relative to "normal" healthy aging, and (ii) "accelerated aging" referring to those who experience a faster rate of age-related decline and changes in brain structure and function [52]. The trajectory of brain aging and cognition likely depends on a variety of genetic, health, environmental, and lifestyle factors [53,54]. This chapter focuses on engagement in exercise as a modifiable protective factor to promote "successful aging" of cognition.

Benefits of exercise for cognition in aging

Regular engagement in exercise appears to be a protective lifestyle factor against cognitive decline in aging [55]. Epidemiological studies show that higher rates of exercise in early and middle adulthood reduce the risk of cognitive decline in later life [56,57]. Exercise is also associated with a reduced risk of developing age-related neurodegenerative diseases such as Alzheimer's disease [58], which are addressed in other chapters of this book. Notwithstanding these associations, the majority of adults become more sedentary as they age. For example, in Australia in 2017–18, 69% of men and 75% of women aged 65 years and older were considered insufficiently active [59].

While exercise across the entire lifespan appears to be advantageous, there are also benefits for those who start a regular exercise routine later in life. In older adults, studies have found that higher levels of cardiorespiratory fitness, as well as engagement in exercise programs, have been associated with reductions in age-related neural changes, and superior performances across various cognitive tasks [55,60–62]. It is important to note that causality between exercise and improvements in cognition and brain function is difficult to establish with the research designs typically used in epidemiological studies. As such, here we focus on experimental studies. In this section, we will discuss current evidence surrounding the benefits of exercise programs for cognitive health in normal aging, as well as potential underlying mechanisms.

Cardiorespiratory exercise programs

In older adults, the link between cardiorespiratory exercise, which relies on oxygen-dependent energy-generating processes such as sustained running, swimming, and aerobics, and improvements in cognitive skills is well established [55,63]. Specifically, regular cardiorespiratory exercise leads to improvements in a variety of executive function skills including inhibitory control [64], idea generation [65], task switching, and working memory [66], when compared to nonexercise control groups. Executive function skills are frequently highlighted as key areas of decline in older age. In a study by Albinet and colleagues [67], older adults who completed a 21-week aqua aerobics program demonstrated significant improvements in inhibitory control and working memory. These improvements were not observed for participants who completed a stretching program of the same duration. Promisingly, similar improvements across executive function skills have been observed following various cardiorespiratory exercise paradigms, including brisk walking [68], and cycling [69]. Accumulating evidence suggests that age-related declines in executive function skills may be somewhat ameliorated or prevented by regular participation in various forms of cardiorespiratory exercise.

Memory is another key area of concern for older adults experiencing cognitive decline and maybe improved by cardiorespiratory exercise [70,71]. One recent study by Kovacevic et al. [72] compared the cognitive benefits of different exercise programs in older adults, (i) a high-intensity interval training (HIIT) program (i.e., burst of high-intensity exercise alternated with periods of moderate-intensity exercise), (ii) a continuous moderate-intensity exercise program, and (iii) a stretching control program. Results showed that participants performed better on a memory task following the HIIT program than those who completed a continuous moderate-intensity exercise program or the control program. Although the optimal exercise parameters for improving memory are still to be determined [73], this finding is generally consistent with the broader literature [74] and suggests that high-intensity cardiorespiratory exercise may be of particular benefit to memory performance for older adults.

In addition to engagement in specific exercise programs, research also suggests a link between overall cardiorespiratory *fitness* and cognitive skills in aging. Higher overall fitness levels have been linked to better executive function skills in older adults [75,76]. Results from the exercise intervention study by Kovacevic and colleagues [72] also found that greater improvements in memory following the exercise intervention were correlated with greater increases in fitness in older adults. These improvements may relate to structural changes, such as protection from age-related brain atrophy, and in some cases increased brain volume [77–79], as well as increased cerebral blood flow [65,70]. However, across studies, changes in cardiorespiratory fitness are not consistently correlated with improvements in cognitive skills following exercise interventions [67]. These conflicting findings may relate to additional factors that contribute to cardiorespiratory fitness levels (e.g., genetics [80]), however, it remains unclear whether improving cardiorespiratory fitness is critical in improving cognitive skills, or whether regular engagement in activity (regardless of fitness) is sufficient.

Resistance training programs

In addition to cardiorespiratory exercise, it has been suggested that resistance training programs may also pose benefits for cognition in older adults. Resistance training involves strengthening exercises that require muscles to contract against an external force such as weights, resistance bands, bodyweight, or water. Many studies exploring the efficacy of resistance training programs in older adults have shown promising results for improving cognitive outcomes (e.g., [75,81,82]). For example, one study indicated significant improvements in response inhibition in a group of older adults following a 6-month home-based resistance training involving 30-min of resistance exercises, three times per week [81]. A recent meta-analysis of 24 studies examining the benefits of long-term (minimum 4 weeks) resistance training programs compared to an active (i.e., stretching or balance exercises) or passive (no change to lifestyle) control group found an overall positive effect of resistance training on cognition, particularly within the domain of executive function [83]. Furthermore, a systematic review of 12 randomized controlled trials, which compared resistance training interventions to active or passive control groups [84], also reported beneficial effects on executive functioning and measures of global cognition. In contrast to these findings showing benefits to executive function, evidence was weak for improvements in memory, and there was no evidence that attention improved. Despite these promising findings, the literature is inconsistent, with some studies observing no change in cognitive outcomes following resistance training interventions [85,86].

The durability of cognitive benefits following resistance training interventions remains unclear because long-term follow-up postintervention has been very limited. One randomized controlled trial that did include long-term follow-up undertook a 52-week program of either once or twice weekly resistance training in older women aged 65–75 years [87]. At both session frequencies, the women showed sustained improvements in cognitive inhibition 1 year after the intervention ended. For the women who engaged in the twice-weekly program, memory improvements and a reduction in cortical white matter atrophy were additional benefits maintained at long-term follow-up. Importantly, based on monthly self-report measures, physical activity levels did not differ between the study groups throughout the 1-year follow-up period after the intervention finished [88]. Therefore, there may be some sustained improvements in cognition following a resistance training program.

Due to inconsistent findings in this area of research, the cognitive benefits of resistance training alone remain equivocal. However, based on the current research there is reason to be optimistic about the role of resistance training in promoting brain and cognitive health. Interestingly, evidence indicates that resistance training *combined with* an aerobic exercise program produces greater improvements in cognitive skills such as memory and executive function, compared to either of these interventions alone [55,89,90]. Additional benefits of resistance training include the provision of an alternative exercise option for older adults with disabilities or conditions that make many forms of aerobic exercise more challenging [91,92].

Acute exercise

Even a single bout of exercise (referred to as acute exercise) can have cognitive benefits for older adults [93,94]. Researchers have found acute exercise-related improvements across various cognitive skills [95–97]. While many of these studies focus on younger populations, increasing evidence indicates that older adults may benefit at least as much from acute exercise compared to young adults [97,98]. For example, in older adults, improvements in fluid intelligence skills such as processing speed [99], executive function [97,98], and working memory have been observed following a bout of acute exercise [100,101,102]. These effects appear to be time-dependent, with maximal improvement observed 15 min following exercise [103], although effects may endure for many hours (e.g., see [100]).

Though much of the research on acute exercise to date has focused on cardiorespiratory exercise, recent studies looking at a single resistance training session have also found benefits for cognition in healthy older adults [104]. For example, when compared to a control group, community-dwelling older adults who completed 45 min of resistance training showed better performance on executive function tasks, specifically in the domains of response inhibition, working memory, and mental flexibility [105]. Interestingly, some evidence indicates that improvements in working memory may be larger in older adults than in younger adults [106]. Taken together, this evidence suggests that cognitive performance in older adults can be improved, at least in the short term, by several different exercise approaches.

Research into the optimal timing, intensity, and duration of acute exercise in older adults is ongoing, and currently, much of our understanding is based on research in young adults. Regarding exercise intensity, some researchers propose an inverted-U-shaped relationship, with the greatest improvements seen following moderate-intensity exercise, and reduced benefits from lower and higher intensities [107,108]. However, findings are inconsistent across the literature and some basic cognitive processes such as attention and processing speed appear to be more enhanced following high-intensity exercise [109]. A wide variation in study designs has contributed to these conflicting findings and has made it difficult to compare how the specifics of certain exercise tasks may influence cognition over a short time frame. Studies that used longer exercise bouts may have resulted in fatigue, which could mask any immediate cognitive benefits, particularly for less fit individuals [97,110]. However, when viewing the present literature overall, meta-analyses typically indicate some form of cognitive enhancement following light, moderate, and high-intensity acute exercise [109,111]. While the cognitive benefits of acute exercise tend to be relatively short-lived, it is important to understand the mechanisms that lead to these benefits. Understanding the neurochemical processes that occur immediately following exercise is critical to understanding how repeated exposure to exercise results in longer term benefits to cognition.

Acute mechanisms of exercise on cognition

The physiological processes that benefit from acute exercise remain unclear, but some authors have suggested that increases in arousal following exercise may benefit cognition. Arousal, in this context, refers to higher overall alertness, and greater activation of cognitive and physiological processes, e.g., increased blood flow [112,113]. However, acute exercise is also associated with increases in the production, availability, and absorption of various neurotransmitters, such as norepinephrine and dopamine [103], and with upregulation of trophic factors such as BDNF and IGF-1 [37,114,115]. Enhancements of these neurochemical processes may explain some of the benefits observed in older adults following acute exercise.

A limitation of this literature is the reliance on animal models and young adults, which complicates generalization to older adults. Although inferences can be made from this literature about how acute exercise may benefit older adults, within the older adult population these mechanisms have not yet been thoroughly tested. To understand how exercise can improve cognition in the longer term, a mechanistic account must be developed to link specifics of exercise to particular physiological and neural outcomes. Once these mechanisms are understood, exercise strategies can be developed to specifically target the desired outcomes.

Long-term mechanisms of exercise on cognition

Long-term engagement in exercise has the capacity to mitigate age-related brain and cognitive decline (Fig. 1). For example, regular exercise reduces age-related brain atrophy [19,60,116,117] in areas particularly susceptible to decline including the hippocampus and prefrontal cortex. A seminal study by Erickson and colleagues demonstrated a 2% increase in hippocampal volume following a 12-month intervention. Impressively, these changes equated to a reversal of age-related volume loss by 1–2 years [71], and were also associated with improved short-term memory. By comparison, a control group that completed a 12-month stretching and toning program showed a 1.4% decrease in hippocampal volume over the same

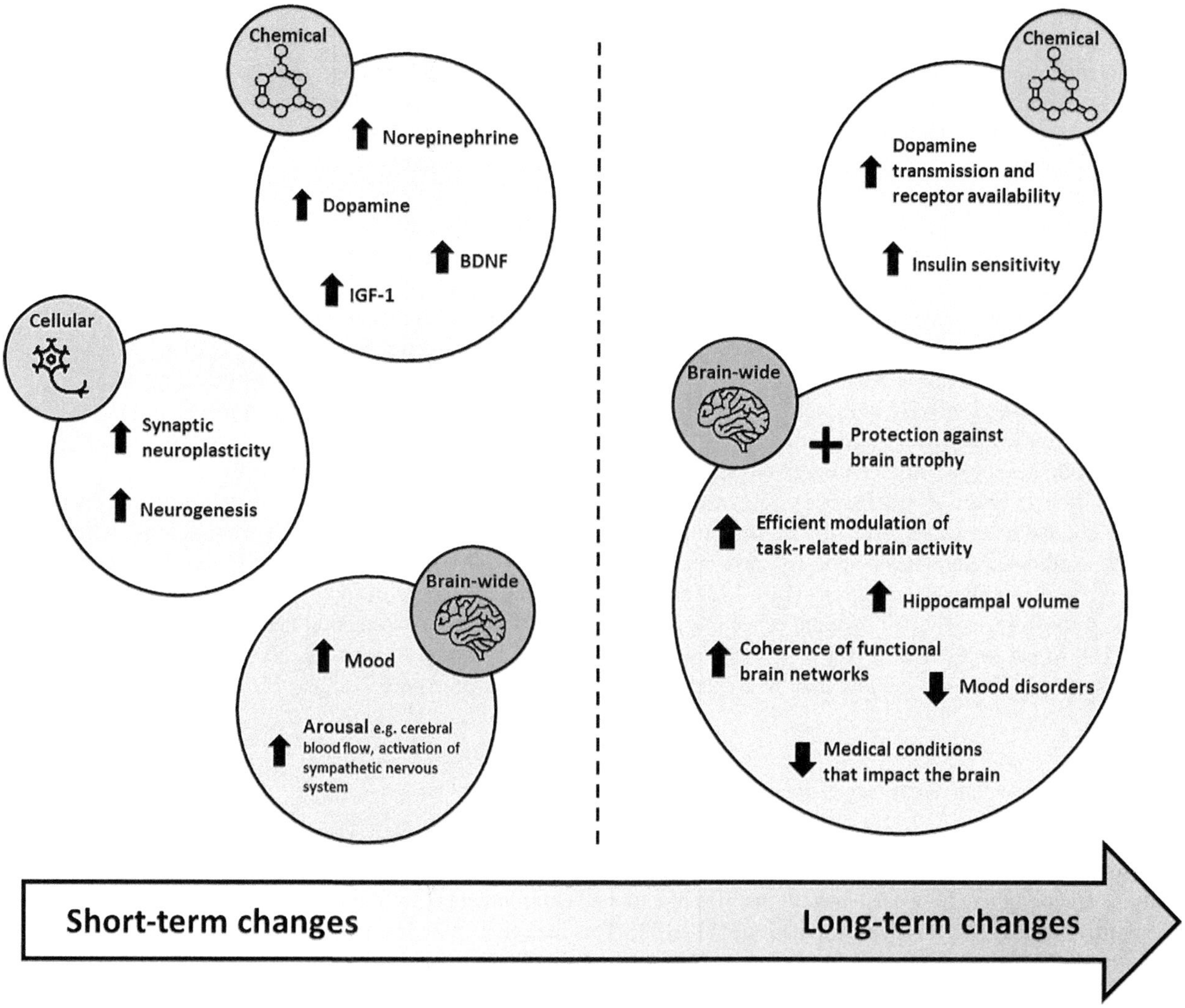

FIG. 1 Proposed mechanisms underlying the cognitive benefits of exercise in older adults. Repeated activation of short-term mechanisms induced by regular exercise likely results in the long-term brain changes observed, although the causative nature of this relationship requires further supporting evidence.

interval. This suggests exercise is not only protective against age-related brain volume loss, but can also reverse age-related atrophy in structures crucial for cognition in older adults. In another study, higher levels of physical activity in older adults were associated with greater volume of the frontal occipital and temporal cortices 9 years later, suggestive of long-term sustained benefits [60]. In contrast, at least two studies have failed to show protective effects of exercise on the structural integrity of cortical brain regions [118,119]. However, these studies did find that improvements in cardiorespiratory fitness were associated with increased white matter integrity [119], and improvements in measures of cognition [118], suggesting some exercise-related benefits.

At a functional level, evidence indicates that engagement in exercise interventions and better cardiorespiratory fitness are associated with selective, region-specific changes in recruitment of brain regions involved in more complex cognitive functions in older adults [120,121]. For example, both higher levels of cardiorespiratory fitness, and 6-month aerobic exercise intervention, were associated with upregulated activity in the dorsolateral prefrontal cortex and parietal regions, and down-regulated activity in the anterior cingulate cortex during a task requiring executive functioning to inhibit or filter misleading information [120]. These changes appear more pronounced during more effortful cognitive tasks [121], and may endure for many years postintervention [122]. These changes are thought to represent both an improvement in top-down cognitive processes, and better regulation of bottom-up processes with increasing task demands during executive function tasks [120,121,123]. This may be a particularly relevant mechanism for older adults, given that there is evidence of reduced modulation of neural recruitment in older adults during demanding tasks [25,26].

Extending on the functional MRI literature discussed above, evidence suggests that exercise may be associated with increased coherence, or synchrony of brain activity, across functional brain networks [124,125]. Coherence is important for efficient communication between brain regions that work together to perform cognitive tasks. After a 1-year aerobic exercise program, Voss and colleagues [125] found improved coherence in two functional networks known to be negatively affected by aging; the default-mode network and the frontal executive network. The hippocampus, a key brain region in the default mode network, is among the brain regions that show the highest exercise-related boost in connectivity when compared to controls [124,126,127].

Regular exercise may also lead to long-term changes in neurochemicals and neurotransmitters important for cognition. Dopamine transmission and insulin sensitivity in the brain have both been shown to diminish with aging, and reductions in these factors are associated with cognitive decline [42,43,128]. Higher levels of cardiovascular fitness in older adults have been associated with longer-term increases in dopamine transmission and receptor availability [129], corroborating an earlier exercise intervention study in mice [130]. Higher levels of dopamine transmission may help to enhance synaptic neuroplasticity [40] and have been associated with superior cognitive skills, particularly for working memory and motor learning [131–134]. Evidence indicates that an active lifestyle can boost insulin sensitivity in the brain [44,135]. Better insulin sensitivity may lead to more efficient uptake of glucose, which is the brain's primary energy source, and also supports synaptic neuroplasticity, neurogenesis, and protection from neuronal cell death [136–140].

In addition to the direct benefits of exercise on brain processes important for cognition, exercise may also reduce the risk of medical conditions which adversely impact brain health, thereby affecting cognition. These conditions include Type 2 diabetes, cardiovascular disease, obesity, and chronic inflammation [141–145]. Furthermore, exercise-related benefits to mood and mental health conditions [146–148] are well established. Feelings of low mood, chronic stress, and anxiety are linked to poorer cognitive function, especially at the severity experienced in mental health disorders like depression [149–151]. Mood improvement following exercise may be particularly important in an older population where rates of loneliness, depression, and anxiety may be higher than in the younger population [152,153].

Combining exercise with cognitive simulation

Combining exercise with a cognitively demanding activity may have superior benefits for cognition in older adults compared to either of these activities alone [154–157]. The hypothesized mechanisms behind the benefits of combined interventions are thought to be consistent with theories of enriched environments (i.e., environments that contain a combination of cognitive, social, and physical stimulation) [119,158]. This idea first came from observations in rodents which showed anatomical and chemical benefits for the brain (e.g., increased brain size and facilitation of neurogenesis and neuroplasticity) when animals were housed in an environment enriched with sensory stimuli compared to a basic cage [159–161]. The parallel to environmental enrichment in rodent studies may be, for humans, the combination of exercise with a cognitive challenge. Research from the animal literature suggests that adding complexity to the exercise environment with cognitively challenging activities may capitalize on the optimal conditions for synaptic neuroplasticity created by exercise, and ensure that newly formed neurons survive [162].

Many studies have explored the efficacy of interventions combining exercise and cognitive tasks on cognitive outcomes for older adults. One meta-analysis conducted by Zhu and colleagues [156] found that these combined interventions led to greater improvements in measures of global cognition and visuospatial ability (i.e., ability to perceive and work with visual stimuli) compared to exercise-only interventions. These findings, however, failed to be sustained in studies that included a long-term follow-up. Of note, the majority of studies included in the analysis used controlled "artificial" cognitive tasks such as learning and memorizing a novel list of words while exercising. These intervention designs can be boring for participants and may reduce willingness to engage. Similarly, another meta-analysis by Gheysen and colleagues [163] found that cognitive-exercise combined interventions yielded better outcomes for cognition compared to exercise-only interventions. Importantly, this analysis included both interventions with artificial cognitive tasks, and those that involved relatively more engaging "real world" activities such as dance and tai chi. These activities are inherently cognitively challenging, with executive functioning demands, learning and memory requirements for step sequences and patterns of movement, and attentional control and visuospatial processing demands [164–167]. Another common form of combined intervention involved "exergaming"—a novel type of intervention that combines exercise and video games with different cognitive demands [168]. The success of these intervention programs is encouraging as they are likely to be more engaging. Older adults often highlight boredom or lack of motivation as reasons not to exercise [169], therefore exercise programs that include cognitive demands may have greater potential for uptake as an exercise option for older adults.

Despite the promising results of many combined intervention studies, several outstanding challenges still need to be considered. One potential limitation of simultaneous programs is that cognitive demands may limit exercise intensity due to competition between cognitive, effort, or physical resources. The considerable heterogeneity across combined intervention designs also adds complexity to this area of research, making it difficult to establish reliable effect sizes and assess the superiority of these interventions beyond regular exercise protocols [163]. For example, studies often involve different modes, intensities, and durations of exercise, and some incorporate single-domain cognitive tasks (e.g.,[170,171]), whereas others use multidomain tasks (e.g.,[172,173]). Therefore, determining the optimal intervention design is challenging and will need to be addressed with carefully designed studies that parametrically vary both cognitive and physical demands across a variety of combinations of component tasks. The lack of combined intervention studies that include long-term interventions or follow-up also limits conclusions about long-term benefits. Although we may not yet be able to say with certainty that combined interventions are superior, no evidence has suggested their inferiority to exercise-alone interventions. In fact, the added cognitive stimulation required during cognitively demanding exercise activities may serve another benefit, which is to boost interest and enjoyment during an exercise program. Hence, combined interventions such as dance, tai chi, or exergaming may be a preferable option for many older adults.

Clinical practice recommendations for exercise prescribers

The current World Health Organization (WHO) recommendations for physical activity prescription in healthy older adults 65 years and over are presented in Fig. 1. The term "physical activity" refers to both planned exercise activities and the incidental exercise achieved when participating in other daily tasks such as walking, occupational activities if the individual is still working, household chores, gardening, and family, leisure, and community activities. The WHO physical activity recommendations are designed to support overall health in older adults [174]. Based on the literature reviewed above, Fig. 2 also includes considerations specific to cognition in older adults. These guidelines are intended for individuals who do not have specific medical conditions that would preclude engaging in physical activity at the recommended duration or intensity. Nevertheless, with appropriate supervision and adaptations, exercise is safe for many older adults with chronic health conditions, injuries, or disabilities, particularly if the volume and/or intensity of exercise is gradually increased if one is beginning from a deconditioned/sedentary state. Safety considerations are discussed in the next section of this chapter.

Evidence suggests that more exercise leads to better outcomes for cardiovascular and respiratory health [175,176]. However, in the context of improving cognition, evidence indicates that long and/or frequent bouts of exercise may not be necessary. A meta-analysis that showed best results for cognition are observed in exercise programs that involve a medium duration of exercise sessions (45–60 min) compared to those with both shorter and longer session durations [89]. In contrast, another meta-analysis showed that exercise programs with session lengths of short duration (<30 min) did not show beneficial effects for cognition above that of controls [55]. However, the intensity of exercise sessions was not reported in this meta-analysis. Based on more recent studies that demonstrate the benefit of acute 20-min bouts of HIIT on mechanisms underlying cognitive skills in younger adults, [177–179], it is reasonable to speculate that session durations shorter than 30 min may be effective in older adults if exercise can be performed at a high intensity. Furthermore, meta-analyses demonstrate that combined exercise/cognitive training programs delivered less frequently had equal or greater efficacy than those delivered more frequently [156,163]. Therefore, even a moderate increase in exercise engagement can be highly valuable for supporting brain health and cognition in older adults.

The best mode and intensity of exercise sessions for supporting cognition in the elderly remains contentious. Some studies show greater cognitive improvements after HIIT programs compared to moderate-intensity programs in healthy older adults (e.g., [72,180]). However, other studies have found no difference between programs that included high-intensity exercise compared to moderate-intensity exercise for cognitive outcomes [89]. In terms of mode of exercise, a combination of aerobic exercise and resistance training may produce superior results [55,89,90], and combining exercise with a cognitively stimulating activity may also benefit cognition [156,163]. Given the limited consensus of evidence for optimal exercise parameters, exercise prescribers should in practice consider prescribing: (i) a variety of modes of exercise with a focus on what the individual enjoys; (ii) exercise sessions of at least moderate intensity for at least 30 min; and (iii) the inclusion of some sessions of HIIT exercise for at least 20 min where there are no contraindications to higher intensity forms of exercise.

In addition to appropriate program design, it is important to build motivation and to identify and address potential barriers to exercise to increase adherence [169,181]. Common self-reported barriers to exercise in inactive older adults include worrying about the potential for injury, difficulty prioritizing time, difficulties with self-discipline and motivation, boredom, and feeling intimidated by gym facilities and exercise groups [169]. To address these barriers, the exercise provider should aim to provide education on the benefits of exercise for cognition, how to exercise safely, how to exercise

FIG. 2 Recommendations for the prescription of exercise to support cognition in older adults. These guidelines are based on the World Health Organization's physical activity recommendations for individuals aged 65 and over [174], with the incorporation of evidence specifically pertaining to improving cognitive outcomes presented throughout this chapter. *Or an equivalent combination of moderate-intensity and high-intensity physical activity. **Heart rate reserve is defined as the difference between an individual's resting heart rate and maximum heart rate and is commonly used to determine target heart rate zones for exercise.

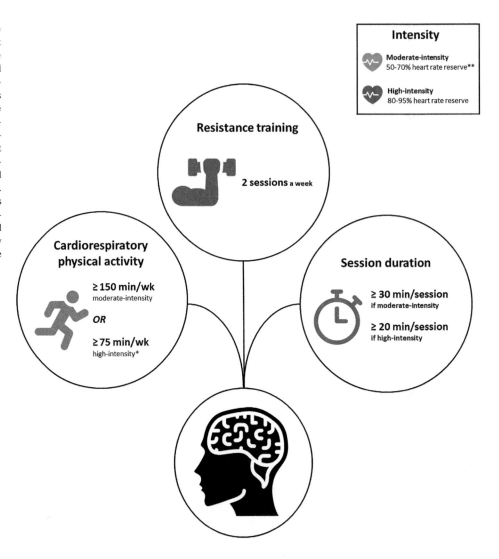

outside the gym setting, and what facilities might be available that specifically cater to older adults [182]. Achievable, personally relevant goals should be set at the beginning of the exercise program, and should be periodically revised, to sustain motivation. These goals could include a focus on supporting their brain health and cognition, as well as goals to improve physical functioning and health management, increasing fitness, and preventing falls [169,183]. Some people may benefit from working with an experienced fitness trainer, who can facilitate exercise sessions and monitor progress. Working with a trainer has also been shown to reduce the fear of injury [184]. Taking the time to identify motivations and address these barriers will be paramount to the success of an exercise prescription.

Safety and precautions

Along with changes in cognition, aging is associated with reduced functional capacity, strength, flexibility, and balance, as well as degenerative musculoskeletal conditions [182,185]. A broad range of medical conditions common in an elderly population may affect their ability to perform exercise. These include, but are not limited to, cardiovascular and pulmonary conditions (e.g., high cholesterol and hypertension), obesity, and Type 2 diabetes [182,186,187]. The prescription of an exercise program to older adults needs to be made in consideration of a person's current capabilities and medical conditions to limit the risk of injury or adverse medical events [182]. See Table 1 for a list of general safety recommendations and precautions when prescribing exercise for older adults.

TABLE 1 Safety recommendations for exercise prescription in older adults.

Gradual introduction of exercise	When prescribing an increase in exercise engagement, a program should start at a level the individual can comfortably achieve. The intensity, duration, and frequency of exercise sessions should be increased slowly, especially for older adults who are deconditioned or have functional or medical limitations. The progression of the exercise program should be carefully tailored to the individual's tolerance and abilities. However, it should be noted that an approach that is too conservative may not confer many benefits [182]
Clearing exercise with a health professional	For those with medical conditions or musculoskeletal injuries, it is recommended that the individual check with an appropriate health professional before increasing their exercise load. Some medical conditions, such as decompensated congestive heart failure or severe aortic stenosis, may mean that exercise is contraindicated or significantly restricted [188]. Advice should be sought on any limitations to the type or intensity of exercise that person can safely engage in. Cardiovascular stress testing prior to engagement in an exercise program that involves high-intensity exercise may be required for individuals with some chronic health conditions [189].
Selecting an appropriate mode of exercise	An appropriate mode of exercise should be selected in line with the individual's ability, interests and health status. Low-impact forms of exercise such as walking, cycling, swimming, tai chi, and aqua aerobics, may be more appropriate for some older adults and can help to decrease stress on joints and reduce pain [190,191].
Supervision of exercise	Supervision of exercise sessions by a qualified trainer, physiotherapist, or exercise physiologist may be necessary for some individuals to ensure safety [182].
Conditioning exercises	Preceding the introduction of cardiorespiratory exercise with muscle strengthening and balance training should be considered among very frail individuals or those at risk of falls [189]. Static flexibility exercises may also assist in improving range of motion and injury prevention [192].

Conclusion

Engagement in exercise is a safe and effective strategy to support the aging brain, and to combat age-related cognitive decline. Evidence posits numerous benefits of exercise for the aging brain, including protecting, or even reversing, age-related brain atrophy; increasing activation and connectivity across brain networks; boosting neurochemicals and transmitters that support processes such as synaptic neuroplasticity and neurogenesis; and improving mood and other health conditions known to have secondary adverse effects on cognition. These exercise-related brain changes are corroborated by studies showing that exercise programs can improve many areas of cognition, particularly complex cognitive skills such as executive functioning skills and memory. However, the optimal exercise mode, intensity, frequency, and duration for supporting cognition in aging are still to be determined. Considering the available evidence to date, exercise programs that combine sessions of cardiorespiratory exercise and resistance training may be particularly beneficial. Evidence also suggests that exercise sessions should involve at least moderate-intensity for at least 30 min, with the inclusion of some sessions of HIIT exercise for a minimum of 20 min for capable individuals. Employing an exercise mode that is cognitively stimulating, such as dance, exergaming or tai chi, could also be considered for its potential benefits on cognition, as well as for improving the enjoyability of exercise. Exercise prescribers should take care to identify personally-relevant goals and to address barriers to exercise prior to the beginning of an exercise program to improve motivation and adherence. Finally, safety considerations should be incorporated in the design of an exercise program such as modifying the program to account for existing medical or musculoskeletal conditions, a gradual build-up of exercise intensity/duration and/frequency, and assessing the need for exercise supervision with a professional.

Acknowledgments

T.C. is supported by the Australian Research Council (DP 180102383, DE 180100389).

J.S. is supported by an Australian National Health and Medical Research Council (NHMRC) Investigator Grant 1173472.

J.C. supported by an Australian Research Council Discovery project (DP 200100234).

References

[1] Anon. World population prospects 2019: highlights. United Nations, Department of Economic and Social Affairs, Population Division; 2019.

[2] Anon. Disability, ageing and carers, australia: summary of findings. Australian Bureau of Statistics; 2019.

[3] McPake B, Mahal A. Addressing the needs of an aging population in the health system: the Australian case. Health Syst Reform 2017;3(3):236–47.

[4] Dodge HH, et al. Cognitive domains and trajectories of functional independence in nondemented elderly persons. J Gerontol A Biol Sci Med Sci 2006;61(12):1330–7.

[5] Horn JL, Cattell RB. Age differences in fluid and crystallized intelligence. Acta Psychol (Amst) 1967;26:107–29.

[6] McArdle JJ, et al. Comparative longitudinal structural analyses of the growth and decline of multiple intellectual abilities over the life span. Dev Psychol 2002;38(1):115–42.

[7] Tucker-Drob EM, et al. Structure and correlates of cognitive aging in a narrow age cohort. Psychol Aging 2014;29(2):236–49.

[8] Salthouse TA. When does age-related cognitive decline begin? Neurobiol Aging 2009;30(4):507–14.

[9] Zimprich D, Martin M. Can longitudinal changes in processing speed explain longitudinal age changes in fluid intelligence? Psychol Aging 2002; 17(4):690.

[10] Svennerholm L, Boström K, Jungbjer B. Changes in weight and compositions of major membrane components of human brain during the span of adult human life of Swedes. Acta Neuropathol 1997;94(4):345–52.

[11] Scahill RI, et al. A longitudinal study of brain volume changes in normal aging using serial registered magnetic resonance imaging. Arch Neurol 2003;60(7):989–94.

[12] Fjell AM, Walhovd KB. Structural brain changes in aging: courses, causes and cognitive consequences. Rev Neurosci 2010;21(3):187–221.

[13] Raz N, et al. Regional brain changes in aging healthy adults: general trends, individual differences and modifiers. Cereb Cortex 2005;15(11): 1676–89.

[14] Tamnes CK, et al. Brain development and aging: overlapping and unique patterns of change. Neuroimage 2013;68:63–74.

[15] Grahn JA, Parkinson JA, Owen AM. The cognitive functions of the caudate nucleus. Prog Neurobiol 2008;86(3):141–55.

[16] Grazioplene RG, et al. Subcortical intelligence: caudate volume predicts IQ in healthy adults. Hum Brain Mapp 2015;36(4):1407–16.

[17] Cardenas VA, et al. Brain atrophy associated with baseline and longitudinal measures of cognition. Neurobiol Aging 2011;32(4):572–80.

[18] Schmidt R, et al. White matter lesion progression, brain atrophy, and cognitive decline: the Austrian stroke prevention study. Ann Neurol 2005; 58(4):610–6.

[19] Colcombe SJ, et al. Aerobic exercise training increases brain volume in aging humans. J Gerontol A Biol Sci Med Sci 2006;61(11):1166–70.

[20] Gunning-Dixon FM, et al. Aging of cerebral white matter: a review of MRI findings. Int J Geriatr Psychiatry 2009;24(2):109–17.

[21] Ziegler DA, et al. Cognition in healthy aging is related to regional white matter integrity, but not cortical thickness. Neurobiol Aging 2010; 31(11):1912–26.

[22] Ritchie SJ, et al. Brain volumetric changes and cognitive ageing during the eighth decade of life. Hum Brain Mapp 2015;36(12):4910–25.

[23] Kerchner GA, et al. Cognitive processing speed in older adults: relationship with white matter integrity. PLoS One 2012;7(11), e50425.

[24] Geerligs L, et al. A brain-wide study of age-related changes in functional connectivity. Cereb Cortex 2014;25(7):1987–99.

[25] Prakash RS, et al. Age-related differences in the involvement of the prefrontal cortex in attentional control. Brain Cogn 2009;71(3):328–35.

[26] Reuter-Lorenz PA, Mikels JA. The aging mind and brain: implications of enduring plasticity for behavioral and cultural change. In: Lifespan development and the brain: the perspective of biocultural co-constructivism; 2006. p. 255–76.

[27] Bear MF, Malenka RC. Synaptic plasticity: LTP and LTD. Curr Opin Neurobiol 1994;4(3):389–99.

[28] Spriggs M, et al. Age-related alterations in human neocortical plasticity. Brain Res Bull 2017;130:53–9.

[29] Freitas C, et al. Changes in cortical plasticity across the lifespan. Front Aging Neurosci 2011;3:5.

[30] Müller-Dahlhaus JFM, et al. Interindividual variability and age-dependency of motor cortical plasticity induced by paired associative stimulation. Exp Brain Res 2008;187(3):467–75.

[31] Tecchio F, et al. Age dependence of primary motor cortex plasticity induced by paired associative stimulation. Clin Neurophysiol 2008;119(3): 675–82.

[32] Eriksson PS, et al. Neurogenesis in the adult human hippocampus. Nat Med 1998;4(11):1313–7.

[33] Gonçalves JT, Schafer ST, Gage FH. Adult neurogenesis in the hippocampus: from stem cells to behavior. Cell 2016;167(4):897–914.

[34] Kuhn HG, Dickinson-Anson H, Gage FH. Neurogenesis in the dentate gyrus of the adult rat: age-related decrease of neuronal progenitor proliferation. J Neurosci 1996;16(6):2027–33.

[35] Siwak-Tapp CT, et al. Neurogenesis decreases with age in the canine hippocampus and correlates with cognitive function. Neurobiol Learn Mem 2007;88(2):249–59.

[36] Lommatzsch M, et al. The impact of age, weight and gender on BDNF levels in human platelets and plasma. Neurobiol Aging 2005;26(1):115–23.

[37] Tapia-Arancibia L, et al. New insights into brain BDNF function in normal aging and Alzheimer disease. Brain Res Rev 2008;59(1):201–20.

[38] Al-Delaimy WK, Von Muhlen D, Barrett-Connor E. Insulinlike growth factor-1, insulinlike growth factor binding protein-1, and cognitive function in older men and women. J Am Geriatr Soc 2009;57(8):1441–6.

[39] van Dam PS, Aleman A. Insulin-like growth factor-I, cognition and brain aging. Eur J Pharmacol 2004;490(1–3):87–95.

[40] Gurden H, Takita M, Jay TM. Essential role of D1 but not D2 receptors in the NMDA receptor-dependent long-term potentiation at hippocampal-prefrontal cortex synapses in vivo. J Neurosci 2000;20(22):RC106.

[41] McGeer PL, McGeer EG, Suzuki JS. Aging and extrapyramidal function. Arch Neurol 1977;34(1):33–5.

[42] Kanaya AM, et al. Change in cognitive function by glucose tolerance status in older adults: a 4-year prospective study of the rancho Bernardo study cohort. Arch Intern Med 2004;164(12):1327–33.

[43] Strachan MW, et al. Cognitive function, dementia and type 2 diabetes mellitus in the elderly. Nat Rev Endocrinol 2011;7(2):108–14.

[44] Baker LD, et al. Aerobic exercise improves cognition for older adults with glucose intolerance, a risk factor for Alzheimer's disease. J Alzheimers Dis 2010;22(2):569–79.

[45] Burke SN, Barnes CA. Neural plasticity in the ageing brain. Nat Rev Neurosci 2006;7(1):30–40.

[46] Mattson MP, Magnus T. Ageing and neuronal vulnerability. Nat Rev Neurosci 2006;7(4):278–94.

[47] Newcomer JW, Farber NB, Olney JW. NMDA receptor function, memory, and brain aging. Dialogues Clin Neurosci 2000;2(3):219.

[48] Foster TC, Norris CM. Age-associated changes in Ca(2+)-dependent processes: relation to hippocampal synaptic plasticity. Hippocampus 1997;7 (6):602–12.

[49] Small SA, et al. Imaging correlates of brain function in monkeys and rats isolates a hippocampal subregion differentially vulnerable to aging. Proc Natl Acad Sci U S A 2004;101(18):7181.

[50] Tucker-Drob E, Salthouse T. Individual differences in cognitive aging. In: Chamorro-Premuzic T, von Stumm S, Furnham A, editors. The Wiley-Blackwell handbook of individual differences. Malden, MA: Wiley-Blackwell; 2011. p. 242–67.

[51] Hayden KM, et al. Cognitive decline in the elderly: an analysis of population heterogeneity. Age Ageing 2011;40(6):684–9.

[52] Thielke S, Diehr P. Transitions among health states using 12 measures of successful aging in men and women: results from the cardiovascular health study. J Aging Res 2012;2012, 243263.

[53] Anstey K, Christensen H. Education, activity, health, blood pressure and apolipoprotein E as predictors of cognitive change in old age: a review. Gerontology 2000;46(3):163–77.

[54] Lipnicki DM, et al. Risk factors for late-life cognitive decline and variation with age and sex in the Sydney memory and ageing study. PLoS One 2013;8(6), e65841.

[55] Colcombe S, Kramer AF. Fitness effects on the cognitive function of older adults: a meta-analytic study. Psychol Sci 2003;14(2):125–30.

[56] Middleton LE, et al. Physical activity over the life course and its association with cognitive performance and impairment in old age. J Am Geriatr Soc 2010;58(7):1322–6.

[57] Yaffe K, et al. Predictors of maintaining cognitive function in older adults: the health ABC study. Neurology 2009;72(23):2029–35.

[58] Buchman AS, et al. Total daily physical activity and the risk of AD and cognitive decline in older adults. Neurology 2012;78(17):1323–9.

[59] Anon. Insufficient physical activity. Canberra: Australian Institute of Health; 2020.

[60] Erickson KI, et al. Physical activity predicts gray matter volume in late adulthood: the cardiovascular health study. Neurology 2010;75(16): 1415–22.

[61] Freudenberger P, et al. Fitness and cognition in the elderly. The Austrian stroke prevention study. Neurology 2016;86(5):418–24.

[62] Falck RS, et al. Impact of exercise training on physical and cognitive function among older adults: a systematic review and meta-analysis. Neurobiol Aging 2019;79:119–30.

[63] Hillman CH, Erickson KI, Kramer AF. Be smart, exercise your heart: exercise effects on brain and cognition. Nat Rev Neurosci 2008;9(1):58.

[64] Boucard GK, et al. Impact of physical activity on executive functions in aging: a selective effect on inhibition among old adults. J Sport Exerc Psychol 2012;34(6):808–27.

[65] Guadagni V, et al. Aerobic exercise improves cognition and cerebrovascular regulation in older adults. Neurology 2020;94(21):e2245–57.

[66] Guiney H, et al. Benefits of regular aerobic exercise for executive functioning in healthy populations. Psychon Bull Rev 2013;20(1):73–86.

[67] Albinet CT, et al. Executive functions improvement following a 5-month aquaerobics program in older adults: role of cardiac vagal control in inhibition performance. Biol Psychol 2016;115:69–77.

[68] Baniqued PL, et al. Brain network modularity predicts exercise-related executive function gains in older adults. Front Aging Neurosci 2017;9:426.

[69] Leyland L-A, et al. The effect of cycling on cognitive function and well-being in older adults. PLoS One 2019;14(2):e0211779.

[70] Chapman SB, et al. Distinct brain and behavioral benefits from cognitive vs. physical training: a randomized trial in aging adults. Front Hum Neurosci 2016;10:338.

[71] Erickson KI, et al. Exercise training increases size of hippocampus and improves memory. Proc Natl Acad Sci U S A 2011;108(7):3017.

[72] Kovacevic A, et al. The effects of aerobic exercise intensity on memory in older adults. Appl Physiol Nutr Metab 2020;45(6):591–600.

[73] Bherer L, Erickson KI, Liu-Ambrose T. A review of the effects of physical activity and exercise on cognitive and brain functions in older adults. J Aging Res 2013;**2013**:657508.

[74] Déry N, et al. Adult hippocampal neurogenesis reduces memory interference in humans: opposing effects of aerobic exercise and depression. Front Neurosci 2013;7:66.

[75] Brown AK, et al. The effect of group-based exercise on cognitive performance and mood in seniors residing in intermediate care and self-care retirement facilities: a randomised controlled trial. Br J Sports Med 2009;43(8):608–14.

[76] Netz Y, et al. Aerobic fitness and multidomain cognitive function in advanced age. Int Psychogeriatr 2011;23(1):114.

[77] Benedict C, et al. Association between physical activity and brain health in older adults. Neurobiol Aging 2013;34(1):83–90.

[78] Flöel A, et al. Physical activity and memory functions: are neurotrophins and cerebral gray matter volume the missing link? Neuroimage 2010;49 (3):2756–63.

[79] Jonasson LS, et al. Aerobic exercise intervention, cognitive performance, and brain structure: results from the physical influences on brain in aging (PHIBRA) study. Front Aging Neurosci 2017;8, 00336.

[80] Levine BD. What do we know, and what do we still need to know? J Physiol 2008;586(1):25–34.

[81] Liu-Ambrose T, et al. Otago home-based strength and balance retraining improves executive functioning in older fallers: a randomized controlled trial. J Am Geriatr Soc 2008;56(10):1821–30.

[82] Cassilhas RC, et al. The impact of resistance exercise on the cognitive function of the elderly. Med Sci Sports Exerc 2007;39(8):1401.

[83] Landrigan J-F, et al. Lifting cognition: a meta-analysis of effects of resistance exercise on cognition. Psychol Res 2020;84(5):1167–83.

[84] Li Z, et al. The effect of resistance training on cognitive function in the older adults: a systematic review of randomized clinical trials. Aging Clin Exp Res 2018;30(11):1259–73.

[85] Kimura K, et al. The influence of short-term strength training on health-related quality of life and executive cognitive function. J Physiol Anthropol 2010;29(3):95–101.

[86] Venturelli M, et al. Positive effects of physical training in activity of daily living–dependent older adults. Exp Aging Res 2010;36(2):190–205.

[87] Best JR, et al. Long-term effects of resistance exercise training on cognition and brain volume in older women: results from a randomized controlled trial. J Int Neuropsychol Soc 2015;21(10):745–56.

[88] Best JR, Nagamatsu LS, Liu-Ambrose T. Improvements to executive function during exercise training predict maintenance of physical activity over the following year. Front Hum Neurosci 2014;8:353.

[89] Northey JM, et al. Exercise interventions for cognitive function in adults older than 50: a systematic review with meta-analysis. Br J Sports Med 2018;52(3):154.

[90] Smith PJ, et al. Aerobic exercise and neurocognitive performance: a meta-analytic review of randomized controlled trials. Psychosom Med 2010;72(3):239.

[91] Ouellette MM, et al. High-intensity resistance training improves muscle strength, self-reported function, and disability in long-term stroke survivors. Stroke 2004;35(6):1404–9.

[92] Yerokhin V, et al. Neuropsychological and neurophysiological effects of strengthening exercise for early dementia: a pilot study. Aging Neuropsychol Cognit 2012;19(3):380–401.

[93] Jaffery A, Edwards MK, Loprinzi PD. The effects of acute exercise on cognitive function: Solomon experimental design. J Prim Prev 2018;39(1):37–46.

[94] Winter B, et al. High impact running improves learning. Neurobiol Learn Mem 2007;87(4):597–609.

[95] Audiffren M, Tomporowski PD, Zagrodnik J. Acute aerobic exercise and information processing: modulation of executive control in a random number generation task. Acta Psychol (Amst) 2009;132(1):85.

[96] Charalambous CC, et al. A single high-intensity exercise bout during early consolidation does not influence retention or relearning of sensorimotor locomotor long-term memories. Exp Brain Res 2019;237(11):2799–810.

[97] McSween M-P, et al. The immediate effects of acute aerobic exercise on cognition in healthy older adults: a systematic review. Sports Med 2019;49(1):67–82.

[98] Ludyga S, et al. Acute effects of moderate aerobic exercise on specific aspects of executive function in different age and fitness groups: a meta-analysis. Psychophysiology 2016;53(11):1611–26.

[99] Barella LA, Etnier JL, Chang Y-K. The immediate and delayed effects of an acute bout of exercise on cognitive performance of healthy older adults. J Aging Phys Act 2010;18(1):87–98.

[100] Wheeler MJ, et al. Distinct effects of acute exercise and breaks in sitting on working memory and executive function in older adults: a three-arm, randomised cross-over trial to evaluate the effects of exercise with and without breaks in sitting on cognition. Br J Sports Med 2020;54(13):776–81.

[101] Kimura K, Hozumi N. Investigating the acute effect of an aerobic dance exercise program on neuro-cognitive function in the elderly. Psychol Sport Exerc 2012;13(5):623–9.

[102] O'Brien J, et al. One bout of open skill exercise improves cross-modal perception and immediate memory in healthy older adults who habitually exercise. PLoS One 2017;12(6), e0178739.

[103] Pontifex MB, et al. A primer on investigating the after effects of acute bouts of physical activity on cognition. Psychol Sport Exerc 2019;40:1–22.

[104] Wilke J, et al. Acute effects of resistance exercise on cognitive function in healthy adults: a systematic review with multilevel meta-analysis. Sports Med 2019;49(6):905–16.

[105] Naderi A, et al. Effects of low and moderate acute resistance exercise on executive function in community-living older adults. Sport Exerc Perform Psychol 2019;8(1):106–22.

[106] Hsieh S-S, et al. The effects of acute resistance exercise on young and older males' working memory. Psychol Sport Exerc 2016;22:286–93.

[107] Bender V, McGlynn G. The effect of various levels of strenuous to exhaustive exercise on reaction time. Eur J Appl Physiol Occup Physiol 1976;35(2):95–101.

[108] Weingarten G, Alexander JF. Effects of physical exertion on mental performance of college males of different physical fitness level. Percept Mot Skills 1970;31(2):371–8.

[109] McMorris T. Exercise-cognition interaction : neuroscience perspectives. London: Elsevier; 2016.

[110] Labelle V, et al. Decline in executive control during acute bouts of exercise as a function of exercise intensity and fitness level. Brain Cogn 2013;81(1):10–7.

[111] Lambourne K, Tomporowski P. The effect of exercise-induced arousal on cognitive task performance: a meta-regression analysis. Brain Res 2010;1341:12–24.

[112] Eysenck HJ, Nias D, Cox D. Sport and personality. Adv Behav Res Ther 1982;4(1):1–56.

[113] Duffy E. The psychological significance of the concept of" arousal" or" activation.". Psychol Rev 1957;64(5):265.

[114] Gómez-Pinilla F, Feng C. Molecular mechanisms for the ability of exercise supporting cognitive abilities and counteracting neurological disorders. In: Functional neuroimaging in exercise and sport sciences. Springer; 2012. p. 25–43.

[115] Huang EJ, Reichardt LF. Neurotrophins: roles in neuronal development and function. Annu Rev Neurosci 2001;24(1):677–736.

[116] Rovio S, et al. The effect of midlife physical activity on structural brain changes in the elderly. Neurobiol Aging 2010;31(11):1927–36.

[117] Torres ER, et al. Physical activity and white matter hyperintensities: a systematic review of quantitative studies. Prev Med Rep 2015;2:319–25.

[118] Ruscheweyh R, et al. Physical activity and memory functions: an interventional study. Neurobiol Aging 2011;32(7):1304–19.

[119] Voss MW, et al. Bridging animal and human models of exercise-induced brain plasticity. Trends Cogn Sci 2013;17(10):525–44.

[120] Colcombe SJ, et al. Cardiovascular fitness, cortical plasticity, and aging. Proc Natl Acad Sci U S A 2004;101(9):3316–21.

[121] Prakash RS, et al. Cardiorespiratory fitness and attentional control in the aging brain. Front Hum Neurosci 2011;4:229.

[122] Rosano C, et al. Psychomotor speed and functional brain MRI 2 years after completing a physical activity treatment. J Gerontol Ser A Biolmed Sci Med Sci 2010;65(6):639–47.

[123] Prakash RS, et al. Physical activity and cognitive vitality. Annu Rev Psychol 2015;66:769–97.

[124] Burdette JH, et al. Using network science to evaluate exercise-associated brain changes in older adults. Front Aging Neurosci 2010;2:23.

[125] Voss MW, et al. Plasticity of brain networks in a randomized intervention trial of exercise training in older adults. Front Aging Neurosci 2010;2:32.

[126] Andrews-Hanna JR, et al. Functional-anatomic fractionation of the brain's default network. Neuron 2010;65(4):550–62.

[127] Buckner RL, Andrews-Hanna JR, Schacter DL. The brain's default network: Anatomy, function, and relevance to disease; 2008.

[128] Volkow ND, et al. Association between decline in brain dopamine activity with age and cognitive and motor impairment in healthy individuals. Am J Psychiatry 1998;155(3):344.

[129] Jonasson LS, et al. Higher striatal D2-receptor availability in aerobically fit older adults but non-selective intervention effects after aerobic versus resistance training. Neuroimage 2019;202, 116044.

[130] Petzinger G, Jakowec MW. Pharmacological and Behavioral enhancement of neuroplasticity in the MPTP Lesioned mouse and nonhuman primate, A. University Of Southern California Los; 2007.

[131] Bäckman L, et al. The correlative triad among aging, dopamine, and cognition: current status and future prospects. Neurosci Biobehav Rev 2006;30(6):791–807.

[132] Cools R, D'Esposito M. Inverted-U–shaped dopamine actions on human working memory and cognitive control. Biol Psychiatry 2011;69(12):e113–25.

[133] Garrett DD, et al. Amphetamine modulates brain signal variability and working memory in younger and older adults. Proc Natl Acad Sci 2015;112(24):7593.

[134] Hosp JA, Luft AR. Dopaminergic meso-cortical projections to M1: role in motor learning and motor cortex plasticity. Front Neurol 2013;4:145.

[135] Seals D, et al. Glucose tolerance in young and older athletes and sedentary men. J Appl Physiol 1984;56(6):1521–5.

[136] Schubert M, et al. Insulin receptor substrate-2 deficiency impairs brain growth and promotes tau phosphorylation. J Neurosci 2003;23(18):7084–92.

[137] Recio-Pinto E, Ishii DN. Effects of insulin, insulin-like growth factor-II and nerve growth factor on neurite outgrowth in cultured human neuroblastoma cells. Brain Res 1984;302(2):323–34.

[138] Dickson BJ. Wiring the brain with insulin. Science 2003;300(5618):440–1.

[139] Yu SW, et al. Autophagic death of adult hippocampal neural stem cells following insulin withdrawal. Stem Cells 2008;26(10):2602–10.

[140] Schulingkamp R, et al. Insulin receptors and insulin action in the brain: review and clinical implications. Neurosci Biobehav Rev 2000;24(8):855–72.

[141] Messier C. Impact of impaired glucose tolerance and type 2 diabetes on cognitive aging. Neurobiol Aging 2005;26(1):26–30.

[142] Sigal RJ, et al. Physical activity/exercise and type 2 diabetes. Diabetes Care 2004;27(10):2518–39.

[143] Spyridaki EC, Avgoustinaki PD, Margioris AN. Obesity, inflammation and cognition. Curr Opin Behav Sci 2016;9:169–75.

[144] Houston DK, Nicklas BJ, Zizza CA. Weighty concerns: the growing prevalence of obesity among older adults. J Am Diet Assoc 2009;109(11):1886–95.

[145] Fratiglioni L, Paillard-Borg S, Winblad B. An active and socially integrated lifestyle in late life might protect against dementia. Lancet Neurol 2004;3(6):343–53.

[146] Wipfli BM, Rethorst CD, Landers DM. The anxiolytic effects of exercise: a meta-analysis of randomized trials and dose–response analysis. J Sport Exerc Psychol 2008;30(4):392–410.

[147] Anderson EH, Shivakumar G. Effects of exercise and physical activity on anxiety. Front Psych 2013;4:27.

[148] Rethorst CD, Wipfli BM, Landers DM. The antidepressive effects of exercise. Sports Med 2009;39(6):491–511.

[149] Beaudreau SA, O'Hara R. The association of anxiety and depressive symptoms with cognitive performance in community-dwelling older adults. Psychol Aging 2009;24(2):507.

[150] Potvin O, et al. Anxiety disorders, depressive episodes and cognitive impairment no dementia in community-dwelling older men and women. Int J Geriatr Psychiatry 2011;26(10):1080–8.

[151] Yochim BP, Mueller AE, Segal DL. Late life anxiety is associated with decreased memory and executive functioning in community dwelling older adults. J Anxiety Disord 2013;27(6):567–75.

[152] Teachman BA. Aging and negative affect: the rise and fall and rise of anxiety and depression symptoms. Psychol Aging 2006;21(1):201.

[153] Aartsen M, Jylhä M. Onset of loneliness in older adults: results of a 28 year prospective study. Eur J Ageing 2011;8(1):31–8.

[154] Law LL, et al. Effects of combined cognitive and exercise interventions on cognition in older adults with and without cognitive impairment: a systematic review. Ageing Res Rev 2014;15:61–75.

[155] Lauenroth A, Ioannidis AE, Teichmann B. Influence of combined physical and cognitive training on cognition: a systematic review. BMC Geriatr 2016;16(1):1–14.

[156] Zhu X, et al. The more the better? A meta-analysis on effects of combined cognitive and physical intervention on cognition in healthy older adults. Ageing Res Rev 2016;31:67–79.

[157] Stanmore E, et al. The effect of active video games on cognitive functioning in clinical and non-clinical populations: a meta-analysis of randomized controlled trials. Neurosci Biobehav Rev 2017;78:34–43.

[158] Hannan A. Environmental enrichment and brain repair: harnessing the therapeutic effects of cognitive stimulation and physical activity to enhance experience-dependent plasticity. Neuropathol Appl Neurobiol 2014;40(1):13–25.

[159] Van Praag H, Kempermann G, Gage FH. Neural consequences of enviromental enrichment. Nat Rev Neurosci 2000;1(3):191–8.

[160] Diamond MC, et al. Effects of environment on morphology of rat cerebral cortex and hippocampus. J Neurobiol 1976;7(1):75–85.

[161] Rosenzweig MR, Bennett EL. Effects of differential environments on brain weights and enzyme activities in gerbils, rats, and mice. Dev Psychobiol 1969;2(2):87–95.

[162] Fabel K, et al. Additive effects of physical exercise and environmental enrichment on adult hippocampal neurogenesis in mice. Front Neurosci 2009;3:2.

[163] Gheysen F, et al. Physical activity to improve cognition in older adults: can physical activity programs enriched with cognitive challenges enhance the effects? A systematic review and meta-analysis. Int J Behav Nutr Phys Act 2018;15(1):63.

[164] Taylor-Piliae RE, et al. Effects of tai chi and Western exercise on physical and cognitive functioning in healthy community-dwelling older adults. J Aging Phys Act 2010;18(3):261–79.

[165] Wayne PM, et al. Effect of Tai Chi on cognitive performance in older adults: systematic review and meta-analysis. J Am Geriatr Soc 2014;62(1):25–39.

[166] Kattenstroth J-C, et al. Six months of dance intervention enhances postural, sensorimotor, and cognitive performance in elderly without affecting cardio-respiratory functions. Front Aging Neurosci 2013;5:5.

[167] Dhami P, Moreno S, DeSouza J. New framework for rehabilitationfusion of cognitive and physical rehabilitation: the hope for dancing. Front Psychol 2014;5:1478.

[168] Ogawa EF, You T, Leveille SG. Potential benefits of exergaming for cognition and dual-task function in older adults: a systematic review. J Aging Phys Act 2016;24(2):332–6.

[169] Costello E, et al. Motivators, barriers, and beliefs regarding physical activity in an older adult population. J Geriatr Phys Ther 2011;34(3):138–47.

[170] Schättin A, et al. Adaptations of prefrontal brain activity, executive functions, and gait in healthy elderly following exergame and balance training: a randomized-controlled study. Front Aging Neurosci 2016;8:278.

[171] Falbo S, et al. Effects of physical-cognitive dual task training on executive function and gait performance in older adults: a randomized controlled trial. Biomed Res Int 2016;2016, 5812092.

[172] Fabre C, et al. Improvement of cognitive function by mental and/or individualized aerobic training in healthy elderly subjects. Int J Sports Med 2002;23(06):415–21.

[173] Oswald WD, et al. Differential effects of single versus combined cognitive and physical training with older adults: the SimA study in a 5-year perspective. Eur J Ageing 2006;3(4):179.

[174] World Health Organisation, editor. Global recommendations on physical activity for health; 65 years and above; 2011.

[175] Health, U.D.o. and H. Services, US Department of Health and Human Services. Physical activity guidelines for Americans. Hyattsville, MD: Author, Washington, DC, 2008; 2008. p. 1–40.

[176] Warburton DE, Nicol CW, Bredin SS. Health benefits of physical activity: the evidence. CMAJ 2006;174(6):801–9.

[177] Mang CS, et al. A single bout of high-intensity aerobic exercise facilitates response to paired associative stimulation and promotes sequence-specific implicit motor learning. J Appl Physiol 2014;117(11):1325–36.

[178] Stavrinos EL, Coxon JP. High-intensity interval exercise promotes motor cortex disinhibition and early motor skill consolidation. J Cogn Neurosci 2017;29(4):593–604.

[179] Andrews SC, et al. Intensity matters: high-intensity interval exercise enhances motor cortex plasticity more than moderate exercise. Cereb Cortex 2020;30(1):101–12.

[180] Mekari S, et al. High-intensity interval training improves cognitive flexibility in older adults. Brain Sci 2020;10(11):796.

[181] Litt MD, Kleppinger A, Judge JO. Initiation and maintenance of exercise behavior in older women: predictors from the social learning model. J Behav Med 2002;25(1):83–97.

[182] Lee PG, Jackson EA, Richardson CR. Exercise prescriptions in older adults. Am Fam Physician 2017;95(7):425–32.

[183] Gillespie LD, et al. Interventions for preventing falls in older people living in the community. Cochrane Database Syst Rev 2012;2012(9), CD007146.

[184] Dauenhauer JA, Podgorski CA, Karuza J. Prescribing exercise for older adults: a needs assessment comparing primary care physicians, nurse practitioners, and physician assistants. Gerontol Geriatr Educ 2006;26(3):81–99.

[185] Kaye JA, et al. Neurologic evaluation of the optimally healthy oldest old. Arch Neurol 1994;51(12):1205–11.

[186] Colberg SR, et al. Exercise and type 2 diabetes: the American College of Sports Medicine and the American Diabetes Association: joint position statement. Diabetes Care 2010;33(12):e147–67.

[187] Ries AL, et al. Pulmonary rehabilitation: joint ACCP/AACVPR evidence-based clinical practice guidelines. Chest 2007;131(5):4S–42S.

[188] Anon. PARmed-X: Physical activity readiness medical examination. [cited 2021 March 3]; Available from: https://www.chp.gov.hk/archive/epp/files/PARmed-X.pdf 2002.

[189] Chodzko-Zajko WJ, et al. Exercise and physical activity for older adults. Med Sci Sports Exerc 2009;41(7):1510–30.

[190] Fransen M, McConnell S, Harmer AR, et al. Exercise for osteoarthritis of the knee: a Cochrane systematic review. Br J Sports Med 2015;49:1554–7.

[191] Bartels EM, et al. Aquatic exercise for the treatment of knee and hip osteoarthritis. Cochrane Database Syst Rev 2016;2016(3), CD005523.

[192] Garber CE, et al. American college of sports medicine position stand. Quantity and quality of exercise for developing and maintaining cardiorespiratory, musculoskeletal, and neuromotor fitness in apparently healthy adults: guidance for prescribing exercise. Med Sci Sports Exerc 2011;43(7):1334–59.

Index

Note: Page numbers followed by f indicate figures t indicate tables and b indicate boxes.